Lecture Notes in Computer Science   12068

More information about this series at http://www.springer.com/series/7412

Yue Lu · Nicole Vincent ·
Pong Chi Yuen · Wei-Shi Zheng ·
Farida Cheriet · Ching Y. Suen (Eds.)

Pattern Recognition and Artificial Intelligence

International Conference, ICPRAI 2020
Zhongshan, China, October 19–23, 2020
Proceedings

 Springer

Editors
Yue Lu
East China Normal University
Shanghai, China

Nicole Vincent
Paris Descartes University
Paris, France

Pong Chi Yuen
Hong Kong Baptist University
Kowloon, Hong Kong

Wei-Shi Zheng
Sun Yat-sen University
Guangzhou, China

Farida Cheriet
Polytechnique Montréal
Montreal, QC, Canada

Ching Y. Suen
Concordia University
Montreal, QC, Canada

ISSN 0302-9743 ISSN 1611-3349 (electronic)
Lecture Notes in Computer Science
ISBN 978-3-030-59829-7 ISBN 978-3-030-59830-3 (eBook)
https://doi.org/10.1007/978-3-030-59830-3

LNCS Sublibrary: SL6 – Image Processing, Computer Vision, Pattern Recognition, and Graphics

This Springer imprint is published by the registered company Springer Nature Switzerland AG
The registered company address is: Gewerbestrasse 11, 6330 Cham, Switzerland

Preface

Message from the Honorable Chair and General Chair

Welcome to the proceedings of the Second International Conference on Pattern Recognition and Artificial Intelligence (ICPRAI 2020). This conference follows the successful ICPRAI 2018, held in Montreal, Canada, in May 2018. It was organized by CENPARMI (Centre for Pattern Recognition and Machine Intelligence) at Concordia University, Montreal, Canada, with the co-operation of the Chinese Association of Science and Technology, China, and other universities from China and abroad. The Organizing Committee of ICPRAI 2020 consisted of well-known experts from all six continents of the world with a dream team of eight world-renowned keynote speakers to cover the new PRAI frontiers in depth and in breadth with insights. The technical program included 74 presentations in 5 tracks related to handwriting and text processing, computer vision and image processing, machine learning and deep networks, medical applications, forensic science and medical diagnosis, features and classification techniques, and various applications. In addition, a workshop on Social Media Processing was also organized.

Pattern Recognition and AI techniques and systems have been applied successfully to solve practical problems in many domains. In co-operation with CENPARMI members and colleagues at other institutions, ICPRAI 2020 is co-sponsored by the Chinese Association of Science and Technology and numerous universities and industrial companies. It brings together a large number of scientists from all over the world to present their innovative ideas and report on their latest findings, as well as the fruitful results of numerous CENPARMI members and their students in handwriting recognition, forensic studies, face recognition, medical imaging, deep learning, and classification techniques. In addition to the rich technical program, our conference also featured practical sessions to encourage author-editor and inventor-investor interactions, and social events for all conference participants.

For all the above activities, we would like to thank all those involved for a superb technical program and conference proceedings: conference co-chairs, Edwin Hancock and Patrick Wang; track co-chairs, Yue Lu, Nicole Vincent, Pong Chi Yuen, Farida Cheriet, and Wei-Shi Zheng; workshop co-chairs, Kam-Fai Wong and Binyang Li; special session chair, Olga Ormandjieva; as well as the publication and publicity chairs and their committee members.

We would like to express our gratitude to the numerous committee members for taking care of financial and sponsorship matters (Louisa Lam, Wen Song, Marleah Blom, Mounîm El Yacoubi, and Jun Tan). Special thanks go to local arrangements of the Chinese Association of Science and Technology and the government officials of Zhongshan City, and Nicola Nobile who dedicated himself to a great variety of duties, ranging from conference secretariat business, publications, software for handling paper submissions, reviews to selections, plus numerous other chores. They, together with

CENPARMI members and students, created a very warm and comfortable environment to work in.

Thanks are also due to the organizations listed in the proceedings and those of the organizers of ICPRAI 2020, and the administration of Zhongshan City.

Finally, I hope you found this conference to be a rewarding and memorable experience. We hope you enjoyed your stay in the beautiful Zhongshan City, a hub in the Guangdong, Hong Kong, Macau Greater Bay Area, and the birthplace of great historical figure Dr. Sun Yat-sen.

August 2020   Ching Y. Suen
                                                                                    Yuan Y. Tang

Organization

Organizing Committees

ICPRAI 2020 was hosted by Zhongshan City and organized by CENPARMI (Centre for Pattern Recognition and Machine Intelligence) of Concordia University, Canada, and journal editors and scientists:

Honorary Chair

Ching Y. Suen, Canada

General Chair

Yuan Y. Tang, Hong Kong, China

Conference Co-chairs

Edwin Hancock, UK
Patrick S. Wang, USA

Special Sessions and Workshops Committee

Adam Krzyzak, Chair, Canada
Camille Kurtz, France
Cheng-Lin Liu, China
Khoa Luu, USA
Takeshi Masuda, Japan
Kam-Fai Wong, Hong Kong, China
Vera Yashina, Russia

Program Chairs

Farida Cheriet, Canada
Yue Lu, China
Nicole Vincent, France
Pong C. Yuen, Hong Kong, China
Wei-Shi Zheng, China

Competition Judges

Mohamed Cheriet, Chair, Canada
Donato Impedovo, Italy
Jay Liebowitz, USA
Jean-Marc Ogier, France
Lihua Yang, China

Exhibitions and Industrial Liaisons

Jun Tan, Chair, China
Alexandre Cornu, Canada
Yufang Tang, China

Publication Chairs

Xiaoyi Jiang, Germany
Muna Khayyat, Canada
Qin Lu, Hong Kong, China

Sponsorship Chairs

Marleah Blom, Canada
Mounîm El Yacoubi, France
Wen Song, China

Financial and Local Arrangements

Louisa Lam, Finance, Canada
Shankang Huang, Liaison, Canada
Connie Chan Kwok, Hong Kong, Canada
Zhongshan Committee, China

Publicity Chairs

Veronique Eglin, France
Soo-Hyung Kim, South Korea
Hongjun Li, China
Li Liu, China
Xiaoqing Lu, China
Jose F Martinez-Trinidad, Mexico
Umapada Pal, India
Alun Preece, UK
Javad Sadri, Canada
Dit-Yan Yeung, Hong Kong, China

Jia You, HK, China
Bill Wang, Canada
David Zhang, China
Xingming Zhao, China
Jie Zhou, USA
Jun Zhou, Australia

Conference Secretariats and Web Design

Nicola Nobile, Secretariat, Canada
Phoebe Chan, Canada
S. Ling Chan Suen, Canada
Yan Xu, China

Keynote Speakers

ICPRAI 2020 was proud to host the following keynote speakers:

Hironobu Fujiyoshi	Chubu University, Japan	"Visualizing Direction of Attention to Understand AI Decisions"
Seong-Whan Lee	Korea University, South Korea	"Recent Advances and Future Prospects of Brain-Computer Interface"
Cheng-Lin Liu	Institute of Automation of Chinese Academy of Sciences, China	"New Frontiers of Document Image Analysis and Recognition"
Farida Cheriet	Polytechnique Montreal, Canada	"PRAI in Multimodal Medical Image Analysis: Applications and Challenges"
Nicole Vincent and Camille Kurtz	University of Paris, France	"Optimal Choices of Features in Image Analysis"
Sheila Lowe	American Handwriting Analysis Foundation, USA	"Revealing Human Personality Through Handwriting Analysis"
Bidyut B. Chaudhuri	Techno India University, India	"Improvement of Deep Neural Net for Machine Learning"
David Zhang	The Chinese University of Hong Kong, Hong Kong, China	"Advanced Biometrics: Research and Development"

Sponsors

Sponsored by CENPARMI and other scientific/technological/industrial partners.

 **GD-ZS CAST**

Contents

Features and Classifiers

Deep Learning

Computer Vision and Image Processing

Handwriting and Text Processing

Gender Detection from Handwritten Documents Using Concept of Transfer-Learning

Najla AL-Qawasmeh[(✉)] [iD] and Ching Y. Suen[(✉)] [iD]

Department of Computer Science and Software Engineering, Concordia University,
CENPARMI, Montreal, Canada
n_alqawa@encs.concordia.ca, suen@cse.concordia.ca
http://www.concordia.ca/research/cenparmi.html

Abstract. Offline gender detection from Arabic handwritten documents is a very challenging task because of the high similarity between an individual's writings and the complexity of the Arabic language as well. In this paper, we propose a new way to detect the writer gender from scanned handwritten documents that mainly based on the concept of transfer-learning. We used a pre-trained knowledge from two convolution neural networks (CNN): GoogleNet, and ResNet, then we applied it on our data-set. We use this two CNN architectures as fixed feature extractors. For the analysis and the classification stage, we used a support vector machine (SVM). The performance of the two CNN architectures concerning accuracy is 80.05% for GoogleNet, 83.32% for ResNet.

Keywords: Handwriting analysis · Transfer-learning · Gender classification · Deep learning

1 Introduction

Gender detection is considered as a biometrical application; hence, the term gender is related to the physical and social condition of being female or male, while gender can be detected from different biometric features such as physiological traits (face) or behavioral traits(handwriting) [1, 2]. Despite the high similarity between individual writings, each person has his/her own writing preferences. Many researchers validated the relation between handwriting and gender detection what makes handwriting analysis for gender detection a vital research area in many sectors, including psychology, document analysis, paleography, graphology, and forensic analysis [3–5]. Computer scientists have also expressed their interest in the handwriting analysis science by developing automatic handwriting analysis systems to detect the writer's gender from scanned images using different machine learning techniques.

The traditional machine learning techniques consist of several complex tasks, including pre-processing, segmentation, feature extraction which degrade the

© Springer Nature Switzerland AG 2020
Y. Lu et al. (Eds.): ICPRAI 2020, LNCS 12068, pp. 3–13, 2020.
https://doi.org/10.1007/978-3-030-59830-3_1

performance of the systems regarding efficiency and accuracy [6]. The concept of CNN was first introduced in 1980 and 1989 to solve these problems [7,8]. However, it reached impressive results only in 2012, when Krizhevsky et al. [9] applied a deep convolution network on the ImageNet [10] to classify 1000 different classes.

Deep CNN becomes a very powerful tool for image classification tasks because it automatically extracts the features from raw data, without the need of the traditional ways of machine learning techniques. This makes the feature extraction process easier and faster. On the other hand, CNN does not produce a good performance when the dataset size is small, as the size of the data must be large to get good results [11].

Collecting a large dataset of handwriting is not an easy task, so to solve this issue, the concept of transfer-learning was introduced which mainly depends on using a pre-trained network as a starting point to train a new task. This concept helps in making the process of training using a small number of samples more effective than training a network from scratch [12,13].

In this proposed work, we used the transfer-learning concept to classify the writer's gender from female and male handwriting images. The main contributions of this paper can be summarized as follow:

- We are proposing a new Arabic handwriting dataset for different classification and recognition tasks.
- We are using the concept of transfer-learning using two different deep-learning architectures.
- We compared the performance of two CNN architectures at the accuracy level.

The rest of the paper is organized as follows: Sect. 2 explores the related works of some recent automatic handwriting gender detection systems. Section 3 explains the details of the proposed approach. Section 4 describes the experiments and results obtained after applying the proposed method on the target dataset. Section 5 concludes the paper and gives the future directions.

2   Related Works

Interest in gender detection began many years ago, when in 1999, Huber and Hadrick [14] presented the correlation between handwriting and several categories, including age, handedness, gender, and ethnicity. However, unfortunately, the old-fashioned way of gender classification suffers from some drawbacks, such as the accuracy of the analysis depends on the analyst's skills, and it is costly and prone to fatigue. These disadvantages created the need to establish an automatic tool, which analyzes the handwriting and detect the writer's gender faster and more precisely with the help of a computer without human intervention [15]. In the following section we are going to explore some recent automatic Handwriting gender classification systems. Table 1 summaries the gender detection systems reviewed in literature.

In 2014, Almadeed and Hussain [16] proposed an offline automatic handwriting prediction system of age, gender, and nationality. They extracted a set of geometric features from the handwriting images, and then they combined these features using random forests and kernel discriminative analysis. Their system classification rates were 74.05%, 55.76%, and 53.66% for gender, age, and nationality classification, respectively. Seddigi et al. [3] proposed an offline prediction system to classify the writer's gender from handwriting images by extracting a set of features such as slant, curvature, texture, and legibility. Artificial neural network(ANN) and support vector machine(SVM) were used for the classification task. They applied their method on two different datasets, QUWI [17] and MSHD [18]. Their approach achieved a classification rate of 68.75 and 67.50 when applying SVM and ANN on QUWI, respectively. Moreover, they produced a classification accuracy of 73.02% and 69.44% when applying SVM and ANN on MSHD dataset, respectively.

In 2015, Maji et al. [19] used handwriting signatures to detect the writer's gender. They extracted several features from the signature's images, including roundness, skewness, kurtosis, mean, standard deviation area, Euler number, the distribution density of black pixels, and the connected components. They used a back-propagation neural network(BPNN) to analyze these features and recognize the writer's gender. Their system achieved an 80.7% accuracy with Euler number feature and a 77% accuracy without the Euler number feature. Nogueras et al. [1] presented a gender classification schema based on online handwriting. The authors used a digital tablet to extract online features related to the pen trajectory, such as pen-up, pen-down, pen pressure. They achieved a classification accuracy rate of 72.6%.

In 2016, Mirza et al. [2] used Gabor-filter to extract the textural features from the handwriting images, to distinguish between male and female writings. The classification task in this work was carried using a feed-forward neural network. They also used QUWI dataset to train and evaluate their method, where they divided the experiments into four tasks. In the first one, they used only Arabic samples for training and testing, while in the second task, they used only English writings for training and testing. But, in the third task, they used Arabic samples for training and English samples for testing. The fourth task was the opposite of the third task, where they used English samples for training and Arabic samples for testing. Their method achieved accuracy rates equal to 70% for task 1, 67% for task 2, 69% for task 3, and 63% for task 4, respectively.

In 2017, Akbari et al. [20] proposed a gender detection system which relies on a global approach that considered the writing images as textures. A probabilistic finite-state automata was used to generate the feature vectors from a series of wavelet sub-bands. They used SVM and ANN for the classification phase, and they used QUWI and MSHD to evaluate their work. Up to 80% accuracy of classification was achieved in this work.

In 2018, Abbas et al. [21] presented a comparative study to evaluate different local-binary methodologies in the classification of gender from handwritings. They proposed a system that presents the images using low- pass filtering features, and then they analyzed these features using SVM technique. The authors evaluated their method using the QUWI dataset. They achieved 72% of classification accuracy. Navya et al. [22] proposed an offline gender detection system which relies on extracting multi-gradient features from the scanned handwritten documents. They evaluated their method using IAM [23] and QUWI datasets. They achieved 75.6% of classification accuracy. Gattal et al. [15] proposed a gender detection system that uses a combination of different configurations of basic oriented image features to extract the texture features from the handwritten samples. They used an SVM to analyze the extracted features and reveal the writer's gender. Their proposed method achieved classification rates of 71.37%, 76%, and 68% for the ICDAR2013, ICDAR2015, and ICFHR2016 experimental settings, respectively. Illouz et al. [24] proposed a novel deep learning approach for an offline handwritten gender classification system. They used a convolution neural network (CNN) which extracted the features automatically from the handwriting images and classify the writer's gender. The authors created their own dataset, which consists of Hebrew and English handwriting samples of 405 participants. They evaluated their system in two categories: first, Hebrew-English classification, where they trained the system on Hebrew samples and tested it on English samples, and they got a 75.65% of classification accuracy. While in the second category, they trained the system on English samples and tested it on Hebrew samples. The accuracy rate for this category was 58.29%. Morera et al. [25] also used a convolution neural network to detect gender and handedness from offline handwritings. They created their own CNN architecture to classify the handwriting into four multi-classes, including left-handed woman, right-handed woman, left-handed male, and right-handed male. The system achieved an accuracy of 80.72% for gender classification on IAM dataset and accuracy of 68.90% on khatt [26] dataset.

In 2019, Maken et al. [27] provided a comprehensive review of various techniques used for gender detection systems. Bi et al. [28] presented a new approach for gender detection from handwriting named kernel mutual information (KMI) which focuses on using multi-features to identify the writer's gender. They extracted geometrical features including, slant orientation, curvatures, and transformed features including, Fourier features and WD-LBP features. They used SVM to classify the gender of the writer. Their system achieved an accuracy of 66.3% on ICDAR2013 dataset and accuracy of 66.7% on the registration-document-form(RDF) dataset.

Table 1. Summary of gender detection systems reviewed in literature

	Author	Features extracted	Database used	Classification rate
1.	Almadeed and Hussain [16]	Geometric features	QUWI	74.06% of gender classification accuracy
2.	Seddigi et. al. [3]	Slant, curvature, legibility	QUWI and MSHD	73.02% and 69.44% of accuracy when they applied SVM and ANN on QUWI and MSHD datasets, resepectively
3.	Maji et al. [19]	Signature features: roundness, skewness, mean, standard deviation, euler number, black pixels distribution density	They used their own database	80.7% of accuracy with euler feature and 77% of accuracy wiyhout euler feature
3.	Nogueras et al. [1]	Online features related to pen-trajectory	They collected it using a tablet and stylus	72.6% of accuracy
3.	Mirza et al. [2]	Textural features	QUWI	They got 70% of classification accuracy, when they used only Arabic samples for training and testing
4.	Akbari et al. [20]	Textural features	QUWI and MSHD	Up to 80% of classification rate
5.	Abbas et al. [21]	Low-pass filtering features	QUWI	75.6% of classification rate
6.	Navya et al. [22]	Multi-gradient features	IAM and QUWI datasets	75.6% of classification accuracy
7.	Gattal et al. [15]	Textural features	ICDAR2013, ICDAR2015 and ICFHR2016	They got a classification rates of 71.37%, 76% and 68% when they applied their method on the aforementioned datasets, respectively
8.	Illouz et al. [24]	Automatically extracted deep features	They used their own dataset	They got 75.65% of classification accuracy, when they trained and tested their method on Hebrew and English samples, respectively
9.	Morera et al. [25]	Automatic deep features	IAM and Khatt datasets	They got an accuracy rate of 80.72% and 68.90%, when they applied their method on IAM and KHATT datasets, respectively
9.	Morera et al. [25]	Geometrical features including, slant orientation, curvatures, and transformed features including, Fourier features and WD-LBP features	Registration-document-form(RDF) and ICDAR2013 dataset	Their system achieved an accuracy of 66.3% on ICDAR2013 dataset and accuracy of 66.7% on the registration-document-form(RDF) dataset

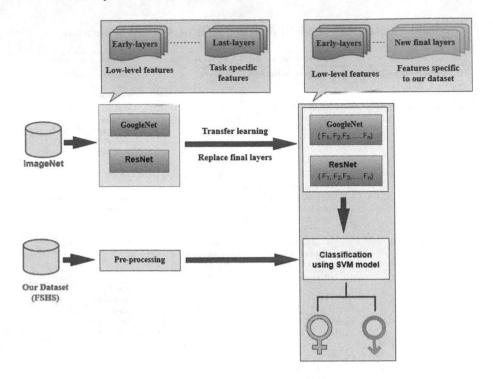

Fig. 1. General structure of the proposed method

3 Methodology

In our proposed method, we used two well-known CNN architectures GoogleNet [29] and ResNet [30] for the detection and the classification of the writer's gender from his/her handwriting. We used the extracted features from each network individually, and then we fed them to an SVM classifier. Figure 1 shows the general structure of our proposed method. The details of each step of our proposed work are given in the following subsections.

3.1 Pre-processing

The pre-processing step in our proposed approach is mainly the resizing of the images to the appropriate size related to each pre-trained network. For ResNet, we resized the images to $(223 \times 223 \times 3)$, and for GoogleNet, we resized the images to $(227 \times 227 \times 3)$.

3.2 Transfer-Learning

We applied the concept of transfer learning using two CNN architectures, as mentioned earlier as fixed feature extractors from the handwritten images. These

CNN architectures have been trained over a million images from ImageNet, and they can classify images into 1000 categories.

In our proposed work, we replace the last layers of each pre-trained network (GoogleNet and ResNet) with new fully connected layers adapted to our dataset. The new final layers can extract features that are more relevant to our target task. Then we specify the number of classes in our dataset, which are female and male. We also kept the weights of the transferred layers (early layers) of the pre-trained networks unchanged (frozen) by setting their learning rate to zero, but we increased the learning rate of the new fully connected layers to speed up training and to prevent overfitting to our dataset.

4 Experiments and Results

4.1 Dataset

As far as we know there is no well-designed and annotated public Arabic hand-written datasets, so we created our own Arabic handwritten dataset to train and test our proposed approach. We called it free-style handwritten samples (FSHS). Hence, datasets should be large enough to cover all the different hand-written styles; we asked 2200 volunteers to write a letter for someone they love. We also asked a portion of the volunteers to copy some paragraphs for future research purposes.

Our dataset in total consists of 2700 handwritten samples which were written by 2200 writers. The writer's ages range from 15 to 75 years old. Men wrote 45% of our dataset, and women wrote the rest. Most of the writer are right-handed, and they are university and school students. Part of the volunteers are employees at some private and governmental companies.

To ensure the convenience of the writers, we did not put any conditions on the type of tool used for writing. However, we provided them with a white sheet of paper to write their letter on it. Volunteers were asked to fill up an information page about their gender, age, handedness, and their work position. We then labeled the dataset concerning the age and the gender of the writer as provided in the information page.

What is significant about our dataset is its size and the number of writers. The variety in age and the large number of writers caused great diversity in the handwriting styles. Table 2 shows a comparison between our dataset and other available datasets. The dataset was mainly written in Arabic, although some writers used the English language, the number of these did not exceed 15 samples. We digitized the handwritten samples using 600 dpi scanner. Our dataset can be used in many research areas related to human interaction, such as handwriting recognition, gender, and age classification. Figure 2 shows some samples of our dataset.

Table 2. Comparison between our dataset and other datasets

	Name of the dataset	Language	Writers	Documents
1.	IFN/ENIT [31]	Arabic	411	2200
2.	AHDB [32]	Arabic	105	315
3.	AWIC2011 [33]	Arabic	54	161
4.	QUWI [17]	Arabic + English	1017	4080
5.	Our dataset(FSHS)	Arabic	2200	2700

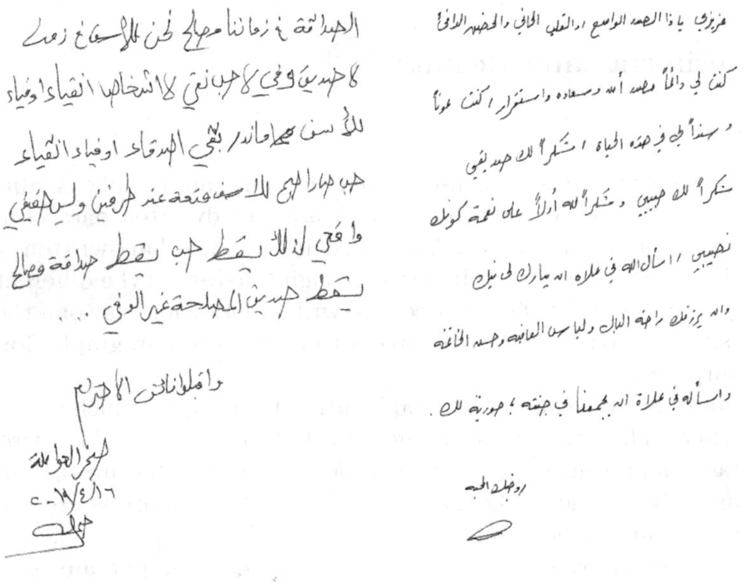

(a) Male   (b) Female

Fig. 2. Samples of the dataset (FSHS)

4.2   Image Classification Results

We evaluated our proposed approach using 75% of our dataset for training and 25% for testing. Table 3 shows the evaluation of the SVM classifier with the Gaussian kernel on the extracted features from each pre-trained network (GoogleNet, ResNet) individually in terms of accuracy and precision metrics.

Table 3. Accuracy and precision rates of the proposed method

	CNN architectures	Accuracy rates	Precision
1	ResNet	83.32%	82.85%
2	GoogleNet	80.05%	81.18%

5   Conclusions and Future Work

In this research, the concept of transfer learning was used using two CNN architectures ResNet and GoogleNet as fixed feature extractors from handwriting samples. Then SVM was used to analyze the extracted features and classify the writer's gender into female and male. We got a high performance of 83.32% of accuracy when we used the ResNet with its powerful shortcut connections. While the performance of the GoogleNet was equal to 80.05% as the use of a bottleneck layer (1×1 convolutions) helped in the reduction of the computation requirements. For future works, we will enlarge the size of our data samples and add new specific features related to the classification of male and female handwriting to improve the performance of the system.

References

1. Sesa-Nogueras, E., Faundez-Zanuy, M., Roure-Alcobé, J.: Gender classification by means of online uppercase handwriting: a text-dependent allographic approach. Cogn. Comput. **8**, 15–29 (2016)
2. Mirza, A., Moetesum, M., Siddiqi, I., Djeddi, C.: Gender classification from offline handwriting images using textural features (2016)
3. Siddiqi, I., Djeddi, C., Raza, A., Souici-meslati, L.: Automatic analysis of handwriting for gender classification. Pattern Anal. Appl. **18**(4), 887–899 (2014). https://doi.org/10.1007/s10044-014-0371-0
4. Hamid, S., Loewenthal, K.M.: Inferring gender from handwriting in Urdu and English. J. Social Psychol. **136**(6), 778–782 (1996). PMID: 9043207
5. Hartley, J.: Sex differences in handwriting: a comment on spear. Brit. Educ. Res. J. **17**(2), 141–145 (1991)
6. Szegedy, C., Ioffe, S., Vanhoucke, V.: Inception-v4, inception-resnet and the impact of residual connections on learning. In: AAAI (2016)
7. Fukushima, K., Miyake, S.: Neocognitron: a self-organizing neural network model for a mechanism of visual pattern recognition. In: Amari, S., Arbib, M.A. (eds.) Competition and Cooperation in Neural Nets. Lecture Notes in Biomathematics, vol. 45, pp. 267–285. Springer, Heidelberg (1982). https://doi.org/10.1007/978-3-642-46466-9_18
8. LeCun, Y., et al.: Backpropagation applied to handwritten zip code recognition. Neural Comput. **1**(4), 541–551 (1989)
9. Krizhevsky, A., Sutskever, I., Hinton, G.E.: Imagenet classification with deep convolutional neural networks. Commun. ACM **60**, 84–90 (2012)
10. Deng, J., Dong, W., Socher, R., Li, L.J., Li, K., Fei-Fei, L.: Imagenet: a large-scale hierarchical image database, pp. 248–255 (2009)

11. Salaken, S.M., Khosravi, A., Nguyen, T., Nahavandi, S.: Extreme learning machine based transfer learning algorithms: A survey. Neurocomputing **267**, 516–524 (2017)
12. Tan, C., Sun, F., Kong, T., Zhang, W., Yang, C., Liu, C.: A survey on deep transfer learning. In: Kůrková, V., Manolopoulos, Y., Hammer, B., Iliadis, L., Maglogiannis, I. (eds.) ICANN 2018. LNCS, vol. 11141, pp. 270–279. Springer, Cham (2018). https://doi.org/10.1007/978-3-030-01424-7_27
13. Pratt, L.Y., Mostow, J., Kamm, C.A., Kamm, A.A.: Direct transfer of learned information among neural networks. In: Proceedings of AAAI-91, pp. 584–589 (1991)
14. Huber, R., Headrick, A.: Handwriting Identification: Facts and Fundamentals. CRC Press, Boca Raton (1999)
15. Gattal, A., Djeddi, C., Siddiqi, I., Chibani, Y.: Gender classification from offline multi-script handwriting images using oriented basic image features (obifs). Exp. Syst. Appl. **99**, 155–167 (2018)
16. Al Maadeed, S., Hassaine, A.: Automatic prediction of age, gender, and nationality in offline handwriting. EURASIP J. Image Video Process. **2014**(1), 1–10 (2014). https://doi.org/10.1186/1687-5281-2014-10
17. Al-ma'adeed, S., Ayouby, W., Hassaïne, A., Jihad, M.: Quwi: an Arabic and English handwriting dataset for offline writer identification (2012)
18. Djeddi, C., Gattal, A., Souici-Meslati, L., Siddiqi, I.: Lamis-mshd: a multi-script offline handwriting database, vol. 2014 (2014)
19. Maji, P., Chatterjee, S., Chakraborty, S., Kausar, N., Dey, N., Samanta, S.: Effect of Euler number as a feature in gender recognition system from offline handwritten signature using neural networks (2015)
20. Akbari, Y., Nouri, K., Sadri, J., Djeddi, C., Siddiqi, I.: Wavelet-based gender detection on off-line handwritten documents using probabilistic finite state automata. Image Vis. Comput. **59**, 12 (2016)
21. Gudla, B., Chalamala, S.R., Jami, S.K.: Local binary patterns for gender classification. In: 2015 3rd International Conference on Artificial Intelligence, Modelling and Simulation (AIMS), pp. 19–22 (2015)
22. Navya, B., et al.: Adaptive multi-gradient kernels for handwritting based gender identification, pp. 392–397 (2018)
23. Marti, U.-V., Bunke, H.: The IAM-database: an English sentence database for offline handwriting recognition. Int. J. Doc. Anal. Recogn. **5**, 39–46 (2002)
24. Illouz, E., (Omid) David, E., Netanyahu, N.S.: Handwriting-based gender classification using end-to-end deep neural networks. In: Kůrková, V., Manolopoulos, Y., Hammer, B., Iliadis, L., Maglogiannis, I. (eds.) ICANN 2018. LNCS, vol. 11141, pp. 613–621. Springer, Cham (2018). https://doi.org/10.1007/978-3-030-01424-7_60
25. Morera, A., Sánchez, N., Vélez, J., Moreno, A.: Gender and handedness prediction from offline handwriting using convolutional neural networks. Complexity **2018**, 1–14 (2018)
26. Mahmoud, S.A., et al.: Khatt: an open Arabic offline handwritten text database. Pattern Recogn. **47**(3), 1096–1112 (2014). Handwriting Recognition and other PR Applications
27. Maken, P., Gupta, A.: A study on various techniques involved in gender prediction system: a comprehensive review. Cybern. Inf. Technol. **19**, 51–73 (2019)
28. Bi, N., Suen, C.Y., Nobile, N., Tan, J.: A multi-feature selection approach for gender identification of handwriting based on kernel mutual information. Pattern Recogn. Lett. **121**, 123–132 (2019). Graphonomics for e-citizens: e-health, e-society, e-education

29. Szegedy, C., et al.: Going deeper with convolutions. In: 2015 IEEE Conference on Computer Vision and Pattern Recognition (CVPR), pp. 1–9 (2015)
30. He, K., Zhang, X., Ren, S., Sun, J.: Deep residual learning for image recognition, pp. 770–778 (2016)
31. Pechwitz, M., Snoussi, S., Märgner, V., Ellouze, N., Amiri, H.: Ifn/enit-database of handwritten arabic words (2002)
32. Ramdan, J., Omar, K., Faidzul, M., Mady, A.: Arabic handwriting data base for text recognition. In: Procedia Technology, 4th International Conference on Electrical Engineering and Informatics, ICEEI 2013, vol. 11, pp. 580–584 (2013)
33. Hassaïne, A., Al-ma'adeed, S., Jihad, M., Jaoua, A., Bouridane, A.: The icdar2011 Arabic writer identification contest, pp. 1470–1474 (2011)

Split and Merge: Component Based Segmentation Network for Text Detection

Pan Cao⬤, Qi Wan⬤, and Linlin Shen$^{(\boxtimes)}$⬤

Computer Vision Institute, School of Computer Science and Software Engineering,
Shenzhen University, Shenzhen, China
{gaopan2017,wanqi2019}@email.szu.edu.cn,llshen@szu.edu.cn

Abstract. This paper presents a novel component-based detector to locate scene texts with arbitrary orientations, shapes and lengths. Our approach detects text by predicting four components like text region (TR), text skeleton (TS), text sub-region (TSR) and text connector (TC). TR and TS can well separate adjacent text instance. TSR are merged by TC to form a complete text instance. Experimental results show that the proposed approach outperforms state-of-the-art methods on two curved text datasets, i.e. 82.42% and 82.63% F-measures were achieved for the Total-Text and CTW1500, respectively. Our approach also achieves competitive performance on multi-oriented dataset, i.e. 85.86% f-measure for the ICDAR2015 was achieved.

Keywords: Text detection · Scene image · Arbitrary shape

1 Introduction

Text is the carrier of human beings to communicate and inherit knowledge, and appears almost in every place of daily life. Recently, due to its wide range of practical applications such as image retrieval, automatic driving, etc., scene text detection has attracted the attention of many scholars. Although the academic community has made great progress over the years, scene text detection remains a challenging task due to complex backgrounds, variations of font, size, color, shape, orientation, language and illumination condition.

With the development of deep learning, the performance of text detection has been improved a lot. The focus of the research has evolved from horizontal text to arbitrary oriented text and more challenging curved or arbitrary shaped text. Recently, the deep learning based method becomes the mainstream and can be roughly divided into two categories, regression based methods and segmentation based methods.

Regression based methods usually detect texts by adapting the general object detection frameworks to the characteristics of texts. TextBoxes [7] utilizes SSD (Single Shot Detection) [10] and try to handle the variation of aspect ratios of text instances by modifying the shape of default anchor and convolution kernels. RRPN (Rotated Region Proposal Network) [15] generate rotated anchor

© Springer Nature Switzerland AG 2020
Y. Lu et al. (Eds.): ICPRAI 2020, LNCS 12068, pp. 14–27, 2020.
https://doi.org/10.1007/978-3-030-59830-3_2

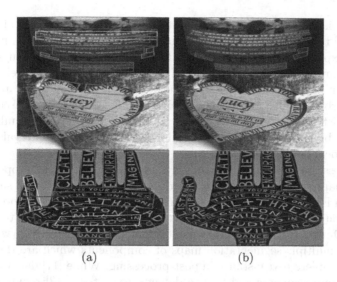

Fig. 1. Example of Curved texts [25] represented using quadrangle (a) and polygon (b).

rectangles in region proposal step to detect arbitrary oriented text in natural images. RRD (Rotation Sensitive Regression Detector) [8] classifies text line and regresses its location with rotation-invariant feature and rotation-sensitive feature respectively, which achieves significant improvement on oriented text line. EAST (Efficient and Accurate Scene Text Detector) [27] generates the segmentation map of shrunk text line and then regresses text boundaries or vertexes on each pixel in that map. NMS (Non-Maximum Suppression) is often required in post processing. In general, regression based methods perform well for horizontal and multi-orientation texts. However, as shown in Fig. 1, it is defective to represent curved text with quadrangle. For example, there are lots of background in the rectangles representing different text fields and lots of overlap among them are available. In addition, due to the limited receptive field, regression-based methods are often unable to obtain accurate boundaries when detecting long text.

Segmentation based methods cast text detection as a semantic segmentation problem and FCN[12] is often taken as the backbone framework. Zhang et al. [26] use FCN to extract text blocks and then search character candidates from these blocks with MSER (Maximally Stable Extremal Regions). Yao et al. [24] modified FCN to produce multiple heat maps corresponding to various properties of text, such as text region and orientation. Pixel-Link [2] defines border on pixel level with 8 directions and uses their connected direction to separate and group text lines. Lyu et al. [14] adopts corner detection to propose text boxes and calculates each box's confidence score on position-sensitive segmentation map. TextField [22] and MSR (Multi-scale Shape Regression network) [23] learn a direction field pointing from the text area to the nearest point on text boundary, the direction information is used to separate adjacent text instances. Wang et al. proposes PSENet [21] to generates various scales of shrunk text segmentation maps, then gradually expands kernels to generate final instance segmentation map. The approach can well handle dense curved text lines. As a pixel-level approach, segmentation based methods generally outperform regression based methods for texts with arbitrary shape. The main problem with segmentation-based methods is how to distinguish two adjacent text instances. Furthermore, multiple false boxes could be generated for extremely long text with space intervals and text with large local curvature.

To address the above problems, mainly inspired by [13,17], we propose a novel arbitrary shape scene text detector based on the merge of detected sub-regions. As shown in Fig. 2, we divide each text instance into four local components such as text region (TR), text skeleton (TS), text sub-region (TSR) and text connector (TC). During testing, a fully convolutional network (FCN)[12] was trained to generate multiple segmentation maps of components which are then merged to form a complete text instance in post-processing. While [17] demands default boxes to obtain segments and fails to detect curved text, [13] requires complex and time consuming post processing including centralizing, striding and sliding. Different from [13,17], the proposed method use multiple segmentation maps to represent local component. The post processing of our approach only requires simple intersection and union. The details of the method will be explained in Sect. 2. The main contributions of this paper are two folds:

- We propose a novel method, namely components based segmentation network, for text detection. The approach firstly detects the local components like TR, TS, TSR and TC, and then merge them into a complete text instance. While TR and TS can well separate adjacent text lines, TSR is easier to be detected and can be merged to generate a complete text instance by TC.
- The proposed method is fully tested on public datasets like IC15, Total-Text and CTW-1500, the results show that the approach can well and efficiently handle scene text with long, arbitrary shapes and orientations.

2   Method

2.1   Overview

The key idea of our method is to split the text field into four local components like TR, TS, TSR and TC, and train an FCN network to segment these component. While TR and TS represent the exact and rough areas coving the text field, TSR represent the left half and right half regions of the text field. TC denotes the central region that connects the two disjoint sub-regions. Since adjacent text instances are not easily distinguished by TR alone, TR and TS are used to distinguish text instances. Then, for each text instance, we combine the text sub-regions through the text connector to form the complete text. The proposed strategy is general and can be applied to different segmentation based text detection methods. In our experiments, PSENet [21] is chosen as the baseline.

Fig. 2. Illustration of the text region (TR), text skeleton (TS), text sub-region (TSR) and text connector (TC).

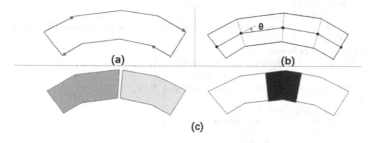

Fig. 3. Illustration of local components generation. (a) the short (in blue color) and long (in red color) boundary lines; (b) segments, and the polyline; (c) TSR and TC obtained by cutting directly from the middle. (Color figure online)

2.2 Generation of Local Components

As shown in Fig. 2, a text instance can be represented by a group of vertexes $\{v_1, v_2, ..., v_n\}$ in clockwise or counterclockwise. TR is exactly the polygon area surrounded by the vertices. TS is obtained by shrinking the original polygon using the Vatti clipping algorithm [20]. The bounding box of the text instance can then be approximated by two long boundary lines (i.e. Edge 1–5, Edge 6–10) and two short boundary lines (i.e. Edge 5–6, Edge 10–1). The two edges (e.g. Edge 4–5, Edge 6–7) near the short boundary lines (e.g. Edge 5–6) are running parallel, but in opposite direction. We firstly find two short boundary lines and two long boundary lines shown in Fig. 3(a). According to the symmetry of the text, we connect the corresponding points on the two long boundaries to divide the entire text area into segments. To measure the local curvature of the text, we connect the midpoints of the connecting lines including the two short border to form a polyline. As illustrated in Fig. 3(b), the angle between adjacent lines on the polyline represents the local curvature of the text.

The length and curvature of the text are the main factors we consider. As illustrated in Fig. 3(c), TSR is obtained by cutting directly from the middle of the text in all our experiments. TC is the region near the dividing line, and its area is about one third of the total text area. The details of local component generation is summarized in Algorithm 1.

Algorithm 1. Local Component Generation

Require: Scene image I and *Texts*. *Texts* is a list of labeled text instance in the given image. $Texts = [text_1, text_2, ..., text_m]$, m is the number of text instance in image. Each text instance is represented by a group of vertexes. $text_i = \{v_1, v_2, ..., v_{ni}\}$, n_i is the number of vertexes in $text_i$
Ensure: TR, TS, TSR, TC. Binary masks with the same shape as the image I
 1: Function Local_component_Generation(I, *Texts*):
 2: **for** *Text* in *Texts* **do**
 3: Initial TR, TS, TSR, TC with all 0
 4: // TR and TS
 5: $TR_{i,j} = 1$, **if** $pixel_{i,j} \in Text$
 6: TS is obtained using Vatti clipping Algorithm based on TR
 7: // TSR and TC
 8: Find the short boundaries B_s and the long boundaries B_l
 9: Generate *Segs* by connecting points on the two long boundaries B_l in order
10: $index_{mid} = \frac{Lenth of Segs}{2}$
11: **if** LenthofSegs is even **then**
12: $TSR1_{i,j} = 1$, **if** $pixel_{i,j} \in Segs[: index_{mid}]$
13: $TSR2_{i,j} = 1$, **if** $pixel_{i,j} \in Segs[index_{mid} :]$
14: Find the split line SL between TSR_1 and TSR_2
15: **else**
16: midseg $= Segs[index_{mid}]$
17: pt_1, pt_2, pt_3, pt_4 = four vertexes of midseg
18: $pt_{midtop} = (pt1 + pt2)/2$
19: $pt_{middown} = (pt3 + pt4)/2$
20: $half_1$ = Region in the quadrangle of
     $[pt1, pt_{midtop}, pt_{middown}, pt4]$
21: $half_2$ = Region in the quadrangle of
     $[pt_{midtop}, pt2, pt3, pt_{middown}]$
22: $TSR1_{i,j} = 1$, **if** $pixel_{i,j} \in Segs[: index_{mid}] \cup half_1$
23: $TSR2_{i,j} = 1$, **if** $pixel_{i,j} \in Segs[index_{mid} + 1 :] \cup half_2$
24: Find the split line SL between TSR_1 and TSR_2
25: **end if**
26: $TSR = TSR_1 \cup TSR_2$
27: $RegionofSL$ is the region after extending SL to 1/3 area of TR
28: $TC_{i,j} = 1$, **if** $pixel_{i,j} \in RegionofSL$
29: **end for**
30: **Return** TR, TS, TSR, TC

2.3 Local Components Segmentation

Segmentation Network. Our network structure is based on feature pyramid network (FPN) [9], which is widely used in object detection and semantic segmentation tasks. ResNet50 is adopted as the network backbone. As illustrated in Fig. 4, after feature extraction, multi-level features are fused by concatenating feature maps from low to high levels. Finally, the fused feature is fed into a 3×3 and 1×1 convolution and produces multichannel segmentation maps consisting of TR, TS, TSR and TC.

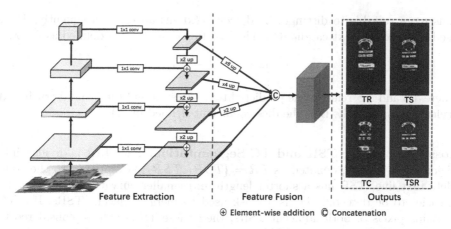

Fig. 4. Illustration of the pipeline of our framework. The left part is feature extraction based on FPN [9]. The middle part is feature fusion from low to high levels. The right part is the output of the network.

Loss Function for TR Segmentation. TR is a single channel segmentation map that classifies each pixel as text or non-text. Pixels inside a text polygon are labeled as positive samples and others are labeled as negative samples. As TS, TSR and TC are part of TR, they shall locate inside TR and those located outside of TR shall be removed. In scene images, text instances usually occupy a small area of the image. Thus, the numbers of text pixels and non-text pixels are rather imbalanced. Instead of binary cross entropy, we adopt Dice Coefficient loss in our paper. The dice coefficient $D(P, G)$ is defined as:

$$D(P, G) = \frac{2 \times |P \cap G|}{|P| + |G|} \tag{1}$$

where $|X|$ represents the number of pixels in X. There are some elements like signs, fences, bricks, and grasses virtually undistinguishable from true text. To further alleviate the class-imbalance problem and better distinguish these elements, Online Hard Example Mining (OHEM) [18] is also adopted. The loss function of TR is defined as follows:

$$L_{tr} = 1 - D(TR \cap M, \widehat{TR} \cap M) \tag{2}$$

where TR and $\widehat{TR}$ refer to the predicted text region and the corresponding ground-truth text region, respectively, and M is the training mask provided by OHEM.

Loss Function for TS Segmentation. TS is obtained after shrinking the text region. TS retains the position of the text, as well as the general shape. Moreover, the space between TS of different texts is larger than that between TR. TS is thus more suitable to distinguish adjacent text instances. Consequently, TS is used to represent text candidates. The loss function of TS is defined as follows:

$$L_{ts} = 1 - D(TS, \widehat{TS}) \tag{3}$$

where TS and $\widehat{TS}$ refer to the predicted text skeleton and the ground-truth text skeleton, respectively. D is the dice coefficient defined in (1).

Loss Function for TSR and TC Segmentation. Each text is divided into disjoint sub-regions, denoted as $TR = \{TSR_1, TSR_2\}$. Compared to the complete text line, TSR has a shorter length and smaller curvature, which makes it easier to be detected. TC is a mask used to merge adjacent TSRs. If a TC contains pixels of both adjacent TSRs, then these TSRs will be considered to belong to the same text. PSENet generates various scales of shrunk text segmentation maps $S_1, S_2, ..., S_n$, then gradually expands kernels to generate final instance segmentation map. We use the same multi-scale method to obtain TSRs and TCs. The loss function can be formulated as:

$$L_p = \lambda L_c + (1 - \lambda) L_s \tag{4}$$

$$L_c = 1 - D(P_n, G_n) \tag{5}$$

$$L_s = 1 - \frac{\sum_{i=1}^{n-1} D(P_i, G_i)}{n - 1} \tag{6}$$

where L_c and L_s represent the losses for the complete region and the shrunk ones respectively, and λ balances the importance between L_c and L_s. D is the dice coefficient, P_i and G_i refer to the predicted segmentation result and the corresponding ground-truth of S_i. The loss functions of TSR and TC are defined as follows:

$$L_{tsr} = L_p(TSR, \widehat{TSR}) \tag{7}$$

$$L_{tc} = L_p(TC, \widehat{TC}) \tag{8}$$

The Full Objective. There are four segmentation tasks in the proposed method, i.e. TR, TS, TSR and TC. The complete loss function can be formulated as:

$$L = \lambda_1 L_{tr} + \lambda_2 L_{ts} + \lambda_3 L_{tsr} + \lambda_4 L_{tc} \tag{9}$$

where L_{tr}, L_{ts}, L_{tsr} and L_{tc} are the loss function of TR, TS, TSR and TC, respectively. Parameters λ_1, λ_2, λ_3 and λ_4 are the balancing factors of the four tasks.

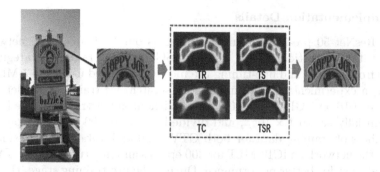

Fig. 5. Illustration of the post-processing.

2.4 Merge of Sub-regions

Given an image, our network outputs TR, TS, TSR and TC. A simple post-processing method is then used to produce the final polygon text boxes from these predictions.

The post-processing for finding the bounding box is summarized as follows. First, binary maps of TR, TS, TSR and TC are obtained by setting a threshold γ_1. The intersection of TR and TS is considered as candidate set of text instances. Subsequently, TSRs are merged by TC at the location corresponding to each text candidate instance to form a complete text box. Finally, the text boxes with the average score lower than the threshold γ_2 or with extreme small area are filtered.

3 Experimental Results

3.1 Dataset

The CTW1500 [25] dataset contains 500 test images and 1000 training images, which contain multi-oriented texts, curved texts and irregular shape texts. Text regions in this dataset are labeled with 14 boundary points at line level.

The Total-Text [1] dataset consists of 300 test images and 1255 training images, the texts in the images present more than 3 different orientations, i.e. horizontal, multi-oriented, and curved. The texts in these images are labeled at word level with adaptive number of corner points.

The ICDAR2017-MLT (IC17-MLT) [16] is a large scale multi-lingual text dataset, which includes 7200 training images, 1,800 validation images and 9,000 test images. The dataset consists of scene text images from 9 languages. All text are annotated by quadrangles at word-level.

The ICDAR2015 (IC15) [6] is a dataset proposed in the Challenge 4 of the 2015 Robust Reading Competition for incidental scene text detection. There are 1,000 images and 500 images for training and testing, respectively. The text instances are annotated by word-level quadrangles.

Figure 6 shows the example images in IC15, CTW1500 and Total-Text. CTW1500 and Total-Text mainly contain curved texts, while IC15 mainly contains oriented texts.

3.2   Implementation Details

We use ResNet-50 pre-trained on ImageNet as our backbone. The network is trained with stochastic gradient descent (SGD). Two training strategies, i.e. training from scratch and fine-tuning the model pretrained using IC17-MLT, are adopted in experiments. For training from scratch, we train our model with a batch size of 16 on 4 GPUs for 600 epochs, without any extra data. The learning rate is initially set to 1×10^{-3}, and divided by 10 at 200 and 400 epochs. As the numbers of training data for both CTW1500 and Total-Text are small, we pretrain the network on IC17-MLT for 300 epochs and fine-tuned it on CTW1500 and Total-Text for better performance. During the pre-training stage, the initial learning rate is set to 1×10^{-3}, and divided by 10 at 100 and 200 epochs. During fine-tuning, the learning rate is initially set to 1×10^{-4}, and divided by 10 at 100 epochs. Online Hard Negative Mining is applied at a ratio of 1:3. By default, λ_1 and λ_2 are set to 0.2, λ_3 and λ_4 are set to 0.3. Data augmentation is adopted to prevent overfitting. We first rescale images with a random ratio of {0.5, 1.0, 2.0, 3.0}. After that, random rotation in range $[-10°, 10°]$ is applied. Next, we randomly crop a 640×640 image from rotated images without crossing texts. In training, unclear text lines are labeled as "DO NOT CARE", and excluded from training by setting their loss weight to zero. Our method is implemented in Pytorch and we conduct all experiments on a GPU workstation with 4 NVIDIA Titan X GPUs and two Intel(R) Xeon(R) E5-2690 v4@ 2.60 GHz CPU.

3.3   Results

The proposed technique was evaluated quantitatively and qualitatively on three public dataset consisting of text instances of different types.

Texts with Arbitrary Shapes. We firstly use CTW1500 and Total-Text to evaluate the performance of our approach on scene texts with arbitrary shapes. CTW1500 dataset mainly contains curved and multi-oriented texts. For each image, the annotation is given at line level. Total-Text mostly contains curved texts labeled at word level. In testing stage, the longer side of images are scaled to 1280 and the threshold γ_1 is set to 0.68 in post-processing. Detection results are shown in Fig. 6 (b), (c). As seen in the figure, While PSENet missed words "To" and "Of", and wrongly detect the words at two lines as an instance, our approach accurately detected all text instances.

Table 1 and 2 lists the precision, recall and F-mesaure of our approach, when external data, i.e. IC17-MLT, is used or not. State of the art approaches like CTD + TLOC [25], TextSnake [13] and PSENet [21] are also included for comparison. The results in Table 1 and 2 suggest that the adoption of external data, i.e. IC17-MLT, can greatly improve the performance of text detection. Take PSENet-1s for example, the recall, precision and F-measure of fine-tuning strategy is around 4–5% higher than that of training from scratch, when CTW-1500 is considered. Similar observation can also be found for our approach. For CTW-1500 dataset, our approach achieves the highest precision (85.26%) and F-measure (82.63%), which is about 1.4% and 0.4% higher than that of runner-up, i.e. PSNet-1s. When IC17-MLT is not used for pre-training, the F-measure (80.03%) of our approach is about 2% higher than that of PSNet-1s (78.0%). For Total-Text dataset, our approach achieves the highest precision (86.84%), recall (78.43%) and F-measure (82.42%), which is about 2.8%, 0.5% and 1.6% higher than that of runner-up,

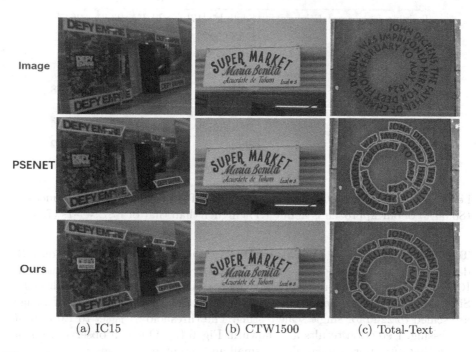

(a) IC15   (b) CTW1500   (c) Total-Text

Fig. 6. Sample images and detection results with PSENet and our approach on IC15 in (a), CTW-1500 in (b) and Total-Text in (c).

Table 1. Single-scale result on CTW-1500. "P", "R" and "F" represent the precision, recall and F-measure respectively. * indicates the results from [25].

Method		CTW-1500		
		P	R	F
Training from scratch	CTPN* [19]	60.4*	53.8*	56.9*
	Seglink* [17]	42.3*	40.0*	40.8*
	EAST* [27]	78.7*	49.1*	60.4*
	CTD+TLOC [25]	77.4	69.8	73.4
	PSENet-1s [21]	80.57	75.55	78.0
	Ours	**83.75**	**76.63**	**80.03**
Pre-trained on external dataset	TextSnak [13]	67.9	**85.3**	75.6
	PSENet-1s [21]	84.8	79.7	82.2
	Ours	**85.26**	80.15	**82.63**

Table 2. Single-scale result on Total-Text. "P", "R" and "F" represent the precision, recall and F-measure respectively. * indicates the results from [1]. Note that EAST and SegLink were not fine-tuned on Total-Text, therefore their results are included only for reference.

Method		Total-Text		
		P	R	F
Training from scratch	Seglink* [17]	30.3*	23.8*	26.7*
	EAST* [27]	50.0*	36.2*	42.0*
	DeconvNet [1]	33.0	40.0	36.0
	PSENet-1s [21]	81.77	75.11	78.3
	Ours	**85.56**	**75.26**	**80.08**
Pre-trained on external dataset	TextSnake [13]	82.7	74.5	78.4
	PSENet-1s [21]	84.02	77.96	80.87
	Ours	**86.84**	**78.43**	**82.42**

i.e. PSENet-1s. When IC17-MLT is not used for pre-training, the F-measure (80.08%) of our approach is about 1.7% higher than that of PSENet-1s (78.3%).

Texts with Arbitrary Orientations. We now validate our method on IC15 dataset to evaluate its performance to detect texts with arbitrary orientations. IC15 mainly contains oriented texts annotated with word-level quadrangles. Since there are many small texts in these images, the longer side of images are scaled to 2240 for single scale testing. The threshold γ_1 is set to 1.0 in post-processing. Detection results are shown in Fig. 6 (a). One can observe from the figure that our approach can detect small and oriented texts vey well.

Table 3 list the precision, recall and F-measure of our approach, when external data, i.e. IC17-MLT, is used or not. State of the art approaches like CTPN (Connectionist Text Proposal Network) [19], Seglink [17], EAST [27], RRD [8], TextSnake [13], PSENet [21] and FOTS (Fast Oriented Text Spotting) [11] et al. are also included for comparison. The adoption of external data can improve the performance of text detection. When IC17-MLT is not used for pre-training, our approach achieves the best precision (87.91%) and F-measure (83.74%), which is about 4.34% and 3.17% higher than that of runner-up, i.e. EAST and PSENet-1s. When external dataset is adopted, regression based method i.e. FOTS achieve the highest F-measure of 87.99%, while our approach achieve competitive F-measure of 85.86%.

As the texts in IC15 were labelled with quadrangles, regression based methods like FOTS achieve quite competitive performance. While FOTS achieve the best results when external dataset is adopted, our approach achieves the best F-measure among segmentation based methods. Although the performance of our method is not as good as FOTS, our approach can handle both curved and oriented texts.

Table 3. Single-scale result on IC15. "P", "R" and "F" represent the precision, recall and F-measure respectively.

Method			IC15		
			P	R	F
Training from scratch	Regression based methods	RRPN [15]	82.0	73.0	77.0
		EAST [27]	**83.57**	73.47	78.2
		DeepReg [4]	82.0	**80.0**	**81.0**
	Non-regression based methods	CTPN [19]	82.0	73.0	77.0
		PSENet-1s [21]	81.49	79.10	80.87
		PixelLink [2]	82.9	**81.7**	82.3
		Ours	**87.91**	79.10	**83.74**
Pre-trained on external dataset	Regression based methods	SSTD [3]	82.23	73.86	76.91
		RRD [8]	85.6	79.0	82.2
		FOTS [11]	**91.0**	**85.17**	**87.99**
	Non-regression based methods	SegLink [17]	73.1	76.8	75.0
		WordSup [5]	79.3	77.0	78.2
		Lyu et al. [14]	**94.1**	70.7	80.7
		TextSnake [13]	84.9	80.4	82.6
		PSENet-1s [21]	86.92	**84.5**	85.69
		Ours	87.74	84.06	**85.86**

4 Conclusion

In this paper, we propose a novel text detector to detect text with arbitrary shape. The proposed method achieves excellent performance on polygon labeled curved dataset (i.e. Total-Text, CTW-1500) and competitive performance on the dataset with arbitrary orientation (i.e. IC15). Even with a small amount of training data, our approach has achieved very competitive performance.

References

1. Ch'ng, C.K., Chan, C.S.: Total-text: a comprehensive dataset for scene text detection and recognition. In: 2017 14th IAPR International Conference on Document Analysis and Recognition (ICDAR), vol. 1, pp. 935–942. IEEE (2017)
2. Deng, D., Liu, H., Li, X., Cai, D.: Pixellink: detecting scene text via instance segmentation. In: Thirty-Second AAAI Conference on Artificial Intelligence (2018)
3. He, P., Huang, W., He, T., Zhu, Q., Qiao, Y., Li, X.: Single shot text detector with regional attention. In: Proceedings of the IEEE International Conference on Computer Vision, pp. 3047–3055 (2017)
4. He, W., Zhang, X.Y., Yin, F., Liu, C.L.: Deep direct regression for multi-oriented scene text detection. In: Proceedings of the IEEE International Conference on Computer Vision, pp. 745–753 (2017)

5. Hu, H., Zhang, C., Luo, Y., Wang, Y., Han, J., Ding, E.: Wordsup: exploiting word annotations for character based text detection. In: Proceedings of the IEEE International Conference on Computer Vision, pp. 4940–4949 (2017)
6. Karatzas, D., et al.: Icdar 2015 competition on robust reading. In: 2015 13th International Conference on Document Analysis and Recognition (ICDAR), pp. 1156–1160. IEEE (2015)
7. Liao, M., Shi, B., Bai, X., Wang, X., Liu, W.: Textboxes: a fast text detector with a single deep neural network. In: Thirty-First AAAI Conference on Artificial Intelligence (2017)
8. Liao, M., Zhu, Z., Shi, B., Xia, G.S., Bai, X.: Rotation-sensitive regression for oriented scene text detection. In: Proceedings of the IEEE Conference on Computer Vision and Pattern Recognition, pp. 5909–5918 (2018)
9. Lin, T.Y., Dollár, P., Girshick, R., He, K., Hariharan, B., Belongie, S.: Feature pyramid networks for object detection. In: Proceedings of the IEEE Conference on Computer Vision and Pattern Recognition, pp. 2117–2125 (2017)
10. Liu, W., et al.: SSD: single shot multibox detector. In: Leibe, B., Matas, J., Sebe, N., Welling, M. (eds.) ECCV 2016. LNCS, vol. 9905, pp. 21–37. Springer, Cham (2016). https://doi.org/10.1007/978-3-319-46448-0_2
11. Liu, X., Liang, D., Yan, S., Chen, D., Qiao, Y., Yan, J.: Fots: fast oriented text spotting with a unified network. In: Proceedings of the IEEE Conference on Computer Vision and Pattern Recognition, pp. 5676–5685 (2018)
12. Long, J., Shelhamer, E., Darrell, T.: Fully convolutional networks for semantic segmentation. In: Proceedings of the IEEE Conference on Computer Vision and Pattern Recognition, pp. 3431–3440 (2015)
13. Long, S., Ruan, J., Zhang, W., He, X., Wu, W., Yao, C.: Textsnake: a flexible representation for detecting text of arbitrary shapes. In: Proceedings of the European Conference on Computer Vision (ECCV), pp. 20–36 (2018)
14. Lyu, P., Yao, C., Wu, W., Yan, S., Bai, X.: Multi-oriented scene text detection via corner localization and region segmentation. In: Proceedings of the IEEE Conference on Computer Vision and Pattern Recognition, pp. 7553–7563 (2018)
15. Ma, J., et al.: Arbitrary-oriented scene text detection via rotation proposals. IEEE Trans. Multimedia **20**(11), 3111–3122 (2018)
16. Nayef, N., et al.: Icdar 2017 robust reading challenge on multi-lingual scene text detection and script identification-rrc-mlt. In: 2017 14th IAPR International Conference on Document Analysis and Recognition (ICDAR), vol. 1, pp. 1454–1459. IEEE (2017)
17. Shi, B., Bai, X., Belongie, S.: Detecting oriented text in natural images by linking segments. In: Proceedings of the IEEE Conference on Computer Vision and Pattern Recognition, pp. 2550–2558 (2017)
18. Shrivastava, A., Gupta, A., Girshick, R.: Training region-based object detectors with online hard example mining. In: Proceedings of the IEEE Conference on Computer Vision and Pattern Recognition, pp. 761–769 (2016)
19. Tian, Z., Huang, W., He, T., He, P., Qiao, Yu.: Detecting text in natural image with connectionist text proposal network. In: Leibe, B., Matas, J., Sebe, N., Welling, M. (eds.) ECCV 2016. LNCS, vol. 9912, pp. 56–72. Springer, Cham (2016). https://doi.org/10.1007/978-3-319-46484-8_4
20. Vatti, B.R.: A generic solution to polygon clipping. Commun. ACM **35**(7), 56–63 (1992)
21. Wang, W., et al.: Shape robust text detection with progressive scale expansion network. In: Proceedings of the IEEE Conference on Computer Vision and Pattern Recognition, pp. 9336–9345 (2019)

22. Xu, Y., Wang, Y., Zhou, W., Wang, Y., Yang, Z., Bai, X.: Textfield: learning a deep direction field for irregular scene text detection. IEEE Trans. Image Process. **28**(11), 5566–5579 (2019)
23. Xue, C., Lu, S., Zhang, W.: MSR: multi-scale shape regression for scene text detection (2019). arXiv preprint arXiv:1901.02596
24. Yao, C., Bai, X., Sang, N., Zhou, X., Zhou, S., Cao, Z.: Scene text detection via holistic, multi-channel prediction (2016). arXiv preprint arXiv:1606.09002
25. Yuliang, L., Lianwen, J., Shuaitao, Z., Sheng, Z.: Detecting curve text in the wild: New dataset and new solution (2017). arXiv preprint arXiv:1712.02170
26. Zhang, Z., Zhang, C., Shen, W., Yao, C., Liu, W., Bai, X.: Multi-oriented text detection with fully convolutional networks. In: Proceedings of the IEEE Conference on Computer Vision and Pattern Recognition, pp. 4159–4167 (2016)
27. Zhou, X., et al.: East: an efficient and accurate scene text detector. In: Proceedings of the IEEE Conference on Computer Vision and Pattern Recognition, pp. 5551–5560 (2017)

Gate-Fusion Transformer for Multimodal Sentiment Analysis

Long-Fei Xie[1,2(✉)] and Xu-Yao Zhang[1,2]

[1] National Laboratory of Pattern Recognition, Institute of Automation of Chinese Academy of Sciences, Beijing, China
xielongfei2017@ia.ac.cn, xyz@nlpr.ia.ac.cn
[2] University of Chinese Academy of Sciences, Beijing, People's Republic of China

Abstract. Computational analysis of human multimodal sentiment is an emerging research area. Fusing semantic, visual and acoustic modalities requires exploring the inter-modal and intra-modal interactions. The first challenge for the inter-modal understanding is to break the heterogeneous gap between different modalities. Meanwhile, when modeling the intra-modal connection of time-series data, we must deal with the long-range dependencies among multiple steps. Moreover, The time-series data usually is unaligned between different modalities because individually specialized processing approaches or sampling frequencies. In this paper, we propose a method based on the transformer and the gate mechanism - the Gate-Fusion Transformer - to address these problems. We conducted detailed experiments for verifying the effectiveness of our proposed method. Because of the flexibility of gate-mechanism for information flow controlling and the great modeling power of the transformer for modeling the inter- and intra-modal interactions, we can achieve superior performance compared with the current state-of-the-art method but more extendible and flexible by stacking multiple gate-fusion blocks.

Keywords: Multimodal learning · Transformer · Sentiment analysis

1 Introduction

Human sentiment analysis is an important research area for understanding the contents in modern media, such as the computational treatment of opinions, sentiment, and subjectivity. In common, human sentiment can be categorized as positive, natural and negative. In this paper, we regard sentiment analysis as a binary classification task [4] to determine the polarity of the input data.

Recently, human affection analysis of unimodality has made progress on the benefit of deep learning [1]. A significant difficulty for computationally modeling the sentiment in a video is the inherent multimodality i.e. usually comprising visual images, acoustic signals and textual subtitles. However, the blooming advances for deep unimodal sentiment analysis provide us a great opportunity to explore multimodal sentiment analysis. The core issue for the unimodal setting is

© Springer Nature Switzerland AG 2020
Y. Lu et al. (Eds.): ICPRAI 2020, LNCS 12068, pp. 28–40, 2020.
https://doi.org/10.1007/978-3-030-59830-3_3

to extract semantic representations richly of the raw data. As speech is inherently time-variant, plenty of works on natural language sentiment analysis focus on modeling the long-range dependencies on different words [14, 20, 35].

The facial expression and gesture movements are sufficiently distinguishable to determine the affection of the person who we face to. Both of them are widely taken as the representation of sentiment in the computer vision community. Recently, due to the development of facial and gesture detection and tracking in the image or video [28, 29] with deep learning methods, we can get more valuable information from raw data. Furthermore, MFCCs [11, 18] is general for acoustic representation in the audio affection recognition scenario.

However, fusing multiple modalities to analyze the sentiment is more difficult than a unimodal scene because of the heterogeneous gaps [25] between heterogeneous modalities. Therefore, the most worthwhile exploration is to modeling the *inter-modal* and *intral-modal* interactions. Thus, the temporal dependencies should be considered more elaborately. Considering the interaction are not limited in a single modality, which is a multi-fold cross-modal reciprocity. Early work about multimodal fusion usually adopted plain concatenation [21], trying to build the joint representation of multiple modalities. Recently, methods for multimodal fusion on feature level typically use more elaborate models [20, 22] or introduce more empirical constraints [23]. However, these methods did not consider the unalignment issue and non-local combinations of multimodalities.

The data collected from different sources are usually unaligned on account of different ways or sampling rates of different modalities [26]. For dealing with this issue, a direct way is to pad the sequences to the same length using zero values or other placeholders[1]. However, these methods introduce a lot of redundant messages. More elaborately, CTC [8] or DTW [3] can be employed to align different sequence to equal length sharing the similar core thoughts as padding methods. Recent methods are asking for alignment by manually forcing the raw input data [19, 31, 32, 34]. Yet the extra auxiliary information is not usually available in real world.

In order to process these problems, we propose a Gate-Fusion Transformer to directly model the dynamic interaction of multimodal time-series data. Also, we perform detailed experiments to evaluate the performance of our proposed method. The results show that we can achieve superior performance compared with the current state-of-the-art method [26] on CMU-MOSI [33] and CMU-MOSEI [34]. Moreover, our proposed method has a better extendibility and flexibility. By stacking the Gate-Fusion Block, we can extract more abstract representations. We could control the information flows more flexibly than MulT thanks to the introduction of the Gate Mechanism.

---

[1] https://pytorch.org/docs/stable/_modules/torch/nn/utils/rnn.html#pad_sequence.

2   Related Work

2.1   Multimodal Pattern Recognition

Following the taxonomy of Baltrušaitis et al. [2], the tasks of multimodal machine learning could be grouped into representation, fusion, alignment, translation, and co-learning. And the most concerned parts are the representation as well as fusion for multimodal pattern recognition. In the recognition scenario, we want a comprehensive representation to describe the raw inputs and an effective fusion to integrate the representations. First, the multimodal raw data are usually with an extreme difference, for example, image - a matrix and audio - a sequence, we can not process them using a unified model. Hence, the majority of multimodal pattern recognition is devoted to narrowing the gap between different modalities. Some work tried to get a shared representation of distinct modalities by forcing them cannot be distinguishable [21,23]. Different from these methods implicitly making the hidden representations similar through loss functions, another kind of approach using the explicit network to map raw inputs into a common space. For instance, Zadeh et al. proposed a memory fusion LSTM [32] to fuse the memories at time $t - 1$ and t with attention-mechanism.

When fusing the hidden representations from different modalities, the most direct method is concatenation [21,32]. But, in [22], they showed that element-wise multiplication is slightly better than concatenation. The fusing procedure and representation are usually undistinguishable for deep-learning methods. However, the gap between these two stages for some traditional methods, such as Multiple Kernel Learning (MKL) [7] or Graphical Models (PGM) [15] are clear. Generally, we can assign a specific kernel for a modality to measure the similarity between different points. Then combine these modal-specific kernels with linear or nonlinear combinations [12,17]. For graphical models, these methods focus on modeling the joint [9] or conditional distribution [13] of different modalities.

2.2   Transformer

The original transformer [27] are proposed for cross-linguistic translation task with the encoder-decoder framework. But for our task, we only introduce the transformer encoder. A transformer encoder is composed of self-attention, residual connection as well as layer normalization and a position-wise feed-forward network. The components of the decoder are similar. An important factor they considered for the translation task is the absolute and relative word position, the original transformer using sin and cos positional encoding. Because the inputs are computing parallelly with no positional information contained as RNN. Recently, the pre-training transformer network - Bert [26] - has achieved great success in various NLP tasks. They use another strategy to encode the positional information by directly learning the positional embedding from the raw inputs. Shaw et al. [24] introduced relative embedding and achieved a better performance, which differs from the absolute embedding employed by the original transformer. And recent transformer-XL [6], they went further, adding global content bias upon relative embedding.

2.3   Gate-Mechanism

In the initial stage for RNN, the simple recursive cell had no capability to store the information far before now. Because the gradients would vanish or explode with the passage of time for Backpropagation Through Time (BPTT). This is generally called the long-range problem. For dealing with this problem, Hochreiter et al. [10] introduced Gated RNN, called Long Short-Term Memory (LSTM) to store the previous states. In LSTM, they used 3 different gates, namely: input gate i_t, forget gate f_t as well as output gate o_t to control the information flows separately.

$$i_t = \sigma \left(W_i x_t + U_i h_{t-1} + b_i \right)$$
$$f_t = \sigma \left(W_f x_t + U_f h_{t-1} + b_f \right)$$
$$o_t = \sigma \left(W_o x_t + U_o h_{t-1} + b_o \right)$$

Afterward, Cho et al. proposed a Gated Recurrent Unit (GRU) [5] with less gate, which can achieve a similar effect as LSTM. Only containing an update gate z_t and a rest gate r_t.

$$z_t = \sigma \left(W_z x_t + U_z h_{t-1} + b_z \right)$$
$$r_t = \sigma \left(W_r x_t + U_r h_{t-1} + b_r \right)$$

3   The Proposed Method

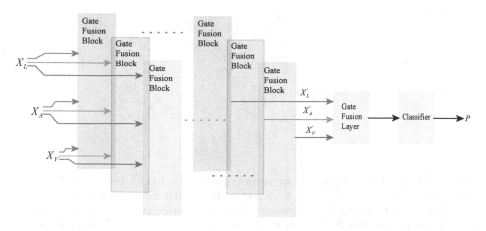

Fig. 1. The overview of our proposed Gate-Fusion Transformer. Given the inputs X_L, X_A, X_V from linguistic, acoustic and visual modality, first they pass several individual Gate-Fusion Block to explore the interactions between different modalities or time steps. Then, with the average pooling and gate-fusion layer, we gain the final representations of all these modalities. Finally, a full-connected neural network is introduced to play the role of a classifier.

For comprehensively utilizing the interactions of different modalities and time steps in the multimodal scene, we proposed the *Gate-Fusion Transformer* (Fig.1) to fuse the information dynamically. Our transformer is based on several stacked *Gate-Fusion Block* which is composed of multiple transformers for each modality. Concretely, a *Self-Transformer* for intra-modal interaction building as a *Cross-Transformer* for inter-modal. And a gate to control the fusion process dynamically.

3.1   Gate-Fusion Block

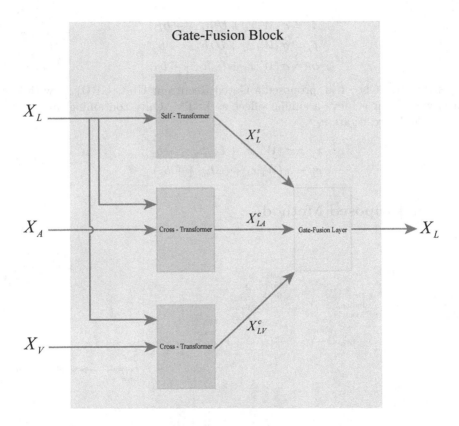

Fig. 2. The detailed structure of our propose gate-fusion block. Taking the language channel as example. X_L, X_A and X_V denotes the corresponding *linguistic, audio* and *visual* inputs for modeling the interactions between language and the other two modalities. Then, passing the *Self-Transformer* and *cross-Transformer*, the corresponding outputs X_L^s, X_{LA}^c and X_{LV}^c model the interactions among different time-steps and modalities. Finally, they are fused in *Gate-Fusion Layer* to obtain the feature outputs X_L for language channel. The other two channels for audio and vision are similar.

A single frame (time steps) is connected to several frames from the same modality and some frames from other modalities. In our setting, we do not have the exact alignment between different modalities because of different sampling frequency or data collecting methods. So we use transformer [27] which is independent of the sequence length to handle the variant length problem (Fig. 2).

The information from different modalities can be divided into complementary and supplymentary [2]. In order to utilize these two kinds of information comprehensively, we proposed a gate mechanism to dynamically control the information fusion process according to the contribution of the different modalities to final classification.

In Sect. 3.1, we first describe the Self-Transformer. Then, in Sect. 3.1, we depict the Cross-Transformer and discuss its relationship with Self-Transformer. Finally, we present the gate-fusion mechanism in Sect. 3.1 for fusing the outputs of Self-Transformer and Cross-Transformer.

For a single video fragment, we can hear the voice ($X_A \in \mathbb{R}^{T_A \times d_A}$) and see the images ($X_V \in \mathbb{R}^{T_V \times d_V}$) as well as the subtitles ($X_L \in \mathbb{R}^{T_L \times d_L}$). We denote our input feature sequence from language, vision, and audio as $\{X_L, X_V, X_A\} \subset X$. We will describe the process of gate-fusion with these notations in subsections.

Self-Transformer. In the pre-processing phase, we use a linear funtion f project all these three inputs to the same dimension d, so the matrix multiplication could proceed for Self-Attention.

$$\hat{X}_i = f(X_i) = X_i W_i, W_i \in \mathbb{R}^{d_i \times d}, i \in \{L, V, A\}. \tag{1}$$

Then following the procedure of the original neural machine translation Transformer [27], we take $\hat{X}_i$ passing through a self-attention block and a position-wise feed-forward neural network with residual connection and layer normalization. A little difference here - we exclude position embedding because for a classification task, the position information will be ignored when putting the final features to the classifier. That is to say, all sequences integrate to 1 feature vector.

The following equations are taking the language channel as example:

$$X_L^1 = \text{softmax}\left(\frac{Q_L K_L^T}{\sqrt{d}}\right) V_L$$

$$= \text{softmax}\left(\frac{\hat{X}_L W_{q_L} W_{k_L}^T \hat{X}_L^T}{\sqrt{d}}\right) \hat{X}_L W_{v_L}, \tag{2}$$

$$X_L^s = \text{FFN}\left(\text{LN}(X_L^1 + \hat{X}_L)\right).$$

where LN means Layer Normalization and FFN is a position-wise feed-forward network. W_{q_L}, W_{k_L} and W_{v_L} are 3 $\mathbb{R}^{d \times d}$ learnable parameter matrixes. Here, we define the *querys* as $Q_L = \hat{X}_L W_{q_L}$, *keys* as $K_L = \hat{X}_L W_{k_L}$ and *values* as $V_L = \hat{X}_L W_{q_L}$. Then a self-attention as well as a FFN is applied to produce the outputs of self-transformer.

Cross-Transformer. In Self-Transformer, the queries, keys, and values are from the same modality. On the contrary, for building the combinations between two different modalities, the information is from 2 different sources. We take the queries from one modality. But the keys, and values are from another modality. Foe example, towards fusing the interactions between language and vision, we use X_L to construct the queries, X_V to construct keys and values, which is named cross-attention in [22,26].

$$
\begin{aligned}
X_L^2 &= \text{softmax}\left(\frac{Q_L K_V^T}{\sqrt{d}}\right) V_V \\
&= \text{softmax}\left(\frac{\hat{X}_L W_{q_L} W_{k_V}^T \hat{X}_V^T}{\sqrt{d}}\right) \hat{X}_V W_{v_V}, \\
X_{LV}^c &= \text{FFN}\left(\text{LN}\left(X_L^2 + \hat{X}_L\right)\right).
\end{aligned}
\tag{3}
$$

As shown in the formula above, except the construction of queries, keys and values, the other processes are same with the Self-Transformer to obtain X_{LV}^c. Another fusing procedure is similar for exploring the interactions between language and audio, resulting in X_{LA}^c.

Gate-Fusion. After the aforementioned two kinds of transformers, we acquire 3 $T_L \times d$ feature sequences for language modality, containing the interactions between language and the other two modalities. Next, we intend to fuse these three information flows as the final representation of language modality.

Inspired by the success of LSTM [10] for fusing information from hidden state h_{t-1} and current input x_t. We introduce a gate mechanism (Eq. 4) to control the contribution of different modalities in the fusing process.

$$
gate_i = \text{sigmoid}(X_L^s W_L^s + X_{LV}^c W_{LV}^c + X_{LA}^c W_{LA}^c).
\tag{4}
$$

where i belongs to $\{s, c_{LV}, c_{LA}\}$ meaning the corresponding gate for X_L^s, X_{LV}^c and X_{LA}^c.

4   Experiments

In this section, we conduct experiments and evaluate the Gate-Fusion Transformer on CMU-MOSI [33] and CMU-MOSEI [34] datasets, which are frequently used for human multimodal affective recognition [2,16,26,30–32]. First, we perform the ablation experiments on CMU-MOSI for determining the influence of distinct components. And then we conduct a contrast experiment on CMU-MOSI and CMU-MOSEI separately under the aligned as well as unaligned settings (Table 1).

Table 1. Comparison of CMU-MOSI and CMU-MOSEI.

Factors	Datasets	
	CMU-MOSI	CMU-MOSEI
Total number of segments	2,199	23,453
Total number of videos	93	3288
Total number of speakers	89	1000
Males/Females	48/41	57/43
Sampling rate Vision	12.5 Hz	15 Hz
Audio	15 Hz	20 Hz

4.1 Datasets

CMU-MOSI. CMU-MOSI [33] is a benchmark dataset for human multimodal sentiment analysis, consisting of 2199 opinion video clips with the annotation in the range $[-3, 3]$. Videos of MOSI were collected from YouTube vlogs, and the subjects were always 1 speaker. Total of 93 videos with 89 different speakers are included in this dataset. Which is approximately gender-balanced - containing 41 females and 48 males. Furthermore, the speakers were all in English when they recorded the vlogs. For comparison, we use the same sampling rate in [26], 12.5 and 15 Hz for vision and audio, respectively.

CMU-MOSEI. CMU-MOSEI [34] is a larger dataset than CMU-MOSI, including more than 23,453 video clips from over 1000 online YouTube speakers. Where the annotation is the same as CMU-MOSI, from -3 (strongly negative) to 3 (strongly positive). The total hours of CMU-MOSEI are more than 65 hours. And the topics contained are very wide, including 250 frequently used topics. Also, it is a gender-balanced dataset, 57% male vs 43% female. For the MOSEI case, the sampling rates 20 Hz for audio 15 Hz for vision.

4.2 Ablation Study

There are three major parts to construct a Gate-Fusion Transformer - a Self-Transformer for modeling the intra-modal combinations among individual time steps, Cross-Transformers used to explore the inter-modal interactions as well as a gate to filter the information flows. To verify the influence of each component, we perform extensive ablation studies on CMU-MOSI. The results are shown in Table 2.

First, we consider the influence of the number of stacked gate-fusion blocks. It can be seen from Table 2 that the performance can improve with blocks adding.

Next, we perform a detailed exploration to investigate the influence of different methods to compute the gate weights. First, with an intuitive consideration that in feature space, the contributions of negative coordinate and positive

Table 2. Ablation studies of our proposed Gate-Fusion Transformer on CMU-MOSI dataset.

Component	Setting	Accuracy
Num of stacked blocks	1	0.757
	2	0.764
	4	0.770
	8	**0.782**
Fusion type	Absolute ratio	0.770
	Time fusion	0.774
	Linear fusion	0.777
	Concat	0.782
	Unimodal sigmoid	0.785
	Addition	0.787
	Feature fusion	**0.792**
Position embedding	Pos-Embedding	0.757
	Avg pooling	**0.792**
Modalities	Language only	0.732
	Audio only	0.509
	Video only	0.523
	L+A	0.744
	L+V	0.750
	A+V	0.525
	L+A+V	**0.792**
d_model	32	0.780
	64	**0.792**
	128	0.765
	256	0.766
Nums of heads	1	0.753
	2	0.755
	4	**0.762**
	8	0.759

should be comparable, we assigned the gate-score as the ratio of separate absolute value to their summation. Different from our proposed methods to combine the $\mathbb{R}^{T_L \times d}$ representations on the feature dimension, time fusion conducts linear transformation on the time dimension. Because of no sequential information containing (without position embedding), the performance is not fairly mediocre. Obviously, direct concatenation and addition are the simplest methods to fuse multiple values to one. They achieve a kind of nice performance. A linear layer is employed on the linear fusion setting to directly add 3 feature sequences together. Another setting is unimodal sigmoid when computing the gate score, we just consider the impact of the single sequence. It achieves a fairly good performance. The last one is our proposed gate-fusion methods, outperforming all other settings.

Then, we evaluate the influence of position embedding. Following [27] we encode the positional information of a sequence via sin and cos functions. Then we select the last hidden states as the representation of the whole sequence. We take the average value of each step in the sequence as representation. Beyond our expectation, a huge performance increase is achieved for average pooling.

After that, we compare the performance among combining different modalities. It is evident that the language modality carries more information about the other modalities. While the other 2 modalities carry lesser messages. As the number of modalities increases, we get more information to depict the affection and a higher accuracy achieved. This phenomenon has been proven by much work on multimodal learning [2].

Finally, following the original transformer [27], we investigate the influence of the embedding dimension (d_model) and the number of heads. Both are important factors in impacting the performance of the seq2seq translation task using the transformer encoder and decoder. Results show that 64 dimensions obtain the best result. In addition, with 4 attention heads, we achieve better performance. Here, the influence of the embedding dimension, as well as the number of the attention heads, is not clear as the original transformer in [27].

4.3 Comparison with State-of-The-Art Methods

Table 3. Comparison with previous state-of-the-art methods on CMU-MOSI and CMU-MOSEI. Where "NaN" means no experiments on the corresponding dataset.

Models	Accuracy	
	CMU-MOSI	CMU-MOSEI
EF-LSTM	0.752	0.782
LF-LSTM	0.767	0.806
TFN [31]	0.771	NaN
MFN [32]	0.774	0.760
Graph-MFN [34]	NaN	0.769
RMFN [16]	0.784	NaN
Ours (aligned)	0.827	0.825
MulT [26]	0.811	0.816
Ours (unaligned)	0.811	0.818

The performance of other previous methods compared with our proposed Gate-Fusion Transformer are shown in Table 3. First, in order to build a baseline for comparasion, we take early-fusing LSTM (EF-LSTM) as well as late-fusing LSTM (LF-LSTM) as our benchmark methods, which are simple concatenation and max-voting separately. Beyond our expectation, the LF-LSTM achieve a fairly high accuracy on CMU-MOSEI, even better than most exquisite model. We argue that this result benefits from the abundant linguistic information of CMU-MOSEI.

The next several methods are Tensor Fusion Network (TFN) [31], Memory Fusion Network (MFN) [32], Graph-Memory Fusion Network (Graph-MFN) [34] and Recurrent Multistage Fusion Network (RMFN) [16]. Both of them are evaluated on word aligned features because their inputs require to be equal-length without the speciality of dealing with unequal sequences. TFN [31] and Graph-TFN [34] (a generalized version of TFN) use the tensor product to fuse the feature sequences. In MFN [32], Zadeh et al. employed 3 LSTM channels separately for extract the hidden representation of individual modality. Then they used an attention-based delta-memory block to fuse every channel. As for RMFN [16], they decompose the fusion into multiple stages to fuse the interactions of temporal multimodal sequences. Due to the power of transformer as well as gate-mechanism, our method outperforms all of those aligned models on CMU-MOSI or CMU-MOSEI in an unaligned manner.

Different from these, multimodal Transformer (MulT) [26] is the most recent state-of-the-art method that can handle unaligned sequences well. Our proposed gate-fusion can achieve comparable performance as it on a small dataset and slightly higher on larger datasets thanks to more input data.

5   Conclusion

In this paper, we present a Gate-Fusion Transformer for human multimodal sentiment analysis. With the great power of transformer [27] for modeling the interaction and the flexibility of the gate mechanism for controlling information flow, we can handle the sequential features with any length and many modalities. Via Gate-Fusion Transformer, we can capture the intra-modal and inter-modal interactions between different time steps through our proposed Gate-Fusion Transformer. In addition, we conducted extensive experiments on different gate mechanisms. The results show that we could achieve a good performance using LSTM like feature-dimensional linear combinations. In the future, we plan to explore the factorization of self-attention for reducing computational complexity. We hope that a well capacity for modeling multimodal interactions with lower consumption of computing power could be achieved using a decomposed transformer.

References

1. Ain, Q.T., et al.: Sentiment analysis using deep learning techniques: a review. Int. J. Adv. Comput. Sci. Appl. 8(6), 424 (2017)
2. Baltrušaitis, T., Ahuja, C., Morency, L.: Multimodal machine learning: a survey and taxonomy. IEEE Trans. Pattern Anal. Mach. Intell. 41(2), 423–443 (2019)

3. Berndt, D.J., Clifford, J.: Using dynamic time warping to find patterns in time series. In: KDD Workshop, Seattle, WA, vol. 10, pp. 359–370 (1994)
4. Cambria, E.: Affective computing and sentiment analysis. IEEE Intell. Syst. **31**(2), 102–107 (2016)
5. Cho, K.,et al.: Learning phrase representations using RNN encoder-decoder for statistical machine translation. In: Proceedings of the 2014 Conference on Empirical Methods in Natural Language Processing (EMNLP), Doha, Qatar, October 2014, pp. 1724–1734. Association for Computational Linguistics (2014)
6. Dai, Z., et al.: Transformer-xl: attentive language models beyond a fixed-length context (2019). arXiv preprint arXiv:1901.02860
7. Gonen, M., Alpaydin, E.: Multiple kernel learning algorithms. J. Mach. Learn. Res. **12**(Jul), 2211–2268 (2011)
8. Graves, A., Fernández, S., Gomez, F., Schmidhuber, J.: Connectionist temporal classification: labelling unsegmented sequence data with recurrent neural networks. In Proceedings of the 23rd International Conference on Machine Learning, pp. 369–376. ACM (2006)
9. Gurban, M., Thiran, J.P., Drugman, T., Dutoit, T.: Dynamic modality weighting for multi-stream hmms inaudio-visual speech recognition. In: Proceedings of the 10th International Conference on Multimodal Interfaces, pp. 237–240. ACM (2008)
10. Hochreiter, S., Schmidhuber, J.: Long short-term memory. Neural Comput. **9**(8), 1735–1780 (1997)
11. Ittichaicharoen, C., Suksri, S., Yingthawornsuk, T.: Speech recognition using MFCC. In: International Conference on Computer Graphics, Simulation and Modeling, pp. 135–138 (2012)
12. Natasha, J., Taylor, S., Sano, A., Picard, R.: Multi-task, multi-kernel learning for estimating individual wellbeing. In: Proceedings NIPS Workshop on Multimodal Machine Learning, Montreal, Quebec, vol. 898, p. 63 (2015)
13. Jiang, X.Y., Wu, F., Zhang, Y., Tang, S.L., Lu, W.M., Zhuang, Y.T.: The classification of multi-modal data with hidden conditional random field. Pattern Recogn. Lett. **51**, 63–69 (2015)
14. Kim, Y.: Convolutional neural networks for sentence classification. In: Proceedings of the 2014 Conference on Empirical Methods in Natural Language Processing, Doha, Qatar, pp. 1746–1751. Association for Computational Linguistics (2014)
15. Koller, D., Friedman, N.: Probabilistic Graphical Models: Principles and Techniques. MIT press, Cambridge (2009)
16. Liang, P.P., Liu, Z., Zadeh, A.A.B., Morency, L.P.: Multimodal language analysis with recurrent multistage fusion. In: Proceedings of the 2018 Conference on Empirical Methods in Natural Language Processing, Brussels, Belgium, October-November 2018, pp. 150–161. Association for Computational Linguistics (2018)
17. Gwen, L., Sikka, K., Bartlett, M.S., Dykstra, K., Sathyanarayana, S.: Multiple kernel learning for emotion recognition in the wild. In: Proceedings of the 15th ACM International Conference on Multimodal Interaction, pp. 517–524 (2013)
18. Logan, B., et al.: Mel frequency cepstral coefficients for music modeling. ISMIR **270**, 1–11 (2000)
19. Tomáš, M., Karafiát, M., Burget, L., Černocky, J., Khudanpur, S.: Recurrent neural network based language model. In: Eleventh Annual Conference of the International Speech Communication Association (2010)
20. Manish, M., Shakya, S., Shrestha, A.: Fine-grained sentiment classification using bert (2019). arXiv preprint arXiv:1910.03474

21. Ngiam, J., Khosla, A., Kim, M., Nam, J., Lee, H., Ng, A.Y.: Multimodal deep learning. In: Proceedings of the 28th International Conference on Machine Learning, pp. 689–696 (2011)
22. Peng, G., et al.: Dynamic fusion with intra- and inter- modality attention flow for visual question answering. In: The IEEE Conference on Computer Vision and Pattern Recognition (2019)
23. Peng, Y., Qi, J.: CM-GANS: cross-modal generative adversarial networks for common representation learning. ACM Trans. Multimedia Comput. Commun. Appl. **15**(1), 22 (2019)
24. Shaw, P., Uszkoreit, J., Vaswani, A.: Self-attention with relative position representations (2018). arXiv preprint arXiv:1803.02155
25. Soleymani, M., Garcia, D., Jou, B., Schuller, B., Chang, S.F., Pantic, M.: A survey of multimodal sentiment analysis. Image Vis. Comput. **65**, 3–14 (2017). Multimodal Sentiment Analysis and Mining in the Wild Image and Vision Computing
26. Tsai, Y.H.H., Bai, S., Liang, P.P., Zico Kolter,J., Morency, L.P., Salakhutdinov, R.: Multimodal transformer for unaligned multimodal language sequences. In: Proceedings of the 57th Annual Meeting of the Association for Computational Linguistics, Florence, Italy, July 2019, pp. 6558–6569. Association for Computational Linguistics (2019)
27. Vaswani, A., et al.: Attention is all you need. In: Advances in Neural Information Processing Systems, pp. 5998–6008 (2017)
28. Wang, J., Fu, J., Xu, Y., Mei, T.: Beyond object recognition: Visual sentiment analysis with deep coupled adjective and noun neural networks. In: IJCAI, pp. 3484–3490 (2016)
29. Wang, N., Gao, X., Tao, D., Yang, H., Li, X.: Facial feature point detection: a comprehensive survey. Neurocomputing **275**, 50–65 (2018)
30. Wang, Y., Shen, Y., Liu, Z., Liang, P.P., Zadeh, A., Morency, L.P.: Dynamically adjusting word representations using nonverbal behaviors. Words can shift. In: Proceedings of the AAAI Conference on Artificial Intelligence, vol. 33, pp. 7216–7223 (2019)
31. Zadeh, A., Chen, M., Poria, S., Cambria, E., Morency, L.P.: Tensor fusion network for multimodal sentiment analysis. In Proceedings of the 2017 Conference on Empirical Methods in Natural Language Processing, Copenhagen, Denmark, September 2017, pp. 1103–1114. Association for Computational Linguistics (2017)
32. Zadeh, A., Liang, P.P., Mazumder, N., Poria, S., Cambria, E., Morency, L.P.: Memory fusion network for multi-view sequential learning. In: Thirty-Second AAAI Conference on Artificial Intelligence (2018)
33. Zadeh, A., Zellers, R., Pincus, E., Morency, L.P.: Mosi: multimodal corpus of sentiment intensity and subjectivity analysis in online opinion videos (2016). arXiv preprint arXiv:1606.06259
34. Zadeh, A.A.B., Liang, P.P., Poria, S., Cambria, E., Morency, L.P.: Multimodal language analysis in the wild: Cmu-mosei dataset and interpretable dynamic fusion graph. In: Proceedings of the 56th Annual Meeting of the Association for Computational Linguistics (Volume 1: Long Papers), pp. 2236–2246 (2018)
35. Zhou, C., Sun, C., Liu, Z., Lau, F.: A c-lstm neural network for text classification (2015). arXiv preprint arXiv:1511.08630

Overview of Mathematical Expression Recognition

Jiashu Huang[✉] ⓘ, Jun Tan ⓘ, and Ning Bi ⓘ

School of Mathematics, Sun Yat-Sen University, Guangzhou, China
hjiashu@mail2.sysu.edu.cn, {mcstj,mcsbn}@mail.sysu.edu.cn

Abstract. Mathematical expression recognition has been one of the most fascinating research among the various researches in field of image processing. This problem typically consists of three major stages, namely, expression positioning, symbol segmentation, symbol recognition, and structural analysis. In this paper, we will review most of the existing work with respect to each of the major stages of the recognition process. Moreover, some important issues in mathematical expression recognition, like handwritten MEs, will be addressed in depth. Finally, point out the future research directions of mathematical expression.

Keywords: Mathematical expression · Symbol recognition · Structural analysis · Handwritten

1 Introduction

Mathematical expression (ME) plays an important role in scientific and technical documents. As an intuitive and easily comprehensible knowledge representation model, MEs are present in various types of literature and could help the dissemination of knowledge in many related domains, including mathematics, physics, economy and many other fields. Mathematical expression recognition is the process of converting scanned images or online handwritten form of mathematical expression into editable text and other forms through related technologies such as image processing, image recognition, semantic analysis, and structural analysis.

With the electronic process of document and the development of online education, the need to identify MEs in pictures or documents, which are based on typesetting systems such as LaTeX and MS Word, has increased recently. It is important to reduce time in converting image-based documents like PDF to text-based documents that are easy to use and edit, but the recognition of MEs is far more difficult than the recognition of traditional text. Since the two-dimensional structure of the mathematical expression is significantly different from the traditional natural language, it is difficult for traditional text OCR technology to be directly applied to mathematical expression recognition.

Meanwhile, benefit from the development and popularization of smart terminal devices based on optical scanning and pen writing input, online handwritten data has become a more common important data type. As one of the branches, people are allowed to use handwritten mathematical expressions (HMEs) as the input data conveniently and

© Springer Nature Switzerland AG 2020
Y. Lu et al. (Eds.): ICPRAI 2020, LNCS 12068, pp. 41–54, 2020.
https://doi.org/10.1007/978-3-030-59830-3_4

naturally, which puts the problem, how to recognize mathematical handwritten expressions, in people perspective. Therefore, the software implementation aspect of the HMEs recognition system has become a key issue. Nevertheless, it is still a difficult problem to recognize HMEs successfully for the reason that the handwritten mathematical expression recognition exhibits three distinct difficulties [1, 2]: i.e., the complex two-dimensional structures, enormous ambiguities in handwriting input and strong dependence on contextual information. Because of the ambiguity present in handwritten input, it is often unrealistic to hope for consistently perfect recognition accuracy.

This paper firstly discusses the three parts of the research status of mathematical expression recognition, including expression detection, symbol recognition, and structural analysis. Then introduces the research status of mathematical expression recognition from document images or handwritten forms. Finally, this paper looks forward to the future development of recognition of mathematical expression, especially handwritten mathematical expression.

2 Major Problems

There are three important directions for the recognition of mathematical expressions today [3, 4]: expression positioning, symbol segmentation, symbol recognition, and structural analysis. There are basically two approaches of mathematical expression recognition: online recognition and offline recognition.

2.1 Symbol Segmentation

In a mathematical expression, there usually exist multiple number, letter, and operator symbols. Before identifying a single symbol, we must first segment the individual symbol from the expression properly.

The mathematical symbol system is a symbol system with a two-dimensional spatial layout, which is different from traditional natural language texts with only a single sequence in the horizontal direction. For example, a fractional expression contains an upper, middle, and lower structure composed of a numerator, a fractional line, and a denominator. A radical expression includes a semi-enclosed structure composed of a root number, a square number, and a squared number. A fixed integrator operation includes an integrator and an integral upper limit. Structures that have both upper and lower relations and left and right relations formed with the integral lower limit, and even some mathematical expressions also contain complex nested structures of these relations. This feature makes the process of segmenting mathematical expression symbols extremely complicated. Figure 1 illustrates the segmentation difficulties mentioned above.

2.2 Expression Positioning

After the symbol segmentation step, we will get a list of objects with some known attribute values, and hope to get the identification of the symbol. In this step, we can apply any symbol recognition method as long as it is designed for the corresponding data

Fig. 1. Some examples of the spatial structure of mathematical expression.

type (i.e., online or offline). Not surprisingly, there are still some objective difficulties in the mathematical symbol recognition.

The features of mathematical symbols which are not conducive to automatic recognition are as shown below.

In the task of identifying mathematical expression, local ambiguity is an inevitable problem. Local ambiguity means that the specific content represented by some strokes is difficult to determine. Especially in the handwriting scene, it is difficult to accurately distinguish many characters from the morphology alone. In some cases, it is difficult for even humans or the writer himself to give unique answers to what the strokes represent. For example, the hand-written English characters "O" and numeric characters "0" are difficult to distinguish, and similar ones include "2" and "z", lowercase letters "x", and multiplication operators "×", "6", and "b" "," 9 "and" q "," B "and" 13 ", etc. Figure 2 illustrates the segmentation difficulties mentioned above.

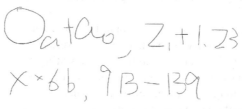

Fig. 2. Some examples of the local ambiguity of mathematical expression. The mathematical expressions are: ① $O_a + a_0$; ② $z_1 + 1.23$; ③ x × 6b; ④ 9B − 13q.

The character set of mathematical symbols is relatively large, including English letters, Roman letters and Greek letters, numbers, and many operators. At the same time, font, size, bold, italics and others make the recognition of mathematical symbols more complicated. For example, in printing scene, the italic English letter "A" and the flower font "$\mathcal{A}$" are difficult to distinguish, which is much harder in handwriting scene. In mathematical expression, implicit multiplication and dot multiplication are quite common, which makes it difficult to distinguish commas, dots, other small symbols, and noise when the sample quality is poor. In addition, the adhesion of characters can also lead to local ambiguity. Figure 3 illustrates the segmentation difficulties mentioned above.

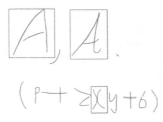

Fig. 3. Some examples of the local ambiguity of mathematical expression. The mathematical expressions are:① the italic English letter "A" and the flower font "$\mathcal{A}$"; ② (p + 2)(y + 6).

The sources of mathematical expression samples are mainly divided into print samples and handwriting samples. For the printing mathematical expression, due to its regular structural shape, the general recognition methods have a good performance in general. However, the recognition of the handwritten mathematical expression requires more consideration of its topological structure. Obviously, the difficulty in recognizing the handwritten mathematical expression is far greater than the recognition of the scanned image of the document.

As mentioned, there are basically two approaches of mathematical expression recognition: online recognition and offline recognition. The available information for online and offline recognition is different. Offline recognition mainly refers to the recognition of the mathematical expression image obtained by the scanner or the document containing the mathematical expression, including the printed matter and the handwriting. Online recognition refers to the recognition of the expression on smart terminal devices based on pen writing input, mainly the handwritten mathematical expression recognition. Compared with offline recognition, online recognition can make more use of handwriting trace information.

For online mathematical expression recognition, the handwriting trace information can help the model to identify segmentation. However, the problems caused by the difference in writing order and separation of handwriting trace from break make it difficult to use the handwriting trace information. For offline mathematical expression recognition, it is often judged by spatial information in binarized pictures. In view of the special circumstances of mathematical expression pictures, such as low pixels, blurred writing, and adhesion, offline mathematical expression recognition also faces many difficulties, and even segmentation ambiguity may occur (see Fig. 3).

2.3 Structure Analysis

To achieve accurate identification of mathematical expression, on the basis of mathematical symbol recognition, it is necessary to assemble, restore, and reconstruct mathematical symbols with the structural characteristics of the mathematical expression itself, to get the result of mathematical expression recognition. Structural analysis is mainly by determining the spatial relationship between the identified symbols, judging their logical relationship and constructive significance. Through structural analysis, the discrete recognition results can be assembled, restored, and reconstructed based on the original structural features, grammatical rules, priorities and other characteristics to complete the mathematical expression recognition of the entire process.

Suppose we are able to correctly segment and recognize mathematical symbols in the previous steps. Then, in the structural analysis stage, we can use the label, size and location of each mathematical symbol to build a parse tree or relationship tree. However, depending on the spatial operator, mathematical expression have the unique two-dimensional structure, which makes the spatial relationship between symbols in mathematical expressions more complicated. Therefore, we should first identify all spatial operators in order to build sub-structures over them and their operands. Using all intermediate substructures and the remaining objects, the final structure can be constructed. It is a pity that in some cases, the spatial relationship between symbols is still very vague, especially the handwritten mathematical expression. And it is difficult to judge some very complicated spatial operators from the simple positional relationship of the context, which requires the contextual semantic information.

The space operators mainly include implicit multiplication, subscripting, or exponentiation, which are widely used in mathematical expressions. However, the spatial relationship of the above space operators may be ambiguous. Figure 4 demonstrates that the same configuration of bounding boxes may reveal different spatial relationships. Furthermore, due to the complexity of large matrices and the structure of equations, it is more difficult to analyze and identify them.

$$\begin{array}{ll} \text{①} & abc \\ \text{②} & a^b c \\ \text{③} & a^{bc} \\ \text{④} & ab_c \\ \text{⑤} & a^{b_c} \end{array}$$

Fig. 4. Implicit multiplication, subscripting, and exponentiation cannot be determined.

3   Document Mathematical Expression Recognition Research

Document mathematical expression recognition, normally, is printed mathematical expression recognition. With many researches in the identification of mathematical symbols at the beginning, many excellent methods have been proposed in the field of document mathematical expression recognition. In recent years, the development of deep learning has also led many researchers try to use neural network as experimental tools.

Ha et al. [5] proposed a recursive X-Y cut segmentation method, which worked well on typeset MEs but not fitted for handwritten cases. In the proposed system, a top-down page segmentation technique known as the recursive X-Y cut decomposes a document image recursively into a set of rectangular blocks. The recursive X-Y cut be implemented using bounding boxes of connected components of black pixels instead of using image pixels.

Kumar et al. [6] try a rule-based approach in the symbol information step. They focus on the recognition of printed MEs and assume connected components (ccs) of a given ME image are labelled. The proposed method comprises three stages, namely symbol formation, structure analysis and generation of encoding form like LATEX. The symbol formation process, where multi-cc symbols (like =, $\equiv$ etc.) are formed, identity of context-dependent symbols (like a horizontal line can be MINUS, OVERBAR, FRACTION etc.) are resolved using spatial relations. Multi-line MEs like matrices and enumerated functions are also handled in this stage. A rule-based approach is proposed for the purpose, where the heuristics based on spatial relations are represented in the form of rules (knowledge) and those rules are fired depending on input data (labelled ccs). As knowledge is isolated from data like an expert system in the approach, it allows for easy adaptability and extensibility of the process (Fig. 5).

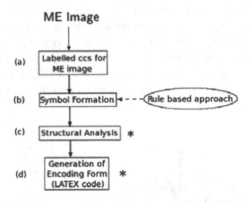

Fig. 5. Various stages of Kumar's approach [6] to ME recognition

Kim et al. [7] point out that the sequence of character segmentation is from left to right, and from top to bottom in case of general character recognition. However, mathematical expression is a kind of two-dimension visual language. Thus, they propose a modified recursive projection profile cutting method of character segmentation in images of mathematical expression, using depth first search for arranging and double linked list for re-arranging.

Murru et al. [8] propose the use of artificial neural networks (ANN) in order to develop pattern recognition algorithms capable of recognizing both normal texts and mathematical expression, and present an original improvement of the backpropagation algorithm. In symbol segmentation step, considering that features selection mainly depends on the experience of the authors, fuzzy logic can be more useful, since it is widely used in applications where tuning of features is based on experience and it can be preferred to a deterministic approach. Thus, a method is proposed that combines, by means of a fuzzy logic based approach, some state–of–the–art features usually exploited one at a time (Fig. 6).

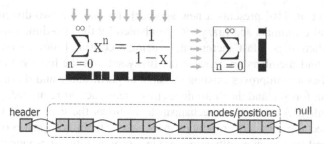

Fig. 6. Kim's [7] modified recursive projection profile cutting method of character segmentation in images of mathematical expression, and the double linked list applied in the paper.

4 Handwritten Mathematical Expression Recognition Research

Handwritten mathematical expression recognition includes both online recognition, such as the input from smart devices like note apps, and offline formula recognition, such as photos of mathematical expression. As mentioned in Sect. 2, due to the freedom of writing, handwritten mathematical expression recognition has greater difficulties than printed ones. Therefore, more diverse and complex technologies are needed, and researchers have invested more interest in it.

Hirata et al. [9] proposes a novel approach, based on expression matching, for generating ground-truthed exemplars of expressions (and, therefore, of symbols). Matching is formulated as a graph matching problem in which symbols of input instances of a manually labeled model expression are matched to the symbols in the model. Pairwise matching cost considers both local and global features of the expression (Fig. 7).

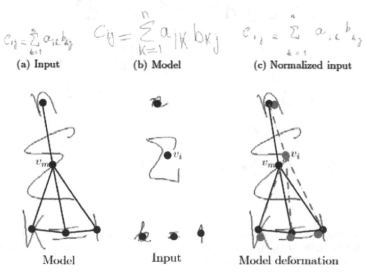

Fig. 7. Normalization of input expressions with respect to its corresponding model and deformation induced to the model graph in Hirata's approach [9].

MacLean et al. [10] presents a new approach for parsing two-dimensional input using relational grammars and fuzzy sets. Motivated by the two-dimensional structure of written mathematics, a fast, incremental parsing algorithm is developed. Then check the similarity and membership between the fuzzy set and the handwritten input. The approach applies and improves existing rectangular partitioning and sharing techniques such as parsing forests, and then introduces new ideas such as relationship classes and interchangeability. A correction mechanism that allows the user to view the parsing results and select the correct interpretation in the case of identifying errors or ambiguities is proposed, and then such modifications are incorporated into subsequent incremental recognition results.

MacLean et al. [11] also uses Bayesian networks for parse tree selection. They proposed a system which captures all recognizable interpretations of the input and organizes them in a parse forest from which individual parse trees may be extracted and reported. If the top-ranked interpretation is incorrect, the user may request alternates and select the recognition result they desire. The tree extraction step uses a novel probabilistic tree scoring strategy in which a Bayesian network is constructed based on the structure of the input, and each joint variable assignment corresponds to a different parse tree. Parse trees are then reported in order of decreasing probability.

Simistira et al. [12] symbolizes the elastic template matching distance based on the directional feature of the pen. The structural analysis is based on extracting the baseline of the mathematical expression, and then classifying symbols into levels above and below the baseline. The symbols are then sequentially analyzed using six spatial relations and a respective two-dimensional structure is processed to give the resulting MathML representation of the ME (Fig. 8).

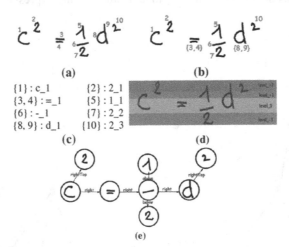

Fig. 8. Processing steps of Simistira's approach [12] to ME recognition. (a) ME with strokes labeled in time order, (b) grouping of strokes into symbols, (c) symbol recognition, (d) assignment of symbols to levels, (e) hierarchical structure of MEs

Álvaro et al. [13] describes a formal model or the recognition of on-line handwritten mathematical expressions using two-dimensional stochastic context-free grammar and hidden Markov models to identify online handwritten mathematical expressions. Hidden Markov models are used to recognize mathematical symbols, and a stochastic context-free grammar is used to model the relation between these symbols. This formal model makes possible to use classic algorithms for parsing and stochastic estimation. The model is able to capture many of variability phenomena that appear in on-line handwritten mathematical expressions during the training process. The parsing process can make decisions taking into account only stochastic information, and avoiding heuristic decisions.

Awal et al. [14] presents an online handwritten mathematical expression recognition system that handles mathematical expression recognition as a simultaneous optimization of expression segmentation, symbol recognition, and two-dimensional structure recognition under the restriction of a mathematical expression grammar. The originality of the approach is a global strategy allowing learning mathematical symbols and spatial relations directly from complete expressions. A new contextual modeling is proposed for combining syntactic and structural information. Those models are used to find the most likely combination of segmentation/recognition hypotheses proposed by a two-dimensional segmentation scheme. Thus, the models are based on structural information concerning the symbol layout (Fig. 9).

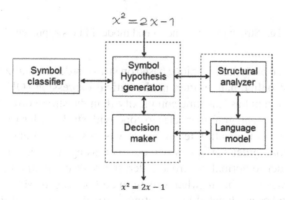

Fig. 9. System architecture of Awal's approach [14] to ME recognition.

Hu et al. [15] proposed a novel framework to analyze the layout and semantic information of handwritten mathematical expressions. The framework includes three steps, namely symbol segmentation, symbol recognition and semantic relationship analysis. For symbol segmentation, a decomposition on strokes is operated, then dynamic programming is adopted to find the paths corresponding to the best segmentation manner and reduce the stroke searching complexity. For symbol recognition, spatial geometry and directional element features are classified by a Gaussian Mixture Model learnt through Expectation-Maximization algorithm. At last, in the semantic relationship analysis module, a ternary tree is utilized to store the ranked symbols through calculating the operator priorities. The motivation for our work comes from the apparent difference in writing styles across western and Chinese populations.

Le et al. [16] present an end-to-end system to recognize Online Handwritten Mathematical Expressions (OHMEs). The system has three parts: a convolution neural network for feature extraction, a bidirectional LSTM for encoding extracted features, and an LSTM and an attention model for generating target LaTex (Fig. 10).

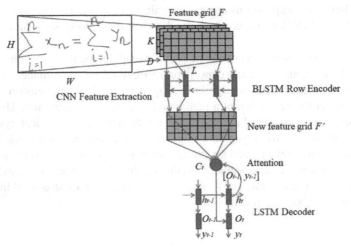

Fig. 10. Structure of the end-to-end model in Le's approach [16].

Le et al. [17] also present a technique based on stroke order normalization for improving recognition of online handwritten mathematical expression (OHME). The stroke order dependent system has less time complexity than the stroke order free system, but it must incorporate special grammar rules to cope with stroke order variations. The presented stroke order normalization technique solves this problem and also the problem of unexpected stroke order variations without increasing the time complexity of ME recognition. In order to normalize stroke order, the X–Y cut method is modified since its original form causes problems when structural components in ME overlap. First, vertically ordered strokes are located by detecting vertical symbols and their upper/lower components, which are treated as MEs and reordered recursively. Second, unordered strokes on the left side of the vertical symbols are reordered as horizontally ordered strokes. Third, the remaining strokes are reordered recursively. The horizontally ordered strokes are reordered from left to right, and the vertically ordered strokes are reordered from top to bottom. Finally, the proposed stroke order normalization is combined with the stroke order dependent ME recognition system (Fig. 11).

Zhang et al. [18] extend the chain-structured BLSTM to tree structure topology and apply this new network model for online math expression recognition. The proposed system addresses the recognition task as a graph building problem. The input expression is a sequence of strokes from which an intermediate graph is derived using temporal and spatial relations among strokes. In this graph, a node corresponds to a stroke and an edge denotes the relationship between a pair of strokes. Then several trees are derived from

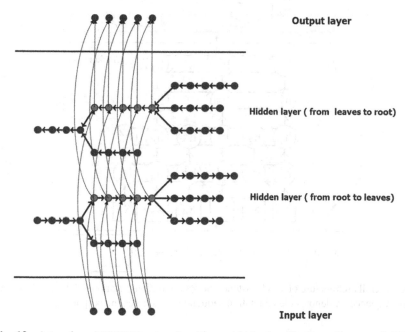

Fig. 11. Horizontal and vertical components and their desired stroke order in Le's approach [17].

the graph and labeled with Tree-based BLSTM. The last step is to merge these labeled trees to build an admissible label graph (LG) modeling two-dimensional expressions uniquely. The proposed system achieves competitive results in online math expression recognition domain (Fig. 12).

Output layer

Hidden layer (from leaves to root)

Hidden layer (from root to leaves)

Input layer

Fig. 12. A tree-based BLSTM network with one hidden level in Zhang's approach [18].

Zhang et al. [19, 20] introduce Track, Attend and Parse (TAP), an end-to-end approach based on neural networks for online handwritten mathematical expression recognition (OHMER). The architecture of TAP consists of a tracker and a parser. The tracker employs a stack of bidirectional recurrent neural networks with gated recurrent units (GRU) to model the input handwritten traces, which can fully utilize the dynamic trajectory information in OHMER. Followed by the tracker, the parser adopts a GRU equipped with guided hybrid attention (GHA) to generate LATEX notations. The proposed GHA is composed of a coverage based spatial attention, a temporal attention and an attention

guider. Moreover, the strong complementarity is demonstrated between offline information with static-image input and online information with ink-trajectory input by blend-
ing a fully convolutional networks based watcher into TAP. Inherently unlike traditional
methods, this end-to-end framework does not require the explicit symbol segmentation
and a predefined expression grammar for parsing (Fig. 13).

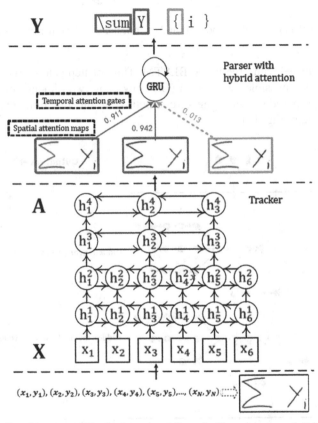

Fig. 13. Overall architecture of Track, Attend and Parse in Zhang's approach [19, 20]. X denotes
the input sequence, A denotes the annotation sequence, Y denotes the output sequence.

5 Conclusion

Through the efforts of researchers, recognition technology is becoming more and more
mature. As mentioned in Sect. 2, mathematical expression recognition mainly includes
three questions. These three tasks can be solved sequentially or jointly. The proposed
solutions can be roughly divided into sequential solutions and integrated solutions. In
addition, with recent advances in deep learning, several end-to-end deep-learning based
systems were proposed for ME recognition.

In the past, due to technical and computational limitations, researchers performed separate calculations for each step. With the development of computing power and algorithms, more and more research attempts to solve problems from a global perspective. In the beginning, researchers attempted to apply much prior knowledge to the three steps of recognition models in order to obtain higher efficiency without a research foundation. This method makes it easier for people to understand the logic of recognition, and it is easier to summarize the reasons for the wrong results.

On the other hand, considering that prior knowledge depends on the experience of the authors, modern researches are more inclined to use of machine learning or deep learning algorithms to allow computers to learn recognition logic from samples in order to discover recognition logic that is not covered by prior knowledge. In addition, since the recognition of mathematical expression is also affected by contextual information, modern researches attempt to recognize the mathematical expression globally to improve the recognition efficiency. The continuous improvement of experimental results proves the advantages of this method, but it may be difficult for people to directly understand the recognition logic obtained by machine learning.

In the future, the work of mathematical expression recognition may include the following aspects. When completing the three steps independently, improve the level of segmentation algorithms and structural analysis, especially in terms of spatial operators and stroke adhesion whether printed or handwritten. Improve the reliability of structural analysis by using more reliable symbolic structural models. In the global recognitions, the algorithm can try to make full use of the context information, and tt can be continuously improved in network design and loss function. Depending on the full development of computing power, the application of deep learning for research may have good prospects. At the same time, how to interpret the results obtained by deep learning is also a potential problem. There are a variety of methods for identifying system integration, especially in terms of syntax and dynamic processing. In addition, more usage scenarios can be considered, such as mathematical expression recognition on the blackboard in teaching scenarios, or the conversion of voice input.

Acknowledgments. This work was supported by the Guangdong Provincial Government of China through the "Computational Science Innovative Research Team" program and the Guangdong Province Key Laboratory of Computational Science at the Sun Yat-Sen University, and the National Science Foundation of China (grant no. 11471012).

References

1. Anderson, R.H.: Syntax-directed recognition of hand-printed two-dimensional mathematics. In: Symposium on Interactive Systems for Experimental Applied Mathematics: Proceedings of the Association for Computing Machinery Inc. Symposium, pp. 436–459. ACM (1967)
2. Belaid, A., Haton, J.-P.: A syntactic approach for handwritten mathematical formula recognition. IEEE Trans. Pattern Anal. Mach. Intell. **PAMI-6**(1), 105–111 (1984)
3. Chan, K.-F., Yeung, D.-Y.: Mathematical expression recognition: a survey. Int. J. Doc. Anal. Recogn. **3**(1), 3–15 (2000). https://doi.org/10.1007/PL00013549
4. Zanibbi, R., Blostein, D.: Recognition and retrieval of mathematical expressions. Int. J. Doc. Anal. Recogn. **15**(4), 331–357 (2012)

5. Ha, J., Haralick, R.M., Phillips, I.T.: Recursive X-Y cut using bounding boxes of connected components. In: International Conference on Document Analysis & Recognition. IEEE (1995)
6. Kumar, P.P., Agarwal, A., Bhagvati, C.: A rule-based approach to form mathematical symbols in printed mathematical expressions. In: Sombattheera, C., Agarwal, A., Udgata, S.K., Lavangnananda, K. (eds.) MIWAI 2011. LNCS (LNAI), vol. 7080, pp. 181–192. Springer, Heidelberg (2011). https://doi.org/10.1007/978-3-642-25725-4_16
7. Yoo, Y.H., Kim, J.H.: Mathematical formula recognition based on modified recursive projection profile cutting and labeling with double linked list. In: Kim, J.H., Matson, E., Myung, H., Xu, P. (eds.) Robot Intelligence Technology and Applications 2012. AISC, vol. 208, pp. 983–992. Springer, Heidelberg (2013). https://doi.org/10.1007/978-3-642-37374-9_95
8. Farulla, G.A., Armano, T., Capietto, A., Murru, N., Rossini, R.: Artificial Neural Networks and Fuzzy Logic for Recognizing Alphabet Characters and Mathematical Symbols. In: Miesenberger, K., Bühler, C., Penaz, P. (eds.) ICCHP 2016. LNCS, vol. 9758, pp. 7–14. Springer, Cham (2016). https://doi.org/10.1007/978-3-319-41264-1_1
9. Hirata, N.S.T., Honda, W.Y.: Automatic labeling of handwritten mathematical symbols via expression matching. In: Jiang, X., Ferrer, M., Torsello, A. (eds.) GbRPR 2011. LNCS, vol. 6658, pp. 295–304. Springer, Heidelberg (2011). https://doi.org/10.1007/978-3-642-20844-7_30
10. MacLean, S., Labahn, G.: A new approach for recognizing handwritten mathematics using relational grammars and fuzzy sets. Int. J. Doc. Anal. Recogn. 16, 139–163 (2013). https://doi.org/10.1007/s10032-012-0184-x
11. Maclean, S., Labahn, G.: A Bayesian model for recognizing handwritten mathematical expressions. Pattern Recogn. 48(8), 2433–2445 (2015)
12. Simistira, F., Papavassiliou, V., Katsouros, V., et al.: A system for recognition of on-line handwritten mathematical expressions. In: International Conference on Frontiers in Handwriting Recognition. IEEE Computer Society (2012)
13. Álvaro, F., Sánchez, J.-A., Benedí, J.-M.: Recognition of on-line handwritten mathematical expressions using 2D stochastic context-free grammars and hidden Markov models. Pattern Recogn. Lett. 35(1), 58–67 (2014)
14. Awal, A.M., Mouchère, H., Viard-Gaudin, C.: A global learning approach for an online handwritten mathematical expression recognition system. Pattern Recogn. Lett. 35, 68–77 (2014)
15. Hu, Y., Peng, L., Tang Y.: On-line handwritten mathematical expression recognition method based on statistical and semantic analysis. In: 2014 11th IAPR International Workshop on Document Analysis Systems (DAS). IEEE Computer Society (2014)
16. Le, A.D., Nakagawa, M.: Training an end-to-end system for handwritten mathematical expression recognition by generated patterns. In: ICDAR 2017. IEEE (2017)
17. Le, A.D., Nguyen, H.D., Indurkhya, B., Nakagawa, M.: Stroke order normalization for improving recognition of online handwritten mathematical expressions. Int. J. Doc. Anal. Recogn. 22(1), 29–39 (2019). https://doi.org/10.1007/s10032-019-00315-2
18. Zhang, T., Mouchère, H., Viard-Gaudin, C.: A tree-BLSTM-based recognition system for online handwritten mathematical expressions. Neural Comput. Appl. 32(9), 4689–4708 (2018). https://doi.org/10.1007/s00521-018-3817-2
19. Zhang, J., Jun, D., Lirong, D.: Track, Attend and Parse (TAP): an end-to-end framework for online handwritten mathematical expression recognition. IEEE Trans. Multimedia 21(1), 221–233 (2019)
20. Zhang, J., Jun, D., Lirong, D.: A GRU-based encoder-decoder approach with attention for online handwritten mathematical expression recognition. In: International Conference on Document Analysis and Recognition, pp. 902–907 (2017)

Handwritten Mathematical Expression Recognition: A Survey

Fukeng He$^{(\boxtimes)}$ (ID), Jun Tan (ID), and Ning Bi (ID)

School of Mathematics, Sun Yat-sen University, Guangzhou, China
hefk@mail2.sysu.edu.cn,
{mcstj,mcsbn}@mail.sysu.edu.cn

Abstract. While the handwritten character recognition has reached a point of maturity, the recognition of handwritten mathematics is still a challenging problem. The problem usually consists of three major parts: strokes segmentation, single symbol recognition and structural analysis. In this paper, we present a review on handwritten mathematical expression recognition to show how the recognition technique is developed. In particular, we put emphasis on the differences between systems.

Keywords: Handwritten mathematics recognition · Template matching · Statistical methods · Neural network

1 Introduction

Optical character recognition (OCR) is a popular topic in computer vision in recent years. Among OCR techniques, handwritten mathematical expression recognition (HMER) is one of the most difficult topic due to the ambiguity of handwritten mathematical symbols and mathematical structure [1,2].

Since mathematical expressions play an important role in scientific and engineering documentation, HMER has a board application, including documentation editing, system entry of mathematical expressions. In addition, the recognition system can connect to mathematical software like MATLAB or Mathematica for algebraic operation.

HMER can be divided into two types [3]: online handwritten mathematics recognition (OHMER) and offline handwritten mathematics recognition (OFHMER). Online recognition deals with the points series produced by a finger or a digital pen in touch screen devices while offline handwritten mathematics recognition deals with the static figures written in paper.

The targets of OHMER and OFHMER are performing a mathematical expression recognition systems to transform the input points series and static figures respectively into MathML or LaTeX [5] format.

HMER typically includes three major steps [4]: strokes segmentation, single symbol recognition and structural analysis. Strokes segmentation is to segment all the strokes into several separated symbols, followed by the single symbol

© Springer Nature Switzerland AG 2020
Y. Lu et al. (Eds.): ICPRAI 2020, LNCS 12068, pp. 55–66, 2020.
https://doi.org/10.1007/978-3-030-59830-3_5

recognition to determinate which type of the symbol belongs to. After that, structural analysis is carried out to analyze the spatial relationship between symbols and parse them into a complete mathematical expression.

However, papers that provide literature survey of the area of handwritten mathematical expression recognition research are rare. In this paper, we present a review on the development path of handwritten mathematical expression recognition including OHMER and OFHMER.

2   Difficulties of HMER

Firstly, In mathematical expressions, symbols and characters are arranged in a complex two-dimension structure. Grouping the strokes into several separated single symbols is not trivial because the strokes segmentation is not necessarily equal to the spatial segmentation, i.e., spatial connected strokes are not necessarily belongs to the same symbol and a symbol may contain different strokes that are not spatially connected.

Secondly, recognition of single symbols is difficult due to the enormous ambiguities in handwritten input and strong dependency of contextual information, e.g., the probability of $a * b * c$ is larger than that of $a * 6 * c$.

Thirdly, structural analysis is also a difficult step because of the ambiguities in the spatial relationship of each symbol and difficulties in arranging the grammar symbols like binding, fence.

3   Symbol Analysis

3.1   Symbol Types

There are four types of mathematical symbols used most often in mathematical expressions [4].

- Digital or character symbols, such as a, b, 1, 2, etc. These symbols are the major part of mathematical expressions.
- Binding symbols, such as fraction line, summation symbol $\sum$, square root $\sqrt{}$, etc. These symbols can determine the spatial relationship between its neighbour symbols.
- Fence symbols, such as (, {,), }, etc. These symbols can group the whole mathematical expression into several units.
- Operator symbols, such as +, −, etc. These symbols represent a operation between their neighbour symbols.

The types above except the first type can invoke some grammar effects and relate to the structure of the mathematical expression.

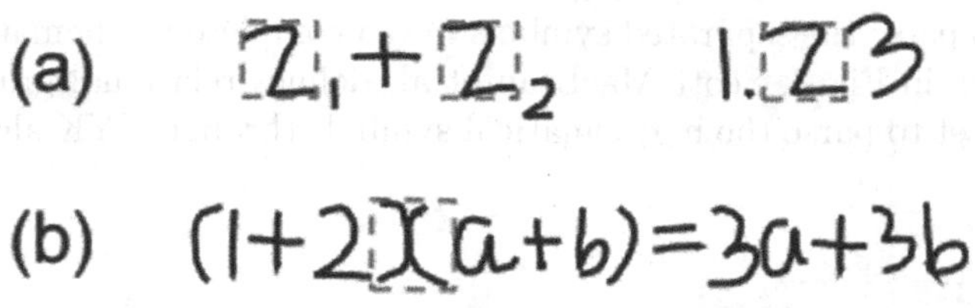

(a) $\overline{Z}_1 + \overline{Z}_2 \quad 1.\overline{Z}3$

(b) $(1 + 2)(a+b) = 3a + 3b$

Fig. 1. The ambiguity of mathematical expressions [6]: (a) it is difficult to distinguish Z and 2 directly, but it can be inferred according to the context. In the blue dotted box in (b), there are two kinds of judgments, one is x, the other is)(, and both of them are likely to occur in the actual scene, so it is difficult to get accurate results even according to the context.

4 Traditional Methods

After several decades of research, many existing recognition techniques are able to achieve satisfactory results (Fig. 1).

In this paper, we divide the development of HMER into two parts: traditional methods and neural network based methods. In this section, we present the development of traditional methods in decades.

Traditional methods includes Sequential solution and global solution.

4.1 Sequential Solutions

Sequential solutions [8,9] first segment input strokes or input static figures into single mathematical symbols and recognize them separately. The analysis of two-dimension structure is then carried out based on the best symbol segmentation model and the best symbol recognition model.

Symbol Segmentation and Symbol Recognition. Traditional methods include template matching techniques, structural approaches and statistical approaches.

Template matching techniques use traditional template matching methods using Hausdorff distance or other distance [10–12].

Structural approaches include Belaid and Haton [13], and Chan and Yeung [14].

Chen and Yin [15], Fateman and Tokuyasu [12] and Lee *et al.* [16] solve the problem of symbol segmentation and recognition using traditional statistical approaches. Besides, HMM (Hidden Markov models) [17,30] is used by Winkler *et al.* to find one or more symbol sequences from *symbol hypothesis net* (SHN) [18] and Scott MacLean [19] et al. uses a probabilistic tree scoring strategy in which a Bayesian network is built.

Fotini Simistira et al. [3] and Anh Duc Le et al. [6] define stochastic context free grammars and use probabilistic SVMs to predict the spatial relation between symbols, followed by CYK parsing algorithm (Cocke-Younger-Kasami algorithm) to parse the separated symbols into a complete mathematical expression, as shown in Fig. 3. Scott MacLean et al. defines relational grammars and utilize fuzzy set to parse the mathematical symbols through CYK algorithm [20] (Fig. 2).

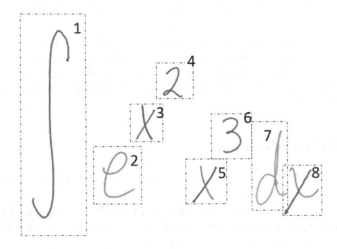

Fig. 2. The segmentation of strokes in CROHME2013 [7,40] train set: different colors represent different strokes, the dashed box contains the strokes combination of the same mathematical symbol, and the numbers beside the dashed box represent the order of writing mathematical symbols.

4.2   Global Solutions

Global solutions integrate symbol segmentation, symbol recognitionand structural analysis but are computationally expensive.

Taik Heon Rhee et al. [21] proposes a layered search framework for handwritten mathematical recognition, with a hope that they may be resolved by later global analysis. This paper formulates the handwritten mathematical expressions recognition as a problem of searching for the most likely interpretation for a given handwritten set of input strokes.

Francisco Alvaro et al. [22] utilizes a efficient search strategy in structure analysis for handwritten mathematical expression recognition. This paper defines stochastic context free grammars as above and present an integrated approach that combines several stochastic sources of information and can globally determine the most likely expression.

4.3   Neural Network Based Methods

Neural network has achieved high performance in many areas of OCR like scene text recognition [31], handwritten text recognition [32,33]. Not surprisingly, neural network can be applied to mathematical expressions recognition.

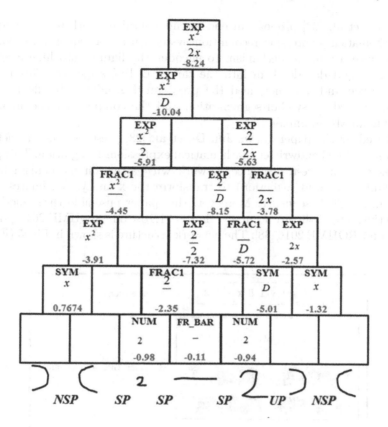

Fig. 3. CYK table of $\frac{x^2}{2x}$ [6].

Sequence-to-Sequence Methods. Encoder-decoder [39], has been exploited specifically to address sequence to sequence learning [23]. The encoder-decoder is designed to handle variable-length input and output sequences. [23] The encoder and decoder are typically recurrent neural networks and they can be convolution neural network in some cases. The encoder learns to encode the input sequence into a fixed-length vector and the decoder uses this vector to produce the variable-length output sequence. In the training stage, to generate the next predicted word, the model provides the ground truth labels from previous words as inputs to the decoder and minimizes the log-loss [34] using stochastic gradient descent [35]. In the inference stage, the model generates the predicted sequence using beam search algorithm [36]. Such a framework has been applied extensively to many applications including machine translation [24,25] and speech recognition [26].

Handwritten mathematical expressions recognition can be also seen as a sequence-to-sequence problem.

Jun Du et al. [27] propose an end-to-end neural network based handwritten mathematical expression recognition system, using a sequence-to-sequence network with attention mechanism to encoder the figures into high-level representations and decode them into the target LaTeX sequence. This method is data-driven and does not need the pre-defined grammar. It alleviates the problems caused by symbol segmentation and the computational demands of employing an ME grammar.

Followed by the paper above, Jun Du et al. [28] presents a neural network method for online handwritten mathematical expressions recognition. This paper proposes an sequence-to-sequence network with attention mechanism named GHA (Guided Hybrid Attention) to transform the x and y coordinates of the strokes into LaTeX sequence. Moreover, this paper ensemble three models to achieve the state-of-the-art accuracy, that is 61.16% on CROHME 2014 [37] and 57.02% on CROHME 2016 [38]. The network structure is shown in Fig. 5 (Fig. 4).

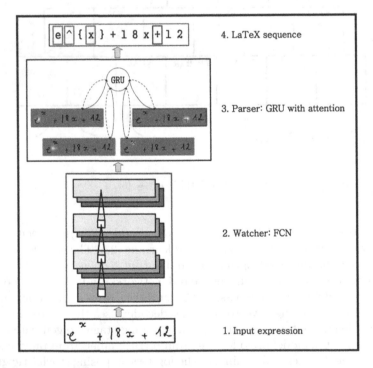

Fig. 4. The network structure of WAP [27].

Tree-Based Recurrent Methods. Since any mathematical expression can be expressed as a tree form, as shown in Fig. 6 [29]. Tree-based recurrent methods [41] can be utilized to mathematical expressions recognition.

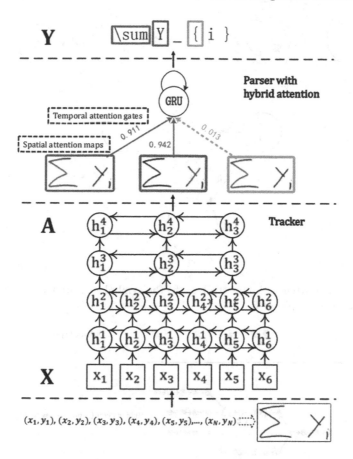

Fig. 5. The network structure of TAP [28].

Ting Zhang et al. [29] proposes a method based on this method. It has no explicit segmentation, recognition, and layout extraction steps but a unique trainable system that produces directly a stroke label graph describing a mathematical expression. In this paper, tree-based BLSTM framework [42] is used to transform the input strokes into the target latex sequence and this network is trained with CTC loss [43], as presented in Fig. 7 (Tables 1 and 2).

4.4 Categorization of the Methods Above

In this section we list the above methods and categorize them into two types: traditional methods and neural network based methods.

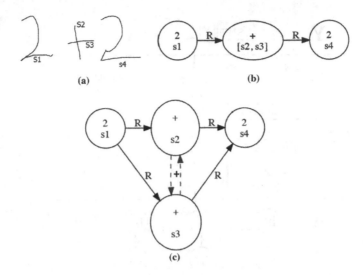

(a)   (b)

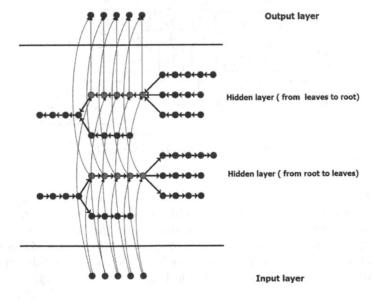

Fig. 6. (a) $2+2$ written with four strokes, (b) the symbol relation tree of $2+2$, (c) the SLG (stroke label graph) of $2+2$.

Fig. 7. A tree-based BLSTM network with one hidden level. We only draw the full connection on one short sequence (red) for a clear view (Color figure online).

Table 1. Traditional methods

Method	Example
Structural approaches	Belaid and Haton, and Chan and Yeung
Template matching	Chou, Nakayama, Okamoto et al.
Traditional statistical approaches	Chen and Yin, Fateman and Tokuyasu and Lee et al.
Hidden Markow models (HMMs)	L.R. Rabiner et al.
Symbol Hypothesis net	Winkler et al.

Table 2. Neural network based methods

Method	Example
Traditional neural networks	Dimitriadis and Coronado et al. and Ha et al.
Sequence-to-sequence based networks	Jun Du et al.
Tree-based recurrent neural networks	Ting Zhang et al.

5 Conclusion

With the development of the pen-based device, we already have all the necessary hardware for entering the mathematical expressions into the computer in offline form or online form. However, the recognition accuracy of mathematical expressions is not satisfactory, with the state-of-the-art accuracy less than 70%. Therefore, this field is potential in OCR and more research is needed.

Moreover, with development of neural networks methods like convolution neural networks and recurrent neural networks, deep learning methods is utilized in both the offline and online mathematical expressions recognition. These methods are end-to-end methods, without explicit segmentation, recognition and layout extraction steps. Besides, they are data-driven, without predefining the grammar of the mathematical expressions.

Acknowledgments. This work was supported by the Guangdong Provincial Government of China through the "Computational Science Innovative Research Team" program and the Guangdong Province Key Laboratory of Computational Science at the Sun Yat-sen University ,and the National Science Foundation of China (grant no. 11471012).

References

1. Anderson, R.H.: Syntax-directed recognition of hand-printed two-dimensional mathematics. In: Symposium on Interactive Systems for Experimental Applied Mathematics: Proceedings of the Association for Computing Machinery Inc., Symposium. ACM, pp. 436–459 (1967)
2. Belaid, A., Haton, J.-P.: A syntactic approach for handwritten mathematical formula recognition. IEEE Trans. Pattern Anal. Mach. Intell. 1, 105–111 (1984)

3. Simistira, F., Katsouros, V., Carayannis, G.: Recognition of online handwritten mathematical formulas using probabilistic SVMs and stochastic context free grammars. Pattern Recogn. Lett. **53**, 85–92 (2015)

4. Chan, K.-F., Yeung, D.-Y.: Mathematical expression recognition: a survey. Int. J. Doc. Anal. Recogn. **3**(1), 3–15 (2000)

5. Lamport, L.: LATEX: A Document Preparation System: User's Guide and Reference Manual. Addison-Wesley, Boston (1994)

6. Le, A.D., Van Phan, T., Nakagawa, M.: A system for recognizing online handwritten mathematical expressions and improvement of structure analysis. In: 11th International Workshop on Document Analysis Systems (DAS 2014). IEEE (2014)

7. Mouchère, H., Zanibbi, R., Garain, U., et al.: Advancing the state of the art for handwritten math recognition: the CROHME competitions, 2011–2014. Int. J. Doc. Anal. Recogn. (IJDAR) **19**(2), 173–189 (2016)

8. Zanibbi, R., Blostein, D., Cordy, J.R.: Recognizing mathematical expressions using tree transformation. IEEE Trans. Pattern Anal. Mach. Intell. **24**(11), 1455–1467 (2002)

9. Álvaro, F., Sánchez, J.-A., Bened'ı, J.-M.: Recognition of on-line handwritten mathematical expressions using 2D stochastic context-free grammars and hidden Markov models. Pattern Recogn. Lett. **35**, 58–67 (2014)

10. Chou, P.A.: Recognition of equations using a two-dimensional stochastic context-free grammar. Proc. SPIE **1199**, 852–863 (1989)

11. Nakayama, Y.: Mathematical formula editor for CAI. ACM SIGCHI Bull. **20**(SI), 387–392 (1989)

12. Fateman, R.J., Tokuyasu, T., Berman, B.P., et al.: Optical character recognition and parsing of typeset mathematics1. J. Vis. Commun. Image Representation **7**(1), 2–15 (1996)

13. Belaid, A., Haton, J.P.: A syntactic approach for handwritten mathematical formula recognition. IEEE Trans. Pattern Anal. Mach. Intell. **1**, 105–111 (1984)

14. Chan, K.F., Yeung, D.Y.: Recognizing on-line handwritten alphanumeric characters through flexible structural matching. Pattern Recogn. **32**(7), 1099–1114 (1999)

15. Chen, L.: A system for on-line recognition of handwritten mathematical expressions. Comput. Process. Chin. Orient. Lang. **6**(1), 19–39 (1992)

16. Lee, H.J., Lee, M.C.: Understanding mathematical expressions in a printed document. In: Proceedings of 2nd International Conference on Document Analysis and Recognition (ICDAR 1993), pp. 502–505. IEEE (1993)

17. Rabiner, L.R.: A tutorial on hidden Markov models and selected applications in speech recognition. Proc. IEEE **77**(2), 257–286 (1989)

18. Koschinski, M., Winkler, H.J., Lang, M.: Segmentation and recognition of symbols within handwritten mathematical expressions. In: 1995 International Conference on Acoustics, Speech, and Signal Processing, vol. 4, pp. 2439–2442. IEEE (1995)

19. MacLean, S., Labahn, G.: A Bayesian model for recognizing handwritten mathematical expressions. Pattern Recogn. **48**(8), 2433–2445 (2015)

20. Chappelier, J.C., Rajman, M.: A generalized CYK algorithm for parsing stochastic CFG. TAPD **98**(133–137), 5 (1998)

21. Rhee, T.H., Kim, J.H.: Efficient search strategy in structural analysis for handwritten mathematical expression recognition. Pattern Recogn. **42**(12), 3192–3201 (2009)

22. Álvaro, F., Sánchez, J.A., Benedí, J.M.: Recognition of on-line handwritten mathematical expressions using 2D stochastic context-free grammars and hidden Markov models. Pattern Recogn. Lett. **35**, 58–67 (2014)

23. Sutskever, I., Vinyals, O., Le, Q.V.: Sequence to sequence learning with neural networks. In: Advances in Neural Information Processing Systems, pp. 3104–3112 (2014)
24. Luong, M.-T., Sutskever, I., Le, Q.V., Vinyals, O., Zaremba, W.: Addressing the rare word problem in neural machine translation, arXiv preprint arXiv: 1410.8206 (2014)
25. Bahdanau, D., Chorowski, J., Serdyuk, D., Brakel, P., Bengio, Y.: End-to-end attention-based large vocabulary speech recognition. In: 2016 IEEE International Conference on Acoustics, Speech and Signal Processing (ICASSP), pp. 4945–4949. IEEE (2016)
26. Chan, W., Jaitly, N., Le, Q.V., Vinyals, O.: Listen, attend and spell, arXiv preprint arXiv: 1508.01211 (2015)
27. Zhang, J., Du, J., Zhang, S., et al.: Watch, attend and parse: an end-to-end neural network based approach to handwritten mathematical expression recognition. Pattern Recogn. **71**, 196–206 (2017)
28. Zhang, J., Du, J., Dai, L.: Track, attend, and parse (TAP): an end-to-end framework for online handwritten mathematical expression recognition. IEEE Trans. Multimed. **21**(1), 221–233 (2018)
29. Zhang, T., Mouchère, H., Viard-Gaudin, C.: A tree-BLSTM-based recognition system for online handwritten mathematical expressions. Neural Comput. Appl. 1–20 (2018)
30. Beal, M.J., Ghahramani, Z., Rasmussen, C.E.: The infinite hidden Markov model. In: Advances in neural information processing systems, pp. 577–584 (2002)
31. Saidane, Z., Garcia, C.: Automatic scene text recognition using a convolutional neural network. In: Workshop on Camera-Based Document Analysis and Recognition, vol. 1 (2007)
32. Kim, G., Govindaraju, V., Srihari, S.N.: An architecture for handwritten text recognition systems. Int. J. Doc. Anal. Recogn. **2**(1), 37–44 (1999)
33. Espana-Boquera, S., Castro-Bleda, M.J., Gorbe-Moya, J., et al.: Improving offline handwritten text recognition with hybrid HMM/ANN models. IEEE Trans. Pattern Anal. Mach. Intell. **33**(4), 767–779 (2010)
34. Bottou, L.: Large-scale machine learning with stochastic gradient descent. In: Proceedings of COMPSTAT'2010. Physica-Verlag HD, pp. 177–186 (2010)
35. Bottou, L.: Stochastic gradient descent tricks. In: Montavon, G., Orr, G.B., Müller, K.-R. (eds.) Neural Networks: Tricks of the Trade. LNCS, vol. 7700, pp. 421–436. Springer, Heidelberg (2012). https://doi.org/10.1007/978-3-642-35289-8_25
36. Tillmann, C., Ney, H.: Word reordering and a dynamic programming beam search algorithm for statistical machine translation. Comput. Linguist. **29**(1), 97–133 (2003)
37. Mouchere, H., Viard-Gaudin, C., Zanibbi, R., et al.: ICFHR 2014 competition on recognition of on-line handwritten mathematical expressions (CROHME 2014). In: 2014 14th International Conference on Frontiers in Handwriting Recognition, pp. 791–796. IEEE (2014)
38. Mouchère, H., Viard-Gaudin, C., Zanibbi, R., et al.: ICFHR2016 CROHME: competition on recognition of online handwritten mathematical expressions. In: 2016 15th International Conference on Frontiers in Handwriting Recognition (ICFHR), pp. 607–612. IEEE (2016)
39. Cho, K., Van Merriënboer, B., Gulcehre, C., et al.: Learning phrase representations using RNN encoder-decoder for statistical machine translation. arXiv preprint arXiv:1406.1078 (2014)

40. Mouchere, H., Viard-Gaudin, C., Zanibbi, R., et al.: ICDAR 2013 CROHME: third international competition on recognition of online handwritten mathematical expressions. In: 2013 12th International Conference on Document Analysis and Recognition, pp. 1428–1432. IEEE (2013)
41. Dam, H.K., Pham, T., Ng, S.W., et al.: A deep tree-based model for software defect prediction. arXiv preprint arXiv:1802.00921 (2018)
42. Sankaran, N., Jawahar, C.V.: Recognition of printed Devanagari text using BLSTM Neural Network. In: Proceedings of the 21st International Conference on Pattern Recognition (ICPR2012), pp. 322–325. IEEE (2012)
43. Hannun, A.: Sequence modeling with CTC. Distill **2**(11), e8 (2017)

Commodity Classification Based on Multi-modal Jointly Using Image and Text Information

Yan Xu[1,2] , Yufang Tang[1,2(✉)] , and Ching Y. Suen[2]

[1] Shandong Normal University, Jinan 250014, Shandong, China
tangyufang@outlook.com
[2] Concordia University, Montreal, QC H3G 1M8, Canada

Abstract. Considering that there exists image and text information almost on every commodity web page, although these two kinds of information belong to different modals, both of them describe the same commodity, so there must be a certain relationship between them. We name this relationship "symbiosis and complementary", and propose a multi-modal based on image and text information for commodity classification algorithm (MMIT). Firstly, we use $\ell_{2,0}$ mixed norm to optimize sparse representation method for image classification, and then employ Bayesian posterior probability to optimize k-nearest neighbor method for text classification. Secondly, we fuse two modal classification results, and build MMIT mathematical model. Finally, we utilize a dataset to train MMIT model, and then employ trained MMIT classifier to classify different commodities. Experimental results show that our method can achieve better classification performance than other state-of-the-art methods, which only exploit image information.

Keywords: Multi-modal · $\ell_{2,0}$ Mixed norm · Sparse representation · Bayesian posterior probability · K-Nearest Neighbor

1 Introduction

When labeling a commodity, most researchers use image information for classification, such as color, shape and texture [1]. Among image classification methods, remarkable results can be obtained by sparse representation classification method(SRC), such as face recognition [2], because discriminate features from images can be extracted by this method [3]. While recommending a commodity, text information always be employed for classification in most recommendation systems, and K-Nearest Neighbor method (KNN) is often used for natural language classification [4]. In fact, consumers pay more attention to image information of commodity. However, commodity can only be depicted from a certain feature space using either image or text information. Some researchers have noticed this problem and started to use the relationship between image and text information for image classification. In recent years, some classification methods

© Springer Nature Switzerland AG 2020
Y. Lu et al. (Eds.): ICPRAI 2020, LNCS 12068, pp. 67–79, 2020.
https://doi.org/10.1007/978-3-030-59830-3_6

based on multi-modal have been proposed, with the idea of fusing multi-modal information for classification [5]. Based on the above discussion, we propose a new method MMIT to classify commodity based on image and text information, making full use of their respective advantages.

The rest of this paper is organized as follows. In Sect. 2, we introduce the concept of "symbiosis and complementary", and discuss the problem of how to select fusion level. In Sect. 3, we propose a new method MMIT. In Sect. 4, we describe experimental results to validate our method. In Sect. 5, we conclude this paper.

2   Background

2.1   "Symbiosis and Complementary" Between Image and Text Information

After analyzing some commodity web pages of different e-commerce sites, we find out that image and text information are almost contained in all commodity web pages. And both of them describe the same commodity from different feature spaces. We name this relationship "symbiosis and complementary". (i) "Symbiosis" means that a commodity can be described by both image and text information, so image and text on the same page belong to the same category. (ii) "Complementary" means that image and text information describe the same commodity from visual feature space and text feature space respectively. Thus, we can utilize this relationship for commodity classification.

2.2   How to Choose Fusion Level

Multi-Modal fusion method can be categorized into four types: sensor level, feature level, classifier level and decision level. Khaleghi [6] proved that only if the fused information belongs to the same modal, different types of information can be fused at sensor or feature level. Wozniak [7] revealed that if we want to fuse different modal information at decision level, the performance of $Classifier_i(i = 1, 2, ..., n)$ should be the same. Moreno [8] demonstrated that multi-modal information can be fused at classifier level.

As mentioned above, because the image and text information belong to different modals, and their performance of classifiers is different, we can exclude the other three types, and choose the classifier level as our fusion level.

3   Multi-modal Based on Image and Text Information for Commodity Classification

3.1   Three Key Issues in Our Method

Suppose that $(\mathbf{x}, \mathbf{t}, y)$ is a commodity to be classified, $\mathbf{x}$ is commodity image, $\mathbf{t}$ is commodity text, y (unknown) is the class label of commodity. There are three key issues to be resolved - (i) How to classify a commodity image to obtain a image classification confidence $Conf(\mathbf{x})$. (ii) How to classify a commodity text to gain a text classification confidence $Conf(\mathbf{t})$. (iii) How to fuse two different modal classification confidences $Conf(\mathbf{x})$ and $Conf(\mathbf{t})$ to get multi-modal classification confidence $Conf(\mathbf{x}, \mathbf{t})$. Finally we can employ classifier $classifier(\cdot) = \arg\min \{Conf(\mathbf{x}, \mathbf{t})\}$ to classify commodity.

3.2 Optimizing Sparse Representation Algorithm for Commodity Image Classification via $\ell_{2,0}$ Mixed Norm

In order to classify commodity image to obtain image classification confidence $Conf(\mathbf{x})$, we optimize traditional sparse representation classification algorithm via $\ell_{2,0}$ mixed norm, shown in Eq. (1).

$$\arg\min_{\alpha} \frac{1}{2}\|\mathbf{x} - \mathbf{D}\alpha\|_2^2 + \lambda\|\alpha\|_0 \tag{1}$$

where $\mathbf{x}$ is a commodity image to be classified, $\mathbf{D}$ is sparse dictionary, α is sparse representation of $\mathbf{x}$.

By solving Eq. (1), we can obtain α, then employ α instead of $\mathbf{x}$ to classify commodity image. There are two important questions in sparse representation classification - (i) how to learn parse dictionary $\mathbf{D}$, and (ii) how to solve sparse representation α.

In order to build sparse dictionary $\mathbf{D}$, Eq. (2) is employed, which is a sparse dictionary learning algorithm proposed in our paper [9].

$$\arg\min_{\mathbf{D}} \frac{1}{2}\|\mathbf{X} - \mathbf{D}\mathbf{A}\|_F^2 + \frac{\lambda_1}{2}\sum_{i\neq j}\|\mathbf{D}_i^T\mathbf{D}_j\|_F^2$$
$$+ \frac{\lambda_2}{2}\|\mathbf{H} - \mathbf{W}\mathbf{A}\|_F^2 \tag{2}$$
$$\text{s.t.} \quad \|\alpha_i\|_0 \leq S, \quad \|\mathbf{D}(j)\|_2^2 = 1$$

where $\mathbf{D}_i$ is the ith sub-dictionary in $\mathbf{D}$, $\mathbf{D}(j)$ is the jth column in $\mathbf{D}$, $\mathbf{X}$ is training data set consisting of commodity image $\mathbf{x}_i$, $\mathbf{H}$ is the class label set of $\mathbf{x}_i$, $\mathbf{W}$ is a linear classifier to be trained, and S is sparse degree.

In order to solve α, Eq. (3) is employed, which is a sparse encoding algorithm proposed in our paper [10].

$$\arg\min_{\alpha} \frac{1}{2}\|\mathbf{x} - \mathbf{D}\alpha\|_2^2 + \lambda\|\mathcal{G}(\alpha)\|_{\ell_{2,0}} \tag{3}$$

where $\|\mathcal{G}(\alpha)\|_{\ell_{2,0}} = \sum_{i=1}^{m}\|\alpha \in C_i\|_2$ is a constraint term of mixed norm $\ell_{2,0}$, λ is a tuning coefficient, which controls the constraint degree of $\ell_{2,0}$. ℓ_2 norm constraint is imposed on sparse sub vector α_i by $\mathcal{G}(\alpha)$, $\alpha \in C_i$, $\alpha = [\alpha_1, \ldots, \alpha_m]$, and C_i is the ith class.

After gaining sparse representation α, we can exploit Eq. (4) to obtain reconstruction residual e_i.

$$e_i = \|\mathbf{x} - \mathbf{D}_i\alpha_i\|_2^2 \tag{4}$$

where $e_i = \|\mathbf{x} - \mathbf{D}_i\alpha_i\|_2^2$ is the reconstruction residual between original image $\mathbf{x}$ and its reconstructed image $\hat{\mathbf{x}} = \mathbf{D}_i\alpha_i$, and α_i is sparse sub vector corresponding to the ith sub-dictionary $\mathbf{D}_i$.

If we only use image $\mathbf{x}$ to classify commodity $\mathbf{t}$, we can obtain label $y = i$ by Eq. (5).

$$y = \arg\min_i \|\mathbf{x} - \mathbf{D}_i\alpha_i\|_2^2 \tag{5}$$

In order to fuse image and text information for commodity classification, we define an equation to calculate image classification confidence $Conf(\mathbf{x})$, as shown in Eq. (6).

$$Conf(\mathbf{x}) = \frac{\|\mathbf{x} - \mathbf{D}_i\alpha_i\|_2^2}{\max\left(\|\mathbf{x} - \mathbf{D}_j\alpha_j\|_2^2\right)}, (1 < i, j \le m) \tag{6}$$

where $Conf(\mathbf{x})$ is the confidence level of $\mathbf{x}$ belonging to the ith class, m is the number of classes.

3.3 Optimizing K-Nearest Neighbor Algorithm for Commodity Text Classification via Bayesian Posterior Probability

In commodity data set $\mathbf{P}$, commodity text $\mathbf{t}_i = \{(t_i, y_i)\} \in \mathbf{P}$, t_i is the text description of the ith commodity, y_i is its label. That is, if $\mathbf{t}_i$ belongs to the lth class, then $y_i = l$. There are m different words $t_i{}^j (1 \le j \le m)$ in text t_i, we can define as $t_i = (t_i^1, t_i^2, \ldots, t_i^j, \ldots, t_i^m)$, with the statistical frequency f_i^j of word t_i^j in t_i, we can rewrite $t_i = (f_i^1, f_i^2, \ldots, f_i^j, \ldots, f_i^m)$.

We employ KNN algorithm to classify a commodity $\mathbf{t}_t = (t_t, y_t)$, its process is as follows: Firstly, we use Eq. (7) to calculate the similarity between $\mathbf{t}_t$ and $\mathbf{t}_i$, then select k most similar $\mathbf{t}_i$ to form data set N_k as k nearest neighbors for $\mathbf{t}_t$.

$$S(\mathbf{t}_t, \mathbf{t}_i) = \frac{\mathbf{t}_t \cdot \mathbf{t}_i}{|\mathbf{t}_t| \times |\mathbf{t}_i|} = \frac{\sum\limits_{j=1}^{m}\left(f_t{}^j \times f_i^j\right)}{\sqrt{\sum\limits_{j=1}^{m}\left(f_t^j\right)^2} \times \sqrt{\sum\limits_{j=1}^{m}\left(f_i^j\right)^2}} \tag{7}$$

Secondly, after obtaining k-nearest neighbor data set N_k, we exploit Eq. (8) to compute the proportion $Prop^l$ of each class in N_k.

$$Prop^l = \frac{\sum\limits_{t_i \in N_k} sign(\mathbf{t}_i)}{k} \quad (1 \le l \le m, 1 \le i \le k)$$

s.t. $\tag{8}$

$$sign(\mathbf{t}_i) = \begin{cases} 1, & y_i = l \\ 0, & y_i \ne l \end{cases}, \quad \sum\limits_{l=1}^{m} Prop^l = 1$$

where $Prop^l$ is the proportion of $t_i \in N_k$ belonging to the lth class.

Finally, after calculating the proportion of each class in N_k, we use Eq. (9) to determine label y_t.

$$y_t = \arg\max_l \left\{ Prop^l \right\} \quad (1 \leq l \leq m) \tag{9}$$

However, there exist two problems in KNN algorithm - (i) when using absolute frequency f_i^j as the contribution degree of word t_i^j for text classification, it can causes high-frequency words to submerge low-frequency words, and (ii) in training data set, the number of samples from different classes is not distributed uniformly, which leads to classification based on Eq. (9) unfair.

In order to solve above two problems, we propose a Weighted Bayesian K-Nearest Neighbor algorithm(WB-KNN). The idea is to employ relative frequency tf_i^j to quantify the contribution degree of word t_i^j, calculate Bayesian posterior probability $P\left(y_t^l\right)$, and use it for text classification, instead of $Prop_i^j$.

The process of WB-KNN is as follows: Firstly, Eq. (10) is employed to calculate the relative frequency tf_i^j of word t_i^j.

$$tf_i^j = \frac{f_i^j}{\sum\limits_l^m f_i^j} \times \log\left(\frac{m+1}{\sum\limits_i^{m+1} sign(\mathbf{t}_i)} \right) \tag{10}$$

$$\text{s.t.} \quad sign\left(\mathbf{t}_i\right) = \begin{cases} 0, & t_i \in \mathbf{t}_i \\ 1, & t_i \notin \mathbf{t}_i \end{cases}$$

where $\sum\limits_i^n f_i$ is the sum of word frequency of t_i in $\mathbf{t}_i$. There are m samples in training data set. Because different text $\mathbf{t}_i$ have different number words, we can normalize f_i^j by $\frac{f_i^j}{\sum\limits_i^m f_i^j}$. With this method, we can obtain the relative frequency tf_i^j of t_i^j.

Secondly, after obtaining relative frequency tf_i^j, we put tf_i^j into Eq. (7) to form N_k, k-nearest neighbor data set. Thirdly, we use N_k as prior knowledge of $\mathbf{t}_t$ to calculate Bayesian posterior probability of $\mathbf{t}_t$ when it belongs to class l. Given a text $\mathbf{t}_t$ to be classified, we can define event as $E_{\mathbf{t}_t}^l$ that $\mathbf{t}_t$ belongs to class l or not. If $\mathbf{t}_t$ belongs to class l, then $y_{\mathbf{t}_t}^l = 1$, otherwise $y_{\mathbf{t}_t}^l = 0$. It is known that N_k is k-nearest neighbors of $\mathbf{t}_t$. When the number of samples in N_k belonging to class l is N_k^l, and $y_{\mathbf{t}_t}^l = 1$, we can define this event as $E_{N_k}^l$, so Bayesian prior probability of $\mathbf{t}_t$ with class l is $P\left(E_{\mathbf{t}_t}^l | E_{N_k}^l\right)$. When $\mathbf{t}_t$ belongs to class l, and the number of samples in N_k belonging to the lth class is N_k^l, the probability of this event is $P\left(E_{N_k}^l | E_{\mathbf{t}_t}^l\right)$.

Based on Bayesian theorem, rewrite the relationship between $P\left(E_{\mathbf{t}_t}^l | E_{N_k}^l\right)$ and $P\left(E_{N_k}^l | E_{\mathbf{t}_t}^l\right)$ is as shown in Eq. (11).

$$P\left(E_{\mathbf{t}_t}^l | E_{N_k}^l\right) = \frac{P\left(E_{\mathbf{t}_t}^l\right) P\left(E_{N_k}^l | E_{\mathbf{t}_t}^l\right)}{P\left(E_{N_k}^l\right)} \tag{11}$$

Finally, with Eq. (12), we can calculate maximum posterior probability $P(y_{\mathbf{t}}^l)$ of $\mathbf{t}_t$ belonging to the lth class, that is $y_{\mathbf{t}_t}^l = 1$.

$$
\begin{aligned}
P(y_{\mathbf{t}_t}^l) &= \arg\max_l \ P(E_{\mathbf{t}_t}^l | E_{N_k}^l) \\
&= \arg\max_l \ \frac{P\left(E_{\mathbf{t}_t}^l\right) P\left(E_{N_k}^l | E_{\mathbf{t}_t}^l\right)}{P\left(E_{N_k}^l\right)} \\
&= \arg\max_l \ P\left(E_{\mathbf{t}_t}^l\right) P\left(E_{N_k}^l | E_{\mathbf{t}_t}^l\right)
\end{aligned}
\tag{12}
$$

where $P\left(E_{\mathbf{t}_t}^l\right)$ and $P\left(E_{N_k}^l | E_{\mathbf{t}_t}^l\right)$ can be derived from training data set.

In order to fuse image and text information for commodity classification, we define an equation to calculate text classification confidence $Conf\,(\mathbf{t})$ when $y_{\mathbf{t}}^i = 1$ as shown in Eq. (13).

$$
Conf\,(\mathbf{t}) = 1 - P(y_{\mathbf{t}}^i)
$$
$$
\text{s.t.} \quad \sum_{i=1}^{m} P(y_{\mathbf{t}}^i = 1).
\tag{13}
$$

where $Conf\,(\mathbf{t})$ is the confidence level of $\mathbf{t}$ belonging to the ith class, m is the number of classes.

3.4 Fusing Two Different Modal Classification Confidences to Get Multi-modal Classification Confidence via Global Normalization

The relationship between multi-modal classifier $classifier_{fusion}\,(\cdot)$ and single-modal classifier $classifier_i\,(\cdot)$ is shown in Eq. (14).

$$
classifier_{fusion}\,(\cdot) = \sum_{i=1}^{n} w_i \times classifier_i\,(\cdot)
\tag{14}
$$

where w_i is weighted coefficient to be found.

Because the output of $classifier\,(\cdot)$ is $Conf\,(\cdot)$, Eq. (14) can be rewritten as Eq. (15).

$$
Conf_{fusion}^l\,(\cdot) = \sum_{i=1}^{n} w_i \times Conf_i^l\,(\cdot)
\tag{15}
$$

After putting $Conf^l(\mathbf{x})$ and $Conf^l(\mathbf{t})$ into Eq. (16), we can obtain multi-modal fusion classification confidence $Conf^l(\mathbf{x}, \mathbf{t})$.

$$
Conf^l(\mathbf{x}, \mathbf{t}) = w_1 \times Conf^l(\mathbf{x}) + w_2 \times Conf^l(\mathbf{t})
\tag{16}
$$

Combining Eq. (6) with Eq. (13), we can rewrite Eq. (16) as Eq. (17).

$$Conf^l(\mathbf{x}, \mathbf{t}) = w_1 \times Conf^l(\mathbf{x}) + w_2 \times Conf^l(\mathbf{t})$$

$$= w_1 \times \frac{\|\mathbf{x} - \mathbf{D}_i \alpha_i\|_2^2}{\max\left(\|\mathbf{x} - \mathbf{D}_j \alpha_j\|_2^2\right)} + w_2 \times (1 - P(y_\mathbf{t}^i)) \tag{17}$$

$$\text{s.t.} \quad \sum_{l=1}^m P\left(y_\mathbf{t}^l\right) = 1, 1 \le i \le n, 1 \le j \le m$$

Because $\|\mathbf{x} - \mathbf{D}_i \alpha_i\|_2^2$ and $1 - P(y_\mathbf{t}^l)$ are two different physical quantities, they cannot be fused directly, and we have to normalize them as shown in Eq. (18).

$$Conf^l(\mathbf{x}, \mathbf{t}) = w_1 \frac{\|\mathbf{x} - \mathbf{D}_l \alpha_l\|_2^2}{\max\left(\|\mathbf{x} - \mathbf{D}_i \alpha_i\|_2^2\right)} + w_2 \frac{\left(1 - P\left(y_\mathbf{t}^l\right)\right)}{\max\left(1 - P\left(y_\mathbf{t}\right)\right)} \tag{18}$$

$$\text{s.t.} \quad \sum_{l=1}^m P\left(y_\mathbf{t}^l\right) = 1, 1 \le i, l \le n$$

After obtaining normalized multi-modal classification confidence $Conf^l(\mathbf{x}, \mathbf{t})$, we use Eq. (19) to classify commodity $\mathbf{t}_t = (t_t, y_t)$.

$$y_t = \arg\min_l \left\{Conf^l(\mathbf{x}, \mathbf{t})\right\} \tag{19}$$

4   Experimental Results and Analysis

Our experiment is carried out in three cases - (i) comparing our SRC-$\ell 2, 0$ algorithm with other algorithms, all algorithms only employ commodity image information, (ii) comparing our WB-KNN algorithm with KNN algorithm, both of them only exploit commodity text information, and (iii) comparing our MMIT algorithm with other algorithms, our algorithm utilizes both commodity image and text information, while others only use commodity image.

We choose two data sets, Swarthmore[1] and IT. Swarthmore is a commodity image data set built by Swarthmore college. Because most public commodity data sets only contain images, we built IT data set by ourselves, which contains both commodity image and text information. With cross validation, we choose average classification accuracy for 10 times as final result.

4.1   Only Using Commodity Image Information for Classification

When we only employ image information for commodity classification, the average classification accuracy of each algorithm in Swarthmore and IT data set are presented in Table 1 and Table 2 respectively, which show that the performance of our algorithm can achieve a higher classification accuracy than other

---

[1] http://www.sccs.swarthmore.edu/users/09/btomasi1/images.zip.

algorithms. Note that all algorithms in the table are based on sparse representation method, and we can obtain better results when using $\ell_{2,0}$ mixed norm to optimize it.

Table 1. Comparison of SRC-$\ell2,0$ with other algorithms, only using image information in Swarthmore data set.

Algorithms	Size of sparse sub-dictionary			
	5	10	15	20
SRC	64.3%	69.7%	75.2%	78.1%
CRC-RLS	67.0%	74.6%	79.4%	82.5%
K-SVD	69.1%	75.8%	81.1%	83.0%
D-KSVD	70.4%	78.7%	82.7%	84.2%
LC-KSVD	71.1%	79.1%	84.0%	85.7%
SRC-$\ell2,0$	**77.3%**	**85.2%**	**90.1%**	**92.5%**

Table 2. Comparison of SRC-$\ell2,0$ with other algorithms, only using image information in IT data set.

Algorithms	Size of sparse sub-dictionary			
	5	10	15	20
SRC	67.1%	71.3%	77.5%	83.2%
CRC-RLS	69.7%	78.6%	82.1%	87.5%
K-SVD	69.8%	79.1%	85.4%	88.6%
D-KSVD	71.5%	81.7%	87.6%	89.4%
LC-KSVD	73.1%	83.3%	88.2%	91.2%
SRC-$\ell2,0$	**83.0%**	**91.7%**	**94.2%**	**95.4%**

4.2 Only Using Commodity Text Information for Classification

When we only employ text information for commodity classification, the performances of algorithms are expressed in Fig. 1, which indicate that our algorithm WB-KNN, using Bayesian posterior probability to optimize KNN algorithm, can achieve a better performance than KNN. Furthermore, it can overcome two problems in the KNN algorithm mentioned above in Sect. 3.3.

4.3   Using both Commodity Image and Text Information for Classification

When we use both image and text information for commodity classification, the average classification accuracy of each algorithm in IT data set is displayed in Table 3. The results demonstrate that with the help of image and text information, the accuracy of MMIT is higher than other algorithms, which only use image information for commodity classification.

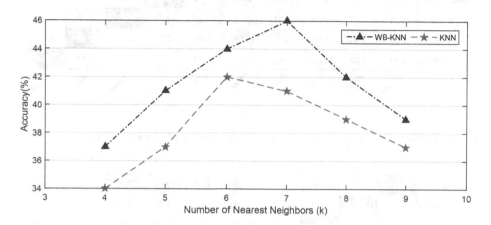

Fig. 1. Comparing the accuracy of WB-KNN with that of KNN, both of them only using text information.

Table 3. Comparison of MMIT with other algorithms, using both image and text information in MMIT, and only using image information in other algorithms.

Algorithms	Size of sub-dictionary			
	5	10	15	20
SRC	67.1%	71.3%	77.5%	83.2%
CRC-RLS	69.7%	78.6%	82.1%	87.5%
K-SVD	69.8%	79.1%	85.4%	88.6%
D-KSVD	71.5%	81.7%	87.6%	89.4%
LC-KSVD	73.1%	83.3%	88.2%	91.2%
MMIT	**88.3%**	**96.1%**	**96.6%**	**97.2%**

It can be seen that - (i) with the increase of the size of sparse sub-dictionary, the accuracy of all algorithms improves gradually, which is consistent with the idea of sparse representation classification, that is, the bigger size of sub-dictionary is, the more powerful ability of each class is expressed by expanded

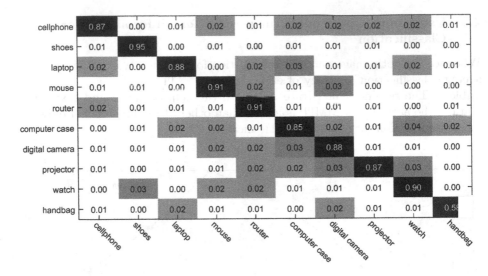

Fig. 2. Performance of SRC algorithm on each class in IT data set, only using image information.

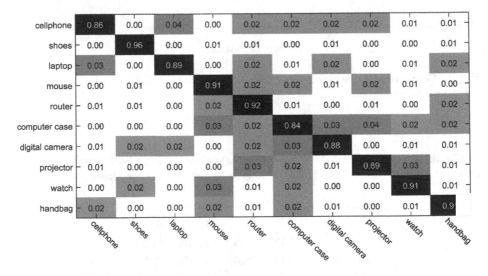

Fig. 3. Performance of CRC-RLS algorithm on each class in IT data set, only using image information.

feature space, (ii) when the size of sub-dictionary is 20, all algorithms can achieve their best performance. It is better than others because our algorithm employs both image and text information for classification, but others only use image information, we can draw conclusion that there is the relationship of "symbiosis and complementary" between commodity image and text information to improve classification performance.

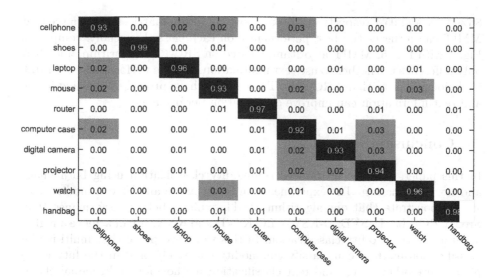

Fig. 4. Performance of SRC-$\ell 2, 0$ algorithm on each class in IT data set, only using image information.

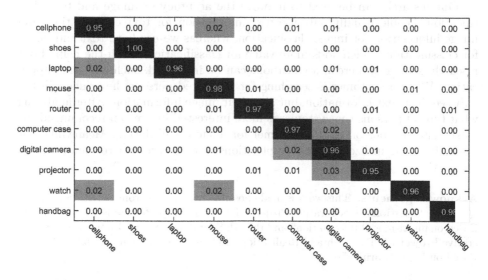

Fig. 5. Performance of MMIT algorithm on each class in IT data set, using both image and text information.

When the size of sub-dictionary is 20, Fig. 2, 3, 4 and 5 reveal the performance of every algorithm on each class in IT data set. For class "cellphone", Fig. 2 shows that the accuracy of SRC is 87%, and 1% images are classified into class "laptop", "router" and "handbag" respectively, and 2% of them are labeled as "mouse", "computer case", "digital camera", "projector" and "watch". Meanwhile the classification accuracy of CRC-RLS, our algorithm SRC-$\ell 2, 0$ and MMIT are

86%, 93% and 95% respectively. From above confusion matrices, we can see that MMIT can achieves the best performance on each class. Especially, comparing Fig. 4 with Fig. 5, MMIT algorithm is based on SRC-$\ell2,0$ algorithm, and their only difference is whether to use text information, but the performance of MMIT is obviously better than SRC-$\ell2,0$. These results have proved that fusing image and text information can improve classification performance.

5   Conclusion

In this paper, we have introduced commodity classification using both image and text information. The experiments on Swarthmore and IT commodity data sets demonstrate that our algorithm MMIT exhibits better performance than SRC, CRC-RLS, K-SVD, D-KSVD, LC-KSVD and KNN. It can achieve a higher accuracy of commodity classification. In this work, we prove that multi-modal classification method can classify commodity effectively, and in the future, we plan to use other image and text classification method for multi-modal classification, such as Convolutional Neural Network (CNN) over a larger data set ImageNet.

Our research can be used to improve the accuracy of image and text classification. People usually only use text information for text classification, and image information for image classification, whereas we use both text and image for classification, which adds more value for classification. With the help of our research, some commercial applications can be improved, such as dating application. When a customer is searching for interested persons, he or she will not only see their text information, but also their image information. After analyzing what type of persons the customer is more interested in, we can recommend the most matching person to the customer for a successful dating. Meanwhile, our research can be utilized in other applications, such as product recommendation application.

Acknowledgements. This work is made possible by support from the 4th Shandong-Quebec International Cooperative Project of "Commodity Recommendation System Based on multi-modal Information" and the 5th Shandong-Quebec International Cooperative Project of "Research and Realization of Commodity Recommendation System Based on Deep Learning".

References

1. Xie, J., Zheng, Z., Gao, R., Wang, W., Zhu, S.C., Nian Wu, Y.: Learning descriptor networks for 3D shape synthesis and analysis. In: Proceedings of the IEEE Conference on Computer Vision and Pattern Recognition, pp. 8629–8638 (2018)
2. Allagwail, S., Gedik, O.S., Rahebi, J.: Face recognition with symmetrical face training samples based on local binary patterns and the Gabor filter. Symmetry **11**(2), 157 (2019)
3. Shen, C., Chen, L., Dong, Y., Priebe, C.: Sparse representation classification via screening for graphs. arXiv preprint arXiv:1906.01601 (2019)

4. Gou, J., Ma, H., Weihua, O., Zeng, S., Rao, Y., Yang, H.: A generalized mean distance-based k-nearest neighbor classifier. Expert Syst. Appl. **115**, 356–372 (2019)
5. Wang, H., Peng, J., Xianping, F.: Co-regularized multi-view sparse reconstruction embedding for dimension reduction. Neurocomputing **347**, 191–199 (2019)
6. Khaleghi, B., Khamis, A., Karray, F.O., Razavi, S.N.: Multisensor data fusion: a review of the state-of-the-art. Inf. Fusion **14**(1), 28–44 (2013)
7. Woźniak, M., Graña, M., Corchado, E.: A survey of multiple classifier systems as hybrid systems. Inf. Fusion **16**, 3–17 (2014)
8. Moreno-Seco, F., Iñesta, J.M., de León, P.J.P., Micó, L.: Comparison of classifier fusion methods for classification in pattern recognition tasks. In: Yeung, D.-Y., Kwok, J.T., Fred, A., Roli, F., de Ridder, D. (eds.) SSPR /SPR 2006. LNCS, vol. 4109, pp. 705–713. Springer, Heidelberg (2006). https://doi.org/10.1007/11815921_77
9. Tang, Y., Li, X., Liu, Y., Wang, J., Xu, Y.: Sparse dimensionality reduction based on compressed sensing. In: IEEE WCNC, pp. 3373–3378 (2014)
10. Yufang, T., Xueming, L., Yan, X., Shuchang, L.: Group lasso based collaborative representation for face recognition. In: 2014 4th IEEE International Conference on Network Infrastructure and Digital Content, pp. 79–83. IEEE (2014)

A New DCT-FFT Fusion Based Method for Caption and Scene Text Classification in Action Video Images

Lokesh Nandanwar[1]([✉]) [iD], Palaiahnakote Shivakumara[1] [iD], Suvojit Manna[2] [iD], Umapada Pal[3] [iD], Tong Lu[4] [iD], and Michael Blumenstein[5] [iD]

[1] Faculty of Computer Science and Information Technology, University of Malayasia, Kuala Lumpur, Malaysia
lokeshnandanwar150@gmail.com, shiva@um.edu.my
[2] Department of Computer Science and Engineering, Jalpaiguri Government Engineering College, Jalpaiguri, India
davsuvo@gmail.com
[3] Computer Vision and Pattern Recognition Unit, Indian Statistical Institute, Kolkata, India
umapada@isical.ac.in
[4] National Key Lab for Novel Software Technology, Nanjing University, Nanjing, China
lutong@nju.edu.cn
[5] University of Technology Sydney, Ultimo, Australia
michael.blumenstein@uts.edu

Abstract. Achieving better recognition rate for text in video action images is challenging due to multi-type texts with unpredictable backgrounds. We propose a new method for the classification of captions (which is edited text) and scene texts (which is part of an image in video images of Yoga, Concert, Teleshopping, Craft, and Recipe classes). The proposed method introduces a new fusion criterion-based on DCT and Fourier coefficients to extract features that represent good clarity and visibility of captions to separate them from scene texts. The variances for coefficients of corresponding pixels of DCT and Fourier images are computed to derive the respective weights. The weights and coefficients are further used to generate a fused image. Furthermore, the proposed method estimates sparsity in Canny edge image of each fused image to derive rules for classifying caption and scene texts. Lastly, the proposed method is evaluated on images of five above-mentioned action image classes to validate the derived rules. Comparative studies with the state-of-the-art methods on the standard databases show that the proposed method outperforms the existing methods in terms of classification. The recognition experiments before and after classification show that the recognition performance rate improves significantly after classification.

Keywords: Caption text · Scene text · Fusion · Caption and scene text classification · Action image recognition

© Springer Nature Switzerland AG 2020
Y. Lu et al. (Eds.): ICPRAI 2020, LNCS 12068, pp. 80–92, 2020.
https://doi.org/10.1007/978-3-030-59830-3_7

1 Introduction

Understanding text in action video images is a hot topic for researchers in the fields of computer vision and image processing [1–3]. This is because of its several real world applications such as surveillance, human-computer interaction, robotics and indexing videos [2, 3]. There are powerful methods for recognizing the action in the video proposed in literature [2, 3]. However, due to the difficulty in defining the exact relationship between coherent objects in images with the help of content-based features, achieving better recognition is still considered as elusive goal. The difficulty is further aggravated for still images without temporal frames [4]. Therefore, in this work, we propose to make an attempt to find a solution to the above complex issues based on text information if text is present in the video images. It is true that either caption, which is edited text or scene text, which is part of an image, play a vital role in action recognition. Since text in the action video images is closer and it provides semantic information about the content of the video image, achieving better recognition results improve performance of the action recognition in the video. Sometimes, a few applications may require only caption text as it describes the content of the video while scene text does not. For these applications, caption text separation from scene text is important to obtain accurate results [5].

(a) Caption: Craft   Yoga   Recipe

(b) Scene: Concert   Teleshopping

Fig. 1. Examples of scene and caption texts in action images.

For recognizing text in video and natural scene images, the methods are developed in the past [6, 7] addresses several challenges of natural scene text. However, when the images contain both caption and scene text, the performance of the method degrades. This is due to different nature and characteristics of caption and scene text. Since caption is edited text, one can expect good contrast and visibility while scene is natural text, one cannot predict the nature of text [8]. Sample caption and scene texts of different action images are shown in Fig. 1(a) and Fig. 1(b), respectively, where texts are marked by red colored rectangles. For the text in the images, the proposed method tested recognition method [7], which is state-of-the-art method before and after classification. Before classification, caption and scene texts are considered as input for recognition while after classification individual text is considered as input for recognition. The recognition results of the method reported in Table 1 show that there is incorrect recognition results before classification and correct recognition results after classification. Therefore, it is noted that classification is necessary to improve performance of recognition of the text in video irrespective of applications. However, in this work, we consider action video images as case study for experimentation. In this work, we consider five video images of action classes, namely, Concert which describes music, Teleshopping which describes marketing of goods, Craft which describes art making, Yoga which describes particular exercise, and Recipe which describes food-making as shown in Fig. 1, where both captions and scene texts are present in these images. The main reason to choose the above-mentioned action classes is that they are quite popular for indexing and retrieval in the field of computer vision.

Table 1. Recognition results of the method [7] before and after classification of action images

Action classes	Before classification	After classification
Craft	cut the wnwarted	cut the unwanted
Yoga	chatvarga nianiasedu	chaturanga dandasana
Recipe	of an inch orite	of an inch or like
Concert	Aloop	Alaap
Teleshopping	orals	oralb

There are methods developed in the literature [8–12] for the classification of scene and caption texts in video images. However, the methods may not give satisfactory results for texts in video images of action classes because the presence of indoor and outdoor scenes in different action classes makes the problem much more complex. It is evident from the images shown in Fig. 1(a) and Fig. 1(b), where one can clearly see the effects of indoor and outdoor scenes on text information. Hence, caption and scene texts separation for video images of action classes is challenging and interesting compared to existing methods.

2   Related Work

In this work, we review the methods which focus on the classification of caption and scene texts in video. Raghunandan et al. [8] proposed a method for classifying video, natural scene, mobile and born digital images based on text information. The methods work well when images have different qualities according to classes. This is not necessarily true for texts in video images of actions, where one can expect texts may share the same quality. Shivakumara et al. [9] proposed the separation of graphics and scene texts in video frames based on the combination of Canny and Sobel edge images along with the ring radius transform. Since the method works based on edge quality, it may not perform well for complex images of video. Roy et al. [10] proposed tampered features for the classification of captions and scene text. In this method, DCT coefficients are explored to detect tampered information in caption images such that it can be separated from scene texts. It is noted that DCT coefficients alone may not be sufficient to classify complex caption and scene texts in action images. In a similar way, Bhardwaj et al. [11] explored tampered features which represent superimposed texts in video for text detection by separating scene texts from captioned ones. The method is good for images with high quality and contrast.

Recently, Roy et al. [12] proposed temporal integration for word-wise caption and scene text identification. This method may not work well for still images as it requires temporal frames for achieving accurate results. Later, Roy et al. [13] proposed rough-fuzzy based scene categorization for text detection and recognition in video. The method defines shapes based on edge components for the classification of videos of different text types. However, the scope of this method is to categorize video frames but not caption and scene texts in video. Since use of convolutional network is popular and it solves complex issue, Ghosh et al. [5] proposed identifying the presence of graphical text in scene images using CNN. The method considers text which is edited and text in natural scene images as graphical text for classification. To achieve this, the method explores CNN. However, the method does not consider caption and scene text in video images for classification. From the above discussions, it is noted that none of the methods are perfect for video images of action classes, where one can expect texts with more unpredictable background variations.

Hence, this work focuses on the classification of caption and scene texts in video images of different action classes. As mentioned in the Introduction Section, caption text exhibits good quality, contrast and uniform color, while the nature of scene text is unpredictable. To extract such an observation, inspired by the method in [10] where it is shown that DCT coefficients help in extracting distinct features for the differentiation of caption texts from scene ones via classification, we explore the DCT domain in a different way in this work. Since the problem being considered is a complex one, we are inspired by the method in [14] where Fourier coefficients are used for separating coefficients that represent actual edge pixels from those that represent noise ones for identifying printers. We explore the Fourier coefficients in a different way for the classification of scene and caption texts in this work. The main contributions of the work are as follows: (1) the way in which the advantages of both DCT and Fourier coefficients are integrated and utilized to obtain fused images is novel, and (2) exploring sparsity at the component level for Canny edge images of fused ones for classifying caption and scene texts is new.

3 Proposed Method

This work considers detected texts from video images of action classes as the input for the classification of scene and caption texts. This is because there are methods which detect texts well irrespective of text types in videos [15]. As mentioned in the previous section, DCT and Fourier coefficients are useful in extracting distinct properties of caption and scene texts. We exploit the above coefficients as motivated by the method in [16], where the fusion concept is introduced in the wavelet domain to enhance degraded medical images. In other words, the method in [16] fuses coefficients that represent edge pixels in different wavelet sub-bands. This observation motivated us to propose a new fusion approach by combining coefficients in the DCT and Fourier domains. This is justifiable because we believe that if caption text has good quality and contrast, the same thing should be reflected in the fusion image. When the nature of scene text is unpredictable, one cannot expect regular patterns in the fused image represented by coefficients. As a result, the fused image of a scene text image results in sparsity, which is opposite to caption text.

Therefore, we can conclude that a fusion image of caption text provides edge information, while a fusion image of scene text provides sparse information. To extract such an observation for classifying caption and scene texts, we perform the inverse Fourier transform on fused images of both scene and caption texts. If the reconstructed image of scene text provides sparse information, the Canny edge operator gives nothing; while for caption text, the Canny edge operator gives edges of text information. In this way, the proposed method explores sparsity for classifying caption and scene texts in this work.

3.1 DCT and FFT Coefficients for Fusion

For the input caption and scene text line images shown in Fig. 2(a), the proposed method obtains Fourier and DCT coefficients, respectively using Eq. (1) and Eq. (2) as shown in Fig. 1(b) and Fig. 1(c). It is observed from Fig. 1(b) and Fig. 1(c) that the caption text of Fourier and DCT coefficients is brighter than the scene text of Fourier and DCT coefficients. This is understandable because caption text has good clarity, contrast and clear differences between the text and its background, while scene text does not. As a result, the Fourier and DCT coefficients of scene text images appear darker compared to caption text images as shown in Fig. 2(b) and Fig. 2(c). With this cue, we propose to compute the variance of each pixel in the input image by defining a local window as defined in Eq. (3) for the respective Fourier and DCT coefficients images. For variance computations, the proposed method considers the respective coefficients corresponding to pixels of the local window defined over the input image. In other words, the variance is computed using frequency coefficients. However, the local window is moved pixel by pixel by referring to the input image. The variances of the respective Fourier and DCT coefficients are used to derive weights as defined in Eq. (4) and Eq. (5). Finally, the derived weights are combined with the respective Fourier and DCT coefficients as defined in Eq. (6), which results in a fused image. The effect of fusion can be seen in Fig. 2(d) for both caption and scene texts, where we can see clearly that the fused image of caption text appears brighter than that of the scene text. This is true because the variance of coefficients of Fourier and DCT matches for caption text, while it mismatches for scene text. As a result, one can expect the fused image of caption text must contain high frequency coefficients that represent edge pixels, while that of scene text contains more low frequency coefficients that represent zero.

$$I_{FFT}(\text{u, v}) = \frac{1}{MN} \sum_{x=0}^{M-1} \sum_{y=0}^{N-1} I(x, y)e^{-2\pi i\left(\frac{ux}{M} + \frac{vy}{N}\right)} \tag{1}$$

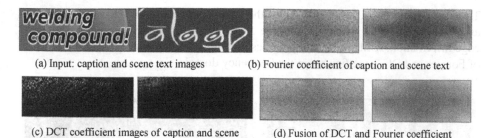

(a) Input: caption and scene text images   (b) Fourier coefficient of caption and scene text

(c) DCT coefficient images of caption and scene   (d) Fusion of DCT and Fourier coefficient

Fig. 2. Examples of obtaining a fused image for caption and scene text images.

$$I_{DCT}(u,\ v) = \frac{2}{\sqrt{MN}} \sum_{x=0}^{M-1} \sum_{y=0}^{N-1} \lambda(x).\lambda(y).\cos\left[\frac{u\pi}{2M}(2x+1)\right].\cos\left[\frac{v\pi}{2N}(2y+1)\right].I(x,y)$$

$$(2)$$

$$Var(x,\ y) = \frac{1}{M \times M} \sum_{x=0}^{M-1} \sum_{y=0}^{N-1} \left[abs(p(i,\ j) - \mu)\right]^2 \tag{3}$$

where $Var(x, y)$ denotes the variance of a local window of size 3×3, $p(i, j)$ is the coefficient value at position (i, j), and μ is the mean of the coefficients of the local window.

$$\omega_{I_{FFT}} = \frac{Var_{I_{FFT}}}{Var_{I_{FFT}} + Var_{I_{DCT}}} \tag{4}$$

$$\omega_{I_{DCT}} = \frac{Var_{I_{DCT}}}{Var_{I_{FFT}} + Var_{I_{DCT}}} \tag{5}$$

Where $\omega_{I_{FFT}}$ denotes the weight with respect to Fourier coefficients, $\omega_{I_{DCT}}$ denotes the weight with respect to DCT coefficients, and I denotes an image.

$$F_{coeff} = I_{FFT} \circ \omega_{I_{FFT}} + I_{DCT} \circ \omega_{I_{DCT}} \tag{6}$$

where $\circ$ is the hadamard product between two given matrices. F_{coeff} denotes the fused image, I_{IFFT} denotes the Inverse Fourier transform, and I_{DCT} denotes the Inverse Discrete Cosine transform.

3.2 Classification of Caption and Scene Text

It is noted from the fused image obtained from the previous step that a caption text image contains vital information, while a scene text image results in sparsity. To extract such an observation, the proposed method applies the Inverse Fourier transform for both the fused images of caption and scene texts using Eq. (7). This step outputs reconstructed images of caption and scene texts as shown in Fig. 3(a). When we look at the reconstructed images of caption and scene texts, it is non-trivial to notice sparsity and non-sparsity.

Therefore, the proposed method applies the Canny edge operator on the fused images as shown in Fig. 3(b), where one can see clearly that caption text provides significant edge information, while scene text provides nothing. This is the advantage of the fusion of Fourier and DCT coefficients in the frequency domain.

(a).   (b)

Fig. 3. Sparse estimation at the component level for classification of caption and scene texts. (a) The result of the inverse Fourier transform on the fused image of caption and scene text images. (b) Canny edge images of a fused image of caption and scene text images

To extract the sparsity property in edge images of caption and scene text images, the proposed method considers edge components as defined by the bounding box shown in Fig. 3(c). For each pixel of each edge component, we define the window of size 3×3 to check whether it contains more than one white pixel. If the window satisfies the above condition, the proposed method counts it as non-sparsity, else it is counted as sparsity. In this way, the proposed method obtains the number of sparsity and non-sparsity counts for each component in the image. Then it computes the average for sparsity and non-sparsity counts of all the edge components separately in the image. If the average of the non-sparsity count is larger than that of the sparsity count, the image is considered as caption text, else it is scene text. Since most of the time, for scene text images, the edge operator gives nothing. Therefore, scene text is classified with high sparsity.

$$I_{IFFT}(u, v) = \sum_{x=0}^{M-1} \sum_{y=0}^{N-1} F_{coeff}(x, y) e^{2\pi i(\frac{ux}{M} + \frac{vy}{N})} \tag{7}$$

4   Experimental Results

Since our target is to classify caption and scene text in action video images, we create our own dataset by collecting from different sources, such as YouTube, Instagram, the Internet and from our own camera, which comprises 429 video images for Concert, 509 for Recipe, 517 for Craft, 507 for Teleshopping and 524 for Yoga action classes. In total, 2486 video images for all the five action classes are considered. The dataset includes video images of different resolutions ranging from 426×240 to 1980×1080, different contrasts, foregrounds and background complexities. We detect caption and scene text images from the above dataset, which consists of 2814 text images including caption and scene texts. Since the dataset includes text with different varieties, the dataset is considered to be complex and fair for evaluating the performance of the proposed and existing methods. To test the objectiveness of the proposed method, we also conducted experiments on a standard dataset, which is used for caption and scene text classification in video images [10]. This dataset consists of 900 caption and 650 scene texts, which gives a total of 1550 text images for experimentation.

To illustrate the effectiveness of the proposed method, we implement two state-of-the-art methods for comparative studies. Roy et al. [10] proposed new tampered features for scene and caption text classification in video, which explores DCT coefficients for extracting tampered features for classification. Ghosh et al. [5] proposed a method for identifying the presence of graphical text in natural scene images by exploring CNN. The reason to choose the above methods is that the method [10] focuses on caption and scene text in video as the proposed method while the method [5] focuses on use of CNN for detecting the presence of graphical text in the natural scene images. In this work, we modify the CNN for classification of caption and scene texts rather than identifying the presence of graphical text. We consider these two methods for comparative study in this work to show that only DCT used in method [10] is not adequate to achieve better results. The method [5] is considered to show that CNN approach does not perform well compared to the proposed rule based method. This can be justifiable because CNN works well when we have a large number of samples for training and learning while the rule based methods do not. In the same way, to validate the effectiveness of the proposed classification, we implement the following text recognition methods. Luo et al. [7] which proposes Multi-Object Rectified Attention Network (MORAN) for scene text recognition and Shi et al. [6] which proposes an Attention Scene Text Recognizer with Flexible Rectification (ASTER). The reason to choose the above two methods is that the above methods explored deep learning models for addressing several challenge of scene text recognition. In addition, since the codes of the above two methods are available, we prefer to use the same for recognition experiments in this work.

For measuring the performance of the proposed and existing methods for classification and recognition, we calculate classification rate through confusion matrix of caption and scene text classification. For recognition, we calculate character recognition rate before and after classification. As mentioned earlier, before classification both caption and scene text are considered as input for recognition while after classification individual classes are considered as input for recognition. It is expected the recognition methods should report better results for after classification compared to before classification. This is valid because the recognizer is trained with caption and scene text together for before classification. On the other hand, the recognizer is trained separately for each class after classification.

4.1 Evaluating the Proposed Classification Method

Sample qualitative results of the proposed method for each action class are shown in Fig. 4, where we can see that the proposed method classifies caption and scene texts successfully regardless of background complexities. Quantitative results of the proposed and existing methods are reported in Table 2 for our dataset. Table 2 shows that the proposed method is the best at the classification rate for all the five action classes compared to existing methods. It is noted from Table 2 that the proposed method scores the highest average classification rate which is mean of diagonal elements of confusion matrix for the Craft class and the lowest for the Concert class. This is due to texts in the Craft class that are not exposed much to open environments, while texts in the Concert class are exposed to open and closed environments.

Caption and scene text from the Concert and Craft action image classes

Caption and scene text from the Recipe and Teleshopping action image classes

Caption text from the Yoga action image class

Fig. 4. Sample images where successful classification is performed by the proposed method.

When images are exposed to both open and closed environments, they get affected by multiple factors such as uneven illumination, dim lighting, different writing styles, and the effect of rough surfaces, perspective distortion, and occlusion. The main reason for the poor results of the existing methods is that they are developed for classifying texts with different objectives but not video images of action classes. When we compare the results of two existing methods, the method [5] scores better results than the method [10]. This shows the method which explores CNN is better than the method which uses only DCT based features for classification. On the other hand, the proposed method is better than the existing methods because of the fusion criteria that have been introduced, which consider the advantages of the Fourier and DCT coefficients in the frequency domain for extracting sparsity. This demonstrates a difference in comparison to existing methods to achieve the best results. Note that the symbol '–' indicates that no classification results reported in Table 2 and this is because the Yoga action class does not provide any scene text. It is reported classification rate of only caption text. It is noted from Table 2 that the methods including proposed one score the best results at classification rate for the caption text compared to scene text. This justifies that caption text provide good clarity and visibility compared to scene text.

To test objectiveness of the proposed method, we conduct experiments on a standard dataset for classification. Quantitative results of the proposed and existing methods are reported in Table 3. It is observed from Table 3 that the proposed method is better than the existing methods in terms of average classification rate. The main drawback of existing methods is that the existing features do not have sufficient discriminative power when compared to the proposed method. When we compare the results on our dataset and the standard dataset, the proposed method gives almost consistent results if we consider the average classification rate, while the existing methods report inconsistent results. This is mainly because of the differences in complexity of the two datasets.

In order to show the usefulness of the proposed classification method, we conduct experiments for text recognition methods using the ASTER and MORAN method

Table 2. Confusion matrix of the proposed and existing methods on action image classes (in %).

Methods	Classes	Recipe		Concert		Crafts		Teleshopping		Yoga	
		Scene	Caption	Scene	Caption	Scene	Caption	Scene	Caption	Scene	Caption
Proposed	Scene	**71.4**	28.5	**74.3**	25.6	**97**	3	**71.8**	28.1	–	–
	Caption	16.2	**83.8**	27.0	**72.9**	0	**100**	20.4	**79.5**	3.42	**96.5**
	Average	**77.6**		**73.6**		**98.5**		**75.6**		**96.5**	
Roy et al. [10]	Scene	64.2	35.7	62.4	37.5	66.6	33.3	63.8	36.1	–	–
	Caption	32.5	67.4	35.2	64.7	33.5	66.4	34.5	65.4	33.2	66.7
	Average	65.8		63.55		66.5		64.6		66.7	
Ghosh et al. [5]	Scene	60.15	39.85	61.61	38.39	64.01	35.99	61.03	38.97	–	–
	Caption	18.7	81.3	8.1	91.9	7.3	92.7	13.4	86.6	6.98	93.02
	Average	70.7		76.7		78.3		73.8		93.0	

Table 3. Confusion matrix of the proposed and existing methods on Dataset [10] (in %)

Types	Proposed Method		Ghosh et al. [5]		Roy et al. [10]	
	Scene	Caption	Scene	Caption	Scene	Caption
Scene	**79.11**	20.88	69.53	15.9	65.69	34.31
Caption	21.99	**78.01**	26.01	79.67	32.62	67.38
Average	**78.56**		74.6		66.53	

[6, 7] respectively. The recognition results of both the methods for all the five action video image classes before and after classification are reported in Table 4. For before classification, we use pre-trained model with same parameters and values while for after classification, we tune the parameters, namely, "Epochs" and "Batch size" according to data of individual classes. This is the advantage of the classification. For training and testing of the recognition methods, we use 75% data for training and 25% data for testing. Quantitative results of the two recognition methods before and after classification for all the five action video image classes are reported in Table 4, where it is noted that both the recognition methods achieve better results after classification compared to before classification for all the five classes on both caption and scene text. It can also be seen in Table 4 that the MORAN is better than ASTER before and after classification. This is due to the MORAN focusses on irregular shaped text along with distortion but the ASTER focusses on distorted images. However, in case of the proposed work, one can expect both irregular shaped text and text affected by distortion. Therefore, we can conclude that the classification is useful and effective to improve recognition performance for video irrespective of applications.

Table 4. Recognition performance of the methods before and after classification (in %)

Methods	ASTER [6]			MORAN [7]		
	Before classification	After classification		Before classification	After classification	
Classes	Captio+Scene	Caption	Scene	Caption + Scene	Caption	Scene
Concert	71.2	73.6	77.8	74.4	73.3	77.2
Recipe	72.4	89.0	63.0	82.2	90.2	76.2
Crafts	74.2	80.8	70.0	75.0	81.9	70.5
Teleshopping	66.8	71.8	67.2	72.9	76.8	72.3
Yoga	79.0	80.4	–	81.2	82.0	–
Dataset [10]	61.8	69.6	66.0	63.0	63.2	69.3

Though the proposed classification method works well for several cases, sometimes, if the images suffer from very low contrast where we cannot differentiate between the foreground and background, the proposed method misclassifies them as per the samples of misclassifications shown between scene and caption text images in Fig. 5. Therefore, there is scope for undertaking improvements in the future.

(a) Scene texts are misclassified as captions due to low contrast between the foreground and background and as well as blur.

(b) Caption texts are misclassified as scene text due to background and foreground complexity variations.

Fig. 5. Limitations of the proposed method.

5 Conclusion and Future Work

In this paper, we have proposed a novel method for the classification of caption and scene texts in action video images of different classes. The proposed method introduces a new fusion concept to integrate the advantages of Fourier and DCT coefficients to extract sparsity for the classification of scene and caption texts in action images. The variances are computed in the frequency domain for both Fourier and DCT coefficients by referring to pixel positions in the input images. The variances are used to derive weights with respect to the Fourier and DCT coefficients. Furthermore, the proposed method uses weights and coefficients to generate fused images. Finally, sparsity is estimated for reconstructed images of the fused ones for classification. Experimental results on our own dataset and a standard dataset show that the proposed method outperforms the existing methods in terms of average classification rate. Experimental results on recognition show that the proposed classification improves recognition performance significantly after classification compared to before classification. However, when an image suffers from very low resolution and low contrast, the proposed method does not perform well. We will explore such situations in our future work.

References

1. Ullah, A., Ahmad, J., Muhamad, K., Sajiad, M., Baik, S.W.: Action recognition in video sequences using deep Bi-directional LSTM with CNN features. IEEE Access **6**, 1155–1166 (2018)

2. Qi, T., Xu, Y., Quain, Y., Wang, Y., Ling, H.: Image based action recognition using hint enhanced deep neural networks. Neurocomputing **267**, 475–488 (2017)
3. Taniski, G., Zalluhoglu, C., Cinbis, N.L.: Facial descriptors for human interaction recognition in still images. Pattern Recogn. Lett. **73**, 44–51 (2016)
4. Yuan, S., Smith, J.S., Zhang, B.: Action recognition from still images based on deep VLAD spatial pyramids. Signal Process. Image Commun. **54**, 118–129 (2017)
5. Ghosh, M., Mukherjee, H., Obaidullah, S. M., Santosh, K. C., Das, N., Roy, K.: Identifying the presence of graphical texts in scene images using CNN. In: Proceedings of the ICDARW, pp. 86–91 (2019)
6. Shi, B., Yang, M., Wang, X., Lyu, P., Yao, C., Bai, X.: ASTER: an attentional scene text recognizer with flexible rectification. IEEE Trans. PAMI **41**, 2035–2048 (2019)
7. Luo, C., Jin, L., Sun, Z.: MORAN: a multi-object rectified attention network for scene text recognition. Pattern Recogn. **90**, 109–118 (2019)
8. Raghunandan, K.S., Shivakumara, P., Kumar, G.H., Pal, U., Lu, T.: New sharpness features for image type classification based on textual information. In: Proceedings of the DAS, pp 204–209 (2016)
9. Shivakumara, P., Kumar, N.V., Guru, D.S., Tan, C.L.: Separation of graphics (superimposed) and scene text in videos. In: Proceedings of the DAS, pp. 344–348 (2014)
10. Roy, S., Shivakumara, P., Pal, U., Lu, T., Tan, C.L.: New tampered features for scene and caption text classification in video frame. In: Proceedings of the ICFHR, pp. 36-41 (2016)
11. Bhardwaj, D., Pankajakshan, V.: Image overlay text detection based on JPEG truncation error analysis. IEEE Signal Process. Lett. **8**, 1027–1031 (2016)
12. Roy, S., Shivakumara, P., Pal, U., Lu, T., Wahab, A.W.B.A.: Temporal integration of word-wise caption and scene text identification. In: Proceedings of the ICDAR, pp. 350–355 (2017)
13. Roy, S., et al.: Rough-fuzzy scene categorization for text detection and recognition in video. Pattern Recogn. **80**, 64–82 (2018)
14. Wang, Z., Shivakumara, P., Lu, T., Basavanna, M., Pal, U., Blumenstein, M.: Fourier-residual for printer identification. In: Proceedings of the ICDAR, pp. 1114–1119 (2017)
15. Liang, G., Shivakumara, P., Lu, T., Tan, C.L.: Multi-spectral fusion based approach for arbitrarily-oriented scene text detection in video image. IEEE Trans. Image Process. **24**, 4488–4500 (2015)
16. Xu, X., Wang, Y., Chen, S.: Medical image fusion using discrete fractional wavelet transform. Biomed. Signal Process. Control **27**, 103–111 (2016)

A New Method for Detecting Altered Text in Document Images

Lokesh Nandanwar[1] , Palaiahnakote Shivakumara[1(✉)] , Umapada Pal[2] ,
Tong Lu[3] , Daniel Lopresti[4] , Bhagesh Seraogi[2] , and Bidyut B. Chaudhuri[2]

[1] Faculty of Computer Science and Information Technology, University of Malaya,
Kuala Lumpur, Malaysia
lokeshnandanwar150@gmail.com, shiva@um.edu.my
[2] Computer Vision and Pattern Recognition Unit, Indian Statistical Institute, Kolkata, India
umapada@isical.ac.in, to.bhagesh.sr@gmail.com,
bbcisical@gmail.com
[3] National Key Lab for Novel Software Technology, Nanjing University, Nanjing, China
lutong@nju.edu.cn
[4] Computer Science and Engineering, Lehigh University, Bethlehem, PA, USA
lopresti@cse.lehigh.edu

Abstract. As more and more office documents are captured, stored, and shared in digital format, and as image editing software becomes increasingly more powerful, there is a growing concern about document authenticity. For example, texts in property documents can be altered to make an illegal deal, or the date on an airline ticket can be altered to gain entry to airport terminals by breaching security. To prevent such illicit activities, this paper presents a new method for detecting altered text in a document. The proposed method explores the relationship between positive and negative coefficients of a DCT to extract the effect of distortions caused by tampering operations. Here we divide DCT coefficients into positive and negative classes, then reconstructs images from the inverse DCT of the respective positive and negative coefficients. Next, we perform Laplacian filtering over reconstructed images for widening the gap between the values of text and other pixels. Then filtered images of positive and negative coefficients are fused by an average operation. For a fused image, we generate Canny and Sobel edge images in order to investigate the effect of distortion through quality measures, namely, MSE, PSNR and SSIM used as features. In addition, for the fused image, the proposed method extracts features based on histograms over the residual images. The features are then passed on to a deep Convolutional Neural Network for classification. The proposed method is tested on our own dataset as well as two standard datasets, namely IMEI and the ICPR 2018 Fraud Contest dataset. The results show that the proposed method is effective and outperforms existing methods.

Keywords: Document digitization · DCT coefficients · Fused image · Altered text detection · Fraud document

© Springer Nature Switzerland AG 2020
Y. Lu et al. (Eds.): ICPRAI 2020, LNCS 12068, pp. 93–108, 2020.
https://doi.org/10.1007/978-3-030-59830-3_8

1   Introduction

Due to increased automated generation and processing of legal documents and records, the importance of document verification has received special attention from researchers [1, 2]. This is because professional and skilled fraudsters may take advantage of advanced tools such as Photoshop and Gimp to alter documents for various purposes at different levels. For instance, in airports, security checks require the verification of travel date and the name of the passenger an a boarding pass or identity card before allowing him/her to enter. There have been a number of cases where individuals have fooled security and gained access to an airport by alternates and names on documents [3]. To alter text in a document, in general, two common operations are used, namely, *copy-paste* and *insertion*. In copy-paste operation, the fraudster extracts desired text from a different document or a different location in the same document to be altered, while in the case of insertion, software tools are used to insert words to change the original text. In such a case, these tools can be used to create text in the same font, size and style as in the original (insertion or imitation) [4]. There are existing methods for forgery detection in document images based on printer identification [5–7]. Such work is based on studying characteristics of connected components in the document image. If the methods find abrupt changes in the shapes of characters or text of a document due to distortion introduced by a different printing device, the document is deemed to be forged. However, such approaches are not robust to altered text detection in document images and this is due to distortion introduced by forgery operations may not be prominent.

Hence, it is challenging to detect the altered texts in documents. It is evident from the illustration in Fig. 1, where the naked eye cannot find any obvious differences between the altered text (marked by a rectangle) and genuine text. However, when adversaries use such tools for altering text, there must be some disturbance in the content, which may be in the form of distortions, loss in pattern regularity, or misplacement of content at the pixel level. These clues lead us to propose several methods to detect altered document images.

Copy-paste operation   Insertion (imitation) operation

Fig. 1. Illustration of sample forged PDF document images by copy-paste and insertion operations. Note: altered texts are enclosed by bounding boxes, which appear to be genuine text in terms of font, color and size.

Several methods related to altered text detection, including forged document identification have been investigated in the literature. Most of those are based on printer identification, ink quality verification, character shape verification, and distortion identification. For example, Khan et al. [5] proposed automatic ink mismatch detection for forensic document analysis. This method analyzes the ink of different pens to find fraudulent documents, which is effective for handwritten documents but not for printed document images. Luo et al. [6] proposed localized forgery detection in hyperspectral document images. This is an improved version of the previous method, which explores ink quality in the hyperspectral domain for fraud document identification. The method is often not effective on printed texts since when digitized, the quality change becomes very low. Khan et al. [7] proposed automated forgery detection in multispectral document images using fuzzy clustering. The method explores ink matching based on fuzzy c-means clustering to partition the spectral responses of ink pixels in handwritten notes into different clusters. The scope of the method is forgery detection in 'handwritten documents'. Raghunandan et al. [8] proposed Fourier coefficients for fraudulent handwritten document classification through age analysis. This approach studies positive and negative coefficients for analyzing image quality and identifying an image as old or new. The method may not work at the text line or word levels and requires the full document.

Most of the above methods target handwritten documents but not the typeset documents. However, Wang et al. [9] proposed a Fourier-residual for printer identification from document images. This method extracts features from residuals given by the Fourier transform for printer identification. The primary goal of this method is to identify printers rather than altered document images. Shivakumara et al. [3] proposed a method for forged IMEI (International Mobile Equipment Identity) in mobile images. The method explores color spaces of the images and it performs fusion operation for combining the different color spaces, which results in fused image. The features based on connected components are extracted from the fused image. The forged IMEI number is detected by comparing the features of template and the input image. However, the performance the method depends on the fusion operation.

It is observed from the above review that most methods extract features at the block level or connected component level for forgery detection in handwritten and typeset images. These methods are good when there are clear differences between forged and genuine text. If there is a minute difference at the pixel level or at character levels, as expected in typeset documents, the methods may not perform well. Therefore, there is a gap between altered text detection in printed documents and other documents.

Hence, in this work, we aim at detecting altered text detection in printed document images. Inspired by the knowledge that Discrete Cosine Transform (DCT) Coefficients have the ability to identify minute changes in image content [4], in this work, we propose to explore DCT coefficients for detecting altered text in PDF document images. Motivated by the divide and conquer method presented in [8] for fraudulent document identification based on the Fourier transform, we explore the idea for extracting features using DCT coefficients to study the differences between altered and genuine texts. Therefore, the main contribution here is to explore DCT coefficients for extracting features, namely quality measures and histogram based features for detecting altered text in document images.

2 Proposed Method

This work considers altered text lines, words or a short paragraph as input for detection. It is true that when the alteration is done using a copy-paste or insertions, some amount of distortion is introduced compared to genuine. Since the text is typeset, the changes may not be visibly apparent. Therefore, inspired by the property of DCT coefficients mentioned in [4] that it has the ability to extract minute changes at the pixel level, we classify DCT coefficients of an input image into positive and negative coefficients. It is expected that if an image is genuine, there is no misclassification of 'positive' as 'negative' or 'negative' as 'positive'. If the input image is altered, one can expect misclassification at the positive and negative coefficient levels due to the distortion in the content. To extract such a distortion, in the proposed method we reconstruct images by applying the inverse DCT on the respective positive and negative coefficients, resulting in two different images.

For widening the gap between text and other pixels, we then perform a Laplacian filtering over each reconstructed image. Then we fuse these two filtered images by averaging operation. It is expected that for genuine texts, the reconstructed image of the original texts must have better quality compared to the reconstructed image of the altered texts. With this cue, the proposed method extracts three quality measures as features, namely, Mean Square Error (MSE), Peak Signal to Noise Ratio (PSNR), and the Structure Similarity Index Metric (SSIM). Note that, for estimating quality measures, we consider Canny and Sobel operator of input images as ground for respective Canny and Sobel image of fused images. Since the problem is complex due to less distortion during alteration, the proposed method extracts more features based on histogram operation over fused and residual images. The residual image is the difference between the fused and input images. When the fused image of altered text is affected by quality, it is reflected in the intensity values compared to intensity values in the input images. So, for extracting more features, we use histogram operation. In total, our method extracts 76 feature vectors including a vector containing 6 quality measures. Then the extracted features are passed to a deep convolutional neural network for detecting altered text from the genuine text. The pipeline of the proposed method can be seen in Fig. 2.

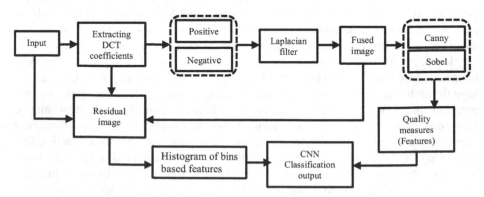

Fig. 2. Workflow of the proposed method

2.1 DCT Coefficient Analysis to Detect Altered Text

For the input original and altered images shown in Fig. 3(a), the proposed method applies the DCT, as defined in Eq. (1) and classifies the coefficients as positive, negative and zero as shown in Fig. 3(b), (c) and (d), respectively. It is noted from Fig. 3(b) that for the original image, there are more bright pixels at the top-left corner, while the brightness reduces gradually towards the right-bottom corner. On the other hand, for the altered one, bright pixels are scattered compared to the original. It is true that high frequency coefficients represent edge pixels, low frequency coefficients represent non-edge pixels, while zero coefficients represent background pixels. If there is no disturbance in the content (original) of the input image, according to DCT, the left-top corner should get high values that represent edge pixels, and the values get reduced gradually towards the right-bottom corner. When there is a disturbance in the content (i.e., altered), one cannot expect the same coefficient distribution as the original image. These observations holds for negative coefficients of the original and altered text images as shown in Fig. 3(c). However, the distribution of zero coefficients of the original and altered images is opposite as shown in Fig. 3(d), where we can see more pink colored pixels at the right-bottom corner for the original, and the pixels decrease gradually towards the top-left corner. Note that this is not true for altered document images. The distribution of zero coefficients is evident from the above observations made with respect to positive and negative coefficients.

$$F[u, v] = \frac{1}{N^2} \sum_{m=0}^{N-1} \sum_{n=0}^{N-1} f[m, n] \cos\left[\frac{(2m+1)u\pi}{2N}\right] \cos\left[\frac{(2n+1)v\pi}{2N}\right] \quad (1)$$

Here, u and v are discrete frequency variables $(0, 1, ..., N-1)$, $f[m, n]$ represent gray value at (m, n) position of $N \times N$ image matrix $(m, n = 0, 1, ..., N-1)$, and $F[u, v]$ is the (u, v) component of resultant DCT frequency matrix.

The proposed method reconstructs images by applying the inverse DCT on positive and negative coefficients of the original and altered images respectively, as in Fig. 4(a) and (b). It is observed from Fig. 4(a) and (b) that the original images reconstructed with respect to positive and negative coefficients appear brighter than the reconstructed altered image made in the same way. Since the alterations caused by copy-paste and insertion operations for printed texts do not create significant changes in content, it is hard to find the differences between the original and altered images. Therefore, to widen the gap between the original and altered text, the proposed method performs Laplacian filtering over the reconstructed images of positive and negative coefficients separately. This eliminates those pixels that represent unwanted noise as shown in Fig. 4(c) and (d), where it is noted that text pixels are enhanced and there are clear differences between the outputs of Laplacian filtering over positive and negative coefficients of the original and forged images.

To take advantage of this observation, we propose to combine the output of Laplacian filtering of positive and negative coefficients with a simple operation called averaging, which results in a fused image for the original and altered images as shown in Fig. 5(a).

98   L. Nandanwar et al.

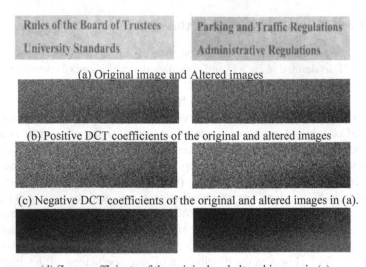

(a) Original image and Altered images

(b) Positive DCT coefficients of the original and altered images

(c) Negative DCT coefficients of the original and altered images in (a).

(d) Zero coefficients of the original and altered images in (a)

Fig. 3. Positive, negative and zero coefficient distributions of the original and altered images.

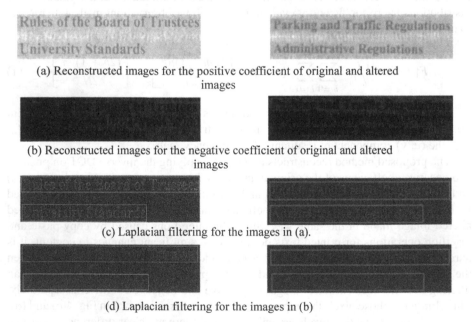

(a) Reconstructed images for the positive coefficient of original and altered images

(b) Reconstructed images for the negative coefficient of original and altered images

(c) Laplacian filtering for the images in (a).

(d) Laplacian filtering for the images in (b)

Fig. 4. Widening the difference between original and altered images using Laplacian filtering

2.2 Altered Text Detection in Document Images

As mentioned in the previous section, due to the operation for alternating text in the images, the image quality affects for the fused images as shown in Fig. 5(a) where we can see the clear difference between fused original and altered images in terms of brightness. It is evident from the Canny and Sobel edge images of original and altered fused images as shown in Fig. 5(b) and (c), respectively. It is observed from Fig. 5(b) and (c) that for the original fused images Canny and Sobel preserve structure of characters while for the altered fused image, the structure of the character lost compared to original fused image. This shows that the original fused image does not suffer from alteration of text while altered image suffer from the operation. To extract such difference, we estimate the quality measures, namely, Mean Square Error (MSE), Peak Signal Noise Ratio (PSNR) and the Structure Similarity Index (SSIM) for the fused and input images. Note: we consider Canny and Sobel of input images as ground truth for respective Canny and Sobel of fused images to estimate the above three quality measures. As a result, this process outputs 6 features, which includes 3 from Canny and 3 more from Sobel for the input image for detecting altered text in the document image. It is a vector containing 6 features.

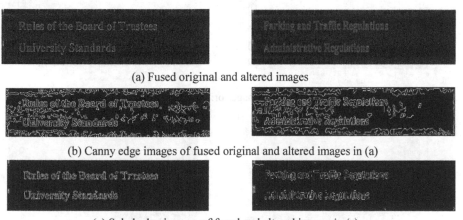

(a) Fused original and altered images

(b) Canny edge images of fused original and altered images in (a)

(c) Sobel edge images of fused and altered images in (a)

Fig. 5. The quality effect can be seen in Canny and Sobel edge images of input and fused images of original and altered.

The problem we are considering here is challenging, so quality measures alone may not be sufficient for achieving better results for altered text detection in document images. The same clue of quality difference in fused images leads us to extract the difference in intensity values in the fused images of original and altered images. This is because when the quality changes, it affects intensity values in the images. It is evident from the histograms plotted over intensity values of input original and altered images with 25 bins as shown in Fig. 6(a), where it is noted that low values are increasing in case of altered image compared to the original images. In the same way, when we plot histogram with bin size for the fused original and altered images as shown in Fig. 6(b), it can be noticed that loss of high values in case of altered image compared to original image. Therefore, to extract the above observation, the proposed method finds difference between the fused and the input images, which results in Residual Image (RI) for respective original and altered images as defined in Eq. (2). The histogram with same the 25-bin size plotted for residual images of original and altered images show that there is a significant difference in distribution of intensity values. Since we consider each bin in the histogram as a feature vector (all the values that contributes to bin), the proposed method gets $25 + 25 = 50$ feature vectors. To strengthen the above feature extraction, since there is difference between histograms of input and fused images of original, altered images, the proposed method further finds the difference between two histograms (CF) at frequency level as defined in Eq. (3), which results in 25 more feature vectors. In total, the proposed method extracts $50 + 25 + 1 = 76$ feature vectors for detecting altered text in document images. Each vector contains number of features according to histogram bin operation.

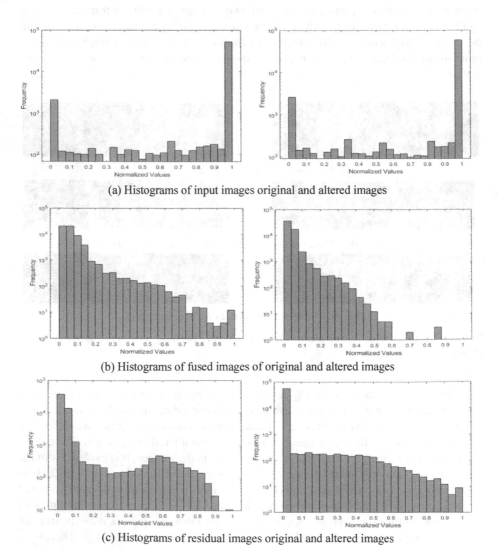

(a) Histograms of input images original and altered images

(b) Histograms of fused images of original and altered images

(c) Histograms of residual images original and altered images

Fig. 6. Feature extraction using histogram of fused and residual images of original and altered images.

$$RI = \sum_{i=0}^{m} \sum_{j=0}^{m} |I(i,j) - FI(i,j)| \tag{2}$$

Where m denotes the size of the Input (I) and Fused Image (FI).

$$CF = \sum_{k=0}^{N} |HI(k) - HFI(k)| \tag{3}$$

where N denotes the bin size of the histogram which is 25 in this work. *HI* and *HFI* denote histograms of input image and fused images, respectively. The value of 25 is determined experimentally for choosing random samples from across datasets. It is justified in experimental section.

Motivated by the strength of discriminative power of deep convolutional neural networks, the extracted 76 feature vectors are passed to the neural network which is shown in Fig. 7 for detecting altered text in document images. In this architecture, we use 'ReLU' activation function for all the layers except the final layer where we use 'Sigmoid' [10] activation. With 'Adam' [11] as optimizer and learning rate of 0.01 and 'binary cross entropy' loss function, the proposed architecture is trained for 100 epochs with the batch size of 8. The loss function as defined in Eq. (4), which is binary cross entropy loss [12] used for classification of altered text in the PDF document images.

$$BCE = -\frac{1}{N} \sum_{i=1}^{N} (y_i . \log(p(y_i)) + (1 - y_i).\log(1 - p(y_i))) \tag{4}$$

where y is the label and $p(y)$ is the estimated probability for total number of N samples. The dropout rate, 0.2 is added in between the convolutional layers to reduce overfitting and more generalization of the results. For all the experiments, we use the system with Nvidea Quadro M5000 GPU for training and testing of the architecture and python framework Keras for this application. The dataset is divided into 80% and 20% for training and testing for all the experiments in this work.

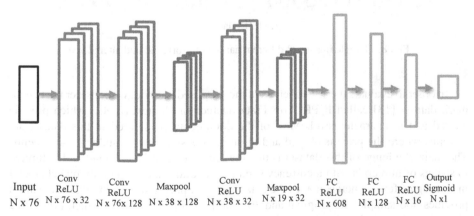

Fig. 7. Architecture of Convolutional Neural Network for Classification

3 Experimental Results

To evaluate our approach, we created our own dataset with the help of copy-paste and insertion operation using Paint Brush Tool, tested on documents drawn from a variety of sources, including property documents, insurance documents and air tickets. Our dataset contains 110 altered text line images and the same number of original text line images, which gives a total of 220 text line images for experimentation. Sample images of our dataset are shown in Fig. 8(a), where it can be seen that it is hard to notice the difference between the altered text and original text line images. Our dataset consists of altered text at line level or words levels. In other words, our dataset does not include character level additions or deletions.

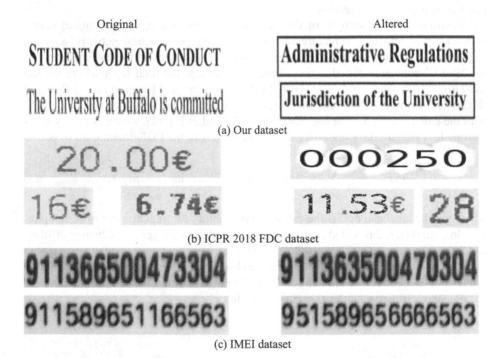

(a) Our dataset

(b) ICPR 2018 FDC dataset

(c) IMEI dataset

Fig. 8. Sample images of different datasets for original and altered text

In order to evaluate the effectiveness of the proposed method, we consider the benchmark dataset [13] called ICPR 2018 Fraud Detection Contest (FDC), which provide altered text at character level. Most of the documents considered in this dataset are receipts, where the price is altered and those are considered as fraudulent documents. The main challenge of this dataset is that the length of text is too small and is limited to strings of numerals with a currency symbol. In addition, altering only one character in a string of few numerals makes the data more complex and challenging. The dataset provides 300 samples for original and 302 samples for altered text, which provides a total of 602 images for experimentation. For our experiment, since the proposed method requires text line of original and altered documents, we segment the altered part from the documents and original documents. The ground truth for the altered region is marked by a rectangle in the dataset. We segment the rectangular region from the altered documents automatically. For original documents, since almost all of the documents are receipts and have the prices in the location where the rectangle is drawn in the altered documents, we use the same segmentation step for extracting price information from original documents. The sample images of altered text and original text are shown in Fig. 8(b), where the clear difference is noticeable. This makes problem easier compared to our own dataset and the IMEI dataset.

In the same way, to test the robustness of the proposed method, we also consider standard dataset of forged IMEI number detection [3], which provides 500 forged and 500 original images. This dataset is challenging because the forgery is done at the character level but not at the word level. In each forged image, we can expect that one or two characters, especially numerals, are forged using copy-paste and insertion operations. This dataset also includes images affected by blur and noise some sample images of which are shown in Fig. 8(c), which are poor quality images. When the image is of poor quality, it is more challenging for forgery detection, because the distortion created by alteration and the poor quality may accentuate each other. The background is complex for these images compared to ours and ICDAR 2018 FDC datasets. Overall, 1822 images are considered for our experiment.

To show the superiority of our proposed technique, we implement two recent methods for comparative study. Wang et al. [9] proposed a method for printer identification by studying the print of different printers. The main basis of the method is that the print of different printers are affected by unique noise/distortion introduced by the printers. The same basis is used for the proposed method. In addition, the text in the print document is printed type as the proposed work. In the same way, we implement Shivakumara et al. [3] method which explores color spaces obtaining the fused image and computing connected based features are extracted for forged IMEI number detection. The objective of the method is same as the proposed method and idea of fusion concept is similar to the proposed method. Furthermore, we run Convolutional Neural Network Inception V3 network [14] (CNN) by passing directly images as input for classification with transfer learning on pertained weights. This is to show that the combination of features and CNN is better than CNN alone especially for two classes with small sized dataset, In order to show that the above methods are not adequate for achieving the results for altered text detection in the document images, we use the above methods for comparative study. For measuring the performance of the proposed and existing methods, we use standard measure average classification rate calculated through confusion matrix. The classification rate is defined as the number of images detected correctly divided by the total number of images. The average classification rate is the mean of diagonal elements of confusion matrix.

As mentioned in the proposed methodology section, the size of the histogram bin is 25. To determine this value, we conduct experiments on random sample chosen from across datasets by varying the size of the bin vs average classification rate as shown in Fig. 9. It is observed from Fig. 9 that as bin size increases the average classification rate

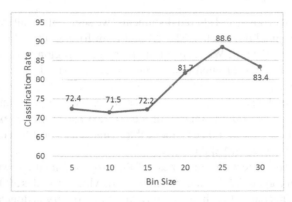

Fig. 9. Determining the value of 25 for achieving better results

increases gradually. This is valid because as the bin size increases, the proposed features acquire more local information and hence the average classification rate increases gradually up to 25 bin size. At the same time, if we continue in increasing bin size, at some point, the proposed features lose vital information in the images due to sparsity. However, when the bin size is 25, the average classification rate reaches highest score, 88.6% and then the average classification rate decreases after bin size 25. This shows that the value of bin size, 25 is the optimal value for achieving better results. The same value are used for all experiments in this work.

In the proposed method, we extract the features, namely, quality measures, histogram based features from fused images, residual images and from the histogram difference between input and fused images. In order to assess the contribution of each feature type, we calculate classification rate for each feature type and the results are reported in Table 1. It is observed from Table 1 that each feature type scores more than 65% average classification rate. Therefore, one can understand that each feature type contributes for achieving better results. At the same time, one can justify that feature type alone is not sufficient to achieve better results as the proposed method. The lowest average classification rate is reported by Quality measures and highest average classification rate is reported by features extracted from histogram of difference between input and fused images. This indicates that the step which classifies DCT coefficients into positive and negative to obtain fused image play a significant role in widening the gap between input and fused images. Since the proposed method uses Canny and Sobel edges of the input and fused images for estimating quality measures, the quality measures do not contribute much for altered text detection because Canny method introduces erratic edges and Sobel operator lose vital information for the fused images of original and altered images, as shown in Fig. 5(b). When we compare the results of fused and residual, the residual contributes more for detecting the altered text. This is because the both input and fused images contributes for the Residual. Overall, to achieve 88.6% average classification rate for the proposed method, four different feature types are employed.

Quantitative results of the proposed and existing methods for our dataset and standard dataset are reported in Table 2, where it is noted that the proposed method is the best in terms of average classification rate for all three datasets compared to the existing

Table 1. Confusion matrix and average classification rate of different features and the proposed method on our dataset (in %).

Features	Quality measures		Fused		Residual		Difference -Hist		Proposed	
Classes	Original	Altered	Original	Altered	Original	Altered	Original	Altered	Original	Altered
Original	61	39	62.8	37.2	75.2	24.8	82.6	17.4	87.4	12.6
Altered	27.5	72.5	29	71	18.4	81.6	20.2	79.8	10.2	89.8
Average	66.75		66.9		78.4		81.2		88.6	

methods. Interestingly, the proposed method scores its highest average classification rate for the ICPR 2018 FDC dataset compared to our and IMEI datasets. This is because the distortions introduced due to forgery operation are noticeable and visible compared to the original image. However, this is not true for our and IMEI datasets. It is evident from the sample image shown in Fig. 8(b), where we can see clearly the changes due to the forgery operation. It is confirmed from the results scored by two existing methods [9, 14] as the methods score high results for ICPR 2018 FDC dataset compared to our and IMEI dataset. Similarly, the proposed method returns its lowest results for the IMEI dataset compared to our own dataset and the ICPR 2018 FDC dataset. The reason is that the images of IMEI dataset are challenging because they were captured by mobile devices, while the images in the other two datasets are captured by scanning the documents. When we compare the results of the three existing methods, Wang et al. is better than the other two existing methods. This is because it extracts spatial, texture and gradient properties of text in the images while the method [13] extracts only the connected component based properties from the edge images and the method [14] does not consider advantage of handcrafted features. In other words, the features extracted in [13, 14] are not good enough to cope with the challenges of different datasets. It is noted from Table 2 that the method [13] achieves highest average classification rate for IMEI dataset compared to our and ICPR 2018 FDC dataset. This is understandable because the method [13] is developed for IMEI forged number detection. Table 2 shows that the CNN [14] reports lowest average classification rate compared to other existing methods for all the three datasets. This shows that CNN does not work well when the dataset size is small. Therefore, one can infer that, in this situation, the combination of handcrafted features with classifier is better than CNN alone. However, the accuracies of the existing methods are lower than the proposed method. Thus, we can conclude that the combination of frequency domain, spatial domain and deep convolutional neural network of the proposed method make difference compared to existing methods.

Although the proposed method achieves better results, there are some cases for which it does not work. One such case is alterations at the text line level, where the proposed method fails to detect the altered text. This is because when we copy and paste the whole text line, the effect of altered operation is minimal compared to character and word levels. One such sample is shown in Fig. 10, where it is hard to see the difference between the original and altered image. Similarly, when the images are degraded due to aging and distortion, the distortion introduced by alteration overlap with the distortion of the image, which is beyond the scope of the work. In this case, it is necessary to introduce context-based features for detecting altered information. This requires image processing and natural language processing to find a solution.

Table 2. Confusion matrix and average classification rate of the proposed and existing methods on our and benchmark datasets (in %)

Methods	Dataset	Own Dataset		IMEI dataset [3]		ICPR 2018 [13]	
		Original	Altered	Original	Altered	Original	Altered
Proposed	Original	**87.4**	12.6	**84.4**	15.6	**84.0**	16.0
	Altered	10.2	**89.8**	14.2	**85.8**	4.50	**95.50**
	Average	**88.6**		**85.1**		**89.41**	
Wang et al. [9]	Original	80	20	83.2	16.8	84.66	8.67
	Altered	13.4	86.6	25.6	74.4	7.95	89.33
	Average	83.3		78.8		86.99	
Shivakumara et al. [3]	Original	60	40	82.2	17.8	92	8
	Altered	35	65	18	82	49.44	50.66
	Average	62.5		82.1		71.33	
Szegedy et al. [14]	Original	34.8	65.2	53.8	46.2	81.8	18.2
	Altered	16.2	83.8	33.7	66.3	47.8	52.2
	Average	59.3		60.1		67.0	

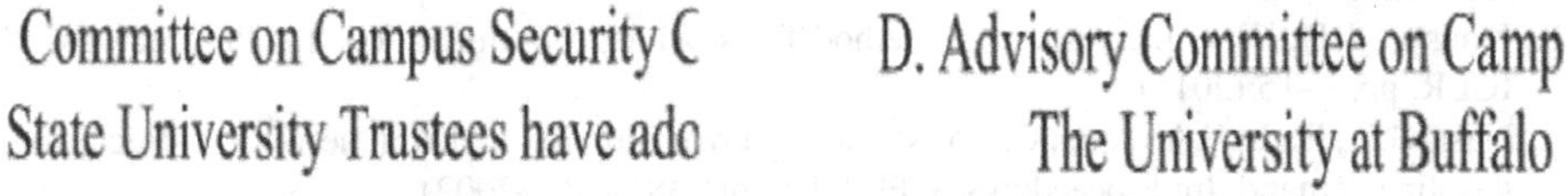

Committee on Campus Security C
State University Trustees have ado

D. Advisory Committee on Camp
The University at Buffalo

(a) Altered text classified as Original   (b) Original text classified as Altered

Fig. 10. Limitation of the proposed method

4 Conclusion and Future Work

In this work, we have proposed a new method for detecting altered text in document images. The proposed method explores applying DCT coefficients in different way for obtaining fused image for the input image. The proposed method extracts features from the fused images based on quality measures and histogram-based features. The extracted features are then passed to deep neural network classifier for detecting altered text in the document images. Experimental results on our own dataset and two standard datasets show that the proposed method outperforms existing methods in terms of average classification rate. In addition, the results show that the proposed method is robust to images which are altered at the character level. However, when the alteration is done at the text line level instead of at the word or character levels, the performance of the proposed method degrades. In addition, when the image is affected by degradations and poor quality, the proposed method may not perform well. This provides scope for future work where we plan to introduce context information through the use of natural language models to find solutions to these remaining problems.

References

1. Alaei, F., Alaei, A., Pal, U., Blumenstein, M.: A comparative study of different texture features for document mage retrieval. Expert Syst. Appl. **121,** 97–114 (2019)
2. Beusekom, J.V., Shafait, F., Breuel, T.: Text line examination for document forgery detection. Int. J. Doc. Anal. Recogn. (IJDAR) **16**(2), 189–207 (2013)
3. Shivakumara, P., Basavaraja, V., Gowda, H.S., Guru, D.S., Pal, U., Lu, T.: A new RGB based fusion for forged IMEI number detection in mobile images. In: Proceedings of ICFHR, pp. 386–391 (2018)
4. Wang,W., Dong, J., Tan, T.: Exploring DCT coefficients quantization effect for image tampering localization. In: Proceedings of IWIFS, pp. 1–6 (2011)
5. Khan, Z., Shafait, F., Mian, A.: Automatic ink mismatch detection for forensic document analysis. Pattern Recogn. **48**, 3615–3626 (2015)
6. Luo, Z., Shafait, F., Mian, A.: Localized forgery detection in hyperspectral document images. In: Proceedings of ICDAR, pp. 496–500 (2015)
7. Khan, M.J., Yousaf, A., Khurshidi, K., Abbas, A., Shafait, F.: Automated forgery detection in multispectral document images using fuzzy clustering. In: Proceedings of DAS, pp. 393–398 (2018)
8. Raghunandan, K.S., et al.: Fourier coefficients for fraud handwritten document classification through age analysis. In: Proceedings of ICFHR, pp. 25–30 (2016)
9. Wang, Z., Shivakumara, P., Lu, T., Basavanna, M., Pal, U., Blumenstein, M.: Fourier-residual for printer identification. In: Proceedings of ACPR, pp. 1114–1119 (2017)
10. Narayan, S.: The generalized sigmoid activation function: competitive supervised learning. Inf. Sci., 69–82 (1997)
11. Kingma, P.D., Bai, J.L.: Adam: a method for stochastic optimization. In: Proceedings of ICLR, pp. 1–15 (2015)
12. Nasr, G.E., Badr, E.A., Joun, C.: Cross entropy error function in neural networks: forecasting gasoline demand. In: Proceedings of FLAIRS, pp. 381–384 (2002)
13. Artaud, C., Sidère, N., Doucet, A., Ogier, J., Yooz, V.P.D.: Find it! Fraud detection contest report. In: Proceedings of ICPR, pp. 13–18 (2018)
14. Szegedy, C., Vanhoucke, V., Ioffe, S., Shlens, J., Wojna, Z.: Rethinking the inception architecture for computer vision. In: Proceedings of CVPR, pp. 2818–2826 (2016)

Hand-Drawn Object Detection for Scoring Wartegg Zeichen Test

Nam Tuan Ly[1]([⊠]) [iD], Lili Liu[2] [iD], Ching Y. Suen[2] [iD], and Masaki Nakagawa[1] [iD]

[1] Tokyo University of Agriculture and Technology, Tokyo, Japan
namlytuan@gmail.com, nakagawa@cc.tuat.ac.jp
[2] Concordia University, Montreal, Canada
li_lil@encs.concordia.ca, suen@cse.concordia.ca

Abstract. The Wartegg Zeichen Test (WZT) is a method of personality evaluation developed by the psychologist Ehrig Wartegg. Three new scoring categories for the WZT consist of Evocative Character, Form Quality, and Affective Quality. In this paper, we present the object detection model in scoring the Affective Quality of WZT. Our works consist of two main parts: 1) using the YOLO (You Only Look Once) model to detect the hand-drawn square boxes from the WZT form, and 2) using YOLO to detect the object in the hand-drawn square box. In the experiments, YOLOv3 achieved 88.94% of mAP for hand-drawn square box detection and 46.90% of mAP for hand-drawn object detection.

Keywords: Wartegg zaichen test · Object detection · YOLO · Human sketches

1 Introduction

The Wartegg Zeichen Test (WZT, or Wartegg Drawing Completion Test) is a method of personality evaluation developed by the psychologist Ehrig Wartegg [1]. The standard WZT form consists of an A4-sized paper sheet with eight white, 4 cm × 4 cm squares in two rows (see Fig. 1). Each square has a simple printed sign, such as a dot or a line. The test person is required to make a complete drawing using the printed sign as the starting point and then give a short explanation or title of each drawing. In 2016, Crisi et al. [2] introduced three new scoring categories of the WZT that present specific personality characteristics: Evocative Character (EC) that is related to social adjustment; Form Quality (FQ) that is connected to reality testing ability and Affective Quality (AQ) that is linked to general mood state. The AQ is simply the assessment of the emotional connotation of each drawing and scored based on the object content of each drawing. The object contained in the WZT drawing can be detected and recognized by the object detection methods.

Object detection is one of the most basic and important tasks in the field of computer vision. In recent years, Deep Neural Networks (DNNs) based object detection models have produced dramatic improvements in both accuracy and speed. The DNNs based object detection methods usually can be divided into two categories, the one-stage methods which achieve high inference speed and the two-stage methods which achieve high

Y. Lu et al. (Eds.): ICPRAI 2020, LNCS 12068, pp. 109–114, 2020.
https://doi.org/10.1007/978-3-030-59830-3_9

Fig. 1. The eight squares of WZT form.

accuracy. The one-stage methods which are popularized by R-CNN family models [3–5] first generate a set of region proposals and then refine them by DNNs. On the other hand, one-stage methods such as YOLO family models [6–8] directly classify the pre-defined anchors and further refining them using DNNs without the proposal generation step.

In this paper, we employ YOLO in scoring the AQ of WZT. Our works consist of two main parts: 1) using YOLO to detect the square boxes from the WZT page, and 2) using YOLO to detect the object in the WZT square box.

The rest of this paper is organized as follows. Section 2 describes the datasets. Section 3 presents an overview of YOLO. The experiments are described in Sect. 4. Finally, conclusions are presented in Sect. 5.

2 Datasets

2.1 CENPARMI WZT Datasets

CENPARMI WZT dataset is a dataset of the Wartegg Zeichen Test produced by Centre for Pattern Recognition and Machine Intelligence, Concordia University which consists of two sub-datasets: CENPARMI WZT page dataset and CENPARMI WZT square box dataset. The CENPARMI WZT page dataset consists of 211 scanned WZT pages collected from 211 test subjects. Each scanned WZT page consists of eight hand-drawn square boxes and was ground-truthed by eight square boxes as shown in Fig. 2. The CENPARMI WZT square box dataset consists of 1,688 hand-drawn square boxes that were cropped from the 211 scanned WZT pages.

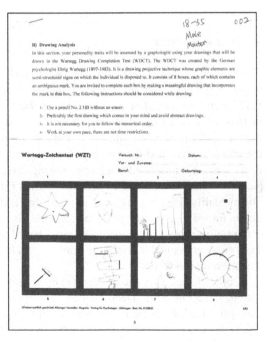

Fig. 2. A sample of the WZT page.

2.2 Sketchy Dataset

The Sketchy dataset [9] is the first large-scale collection of sketch-photo pair dataset which spans 125 categories and consists of 12,500 unique photographs of objects and 75,471 human sketches of objects inspired by those photographs. In this work, we employ the human sketches of the Sketchy dataset to train the hand-drawn object detector. The above sketches were randomly split into two parts, with approximately 90% for training and the remainder for validation.

3 Overview of the YOLO

R. Joseph et al. proposed an object detection framework called YOLO in 2015 [6]. It is the first one-stage object detector based on DNNs which makes use of the whole topmost feature map to predict both confidences of categories and bounding boxes. YOLO splits an input image into an $S \times S$ grid, where each grid cell predicts B bounding boxes and their corresponding confidence scores as shown in Fig. 3. YOLO can process images in real-time with better results than other real-time models.

YOLOv2 [7] is an improved version of YOLO which adopts several impressive strategies, such as convolutional layers with anchor boxes and multi-scale training. YOLOv3 [8] is the latest version of YOLO which has a few incremental improvements on YOLOv2 such as DarkNet-53 as well as feature map upsampling and concatenation. This paper employs YOLOv3 in scoring the AQ of WZT.

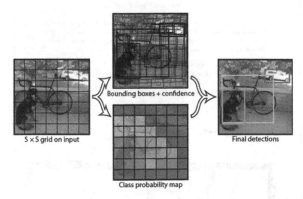

Fig. 3. The basic idea of YOLO [6].

4 Experiments

4.1 Hand-Drawn Square Box Detection

In this section, we present the experiment of YOLOv3 for detecting hand-drawn square boxes in a scanned WZT page. We use the CENPARMI WZT page dataset to train the

YOLOv3 model. The 211 scanned WZT page images were divided into three groups: the training set of 171 images, the validation set of 19 images, and the testing set of 21 images.

Table 1 shows the mAP, Recall and Precision of YOLOv3 on the test set of the CENPARMI WZT page dataset with the IoU threshold of 0.9, 0.8, and 0.7. The YOLOv3 with the IoU threshold of 0.8 achieved 88.94% of mAP on the test set. This result implies that YOLOv3 can assist the detecting WZT square boxs on a WZT page with significant accuracy. Figure 4 shows an example of hand-drawn square box detection results in the test set of the CENPARMI WZT page dataset.

Table 1. Results on the CENPARMI WZT page dataset (%).

Model	Test set		
	mAP(%)	Recall(%)	Prec(%)
YOLOv3 (threshold = 0.9)	63.17	68.45	7.3
YOLOv3 (threshold = 0.8)	88.94	90.48	9.65
YOLOv3 (threshold = 0.7)	97.67	98.21	10.48

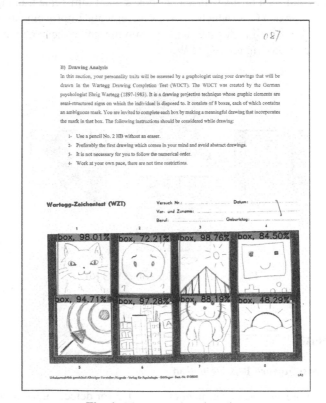

Fig. 4. The square box detection.

4.2 Hand-Drawn Object Detection

In this section, we conducted an experiment of YOLOv3 for detecting the hand-drawn object in the WZT hand-drawn square box. We use the human sketches of the Sketchy dataset to train and validate the YOLOv3 model. The trained YOLOv3 model is tested in the CENPARMI WZT square box dataset. Since the Sketchy dataset spans only 125 object categories, then we exclude the WZT square box which has the object that does not belong to 125 categories. Consequence, we have 299 WZT square box images for testing the trained YOLOv3 model.

Table 2. Results on CENPARMI WZT square box dataset (%).

Model	Test set		
	mAP(%)	Recall(%)	Prec(%)
YOLOv3 (threshold = 0.5)	46.90	55.50	6.4

Table 2 shows the mAP, Recall, and Precision of YOLOv3 on the test set of the CENPARMI WZT square box dataset with the IoU threshold of 0.5. With 46.90% of mAP on the WZT square box dataset, YOLOv3 can assist the detecting hand-drawn objects in scoring the AQ of WZT. Figure 5 shows some examples of hand-drawn object detection results in the CENPARMI WZT square box dataset.

Fig. 5. The hand-drawn object detection.

5   Conclusion

In this paper, we presented the use of YOLOv3 for hand-drawn object detection and hand-drawn square box detection in a WZT page image. Following the experiments, the hand-drawn square box detector by YOLOv3 achieved 88.94% of mAP with the IoU threshold of 0.8 and the hand-drawn object detector by YOLOv3 achieved 46.90% of mAP with the IoU threshold of 0.5. The object detection methods can assist in scoring the Affective Quality of WZT.

In future works, we will expand the hand-drawn object detector to detect multiple objects as well as add more object categories. We also plan to collect more WZT page images to improve the performance of the hand-drawn square box detector. The experiments of other object detectors such as Faster R-CNN also will be considered.

References

1. Wartegg, E.: Gestaltung und Charakter [Formation of gestalts and personality]. Zeitschrift für Angewandte Psychologie und Charakterkunde 84, Beiheft 2
2. Crisi, A., Dentale, F.: The wartegg drawing test: inter-rater agreement and criterion validity of three new scoring categories. Int. J. Psychol. Psychol. Ther. **16**, 83–90 (2016)
3. Girshick, R., Donahue, J., Darrell, T., Malik, J.: Rich feature hierarchies for accurate object detection and semantic segmentation. In: CVPR (2014)
4. Girshick, R.: Fast r-cnn. In: ICCV (2015)
5. Ren, S., He, K., Girshick, R., Sun, J.: Faster r-cnn: towards realtime object detection with region proposal networks. In: NIPS, pp. 91–99 (2015)
6. Redmon, J., Divvala, S., Girshick, R., Farhadi, A.: You only look once: unified, real-time object detection. In: CVPR (2016)
7. Redmon, J., Farhadi, A.: Yolo9000: better, faster, stronger (2016). arXiv:1612.08242
8. Redmon, J., Farhadi, A.: Yolov3: an incremental improvement (2018). arXiv preprint arXiv: 1804.02767
9. Sangkloy, P., Burnell, N., Ham, C., Hays, J.: The sketchy database: learning to retrieve badly drawn bunnies. ACM Trans. Graph. **35**(4), 119:1–119:12 (2016)

Application of Deep Learning for Online Handwritten Mathematical Expression Recognition: A Review

Danfeng Zhang$^{(\boxtimes)}$ (iD) and Jun Tan (iD)

School of Mathematics, Sun Yat-Sen University, Guangzhou, China
zhangdf3@mail2.sysu.edu.cn

Abstract. Handwritten mathematical expression recognition, which can be classified into offline handwritten mathematical expression recognition and online handwritten mathematical expression recognition (OHMER) is an important research filed of pattern recognition, which has attracted extensive studies during the past years. With the emergence of deep learning, new breakthrough progresses of handwritten mathematical expression recognition have been obtained in recent years. In this paper, we review the applications of deep learning models in the field of OHMER. First, the research background and current state-of-the-art OHMER technologies as well as the major problems are introduced. Then, OHMER systems based on different deep learning methods such as CNN-based system, RNN-based system, encoder-decoder approach and so on are introduced in detail including the summaries of technical details and experiment results. Finally, further research directions are discussed.

Keywords: Handwritten mathematical expression recognition (OHMER) · Pattern recognition · Deep learning · CNN · RNN

1 Introduction

Mathematical expressions play an import role in describing problems and theories in math, physics and many other fields. However, how to successfully recognize handwritten mathematical expressions is difficult and immature. Therefore handwritten mathematical expression recognition is an import field in pattern recognition. Based on the data collection mode, handwritten mathematical expression recognition can be divided into two recognition system, offline handwritten mathematical expression recognition and online handwritten mathematical expression recognition.

Offline is when two dimensional pictures of handwritten mathematical expression are collected and pass to the recognition system. Online recognition system traces the written track and records the sequential information as the input by physical devices (such as digital pen, digital tablet or touch screen). Because of the different recognition objects, the methods and strategies used in these two kinds of handwriting recognition technologies are different. The former is to recognize a two-dimensional picture while

Y. Lu et al. (Eds.): ICPRAI 2020, LNCS 12068, pp. 115–124, 2020.
https://doi.org/10.1007/978-3-030-59830-3_10

the latter is to recognize series of sampling point information arranged in mathematical grammar. Due to the different illumination, resolution, writing paper and other strips in the photographing and scanning equipment, digitization will bring some noise interference which makes offline handwritten mathematical expression recognition is more difficult than online handwritten mathematical expression recognition in general. This paper focus on online hand written mathematical expression recognition (OHMER).

In the past years, pattern recognition has done rapid development and the performance of OHMER has improved. But OHMER still consists of two major problems [1], symbol recognition and structural analysis. Symbol recognition can be solved sequentially and structural analysis globally. To solve the problem sequentially, it first makes a segmentation on handwritten mathematical expressions and recognizes the symbols separately [2, 3]. Then implement structural analysis by using the global information of the mathematical expression, which allows symbol recognition and structural relations to be learned from the whole handwritten expression [4, 5].

In view of the related academic and technical progress, this paper mainly review and analyze the latest development of OHMER based on deep learning.

2   OHMER Based on Deep Learning

Traditional handwritten mathematical expression recognition generally includes preprocessing (such as normalization), feature extraction, feature dimensionality reduction, classifier design and so on. However, the convolution neural network (CNN) makes it possible to design an end-to-end handwritten mathematical expression recognition system without the process mentioned above. In addition, making use of deep learning can achieve better performance than the traditional method. For OHMER, many improved deep learning methods have been proposed in recent years.

2.1   CNN-Based OHMER System

Convolutional neural network (CNN) was came up in 1990s. It is a feedforward neural network with deep structure and convolution computation, which has great performance in picture processing. In English handwriting recognition, the LeNet5 model [6] achieves a reported result with an expression recognition accuracy of 99.05% and it can be improved to 99.20% with transformed training set, which is greatly ahead of traditional machine learning methods such as SVM, boosting, multi-layer perceptron and so on.

Inspired by the great achievement of CNN-based English handwriting recognition, researchers applied CNN in OHMER and got inspiring experiment results. Irwansyah Ramadhan et al. built a CNN for mathematical symbol recognition with the following architecture as showed in Fig. 1, input - convolutional - pooling - convolutional - pooling -full connected - softmax [7]. Each convolutional layers of their model contains 5×5 size of the receptive field and 2×2 size of the receptive field for pooling layers with the max pooling operation, while performing sigmoid function for non-linearity in every convolutional and fully-connected layer. Their model achieved 87.72% accuracy using CROHME 2014 dataset, which outperforms the previous work.

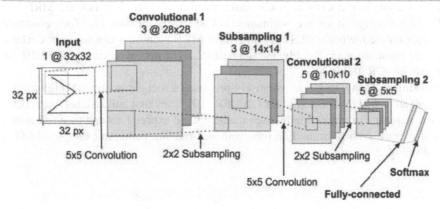

Fig. 1. Architecture of CNN proposed by Irwansyah Ramadhan et al. [7].

Although the work of Irwansyah Ramadhan et al. makes improvement, it is applied for handwritten mathematical symbol recognition, which is am easier task tan OHMER. Tran et al. use a SSD model (Fig. 2) for symbol recognition and DRACULAE for structural analysis to generate LaTeX string [8]. They use mAP score to evaluate their model and it gets 0.65 in the best model.

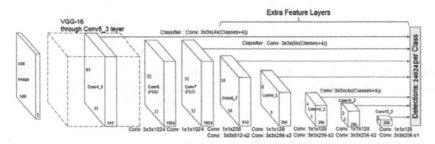

Fig. 2. Architecture of SSD model [8].

2.2 RNN-Based OHMER System

Though CNN has a good performance in picture processing, for OHMER, the input is a sequential data, making recurrent neural network (RNN) a more effective method.

RNN is a class of artificial neural networks where connections between nodes form a directed graph along a temporal sequence. This allows it to exhibit temporal dynamic behavior. It can use internal state (memory) to process sequences of inputs. This makes it applicable to tasks like connected handwriting recognition. Furthermore, the LSTM advanced RNN allows that cells can access to context information over long periods of time [9].

Dai Nguyen et al. apply max-out-based CNN and BLSTM to image patterns created from online patterns and to the original online patterns, respectively and combine them.

They also compare models with traditional recognition methods which are MRF and MQDF by carrying out some experiments on CROHME database [9]. They conclude that, as for online methods, BLSTM outperform MRF because it can access the whole context of input sequence flexibly and the best BLSTM gets an accuracy of 87.7% on testing set TestCROHME2014.

In addition, Zhang Ting et al. creatively proposed a merged multiple 1D interpretations with advanced RNN BLSTM for OHMER [10]. They first build a label graph (LG) from the sequence of online handwritten mathematical expressions and then select two types of merged path from the LG, time path and random path. An example of LG is showed in Fig. 3.

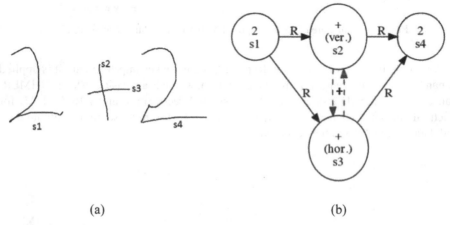

(a)   (b)

Fig. 3. (a) '2 + 2' written with 4 strokes; (b) the label graph of '2 + 2'. The four strokes are indicated as s1, s2, s3, s4 in writing order. (ver.) and (hor.) are added to differentiate the vertical and the horizontal strokes for '+'. 'R' is for left-right relationship [10].

Time path covers all the nodes but could miss some edges while random path is used to catch more edges. Followed, they train a BLSTM to label these two paths, using connectionist temporal classification (CTC) to post-processing. The experiment shows that the Rec. of Segments and Seg + Class are 92.77% and 85.17% testing in CROHME 2014. However, at the global expression level, the recognition rate with no error is low, which is only 13.02% in CROHME.

2.3 GRU-Based Encoder-Decoder Approach

Methods mentioned above need to segment the handwritten mathematical symbols and rely on a predefined grammar to do the structural analysis. Encoder-decoder architecture [11] can unify the symbol segmentation, symbol recognition, and structural analysis in one data-driven framework to output a character sequence in LaTeX format [12].

In 2017, Jianshu Zhang, Jun Du, and Lirong Dai proposed a gated recurrent unit based recurrent neural network (GRU-RNN), in which the input two-dimensional ink trajectory information of handwritten expression is encoded to the high level representations via the stack of bi-directional GRU-RNN and the decoder is implemented by a unidirectional GRU-RNN with a coverage-based attention model [13]. Figure 4 is the architecture of the GRU-RNN.

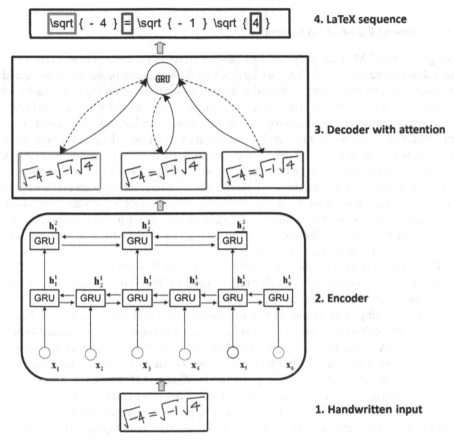

Fig. 4. The overall architecture of attention based encoder-decoder model proposed by Jianshu Zhang, Jun Du, and Lirong Dai [13].

In encoder, since simple RNN has the problems of the vanishing gradient and exploding gradient, Zhang, Du and Dai utilize GRU as a improved version of simple RNN in encoder. To use the future information, they pass the input vectors through two GRU layers running in opposite directions and concatenate their hidden state vectors.

In decoder, given a context vector, a unidirectional GRU is utilized to produce the LaTeX sequences symbol by symbol. And the unidirectional GRU also takes the previous target symbol as input. To refine the information of the input sequence, the decoder can adopt an attention mechanism, which is improved by appending a coverage vector to the computation attention on the purpose of addressing the problem of the lack of convergency [13], to link to the related part of input sequence and then assign a higher weight to the corresponding annotation vector.

The experiments show that the accuracy of recognition is 52.43% in competition dataset CROHME 2014. Even though the accuracy of recognition is not satisfactory enough, [13] proposed a model that is capable of performing symbol segmentation automatically and learning to grasp a math grammar without prior knowledge.

2.4 Advanced End-to-End System

Though the BLSTM model proposed by Zhang et al. is creative, the performance is still lower than the previous work. Anh Duc Le and Masaki Nakagawa presented an advanced end-to-end system employing convolution neural network based on the attention-based encoder-decoder model [14]. The overall architecture of the model is showed in Fig. 5.

As for the end-to-end system proposed by Anh Duc Le and Masaki Nakagawa, it has three parts: a convolution neural network for feature extraction, a BLSTM for encoding the extracted features and a LSTM with an attention model for generating the LaTeX sequence. The CNN for feature extraction is a standard CNN without new technique. To extend training set because of the difficulty of collecting handwritten mathematical expressions, they also adopt local and global distortion models for data generation. Local distortion consist of shearing, shrinking, shrinking plus rotation, perspective and perspective plus rotation. Global distortion distorts an online handwritten mathematical expression with rotation and scaling models.

The composed model achieves 28.09%, 34.99% and 35.19% by using distortion on the CROHME training set, G_CROHME1, and G_CROHME2, respectively which shows that the end-to-end systems is potential.

In 2018, Zhang et al. proposed a novel end-to-end based approach named Track, Attend and Parse (TAP), consisting of a tracker and a parser equipped with guided hybrid attention (GHA) [15]. The two typical problems of symbol recognition and structural analysis can be optimized jointly through this end-to-end based approach. Figure 6 is the overall architecture of TAP model proposed by Zhang et al.

They employed a GRU as the tracker which takes handwritten expression traces as input and maps the trajectory information to high-level representations. The parser consists of a unidirectional GRU and a GHA where GHA is composed of a coverage based spatial attention, a temporal attention and an attention guider. More specifically, for each predicted symbol, the spatial attention teaches the parser where to attend for describing a math symbol or an implicit spatial operator. The temporal attention informs the parser when to rely on the product of spatial attention or to just rely on the language model built in the parser. In the procedure of training, the learning of spatial attention is also guided by an attention guider which performs as a regularization term.

The TAP model achieves a recognition accuracy of 61.16% on CROHME 2014 and 57.02% on CROHME 2016, outperforming the state-of-the-art methods published by the CROHME competition, which makes a improvement of online handwritten mathematical expression recognition.

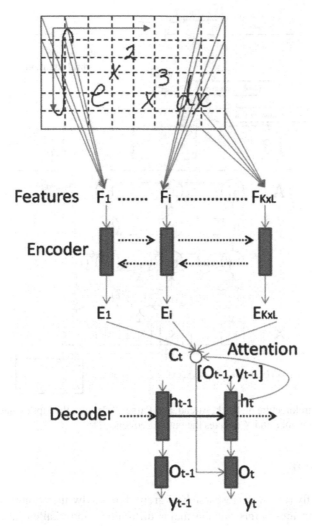

Fig. 5. Structure of the end-to-end model. $F_i (i = 1, \dots, K \times L)$ are the features of handwritten mathematical expression and $E_i (i = 1, \dots, K \times L)$ are the outputs generated by the encoder. At each time step t, the decoder predict symbol y_t based on the current output O_t, and the context vector C_t. O_t is calculated from the previous hidden state of the decoder h_{t-1}, the previous decoded vector O_{t-1}, and the previous symbol y_{t-1} [14].

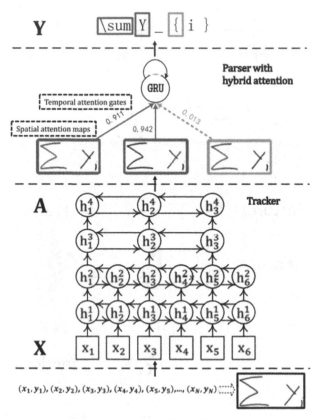

Fig. 6. Overall architecture of Track, Attend, and Parse. X denotes the input sequence, A denotes the annotation sequence and Y denotes the output sequence [15].

3 Conclusion

This paper briefly review the research background of handwritten mathematical expression recognition and points out the major difficulties and challenges in the current research. This paper summarizes and analyzes the latest development of online handwritten mathematical expression recognition based on depth learning in recent years, and summarizes the research status of various depth learning models in handwritten mathematical expression recognition.

In recent years, online homework assistant platform becomes a promising business model, which requires a handwritten mathematical expression recognition system with a satisfactory recognition accuracy. In addition, because of the rapid development of touch-screen smart devices such as smart cell phone and tablet computer, handwriting input is becoming a more and more popular touch-screen interactive application. Therefore, handwritten mathematical expression recognition technology still deserves more attention and further research.

Both offline and online handwritten mathematical expression recognition have got improvement with the application of deep learning. However, there are still many problems to be solved.

First, existing work on OHMER can only achieve high recognition accuracy on handwritten mathematical symbol recognition. On long handwritten mathematical expression with complex structure, there is still a lot of room for improvement. Most of the proposed model can only achieve recognition accuracy of about 50% to 60% on standard competition dataset, which is far away being applied on realistic task.

Moreover, at present, the deep learning model which can achieve significantly better than the traditional method in OHMER field is mainly based on RNN and various improved methods. Other deep learning models like GAN model has not been applied on OHMER. The research on the interrelation and fusion application of various deep learning models is not in-depth enough. We are looking forward to novel and better-performed deep learning models are proposed for OHMER in the future and make breakthroughs to promote the research and development in this field.

In conclusion, deep learning provides a new idea and technology for OHMER and a lot of achievements have been made in this filed in recent years. But there are still many problems worthy of further research. Through the review and discussion of the research in OHMER, this paper is aimed at bringing new information and ideas to researchers and jointly promote the further development and prosperity of OHMER and related document analysis and recognition fields.

References

1. Chan, K.-F., Yeung, D.-Y.: Mathematical expression recognition: a survey. Int. J. Doc. Anal. Recog. **3**(1), 3–15 (2000)
2. Zanibbi, R., Blostein, D., Cordy, J.R.: Recognizing mathematical expressions using tree transformation. IEEE Trans. Pattern Anal. Mach. Intell. **24**(11), 1455–1467 (2002)
3. Alvaro, F., Sánchez, J.-A., Benedí, J.-M.: Recognition of online handwritten mathematical expressions using 2D stochastic contextfree grammars and hidden markov models. Pattern Recog. Lett. **35**, 58–67 (2014). Author, F.: Contribution title. In: 9th International Proceedings on Proceedings, pp. 1–2. Publisher, Location (2010)
4. Awal, A., Mouchere, H., Viardgaudin, C.: A global learning approach for an online handwritten mathematical expression recognition system. Pattern Recogn. Lett. **35**(1), 68–77 (2014)
5. Alvaro, F., Sanchez, J., Benedi, J.: An integrated grammarbased approach for mathematical expression recognition. Pattern Recogn. **51**, 135–147 (2016)
6. LeCun, Y., Bottou, L., Bengio, Y., Haffner, P.: Gradient-based learning applied to document recognition. Proc. IEEE **86**(11), 2278–2324 (1998)
7. Ramadhan, I., Purnama, B., Al Faraby, S.: Convolutional neural networks applied to handwritten mathematical symbols classification. In: 2016 4th International Conference on Information and Communication Technology (ICoICT). IEEE (2016)
8. Tran, G.S., et al.: Handwritten mathematical expression recognition using convolutional neural network. In: 2018 3rd International Conference on Control, Robotics and Cybernetics (CRC). IEEE (2018)
9. Dai Nguyen, H., Le, A.D., Nakagawa, M.: Deep neural networks for recognizing online handwritten mathematical symbols. In: 2015 3rd IAPR Asian Conference on Pattern Recognition (ACPR). IEEE (2015)

10. Zhang, T., Mouchère, H., Viard-Gaudin, C.: Online handwritten mathematical expressions recognition by merging multiple 1D interpretations. In: 2016 15th International Conference on Frontiers in Handwriting Recognition (ICFHR). IEEE (2016)
11. Bahdanau, D., Cho, K., Bengio, Y.: Neural machine translation by jointly learning to align and translate. arXiv preprint arXiv:1409.0473 (2014)
12. Lamport, L.: LaTeX: A Document Preparation System: User's Guide and Reference. Illustrations by Duane Bibby. AddisonWesley Professional, Reading (1994). ISBN 0-201-52983-1
13. Zhang, J., Du, J., Dai, L.: A GRU-based encoder-decoder approach with attention for online handwritten mathematical expression recognition. In: 2017 14th IAPR International Conference on Document Analysis and Recognition (ICDAR), vol. 1. IEEE (2017)
14. Le, A.D., Nakagawa, M.: Training an end-to-end system for handwritten mathematical expression recognition by generated patterns. In: 2017 14th IAPR International Conference on Document Analysis and Recognition (ICDAR), vol. 1. IEEE (2017)
15. Zhang, J., Jun, D., Dai, L.: Track, Attend, and Parse (TAP): an end-to-end framework for online handwritten mathematical expression recognition. IEEE Trans. Multimedia 21(1), 221–233 (2018)

Comparison of Persian Handwritten Digit Recognition in Three Color Modalities Using Deep Neural Networks

Abbas Zohrevand[1], Mahdi Sattari[2], Javad Sadri[3(✉)], Zahra Imani[4],
Ching Y. Suen[3], and Chawki Djeddi[5]

[1] Department of Computer Engineering, Kosar University of Bojnord, Bojnord, Iran
Zohrevand86@gmail.com
[2] Department of Computer Engineering, University of Tabriz, Tabriz, Iran
Eng.MahdiSattari@yahoo.com
[3] Department of Computer Science and Software Engineering,
Faculty of Engineering and Computer Science, Concordia University,
Montreal, QC, Canada
{j_sadri,suen}@cs.concordia.ca
[4] Department of Electrical and Computer Engineering,
Babol Noshirvani University of Technology, Babol, Iran
Z.imani13@gmail.com
[5] Department of Mathematics and Computer Science,
Larbi Tebessi University, Tebessa, Algeria
C.djeddi@univ-tebessa.dz

Abstract. Most of the methods on handwritten digit recognition in the literature are focused and evaluated on black and white image databases. In this paper we try to answer a fundamental question in document recognition. Using Convolutional Neural Networks (CNNs), we investigate to see whether color modalities of handwritten digits affect their recognition rate? To the best of our knowledge, so far this question has not been answered due to the lack of handwritten digit databases that have all three color modalities of handwritten digits. To answer this question, we select 13,330 isolated digits from novel Persian handwritten database, which have three different color modalities and are unique in term of size and variety. Our selected dataset are divided into training, validation, and testing sets. Afterward, similar conventional CNN models are trained with the samples of our training set. While the experimental results on the testing set show that CNN on the black and white digit images has a better performance compared to the other two color modalities, in general there are no significant differences for network accuracy in different color modality. Also, comparisons of training times in three color modalities show that recognition of handwritten digits in black and white using CNN is much more efficient.

Keywords: Handwritten digit recognition · Handwritten data bases · Image modalities · Convolutional Neural Networks · Deep learning

© Springer Nature Switzerland AG 2020
Y. Lu et al. (Eds.): ICPRAI 2020, LNCS 12068, pp. 125–136, 2020.
https://doi.org/10.1007/978-3-030-59830-3_11

1 Introduction

Recognition of handwritten characters and digits is one of the most impressing subjects in pattern recognition [1]. Handwritten digit recognition has many applications such as processing of bank cheques, postal mail sorting, reading data from forms, archiving and retrieving medical and business forms. Due to importance of these applications, in the last twenty years many works have been proposed on handwritten digit recognition for different scripts such as Latin, Chinese, Indian, Arabic, and Persian (please see [2–9]).

Besides developing many interesting and novel recognition methods, several handwritten digit databases have been provided to evaluate and compare recognition methods in different scripts. For example, CENPARMI[1] digit dataset [10] which, comprises of 6000 digits prepared from the envelop USPS[2] images. In this database 4000 samples, 400 images per digit, are determined for training and 2000 images for test. The CEDAR[3] digit dataset is another database, which comprises of 18468 and 2711 digits as the training and test sets, respectively [11].The MNIST dataset [12] was another database which extracted from the NIST[4] dataset. In MNIST, images are normalized into 20 * 20 gray scale images, and then the normalized samples are placed in 28 * 28 frames [1]. Another popular database is USPS digit dataset which has 7291 training and 2007 test samples. Similar databases also exist for a few other languages such as Chinese, Arabic and Indian [13,14]. In Persian, HODA [1] is popular database which, their image samples were extracted from 12,000 registration forms. This dataset contains 102,352 digits and partitioned into train, test and evaluation groups. There are other popular Persian handwritten digit dataset [15,16]. Although these datasets are very great works, to the best of our knowledge most of them do not provide all color modalities (True color, gray level, and binary images) of handwritten characters or digits. Providing all color modalities of the same images of handwritten characters or digits can provide a lot of opportunities for new investigations in computer vision and for recognition and time comparison. Recently, some publicly available handwritten databases such as [17,18] have provided all color modalities of the images of their handwritten samples of characters or digits and have paved the way for new research directions in handwritten document recognition.

Recall that humans have color vision due to existence of red, blue and green cones in the retina [19]. However, the retina of cats and dogs possess solely blue and green cones, resulting in a much more muted perception of color, which is similar to color blindness in humans [19]. This paper tries to answer fundamental question in document recognition. Using Deep Convolutional Neural Networks (CNNs), as eye simulators, we investigate to see whether color modalities of handwritten digits effect their recognition accuracy and speed. To the best of

---

[1] Center for Pattern Recognition and Machine Intelligence.
[2] United States Postal Service.
[3] Center of Excellence for Document Analysis and Recognition.
[4] National Institute of Standards and Technology.

our knowledge, so far this question has not been answered due to lack of handwritten digit databases that have all three-color modalities of the handwritten digits. To answer this question, samples of Persian isolated digits provided by a new comprehensive handwritten Persian database [17] in three different color modalities are selected and they are divided into training, validation, and testing sets. Then, similar deep conventional CNN models are created with the samples of our training set. The experimental result on the testing set show that CNN on the black and white digit images has a better performance compared to the other two modalities. In general there is no significant difference for network accuracy in different color models. Also, comparisons of training times in three color modalities show that recognition of handwritten digits in black and white using CNN is much more efficient. Details of experiments are shown in the paper.

The rest of paper organized as follow; In Sect. 2 the architecture of proposed CNNs for all color modulates has been explained. In the Sect. 3 the new database which has three colors modalities has been reviewed. In the Sect. 4 the data augmentation has been explained. Details of experimental results are described in Sect. 5, and finally in Sect. 6 conclusion and future works explained.

2   Methodology

For the first time Convolutional Neural Networks proposed by LeCun et al. [20]. These types of artificial neural networks combine feature extraction and classification tasks together, and the main purpose of their designs is to recognize 2D shapes due to scale, shift and distortions [20]. In comparison of other classical approaches, CNN architecture extract features automatically. Indeed, the network input is the raw image and the winning class presented by the output. The architecture of CNN, which is proposed in this paper shown in Fig. 1. This CNN network contains: one input layer, two convolutional, two sub-sampling layers for extracting features automatically, two fully connected layers as multi-layer perceptron for classification, and one output layer. Detail of our proposed architecture shown in Table 1. In this paper, we apply this architecture to true color, gray level and black and white images, separately, and compare them due to the recognition rates and the training times.

3   Database

There are several databases for the evaluation of Persian handwritten recognition systems [1, 21–23]. Although efforts to produce these databases have been remarkable, none of them provided three image modalities (true color, gray level and black and white) for handwritten characters or digits. To cover this shortcoming, a vast and a comprehensive Persian handwritten database recently has been introduced [17] and used for our experiment. One example of three image color modalities written by the same writer taken from this database is shown in Fig. 2.

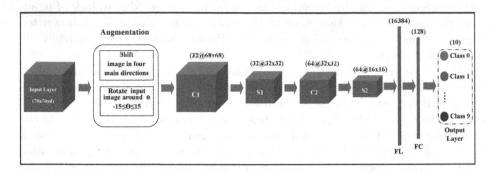

Fig. 1. Convolutional Neural Network (C: convolutional layer, S: subsampling layer, FL: flatten layer, FC: fully connected). In above figure 'd' the input layer indicates the depth of the input images which is (d = 1 for Black and White and Gray level images, d = 3 for True color images). (Color figure online)

Table 1. Details of layers in proposed CNN.

Layer	Layer operation	Feature map no.	Feature map size	Window size	Parameter
C1	Convolution	32	68 * 68	3 * 3	896
S1	Max-Pooling	32	34 * 34	2 * 2	0
C2	Convolution	64	32 * 32	3 * 3	18496
S2	Max-Pooling	64	16 * 16	2 * 2	0
FC	Flatten Layer	N/A	N/A	N/A	0
FC	Fully Connected	N/A	N/A	N/A	2097280
FC	Output	N/A	N/A	N/A	1290

Digit	0	1	2	3	4	5	6	7	8	9
TC										
GL										
BW										

Fig. 2. Samples of isolated handwritten Persian digits [17] (TC: True Color, GL: Gray Level, BW: Black & White Images). (Color figure online)

This database can be used to develop most applications such as general handwriting recognition [24–26], segmentation, word spotting, letter recognition, digit recognition, gender recognition [27,28], and writer identification [29], proposed In [17]. This database includes digits, numeral dates, alphabet letters, worded dates, numeral strings, touching digits, symbols, words, and free texts (paragraphs), and recorded in three image modalities (true color, gray level and black and white).

4 Data Augmentation

In this paper, a subset including of 13,330 samples of isolated digits have been selected from our database and then 80% of them have been selected for training and validating CNNs and 20% remaining images for testing of trained CNN models. Existence many trainable parameters are one of the challenging point in deep neural network [30]. As shown in Table 1, there are many parameters (weights) that should be optimized in CNN, and for training these huge numbers of weight parameters there must be enough image samples. For creating suitable and enough images, original images must be augmented as follows. In this paper we have used efficient methods like shifting (in four main directions) and rotation (with constant probability) for creating new samples. As shown in Fig. 3, each input image firstly shifted in four main directions and then each shifted and original images are rotated around −15 to 15 degrees with fix probability.

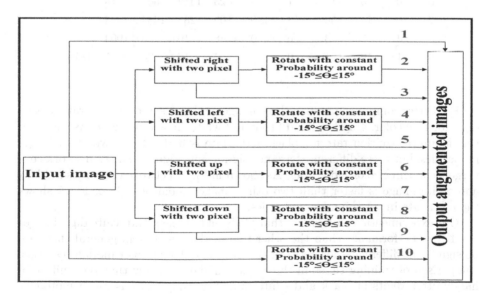

Fig. 3. Procedure of augmenting for each input image. As shown in the above figure, input images and all their shifted images are saved as new images and then each shifted and original image is rotated randomly with a fix probability = 0.9).

Totally after augmentation, there are 100000 images, which 80% (0.8 * 100000 = 80000) of data provided for training and validation of CNN and remaining 20%(0.2 * 100000 = 20000) are used for testing CNN model.

5   Experimental Result

In our experiments we apply similar architecture (see Fig. 1) for recognizing Persian handwritten digit database. In three image modalities, proposed CNN trained by stochastic gradient descent algorithm using empirical optimized parameters. The confusion matrix for true color, gray level and black & white image shown in Table 2, 3 and 4 respectively.

Table 2. Confusion matrix on true color images.

–	0	1	2	3	4	5	6	7	8	9
0	1983	0	5	1	0	2	2	3	1	3
1	0	1972	1	0	1	0	12	10	0	4
2	0	2	1864	78	39	1	10	5	1	0
3	0	0	60	1883	47	7	1	2	0	0
4	0	22	32	87	1822	8	5	15	4	5
5	4	8	1	7	14	1912	8	25	10	11
6	0	5	13	1	8	23	1412	507	14	17
7	0	14	56	13	120	109	288	1384	4	12
8	4	5	3	0	0	48	10	6	1904	20
9	1	10	1	0	5	57	16	25	7	1878

For better comparison among three image modalities, recognition rate of each digit for true color, gray level and black and white shown in Fig. 4. As shown in this figure recognition rate for black and white is better than two other image modalities, but these differences are not significant. Figure 5 shows network convergence in three color modalities. As shown in Fig. 5 network converges in black and white image is faster than two other image modalities. Also, Fig. 6 shows the error rate in three image modalities.

Our experimental result shows that CNN on the black and white digit images has better performance than the other two modalities, but in general there are no significant differences for network accuracy in different color modalities. Also, comparison of training times in three color modalities show that recognition of handwritten digits in black and white modality using CNN is much more efficient. All implementations are based on Python Scripting language and Tensor-Flow library (CPU version). All our experiments are done on a Core i5 processor 2.50 GHz with 4 GB onboard memory under Linux Ubuntu 16.04 Operating system. Table 5 shows running times in training and testing phases based on this

Table 3. Confusion matrix on gray level images.

–	0	1	2	3	4	5	6	7	8	9
0	1992	3	0	0	2	3	0	0	0	0
1	0	1984	0	0	0	0	3	8	3	2
2	0	3	1928	52	8	0	0	4	4	1
3	0	0	55	1924	8	4	0	7	2	0
4	0	3	20	18	1932	3	1	10	4	9
5	0	5	0	0	9	1963	1	16	4	2
6	3	3	5	0	3	1	1521	455	0	9
7	0	26	34	15	13	48	100	1750	4	10
8	0	7	0	1	0	4	3	2	1976	7
9	0	9	0	0	4	42	7	13	1	1924

Table 4. Confusion matrix on black & white images.

–	0	1	2	3	4	5	6	7	8	9
0	1989	2	2	0	3	4	0	0	0	0
1	0	1987	2	0	0	0	4	7	0	0
2	0	2	1937	36	17	0	1	5	2	0
3	0	0	56	1925	10	5	1	3	0	0
4	0	2	10	13	1950	3	0	13	3	6
5	0	5	0	1	7	1974	3	5	2	3
6	0	1	3	0	1	3	1562	424	0	6
7	0	10	7	8	6	29	56	1865	12	16
8	0	7	0	0	1	8	2	1	1972	9
9	0	13	0	0	4	40	3	8	0	1932

platform for our whole image database as mentioned in Sect. 4. As shown in this table, in the training phase recognizing handwritten digits in black and white modality is more efficient than the two other modalities, however in the testing phase, recognizing handwritten digits in three modalities nearly have equal running times.

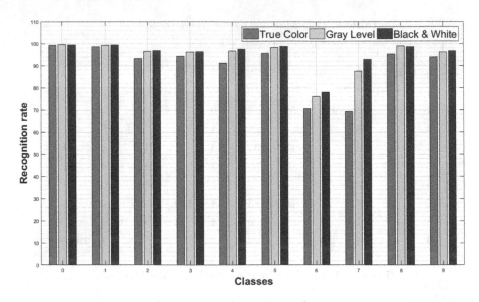

Fig. 4. Overall recognition rate for each digit class in three different image modalities.

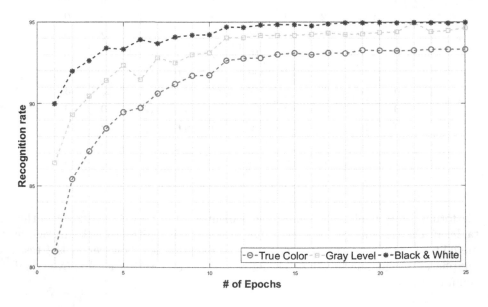

Fig. 5. Overall recognition rate of CNN network versus training epochs in three different image modalities.

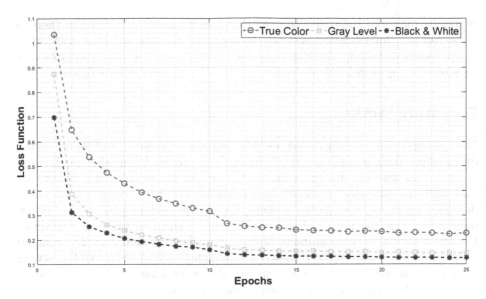

Fig. 6. Overall error rate for each digit class in three different image modalities.

Table 5. Running times for training and testing CNN (in seconds).

Image modality	Training times	Testing times
True color	68016.16	347.7
Gray level	58705.5	329.4
Black & White	52261.0	337.7

6   Conclusion and Discussion

Humans have color vision due to existence of red, blue and green cones in the retina. In other mammals (such as cats and dogs) color vision is limited, resulting in a much more muted perception of color, which is similar to color blindness in humans. In this paper we conducted a case study on Persian handwritten digit recognition in three different color vision modalities. Using Deep Convolutional Neural Networks (CNNs), as eye simulators, we investigate to see whether color modalities of handwritten digits would affect their recognition accuracy and speed. To the best of our knowledge, so far this question has not been answered, quantitatively, due to the lack of handwritten digit databases that have all three-color modalities of the handwritten digits. To answer this question, samples of Persian isolated digits provided by a new comprehensive handwritten Persian database [17] in three different color modalities are selected and they are divided into training, validation, and testing sets. Then, similar deep conventional CNN models are created with the samples of our training set. The experimental results on the testing set show that CNN on the black and white digit images has a better performance compared to the other two modalities. In general there is no significant difference for network accuracy in different color models. Also, comparisons of training times in three color modalities show that recognition of handwritten digits in black and white using CNN is much more efficient.

7   Future Works

In this paper, we investigate to see whether color modalities of handwritten digits affect their recognition rate? To the best of our knowledge, so far this question has not been answered due to the lack of handwritten databases that have all three color modalities of handwritten images. Sadri et al. [17] proposed database which includes digits, numeral dates, alphabet letters, worded dates, numeral strings, touching digits, symbols, words, and free texts (paragraphs). For the creation of this database, 500 individuals (250 men and 250 women) each completed seven special Handwriting Samples Forms (HSF). All HSFs were gathered, scanned (with 300 dpi), segmented, and then recorded in three image modalities (true color, gray level and black and white). Because of more complexity and variety in words and free text,in the next step, we want to investigate same question on word and texts.

References

1. Khosravi, H., Kabir, E.: Introducing a very large dataset of handwritten Farsi digits and a study on their varieties. Pattern Recogn. Lett. **28**(10), 1133–1141 (2007)
2. Qiao, J., Wang, G., Li, W., Chen, M.: An adaptive deep Q-learning strategy for handwritten digit recognition. Neural Netw. **107**, 61–71 (2018)
3. Mayraz, G., Hinton, G.E.: Recognizing hand-written digits using hierarchical products of experts. In: Advances in neural information processing systems, pp. 953–959 (2001)
4. Oliveira, L.S., Sabourin, R., Bortolozzi, F., Suen, C.Y.: Automatic recognition of handwritten numerical strings: a recognition and verification strategy. IEEE Trans. Pattern Anal. Mach. Intell. **24**(11), 1438–1454 (2002)
5. Sarker, S., Sarker, S., Rahman, S., Jabiullah, M.I.: A Lenet-5 based Bangla handwritten digit recognition framework. Adv. Image Process. Pattern Recogn. **2**(3), 1–7 (2019)
6. Ghofrani, A., Toroghi, R.M.: Capsule-based Persian/Arabic robust handwritten digit recognition using EM routing. In: 2019 4th International Conference on Pattern Recognition and Image Analysis (IPRIA), pp. 168–172. IEEE (2019)
7. Alani, A.: Arabic handwritten digit recognition based on restricted Boltzmann machine and convolutional neural networks. Information **8**(4), 142 (2017)
8. Liu, X., Hu, B., Chen, Q., Wu, X., You, J.: Stroke sequence-dependent deep convolutional neural network for online handwritten Chinese character recognition. IEEE Trans. Neural Netw. Learn. Syst. 1–12 (2020)
9. Ghanbari, N.: A review of research studies on the recognition of Farsi alphabetic and numeric characters in the last decade. In: Montaser Kouhsari, S. (ed.) Fundamental Research in Electrical Engineering. LNEE, vol. 480, pp. 173–184. Springer, Singapore (2019). https://doi.org/10.1007/978-981-10-8672-4_13

10. Suen, C.Y., Nadal, C., Legault, R., Mai, T.A., Lam, L.: Computer recognition of unconstrained handwritten numerals. Proc. IEEE **80**(7), 1162–1180 (1992)
11. Liu, C.-L., Nakashima, K., Sako, H., Fujisawa, H.: Handwritten digit recognition: benchmarking of state-of-the-art techniques. Pattern Recogn. **36**(10), 2271–2285 (2003)
12. LeCun, Y., Bottou, L., Bengio, Y., Haffner, P., et al.: Gradient-based learning applied to document recognition. Proc. IEEE **86**(11), 2278–2324 (1998)
13. Bhattacharya, U., Chaudhuri, B.B.: Databases for research on recognition of handwritten characters of Indian scripts. In: Eighth International Conference on Document Analysis and Recognition (ICDAR 2005), pp. 789–793. IEEE (2005)
14. Mihov, S., et al.: A corpus for comparative evaluation of OCR software and post-correction techniques. In: Eighth International Conference on Document Analysis and Recognition (ICDAR 2005), pp. 162–166. IEEE (2005)
15. Haghighi, P.J., Nobile, N., He, C.L., Suen, C.Y.: A new large-scale multi-purpose handwritten Farsi database. In: Kamel, M., Campilho, A. (eds.) ICIAR 2009. LNCS, vol. 5627, pp. 278–286. Springer, Heidelberg (2009). https://doi.org/10. 1007/978-3-642-02611-9_28
16. Mozaffari, S., Faez, K., Faradji, F., Ziaratban, M., Golzan, S.M.: A comprehensive isolated Farsi/Arabic character database for handwritten OCR research. In: Tenth International Workshop on Frontiers in Handwriting Recognition. Suvisoft (2006)
17. Sadri, J., Yeganehzad, M.R., Saghi, J.: A novel comprehensive database for offline Persian handwriting recognition. Pattern Recogn. **60**, 378–393 (2016)
18. Akbari, Y., Jalili, M.J., Sadri, J., Nouri, K., Siddiqi, I., Djeddi, C.: A novel database for automatic processing of Persian handwritten bank checks. Pattern Recogn. **74**, 253–265 (2018)
19. Jacobs, G.H.: Colour vision in animals. Endeavour **7**(3), 137–140 (1983)
20. LeCun, Y., Bengio, Y., et al.: Convolutional networks for images, speech, and time series. In: The Handbook of Brain Theory and Neural Networks, vol. 3361(10) (1995)
21. Ziaratban, M., Faez, K., Bagheri, F.: FHT: an unconstraint Farsi handwritten text database. In: 2009 10th International Conference on Document Analysis and Recognition, pp. 281–285. IEEE (2009)
22. Solimanpour, F., Sadri, J., Suen, C.Y.: Standard databases for recognition of handwritten digits, numerical strings, legal amounts, letters and dates in Farsi language. In: Tenth International workshop on Frontiers in Handwriting Recognition. Suvisoft (2006)
23. Imani, Z., Ahmadyfard, A.R., Zohrevand, A.: Introduction to database FARSA: digital image of handwritten Farsi words. In: 11th Iranian Conference on Intelligent Systems in Persian, Tehran, Iran (2013)
24. Imani, Z., Ahmadyfard, Z., Zohrevand, A.: Holistic Farsi handwritten word recognition using gradient features. J. AI and Data Min. **4**(1), 19–25 (2016)
25. Imani, Z., Ahmadyfard, A., Zohrevand, A., Alipour, M.: Offline handwritten Farsi cursive text recognition using hidden Markov models. In: 2013 8th Iranian Conference on Machine Vision and Image Processing (MVIP), pp. 75–79. IEEE (2013)
26. Mersa, O., Etaati, F., Masoudnia, S., Araabi, B.N.: Learning representations from Persian handwriting for offline signature verification, a deep transfer learning approach. In: 2019 4th International Conference on Pattern Recognition and Image Analysis (IPRIA), pp. 268–273. IEEE (2019)
27. Akbari, Y., Nouri, K., Sadri, J., Djeddi, C., Siddiqi, I.: Wavelet-based gender detection on off-line handwritten documents using probabilistic finite state automata. Image Vis. Comput. **59**, 17–30 (2017)

28. Bi, N., Suen, C.Y., Nobile, N., Tan, J.: A multi-feature selection approach for gender identification of handwriting based on kernel mutual information. Pattern Recogn. Lett. **121**, 123–132 (2019)
29. Aubin, V., Mora, M., Santos-Peñas, M.: Off-line writer verification based on simple graphemes. Pattern Recogn. **79**, 414–426 (2018)
30. Liu, W., Wang, Z., Liu, X., Zeng, N., Liu, Y., Alsaadi, F.E.: A survey of deep neural network architectures and their applications. Neurocomputing **234**, 11–26 (2017)

Application of Deep Learning in Handwritten Mathematical Expressions Recognition

Hong Lin$^{(\boxtimes)}$ ⓘ and Jun Tan ⓘ

School of Mathematics, Sun Yat-Sen University, Guangzhou, China
lh19960327@163.com, mcstj@mail2.sysu.edu.com

Abstract. Handwritten mathematical expression (HME) is a research highlight in pattern recognition area since it has many practical application scenarios such as scientific documents digitalization and online education. However, it can be very challenging due to its complicated two-dimensional structure and great variety among different individuals. Inspired by the excellent performance of deep learning in various recognition problems, some researchers in this field began to use deep learning method to solve the problem of handwriting recognition in order to reach the prediction accuracy that traditional resolutions cannot achieve. In this survey paper, we aim to review the solutions of deep learning (encoder-decoder, GRU, attention mechanism and so on) to various problems of handwritten mathematical expression recognition. In particular, some important methods and issues will be described in depth. Moreover, we will try to discuss the future application of deep learning to HME recognition. All these together will help us have a clear picture about how deep learning methods are applied to in this research area.

Keywords: Handwritten mathematical expression recognition deep learning · Encoder-decoder · Attention mechanism

1 Introduction

1.1 A Subsection Sample

It is well known that writing, which is almost as long as the history of human civilization is a very important skill of human beings. We can store our knowledge and thoughts for a longer time and pass them on to future generations. Nowadays, with the rapid increment of Internet users and machine learning technology, people pursue to automatically recognize and store writings in computers so as to store and deliver information more effectively. Handwritten mathematical expression recognitions, which is an important part of science and engineering literatures, plays a vital rule in transcribing paper documents (both printed forms and written forms) into electronic documents

Research on mathematical expressions recognition can be traced back to 1967 [2] but develop slowly due to lack of efficient methods. At the beginning of the 21th century, the problems of handwritten mathematical expression recognition (HMER) once again attracted researchers' attention because of the development of high performance

© Springer Nature Switzerland AG 2020
Y. Lu et al. (Eds.): ICPRAI 2020, LNCS 12068, pp. 137–147, 2020.
https://doi.org/10.1007/978-3-030-59830-3_12

computing. People haven't been constantly improving their methods to improve the efficiency of writing recognition since then. And the Competition on Recognition of Online Handwritten Mathematical Expressions (CHROME), which is one of the most important competition in HMER area was held for the first time in 2011 [3].

Typically, this research area can be divided into two major parts: online and offline [4], based on the forms of HME. The formers are received from smart electronic devices such as pen-based tablets, while the latter are acquired from paper documents (That is to say, offline mathematical expressions are some images). Different from plain text recognition, mathematical expressions recognition is more complicated. Firstly, HMEs are two-dimensional which means we should consider spatial relationship among mathematical symbols in HMEs since they vary greatly, representing different meanings. For example: $y + \frac{1}{x}$ is totally different from $\frac{y+1}{x}$. Secondly, HMEs can be very ambiguous due to many reasons. To start with, HMEs consist of different kinds of symbols such as Arabic numerals, English letters, Greek letters, special operation symbols and so on. As a result some similar symbols in different categories can be confused [5]. In addition, a particular HME may look very different because different writers have different handwriting styles. Finally, in some cases spatial relationships between symbols can be ambiguous [1], for example, the following picture can be interpreted to η^x or $\eta \cdot x$ (Fig. 1):

Fig. 1. The ambiguity of the formula process of HMER

Traditionally, the process of HMER consists of two major stages: symbols recognition and structural analysis [1]. We can see many papers focus on improving one or more than one aspects in above two stages. Chan and Yeung [1] have already given a survey about traditional solutions of HMER before 2000. However, since then many new approaches applied in this research area have been presented. Besides, deep learning methods are one of the top research highlights in recent years since they have achieved great performance in many specific application scenarios of science and engineering. Researchers begin to make use of those methods to solve the problem of handwritten mathematical expressions recognition (such as WAP [6], TAP [7]). To the best of our knowledge, by now there hasn't been a survey paper which provides an overall review about these promoted HMER methods published since 2000. And not a paper gives a detailed comparison between these new methods including deep learning.

In this paper, we will remedy this missing part by reviewing those important systems applied to HMER. In particular, we will pay more attention to those methods based on deep learning and emphasize their power and differences from other kinds of methods. The rest of the paper is organized as follow. In Sect. 2, we give a brief introduction of the traditional solution on HMER problems. In Sect. 3, Application of deep learning method in handwritten mathematical expressions recognition are described. In Sect. 4, we will try to predict development direction of deep learning method in the research area of HMER.

2 Traditional Solution of HMER

As we have presented above, problems of HMER can be divided into two major parts: symbols recognition and structural analysis. Both of them have been studied for decades and have many mature solutions. But in the research area of HMER, we consider study integration of these two kinds of problems [1]. These problems can be solved consequently or globally [8]. The formers address problems step-by-step. That is to say, consequential solution will firstly try to recognize the mathematical symbols of expressions and then establish the structure of formulas according to the relationship between symbols. Conversely, the latter will consider combining two stages in order to make use of global information of mathematical expressions. More concretely, the flow of solving the problems of HMER is as shown in the Fig. 2. Next, we will introduce how traditional methods solve problems in each part of the process.

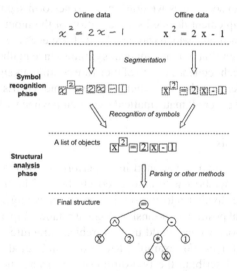

Fig. 2. The process of HMER [1]

2.1 Symbols Recognition

As we know, a mathematical expression contains multiple mathematical symbols, while many mature recognition techniques can only be applied to isolated symbols. So a naive thought is to segment individual symbols from expression before we recognize symbols separately. There are plenty of mathematical expression recognition systems which are based on this thought. On the other hand, other methods consider combine segmentation and recognition of symbols since precisely segment symbols of mathematical expressions can be very difficult. We will discuss different systems separately.

As for segmentation of symbols, Chan and Yeung [1] have summarized several commonly used methods. However, all of them have flaws have defects that seriously reduce the accuracy of identification. Since 2000, some new technologies have emerged. For example, Álvaro et al. computed components of mathematical expressions by using linear interpolation between the points belonging to the same stroke and then grouped connected components into one symbol [12]. What more, Hu and Zanibbi applied two-value AdaBoost classifier to segmenting mathematical symbols. Based on shape context-based features, the system achieved great performance. After the expressions are segmented and combined into symbol sequence, the recognition can be carried out. Chan and Yeung classify recognition methods into three categories: template matching approaches, structural approaches and statistical approaches [1]. Some statistical learning methods, such as support vector machine (SVM) [10], multi-layer perceptron (MLP) [11] can be effectively applied in mathematical symbols recognition and achieved higher recognition accuracy.

Due to the complicated geometry structural, mathematical expressions are very difficult to segment. Besides, data of online handwritten mathematical expressions (OHME) contain temporal information, based on which we can give mathematical expressions rough segmentation. So sometimes we don't have to perform segmentation and recognition separately or implement detailed segmentation of the input sequences. Methods based on hidden markov model which was developed for speech recognition can combine segmentation and recognition of mathematical symbols since mathematical expressions are quite similar to speech sequences (both of them are continuous signal in time domain). But there are also differences between them. For example, HMEs have high degree of freedom, so we cannot cover all mathematical expression when we build HMM models.

2.2 Structural Analysis

In general, when the methods mentioned in A perform recognition stage, we acquire a list of mathematical symbols with associated attributes. The next step is structural analysis which can be divided into two major parts: analyzing and error correction. According to the spatial position relationship, logical relationship and semantic analysis of mathematical symbols, we can build up hierarchical structure of HMEs in the form of a parse or relational trees. For example, Richard Zanibbi et al. construct a Baseline Structure Tree (BST) describing the two-dimensional arrangement of input symbols [13]. What more, Garain and Chaudhuri designed a context-free grammar to explore the spatial relationships among symbols according to the combination of online and offline features [14]. The computer can recover the true mathematical meaning of the formula by specific traversal method.

Sometimes due to the similarity of the handwriting of different symbols, there will be errors in recognition. The analysis of the structure can only help us to construct the corresponding structure tree, and cannot help us to correct the error of symbol recognition. Therefore, we need to correct the unreasonable elements in the structure tree with context and semantics features. And that is what Garain and Chaudhuri did in their paper [14]. What is more, Lee and Wang use four heuristic rules for error detection [15], and Chan and Yeung give some solutions to syntactic errors [16]. However, the errors of symbol recognition and segmentation are subsequently inherited by the structural analysis. Besides, mathematical expressions are complex and changeable, which makes it difficult to correct some lexical and logical errors. Researchers began to study the global approaches to perform these steps simultaneously.

2.3 Global Approaches

Ahmad-Montaser. Awal et al. present a HMER system with global strategy that allow segmenting mathematical expressions recognizing mathematical symbols and 2D structures of mathematical expressions simultaneously. Different from those sequential solutions mentioned above, the model can learn mathematical symbols and spatial relations directly from complete expressions, so it can reduce error propagation of sequential solutions [17]. Besides Alvaro et al. also presented a statistical model that simultaneously optimizing symbol segmentation, symbol recognition and structure searching.

However, systems with global approaches are more computationally than others [8]. When the complexity of HMEs' architecture increases exponentially with the size of predefined grammar, the models need to be modified.

3   Solving HMER Problems by Deep Learning Methods

Into the twenty-first century, deep learning has become a research highlight since the improvement of computer computing performance makes large-scale computing possible. Researches began to apply deep learning framework to handwritten expression recognition and we will introduce them in the rest of this section.

3.1   Commonly Used Deep Learning Framework

Since the systems that apply deep learning to solve the problem of HMER problems often establish a structure that combines different kinds of deep neural networks, we will firstly give a brief introduction of those commonly used deep learning framework.

Convolutional neural network has been widely used in many visual research fields, such as image classification, face recognition, audio retrieval, ECG analysis and so on Convolution neural network is so successful because of its convolution, pooling and other operations. Convolution operation has the characteristics of local connection and weight sharing, which can well retain the spatial information of two-dimensional data, while pooling operation can well meet the translation invariance, which is very important in the classification tasks [19]. In HMER system, CNN is often used to recognize the pictures of offline handwritten mathematical expressions.

Different from traditional feedforward network convolutional neural network, Recursive Neural Network (RNN) introduces states and cycles into neural networks so that such neural networks can model sequences. Figure 3 illustrates a simple RNN structure.

Given an input sequence $(x_1, x_2, x_3 \cdots x_n)$, we can adopt RNN as the encoder to compute the sequence of hidden state $(h_1, h_2, \ldots h_N)$ and the output sequence:

$$h_t = \tanh(W_{xh}x_t + W_{hh}h_{t-1}) \tag{1}$$

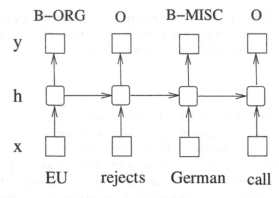

Fig. 3. The Recursive Neural Network architecture [24]

$$y_t = \text{softmax}(Vh_t) \tag{2}$$

Where W_{xh}, W_{hh}, V is the connection weight matrix weight to be trained.

Since natural language is a natural sequential format of data, RNN has become an important tool in many natural language processing (NLP) tasks. For example, in the seq2seq model of machine translation, RNN plays the role of encoder and decoder. Because the data of handwritten mathematical expressions are also sequence, so sequence model can be applied to HEMR research area (Fig. 4).

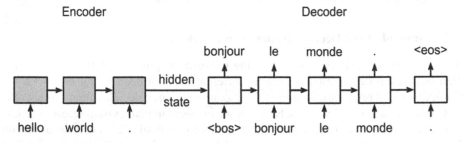

Fig. 4. The Sequence to Sequence model architecture [25]

However, when the sequence is too long, using RNN is prone to gradient disappearance and gradient explosion. Long short term memory (LSTM) network can well address this problem by adding gate mechanism. In LSTM networks, the hidden layer updates are replaced by purpose-built memory cells. As a result, they may be better at finding and exploiting long range dependencies in the data. Figure 5 illustrates the structure of a single LSTM memory cell.

LSTM memory cell is implemented as the following:

$$i_t = \sigma(W_{xi}x_t + W_{hi}h_{t-1} + W_{ci}c_{t-1} + b_i) \tag{3}$$

$$f_t = \sigma\left(W_{xf}x_t + W_{hf}h_{t-1} + W_{cf}c_{t-1} + b_f\right) \tag{4}$$

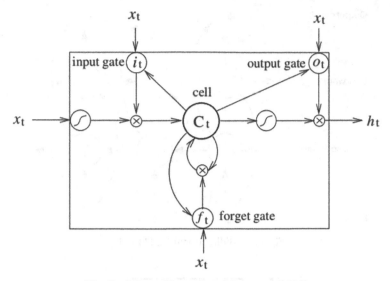

Fig. 5. A Long Short-Term Memory cell [24]

$$c_t = f_t c_{t-1} + i_t \tanh(W_{xc}x_t + W_{hc}h_{t-1} + b_c) \tag{5}$$

$$o_t = \sigma(W_{xo}x_t + W_{ho}h_{t-1} + W_{co}c_t + b_o) \tag{6}$$

$$o_t \tanh(c_t) \tag{7}$$

Where σ is the logistic sigmoid function and i, f, o and c are noted as input gate, forget gate, output gate and cell vectors, all of which are the same size as the hidden vector h. And W_{x*}, $W_{h*}(* = i, f, o, c)$ are the connection weight matrix weights to be trained. $b_*(* = i, f, o, c)$ are bias vector.

Sometimes we have access to both past and future input features for a given time. Bidirectional LSTM can help us address this issue. Input vectors are passed through two LSTM layer running in opposite directions and their hidden state vectors are concatenated together. In doing so, the past features and future features can be efficiently learned by the network for a specific time frame (Fig. 6).

But the LSTM network has too many parameters to converge effectively. Therefore, GRU which is a simplified version of LSTM was presented. It can achieve almost the same performance of LSTM while having fewer parameters (Fig. 7).

A GRU hidden state h_t is implemented as follows:

$$z_t = \sigma(W_{xz}X_t + U_{hz}h_{t-1}) \tag{8}$$

$$r_t = \sigma(W_{xr}X_t + U_{hr}h_{t-1}) \tag{9}$$

$$\tilde{h}_t = \tanh(W_{xh}X_t + U_{rh}(r_t \otimes h_{t-1})) \tag{10}$$

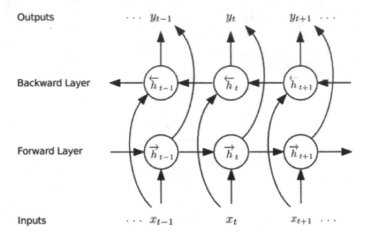

Fig. 6. A Bidirectional LSTM [24]

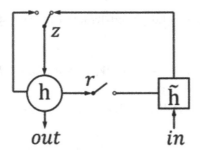

Fig. 7. The GRU architecture [24]

$$h_t = (1 - z_t) \otimes h_{t-1} + z_t \otimes \tilde{h}_t \tag{11}$$

Where σ is the sigmoid function and $\otimes$ is element-wise multiplication operator. z_t, r_t and $\tilde{h}_t$ are the update gate, reset gate and candidate activation respectively. W_{xz}, W_{xh}, W_{xr}, U_{hz}, U_{hr} and U_{rh} are related weight matrices. The GRU hidden state can be simply denoted as $h_t = GRU(X_t, h_{t-1})$.

Just like LSTM, we can also use the bidirectional GRU to capture past and future input features.

Attention mechanism is to select more critical information from a large number of information, and then invest more attention resources in the target areas that need to be focused. Adding attention mechanism to LSTM or GRU neural network can selectively give different weight to different text, and then using context based semantic association information can effectively make up for the lack of deep neural network in obtaining local features. Most of the sequence to sequence models rely on complex RNN and CNN as encoder and decoder. In order to reduce sequential computation, Vaswani et al. propose a new simple network architecture, the Transformer, based solely on attention mechanisms [28]. The Transformer follows the traditional architecture of encoder and decoder using stacked self-attention (Fig. 8).

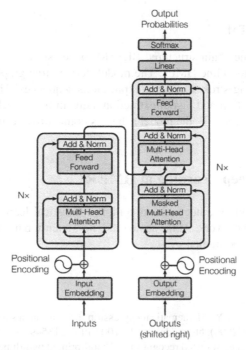

Fig. 8. The Transformer - model architecture [24]

3.2 End-to-End Approach

Zhang et al. present several end-to-end approached to solve HMER problems (including online HMEs and offline HMEs) [19–22]. End-to-end means that the input is the original data and the output is the final result. In 2017, they present an encoder-decoder framework with the attention mechanism for online HMEs. In their model, input data is encoded to the high level representations via the stack of bi-directional GRU, and the decoder was GRU with a coverage-based attention model which can helps complete the process of symbol segmentation [20]. In 2018, they improved their encoder-decoder framework by adding new attention mechanism. To be exact, they presented a guided hybrid attention (GHA) which consist of a coverage based spatial attention, a temporal attention and an attention guider [21]. The system presented above is designed for online HMEs. On the other hand, they present an end-to-end model for offline HMEs in 2017. In the procedure of encoder, they apply deep convolution network to identify two-dimensional expressions images [19]. In 2018, they improved their framework by applying multi-Scale Attention mechanism [22]. Their end-to-end approach avoids problems that stem from symbol segmentation and does not require a predefined expression grammar [19].

Hong et al. further improve feature extraction by employing residual connection in the Bi-RNN encoder layer. In the meantime, they introduce Markovian transition probability matrix in decoder to effectively make use of long-term information [27].

3.3 Tree-Based LSTM

Zhang et al. extend the chain-structured BLSTM to tree structure topology and apply this new framework to online HMEs. The model derived tree graphs by temporal and spatial relations among strokes of input expressions sequences. There are no explicit segmentation, recognition and layout extraction steps in the model, but by producing directly SLGs to describe ME, the model achieves competitive results on symbol level [23].

4 Prospect of Deep Learning in HMER

In the future, it should be a hot trend to combine statistical learning models such as HMM, CRF and deep network. At the same time, new attention mechanism model such as transformer should be studied.

References

1. Chan, K.-F., Yeung, D.-Y.: Mathematical expression recognition: a survey. Int. J. Doc. Anal. Recogn. **3**(1), 3–15 (2000). https://doi.org/10.1007/PL00013549
2. Anderson, R.H.: Syntax-directed recognition of hand-printed two-dimensional mathematics. In: Symposium on Interactive Systems for Experimental Applied Mathematics: Proceedings of the Association for Computing Machinery Inc. Symposium, pp. 436–459. ACM (1967)
3. Mouchère, H., Viardgaudin, C., Garain, U., et al.: CROHME2011: Competition on Recognition of Online Handwritten Mathematical Expressions (2011)
4. Plamondon, R., Srihari, S.N.: On-line and off-line handwriting recognition: a comprehensive survey. IEEE Trans. Pattern Anal. Mach. Intell. (T-PAMI) **22**(1), 63–84 (2002)
5. Blostein, D., Grbavec, A.: Recognition of mathematical notation. In: Handbook of Character Recognition and Document Image Analysis, pp. 557–582 (1997)
6. Zhang, J., Du, J., Zhang, S., et al.: Watch, attend and parse: An end-to-end neural network based approach to handwritten mathematical expression recognition. Pattern Recogn. **71**, 196–206 (2017)
7. Zhang, J., Du, J., Dai, L.: Track, Attend, and Parse (TAP): an end-to-end framework for online handwritten mathematical expression recognition. IEEE Trans. Multimedia **21**(1), 221–233 (2018)
8. Zhang, J., Du, J., Dai, L.: Track, Attend, and Parse (TAP): an end-to-end framework for online handwritten mathematical expression recognition. IEEE Trans. Multimedia **21**(1), 221–233 (2018)
9. Hu, L., Zanibbi, R.: Segmenting handwritten math symbols using AdaBoost and multi-scale shape context features. In: 2013 12th International Conference on Document Analysis and Recognition, pp. 1180–1184. IEEE (2013)
10. Suykens, J.A.K., Vandewalle, J.: Least Squares Support Vector Machine Classifiers. Kluwer Academic Publishers (1999)

11. Gardner, M.W.: Artificial neural networks (the multiplayer perceptron)-a review of applications in the atmospheric sciences. Atmospheric Environ. **32** (1998)
12. Álvaro, F., Sánchez, J.-A., Benedí, J.-M.: Recognition of on-line handwritten mathematical expressions using 2D stochastic context-free grammars and hidden Markov models. Pattern Recogn. Lett. **35**(1), 58–67 (2014)
13. Zanibbi, R., Blostein, D., Cordy, J.R.: Recognizing mathematical expressions using tree transformation. IEEE Trans. Pattern Anal. Machine Intell. **24**(11), 1455–1467 (2002)
14. Garain, U., Chaudhuri, B.B.: Recognition of online handwritten mathematical expressions. IEEE Trans. Syst. Man Cybern. Part B: Cybern. **34**(6), 2366–2376 (2004)
15. Lee, H.J., Wang, J.S.: Design of a mathematical expression recognition system. Pattern Recogn. Lett. **18**(3), 1084–1087 (1997)
16. Chan, K.F., Yeung, D.Y.: PenCalc: a novel application of on-line mathematical expression recognition technology. In: Proceedings of Sixth International Conference on Document Analysis and Recognition, pp. 774–778. IEEE (2001)
17. Awal, A.-M., Mouchère, H., Viard-Gaudin, C.: A global learning approach for an online handwritten mathematical expression recognition system. Pattern Recogn. Lett. **35**, 68–77 (2014)
18. Alvaro, F., Sánchez, J.A., Benedí, J.M.: An integrated grammar-based approach for mathematical expression recognition. Pattern Recogn. **51**, 135–147 (2016)
19. Zhang, J., Du, J., Zhang, S., et al.: Watch, attend and parse: an end-to-end neural network based approach to handwritten mathematical expression recognition. Pattern Recogn. **71**, 196–206 (2017)
20. Zhang, J., Du, J., Dai, L.: A GRU-based encoder-decoder approach with attention for online handwritten mathematical expression recognition. In: 2017 14th IAPR International Conference on Document Analysis and Recognition (ICDAR), vol. 1, pp. 902–907. IEEE (2017)
21. Zhang, J., Du, J., Dai, L.: Track, Attend, and Parse (TAP): an end-to-end framework for online handwritten mathematical expression recognition. IEEE Trans. Multimedia **21**(1), 221–233 (2018)
22. Zhang, J., Du, J., Dai, L.: Multi-scale attention with dense encoder for handwritten mathematical expression recognition. In: 2018 24th International Conference on Pattern Recognition (ICPR), pp. 2245–2250. IEEE (2018)
23. Zhang, T., Mouchère, H., Viard-Gaudin, C.: A tree-BLSTM-based recognition system for online handwritten mathematical expressions. Neural Comput. Appl. **32**, 4689–4708 (2020)
24. Huang, Z., Xu, W., Yu, K.: Bidirectional LSTM-CRF models for sequence tagging. arXiv preprint arXiv:1508.01991 (2015)
25. Zhang, A., Lipton, Z.C., Li, M., et al.: Dive into Deep Learning. Unpublished draft, vol. 3, p. 319 (2019). Accessed
26. Graves, A., Mohamed, A., Hinton, G.: Speech recognition with deep recurrent neural networks. In: 2013 IEEE International Conference on Acoustics, Speech and Signal Processing, pp. 6645–6649. IEEE (2013)
27. Hong, Z., You, N., Tan, J., et al.: Residual BiRNN based Seq2Seq model with transition probability matrix for online handwritten mathematical expression recognition. In: 2019 International Conference on Document Analysis and Recognition (ICDAR), pp. 635–640. IEEE (2019)
28. Vaswani, A., Shazeer, N., Parmar, N., et al.: Attention is all you need. In: Advances in Neural Information Processing Systems, pp. 5998–6008 (2017)

Automating Stress Detection
from Handwritten Documents

Najla AL-Qawasmeh$^{(\boxtimes)}$ and Muna Khayyat$^{(\boxtimes)}$

Department of Computer Science and Software Engineering, Concordia University,
CENPARMI, Montreal, Canada
n_alqawa@encs.concordia.ca,
muna.khayyat@gmail.com
http://www.concordia.ca/research/cenparmi.html

Abstract. Stress is a crucial problem in life, which is growing because of the fast-paced and demanding modern-life. Meanwhile, stress must be detected at early stages to prevent the negative effects on human health. Graphologist who analyze human handwriting have been able to detect stress from human handwriting. This is by extracting some features from the handwriting to detect stress level. The manual stress detection process is expensive, tedious and exhausting. This made the automation of the stress detection system important. Little research has been done on this field. In this research we will focus on automating stress detection from handwritten documents. We are working with graphologist to create a database of handwritten documents for stress detection. Later we will experiment different features to automate stress detection from the person's handwriting.

Keywords: Handwriting analysis · Feature extraction · Stress detection · Machine learning

1 Introduction

Modern life and technology have facilitated many aspects for human being. However, this has increased their levels of stress. Many studies have been done on stress detection since stress affects our life significantly. These studies are grouped into four main aspects: psychological, physiological, behavioral and physical. The manual process of detecting stress is tedious, inaccurate and time consuming. This has motivated many researchers to work on automatic stress detection.

This study is focused on behavioural aspect, which is related to person or group reactions or behavior in a certain situation, for example being more irritated, angry, being very aggressive with technological devices or using a certain aggressive words [1].

© Springer Nature Switzerland AG 2020
Y. Lu et al. (Eds.): ICPRAI 2020, LNCS 12068, pp. 148–155, 2020.
https://doi.org/10.1007/978-3-030-59830-3_13

1.1   Related Works

Benali et al. [2] Developed a computational approach for emotion detection in text. Their work was based on extracting emotional keywords from a gazetteer list in conjunction with syntactic data. A semantics analysis method was then used to enhance the detection accuracy. The system achieved a detection accuracy of 96.43% using Support Vector Machines (SVM) method.

Ezhilarasi and minu [3] proposed an automatic emotion recognition and classification system based on natural language processing. They analyzed and recognized the emotional outlay in English sentences by creating an emotion ontology using Word-net and its constructions. Giakoumis et al. [4] developed a stress detection system based on using a computer system to extract activity-related behavioral features automatically. The main idea of their work is the processing of appropriate video and accelerometer recorded from nineteen participants who followed a stress induction protocol using the Stroop color word test. Their experiments showed a potential correlation between the extracted features and self-reported stress. They obtained the same detection accuracy rate of 92.59% for both when applied a linear regression on the behavioral extracted features or physiological features such as Electrocardiogram and Galvanic Skin Response.

Shahin et al. [5] developed a lexical stress classification system in children's speech. They classified the stress into three categories: strong-weak(sw), weak -strong (ws) and equal-stress (ss/ww), using deep neural networks. The system achieved an accuracy of 83.2% for the classification of unequal lexical stress patterns, and accuracy of 72.3% when used for the classification of equal and unequal lexical stress patterns. Gonzalo et al. [6] proposed an online stress analysis system in mobile biometrics. They applied their system on dynamic handwritten signatures, where the participants were asked to sign on a smart-phone using a stylus. The system achieved an 83% classification accuracy.

Aigrain et al. [7] developed a stress detection system based on using a crowdsourcing platform to collect answers of binary questions as well as collecting a video data to capture facial and body features. They used several classification methods, including SVM, NN (neural network), RF (random forest) and NDF (neural decision forest). The best method in terms of accuracy was NN, where it achieved near 71% of classification accuracy.

Thelwall [8] proposed stress and relaxation detection system based on analyzing social media text messages. The author evaluated his system using texts from a corpus of 3066 English language tweets. Sulem et al. [9] provided a database for emotional state recognition from handwriting and drawing, called EMOTHAW. The dataset was collected from 129 participants, where they were asked to finish seven tasks, including the drawing of pentagons, houses, circles and clocks, as well as copying separate words and sentences. The participant's emotional states were assessed by the depression- anxiety- stress scales (DASS). A set of features related to the pen position, timing and ductus were extracted from the dataset. The system achieved a stress detection accuracy rate of 51.2% and 60.2% from writing and drawing, respectively.

Ayzeren et al. [10] proposed an online emotional states prediction system from handwriting and signatures. A set of online features including pen path and timing-dependent features were extracted. They used K-nearest neighbor and random forest for classification. Their system achieved a stress detection accuracy rate of 75.19% and 84.96%, respectively.

In this study we will focus on detecting stress from handwriting, in which we will proposed a machine learning system to automate stress detection using annotated handwritten documents. The rest of this paper is organized as follows: Sect. 2, discusses the research methodology including the proposed database and the proposed stress detection features to be extracted. Finally, Sect. 3 concludes the paper.

2   Research Methodology

In this research we will present a novel database of Arabic handwritten documents, each of which is annotated to detect if the writer is stressed or not. In addition, we will work on a machine learning based stress detection system, to reveal stress from handwritten document. Extracted features will be used based on a recommendation given by a psychologist together with a graphologist. These features will be fed into different classifiers to detect stress. In addition, we will compare our results with the results generated by Convolutional Neural Network or using other deep learning neural network models such as ResNet.

2.1   Dataset

In the literature of handwriting analysis and recognition, several databases have been used to conduct research. These databases differ in term of the language used, number of writers, number of documents. None of which has ground-truth on stress detection.

We are creating an Arabic handwriting database of free-style handwritten samples (FSHS). The database was written by 2200 volunteers, each of which handwritten a letter for someone he/she loves. We have also asked some of the volunteers to copy some paragraphs for future research purposes. This generated a database of 2700 handwritten samples. The writer's ages range from 15 to 75 years old. Men wrote 45% of our dataset, and women wrote the rest. Most of the writer are right-handed, and they are university and school students. Part of the volunteers are employees at some private and governmental companies.

To ensure the convenience of the writers, we did not put any conditions on the type of tool used for writing. However, we provided them with a white sheet of paper to write their letter on it. Volunteers were asked to fill up an information page about their gender, age, handedness, and their work position, as shown in Fig. 2. The variety in age and the large number of writers resulted in a great diversity in the handwriting styles. The dataset was mainly written in Arabic, although some writers used the English language, the number of these did not exceed 15 samples. we digitized the handwritten samples using a RICOH scanner

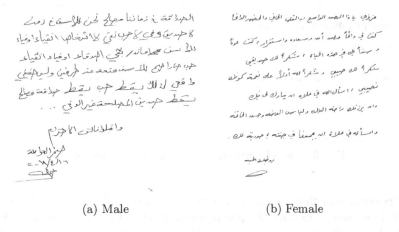

(a) Male   (b) Female

Fig. 1. Samples of the dataset

with a resolution of 600 dpi. Our dataset can be used in many research areas related to human interaction, such as handwriting recognition, gender, and age classification. Figure 1 shows some samples of our dataset.

Currently, we are working with a phycologist and a graphologist on labeling the dataset in which they will indicate if the writer of each of the document has been stressed when the page was written or not, together with other psychological related information. This database is the first database on stress detection, which will help research to continue working on automating stress detection based on human handwriting, since machine learning techniques have shown promising results in many different handwriting analysis studies.

2.2 Feature Extraction Phase

The primary goal of feature extraction is to extract the most relevant information from a raw data and represent them in a reduced set of feature vectors, which makes it easier for the classifier to analyze the extracted features instead of the raw data, since using all candidates will easily lead to over-fitting. It is important to focus on the feature extraction phase because the performance of the recognition system is highly dependent on the extraction features. The primary goals of feature extraction can be summarized as follow:

- Extract most relevant information to increase the discrimination between the classes.
- Extract features to reduce the within class variability meanwhile, increase the between class variability.
- Discard Redundant information.

In our proposed work, four types of features will be extracted, namely: pen pressure, baseline direction, space between words and letters. Experiments will be conducted to determine the best set of features.

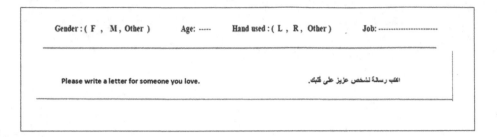

Fig. 2. The information page of the dataset samples.

Pen Pressure can show the mental state of the writer. The standard deviation (SD) will be used to extract the pen pressure feature from the gray-scale hand-written documents using Eq. (1). Then the mean value will be calculated and used as a threshold to evaluate the value of the standard deviation. Where light pressure will be identified if the value of the SD is higher than the threshold. Otherwise, it will be considered as a heavy pressure. The mean value will be calculated using Eq. (2). We will use pen pressure feature to extract both the stress and gender from handwritten documents.

$$SD = \sqrt{\frac{1}{n-1} \times \sum_{i=1}^{n}(X_i - X_i')^2} \tag{1}$$

$$Mean = \frac{\sum_{i=1}^{n}(X_i)}{n} \tag{2}$$

The Direction of the Baseline gives an indication of a stressed writer. If the baseline direction is down, then the writer was under stress when he wrote the document. Otherwise, the writer is considered calm. To calculate the baseline, we first need to find the best fit line of the foreground pixels using least square method to find the equation of the line $Y = mX + b$. For an ordered points $(x_1, y_1), (x_2, y_2) \ldots \ldots (x_n, y_n)$, we need to find the mean of both x-values and y-values using Eq. (2). Then for each i we need to calculate the slop using Eq. (3). Since we have the value of the slope, we can find the Y-intercept of the line using Eq. (4) which is the formula of the best fit line. Figure 3 shows the slope of a text-line.

$$Slope(m) = \frac{\sum_{i=1}^{n}(y_i - Y_{mean})}{\sum_{i=1}^{n}(x_i - X_{mean})} \tag{3}$$

$$b = Y_{mean} - m \times X_{mean} \tag{4}$$

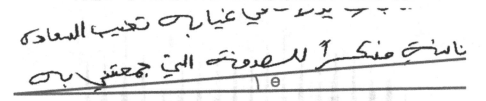

Fig. 3. Slope of text-line

The Space Between Words and Letters can give us an indication of the mental state of the writer, where the wide space means a calm and non-stressed writer. Otherwise, the writer is considered to be a stressful one. The database used consists of Arabic handwritten documents. Arabic writing is a cursive horizontal script whose words consist of sub-words or Pieces of Arabic Words , each of which consists of one or more letters, this can be shown in Fig. 4. Consequently, a connected component (Sub-word) can be used as a word in Latin languages, in which we will be measuring the space between the sub-words.

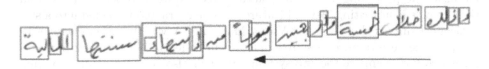

Fig. 4. Arabic handwritten text-line

Handwriting Irregularity. Irregularity is one of the features that reveal stress. There are many aspect of irregularity in handwriting, such as: irregularity in the slant, and the irregularity between handwritten lines. We focused on extracting irregularity between handwritten lines. We measure lines irregularity using the projection profile after dilating the text in a given document [11]. Figure 5 shows the horizontal projection profile of two different documents. The horizontal projection profile in Fig. 5 (a) shows that the lines are irregular and not well separated and words are sparse all over the document. In addition, there are no deep valleys in many parts of the profile, which gives an indication that the lines are skewed and close to each other. The profile of Fig. 5 (b) shows that the lines in the documents are nicely separated, since it has deep valleys. We will be exploring the best way to feed the projection profile results into the classifier.

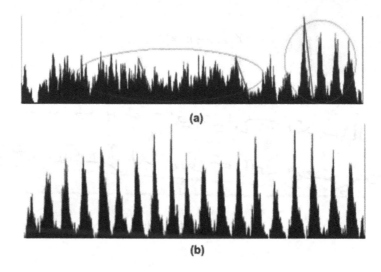

Fig. 5. Projection profiles of two different documents.

3　Conclusion and Future Works

We are proposing a novel and unique database for stress detection based on handwriting. The database includes additional information, other than the stress related labeled document. These information include: age, gender, profession, etc. This allows researchers to use the database for different purposes. Furthermore, we are proposing a machine learning based system to detect stress from the person's handwriting. Several features will be extracted to be fed into a statistical classifier such as Support Vector Machine (SVM), and other classifier to compare the performance. Also row images will be fed into a deep learning neural network to compare the two systems. We will also be working on exploring other features during our research. Finally, this research brings another area or field of handwriting analysis to which other researchers can continue working on and build on top on it.

References

1. Alberdi, A., Aztiria, A., Basarab, A.: Towards an automatic early stress recognition system for office environments based on multimodal measurements: a review. J. Biomed. Inform. **59**, 49–75 (2016)
2. Binali, H., Wu, C., Potdar, V.: Computational approaches for emotion detection in text. In: 4th IEEE International Conference on Digital Ecosystems and Technologies, pp. 172–177, April 2010
3. Minu, R.I., Ezhilarasi, R.: Automatic emotion recognition and classification. Procedia Eng. **38**, 21–26 (2012)
4. Giakoumis, D., et al.: Using activity-related behavioural features towards more effective automatic stress detection. PloS One **7**, e43571 (2012)

5. Shahin, M.A., Ahmed, B., Ballard, K.J.: Classification of lexical stress patterns using deep neural network architecture, December 2014
6. Blanco-Gonzalo, R., Sanchez-Reillo, R., Miguel-Hurtado, O., Bella-Pulgarin, E.: Automatic usability and stress analysis in mobile biometrics. Image Vis. Comput. **32**(12), 1173–1180 (2014)
7. Aigrain, J., Dapogny, A., Bailly, K., Dubuisson, S., Detyniecki, M., Chetouani, M.: On leveraging crowdsourced data for automatic perceived stress detection, October 2016
8. Thelwall, M.: Tensistrength: stress and relaxation magnitude detection for social media texts. Inf. Process. Manag. **53**(1), 106–121 (2017)
9. Likforman-Sulem, L., Esposito, A., Faundez-Zanuy, M., ClémençSon, S., Cordasco, G.: EMOTHAW: a novel database for emotional state recognition from handwriting and drawing. IEEE Trans. Hum.-Mach. Syst. **47**, 273–284 (2017)
10. Bay, Y., Erbilek, M., Celebi, E.: Emotional state prediction from online handwriting and signature biometrics. IEEE Access 1 (2019)
11. Khayyat, M., Lam, L., Suen, C.Y., Yin, F., Liu, C.L.: Arabic handwritten text line extraction by applying an adaptive mask to morphological dilation. In: 10th IAPR International Workshop on Document Analysis Systems (DAS 2012), Gold Coast, Queenslands, Australia, 27–29 March 2012, pp. 100–104 (2012)

Recognition of Cursive Caption Text Using Deep Learning - A Comparative Study on Recognition Units

Ali Mirza$^{(\boxtimes)}$ (iD), Imran Siddiqi (iD), Umar Hayat (iD), Muhammad Atif (iD),
and Syed Ghulam Mustufa (iD)

Bahria University, Islamabad, Pakistan
{alimirza,imran.siddiqi}@bahria.edu.pk,umarhayatuetpeshawar@gmail.com,
atifmaju@gmail.com, ghulam.mustufa31@gmail.com

Abstract. Text appearing in videos carries rich information that can be exploited for content based retrieval applications. An important component in such systems is the development of a video optical character recognition (V-OCR) system that converts images into text. Though mature recognition system have been developed for recognition of text in non-cursive scripts, recognition of cursive text remains a challenging problem. An important factor in developing recognition systems for cursive text is the choice of recognition unit. Unlike words which can be extracted in non-cursive scripts, recognition of cursive text relies either on partial words/ligatures (holistic techniques) or characters (analytical techniques). This paper presents a comparative study on the effectiveness of holistic and analytical recognition using the latest deep learning techniques. More specifically, we employ convolutional neural networks for recognition of ligatures segmented from caption text and a combination of convolutional and recurrent neural networks for recognition of characters from text line images. Experiments are carried out on 16,000 text lines (extracted from 5000 video frames) containing cursive Urdu text extracted from videos of various News channels. The experimental results demonstrate that analytical techniques are more robust as compared to the holistic techniques and the findings can be generalized to other cursive scripts as well.

Keywords: Caption text · Video retrieval · Cursive text recognition · Deep Convolutional Neural Networks (CNNs) · Long-Short-Term Memory Networks

1 Introduction

With the enormous growth in the amount of digital multimedia data, the conventional tag-based search engines are being replaced by intelligent content-based retrieval systems. While traditional retrieval systems rely on matching the

Y. Lu et al. (Eds.): ICPRAI 2020, LNCS 12068, pp. 156–167, 2020.
https://doi.org/10.1007/978-3-030-59830-3_14

queried keyword with user-assigned annotations to videos and images, content-based search exploits the visual (e.g. buildings or objects), audio (spoken keywords) or textual (news tickers for instance) content in data to retrieve the desired images or videos. The recent advancements in deep learning have also resulted in the development of state-of-the-art object detection and recognition systems which contributed to enhancing the research in content-based retrieval systems as well. Among different modalities allowing content-based retrieval, in the present study, we focus on the textual content appearing in videos. Extracting and recognizing text from videos can be exploited to develop a number of useful applications including retrieval of videos containing certain keywords, user alerts, summarization of ticker news and opinion mining systems.

The development of textual content-based retrieval systems is marked by many challenges, recognition of text being the most significant of these. In other words, Video Optical Character Recognition (V-OCR) systems need to be developed to convert regions of interest in video frames into text. The problem of OCR has witnessed more than four decades of extensive research. Thanks to these research endeavors, state-of-the-art recognition systems have been developed for many scripts both for document images (scanned or camera based) and video frames. Recognition of text in cursive scripts, however, still remains a challenging problem. Characters in cursive scripts join to form partial words (ligatures) and shapes of characters vary as a function of position within the sub-words. There are non-uniform inter and intra word spacings and many characters share the same shape of the main body with differences only in the number and position of dots or diacritic marks. Common examples of the cursive script include Arabic, Persian, Urdu, and Pashto, etc. The current study targets recognition of Urdu text but the findings can be generalized to most of the other cursive scripts due to the sharing of major proportions of alphabets as well as the rules of joining characters. In addition to the complexities of the script, the low resolution of text appearing in video frames makes its recognition more challenging as opposed to text on high-resolution digitized document images.

As a function of the recognition unit, recognition techniques for cursive scripts are categorized into holistic and analytical methods. Holistic or segmentation-free techniques employ sub-words (also known as ligatures) as units of recognition while analytical approaches recognize individual characters. While holistic methods avoid the complex segmentation part, a major challenge is a large number of unique ligature classes as the number of unique sub-words in a given language can be substantially huge. On the other hand, analytical methods need to discriminate among a small number of character classes which is equal to the number of unique characters in the alphabet and their various context-dependent shapes. Segmentation of cursive text into characters, however, is itself a highly complex problem. As a result, implicit segmentation based techniques (discussed in Sect. 2) are generally employed to avoid explicit segmentation.

This paper presents a comparative analysis of holistic and analytical recognition techniques for the recognition of cursive video text. More specifically, we compare the challenges and recognition performance using ligatures and characters as recognition units. The ligature based recognition employs Convolutional Neural Networks (CNNs) and extends our recognition framework presented in [7] while the character level recognition is carried out using Long-Short-Term-Memory (LSTM) networks. Experiments on a dataset of 16,000 video text lines with 1790 ligature classes reveal that (implicit) segmentation based recognition outperforms ligature based recognition. More details are presented in the subsequent sections of the paper.

2 Background

Text recognition has been extensively researched for printed documents, handwriting images, scene images and caption text in videos [27,28,35,39]. From the viewpoint of caption text, the problem of low resolution is generally resolved using multiple frame integration [15,17,26] enhancing the resolution of images as a pre-recognition step. Among recent works on recognition of cursive scripts from video, a segmentation-free technique exploiting Long-Short-Term-Memory (LSTM) networks is presented by Zayene et al. [42]. Evaluations on two different datasets (ACTiV [41] and ALIF [40]) with Arabic text in video frames report high recognition rates. In a similar work [9], the hybrid CNN-LSTM combination is investigated to recognize Arabic text appearing in video frames.

Focusing specifically on Urdu text, a number of robust recognition techniques, mainly targeting printed documents, have been proposed in recent years. A major milestone was the development of labeled datasets like UPTI [29] and CLE [1] revitalizing the research on this problem. A major proportion of holistic recognition techniques employed hidden Markov models (HMMs) for classification of ligatures [2,3,11,13]. Features extracted from ligatures using sliding windows are used to train a separate model for each (ligature) class. To reduce the total number of unique classes, a number of techniques rely on separate recognition of main body ligatures and dots [5].

Segmentation of cursive text into characters is highly challenging [18,20], consequently, rather than attempting to explicitly segment the sub-words into characters, techniques based on implicit segmentation have been exploited for recognition of characters [10,21–23]. Such methods rely on feeding the images of text with its transcription to a learning algorithm that is expected to learn the character shapes and segmentation points. Most of these techniques employ different variants of LSTMs with a CTC layer [24,37] reporting high character recognition rates on printed Urdu text. From the viewpoint of video text, the research is still in early days where few preliminary studies have been carried out exploiting analytical [36] as well as holistic recognition techniques [7]. With the huge collections of videos containing textual content in cursive scripts and the increasing number of News channels continuously flashing the News tickers (and other information), the problem of V-OCR for cursive scripts offers an immense research potential.

The next section presents the details of the dataset employed in our study.

3   Dataset

While benchmark datasets have been developed for printed Urdu text, to the best of authors' knowledge, no labeled dataset is publicly available to evaluate the localization and recognition performance of Urdu caption text in videos. To address this issue, we are in the process of developing a comprehensive database of labeled video frames allowing evaluation of text detection and text recognition tasks in multiple scripts. We have collected more than 400 hours of videos from multiple channels and presently, more than 10,000 video frames have been labeled (using our ground truth labeling tool). The ground truth data contains information on the bounding box of each textual region, its transcription and other metadata in XML format. The database will be made publicly available once the labeling process is complete. More details on the database and its labeling can be found in our previous work [7,19] while sample frames from the database are illustrated in Fig. 1.

In the present study, we employ a total of 5000 video frames having 16,000 text line images. Out of these, 13,000 text lines (of 4000 frames) are used in the training set while 3000 text lines (from 1000 frames) in the test set. More details on various experimental settings are provided in Sect. 5 of the paper.

4   Methods

This section presents the details of the analytical and holistic recognition techniques. From the viewpoint of a complete textual content-based indexing and retrieval system, the textual content is detected (localized) from the video frames and is fed to the recognition engine. The focus of the current study, however, is on the recognition of text and not on its localization. Hence, we employ the ground truth information of video frames to extract the text lines which are processed further.

For text in cursive scripts (like Urdu and Arabic etc.), characters may appear in isolated form or are joined with other characters using the joiner rules. The shape of a character, therefore, varies depending upon whether it appears in isolation, or at the start, end or middle of a sequence of joined characters. Since word boundaries are hard to identify in cursive scripts, partial words or ligatures are typically extracted from text lines. Ligatures are further categorized into primary and secondary ligatures. Primary ligatures are the main body component while secondary ligatures correspond to dots and diacritics (Fig. 2).

Fig. 1. Sample video frames with occurrences of textual content

Fig. 2. (a): A complete word (b): Ligatures (c): Main body (primary ligature) (d): Dots and diacritics (secondary ligatures)

Prior to recognition, the textual regions need to be binarized. While simple global thresholding is sufficient in case of high resolution scanned document images, binarization of video text offers a challenging scenario. Text can be in low resolution and on non-homogenous backgrounds. In some cases, the binarization techniques assume the convention of dark text on a bright background. Consequently, as a first step, we ensure that text line images fed to the binarization module follow the same convention. The canny edge detector is applied to the grayscale text line image to find the (rough) text blobs. The median gray value is computed for all blobs ($Text_{med}$) as well as for the entire image region that is not a part of any blob ($Back_{med}$). The polarity of the image is reversed if ($Back_{med} < Text_{med}$). The median value is preferred over mean as applying the edge detector on the grayscale image does not necessarily segment text regions and some parts of background may also be falsely identified as text blobs which may disturb the mean value (median is insensitive to such outliers).

For binarization, we investigated a number of binarization techniques. These include the classical Otsu's global threshodling [25] as well as various local thresholding techniques including Niblack [14], Sauvolo [30], Feng's [6] and Wolf's [38] algorithms. Each of these techniques considers a local neighborhood around each pixel and compute the binarization threshold as a function of statistic computed from the neighboring pixels. Among these, Wolf's binarization technique [38] was specifically designed to work on low-resolution video images and it outperformed other techniques on images in our dataset as well. Binarization results of various algorithms on a sample text line image from a video frame are illustrated in Fig. 3. All text line images in our dataset were binarized using Wolf's algorithm and the binarized images were fed to the (holistic and analytical) recognition engines as discussed in the following.

4.1   Recognition Using Analytical Technique

As discussed earlier, analytical recognition employs characters as units of recognition. Segmentation of cursive text into characters using image analysis techniques has remained a very challenging problem. However, thanks to the recent advancements in deep learning, explicit segmentation into characters is no longer a prerequisite for the recognition task. The learning algorithm can be provided

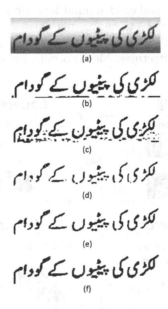

Fig. 3. Binarization results on a sample text line (a): Original image (b): Global thresholding (c): Niblack (d): Sauvola (e): Feng's algorithm (f): Wolf's algorithm

with training text line images along with the corresponding textual transcription. The algorithm not only learns various character shapes but also character boundaries.

In our implementation, we employ the Convolutional Recurrent Neural Network (CRNN) [31] which is a combination of CNN, RNN and CTC (Connectionist Temporal Classification) loss [31] and is summarized in Fig. 4. The height of each (binarized) text line is normalized and the convolutional part of the network extracts features from windows that slide over the image. These activation sequences are fed as input to the bidirectional RNN along with the transcription. The CTC layer serves to align the feature sequences with their corresponding ground truth labels in the transcription. RNN predicts the output of each part of the image (the pre-frame predictions) and these outputs are converted into labels using a look-up table.

4.2   Recognition Using Holistic Technique

The holistic recognition technique employs ligatures as units of recognition. As a first step, ligatures need to be extracted from the binarized text line images. Ligatures are extracted using connected component labeling and the secondary ligatures (dots and diacritics) are associated with their parent primary ligatures by performing morphological dilation (with a vertical structuring element). From the 16,000 text lines in our study, a total of 260,000 ligatures are extracted. In order to prepare the training and test data, these ligatures are organized into classes (clusters) where each class is a collection of images that correspond to a single ligature. The total number of unique ligature classes in our study sums up to 1790 with an average of 145 images per class.

For the recognition of ligatures, we investigated a number of deep convolutional neural network architectures. More specifically, we employed a number of

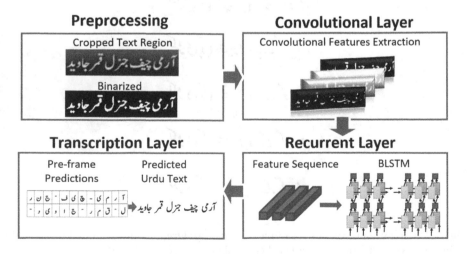

Fig. 4. Text line recognition with C-RNN

pre-trained CNN models using the transfer learning framework. These include the classical AlexNet [16], VGG Nets [32], GoogLeNet [33], InceptionV3 [34] and ResNet101 [8]. Transfer learning allows adapting a pre-trained network for a different problem. A pre-trained model can be used as a feature extractor (using convolutional layers only) and these features can be fed to a separate classifier similar to the traditional machine learning framework. Another common technique is to replace the last fully connected layer of a pre-trained network with class labels of the dataset under study and continue back propagation (either on all or few last layers of the network) to adjust the network weights. In our study, we investigate both the possibilities employ pre-trained models as feature extractors as well as fine-tuned them to our set of ligatures. A summary of the networks considered in our work is presented in Table 1.

Table 1. Summary of pre-trained models employed in our study

Model	Input size	Depth	FC layers
Alex-net [16]	$227 \times 227 \times 3$	8	3
VGG-16 [32]	$224 \times 224 \times 3$	16	3
VGG-19 [32]	$224 \times 224 \times 3$	19	3
Google-net [33]	$224 \times 224 \times 3$	22	1
Inceptionv3 [34]	$299 \times 299 \times 3$	48	1
Resnet-101 [8]	$224 \times 224 \times 3$	101	1

After having discussed the analytical and holistic recognition techniques, we present the experimental setup and the corresponding results in the next section.

5   Experiments and Results

As discussed earlier, for all experiments, we employ 13,000 text lines in the training set and 3000 in the test set. For the holistic recognition technique, the ligature classes corresponding to the same set of 13,000 and 3000 lines are employed in the training and test sets respectively. The analytical recognition outputs transcription of the queried text line and the character recognition rate is computed by calculating the edit distance between the predicted and the ground truth transcription. In the case of holistic recognition, the ligature recognition rate is computed as a fraction of ligatures correctly recognized by the system.

The recognition rates of the two techniques are summarized in Table 2. For holistic recognition, the results are presented for both fine-tuning and feature extraction using multiple pre-trained models. Comparing the performance of various pre-trained models, it can be seen that fine-tuning outperforms feature extraction for all models. The observation is natural as fine-tuning allows adjusting the weights of the network according to images under study hence the extracted features are likely to be more effective. The highest recognition rate is reported by AlexNet reading 83.50%. For analytical recognition, the system reports a recognition rate of 90.50%. When comparing the recognition rates of analytical and holistic techniques, it is important to mention that the former reports character recognition rates while the later reports ligature recognition rates. Although it can be argued the two recognition rates are not directly comparable, the analytical technique outperforms the holistic method by a significant margin validating its robustness.

An interesting observation in the results of holistic recognition is that the highest recognition rate is reported by AlexNet which has the least depth among the investigated pre-trained models. This observation is consistent with our previous findings on the recognition of ligatures in video [7] as well as printed (scanned) documents [12]. This observation can be attributed to the fact that all these models are trained on the ImageNet [4] dataset which contains colored images of objects. We, on the other hand, deal with binary images of ligatures representing a different scenario. Consequently, networks with relatively fewer convolutional layers are able to learn the discriminative features reporting acceptable recognition rates.

In addition to recognition rates, we also present a comparative analysis of the two techniques in terms of the pros and cons of each. Based on the findings of this study, we came up with the following remarks.

– Analytical techniques are more robust as opposed to holistic techniques in terms of training as well as recognition performance.
– No segmentation is required in analytical techniques while holistic techniques require segmented ligatures (along with secondary ligatures associated with their parent primary ligatures). Errors in segmentation add to the overall recognition errors of the system.
– To correctly learn the character shapes and boundaries, analytical techniques require relatively more training data as opposed to holistic techniques that need to learn ligature classes only (and not any boundaries).
– Training an analytical system requires only text lines and the corresponding transcription. On the other hand, holistic methods require ligatures organized into clusters. While these clusters can be prepared for manageable dataset

Table 2. Recognition rates of analytical and holistic techniques

Technique	model	Recognition rate	
Analytical	C-RNN	90.50	
		Feature extraction	Fine-tuning
Holistic	AlexNet	78.27	83.50
	VGG16	76.60	82.95
	VGG19	76.79	82.96
	GoogleNet	79.28	82.60
	InceptionV3	76.91	81.47
	ResNet	67.94	75.15

sizes, manually generating them for larger datasets would be a tedious job. None of the currently available datasets contain ground truth information at the ligature level.

Summarizing, we can conclude that analytical techniques indeed represent the effective and robust choice for recognition of cursive scripts.

6 Conclusion

This paper presented a comparative study of analytical and holistic recognition techniques for the recognition of cursive video text. We employed Urdu text as a case study but the findings can be generalized to other cursive scripts as well. The holistic technique relies on extracting partial words or ligatures from text lines, grouping them into classes and recognizing them using a number of pre-trained convolutional neural networks. The analytical technique employs the combination of convolutional and recurrent neural networks to produce the transcription of input text lines. Using the same experimental setup, we demonstrated the robustness of analytical over holistic recognition.

In our further study on this problem, we intend to evaluate the analytical technique on a larger set of text lines. The impact of training data on recognition performance is also planned to be studied. The training data can be further enhanced by employing data augmentation as well as making printed text (for which benchmark datasets are available) a part of the training set. The recognition engine will also be combined with the detection module to develop a comprehensive system allowing retrieval of videos on queried keywords.

Acknowledgment. This study is supported by IGNITE, National Technology Fund, Pakistan under grant number ICTRDF/TR&D/2014/35.

References

1. Center for language engineering. http://www.cle.org.pk/. Accessed 15 Apr 2019
2. Ahmad, I., Mahmoud, S.A., Fink, G.A.: Open-vocabulary recognition of machine-printed Arabic text using hidden Markov models. Pattern Recogn. **51**, 97–111 (2016)
3. Javed, S.T., Hussain, S., Maqbool, A., Asloob, S., Jamil, S., Moin, H.: Segmentation free nastalique urdu ocr. World Acad. Sci. Eng. Technol. **46**, 456–461 (2010)
4. Deng, J., Dong, W., Socher, R., Li, L. J., Li, K., Fei-Fei, L.: Imagenet: a large-scale hierarchical image database. In: 2009 IEEE Conference on Computer Vision and Pattern Recognition, pp. 248–255. IEEE (2009)
5. Din, I.U., Siddiqi, I., Khalid, S., Azam, T.: Segmentation-free optical character recognition for printed urdu text. EURASIP J. Image Video Process. **2017**(1), 62 (2017)
6. Feng, M.-L., Tan, Y.-P.: Contrast adaptive binarization of low quality document images. IEICE Electron. Express **1**(16), 501–506 (2004)

7. Hayat, U., Aatif, M., Zeeshan, O., Siddiqi, I.: Ligature recognition in urdu caption text using deep convolutional neural networks. In: 2018 14th International Conference on Emerging Technologies (ICET), pp. 1–6. IEEE (2018)
8. He, K., Zhang, X., Ren, S., Sun, J.: Deep residual learning for image recognition. In: Proceedings of the IEEE Conference on Computer Vision and Pattern Recognition, pp. 770–778 (2016)
9. Jain, M., Mathew, M., Jawahar, C.V..: Unconstrained scene text and video text recognition for arabic script. In: 2017 1st International Workshop on Arabic Script Analysis and Recognition (ASAR), pp. 26–30. IEEE (2017)
10. Javed, N., Shabbir, S., Siddiqi, I., Khurshid, K.: Classification of urdu ligatures using convolutional neural networks-a novel approach. In: 2017 International Conference on Frontiers of Information Technology (FIT), pp. 93–97. IEEE (2017)
11. Javed, S.T., Hussain, S.: Segmentation based urdu nastalique OCR. In: Ruiz-Shulcloper, J., Sanniti di Baja, G. (eds.) CIARP 2013. LNCS, vol. 8259, pp. 41–49. Springer, Heidelberg (2013). https://doi.org/10.1007/978-3-642-41827-3_6
12. Khattak, I.U., Siddiqi, I., Khalid, S., Djeddi, C.: Recognition of urdu ligatures- a holistic approach. In: 2015 13th International Conference on Document Analysis and Recognition (ICDAR), pp. 71–75. IEEE (2015)
13. Khémiri, A., Echi, A.K., Belaïd, A., Elloumi, M.: Arabic handwritten words off-line recognition based on HMMs and DBNs. In: 2015 13th International Conference on Document Analysis and Recognition (ICDAR), pp. 51–55. IEEE (2015)
14. Khurshid, K., Siddiqi, I., Faure, C., Vincent, N.: Comparison of Niblack inspired binarization methods for ancient documents. In: Document Recognition and Retrieval XVI, vol. 7247, pp. 72470U. International Society for Optics and Photonics (2009)
15. Kim, D., Sohn, K.: Static text region detection in video sequences using color and orientation consistencies. In: 2008 19th International Conference on Pattern Recognition, ICPR 2008, pp. 1–4. IEEE (2008)
16. Krizhevsky, A., Sutskever, I., Hinton, G.E.: Imagenet classification with deep convolutional neural networks. In: Advances in Neural Information Processing Systems, pp. 1097–1105 (2012)
17. Lee, C.W., Jung, K., Kim, H.J.: Automatic text detection and removal in video sequences. Pattern Recogn. Lett. **24**(15), 2607–2623 (2003)
18. Malik, H., Fahiem, M.A.: Segmentation of printed urdu scripts using structural features. In: 2009 Second International Conference in Visualisation, VIZ 2009, pp. 191–195. IEEE (2009)
19. Mirza, A., Fayyaz, M., Seher, Z., Siddiqi, I.: Urdu caption text detection using textural features. In: Proceedings of the 2nd Mediterranean Conference on Pattern Recognition and Artificial Intelligence, pp. 70–75. ACM (2018)
20. Morales, A., Fierrez, J., Sánchez, J.S., Ribeiro, B. (eds.): IbPRIA 2019. LNCS, vol. 11868. Springer, Cham (2019). https://doi.org/10.1007/978-3-030-31321-0
21. Naseer, A., Zafar, K.: Comparative analysis of raw images and meta feature based urdu OCR using CNN and LSTM. Int. J. Adv. Comput. Sci. Appl. **9**(1), 419–424 (2018)
22. Naz, S., et al.: Offline cursive urdu-nastaliq script recognition using multidimensional recurrent neural networks. Neurocomputing **177**, 228–241 (2016)
23. Naz, S., et al.: Urdu Nastaliq recognition using convolutional-recursive deep learning. Neurocomputing **243**, 80–87 (2017)
24. Naz, S., Umar, A.I., Ahmed, R., Razzak, M.I., Rashid, S.F., Shafait, F.: Urdu Nasta'liq text recognition using implicit segmentation based on multi-dimensional long short term memory neural networks. SpringerPlus **5**(1), 2010 (2016)

25. Otsu, N.: A threshold selection method from gray-level histograms. IEEE Trans. Syst. Man Cybern. **9**(1), 62–66 (1979)
26. Phan, T.Q., Shivakumara, P., Lu, T., Tan, C.L.: Recognition of video text through temporal integration. In: 2013 12th International Conference on Document Analysis and Recognition (ICDAR), pp. 589–593. IEEE (2013)
27. Plamondon, R., Srihari, S.N.: Online and off-line handwriting recognition: a comprehensive survey. IEEE Trans. Pattern Anal. Mach. Intell. **22**(1), 63–84 (2000)
28. Ye, Q., Doermann, D.: Text detection and recognition in imagery: a survey. IEEE Trans. Pattern Anal. Mach. Intell. **37**, 1480–1500 (2015)
29. Sabbour, N., Shafait, F.: A segmentation-free approach to Arabic and urdu OCR. In: IS&T/SPIE Electronic Imaging, pp. 86580N–86580N. International Society for Optics and Photonics (2013)
30. Sauvola, J., Pietikäinen, M.: Adaptive document image binarization. Pattern Recogn. **33**(2), 225–236 (2000)
31. Shi, B., Bai, X., Yao, C.: An end-to-end trainable neural network for image-based sequence recognition and its application to scene text recognition. IEEE Trans. Pattern Anal. Mach. Intell. **39**(11), 2298–2304 (2017)
32. Simonyan, K., Zisserman, A.: Very deep convolutional networks for large-scale image recognition (2014). arXiv preprint arXiv:1409.1556
33. Szegedy, C., et al.: Going deeper with convolutions. In: Proceedings of the IEEE Conference on Computer Vision and Pattern Recognition, pp. 1–9 (2015)
34. Szegedy, C., Vanhoucke, V., Ioffe, S., Shlens, J., Wojna, Z.: Rethinking the inception architecture for computer vision. In: Proceedings of the IEEE Conference on Computer Vision and Pattern Recognition, pp. 2818–2826 (2016)
35. Tao, W., Wu, D.J., Coates, A., Ng, A.Y.: End-to-end text recognition with convolutional neural networks. In: 2012 21st International Conference on Pattern Recognition (ICPR), pp. 3304–3308. IEEE (2012)
36. Tayyab, B.U., Naeem, M.F., Ul-Hasan, A., Shafait, F., et al.: A multi-faceted OCR framework for artificial urdu news ticker text recognition. In: 2018 13th IAPR International Workshop on Document Analysis Systems (DAS), pp. 211–216. IEEE (2018)
37. Ul-Hasan, A., Ahmed, S.B., Rashid, F., Shafait, F., Breuel, T.M.: Offline printed urdu nastaleeq script recognition with bidirectional LSTM networks. In: 2013 12th International Conference on Document Analysis and Recognition, pp. 1061–1065. IEEE (2013)
38. Wolf, C., Jolion, J.-M.: Extraction and recognition of artificial text in multimedia documents. Formal Pattern Anal. Appl. **6**(4), 309–326 (2004)
39. Yin, X.-C., Zuo, Z.-Y., Tian, S., Liu, C.-L.: Text detection, tracking and recognition in video: a comprehensive survey. IEEE Trans. Image Process. **25**(6), 2752–2773 (2016)
40. Yousfi, S., Berrani, S.A., Garcia, C.: Deep learning and recurrent connectionist-based approaches for arabic text recognition in videos. In: 2015 13th International Conference on Document Analysis and Recognition (ICDAR), pp. 1026–1030. IEEE (2015)
41. Zayene, O., Hennebert, J., Touj, S.M., Ingold, R., Amara, N.E.B.: A dataset for Arabic text detection, tracking and recognition in news videos - activ. In: 2015 13th International Conference on Document Analysis and Recognition (ICDAR) (2015)
42. Zayene, O., Touj, S.M., Hennebert, J., Ingold, R., Amara, N.E.B.: Multidimensional long short-term memory networks for artificial Arabic text recognition in news video. In: IET Computer Vision (2018)

Features and Classifiers

A Hybrid Multiple Classifier System Applied in Life Insurance Underwriting

Chun Lei He[1](✉) ⓘ, Dave Keirstead[1] ⓘ, and Ching Y. Suen[2] ⓘ

[1] Advanced Analytics, Canadian Division, Manulife, Toronto, ON, Canada
{chun_lei_he,dave_keirstead}@manulife.ca
[2] CENPARMI (Centre for Pattern Recognition and Machine Intelligence), Computer Science and Software Engineering Department, Concordia University, Montreal, QC, Canada
suen@cenparmi.concordia.ca

Abstract. In an insurance company, manual underwriting is costly, time consuming, and complex. Simulating underwriters with AI is an absolutely time saving and cost fitting solution. As a result, a Hybrid Multiple Classifier System, combining three classifiers with rejection options: XGBoost, Random Forest, and SVM, was designed and applied on production. An optimal rejection criterion on classification, so-called Linear Discriminant Analysis Measurement (LDAM), is applied the first time in industry. This system is the first AI driven underwriting system in Canadian life insurance, and it helps Manulife expand digital capabilities, reorient customer experience focus and grow its business.

Keywords: Underwriting · A hybrid multiple classifier system · Rejection criterion · First AI driven underwriting system in Canadian life insurance · Save the cost · Expand digital capabilities · Reorient customer experience focus

1 Introduction

Underwriting in life insurance is a procedure to measure the risk exposure and determine appropriate rate class on the life insurance. Underwriting is costly and complex. Traditionally, the procedure of making a life insurance underwriting decision may involve in underwriters from different levels (from Junior to Director), medical consultants, risk teams, pricing teams, etc. Underwriters play key roles in assessing the risk factors presented in each case, and determine whether to accept a risk, how to determine appropriate levels of insurance coverages and what premium to charge, the rate-making process and portfolio management. Accordingly, the traditional manual underwriting faces challenges: Time, Cost, and Complexity, etc.

In order to save the workforce and fit the cost, we proposed a novel hybrid multiple classifier system (HMCS) to automatically make underwriting decisions with AI in real time. This system helps to fill the gaps by speeding up the process of the underwriting automatically while avoiding the complex underwriting procedures and bias from underwriters. The model acts as a triage system that classifies all applications into different risk categories. The lowest risk applications are processed automatically, without human

© Springer Nature Switzerland AG 2020
Y. Lu et al. (Eds.): ICPRAI 2020, LNCS 12068, pp. 171–176, 2020.
https://doi.org/10.1007/978-3-030-59830-3_15

intervention. Those identified as higher risk go through traditional underwriting. The target of this underwriting model is to predict three decisions (classes): Low, Medium, and High. A novel efficient rejection criterion is applied to improve the precision on Class Low and the entire system's reliability.

This system is the first Canadian life insurance provider adopting an AI decision algorithm for underwriting [1]. As Karen Cutler, Head of Underwriting, Manulife, said, "Artificial intelligence is helping to free up our valued underwriters to focus on challenging cases, and this sophisticated technology handles the straightforward policies. In addition, advisors are still going to be front and centre, helping the customer determine what kind of product they need" [2].

2   A Hybrid Multiple Classifier System (HMCS)

2.1   Data Pre-processing and Feature Extraction

In life insurance, underwriters consider the collected data from the application forms as main source for decision making. It includes risk related information such as personal & family medical history, applicants' demographic information, financial information, occupation and behavioral information (such as driving history), etc. In this system, we consider all the questions from application forms as input.

There are a few generated features based on their business meanings. For example, AGEBMI is designed by Age* BMI (body mass index). BMI is a measure of body fat based on height and weight that applies to adult men and women. It is obvious that when people are aging, BMI impact the health more and more. Thus, we design this specific feature as one of the factors for prediction.

2.2   Classification

Misclassification is very costly, so cases predicted with low confidence or high risk need to be re-evaluated by the senior underwriters. Thus, the goal of the algorithm is to improve the precision on Class Low Risk while maintaining a reasonable Rejection Rate. Rejection is defined as the cases predicted as low confidence value and high risk and sent to underwriters for manual procedure.

The Decision Algorithm is designed as a Hybrid Multiple Classifier System (HMCS) [3] with a combination of a few classifiers in both serial and parallel. Business expects very high precision in order to replace underwriters, but any single classifier may not perform well enough. Accordingly, a multiple classifier system may help on this approach. Only when all classifiers agree on decisions as Class Low Risk on the cases, the system identifies the cases as Low and allow them to pass through the system. Otherwise, the cases are rejected to underwriters.

Instead of combining classifiers in parallel, we choose a HMCS. We assign the classifier with the best performance as a primary classifier. The advantage of having a primary classifier is speed. When the cases rejected by the primary classifier, they can be sent to manual underwriters and do not need to pass through other classifiers.

Primary Classifier. XGBoost [4] is built as a primary classifier due to its performance and speed. XGBoost is short for "Extreme Gradient Boosting", where the term "Gradient Boosting" is proposed in the paper Greedy Function Approximation: A Gradient Boosting Machine, by Friedman. In training, pre-defined evaluation function was embedded to it. This evaluation function is based on the costs of pricing in underwriting. In addition, as the number of records in each class is unbalanced, subsampling and weights were applied to improve the performance.

Secondary Classifiers. Business expects very high precision, and other "experts" with agreements or disagreements may help XGBoost on improve the precision. Therefore, when XGBoost predicts cases as "Standard", the cases should be classified by some secondary classifiers. If any of these classifiers has a disagreement on classification, them should be rejected to underwriters. These classifiers are: Random Forest (RF) [5], SVM, XGBoost on Probabilities, and RF on Probabilities.

Random Forest. Random forests is an ensemble learning methodology for classification, regression and other tasks, that operate by constructing a multitude of decision trees at training time and outputting the class that is the mode of the classes (classification) or mean prediction (regression) of the individual trees. Random decision forests correct for decision trees' habit of overfitting to their training set. The input of RF is the same as XGBoost in this system.

Support Vector Machine (SVM) [6]. In machine learning, an SVM model is a representation of the examples as points in space, mapped so that the examples of the separate categories are divided by a clear gap that is as wide as possible. New examples are then mapped into that same space and predicted to belong to a category based on which side of the gap they fall. The probabilities and LDAM of each class from each classifier are input for SVM to train and predict on samples.

XGB on Probabilities and RF on Probabilities. When a certain threshold set on FRM XGB/RF outputs, the prediction results may vary to XGB/RF. It provides the flexibility on the rejection options on the classification.

2.3 Rejection Criterion

Even low error rates can be costly in this application. For example, if the system accepted a high-risk person, who should be declined, in underwriting, the financial loss should be extremely high. On the contrary, the financial loss on misclassification of Low Risk is not that high. Therefore, it is the general expectation that this system should achieve a high accuracy as well as high reliability. In this system, Linear Discriminant Analysis Measurement (LDAM) [7] is applied.

LDAM. When rejection is considered as a two-class problem, (acceptable or rejected classification), the outputs at the measurement level can be considered as features for the rejection option. In an output vector, whose components may represent distances or probabilities, we expect the confidence value (measure) of the first rank (most likely class) to be far distant from the confidence values or measures of the other classes. In other words, good outputs should be easily separated into two classes: the confidence value of the first rank and others. Linear Discriminant Analysis (LDA), which is an effective classification method, can be used to optimize the criterion for rejection. In [7], a Linear Discriminant Analysis Measurement (LDAM) is designed to take into consideration the confidence values of the classifier outputs and the relations between them.

To consider the relative difference between the measurements in the first two ranks and all other measurements, LDAM is defined and applied. Since rejection in classification can be considered as a two-class problem (acceptance or rejection), we apply LDA [8] for two classes, to implement rejection. LDA approaches the problem by assuming that the conditional probability density functions of the two classes are both normally distributed. There are $n = n_1 + n_2$ observations with d features in the training set, where $\{x_{1i}\}_{i=1}^{n_1}$ arise from class ω_1 and $\{x_{2i}\}_{i=1}^{n_2}$ arise from class ω_2. Gaussian-based discrimination assumes two normal distributions: $(x|\omega_1) \sim N(\mu_1, \Sigma_1)$ and $(x|\omega_2) \sim N(\mu_2, \Sigma_2)$. In LDA, the projection axis (discriminant vector) w for discriminating between two classes is estimated to maximize the Fisher criterion:

$$J(w) = tr((w^T S_w w)^{-1}(w^T S_B w))$$

where tr($\cdot$) denotes the trace of a matrix, S_B and S_w denote the between-class scatter matrix and within-class scatter matrix respectively, and w is the optimal discriminant vector. For the two classes ω_1 and ω_2, with a priori probabilities p_1 and p_2 (it is often assumed that $p_1 = p_2 = 0.5$), the within-class and between-class scatter matrices can be written as

$$S_w = p_1 \Sigma_1 + p_2 \Sigma_2 = \Sigma_{12}$$
$$S_B = (\mu_1 - \mu_2)(\mu_1 - \mu_2)^T,$$

where Σ_{12} is the average variance of the two classes. It can be shown that the maximum separation occurs when:

$$w = S_w^{-1}(\mu_1 - \mu_2) = (p_1 \Sigma_1 + p_2 \Sigma_2)^{-1}(\mu_1 - \mu_2)$$
$$= \Sigma_{12}^{-1}(\mu_1 - \mu_2)$$

To apply this principle to the outputs for the rejection option as a one-dimensional application, we define the two sets $G^{(1)}(x) = \{p_1(x)\}$, and $G^{(2)}(x) = \{p_2(x), p_3(x), \ldots, p_M(x)\}$.
Then it follows that

$$\mu_1 = p_1(x),$$

$$\mu_2 = \frac{1}{M-1} \sum_{i=2}^{M} p_i(x),$$

$$\Sigma_1 = (p_1(x) - \mu_1)^2 = 0,$$

$$\Sigma_2 = \frac{1}{M-1} \sum_{i=2}^{M} (p_i(x) - \mu_2)^2,$$

and $\Sigma_{12} = \frac{1}{2}\Sigma_2$.

Thus, in LDA,

$$w = \Phi_3(x) = \frac{\sum_{i=2}^{M} ||p_1(x) - p_i(x)||}{(M-1)\cdot \Sigma_{12}}$$

Then the decision function would be based on $\text{sgn}(\Phi_3(x) - T_3)$, where T_3 is a threshold derived from the training set, and all values are scaled to [0, 1].

3   Results

The project leveraged variables from the data derived from a few years' applications. ¾ were randomly selected as Training data, and the rest as Test data. The Precision on Class Low Risk is considered as performance evaluation in this HMCS. As a result, this HMCS enable us to identify those people so we can get them through, to purchase insurance much simpler and faster. The performance on pass through cases are similar to human underwriters. This HMCS is flexible and reliable as the thresholds on FRM and LDAM can be adjusted due to the tolerance of the rejection rate.

4   Conclusion

Artificial intelligence (AI) technology is taking industries by storm, so transferring the existing research innovation in AI to profitable products in business is crucial. This HMCS is a great example of the right way to save the cost and improve the efficiency in industry. It makes underwriting more efficient by automating the "easy pile" – the super-straightforward applications that just need to be pushed through. This system also helps us invest in a stronger future – and help make the insurance company a place where customers want to do business and people want to work.

"Manulife has a proud 130-year history of innovation and leadership in insurance," said Alex Lucas, head of individual Insurance, Manulife, in a statement. "Not only do we expect strong growth from our new par product and our recently expanded Manulife Vitality program, we also expect increased efficiencies as a result of the launch of our new artificial intelligence (AI) tool for underwriting which will dramatically reduce turnaround times for many of our customers' applications." [9].

References

1. Cheung, K.C.: First Canadian Life Insurer to Underwrite Using Artificial Intelligence.https://algorithmxlab.com/blog/first-canadian-life-insurer-to-underwrite-using-artificial-intelligence-2/Jun 21, 2018.

2. Anonymous: Hello data! Look who's changing the insurance game. https://mfcentral.man ulife.com/cms__Main?name=Mission-Control#!/article/hello-data-look-who-s-changing-the-insurance-game-search/a0Qf200000Jq6gNEAR July 9, 2018.

3. He, C.L., Suen, C.Y.: A hybrid multiple classifier system of unconstrained handwritten numeral recognition. Pattern Recognit. Image Anal. **17**(4), 608–611 (2007). https://doi.org/10.1134/S10 54661807040219

4. Chen, T., Guestrin C.: XGBoost: a scalable tree boosting system. Technical report, LearningSys (2015)

5. Ho, T.K.: Random decision forests. In: Proceedings of the 3rd International Conference on Document Analysis and Recognition, pp. 278–282 (1995)

6. Cortes, C., Vapnik, V.N.: Support-vector networks. Mach. Learn. **20**(3), 273–297 (1995). https://doi.org/10.1007/BF00994018

7. He, C.L., Lam, L., Suen, C.Y.: Optimization of rejection parameters to enhance reliability in handwriting recognition. In: Handbook of Pattern Recognition and Computer Vision: 4th, pp. 377–395 (2009)

8. Fisher, R.A.: The use of multiple measurements in taxonomic problems. Ann. Eugen. **7**, 179–188 (1936)

9. Anonymous: Manulife Growing Canadian Insurance Business: Re-enters Participating Whole Life Insurance Market and the first in Canada to underwrite using artificial intelligence. http://manulife.force.com/Master-Article-Detail?content_id=a0Qf200000Jq4krEAB& ocmsLang=en_USJune 19, 2018.

Generative Adversarial-Synergetic Networks
for Anomaly Detection

Hongjun Li[1,2,3,4](✉) , Chaobo Li[1] , and Ze Zhou[1]

[1] School of Information Science and Technology, Nantong University, Nantong 226019, China
lihongjun@ntu.edu.cn
[2] State Key Laboratory for Novel Software Technology, Nanjing University, Nanjing 210023, China
[3] Nantong Research Institute for Advanced Communication Technologies, Nantong 226019, China
[4] TONGKE School of Microelectronics, Nantong 226019, China

Abstract. Anomaly detection is an important and demanding problem in social harmony. However, due to the uncertainty, irregularity, diversity and scarcity of abnormal samples, the performance is often poor. This paper presents Generative Adversarial-Synergetic Networks (GA-SN) to improve the discriminative ability. It is built on the Adversarial Discriminant Network (ADN) for detecting anomaly and the Synergetic Generative Network (SGN) for extracting pivotal and discriminant feature. The ADN takes advantage of adversary structure between discriminator and the asymmetric generator. As an important hub of ADN and SGN, the asymmetric generator is combined with promotor by synergy in SGN. Different from other methods which based on generative adversarial network, our method pays more attention to the discriminant performance rather than the reconstruction performance. Our method was verified on MNIST and UCSD Ped2 dataset, where MNIST obtained higher values on both area under curve and F1-score, and 12.56% equal error rate is obtained on UCSD Ped2. The experimental results show the superiority of proposed method compared to existing methods. So the GA-SN combines synergy and adversarial relationship, which helps to improve discriminative performance in detection.

Keywords: Anomaly detection · Adversarial · Synergetic

1 Introduction

Anomaly detection is one of the most important aspects of intelligent video surveillance, which is an inexhaustible force for sustainable and stable development of society. On one hand, anomalies are happening more and more frequently, and people's attention and sensitivity are getting higher and higher. On the other hand, it is extremely challenging because anomaly are uncertain, irregular, diverse and unpredictable. It is almost infeasible to gather all kinds of anomaly and tackle the problem with a classification method. With challenges and people's urgent needs, lots of efforts have been made for anomaly detection.

© Springer Nature Switzerland AG 2020
Y. Lu et al. (Eds.): ICPRAI 2020, LNCS 12068, pp. 177–190, 2020.
https://doi.org/10.1007/978-3-030-59830-3_16

In recent years, the work of anomaly detection mainly based on traditional methods and deep learning. Traditional methods usually include sparse representation [1, 2], support vector machine [3], gaussian mixture model [4] and others [5, 6]. However, traditional methods need to select features manually where generalization ability is poor. If the feature extraction is not ideal, the performance will be greatly affected. Luckily, features extracted by deep learning tend to be more suitable for representing the diversity of numerous samples. The deep learning methods are usually based on convolutional neural network [7], recurrent neural network [8], auto-encoder [9] and various improved networks [10]. As we all know, a large number of samples are required for training deep network. But compared with normal samples, the abnormal samples are less. Meanwhile it is difficult to collect all types of abnormal samples and construct large abnormal database with annotation, so the abnormal features are difficult to extract effectively.

Many researchers begin using the advantages of image generation to solve the lack of abnormal samples, inspired by the Generative Adversarial Networks (GAN) [11–13]. Similar to some previous relevant researches, some results show that improving performance is not always possible by directly generating samples [14]. Therefore, researchers began to study anomaly detection through reconstruction error [15, 16]. These approaches mainly train networks using normal data to avoid the definition of anomaly. However, these methods are based on an ideal assumption that it is difficult to generate the abnormal samples if the normal samples are used for training. In fact, there are two situations. One is the anomalies are distorted and the normal images are reconstructed inadequately. The other is that the normal samples can be reconstructed well, but anomalies can also be reconstructed roughly due to the capability of deep network in feature representation. Therefore, it is not accurate to judge the anomaly according to the reconstruction error.

To overcome aforementioned problems and be inspired by the cooperative mechanism [17], we propose the Generative Adversarial-Synergetic Networks (GA-SN) for anomaly detection. It consists of the Adversarial Discriminant Network (ADN) and the Synergetic Generative Network (SGN), which includes generator, discriminator and promotor. The discriminator acts as a detector, and generator is an important hub in ADN and SGN, while promotor facilitates the other modules to extract key and discriminant features. Three modules supervise and promote each other. In summary, the main contributions of this paper are as follows:

(1) The ideas of adversarial relationship and synergy are combined. Synergy increases the effectiveness of adversarial relationship, in turn the adversarial relationship improves the ability of synergy.
(2) The GA-SN with more powerful discrimination is built for anomaly detection. The promotor is introduced innovatively so as to assist asymmetric generator and discriminator to extract more pivotal and discriminant features, which improves the detection accuracy.

The remainder of the paper is organized as follows: the generative adversarial network is briefly introduced in Sect. 2. Section 3 discusses the proposed method. In Sect. 4, some experiments and results are described, while Sect. 5 ends the paper with our conclusion.

2 Generative Adversarial Network

Generative adversarial network is a typical generative model. It mainly consists of two parts: generator G and discriminator D. The generator is used to obtain the distribution of the real data and construct the generated data $G(z)$ according to the obtained distribution. The discriminator is equivalent to a binary classifier which used to judge whether the input data is from the real data or the generated data. The generator and discriminator are trained by discussing the minimax game:

$$\min_{G} \max_{D} V(D, G) = E_{x \sim p_{d(x)}}[\log D(x)] + E_{z \sim p_{z(z)}}[\log(1 - D(G(z)))] \qquad (1)$$

Where the generator obtains the distribution p_d from the real data x, the prior distribution of the noise is $p_z(z)$, $D(x)$ represents the probability of x that is from real data. When the number of samples is sufficient, (1) is deformed to:

$$\begin{aligned} \max_{D} V(G, D) &= E_{x \sim p_{data}}[\log D_G^*(x)] + E_{z \sim p_z}[\log(1 - D_G^*(G(z)))] \\ &= E_{x \sim p_{data}}[\log D_G^*(x)] + E_{z \sim p_z}[\log(1 - D_G^*(x))] \\ &= E_{x \sim p_{data}}[\log \frac{p_{data}(x)}{p_{data}(x) + p_g(x)}] + E_{z \sim p_z}[\log \frac{p_g(x)}{p_{data}(x) + p_g(x)}] \qquad (2) \end{aligned}$$

The global optimal solution of minimax problem is transformed into the relationship between real data distribution and generated data distribution, and the Nash equilibrium is achieved. The discriminator can not distinguish the generated data from the real data. That is, $p_g = p_{data}$, $D_G^*(x) = 0.5$.

3 The Generative Adversarial-Synergetic Networks

From the psychology, if the child is often denied by parents, he or she may lose confidence and even eventually produce inferiority. If the child is often praised, he or she may be self-confident and eventually conceited. The child who is given praise and criticism will develop strengths and correct shortcomings. So the parents and child can communicate with each other better [18]. Combined with the generative adversarial network, the generator is equivalent to the child and the discriminator corresponds to the parents. The discriminator denies all of the generator, which may cause the generator negative. Inspired by this thought, we proposed the generative adversarial-synergetic networks, introducing the promotor to give the generator encouragement and help generator and discriminator extract effective information. The overall scheme of proposed method is shown in Sect. 3.1, and the detailed descriptions of each network are given in Sect. 3.2.

3.1 Overall Scheme

The GA-SN is composed of the Adversarial Discriminant Network (ADN) and the Synergetic Generative Network (SGN), as shown in Fig. 1. Since anomaly exists in some area randomly, image partition is carried out to obtain the overlapping patches.

Aim to discriminate the normal and abnormal attributes of the unknown data, the ADN is designed to learn features from the normal frames during training. There are Generator (G) and Discriminator (D) improving each other through adversarial learning. For generator, it takes noisy patches $\hat{x}$ as inputs where real patches x are attached to noise $\eta \sim N(0, \sigma^2\mathbf{I})$, and outputs generated patches $G(\hat{x})$ with the same dimensions of $\hat{x}$ but without noise. For discriminator, the inputs are both the real patches and generated patches, then the probability scores of each patches are obtained. The objective function of ADN is shown as follows:

$$\min_{G} \max_{D} \left(E_{x \sim p_d}[\log(D(x))] + E_{\hat{x} \sim \hat{p}_d}[\log(1 - D(G(\hat{x})))] \right) \tag{3}$$

Where noisy patches $\hat{x} = x + \eta$ and its distribution $\hat{p}_d = p_d + N(0, \sigma^2\mathbf{I})$, real patches x and noise η follow the distribution p_d and $N(0, \sigma^2\mathbf{I})$ respectively, i.e. $x \sim p_d$ and $\eta \sim N(0, \sigma^2\mathbf{I})$. In contrast to the conventional GAN, instead of mapping the noise to these patches with the distribution p_d, G maps noise patches $\hat{x}$ to distribution p_d in ideal, and D explicitly decides whether x and $G(\hat{x})$ follow p_d or not.

In addition, the ADN is trained on normal patches so that generator and discriminator can only learn the normal features. If abnormal samples appear in the test, the generator reconstruct abnormal features poorly and the discriminator naturally distinguish that the abnormal distribution is not p_d. So the discriminator obtains low probability score and discriminate abnormal ones. In order to extract more pivotal features, it is necessary to require the generated patches to obey the distribution p_d of real patches, avoiding the generated patches with different distributions to disturb the judgment of discriminator. The generated patches are incompletely subjected to distribution p_d if only rely on the discriminator to constraint generator [14]. We further propose the SGN to assist the generator to generate patches that comply with distribution p_d.

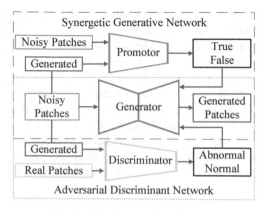

Fig. 1. The framework of GA-SN.

Based on ADN, the SGN is mainly designed to help generator learn the distribution and extract effective features from normal samples, improving the discriminant performance of ADN. Here the effective features refer to the pivotal and discriminant-features. The Generator (G) and Promotor (P) are built simultaneously in SGN through synergetic learning. The generator in SGN is the same as in ADN, which takes noisy patches $\hat{x}$ as inputs and generated patches $G(\hat{x})$ as outputs. Different from ADN, the introduced promotor takes the noisy patches and generated patches as inputs. As the noisy patches contain much interference, the promotor always trusts the ability of generator and believes generated patches are better than the noisy patches. So promotor gives the generator confidence and promotes it perfection. The objective function of SGN is shown as follows:

$$\min_{G,P} \left(E_{\hat{x} \sim \hat{p}_d} [\log(1 - P(G(\hat{x})))] + E_{\hat{x} \sim \hat{p}_d} [\log(P(\hat{x}))] \right) \tag{4}$$

Where the purposes of G and P are minimizing the objective function. Specially, the promotor regards the generated patches and noisy patches as real patches and fake patches respectively, so that the generator is aware that the fake patches have good features and strong competitiveness, and it needs to perfect itself to make the generated patches have more effective features.

In particular, ADN and SGN are combined to form GA-SN sharing the generator. Generally speaking, the GA-SN contains generator, discriminator and promotor. It combines the different features to obtain effective information, which extracted from the noisy patches and the real patches respectively. On the one hand, the redundant information is reduced. On the other hand, even if the input data is corrupted by various noises or other interference, the effective information can be extracted for the discriminator to make judgments.

3.2   Net-Architecture

It is obvious that the good performance corresponds to the correct theory, the appropriate networks, the reasonable optimization and so on. According to the effects of different modules in GA-SN, the structure of each network and the overall optimization are introduced in detail.

The discriminator D is designed mainly to distinguish the normal or abnormal data without any supervision. So the D is mainly consisted of a sequence of convolution. The third row in Fig. 2 shows the detailed architecture of D. There are three Convolutional layers (Conv) and one Fully Connected layer (FC). The Batch Normalization (BN) is used after the last two convolutional layers. Activation functions exist at the end of each layer, where the first three layers use leaky Rectified linear unit (Relu) and the last layer uses Sigmoid function. Because the Sigmoid function can be regarded as a classifier, $D(*)$ is relative to the likelihood of its input following the normal class. Therefore, D outputs a scalar value, which can be considered as a probability score for any given input.

The generator G mainly extracts the effective information from inputs and assists discriminator to detect anomaly without supervision. The second row in Fig. 2 shows the detailed architecture of asymmetric generator. There are four layers in network. The

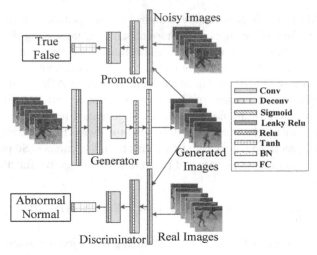

Fig. 2. The detailed architecture of GA-SN.

first two layers are convolutional layers, whose behind are the batch normalization and Leaky Relu. The last two layers are Deconvolutional layers (Deconv) where the former one is activated by Relu function and the latter layer is activated by Tanh function. So G can learn and reconstruct the normal features and then convey the learned knowledge to the discriminator. As a result, G efficiently learn the feature of these samples that share a similar concept with the normal class, and it poorly learns them when inputting abnormal samples. In addition, G makes it easier for the discriminator to separate the anomaly from the normal ones.

In order to urge the generator to learn more effective information directly and encourage the discriminator to make judgments indirectly, the promotor P is designed. The first row in Fig. 2 shows the detailed architecture of P. As we can see, this architecture is similar to the discriminator which contains three convolutional layers and one fully connected layer. However, different from the discriminator, the promotor determines whether the input data are generated patches or noisy patches and produces the probabilities that the inputs belong to the generated patches. These probabilities is further fed back to the generator, which facilitates the G to extract helpful information from the images with different interference.

In addition, we list the structures of discriminator, generator and promotor in detail, as shown in Table 1. Compared with the existing networks [19], our network has fewer layers. Specially, because of redundancy and interference information in images, the convolutional layer of the generator is less than others, and the number of feature maps between convolution and the deconvolution is not symmetrical. So the generator and discriminator learn the pivotal and discriminant features.

It is necessary to consider the optimization after building the network. In ADN, the discriminator and generator are optimized in adversary, while the generator and promotor are optimized in synergetic way for SGN. The overall optimization is shown in Table 2. The generator generates data and inputs them to promotor. Promotor updates and feedback the results to the generator, and the generator improves itself. The generated

Table 1. Detailed structural parameters of each module in GA-SN

Layers	Discriminator	Generator	Promotor
Input	$x, G(\hat{x})$	$\hat{x}$	$\hat{x}, G(\hat{x})$
Conv_1	(5, 5, 2, 64)	–	(5, 5, 2, 64)
Conv_2	(5, 5, 2, 128)	(5, 5, 2, 128)	(5, 5, 2, 128)
Conv_3	(5, 5, 2, 256)	(5, 5, 2, 256)	(5, 5, 2, 256)
Deconv_1	–	(5, 5, 2, 64)	–
Deconv_2	–	(5, 5, 2, 3)	–
FC	(256,1)	–	(256,1)

patches are given to discriminator, and then the discriminator and generator are updated in sequence. Obviously, generator is an important hub of ADN and SGN. Each network share information and promote each other. The discriminator can obtains the probability of each patch when testing, and the mean value of all patches in an image is the final probability score.

Table 2. The overall optimization of GA-SN

Initial all parameters
for the number of iterations do:
   SGN update:
$$\min_{G,P} \left(E_{\hat{x} \sim \hat{p}_d}[\log(1 - P(G(\hat{x})))] + E_{\hat{x} \sim \hat{p}_d}[\log(P(\hat{x}))] \right)$$
   Update the promotor and feedback to generator
   Update the generator
   ADN update:
$$\min_{G} \max_{D} \left(E_{x \sim p_d}[\log(D(x))] + E_{\hat{x} \sim \hat{p}_d}[\log(1 - D(G(\hat{x})))] \right)$$
   Update the discriminator and feedback to generator
   Update the generator
end for

4 Experiments

In this section we evaluate our method using two well-known datasets. The rest of this section describes the implementation and the results which are analyzed in details.

4.1 Implementation

Setup. All the reported results are from our implementation using the Intel Core i7-6800 K 3.40 GHz CPU, NVIDIA GTX 1080 GPU. Our method is mainly implemented in Anaconda3 64-bit, Python3.5 and TensorFlow, an open source framework for GAN training and testing on Windows 10. The detailed structures of networks are explained in Sect. 3. These structures are kept fixed for different tasks. The convolution kernel is 5×5 and the stride is 2×2. The hyperparameters of batch normalization are set as $\varepsilon = 10^{-6}$ and momentum $= 0.9$. The parameter of Leaky Relu activation function is 0.2. The learning rate of the Adaptive moment estimation optimizer is 0.002, and exponential decay rates β_1, β_2 for the moment estimates is 0.5 and 0.999 respectively. Additionally, the standard Gaussian noise is chosen for noisy patches, that is $\sigma = 1$.

Datasets. The MNIST dataset [20] and the UCSD anomaly detection dataset [21] are used. The MNIST dataset contains 70000 binary images of handwritten numbers, 28 $\times$ 28 in size, with 60000 training images and 10000 test images. Many researchers use it to prove the effectiveness of their proposed method. The UCSD dataset is split into two subsets: one is Ped1 which contains 34 training sequences and 16 testing sequences, the other is Ped2 that contains 16 training videos and 12 testing videos. This dataset is challenging due to the low-resolution, different and small moving objects and one or more anomalies in the scene.

Evaluation Metrics. The probability score can be used to distinguish whether the input frame is normal or abnormal. The threshold of probability score is used to identify the abnormal frames, which is manually specified. The optimal value of this parameter is very important since a higher threshold leads to a higher false negative rate, while a lower threshold leads to a higher false positive rate. Thus, the Area Under Curve (AUC) is a more suitable metric [21], which measures the performance by changing different thresholds. In addition, we also evaluate the performance using the Equal Error Rate (EER), which is also used in [22]. For a more comprehensive comparison, we also list the results with F1-score [19].

4.2 Verification Experiment on MNIST Dataset

During the training, the size of the input images is 28 $\times$ 28. Due to the small size of MNIST images, it is difficult to judge the category of digits if object partition is carried out. Therefore, the input images are as input patches which are operated directly. Then the Gaussian noise is added to input patches as noisy patches, and then the batch size is set to 64. The proposed method is compared with the other algorithms, and our experiments are divided into two parts.

On one hand, each of the ten categories is treated as an anomaly, while the rest of the categories are regarded as normal classes. Here 80% of normal samples are selected randomly to constitute the training dataset. The testing dataset is composed of the rest normal images and all abnormal images. Figure 3 shows the results of AUC which are obtained through experimenting several times and the other results were obtained from [23].

As we can see in Fig. 3, the AUC of GA-SN is lower than CVGAD when digit 8 is chosen as anomalous, that may be due to the more irregular writing of 8, which is easily misjudged as normal category. But it is superior to the CVGAD [23], GANomaly [24], EGBAD [25] and AnoGAN [26] on the whole.

On the other hand, each of the ten categories of digits is taken as the normal class, and we simulate anomaly by randomly selecting images from other categories with a

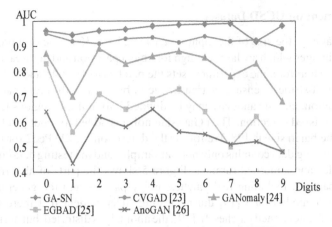

Fig. 3. The AUC results obtained on MNIST.

proportion of 10% to 50%, which is repeated for all of the ten digits. We report the F1-score to evaluate the performance and compare it with others. Figure 4 shows the F1-score of different methods on several proportions of anomaly. As it can be seen, our method operates more efficiently than the Local Outlier Factor (LOF) [27] and Discriminative Reconstruction AutoEncoder (DRAE) [28] obtained from [19]. In Addition, when the abnormal proportions are 10% and 50%, the F1-score are 0.9764 and 0.9211 respectively, which decreased about 5%. It is obvious that our method has only smaller decline, with the ascendant proportion of abnormal samples.

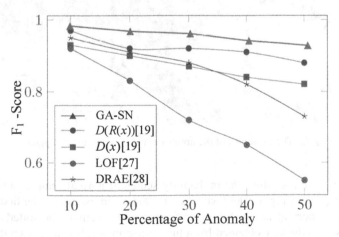

Fig. 4. The F1-score of different methods on several proportions of anomaly.

4.3 Experiment on UCSD Dataset

During the training, the size of the input images is 240 × 360. One or more anomalies may occur in images which are large enough for object partition and the area of anomalies may be small. Therefore, the experiment sets the patches to 45 × 45 and the sliding step is 20. There is a stride to ensure that an image is blocked to different patches with a certain correlation, and the anomaly may be divided into multiple regions to reduce the possibility of missed detection. Then Gaussian noise is added to input patches as noisy patches and the batch size is 128. We follow the dataset on UCSD Ped2 for training and testing. The training data contains only normal samples and the testing data contains both normal samples and abnormal samples. Figure 5 shows the parts of abnormal patches and normal patches, including real patches, noisy patches and the generated patches obtained by our method and [19] during testing. It is evident that there is no noise interference in the generated patches. In [19], the anomaly is distorted, but normal patches are not adequately reconstructed. The normal objects can be reconstructed better and anomaly is not adequately reconstructed by proposed GA-SN. This shows the generator is good enough to extract features under the synergy of promoter, and it is conducive for discriminator to make decision.

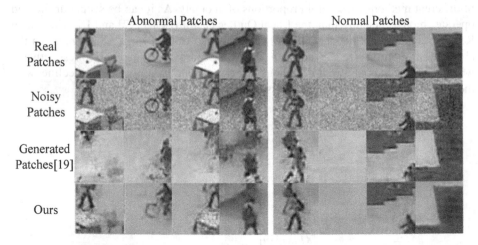

Fig. 5. The examples of the abnormal patches and normal patches

Meanwhile, we also show the probability scores of normal samples and abnormal samples, as shown in Fig. 6. Here we take Test004 as an example. The first 31 frames from dataset are normal and the following frames are abnormal. The dotted curve is the probability scores directly obtained from the discriminator. It can be seen that the first 31 probability scores are relatively larger, while the latter are relatively smaller. Samples with low probability scores are detected as abnormal. It is convenient to distinguish normal samples from abnormal samples according to the probability scores, achieving anomaly detection.

Further more, we compare our proposed method with the other algorithms. Currently, there are two main evaluation methods of frame-level evaluation and pixel-level anomaly

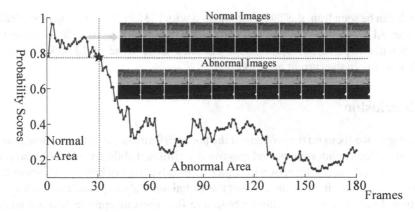

Fig. 6. The probability scores of Test004

localization. In the frame-level evaluation, if one or more pixel is predicted as anomaly, an abnormality label is predicted for the given frame. In this case the abnormality label is assigned based on pixel for the whole frame. For pixel-level evaluation, the predicted abnormal pixels are compared with the ground truth. The test frame is judged as abnormal if the area of the predicted anomaly overlaps with the ground-truth area at least 40%, otherwise the frame is counted as normal. But frame-level evaluation often detects many of the normal scenes as anomalies and pixel-level anomaly localization should compare every pixel in each image, which takes lots of computation. Therefore, we calculate the probability scores of each patch and then take average of them which are from the whole image. In this way, local anomaly and global anomaly are combined to reduce the computation. Table 3 shows the AUC and EER results and the other results were obtained from [19].

Table 3. The AUC and EER results on UCSD Ped2.

Methods	AUC	EER
MPPCA [29]	69.3%	30%
Social force [30]	55.6%	42%
SF + MPPCA [31]	61.3%	36%
MDT [31]	82.9%	25%
TCP [32]	88.4%	18%
AMDN (double fusion) [33]	90.8%	17%
Convolutional AE [34]	90.0%	21.7%
RE [35]	–	15%
Ravanbakhsh et al. [36]	–	14%
Sabokrou et al. [19]	–	13%
Ours	89.6%	12.56%

As it can be seen from the Table 3, compared with [33] and [34], our method obtained the lower AUC. However, There is a better EER value in [19] of 13%. However, our method without parameter optimization obtained EER value of 12.56%, which was 0.44% lower than the current networks.

5 Conclusion

In this paper, we focus on the problem of the poor performance of anomaly detection. To extract the effective information and improve discriminant ability, we propose the generative adversarial-synergetic networks, which takes advantage of both the adversarial and the synergetic ideas. It takes into account the global anomaly to provide the high-quality detection. Further more, the different objective functions of optimization are taken to control pivotal and discriminant features extraction. The experimental results on several datasets demonstrating that the proposed GA-SN is capable of anomaly detection. So our work shows a perspective of designing the network to detection anomaly. In future work, considering that noise is necessary in network design, it is feasible to further study the influence of different noises on network. In addition, the performance can be further improved by designing more excellent and balanced networks, anomaly metrics and so on.

Acknowledgment. This work is supported by National Natural Science Foundation of China (NO.61871241, NO.61976120); Ministry of education cooperation in production and education (NO.201802302115, NO.201901009007, NO.201901009044); Educational Science Research Subject of China Transportation Education Research Association (Jiaotong Education Research 1802-118); the Science and Technology Program of Nantong (JC2018025, JC2018129); Nantong University-Nantong Joint Research Center for Intelligent Information Technology (KFKT2017B04); Nanjing University State Key Lab. for Novel Software Technology (KFKT2019B15); Postgraduate Research and Practice Innovation Program of Jiangsu Province (KYCX19_2056).

References

1. Jin, D., Zhu, S., Wu, S., Jing, X.: Sparse representation and weighted clustering based abnormal behavior detection. In: 24th International Conference on Pattern Recognition, pp. 1574–1579. IEEE, New York (2018)
2. Li, H.J., Suen, C.Y.: Robust face recognition based on dynamic rank representation. Pattern Recogn. **60**(12), 13–24 (2016)
3. Singh, D., Krishna Mohan, C.: Graph formulation of video activities for abnormal activity recognition. Pattern Recogn. **65**, 265–272 (2017)
4. Avinash, R., Vinod, P.: Tucker tensor decomposition-based tracking and Gaussian mixture model for anomaly localisation and detection in surveillance videos. IET Comput. Vision **12**(6), 933–940 (2018)
5. Afiq, A.A., et al.: A review on classifying abnormal behavior in crowd scene. J. Vis. Commun. Image Represent. **58**, 285–303 (2018)
6. Li, H.J., Suen, C.Y.: A novel Non-local means image denoising method based on grey theory. Pattern Recogn. **49**(1), 217–248 (2016)

7. Xu, Y., Lu, L., Xu, Z., He, J., Zhou, J., Zhang, C.: Dual-channel CNN for efficient abnormal behavior identification through crowd feature engineering. Mach. Vis. Appl. **30**(5), 945–958 (2018). https://doi.org/10.1007/s00138-018-0971-6
8. Ko, K.E., Sim, K.B.: Deep convolutional framework for abnormal behavior detection in a smart surveillance system. Eng. Appl. Artif. Intell. **67**, 226–234 (2018)
9. George, M., Jose, B.R., Mathew, J., Kokare, P.: Autoencoder-based abnormal activity detection using parallelepiped spatio-temporal region. IET Comput. Vision **13**(1), 23–30 (2019)
10. Li, H.J., Zhou, Z., Li, C.B., Zhang, S.B.: RCCM: Reinforce Cycle Cascade Model for image recognition. IEEE Access **8**, 15369–15376 (2020)
11. Goodfellow, I.J., Pouget-abadie, J., Mirza, M., Xu, B.: Generative adversarial nets. In: International Conference on Neural Information Processing Systems, pp. 2672–2680. MIT Press, Cambridge (2014)
12. Yu, H., Li, X.: Intelligent corner synthesis via cycle-consistent generative adversarial networks for efficient validation of autonomous driving systems. In: Design Automation Conference, pp. 9–15. IEEE, New York (2018)
13. Zhai, W., Zhu, J., Cao Y., Wang, Z.: A generative adversarial network based framework for unsupervised visual surface inspection. In: 2018 IEEE International Conference on Acoustics, Speech and Signal Processing, pp. 1283–1287. IEEE, New York (2018)
14. Adiga, S., Attia, M. A., Chang W., Tandon, R.: On the tradeoff between mode collapse and sample quality in generative adversarial networks. In: IEEE Global Conference on Signal and Information Processing, pp. 1184–1188. IEEE, New York (2019)
15. Xia, Y., Cao, X.D., Wen, F., Hua, G., Sun, J.: Learning discriminative reconstructions for unsupervised outlier removal. In: IEEE International Conference on Computer Vision, pp. 1511–1519. IEEE, New York (2015)
16. Ravanbakhsh, M., Sangineto, E., Nabi, M., Sebe, N.: Training adversarial discriminators for cross-channel abnormal event detection in crowds. In: 2019 IEEE Winter Conference on Applications of Computer Vision, pp. 1896–1904. IEEE, New York (2019)
17. Zhang, L., Zhao, J.Y., Ye, X.L., Dong, W.: Co-operative generative adversarial nets. Acta Automatica Sin. **44**(5), 804–810 (2018)
18. Friedman, R.: The theory and art of child psychotherapy: a corrective developmental approach. Psychoanal. Rev. **104**(5), 561–593 (2017)
19. Sabokrou, M., Khalooei, M., Fathy M., Adeli, E.: Adversarially learned one-class classifier for novelty detection. In: IEEE Conference on Computer Vision and Pattern Recognition, pp. 3379–3388. IEEE, New York (2018)
20. Lecun, Y., Bottou, L., Bengio, Y., Haffner, P.: Gradient-based learning applied to document recognition. Proc. IEEE **86**(11), 2278–2324 (1998)
21. Luo, W., Liu, W., Gao, S.: A revisit of sparse coding based anomaly detection in stacked RNN framework. In: IEEE International Conference on Computer Vision, pp. 341–349. IEEE, New York (2017)
22. Sabokrou, M., Fayyaz, M., Fathy, M., Moayed, Z., Klette, R.: Deep anomaly: fully convolutional neural network for fast anomaly detection in crowded scenes. Comput. Vis. Image Underst. **172**, 88–97 (2018)
23. Bian, J., Hui, X., Sun, S., Zhao, X., Tan, M.: A novel and efficient cvae-gan-based approach with informative manifold for semi-supervised anomaly detection. IEEE Access **7**, 88903–88916 (2019)
24. Akcay, S., Atapour-Abarghouei, A., Breckon, Toby P.: GANomaly: semi-supervised anomaly detection via adversarial training. In: Jawahar, C.V., Li, H., Mori, G., Schindler, K. (eds.) ACCV 2018. LNCS, vol. 11363, pp. 622–637. Springer, Cham (2019). https://doi.org/10.1007/978-3-030-20893-6_39

25. Zenati, H., Foo, C. S., Lecouat, B., Manek, G., Chandrasekhar, V.R.: Efficient gan-based anomaly detection. https://arxiv.org/pdf/1802.06222.pdf. Accessed 20 Nov 2019
26. Schlegl, T., Seebock, P., Waldstein, S.M., Schmidt-Erfurth, U., Langs, G.: Unsupervised anomaly detection with generative adversarial networks to guide marker discovery. In: Information Processing in Medical Imaging, pp. 146–157. Springer, Switzerland (2017)
27. Breunig, M.M., Kriegel, H.P., Ng, R.T., Sander, J.: Lof: identifying density-based local outliers. ACM SIGMOD Int. Conf. Manage. Data **29**(2), 93–104 (2000)
28. Xia, Y., Cao, X., Wen, F., Hua, G., Sun, J.: Learning discriminative reconstructions for unsupervised outlier removal. In: IEEE International Conference on Computer Vision, pp. 1511–1519. IEEE, New York (2015)
29. Kim J., Grauman, K.: Observe locally, infer globally: a space-time MRF for detecting abnormal activities with incremental updates. In: IEEE Conference on Computer Vision and Pattern Recognition, pp. 2921–2928. IEEE, New York (2009)
30. Mehran, R., Oyama, A., Shah, M.: Abnormal crowd behavior detection using social force model. In: IEEE Conference on Computer Vision and Pattern Recognition, pp. 935–942. IEEE, New York (2009)
31. Mahadevan, V., Li, W., Bhalodia, V., Vasconcelos, N.: Anomaly detection in crowded scenes. In: IEEE Conference on Computer Vision and Pattern Recognition, pp. 1975–1981. IEEE, New York (2010)
32. Ravanbakhsh, M., Nabi, M., Mousavi, H., Sangineto, E., Sebe, N.: Plug-and-play CNN for crowd motion analysis: an application in abnormal event detection. In: 18th IEEE Winter Conference on Applications of Computer Vision, pp. 1689–1698. IEEE, New York (2018)
33. Xu, D., Yan, Y., Ricci, E., Sebe, N.: Detecting anomalous events in videos by learning deep representations of appearance and motion. Comput. Vis. Image Underst. **156**, 117–127 (2017)
34. Hasan, M., Choi, J., Neumann, J., Roy-Chowdhury A.K., Davis, L.S.: Learning temporal regularity in video sequences. In: IEEE Conference on Computer Vision and Pattern Recognition, pp. 733–742. IEEE, New York (2016)
35. Sabokrou, M., Fathy, M., Hoseini, M.: Video anomaly detection and localisation based on the sparsity and reconstruction error of auto-encoder. Electron. Lett. **52**(13), 1122–1124 (2016)
36. Ravanbakhsh, M., Nabi, M., Sangineto, E., Marcenaro, L., Regazzoni, C., Sebe, N.: Abnormal event detection in videos using generative adversarial nets. In: 24th IEEE International Conference on Image Processing, pp. 1577–1581. IEEE, New York (2017)

Multi-features Integration for Speech Emotion Recognition

Hongjun Li[1,2,3,4](✉) ⓘ, Ze Zhou[1] ⓘ, Xiaohu Sun[1] ⓘ, and Chaobo Li[1] ⓘ

[1] School of Information Science and Technology, Nantong University, Nantong 226019, China
lihongjun@ntu.edu.cn
[2] State Key Lab, Novel Software Technology, Nanjing University, Nanjing 210023, China
[3] Nantong Research Institute, Advanced Communication Technologies, Nantong 226019, China
[4] TONGKE School of Microelectronics, Nantong 226019, China

Abstract. Speech not only conveys the content information but also reveals the emotions of speakers. In order to achieve effective speech emotion recognition, a novel multi-features integration algorithm has been proposed. The statistical Mel frequency cepstrum coefficient (MFCC) features are directly evolved from the original speech. To further mine more useful information among statistical features, sparse groups are presented to extract the discriminative features. For enhancing the nonlinearity of features, we map features to nonlinear space to obtain nonlinear features by the orthogonal matrix. Multiple features integrated enable them to work for speech emotion recognition together. Extensive experiments comparison with state-of-the-art algorithms on CASIA dataset confirm that our algorithm can achieve effective and efficient speech emotion recognition. In addition, the analysis of different features indicates multi-features integration is superior than single type of features, where the MFCC features contribute greater in recognition accuracy and at the same time it also takes more time for features extraction.

Keywords: Speech emotion recognition · Multi-features integration · Sparse groups · Nonlinear features · CASIA

1 Introduction

Speech emotion recognition has been widely applied in the field of human-computer interaction [1–3]. It is the simulation of human brain system's processing of speech emotion. It can be defined as extracting appropriate emotional state from human speech signals [4]. At the same time, speech emotion recognition is also an important and challenging recognition technology [5].

In speech emotion recognition, there are mainly two difficulties that how to construct a suitable speech emotion recognition model and how to find effective speech emotion features [6]. To realize speech emotion recognition, substantial research work on recognition model have been proposed, such as traditional algorithms support vector machine (SVM) [7–9], Bayesian classifiers [10, 11], k nearest neighbors [12, 13],

© Springer Nature Switzerland AG 2020
Y. Lu et al. (Eds.): ICPRAI 2020, LNCS 12068, pp. 191–202, 2020.
https://doi.org/10.1007/978-3-030-59830-3_17

and Gaussian mixture model [14, 15], hidden Markov model [16, 17]. In recent years, some networks based on deep learning [18] has been widely used for speech emotion recognition, such as deep neural network (DNN) [19, 20], convolutional neural network (CNN) [21, 22], long short-term memory [23, 24], recurrent neural network [25, 26], where deep networks usually take the spectrum of the speech as input. Ren et al. [27] proposed a multi-modal correlation network for emotion recognition, which combined audio and video features and transmitted the Mel spectrum processed to CNN. Features are essential for speech emotion recognition, Nwe et al. [28] pointed out that acoustic features related to speech emotion can be mainly divided into spectral features, prosodic features, and sound quality features. Many methods of speech emotion recognition are focusing on extracting these statistical features from the original speech. For example, Chen et al. [29] utilized openSMILE to extract fundamental frequency, zero crossing rate in the short-time speech, spectral energy, and MFCC features. Liu et al. [30] proposed a correlation analysis and Fisher's feature selection manner, which extracts the speaker-dependent features (i.e., fundamental frequency, short-time energy, MFCC features, spectral energy dynamic coefficients, etc.) and speaker-independent features (i.e., the average change rate of fundamental frequency, the change rate of short-time energy frequency and so on) to perform speech emotion analysis. Chen et al. [31] used a three-level speech emotion recognition model to solve the speaker independent emotion recognition problem and extracted the energy, zero crossing rate, pitch, the first to third formants, spectrum centroid, spectrum cut-off frequency, correlation density, fractal dimension, and five Mel frequency bands energy. Pan et al. [32] combined MFCC features, Mel-energy spectrum dynamic coefficients with energy to recognize emotions by a SVM classifier.

Statistical features can be considered as the first evolution of raw speech, which can reflect the attributes of speech best. These features are highly correlated and contain massive useful information. However, many algorithms aforementioned always spare no effort to explore more similar statistical features, without further excavating more useful features among them. Here we propose a multi-features integration algorithm for speech emotion recognition, in which we only extract MFCC features as statistical features, and the focus is to mine the correlation between these features through adopting sparse groups to extract discriminative features. To enhance the independence of features, we map the features to nonlinear features so as to refine features which have decisive function of emotion recognition.

The remainder of the paper is organized as follows: the related work is briefly introduced in Sect. 1. Section 2 discusses the proposed method. In Sect. 3, we describe experiments and results, while Sect. 4 ends the paper with our conclusion.

2   Related Work

2.1   MFCC

MFCC [33] was proposed based on the auditory characteristics of the human ear, taking into account the human ear's auditory perception of sound. This feature is mainly used for speech feature extraction and reduction of computational dimensions. It has a nonlinear correspondence with frequency and has a certain improvement on the performance of the speech recognition system.

The auditory characteristics of the human ear are consistent with the increase of Mel frequency, that is, the actual frequency shows a linear distribution below 1000 Hz, and a logarithmic increase above 1000 Hz. The relationship between the MFCC features and the frequency can be approximated by the formula:

$$Mel(f) = 2595 \cdot \lg\left(1 + \frac{f}{700\,\text{Hz}}\right) \tag{1}$$

where f is the frequency in Hz.

2.2 Other Statistical Features

Speech changes slowly over time, which generally can be considered as an expectant steady state process. If the speech in the short-time (10 ms–30 ms) stationary state is processed, its intrinsic features can be remained, so the speech should be framed firstly.

Fundamental frequency is the frequency of the vocal cord vibration, which carries important features that reflect speech emotion. Theoretically, if a signal has periodicity, then its autocorrelation function also has the same period. Therefore the pitch period of the signal can be estimated from the position of the first maximum of the autocorrelation function regardless of the start time.

$$R(\Delta n) = \sum_{m=0}^{N-\Delta n-1} x(n)x(n + \Delta n) \tag{2}$$

where $x(n)$ is a frame of speech, N is the length of it, Δn is a delay. $R(\Delta n)$ reflects the similarity between $x(n)$ and its delay.

Short-time energy is used to measure the energy of a frame of speech, which can distinguish the unvoiced and voiced sounds and discriminate the initials and finals.

$$E_f = \sum_{n=0}^{N-1} |w(n)x(n)|^2 \tag{3}$$

where $w(n)$ is a window function.

Average zero crossing rate in the short-time speech indicates the number of times the speech signal waveform crosses zero-level. The voiced sound has lower zero crossing rate than that of the unvoiced sound usually.

$$ZC = \frac{1}{2} \sum_{n=0}^{N-1} |\text{sgn}(x(n)) - \text{sgn}(x(n - 1))| \tag{4}$$

where $\text{sgn}(\cdot)$ is a sign function.

Formant is the area where the spectral energy of the speech relatively concentrates, which reflects the important parameters of channel information. The common formant extraction methods include spectral envelope method [34], LPC interpolation method [35], LPC root finding method [36], Hilbert transform method [37] and so on.

3 Methodology

Considering that most algorithms of speech emotion recognition focus on extracting statistical features from the original speech, such as energy, zero crossing rate and formant, which remain many more useful information for recognition. We have proposed multi-features integration algorithm to improve the recognition rate of speech emotion and explore more decisive information. The MFCC features are directly evolved from the original speech, which promotes the raw signal to the primary feature level. To mine more useful features, we adopt sparse groups to extract the discriminative features, representing the speech with sparse features. Nonlinear features are further mapped to increase the nonlinearity among the discriminative features. The integration of these features enables them to play a decisive role in emotion recognition. The basic flow chart of our algorithm is shown in Fig. 1.

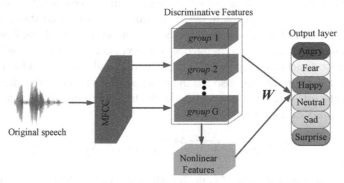

Fig. 1. The framework of our algorithm. The MFCC features are directly transformed from the original speech. Then sparse groups are utilized to further mine the correlations between features, extracting the discriminative features. After that, we map features to nonlinear features to enhance its independence. Finally, the multi-features integration are sent to the output layer for calculating the output weights W.

3.1 Spectral Feature Extraction

Spectral features are considered to be the manifestation of the correlation between channel shape changes and vocal motion [38]. Because the sound level heard by humans is not linearly proportional to the frequency, yet MFCC is based on the auditory characteristics of human ears, it has good robustness and accuracy in speech emotion recognition.

The details of MFCC feature extraction can be represented as:

1) Pre-process consists of pre-emphasis, framing, and windowing.

$$\begin{cases} s_p(n) = s(n) - \partial \cdot s(n-1) \\ x(n) = s_{pf}(n) \otimes w_m(n) \end{cases} \tag{5}$$

where $s(n)$ represents a piece of speech, ∂ is pre-emphasis coefficient, $s_p(n)$ and $s_{pf}(n)$ are the output of pre-emphasis and framing respectively, in which we set 25 ms and 10 ms respectively as the period and time shift of the short-term stationary signal, $w_m(n)$ is used as Hamming window, and N is the length of the window and the short-time speech.

$$w_m(n) = 0.54 - 0.46 \cos\left(\frac{2\pi n}{n-1}\right), \quad 0 \le n \le N-1 \tag{6}$$

2) Perform discrete Fourier transform on each frame of speech $x(n)$ as

$$X(k) = \sum_{n=0}^{N-1} x(n)e^{-j2\pi nk/N}, \quad n \ge 0, \ 0 \le k \le N-1 \tag{7}$$

where $X(k)$ is the frequency domain output.

3) Then we filter the frequency domain by Mel filter bank, which is usually a set triangular filters and conduct logarithmic energy operation.

$$E(m) = \ln\left(\sum_{k=1}^{N-1} |X(k)|^2 H_m(k)\right), \quad 0 \le m \le M \tag{8}$$

where $E(m)$ is the logarithmic energy of each filter bank output, M is the number of triangular filters, $H_m(k)$ is the m th triangular filter,

$$H_m(k) = \begin{cases} \frac{k-f_c(m-1)}{f_c(m)-f_c(m-1)}, & f_c(m-1) \le k \le f_c(m) \\ 0, & others \\ \frac{f_c(m+1)-k}{f_c(m+1)-f_c(m)}, & f_c(m) \le k \le f_c(m+1) \end{cases} \tag{9}$$

where $f_c(m)$ is the center frequency of the m th triangular filter.

4) Obtain the MFCC features by discrete cosine transform.

$$MFCC(l) = \sum_{m=0}^{M-1} E(m) \cdot \cos\left(\frac{\pi l(m-0.5)}{M}\right), \quad l = 1,\ldots,L \tag{10}$$

where L is the dimension of MFCC features.

3.2 Discriminative Feature Extraction

Though MFCC features can represent the original speech signal, a piece of speech usually contains numerous frames, the MFCC features in which have great correlation. In order to explore the relationship of spectral features and reduce feature redundancy further, we adopt sparse groups to extract the discriminative features.

$$Z_i = \phi(VW_{ei} + \beta_{ei}), \quad i = 1,\ldots,G \tag{11}$$

where $V \in \mathbb{R}^{N \times M N \times M}$ are the MFCC features, N, M are the number and the dimension of samples, and Z_i are the discriminative features extracted by the i th sparse group, and the W_{ei}, β_{ei} are the sparse weights and offsets, respectively. The sparse process can be considered as an optimization in Eq. (12),

$$\min_{W_{ei}} : \|Z_i W_{ei} - V\|_2^2 + \lambda \|W_{ei}\|_1 \tag{12}$$

Then optimization process of the sparse weights is as follows:

$$\min_{W_{ei}} : p(W_{ei}) + g(O_{ei}), \quad s.t. \ W_{ei} - O_{ei} = 0 \tag{13}$$

and the above optimization of W_{ei} is described as

$$\begin{cases} W_{ei}^{t+1} = (Z_i^T Z_i + \rho I)^{-1} (Z_i^T V + \rho (O_{ei}^t - Q^t)) \\ O_{ei}^{t+1} = S_\kappa (W_{ei}^{t+1} + Q^t) \\ Q^{t+1} = Q^t + (W_{ei}^{t+1} - O_{ei}^{t+1}) \end{cases} \tag{14}$$

where O_{ei} are the dual matrix of W_{ei}, Q are the residual between them, t is the number of iteration, $\rho > 0$, and $S_\kappa(\cdot)$ is the soft threshold function, which is expressed as

$$S_k(\alpha) = \begin{cases} \alpha - \kappa & \alpha > \kappa \\ 0 & |\alpha| \le \kappa \\ \alpha + \kappa & \alpha < -\kappa \end{cases} \tag{15}$$

We define the discriminative features set Z^G as $[Z_1, \ldots, Z_G]$.

3.3 Nonlinear Feature Extraction and Recognition Decision

In order to enhance the nonlinearity of the discriminative features, we try to map these features into a nonlinear space by using an orthogonal matrix, the vectors in which are independent and orthogonal.

$$H = \xi(Z^G \cdot W_h \cdot \eta \cdot \mu) \tag{16}$$

where H are the nonlinear features, η is shrinkage scale, $\xi(\cdot)$ is the activation function $\tan sig(\cdot)$, which can be represented

$$\tan sig(x) = \frac{2}{1 + e^{-2x}} - 1 \tag{17}$$

Specially, the orthogonal matrix W_h and μ are shared the training and testing phase, in which μ is calculated as

$$\mu = \frac{1}{(\max(\max(Z^G \cdot W_h)))} \tag{18}$$

After feature extraction, we combine multiple features together as $[Z^G \mid H]$. Suppose the target output $Y \in \mathbb{R}^{N \times C}$, where C is the number of emotion. To acquire the output weights W for recognition decision, we consider it as an optimization problem.

$$\arg\min_{W} \|[Z^G \mid H]W - Y\|_2^2 + \lambda \|W\|_2^2 \tag{19}$$

where it is considered as least square optimization problem with $L2$ constraints, which can be fast solved by calculating the pseudo-inverse. Assume that the minimum value of Eq. (19) is $\mathbf{0}$, and then calculate the derivation of W as Eq. (21).

$$||[Z^G \mid H]W - Y||_2^2 + \lambda ||W||_2^2 = \mathbf{0} \tag{20}$$

$$2[Z^G \mid H]^T (Y - [Z^G \mid H]W) - 2\lambda W = \mathbf{0} \tag{21}$$

Therefore,

$$W = ([Z^G \mid H]^T [Z^G \mid H] + \lambda I)^{-1} [Z^G \mid H]^T Y \tag{22}$$

where λ is a positive constant, and I is an identity matrix.

4 Simulation Results

All experiments are conducted on a Windows 10 PC laptop Intel Core i7-8700 3.20 GHz CPU, 8 GB RAM and MATLAB R2016a 64-bit. The speech dataset used in this paper is CASIA, Chinese emotion corpus recorded by the Institute of Automation, Chinese Academy of Sciences [39], which is recorded by four people (i.e., two men and two women) in a clean recording environment (SNR is about 35 dB), and adopting 16 kHz sampling, 16 bit quantified. It consists of Mandarin utterances of six basic emotions (i.e., angry, fear, happy, neutral, sad, and surprise), and each of which contains 200 speeches expressing 50 audios of the same text. In all experiments, we randomly select 30 samples of each speaker with each emotion, i.e., 720 samples of emotional speech in total for training. For the test samples, each speakers' speech samples are utilized as a test set, (i.e., four test sets in total) and there are 20 samples of each speaker with each emotion, i.e., each test set includes 120 samples of emotion speech in total. Comparative experiments with DNN, SVM and CNN have been conducted to demonstrate the superiority of our algorithm in terms of training time and testing accuracy. In addition, we conduct experiments under different features to illustrate the importance of the combination of multiple features.

4.1 Comparison Between Our Algorithm and Others

For realizing efficient speech emotion recognition, we have proposed a multi-features integration algorithm for emotion recognition, and compared with DNN, SVM and CNN to demonstrate the superiority of our algorithm on the same dataset. The experimental results are shown in Table 1.

Where the "Wang (M)", "Liu (W)", "Zhao (M) and Zhao (W)" respectively refer to the last names of two males and two females experimenters (i.e., Zhe Wang, Changhg Liu, Quanyin Zhao, and Zuoxiang Zhao) who contributed all speech data of the CASIA database. It can be seen from Table 1 that our algorithm performs better than DNN, SVM and CNN, regardless of recognition accuracy and training time-consuming. For the four speakers, the testing rate of ours is respectively higher than DNN, SVM and CNN by

Table 1. Comparison of experimental results with DNN, SVM and CNN

Group	DNN		SVM		CNN		Ours	
	Ac.(%)	Time(s)	Ac.(%)	Time(s)	Ac.(%)	Time(s)	Ac.(%)	Time(s)
Wang (M)	68.33	15.12	70.83	7.04	74.3	382.4	**76.66**	**6.88**
Liu (W)	70.83	15.05	73.33	7.12	80.7	388.1	**87.5**	**6.84**
Zhao (M)	71.67	14.96	73.33	7.13	73.8	381.6	**74.44**	**6.81**
Zhao (W)	70.83	15.06	71.67	7.06	79.6	385.5	**89.16**	**6.79**

(8.33%, 5.83%, 2.36%), (16.67%, 14.17%, 6.8%), (2.77%, 1.11%, 0.64%), (18.33%, 17.49%, 10.01%), and our algorithm takes least time to train our algorithm. Thanks to the integration of multiple features, all the refined features have decisive function in speech emotion recognition, and the classification algorithm we used does not involve an iterative process, which results in fast training. In addition, for the three recognition algorithms, we are surprised to find that the accuracy of male emotion recognition is significantly lower than that of female. According to our analysis, the fundamental difference between male and female voices lies in the difference in pitch. The overtone spacing of male voices is smaller than that of female voices, which leads to the lower fundamental frequency and the appearance of "roughness".

4.2   The Importance of Multi-feature Integration

We integrate three types of features to recognize emotion based on speech. In order to verify the importance of multiple features integration, we perform the following experiments, in which the speech of a female called Liu is taken as an example, and the experimental results are illustrated in Fig. 2.

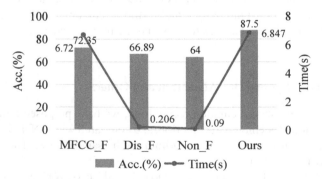

Fig. 2. Comparison of results under different features

Here MFCC_F, Dis_F, Non_F respectively refer to the cases that our algorithm only contains MFCC features, Discriminative features and Nonlinear features. As Fig. 2 shows that the integration of multiple features is more effective for emotion recognition based on speech, in which the recognition accuracy under integration features is respectively greater than MFCC features, discriminative features, and nonlinear features by 15.15%, 20.16%, and 23.5%. Compared with using single feature, our algorithm first converts the original speech into MFCC features, which specializes in representing speech, and then analyzes the correlation of these features and adopts sparse groups to extract discriminative features, and finally extract nonlinear features for enhancing the nonlinearity of features. Thanks to that, the multi-features integration is refined to represent the speaker's emotions. Moreover, in order to explore which type of features in this paper contributes greater to emotion recognition, we conduct the following experiments, and the experimental results are shown in Fig. 3.

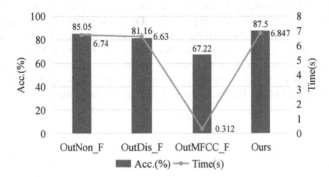

Fig. 3. The contribution analysis of different features

Here, OutNon_F, OutDis_F, and OutMFCC_F mean that the lack of nonlinear features, discriminative features and MFCC features respectively in our algorithm. The visual experimental results in Fig. 3 indicate that if the nonlinear features, the discriminative features, and the MFCC features are respectively missing, the corresponding recognition accuracy will reduce by 2.45%, 6.34%, 20.28%, which illustrates that the MFCC features have more contribution to recognizing emotion. This mainly because MFCC features evolve from the original speech and contains the most information, the integration of any other features with the MFCC features can achieve better recognition accuracy. Without MFCC features, discriminative features and nonlinear features are extracted directly from the raw speech, which can be considered as the phase of data processing, and the extracted features can not reflect the primitive speech professionally. In addition, we also notice that the process of extracting the MFCC features is very time consumption by comparison, and the reason can be attributed that the cumbersome MFCC feature extraction process is applied to the original speech with large size.

5   Conclusion

In this paper, we have proposed multi-features integration for speech emotion recognition. The MFCC features are the primary representation of the original speech, which are directly transformed from it. The discriminative features extracted by sparse groups are considered as advanced features, which further mines the correlation between MFCC features. In order to refine features and improve the independence between features, nonlinear features have been mapped. Multiple features integrated enable them to work for speech emotion recognition together. Extensive comparison experiments with several state-of-the-art algorithms on CASIA dataset show that our algorithm is superior in terms of recognition accuracy and training time. In addition, comparative experiments under different features indicates the importance of multi-features integration and points out that MFCC features contribute more to recognition rate and spend more time on feature learning as well. In the future work, we will explore features that are more generalized and further improve the recognition accuracy on speech datasets with more emotions.

Acknowledgment. This work is supported by National Natural Science Foundation of China (NO. 61871241, NO. 61976120); Ministry of education cooperation in production and education (NO.201802302115, NO. 201901009007, NO. 201901009044); Educational Science Research Subject of China Transportation Education Research Association (Jiaotong Education Research 1802-118); the Science and Technology Program of Nantong (JC2018025, JC2018129); Nantong University-Nantong Joint Research Center for Intelligent Information Technology (KFKT2017B04); Nanjing University State Key Lab. for Novel Software Technology (KFKT2019B15); Postgraduate Research and Practice Innovation Program of Jiangsu Province (KYCX19_2056).

References

1. Albornoz, E.M., Milone, D.H.: Emotion recognition in never-seen languages using a novel ensemble method with emotion profiles. IEEE Trans. Affect. Comput. **8**(1), 43–53 (2017)
2. Kim, J., André, E.: Emotion recognition based on physiological changes in music listening. IEEE Trans. Pattern Anal. Mach. Intell. **30**(12), 2067–2083 (2008)
3. Le, B. V., Lee, S.: Adaptive hierarchical emotion recognition from speech signal for human-robot communication. In: 2014 Tenth International Conference on Intelligent Information Hiding and Multimedia Signal Processing, Kitakyushu, pp. 807–810. IEEE (2014)
4. Tarunika, K., Pradeeba, R.B., Aruna, P.: Applying machine learning techniques for speech emotion recognition. In: 2018 9th International Conference on Computing, Communication and Networking Technologies (ICCCNT), Bangalore, pp. 1–5. IEEE (2018)
5. Li, H., Suen, C.Y.: Robust face recognition based on dynamic rank representation. Pattern Recogn. **60**(12), 13–24 (2016)
6. Sun, L., Fu, S., Wang, F.: Decision tree SVM model with Fisher feature selection for speech emotion recognition. EURASIP J. Audio Speech Music Process. **2019**(1), 1–14 (2019). https://doi.org/10.1186/s13636-018-0145-5
7. Sun, L.H., Zou, B., Fu, S., Chen, J., Wang, F.: Speech emotion recognition based on DNN-decision tree SVM model. Speech Commun. **115**, 29–37 (2019)
8. Schuller, B., et al.: Cross-corpus acoustic emotion recognition: variances and strategies. IEEE Trans. Affect. Comput. **1**(2), 119–131 (2011)
9. Wang, K., An, N., Li, B.N., Zhang, Y., Li, L.: Speech emotion recognition using Fourier parameters. IEEE Trans. Affect. Comput. **6**(1), 69–75 (2015)
10. Xu, X.Z., et al.: A two-dimensional framework of multiple kernel subspace learning for recognizing emotion in speech. IEEE-ACM Trans. Audio Speech Lang. Process. **25**(7), 1436–1449 (2017)

11. Alvarez, A., et al.: Feature subset selection based on evolutionary algorithms for automatic emotion recognition in spoken Spanish and standard Basque language. In: 9th International Conference on Text, Speech and Dialogue, Brno, Czech Republic, pp. 565–572 (2006)

12. Morrison, D., Wang, R., Silva, L.C.D.: Ensemble methods for spoken emotion recognition in call-centres. Speech Commun. **49**, 98–112 (2007)

13. Pao, T.L., Liao, W.Y., Chen, Y.T., Yeh, J.H.: Comparison of several classifiers for emotion recognition from noisy Mandarin speech. In: 3rd International Conference on Intelligent Information Hiding and Multimedia Signal Processing, pp. 23–26. IEEE, Kaohsiung, TaiWan (2007)

14. Shahin, I., Nassif, A.B., Hamsa, S.: Emotion recognition using hybrid Gaussian mixture model and deep neural network. IEEE Access **7**, 26777–26787 (2019)

15. Li, H.J., Suen, C.Y.: A novel non-local means image denoising method based on grey theory. Pattern Recogn. **49**(1), 217–248 (2016)

16. Absa, A.H.A., Deriche, M.: A two-stage hierarchical bilingual emotion recognition system using a hidden Markov model and neural networks. Arab. J. Sci. Eng. **42**(12), 5231–5249 (2017)

17. Nogueiras, A., Moreno, A., Bonafonte, A., Mariño, J.B.: Speech emotion recognition using hidden Markov models. In: 7th European Conference on Speech Communication and Technology, pp. 2679–2682 (2001)

18. Hongjun, L., Ze, Z., Chaobo, L., Shibing, Z.: RCCM: reinforce cycle cascade model for image recognition. IEEE Access **8**, 15369–15376 (2020)

19. Hifny, Y., Ali, A.: Efficient arabic emotion recognition using deep neural networks. In: 44th IEEE International Conference on Acoustics, Speech and Signal Processing (ICASSP), Brighton, United Kingdom, pp. 6710–6714. IEEE (2019)

20. Desai, S., Raghavendra, E.V., Yegnanarayana, B., Black, A.W., Prahallad, K.: Voice conversion using artificial neural networks. In: IEEE International Conference on Acoustics, Speech, and Signal Processing, pp. 3893–3896 (2009)

21. Zhao, J., Mao, X., Chen, L.: Learning deep features to recognise speech emotion using merged deep CNN. IET Sig. Proc. **12**(6), 713–721 (2018)

22. Neumann, M., Vu, N.T.: Attentive convolutional neural network based Speech emotion recognition: a study on the impact of input features, signal length, and acted speech. In: 18th Annu. Conf. Int. Speech Commun. Assoc, pp. 1263–1267 (2017)

23. Xie, Y., Liang, R., Liang, Z., Huang, C., Zou, C., Schuller, B.: Speech emotion classification using attention-based LSTM. IEEE-ACM Trans. Audio Speech Lang. Process. **27**(11), 1675–1685 (2019)

24. Greff, K., Srivastava, R.K., Koutník, J., Steunebrink, B.R., Schmidhuber, J.: LSTM: a search space odyssey. IEEE Trans. Neural Netw. Learn. Syst. **28**(10), 2222–2232 (2017)

25. Zhang, T., Wu, J.: Speech emotion recognition with i-vector feature and RNN model. In: 2015 IEEE China Summit and International Conference on Signal and Information Processing (ChinaSIP), Chengdu, pp. 524–528. IEEE (2015)

26. Keren G., Schuller, B.: Convolutional RNN: an enhanced model for extracting features from sequential data. In: 2016 International Joint Conference on Neural Networks (IJCNN), Vancouver, BC, Canada, pp. 3412–3419 (2016)

27. Ren, M.J., Nie, W.Z., Liu, A.A., Su, Y.T.: Multi-modal correlated network for emotion recognition in speech (in press)

28. Nwe, T.L., Foo, S.W., De Silva, L.C.: Speech emotion recognition using hidden Markov models. Speech Commun. **41**(4), 603–623 (2003)

29. Chen, L.F., Su, W.J., Feng, Y., Wu, M., She, J.H., Hirota, K.: Two-layer fuzzy multiple random forest for speech emotion recognition in human-robot interaction. Inf. Sci. **509**, 150–163 (2020)

30. Liu, Z.T., Wu, M., Cai, W.H., Mao, J.W., Xu, J.P., Tan, G.Z.: Speech emotion recognition based on feature selection and extreme learning machine decision tree. Neurocomputing **273**(17), 271–280 (2018)
31. Chen, L.J., Mao, X., Xue, Y.L., Cheng, L.L.: Speech emotion recognition: features and classification models. Digit. Sig. Proc. **22**(6), 1154–1160 (2012)
32. Pan, Y.X., Shen, P.P., Shen, L.P.: Speech emotion recognition using support vector machine. Int. J. Smart Home **6**(2), 101–108 (2012)
33. Murty, K.S.R., Yegnanarayana, B.: Combining evidence from residual phase and MFCC features for speaker recognition. IEEE Sig. Process. Lett. **13**(1), 52–55 (2006)
34. Kameoka, H., Ono, N., Sagayama, S.: Speech spectrum modeling for joint estimation of spectral envelope and fundamental frequency. IEEE Trans. Audio Speech Lang. Process. **18**(6), 1507–1516 (2010)
35. Levine, A.: The use of LPC coefficient interpolation to improve the naturalness of sentence formation by word concatenation. J. Acoust. Soc. Am. **66**, 1947–1948 (1979)
36. Zapata, J.G., Díaz Martín, J.C., Vilda, P.G.: Fast formant estimation by complex analysis of LPC coefficients. In: 2004 12th European Signal Processing Conference, Vienna, pp. 737–740. IEEE (2004)
37. Mundodu, P.K., Ramaswamy, K.: Single channel speech separation based on empirical mode decomposition and Hilbert Transform. IET Signal Proc. **11**(5), 579–586 (2017)
38. Björn, W.: Speech emotion recognition: two decades in a nutshell, benchmarks, and ongoing trends. Commun. ACM **61**(5), 90–99 (2018)
39. Tao, J., Liu, F., Zhang, M., Jia, H.: Design of speech corpus for mandarin text to speech, pp. 1–4 (2018)

Analysis of Multi-class Classification of EEG Signals Using Deep Learning

Dipayan Das[1](✉)(iD), Tamal Chowdhury[1](iD), and Umapada Pal[2](iD)

[1] National Institute of Technology, Durgapur, India
dipayan.das2010@gmail.com, tgchowdhury101@gmail.com
[2] Indian Statistical Institute, Kolkata, India
umapada@isical.ac.in

Abstract. After promising results of deep learning in numerous fields, researchers have started exploring Electroencephalography (EEG) data for human behaviour and emotion recognition tasks that have a wide range of practical applications. But it has been a huge challenge to study EEG data collected in a non-invasive manner due to its heterogeneity, vulnerability to various noise signals and variant nature to different subjects and mental states. Though several methods have been applied to classify EEG data for the aforementioned tasks, multi-class classification like digit recognition, using this type of data is yet to show satisfactory results. In this paper we have tried to address these issues using different data representation and modelling techniques for capturing as much information as possible for the specific task of digit recognition, paving the way for Brain Computer Interfacing (BCI). A public dataset collected using the MUSE headband with four electrodes (TP9, AF7, AF8, TP10) has been used for this work of categorising digits (0–9). Popular deep learning methodologies like CNN (Convolutional Neural Network) model on DWT (Discrete Wavelet Transform) scalogram, CNN model on connectivity matrix (mutual information of time series against another), MLP (Multilayer Perceptron) model on extracted statistical features from EEG signals and 1D CNN on time domain EEG signals have been well experimented with in this study. Additionally, methodologies like SVC (Support Vector Classifier), Random Forest and AdaBoost on extracted features have also been showcased. Nevertheless, the study provides an insight in choosing the best suited methodology for multi-class classification of EEG signals like digit recognition for further studies.

Keywords: EEG · Deep learning · Classification · Signal processing · Convolutional neural network · LSTM · Discrete wavelet transform

1 Introduction

Electroencephalography (EEG) signal is generated by the spontaneous firings of the numerous neurons present in the brain. The rapid charging and discharging

© Springer Nature Switzerland AG 2020
Y. Lu et al. (Eds.): ICPRAI 2020, LNCS 12068, pp. 203–217, 2020.
https://doi.org/10.1007/978-3-030-59830-3_18

of the neurons causes varying electrostatic potential on the external skin surface, which when measured with respect to a ground potential (generally, the pinna or external ear is used as the reference potential) via a sensitive voltmeter, a voltage (usually in milli Volts) versus time graph is obtained. This graph is like a reflection of the internal processing of the brain because the major frequency components are delivered by the cortical activities that are in close proximity to the external scalp.

In the recent years, EEG signal processing has gained high importance due to the emergence of interest in deciphering the brain's state and brain-computer interfacing mechanisms. It is believed that one can extract features from the EEG signals captured from a subject and readily predict certain aspects of the subject's brain or mind like attention level, seizures, anomalies and so on. But recently, researchers have decided to take a step forward and use EEG signals to predict much more specific aspects like emotion [1–4], motor control intention [5] and so on. Hence the thought arises whether someone could use it to determine the exact thought that a person might be thinking during an EEG recording session.

From the beginning of this decade, researchers have shown some promising results in predicting the exact state of mind (in terms of calm, positive and negative emotions) that a subject was in while recording EEG signals. But the state of the art results show that the prediction or classification task has only been limited to either a binary classification task or a multi-class classification which contains no more than five classes. Since the diversity of human thoughts is much beyond this, the next step towards this movement is to predict thoughts that belong to more classes. Hence, digit recognition (0 to 9) is a great opportunity for the same. But this also means that we would need much more complex and robust methods to decrypt the EEG signals into the thoughts that generated it, and here comes deep learning.

Deep learning has grown exponentially in the recent years in numerous fields for prediction and classification tasks [6]. From cancer detection in medical imaging to detecting globular clusters with black hole sub-systems in astrophysics, deep learning has been a delight for the researchers who are trying to find patterns in the extensive amount of data that they have recorded. Hence it becomes an obvious option for the researchers to use it EEG signal processing too. But till date, the most extensively used methodologies for decrypting EEG signals has been from the field of machine learning or classical statistical learning like Support Vector Machines, K-Nearest Neighbouring and so on, supported by feature extraction policies like Fast Fourier Transforms [7] and Principal Component Analysis. To our best knowledge, deep learning methodologies are yet to be explored for the same purpose. Therefore, in this study we showcase a comparative analysis of using various popular deep learning algorithms on EEG signals to do multi-class classification, for which digit recognition is taken as the example task.

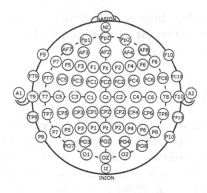

Fig. 1. Electrode positions for Interaxon MUSE headset

2   Dataset

The dataset is taken from a public website 'mindbigdata.com/opendb/'. The International 10–20 System was used as the reference standard for electrode placement. The data is collected using an Interaxon MUSE headband which contains four electrodes – TP9, TP10, FP1, FP2, placed on the scalp as shown in Fig. 1. The data is recorded in the following fashion – the subject is allowed to see a digit from 0 to 9 and the corresponding EEG signal is recorded for exactly 2 s at a sampling rate 220 Hz. There are 40983 data samples in this dataset, where each sample is composed of four time sequences corresponding to four electrodes. For each sequence, the number of time steps vary from 430 to 612. The EEG signal distribution corresponding to stimulus digits 0–9 is shown in Table 1. The peak voltage amplitudes vary between 400 mV to 700 mV, along

Table 1. Distribution table of the EEG samples according to digit stimulation.

Digit stimulation	No. of EEG samples
0	2,976
1	2,908
2	2,980
3	2,958
4	2,884
5	3,013
6	3,092
7	3,020
8	3,052
9	2,997
−1	11,103

with some data samples that shoot up to 1500mV as well. These are most likely due to noise signals that contain little to no information and taken care of during the preprocessing stage.

3   Preprocessing

The file format of the dataset includes - 'id', 'event', 'device name', 'channel', 'label', 'number of time steps' and the 'time series' in each row. For each label, four consecutive rows are dedicated. Likewise, a dataframe management python library is used to extract the time series and the labels from every fourth entry in the data file. Thus a total of 40983 data samples are obtained with varying lengths of the time sequences from around 430 to 612. The general pre-processing steps include:

3.1   Mean Padding

At very beginning each time sequence of four electrodes for individual samples is being padded to maximum time steps using mean-padding in order to bring them to a fixed length. Truncation of sequences is avoided to eliminate the possibility of information loss. This technique considers the mean value of an entire sequence and pad this value to the end of the sequence upto required time steps to get a uniform length for all the sequences.

3.2   Min-Max Scaling

For smooth and effective training of Deep learning model a normalised data distribution is expected at the input. We have used Min-Max scaling as the normalization method to bound it to a range of 0 to 1. The cost of having this bounded range in contrast to standardization is that we will end up with smaller standard deviations, which can suppress the effect of out-liars or electric fluctuations present in the data making the learning algorithm more robust to noises. A Min-Max scaling is typically performed using the following equation:

$$X_{norm} = \frac{X - X_{min}}{X_{max} - X_{min}} \tag{1}$$

3.3   Data Smoothing

To remove the noise signals from EEG data we passed each data sample through a Triangular Moving Average (TMA) filter. It is basically a Low Pass Filter that removes the high frequency noise components of a signal. The low pass filtering effect of the TMA on an EEG signal is shown in Fig. 2. It is an average of an average, of the last N data points(P). As the filter length increases (the parameter N) the smoothness of the output increases, whereas the sharp transitions in the data are made increasingly blunt. This implies that this filter has excellent time domain response but a poor frequency response. The TMA can also be expressed as:

$$TMA = \frac{SMA_1 + SMA_2 + SMA_3 + ...SMA_N}{N} \tag{2}$$

where,

$$SMA = \frac{P_1 + P_2 + P_3 + ...P_N}{N} \tag{3}$$

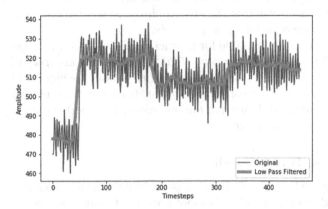

Fig. 2. Smoothing of EEG signal

3.4 Down-Sampling

All the samples are then down-sampled to 40 time steps to remove redundant information from the data and reducing computational load during model training and inference.

After these primary pre-processing steps method wise data preparation is performed for individual modeling techniques. Details of each of those techniques are mentioned in the corresponding sections.

4 Methods

The motive of this study is mainly dedicated to the comparative analysis of EEG signals for multi-class classification using various state of the art deep learning and data representation techniques. Hence a number of traditional machine learning techniques are also studied in order to get a reference for the deep learning approach. The following list shows the techniques that yielded the maximum results in terms of prediction accuracy. Even though the accuracies are quite low with respect to normal deep learning performances, they provide a rough guideline for choosing the required methodology for further work and improvement in the same.

4.1 Multi-class Support Vector Classifier (SVC)

Support Vector Classifier or SVC has been a remarkable machine learning technique for classification and regression problems [8]. In this technique a data point is represented in a n-dimensional hyper-space, by using each feature of the data point as a coordinate axis. So for our problem, the time sequences were represented as vectors in high dimensional feature space where each time step corresponded to one feature or coordinate axis. After the representation, a number of suitable hyperplanes are found according to number of clusters or unique output labels that separates the data points having maximum and minimum inter and intra class variances respectively.

Though SVC is not suitable for multiclass classification problem and overfits quickly yielding a training accuracy of 79.2% and a test accuracy of 27.8%. Nevertheless, the test accuracy showed by support vector classifier for 11 classes show that it has learnt at least some information from the data, even if not remarkably. Further tuning of hyper-parameters along with better data representation might aid to its performance.

4.2 Random Forest Classifier

Random Forest Classifier [9] is an enesemble method in which a set of decision trees are used to classify data. Random Forest is based on the notion that a large number of uncorrelated decision makers make a better decision. Hence each tree in a random forest picks up a random subset of the data features and conditions itself to predict a class based on the features and corresponding feature value chosen. In this study, the random forest classifier is implemented using the python library 'sklearn' and the hyper-parameters were taken as number of estimators = 100 and a maximum depth of 10. Using this configuration the classifier yielded a training accuracy of 35.3% and a test accuracy of 31.7%. This shows a better result than the former SVC method and also shows well fitting over the given data. On further increasing the number of estimators/trees in the ensemble or the maximum allowable depth of the trees, overfitting was observed.

4.3 Gradient Boosting Classifier

Gradient Boosting Classifier is yet another type of ensemble method where a set of weak learners, like decision trees are cascaded to do regression or classification task. This method builds the model in a stage wise fashion, like other boosting methods and then generalizes them by optimizing an arbitrary differentiable loss function. For our task, the hyper-parameters were taken as number of estimators = 50, learning rate = 0.01 and max depth = 5, which yielded a training accuracy of 27.2% and a test accuracy of 27.04%.

4.4 Long Short Term Memory Unit or LSTM

In the study of sequences, recurrent neural networks or RNNs have emerged as a promising methodology. But in the recent years, a better version of RNNs, known as the Long Short Term Memory units or LSTMs [10] have come into play. LSTMs are basically gated RNNs where the gates are key components that learn patterns in the input data sequences. Since EEG signals are nothing but time sequence data, LSTM becomes an obvious option to be experimented with. Using the python library 'keras' as wrapper library and tensorflow as the backend, an LSTM model with 100 units in the first layer and 20 units in the second layer, followed by a fully connected layer with 11 nodes was created. The input data sequence was downsampled to 40 timesteps, specifically for the reason that LSTMs tend to break down time sequences having higher number of timesteps. Additionally it was ensured that the time sequences were low pass filtered to remove spikes and high frequency noise components. The model yielded a training accuracy of 27.47% and a test accuracy of 27%. It can be observed that since EEG signals contains information in its various frequency components, removal of higher frequency components removes a certain amount of information. Nevertheless, the low frequency components do retain some information which causes the LSTM model to learn atleast upto a certain degree.

4.5 Statistical Feature Extraction and Dense Neural Network

In the field of classical machine learning, statistical feature extraction plays a very important role. This involves gathering specific feature values from the given data. In our study, the time sequences are represented as a series of timesteps. So in order to extract some meaningful information from this sequences, certain statistical features were chosen, to interpret each time sequence. The features hence chosen are as follows – Mean, Median, Mode, Standard Deviation, Variance, Skewness, Kurtosis. Then the sequence was divided into two segments equally and the maximum and minimum values of each segment was taken. Then the original sequence was again divided into four segments and the maximums, minimums and means of each segment was also calculated, along with the temporal distances between each of these four segments' maximums, minimums and means. Lastly the entropy and Logen entropy was calculated for the entire sequence. This gives us a total of 43 features for each time sequence. Apart from this, Fast Fourier Transform was applied on the time sequences and the mean amplitudes of the spectrogram at specific band regions (Delta – 0.5 4 Hz, Theta – 4 8 Hz, Alpha – 8 12 Hz, Beta – 12 30 Hz and Gamma – 30 60 Hz) were also taken for each sequence, which added five more features to our feature list. Hence a total of 48 features were calculated for each sequence. So for each data sample, which contains 4 time sequences from 4 electrodes, a total of 192 features were created. Using this format of data representation for each sample, a Dense Neural Network (DNN) [11] model was created with 100, 50 and 11 neurons in 3 layers respectively, which yielded a training accuracy of 32.3% and test accuracy of 31.7%.

4.6 1D Convolutional Neural Network or CNN

Convolutional Neural Network or CNN is a major breakthrough in the field of deep learning in this decade. Although it is mainly used for image classification, CNNs are very versatile to be used in other aspects of data also. In CNNs, a stack of kernel matrices are generated by learning the input data, which produces activation map. This activation map is a result of the convolution of the input data with the kernel matrices. Since image data is nothing but a matrix of pixel values, CNNs are well suited for tasks where the input data is a matrix or a stack of matrices (where, the depth of the matrices is represented as the channel dimension) [12]. In our case, the EEG signals can be represented as a 2D matrix of dimension (number of electrodes x number of timesteps) and the pixel values correspond to the amplitude values in the EEG signals. Hence, a CNN model was created having 25, 12 and 12 kernels in the first, second and third convolutional layer respectively, followed by 80 and 11 fully connected neurons. Note that, we used 1D convolution, wherein the kernels have the shape of a row vector with varying lengths. The motive behind this is that the kernels would extract singular and independent features from each row of a data sample. The model hence created yielded a training accuracy of 36.2% and a test accuracy of 32.7%, on an input data where the maximum timesteps was taken as 40.

4.7 Connectivity Matrix

In the field of information theory, mutual information is defined as the amount of uncertainty reduced regarding a random variable on observing another random variable. Hence this concept was used in this study to calculate the mutual information among each of the four time sequences belonging to a single data sample. This yielded a 4×4 matrix, which is known as the connectivity matrix [13]. Examples of this generated connectivity matrices are shown in Fig. 3. Hence, this representation was used as an input to deep learning models like Dense Neural Network (by flattening the 4×4 matrix into a row vector of length 16) and 2D Convolutional Neural Network, which produced accuracies of 32.2% and 31.7% respectively. The flattened connectivity matrix was also tested using a Random Forest Classifier which yielded an accuracy of 31.09%.

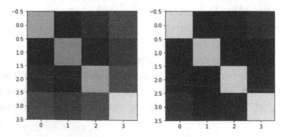

Fig. 3. Connectivity matrices of two distinct EEG signals

4.8   Discrete Wavelet Transform (DWT) and Convolutional Neural Network

EEG signal by its inherent nature is a non-stationary signal. So the frequency components of a EEG signal is not ever present throughout the entire time period of the data sequence. Certain frequency components occur at some interval while its absent in some other interval. This nature makes FFT or Fast Fourier Transform inapplicable to be used for frequency decomposition of EEG signals, which assumes that the input signal is stationary, i.e. all the frequency components are present through out the time sequence. Here comes Wavelet Transform [14,15], a novel frequency decomposition technique, in which along with the frequency components, the time instants are also calculated where each of the frequency components occur. This gives us a frequency versus time versus amplitude graph, which can be interpreted as an image with one axis denoting frequency, another axis denoting time and the pixel values denoting the amplitudes. Examples of DWT of EEG signals are shown in Fig. 4. Using this procedure, each data sample was converted into an image with 4 layers (corresponding to 4 electrodes) and each layer corresponding to the Wavelet Transform of each electrode's time sequence. Since our data sequence is a discrete data, Discrete Wavelet Transform was used to calculate the DWT coefficients and then DWT scalogram images were generated from those coefficients. The DWT denoted by $W(a, b)$ of a signal is given as:

$$W(j, k) = \int_f f(t) 2^{j/2} \psi_{j,k}(2^j t - k) dt \qquad (4)$$

where $\psi_{j,k}(t)$ are the wavelets with $\psi_{0,0}(t)$ as the mother wavelet and j and k are integers representing the scale and translation respectively.

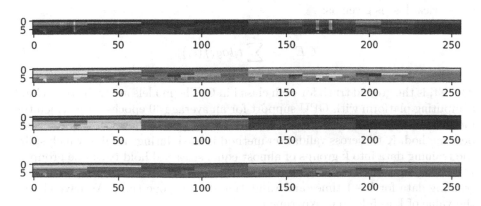

Fig. 4. Discrete wavelet transform scalograms of four different EEG signals

On this generated data, two types of convolutional networks were experimented, 3D CNN and 2D CNN, which yielded accuracies of 33.7% and 34.1%. For compiling and training the network 'categorical crossentropy' was used as

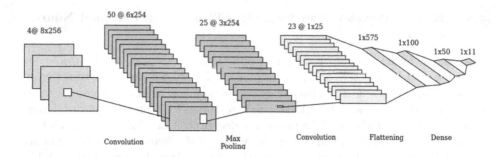

Fig. 5. Block diagram of the Convolutional Neural Network architecture used on the DWT scalograms

the loss function, 'adam' was used as the optimizer and the training process was carried out for 40 epochs. To our best knowledge, this result of 34.1% is by far the maximum accuracy achieved in EEG digit recognition task using specifically deep learning approach. The model architecture for the 2D CNN used is shown in Fig. 5.

5   Loss Function and Training Scheme

We have used categorical crossentropy as the loss function during the training of our Deep Learning models for multi-class classification tasks. Cross-entropy loss, or log loss, measures the performance of a classification model whose output is a probability value between 0 and 1. Cross-entropy loss increases as the predicted probability diverges from the actual label. The equation for calculating categorical loss is given as:

$$CE = - \sum_i^C t_i log(f(s_i))$$ (5)

where t_i is the ground truth for each class i in C. The models are trained on cloud computing platform with GPU support for an average 40 epochs after which the model starts to overfit. An epoch versus accuracy plot is shown in Fig. 6 for our best method. K-fold cross validation method is used during training which splits the training data into k groups of almost equal sizes and hold back one group as validation data while training on rest $k - 1$ groups. Each fold or group is used as training data for $k - 1$ times and validation data for one time. We have chosen the value of k as 5 for our experiments.

6   Experimentation and Comparative Results

Deep learning approaches have reached a golden stage in terms of image recognition to time series predictions. Hence, it is just a matter of time that these

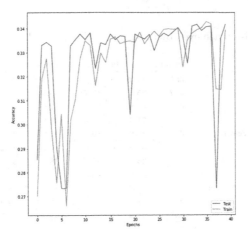

Fig. 6. Accuracy versus epoch plot for the training procedure.

approaches shall open new doors for Brain Computer Interfacing (BCI) too. EEG is a promising approach in the field of BCI, but the amount of data redundancy captured in EEG is what needs to be battled before using EEG as an easy mean of BCI. Through this study we show a comparative analysis of using some of the popular deep learning techniques to predict multi-class labels like numerical digits from EEG signals and the most promising result being shown by the 2D CNN on Discrete Wavelet Transform scalograms, which accounts to 34.1%. This result surpasses the current state of the art result of 31.35%, as stated in [16] and [17] by a marginal amount, which was achieved by using Multilayer Perceptron model (MLP) on the same dataset.

Table 2. 5 fold Accuracy achieved by different methods on a 11 class classification task

Methods	1st fold	2nd fold	3rd fold	4th fold	5th fold	Avg. Accuracy
SVC	27.5	27.9	27.6	28.1	27.9	27.8
Random Forest	31.5	31.6	31.4	32.1	31.9	31.7
Gradient Boost	26.8	27.0	27.2	27.0	27.2	27.04
LSTM	26.8	26.9	27.3	27.0	27.1	27.02
DNN with Statistical Features	31.2	31.5	31.9	32.4	32.0	31.8
Conv1D	32.4	32.3	32.9	32.8	32.8	32.7
RF with connectivity matrix	30.8	30.9	30.9	31.3	31.4	31.07
Conv2D with connectivity matrix	31.5	31.7	31.8	32.3	32.1	31.9
ANN with Connectivity Matrix	32.0	32.2	32.4	32.9	32.5	32.4
Conv2D with DWT	34.0	34.2	34.4	34.9	34.8	34.46
Conv3D with DWT	33.2	33.6	33.8	33.9	33.9	33.7

Table 3. 5 fold Precision achieved by different methods on a 11 class classification task

Methods	1st fold	2nd fold	3rd fold	4th fold	5th fold	Avg. Precision
SVC	32.5	34.2	33.3	35.5	33.7	33.84
Random Forest	35.2	35.4	36.7	38.8	39.0	37.02
Gradient Boost	30.8	31.8	32.5	32.0	32.8	31.98
LSTM	30.0	31.5	32.4	32.6	32.8	31.86
DNN with Statistical Features	35.5	35.9	36.5	37.0	36.5	36.28
Conv1D	36.5	36.8	37.8	37.5	38.2	37.36
RF with Connectivity Matrix	35.5	36.6	36.8	38.2	39.0	37.22
Conv2D with Connectivity Matrix	34.6	35.2	36.8	36.2	37.1	35.98
ANN with Connectivity Matrix	35.9	36.8	37.5	38.2	38.5	37.38
Conv2D with DWT	38.5	37.9	38.2	39.4	39.0	38.6
Conv3D with DWT	36.2	38.4	38.1	38.9	37.5	37.82

Table 4. 5 fold Recall achieved by different methods on a 11 class classification task

Methods	1st fold	2nd fold	3rd fold	4th fold	5th fold	Avg. Recall
SVC	28.0	28.2	28.4	28.6	28.0	28.24
Random Forest	32.5	32.7	33.0	32.8	33.6	32.92
Gradient Boost	28.2	28.4	28.0	28.6	29.2	28.48
LSTM	27.6	27.8	28.4	28.6	28.1	28.1
DNN with statistical features	32.4	32.6	33.5	33.1	32.9	32.9
Conv1D	38.8	38.4	39.0	39.4	39.2	38.96
RF with connectivity matrix	32.5	32.1	33.2	33.4	34.0	33.04
Conv2D with connectivity matrix	32.5	32.8	33.3	33.5	33.8	33.18
ANN with connectivity matrix	33.1	33.3	33.6	33.9	34.0	33.58
Conv2D with DWT	35.5	35.8	35.2	35.6	36.2	35.66
Conv3D with DWT	34.2	34.6.6	34.0	34.2	33.9	34.18

Table 5. 5 fold F1 Score achieved by different methods on a 11 class classification task

Methods	1st fold	2nd fold	3rd fold	4th fold	5th fold	Avg. F1 Score
SVC	27.9	28.3	28.1	28.3	28.7	28.26
Random Forest	32.4	32.5	33.3	32.5	33.1	32.76
Gradient Boost	28.0	28.2	28.7	28.5	29.4	28.56
LSTM	27.1	27.5	28.0	28.9	28.1	27.92
DNN with statistical features	32.1	32.9	33.2	33.0	32.9	32.82
Conv1D	38.3	38.8	39.9	39.5	39.4	38.98
RF with connectivity matrix	32.6	32.5	33.4	33.1	34.2	33.16
Conv2D with connectivity matrix	32.7	32.4	33.2	33.6	32.9	32.96
ANN with connectivity matrix	33.4	33.1	33.5	33.2	34.7	33.58
Conv2D with DWT	35.8	35.8	35.4	35.1	36.0	35.62
Conv3D with DWT	34.0	34.7	34.2	34.6	33.5	34.2

The 5-fold accuracy, precision, recall and F1 score achieved using the various methods is shown in Table 2, 3, 4 and 5 respectively, and the comparison of our result to the state of the art result is presented in Table 6.

Table 6. Comparison with state of the art result

Methodology	Accuracy
State of the art result using Multi-Layer Perceptron Model [16]	31.35%
2D CNN on Discrete Wavelet Transform features (our approach)	**34.46%**

7   Discussion Potential

After performing several experiments with different data representation techniques as well as different learning algorithms it is observed that the accuracy level reached is quite nominal, which can be justified in many ways. Firstly, EEG signals captured by non-invasive devices are extremely noise prone and requires extensive pre-processing to consider it as a good distinguishable information source. Very marginal improvements over different models suggests that the amount of data required for a learning algorithm to approximate the output mapping function is not sufficient. Also the promising performance shown by a recent work for the same task on a different dataset suggests that the data distribution of MUSE dataset is quite poor due to limited number of samples per class and class imbalance. Almost 30% of the total data belongs to a redundant class that contributes little to nothing for our model performance. Additionally, it can also be said that information encoding in EEG signals is also subjected to the mental state of the person recording the signals. A slight deviation from the desired thought would get highly reflected in the EEG signals and since there is no direct correspondence between the recorded signal and the person's thought. The plots in Fig. 7 show that for a single digit class, the variation in EEG signal pattern is quite dissimilar to one another, which makes the classification quite hard in time domain itself.

Apart from this, EEG being a non-invasive method contains information that corresponds to only those events in our brain that involve cortical cluster activities. Hence, tasks which involve much more specific/fine details, might fail to use EEG as an approach. Electrode positions also play an important role in this aspect due to the localization of specific regions in our brain. Owing to all these complexities, the amount of accuracy achieved by our approaches seems justified.

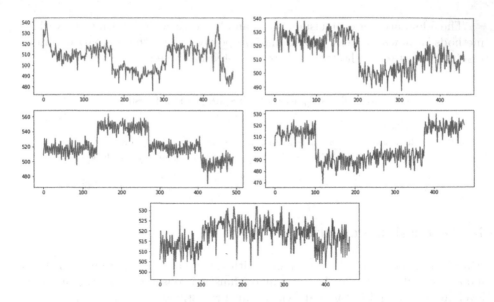

Fig. 7. Different samples of EEG for class digit 0

8   Conclusion

To our best knowledge this the highest accuracy level achieved on MUSE dataset for the task of digit recognition. The accuracy achieved is not satisfactory for practical applications. The MUSE dataset should be balanced properly by adding more number of training samples. Also more emphasis should be given to data processing and representation techniques to capture more information and suppress the vulnerability to noise signals. Overall, EEG has been a promising approach in understanding and monitoring brain activities non-invasively and hence is opening up doors for Brain Computer Interfacing. It is only a matter of time that we reach the stage of using it flawlessly to understand one's thought and execute an action accordingly. People suffering from different ailments like speaking disability or coma can be well aided with EEG monitoring devices. Through this study we put forward a new direction of using deep learning in the field of EEG signal processing. The results give a vivid understanding of choosing the proper direction for further improvements and future work.

References

1. Murugappan, M., et al.: Time-frequency analysis of EEG signals for human emotion detection. In: Abu Osman, N.A., Ibrahim, F., Wan Abas, W.A.B., Abdul Rahman, H.S., Ting, H.N. (eds.) 4th Kuala Lumpur International Conference on Biomedical Engineering 2008, vol. 21, pp. 262–265. Springer, Heidelberg (2008). https://doi.org/10.1007/978-3-540-69139-6_68

2. Mohammadi, Z., Frounchi, J., Amiri, M.: Wavelet-based emotion recognition system using EEG signal. Neural Comput. Appl. **28**, 1985–1990 (2017)
3. Li, Y., Zheng, W.M., Cui, Z., et al.: EEG emotion recognition based on graph regularized sparse linear regression. Neural Process. Lett. **1**, 1–17 (2018)
4. Yang, Y., Wu, Q.M.J., Zheng, W.L., et al.: EEG-based emotion recognition using hierarchical network with sub-network nodes. IEEE Trans. Cogn. Dev. Syst. **10**, 408–419 (2018)
5. Neuper, C., et al.: Motor imagery and EEG-based control of spelling devices and neuroprostheses. Progress Brain Res. **159**, 393–409 (2006)
6. Schirrmeister, R.T., et al.: Deep learning with convolutional neural networks for EEG decoding and visualization. Human Brain Mapp. **38**(11), 5391–5420 (2017)
7. Van Loan, C.: Computational frameworks for the fast Fourier transform, vol. 10. Siam (1992)
8. Subasi, A., Ismail Gursoy, M.: EEG signal classification using PCA, ICA, LDA and support vector machines. Expert Syst. Appl. **37**(12), 8659–8666 (2010)
9. Liaw, A., Wiener, M.: Classification and regression by random Forest. R News **2**(3), 18–22 (2002)
10. Hochreiter, S., Schmidhuber, J.: Long short-term memory. Neural Comput. **9**(8), 1735–1780 (1997)
11. Sarle, W.S.: Neural networks and statistical models (1994)
12. Acharya, U.R., et al.: Deep convolutional neural network for the automated detection and diagnosis of seizure using EEG signals. Comput. Biol. Med. **100**, 270–278 (2018)
13. Chen, H., Song, Y., Li, X.: A deep learning framework for identifying children with ADHD using an EEG-based brain network. Neurocomputing **356**, 83–96 (2019)
14. Daubechies, I.: The wavelet transform, time-frequency localization and signal analysis. IEEE Trans. Inf. Theory **36**(5), 961–1005 (1990)
15. Pattnaik S.: DWT-based feature extraction and classification for motor imaginary EEG signals. Paper presented at: 2016 International Conference on Systems in Medicine and Biology (ICSMB), 4–7 January, Kharagpur, India (2016)
16. Bird, J.J., et al.: A deep evolutionary approach to bioinspired classifier optimisation for brain-machine interaction. Complexity (2019)
17. Jolly, B.L.K., Aggrawal, P., Nath, S.S., Gupta, V., Grover, M.S., Shah, R.R.: Universal EEG encoder for learning diverse intelligent tasks. In: 2019 IEEE Fifth International Conference on Multimedia Big Data (BigMM), pp. 213–218. IEEE, September 2019

A Deep Object Detection Method for Pineapple Fruit and Flower Recognition in Cluttered Background

Chen Wang[1], Jun Zhou[2(✉)], Cheng-yuan Xu[3], and Xiao Bai[1]

[1] School of Computer Science and Engineering, Beihang University, Beijing, China
[2] School of Information and Communication Technology, Griffith University,
Brisbane, Australia
jun.zhou@griffith.edu.au
[3] School of Health, Medical and Applied Sciences, Central Queensland University,
Rockhampton, Australia

Abstract. Natural initiation of pineapple flowers is not synchronized, which yields difficulties in yield prediction and the decision of harvest. Computer vision based pineapple detection system is an automated solution to address this issue. However, it is faced with significant challenges, e.g. pineapple flowers and fruits vary in size at different growing stages, the images are influenced by camera viewpoint, illumination conditions, occlusion and so on. This paper presents an approach for pineapple fruit and flower recognition using a state-of-the-art deep object detection model. We collected images from pineapple orchard using three different cameras and selected suitable ones to create a dataset. The experimental results show promising detection performance, with an mAP of 0.64 and F_1 score of 0.69.

Keywords: Pineapple detection · Deep learning · YOLOv3

1 Introduction

Pineapple is a popular horticulture product of tropical and subtropical Australia and produces the only edible fruit in plants of the family Bromeliaceae. A major challenge faced by pineapple growers in Australia is natural initiation of flowers. As the natural initiation is not synchronized, fruiting can be spread out over two to three months, leading to lots of small harvest rounds with high human labor and financial costs. As pineapple does not have starch as storage, sugar content in fruits would not change during storage, so it is not viable to reduce harvest rounds by picking premature fruits. Currently, farmers make decisions based on experience and there is no technology available for reliable estimation of flowers and fruits in the field to support accurate estimation of profit margin for each harvest round. Figure 1(a) depicts the pineapple harvest. In practice, efficient harvest planning needs a sophisticated detection algorithm in order to overcome

© Springer Nature Switzerland AG 2020
Y. Lu et al. (Eds.): ICPRAI 2020, LNCS 12068, pp. 218–227, 2020.
https://doi.org/10.1007/978-3-030-59830-3_19

challenges such as insufficient or over illumination, heavy occlusion, variation of growing stage and fruit size, camera viewpoint and so on.

Over the years, many researchers have made efforts to develop automatic methods for fruit detection and localization [5,7,11,21]. Most of the techniques [16,21] relied on hand-crafted features such as color, shape, reflectance, etc. In recent years, with the significant increase in computational power and large amount of labeled data, deep convolutional neural network (CNN) based methods are widely used in various areas [1,24,25], including remote sensing [22,23], agricultural [3,19] and unmanned aerial vehicle [12]. By using deep learning techniques, the need for specific handcrafted features was eliminated, so feature extraction and object recognition become an integrated process.

(a) Pineapple Harvest   (b) Pineapple Detection Results

Fig. 1. (a) Harvesting pineapple in field. One harvester usually cover 4–8 rows, with one tractor pulling it and another tractor or forklifter transporting harvested pineapples to processing shed. The whole system includes 8–15 staff and costs 300 to 600 AUD per hour to operate. (b) Pineapple Detection Results. Fruits and flowers with different sizes, viewpoints and occlusions can be well detected.

In this paper, we presented a CNN model for accurate pineapple fruit and flower recognition using a state-of-the-art deep learning based object detector – YOLOv3 model. We collected real images from pineapple orchard using digital cameras mounted on different platforms. Then we selected suitable images and manually annotated them with the ground truth bounding boxes to create a dataset. The deep neural network was pre-trained on COCO dataset [9] and fine-tuned on the pineapple dataset. Experimental results show promising performance on pineapple fruit and flower recognition. Figure 1(b) shows a sample of pineapple detection results.

2 Related Work

Traditional computer vision approaches have been extensively adopted for fruit detection. Early methods [4,14] used fruit color as an import feature. However, these methods were only useful for fruits that have different color from the background. Shape based methods overcome some of the limitations of color-based methods. Bansal *et al.* [2] proposed an approach for detecting immature green citrus fruits, by using the symmetrical shape of smooth spherical fruit. Rabatel and Guizard [15] described a method to detect berry shapes from visible fruit edges using an elliptical contour model. Sengupta and Lee [20] used shape analysis to detect fruit. These methods heavily relied on manually developed features and were very sensitive to changes in viewpoint and occlusion.

In recent years, deep learning methods have generated breakthroughs in fruit detection. CNN models have been used for pixel-wise apple classification [3], followed by Watershed Segmentation (WS) to identify individual apples. Mai *et al.* proposed to incorporate a multiple classifier fusion strategy into a Faster R-CNN network [18] for small fruit detection. Sa *et al.* [19] developed a fruit detection model called Deep Fruits. Faster-RCNN was used as feature extraction frameworks. And then a multi-modal fusion approach that combined the information from RGB and NIR images was used.

In computer vision, research on object detection network has been heavily investigated. The reported methods are usually divided into two main groups: one-stage and two stage detectors. Two-stage detectors, such as Faster R-CNN (Region-based Convolutional Neural Networks) [18] and Mask R-CNN [6], use a region proposal network to generate regions of interests in the first stage and then use the region proposals for object classification and bounding-box regression. On the contrary, one-stage detectors, such as You Only Look Once(YOLO) [17] and Single Shot MultiBox Detector (SSD) [10], treat object detection as a simple regression problem which takes an input image and learns the class probabilities and coordinates of bounding boxes to show the location of objects. In this research, we use YOLOv3 model as the backbone framework for pineapple fruit and flower recognition.

3   Method

In this section, we briefly describe the YOLOv3 network architecture and loss functions. The structure of the YOLOv3 model is illustrated in Fig. 2. The YOLOv3 network takes the full size image as input and does not need any region proposal operation. The feature extraction part is called Darknet-53 because it has 53 convolutional layers in total. The network uses successive 3×3 and 1×1 convolutional layers with some shortcut connections. YOLOv3 extracts features at 3 different scales and then predicts bounding boxes. The last of these feature maps predict a 3D tensor encoding bounding box coordinates, objectness, and class predictions. In our experiments with pineapple dataset, we predict 3 boxes at each scale so the tensor is $N \times N \times [3*(4+1+2)]$ for the setting of 4 bounding offsets, 1 objectness prediction, and 2 class predictions. N means the width and height of feature map. The size of total 9 anchors are: $(10 \times 13), (16 \times 30), (33 \times 23), (30 \times 61), (62 \times 45), (59 \times 119), (116 \times 90), (156 \times 198), (373 \times 326)$.

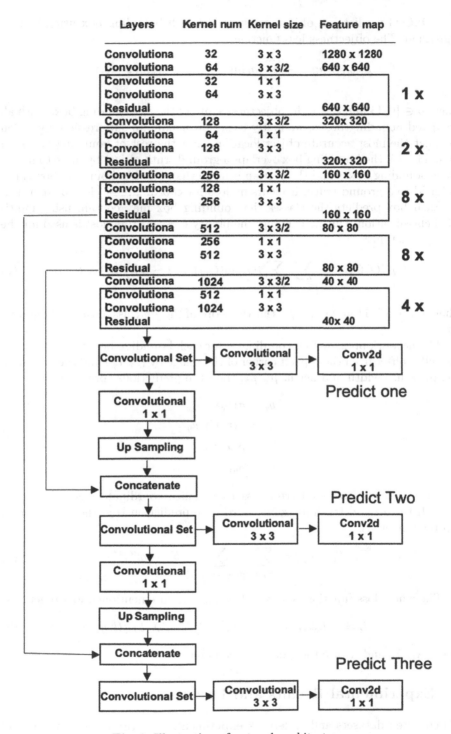

Fig. 2. Illustration of network architecture

YOLOv3 predicts an objectness score for each bounding box using logistic regression. The objectness loss function is:

$$L_{conf}(o,c) = -\sum_i (o_i \ln(c_i) + (1 - o_i) \ln(1 - c_i)) \tag{1}$$

where $o \in \{0,1\}$ and c_i are the objectness score of the ith bounding box. With all predicted bounding boxes, we firstly used non-maximum suppression operation to select the most accurate object locations. For the rest of bounding boxes, o_i should be 1 if the bounding box overlaps a ground truth object by more than any other bounding boxes and the overlap is more than 0.5. If a bounding box is not assigned to a ground truth, it incurs no loss for coordinate or class predictions.

Each box predicts the classes the bounding box may contain using multi-label classification. During training, the binary cross-entropy loss is used for the class predictions:

$$L_{cla}(O,C) = -\sum_{i \in Pos} \sum_{j \in cla} (O_{ij} \ln(C_{ij}) + (1 - O_{ij}) \ln(1 - C_{ij})) \tag{2}$$

where $O_{ij} \in \{0,1\}$ and C_{ij} represent the score of the jth class of the ith bounding box.

The network predicts four coordinates for each bounding box, t_x, t_y, t_w, t_h. If the cell is offset from the top left corner of the image by (c_x, c_y) and the bounding box prior has width and height p_w, p_h, then the predictions correspond to:

$$\begin{aligned} b_x &= \sigma(t_x) + c_x \\ b_y &= \sigma(t_y) + c_y \\ b_w &= p_w e^{t_w} \\ b_h &= p_h e^{t_h} \end{aligned} \tag{3}$$

We use sum of squared error loss to penalize coordinate offset regression error. If the ground truth for some coordinate prediction is g, the loss function can be written as:

$$L_{loc}(b,g) = \sum_{i \in pos} \sum_{m \in \{x,y,w,h\}} (b_i^m - g_i^m)^2 \tag{4}$$

The whole loss function is a weighted sum of aforementioned loss functions:

$$L = \lambda_1 L_{conf}(o,c) + \lambda_2 L_{cla}(O,C) + \lambda_3 L_{loc}(l,g) \tag{5}$$

where λ_1, λ_2 and λ_3 are the balance parameter.

4   Experimental Results and Discussion

We collected datasets and undertook experiments to verify the effectiveness of the proposed method.

4.1   Dataset

Three cameras were used to collect data in different imaging settings from a pineapple orchard near Yandaran, Queensland, Australia during November to December of 2018[1]. The imaging process considered scales, illuminations and viewpoints. Camera 1 took images from the top of pineapples in close range, obtaining images with detailed features. Camera 2 and 3 were mounted on the spray boom, taking pictures from a bird's-eye view. Camera 3 used a wide-angle lens to get wide-angle images. We selected a total of 200 images to form the dataset, and then manually annotated each image. Figure 3 shows three sample images in the dataset. Note that each image may contain multiple pineapple fruits and flowers. There were totally 786 bounding boxes for fruits and 2020 bounding boxes for flowers. In the experiments, we divided them into three groups: 140 images for training, 30 images for validation and the rest 30 images for testing.

Fig. 3. Samples in the pineapple dataset. (a) Image from camera 1 (Huawei P8 Max). (b) Image from camera 2 (DJI Phantom 3 Advanced). (c) Image from camera 3 (GoPro Hero 4). Pineapples are captured in various illumination condition, different sizes and heavy occlusion.

4.2   Implementation Details

The fruit and flower recognition architecture in this research was implemented using PyTorch [13] and was optimized with Adam [8]. The model was trained on a single Nvidia GTX 1080 Ti GPU with 2 images per mini-batch. Each image was resized to 1280×1280. Our model was pre-trained on COCO dataset [9] and then refined on the dataset we collected 40 K iterations. The learning rate was initially set as 0.001, and then was divided by 5 after 20 K iterations. For all experiments, we set $\lambda_1 = \lambda_2 = \lambda_3 = 1$. With all bounding-box coordinates as the output, we used Non-maximum suppression operation to select the most accurate locations. The time cost of refinement for this whole detection network is about 6 h.

[1] We thank John Steemson for permitting image collection in the orchard, Simon White for field help on data collection, and Tham Dong for image annotation.

Table 1. Pineapple fruit and flower recognition results.

	Precision	Recall	mAP	F_1
YOLOv3	0.656	0.741	0.64	0.69

4.3   Results and Discussion

We used four evaluation criteria to assess the performance of the proposed method, including precision, recall, mAP and F_1. The precision and recall are defined as follows:

$$precision = \frac{TP}{TP+FP}, recall = \frac{TP}{TP+FN} \tag{6}$$

where TP is true positive predictions, FP is false positive predictions and FN is false negatives predictions.

By setting IoU as 0.5, mAP can be calculated using precision-recall pairs. We also use the F_1 score which takes consideration of both precision and recall and how well the prediction fits the ground truth:

$$F_1 = \frac{2 \times precision \times recall}{precision + recall} \tag{7}$$

Table 1 shows the prediction results. The average mAP and F_1 score of two classes are 0.64 and 0.69, respectively. Figure 4 visualizes the bounding boxes of the YOLOv3 model on pineapple dataset. From Fig. 4, we can see that YOLOv3 obtains a promising detection performance under various illumination conditions, object sizes, viewpoints and heavy occlusion. When there are other objects with similar color or shape with pineapple, the YOLOv3 algorithm is prone to false detection. For pineapple, the process from flowering to fruiting is gradual. This results in late stage flowers that are very similar to early stage fruits. Therefore, the flowers may be mis-identified as fruits, or vise versa.

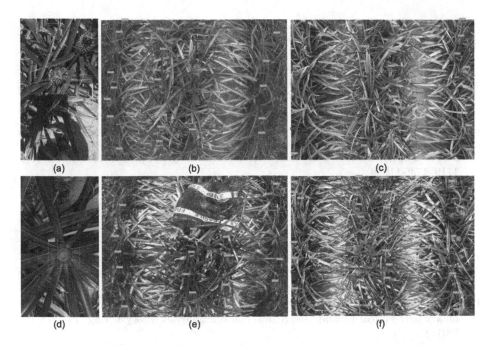

Fig. 4. Example results of pineapple fruit and flower recognition. The red arrows indicate incorrectly predicted bounding boxes. (Color figure online)

5 Conclusion

In this paper, we applied a deep object detection model for pineapple fruit and flowers recognition. To facilitate model training, we collected and annotated a large amount of pineapple images to form a dataset, and then trained the deep network on this dataset. The experimental results show that the proposed model has strong robustness to illumination, object size, viewpoint and occlusion. The similarity of late stage flower and early stage fruit leads to mis-identification, which is the key problem we will study in the future. Through the experimental analysis, we found that the collection of larger dataset and more accurate labeling are conducive to further improve the accuracy.

References

1. Bai, X., Yan, C., Yang, H., Bai, L., Zhou, J., Hancock, E.R.: Adaptive hash retrieval with kernel based similarity. Pattern Recogn. **75**, 136–148 (2018)
2. Bansal, R., Lee, W.S., Satish, S.: Green citrus detection using fast Fourier transform (FFT) leakage. Precis. Agric. **14**(1), 59–70 (2013)
3. Bargoti, S., Underwood, J.P.: Image segmentation for fruit detection and yield estimation in apple orchards. J. Field Rob. **34**(6), 1039–1060 (2017)

4. Dunn, G.M., Martin, S.R.: Yield prediction from digital image analysis: a technique with potential for vineyard assessments prior to harvest. Aust. J. Grape Wine Res. **10**(3), 196–198 (2004)
5. Gongal, A., Amatya, S., Karkee, M., Zhang, Q., Lewis, K.: Sensors and systems for fruit detection and localization: a review. Comput. Electron. Agric. **116**, 8–19 (2015)
6. He, K., Gkioxari, G., Dollár, P., Girshick, R.: Mask R-CNN. In: Proceedings of the IEEE International Conference on Computer Vision, pp. 2961–2969 (2017)
7. Jiménez, A.R., Jain, A.K., Ceres, R., Pons, J.L.: Automatic fruit recognition: a survey and new results using range/attenuation images. Pattern Recogn. **32**(10), 1719–1736 (1999)
8. Kingma, D.P., Ba, J.: Adam: a method for stochastic optimization (2014). arXiv preprint arXiv:1412.6980
9. Lin, T.Y., et al.: Microsoft COCO: common objects in context. In: Fleet, D., Pajdla, T., Schiele, B., Tuytelaars, T. (eds.) ECCV 2014. LNCS, vol. 8693, pp. 740–755. Springer, Cham (2014). https://doi.org/10.1007/978-3-319-10602-1_48
10. Liu, W., et al.: SSD: single shot multibox detector. In: Leibe, B., Matas, J., Sebe, N., Welling, M. (eds.) ECCV 2016. LNCS, vol. 9905, pp. 21–37. Springer, Cham (2016). https://doi.org/10.1007/978-3-319-46448-0_2
11. Lu, J., Lee, W.S., Gan, H., Hu, X.: Immature citrus fruit detection based on local binary pattern feature and hierarchical contour analysis. Biosyst. Eng. **171**, 78–90 (2018)
12. Luo, C., Yu, L., Ren, P.: A vision-aided approach to perching a bioinspired unmanned aerial vehicle. IEEE Trans. Ind. Electron. **65**(5), 3976–3984 (2017)
13. Paszke, A., et al.: Pytorch: an imperative style, high-performance deep learning library. In: Advances in Neural Information Processing Systems, pp. 8024–8035 (2019)
14. Payne, A.B., Walsh, K.B., Subedi, P., Jarvis, D.: Estimation of mango crop yield using image analysis-segmentation method. Comput. Electron. Agric. **91**, 57–64 (2013)
15. Rabatel, G., Guizard, C.: Grape berry calibration by computer vision using elliptical model fitting. In: European Conference on Precision Agriculture, vol. 6, pp. 581–587 (2007)
16. Rahnemoonfar, M., Sheppard, C.: Deep count: fruit counting based on deep simulated learning. Sensors **17**(4), 905 (2017)
17. Redmon, J., Divvala, S., Girshick, R., Farhadi, A.: You only look once: unified, real-time object detection. In: Proceedings of the IEEE Conference on Computer Vision and Pattern Recognition, pp. 779–788 (2016)
18. Ren, S., He, K., Girshick, R., Sun, J.: Faster r-cnn: towards real-time object detection with region proposal networks. In: Advances in Neural Information Processing Systems, pp. 91–99 (2015)
19. Sa, I., Ge, Z., Dayoub, F., Upcroft, B., Perez, T., McCool, C.: Deepfruits: a fruit detection system using deep neural networks. Sensors **16**(8), 1222 (2016)
20. Sengupta, S., Lee, W.S.: Identification and determination of the number of green citrus fruit under different ambient light conditions. In: International Conference of Agricultural Engineering (2012)
21. Song, Y., Glasbey, C., Horgan, G., Polder, G., Dieleman, J., Van der Heijden, G.: Automatic fruit recognition and counting from multiple images. Biosyst. Eng. **118**, 203–215 (2014)

22. Wang, C., Bai, X., Wang, S., Zhou, J., Ren, P.: Multiscale visual attention networks for object detection in VHR remote sensing images. IEEE Geosci. Rem. Sens. Lett. **16**(2), 310–314 (2018)
23. Wang, C., Wang, X., Bai, X., Liu, Y., Zhou, J.: Self-supervised deep homography estimation with invertibility constraints. Pattern Recogn. Lett. **128**, 355–360 (2019)
24. Yan, C., Pang, G., Bai, X., Shen, C., Zhou, J., Hancock, E.: Deep hashing by discriminating hard examples. In: Proceedings of the 27th ACM International Conference on Multimedia, pp. 1535–1542 (2019)
25. Zhou, L., Xiao, B., Liu, X., Zhou, J., Hancock, E.R., et al.: Latent distribution preserving deep subspace clustering. In: 28th International Joint Conference on Artificial Intelligence, New York (2019)

A New Credit Scoring Model Based on Prediction of Signal on Graph

Zhihua Yang[1]⬤, Qian Zhang[2]⬤, Feng Zhou[1]⬤, and Lihua Yang[3](✉)

[1] Information Science School, Guangdong University of Finance and Economics, Guangzhou, China
[2] College of Mathematics and Statistics, Shenzhen University, Shenzhen, China
[3] School of Mathematics, Sun Yat-sen University, Guangzhou, China
mcsylh@mail.sysu.edu.cn

Abstract. Study on high-precision credit scoring model has become a challenging issue. In this paper, a new credit scoring model is proposed. We model the users and their relationship as a weighted undirected graph, and then the credit scoring problem is reduced to a prediction problem of signal on graph. The new model utilizes both the information of the unlabeled samples and the location information of the samples in the feature space, and thus achieves an excellent predictive performance. The experimental results on the open UCI German credit dataset are compared with those of seven classical models that the prediction performance of the proposed model is significantly better than that of the reference models. The Friedman test indicate that the experimental results have a high confidence level.

Keywords: Credit scoring · Classification · Signal processing on graph

1 Introduction

The credit scoring method is to establish the credit scoring model according to some information, and to rate the credit of the loan applicant, so as to assist the lender in making decisions. A good credit scoring model can reduce the cost of lending and the loss of non-performing loans. It is stated that companies could make significant future savings if an improvement of only a fraction of a percent could be made in the accuracy of the credit scoring techniques implemented [1]. Therefore, the study of high-precision credit scoring methods has become a challenging problem in related fields.

The problem of credit scoring is essentially a classification problem. The goal of credit scoring model is to generate classifiers according to the distribution of known good users and bad users in feature space in order to predict the credit of unknown users accurately. In the past few decades, many scholars have put

Supported by National Natural Science Foundations of China (Nos. 11771458, 11431015, 11601346), and Guangdong Province Key grant (No. 2016B030307003).

Y. Lu et al. (Eds.): ICPRAI 2020, LNCS 12068, pp. 228–237, 2020.
https://doi.org/10.1007/978-3-030-59830-3_20

forward many credit scoring models. These models can be divided into two categories: single model and ensemble model. The single model is mainly based on statistical and machine learning methods, including linear discriminant analysis (LDA) [2], logistic regression (LR) [3], decision trees (DT) [4], naive Bayes (NB) [5], neural network (NN) [6], support vector machine (SVM) [7] and random forest (RF) [5] et al. The idea of ensemble models is to combine the results of multiple single classifiers, called a base classifier, to produce a comprehensive score. Therefore, the selection of base classifiers should be as good and different as possible, that is, to improve the accuracy and diversity of base classifiers as much as possible [8]. At present, most ensemble models use logical regression, support vector machine and so on as the base classifiers of ensemble models [3].

From the point of view of machine learning, the existing models are basically based on supervised learning. This means that all unknown samples do not participate in the training of classifiers at all. Recently, semi-supervised learning techniques have made great progress [9]. Researches have found that unlabeled data, when used in conjunction with labeled data, can produce considerable improvement in learning accuracy. This idea motivates us to make use of the information of unknown users to participate the credit scoring prediction.

To take advantage of the information carried by the unknown users, we model the users as well as their relationships as an undirected and weighted graph, $\mathcal{G} = (\mathcal{V}, \mathbf{W})$, in which $\mathcal{V}$ is a set of vertices, one vertex indicates a user and the weight matrix $\mathbf{W}$ represents the connections or similarity between the vertices. With this model, the problem of predicting credit of unknown users becomes a sampling reconstructive problem of signal on the graph. On the other hand, we map a vertex which is represented by a feature vector into a high-dimensional feature space through a feature mapping. According to the manifold assumption [9], the images of all the vertices in the high dimensional feature space do not uniformly fill the entire feature space and are usually distributed on a low dimensional manifold in an implicit and easy-to-classification manner. This means that the signal which is composed of vertex labels is continuous in the feature space, which inspires us to find the reconstruction signal in a reproducing kernel Hilbert space. Finally, the original problem is approximately modeled as a quadratic unconditional optimization problem in a reproducing kernel Hilbert space. The famous Representer Theorem of the Reproducing Kernel Hilbert Space (RKHS) is employed to get the analytical solution. We conduct experiments on the open UCI German credit dataset, and the encouraging results are received.

The rest of the paper is organized as follows: Sect. 2 includes the main contributions of this paper, which discusses the prediction model of signal on graph and its solutions, and gives the new credit scoring model. In Sect. 3, the details of experiment are introduced and the experimental results are exhibited. Finally, a brief conclusion of the paper is given in Sect. 4.

2   The Prediction Model of Signal on Graph

In this section, we discuss how to use the manifold assumption to model the prediction problem as an unconstrained optimization objection, as well as its solving strategies.

2.1   Mathematical Formulation of the Prediction Problem

Let $\mathcal{V}$ be a set of vertices, each of which is described by a two-tuples, i.e., $\mathcal{V} = \{(\mathbf{v}_i, y_i) | \mathbf{v}_i \in \mathcal{R}^m, y_i \in \mathcal{R}, i = 1, \cdots, n\}$, where $\mathbf{v}_i$ is m-dimensional feature vector and y_i is called as the label of the vertex $\mathbf{v}_i$. In our problem, only part of the vertices's label is known. We call these vertices the labeled ones, and others the unlabeled ones. Let $\mathcal{K} = \{(\mathbf{v}_i, y_i), \ i = 1, \cdots, l\} \in \mathcal{V}$ be the set of all the labeled vertices. We want to find a prediction function to estimate the label of the unlabeled vertices. A direct approach of learning the prediction function is to minimize the prediction error on the set of labeled samples, which is formulated as follows:

$$\arg \min_{f \in \mathcal{H}_\kappa} J(f) := \arg \min_{f \in \mathcal{H}_\kappa} \sum_{\mathbf{v}_j \in \mathcal{K}} |f(\mathbf{v}_j) - y_j|^2 + \lambda \|f\|_{\mathcal{H}}^2, \tag{1}$$

where $\mathcal{H}_\kappa$ is the Reproducing Kernel Hilbert Space (RKHS) associated with the kernel κ. It will be the completion of the linear span given by $\kappa(\mathbf{v}_i, \cdot)$ for all $\mathbf{v}_j \in \mathcal{V}$, i.e., $\mathcal{H}_\kappa = span\{\kappa(\mathbf{v}_i, \cdot) | \mathbf{v}_i \in \mathcal{V}\}$.

In the above model, we only use the label data, and the unlabeled data does not participate in the modeling of the prediction function at all. In order to use any of unlabeled data, one must assume some structures for the underlying distribution of data. In this paper, we follow the manifold assumption [9]. We model the vertices as well as their relationships as an undirected and weighted graph, $\mathcal{G} := (\mathcal{V}, \mathbf{W})$, where $\mathbf{W}$ be a matrix of order n whose (i, j)th entry, w_{ij}, is a nonnegative real number, called weight, to measure the affinity between the vertices $\mathbf{v}_i$ and $\mathbf{v}_j$.

According to the manifold assumption, vertices with large weights should have similar values [10]. This means that the function f should be continuous on the graph. This observation induces us to introduce the following continuity measure:

$$\sigma(f) := \frac{1}{2} \sum_{i,j=1}^{n} w_{ij}(f(\mathbf{v}_i) - f(\mathbf{v}_j))^2. \tag{2}$$

Let $\mathbf{L}$ be the Laplacian matrix of the graph defined by $\mathbf{L} := \mathbf{D} - \mathbf{W}$, where $\mathbf{D}$ is the degree matrix of the graph. A simple calculation deduces that the smoothness measurement term $\sigma(f)$ defined by Eq. (2) can be rewritten as [11],

$$\sigma(f) = \mathbf{f}^T \mathbf{L} \mathbf{f}, \quad \text{where } \mathbf{f} := (\mathbf{f}(\mathbf{v_1}), \cdots, \mathbf{f}(\mathbf{v_n}))^T \in \mathbb{R}^n. \tag{3}$$

Combining the smoothness term $\sigma(f)$ with Eq. (1), we obtain the following modified objection:

$$\arg \min_{f \in \mathcal{H}_\kappa} J(f) := \arg \min_{f \in \mathcal{H}_\kappa} \sum_{\mathbf{v}_j \in \mathcal{K}} |f(\mathbf{v}_j) - y_j|^2 + \lambda \|f\|^2 + \gamma \mathbf{f}^T \mathbf{L} \mathbf{f}, \tag{4}$$

where the penalty term $|f(\mathbf{v}_j) - y_j|^2$ is used to ensure that the value of f at the labeled vertex $\mathbf{v}_j$ is approximately equal to the known label, y_j. The regularization term $\mathbf{f}^T \mathbf{L} \mathbf{f}$ makes the solution f be as smooth as possible. The parameter, $\gamma > 0$, controls the importance of the regularization term and the penalty terms.

2.2 Solution of the Optimization Objection Eq. (4)

In order to solve the above optimization problem, we introduce a famous theorem in RKHS, that is the representer theorem [10]:

Theorem 21 (Representer Theorem). *Let $\mathcal{V}$ be a nonempty set and κ a reproducing kernel on $\mathcal{V} \times \mathcal{V}$ with corresponding reproducing kernel Hilbert space $\mathcal{H}$. Given a training sample set including both labeled and unlabeled samples. Without loss of generality, let its first l samples be labeled ones, i.e., $\{(\mathbf{v}_1, y_1), \cdots, (\mathbf{v}_l, y_l)\} \in \mathcal{V} \times \mathcal{R}$. Then for any function $c : (\mathcal{V} \times \mathcal{R})^l \to \mathcal{R} \cup \{\infty\}$ and monotonic increasing function $\Omega : [0, \infty) \to \mathcal{R}$, each minimizer $f \in \mathcal{H}$ of the regularized risk functional*

$$c\left((y_1, f(\mathbf{v}_1)), \cdots, (y_l, f(\mathbf{v}_l))\right) + \Omega\left(\|f\|_{\mathcal{H}}\right) + \gamma \mathbf{f}^T \mathbf{L} \mathbf{f} \tag{5}$$

admits an expansion

$$f(\mathbf{v}) = \sum_{i=1}^{n} a_i \kappa(\mathbf{v}_i, \mathbf{v}) \tag{6}$$

in terms of both labeled and unlabeled samples.

Denoting

$$\mathbf{K} = \begin{bmatrix} \kappa(\mathbf{v}_1, \mathbf{v}_1) & \cdots & \kappa(\mathbf{v}_1, \mathbf{v}_n) \\ \vdots & & \vdots \\ \kappa(\mathbf{v}_n, \mathbf{v}_1) & \cdots & \kappa(\mathbf{v}_n, \mathbf{v}_n) \end{bmatrix} \tag{7}$$

as the kernel gram matrix and

$$\mathbf{f} = \begin{bmatrix} f(\mathbf{v}_1) \\ \vdots \\ f(\mathbf{v}_n) \end{bmatrix}, \quad \mathbf{a} = \begin{bmatrix} a_1 \\ \vdots \\ a_n \end{bmatrix}. \tag{8}$$

then the solution of Eq. (4) can be written as

$$\mathbf{f} = \mathbf{K}\mathbf{a}. \tag{9}$$

Without loss of generality, we assume that $\{\mathbf{v}_j\}_{j \in J}$ for $J := \{1, \cdots, l\}$ to be the labelled vertices and $\{y_j\}_{j \in J}$ to be the corresponding labels. Let $\mathbf{y}_J := (y_1, \cdots, y_l)^T$ and $\mathbf{K}_J$ the sub-matrix consisting of the first l rows of $\mathbf{K}$. It is easy to see that $\mathbf{K}_J \mathbf{a} - \mathbf{y}_J$ is a l-dimensional vector whose ith element is

$$(\mathbf{K}_J \mathbf{a} - \mathbf{y}_J)_i = \sum_{j=1}^{n} \kappa(\mathbf{v}_i, \mathbf{v}_j) a_j - y_i = f(\mathbf{v}_i) - y_i, \ i \in J.$$

Thus the Euclidean norm of $\mathbf{K}_J\mathbf{a} - \mathbf{y}_J$ in $\mathbb{R}^l$ is

$$\|\mathbf{K}_J\mathbf{a} - \mathbf{y}_J\|_2^2 = \sum_{i \in J} |f(\mathbf{v}_i) - y_i|^2. \tag{10}$$

On the other hand, according to [10], the norm of f in $\mathcal{H}$ is

$$\|f\|_{\mathcal{H}}^2 = \sum_{i=1}^{n} \sum_{j=1}^{n} a_i a_j K(\mathbf{v}_i, \mathbf{v}_j) = \mathbf{a}^T \mathbf{K} \mathbf{a}. \tag{11}$$

According to Eq. (9), Eq. (10) and Eq. (11), the optimization problem Eq. (4) can be reduced to the following equivalent problem

$$\min_{f \in \mathcal{H}} J(f) := \|\mathbf{K}_J\mathbf{a} - \mathbf{y}_J\|_2^2 + \gamma \mathbf{a}^T \mathbf{K}^T \mathbf{L} \mathbf{K} \mathbf{a} + \lambda \mathbf{a}^T \mathbf{K} \mathbf{a}. \tag{12}$$

Using $\|\mathbf{K}_J\mathbf{a} - \mathbf{y}_J\|_2^2 = (\mathbf{K}_J\mathbf{a} - \mathbf{y}_J)^T(\mathbf{K}_J\mathbf{a} - \mathbf{y}_J)$, the objective function of Eq. (12) can be further simplified in form of

$$\begin{aligned} J(f) = \mathbf{a}^T \left(\mathbf{K}_J^T \mathbf{K}_J + \gamma \mathbf{K}^T \mathbf{L} \mathbf{K} + \lambda \mathbf{K} \right) \mathbf{a} \\ - 2(\mathbf{K}_J^T \mathbf{y}_J)^T \mathbf{a} + \|\mathbf{y}_J\|_2^2. \end{aligned} \tag{13}$$

Since $\mathbf{L}$ is a semi-positive matrix, if $\mathbf{K}$ is positive definite and $\lambda > 0$, then $\mathbf{K}_J^T \mathbf{K}_J + \gamma \mathbf{K}^T \mathbf{L} \mathbf{K} + \lambda \mathbf{K}$ is positive definite. Therefore, $J(f)$ is the convex quadratic function of $\mathbf{a}$ which implies that the solution of Eq. (12) exists uniquely and can be expressed analytically as:

$$\mathbf{a} = \left(\mathbf{K}_J^T \mathbf{K}_J + \gamma \mathbf{K}^T \mathbf{L} \mathbf{K} + \lambda \mathbf{K} \right)^{-1} \mathbf{K}_J^T \mathbf{y}_J. \tag{14}$$

Once $\mathbf{a}$ has been obtained, we can get the optimal predicted values on all vertices from Eq. (9).

2.3   A New Credit Prediction Algorithm

In this section, we will discuss the application of the prediction model on the graph to the problem of credit scoring. We model a user as a vertex of a weighted undirected graph, in which the user information constitutes the vertex's feature vector and the near neighbor relation of the feature vectors are used to establish the graph Laplace matrix. Therefore, this problem that predicting the unknown users's credit scores by some known credit scores is essentially a prediction problem of signal on graph. Figure 1 shows the diagram of the proposed credit scoring model.

In the new model, the weight matrix $\mathbf{W}$ plays a crucial role. Its construction has become an important issue in the signal processing on graphs [12]. So far, there are two strategies: The first one is by means of prior domain knowledge [13]; The second one is learning from observed data [14]. Since learning methods usually require a lot of computation, this paper adopts the first strategy, i.e., creates a k-nearest neighbors graph from users's positional data in feature space.

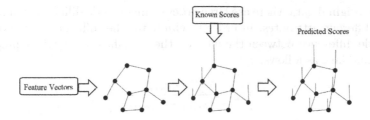

Fig. 1. Diagram of the credit scoring model.

The next issue is to choose the kernel function $\kappa : \mathcal{V} \times \mathcal{V} \to \mathbb{R}$, or equivalently matrix $\mathbf{K} = (\kappa(\mathbf{v}_i, \mathbf{v}_j))_{i,j \in \mathcal{V}}$, where $\mathbf{v}_i$ and $\mathbf{v}_j$ are the feature vectors of the users i and j. In this paper, we choose the most commonly used Gaussian kernel, that is,

$$\kappa(\mathbf{v}_i, \mathbf{v}_j) = \exp(-\|\mathbf{v}_i - \mathbf{v}_j\|^2/\sigma^2), \tag{15}$$

with the constant parameter σ.

Now, we propose the following prediction algorithm.

Algorithm 21. *Given matrix $\mathbf{V}_{n \times m}$ which consists of the feature vectors of all users and whose first l rows indicate the known users, $\mathbf{y}_J = \mathbf{1}_{l \times 1}$ is the labels of the known users, γ, λ are the regularization parameters.*

1. *Compute the adjacent matrix $\mathbf{W}$ and Laplacian matrix $\mathbf{L}$ to form an undirected and weighted graph $\mathcal{G} := \{\mathcal{V}, \mathbf{W}\}$, where $\mathcal{V} := \{\mathbf{v}_1, \cdots, \mathbf{v}_n\}$ is the set of vertices.*
2. *Compute the kernel matrix $\mathbf{K}$ by Eq. (7), and (15).*
3. *Pick out the first l rows of $\mathbf{K}$ to form the sub-matrix $\mathbf{K}_J$.*
4. *Compute $\mathbf{A} := \mathbf{K}_J^T \mathbf{K}_J + \gamma \mathbf{K}^T \mathbf{L} \mathbf{K} + \lambda \mathbf{K}$;*
5. *Solve $\mathbf{A}\mathbf{a} = \mathbf{K}_J^T \mathbf{y}_J$ to get $\mathbf{a}$ and the compute $\mathbf{f} = \mathbf{K}\mathbf{a}$.*

The last $n - l$ items are the predicted labels of the unknown users.

3 Experimental Results and Analysis

In this section, experimental results are presented to demonstrate the superiority of the proposed method. The experiments are conducted on the famous UCI German Credit dataset which is publicly available at the UCI repository (http://kdd.ics.uci.edu/). This dataset contains 1000 instances, of which 70% are good users and 30% are bad users. It provides two files: the file "german.data" is the original dataset which includes 20 attributes (7 numerical, 13 categorical), and the file "german.data-numeric" is the numerical version, in which the original attributes have been edited and several indicator variables have been added to form the final 24 numerical attributes to make it suitable for algorithms. In the experiments, we adopt the numerical version.

In the original data, there are large order of magnitude differences among the data of different attributes. In order to eliminate the influence of the order of magnitude difference between the data on the classification, we first preprocess the original data as follows:

$$v_{i,k} = \frac{v_{i,k} - v_{kmax}}{v_{kmax} - v_{kmin}}$$

where $v_{i,k}$ indicates the ith user's the kth attribute, v_{kmax} and v_{kmin} are the maximum and the minimum of the kth attribute of all the users, respectively.

We chose LDA, LR, DT, NB, NN, SVM and RF as the reference models to compare the performance of the proposed model. Since the area under the receiver operating characteristic curve (AUC) is independent of the threshold and the error, we use it as the performance metric. The value of AUC is $0 \sim 1$, and the greater the value, the better the performance of the model.

At first, we conduct 5-folds cross-validation on the original data set to find the good parameters. The final parameters are selected as follows: The number of nearest neighbors is $20, \gamma = 0.0001, \lambda = 0.5$, the σ in Eq. (15) is set to 10. Seven MATLAB library functions that are fitcdiscr, fitglm, fitctree, fitcnb, feedforwardnet, fitcsvm and TreeBagger which correspond to seven reference models are called with 10 hidden nodes for feedforwardnet, 90 trees for TreeBagger and the default options for other functions.

We use Friedman's test [16] to compare the AUCs of the different models. The Friedman test statistic is based on the average ranked (AR) performances of the models on each data set, and is calculated as follows:

$$\chi_F^2 = \frac{12D}{K(K+1)} \left[\sum_{j=1}^{K} AR_j^2 - \frac{K(K+1)^2}{4} \right], \tag{16}$$

where $AR_j = \frac{1}{D} \sum_{i=1}^{D} r_i^j$, D denotes the number of data sets used in the study, K is the total number of models and r_i^j is the rank of model j on data set i. χ_F^2 is distributed according to the Chi-square distribution with $K - 1$ degrees of freedom. If the value of χ_F^2 is large enough, then the null hypothesis that there is no difference between the models can be rejected.

The 5-folds cross-validation on the original data set is done by using the above parameters. The results are list in Table 1, in which the AUCs of all eight models on each fold are listed. The ARs for the different models and the Friedman test statistic are also shown. The model achieving the highest AUC on each fold is underlined as well as the overall highest ranked model. Table 1 shows that the proposed model has the highest Friedman score (average rank (AR)), and the average of AUCs on five folds wins out over all the benchmark models.

In order for the percentage reduction in the bad observations, we under-sampling the bad observations that only $233, 175, 124, 78, 37, 18$ ones from a total of 300 bad observations have been used to give $75/25, 80/20, 85/15, 90/10, 95/5, 97.5/2.5$ class distributions. For each split, the

Table 1. Area under the receiver operating characteristic curve (AUC) results.

Model	Friedman test statistic: $\chi_F^2 = 30.87, Prob > \chi_F^2 : 6.579e^{-05}$						
	Fold 1	Fold 2	Fold 3	Fold 4	Fold 5	Avg.	AR
Graph	0.7796	0.8089	0.8515	0.8394	0.8523	0.8263	1.0
LDA	0.7643	0.8014	0.8238	0.8177	0.8345	0.8084	3.0
LR	0.7649	0.8014	0.8255	0.8121	0.8303	0.8068	3.2
DT	0.6395	0.7502	0.7234	0.7025	0.7139	0.7059	7.8
NB	0.7412	0.7685	0.7989	0.7875	0.7641	0.7721	6.2
NN	0.7407	0.7027	0.7947	0.7971	0.7999	0.7670	6.8
SVM	0.7723	0.8085	0.8157	0.8101	0.8330	0.8079	3.2
RF	0.7439	0.7935	0.7987	0.8161	0.8296	0.7964	4.8

Table 2. Average of AUCs on five folds for the different percentage class splits.

Model	Percentage class splits						
	70/30	75/25	80/20	85/15	90/10	95/5	97.5/2.5
Graph	0.8263	0.7894	0.7817	0.7601	0.7517	0.7758	0.6511
LDA	0.8084	0.7837	0.7865	0.7595	0.7272	0.7820	0.7407
LR	0.8068	0.7840	0.7816	0.7569	0.7307	0.7811	0.6588
DT	0.7059	0.6591	0.6328	0.6483	0.5449	0.5251	0.4993
NB	0.7721	0.7536	0.7600	0.7362	0.7047	0.7482	0.6908
NN	0.7670	0.7295	0.7443	0.6642	0.6677	0.5636	0.7573
SVM	0.8079	0.7788	0.7738	0.6971	0.6417	0.6616	0.5386
RF	0.7964	0.7690	0.7555	0.7353	0.6789	0.6991	0.5216

Table 3. ARs on the different percentage class splits.

Model	Percentage class splits						
	70/30	75/25	80/20	85/15	90/10	95/5	97.5/2.5
Graph	1.0	2.6	3.2	2.6	1.6	3.6	3.4
LDA	3.0	3.2	2.4	3.0	3.4	2.6	3.0
LR	3.2	3.2	3.4	2.8	2.2	2.2	4.4
DT	7.8	7.8	8.0	6.0	8.0	7.4	6.2
NB	6.2	5.6	4.4	4.0	4.6	3.8	4.0
NN	6.8	5.8	5.2	7.0	5.2	7.0	3.0
SVM	3.2	3.2	4.0	6.0	6.2	5.2	5.8
RF	4.8	4.6	5.4	4.6	4.8	4.2	6.2
χ_F^2	30.87	18.73	17.6	16.17	25.53	21.2	10.58
$Prob > \chi_F^2$	$6.579e^{-05}$	0.0091	0.0139	0.0212	0.0006	0.0035	0.1578

5-folds cross-validation is conducted, AR and the average of AUCs are calculated, the results are displayed in Table 2 and Table 3, in which the best results have been underlined. It shows that the proposed model is superior to the benchmark models on almost all the class splits. The averages of AUC are larger and the ARs are less than those of the benchmark models.

4 Conclusion

In this paper, a new credit scoring model is proposed. Unlike the traditional methods, first, the proposed model makes full use of the information of unknown users to participate in the decision, while in the traditional methods, they are completely abandoned; Second, the solution of the problem is restricted to a suitable RKHS so that the location information of the samples in the feature space can be fully utilized. These form the main innovations of this paper. Based on the above idea, a new prediction algorithm is given. Experiments confirm these novel strategies have generated a very high decisional accuracy. We carry out experiments on the open UCI German credit dataset and compare them with seven classical models. The results show that the prediction performance of the proposed model is obviously better than that of reference models.

References

1. Henley, W.E., Hand, D.J.: Construction of a k-nearest neighbor credit-scoring system. IMA J. Manag. Math. **8**(4), 305–321 (1997)
2. Baesens, B., Van Gestel, T., Viaene, S., Stepanova, M., Suykens, J., Vanthienen, J.: Benchmarking state-of-the-art classification algorithms for credit scoring. J. Oper. Res. Soc. **54**(6), 627–635 (2003)
3. Brown, I., Mues, C.: An experimental comparison of classification algorithms for imbalanced credit scoring datasets. Expert Syst. Appl. **39**(3), 3446–3453 (2012)
4. Arminger, G., Enache, D., Bonne, T.: Analyzing credit risk data: a comparison of logistic discrimination, classification tree analysis, and feed forward networks. Comput. Stat. **12**, 293–310 (1997)
5. Lessmann, S., Baesens, B., Seow, H.V.: Benchmarking state-of-the-art classification algorithms for credit scoring: an update of research. Eur. J. Oper. Res. **247**(1), 1–32 (2015)
6. West, D.: Neural network credit scoring models. Comput. Oper. Res. **27**(11–12), 1131–1152 (2002)
7. Baesens, B., Vanthienen, J.: Benchmarking state-of-the-art classification algorithms for credit scoring. J. Oper. Res. Soc. **54**(6), 627–635 (2003)
8. Nani, L., Lumini, A.: An experimental comparison of ensemble classifiers for bankruptcy prediction and credit scoring. Expert Syst. Appl. **36**(2), 3028–3033 (2009)
9. Chapelle, O., Schölkopf, B., Zien, A.: Semi-supervised Learning. MIT Press, Cambridge (2013)
10. Belkin, M., Niyogi, P., Sindhwani, V.: Manifold regularization: a geometric framework for learning from examples. J. Mach. Learn. Res. **7**, 2399–2434 (2006)

11. Liu, X., Zhai, D., Zhao, D., Zhai, G., Gao, W.: Progressive image denoising through hybrid graph Laplacian regularization: a unified framework. IEEE Trans. Image Process. **23**(4), 1491–1503 (2014)
12. Shuman, D., Narang, S., Frossard, P., Ortega, A., Vandergheynst, P.: The emerging field of signal processing on graphs: extending high- dimensional data analysis to networks and other irregular domains. IEEE Signal Process. Mag. **30**(3), 83–98 (2013)
13. Sandryhaila, A., Moura, J.M.F.: Discrete signal processing on graphs. IEEE Trans. Signal Process. **61**(7), 1644–1656 (2013)
14. Dong, X., Thanou, D., Frossard, P., Vandergheynst, P.: Laplacian matrix learning for smooth graph signal representation. In: 2015 IEEE International Conference on Acoustics, Speech and Signal Processing (ICASSP), pp. 3736–3740 (2015)
15. Bradley, A.P.: The use of the area under the ROC curve in the evaluation of machine learning algorithms. Pattern Recogn. **30**(7), 1145–1159 (1997)
16. Friedman, M.: A comparison of alternative tests of significance for the problem of m rankings. Ann. Math. Stat. **11**(1), 86–92 (1940)

Characterizing the Impact of Using Features Extracted from Pre-trained Models on the Quality of Video Captioning Sequence-to-Sequence Models

Menatallh Hammad⑩, May Hammad⑩, and Mohamed Elshenawy⁽✉⁾⑩

Zewail City of Science and Technology, Giza, Egypt
{s-menah.hammad,s-maymahmod,melshenawy}@zewailcity.edu.eg

Abstract. The task of video captioning, that is, the automatic generation of sentences describing a sequence of actions in a video, has attracted an increasing attention recently. The complex and high-dimensional representation of video data make it difficult for a typical encoder-decoder architectures to recognize relevant features and encode them in a proper format. Video data contains different modalities that can be recognized using a combination of image, scene, action and audio features. In this paper, we characterize the different features affecting video descriptions and explore the interactions among these features and how they affect the final quality of a video representation. Built on existing encoder-decoder models that utilize limited range of video information, our comparisons show how the inclusion of multi-modal video features can make a significant effect on improving the quality of generated statements. The work is of special interest to scientists and practitioners who are using sequence-to-sequence models to generate video captions.

Keywords: Neural networks · Feature extraction · Machine intelligence · Pattern analysis · Video signal processing.

1 Introduction

The problem of generating accurate descriptions for videos has received growing interest from researchers in the past few years [1–4]. The different modalities contained in a video scene and the need to generate coherent and descriptive multi-sentence descriptions have imposed several challenges on the design of effective video captioning models. Early attempts focus on representing videos as a stack of images frames such as in [5] and [6]. Such representation, however, ignores the multi-modal nature of video scenes and has a limited ability to generate accurate descriptions that describe temporal dependencies in video sequences. Also, it ignores some important video features such as activity flow,

Supported by Zewail City of Science and Technology, Giza, Egypt.

ⓒ Springer Nature Switzerland AG 2020
Y. Lu et al. (Eds.): ICPRAI 2020, LNCS 12068, pp. 238–250, 2020.
https://doi.org/10.1007/978-3-030-59830-3_21

sound and emotions. This problem was addressed in more recent approaches [7] and [8], in which videos are represented using multi-modal features that are capable of capturing more detailed information about scenes and their temporal dynamics.

While multi-modal approaches have shown promising results improving the efficiency of the video captioning techniques in general [9,12], their implementation has several challenges that hamper their use in video captioning applications. Challenges include:

- Choosing the right combination of features describing a scene: Humans have the ability to extract the most relevant information describing a movie scene. Such information describes the salient moments of a scene rather than regular dynamics (e.g. regular crowd motion in a busy street). The ability of a machine learning model to make such distinction is limited by the proper choice of features, whether 2D or 3D, that encodes the activity of each object in a scene.
- Choosing the right concatenation of features: presented models adopt several different concatenation techniques to merge various combinations of input feature vectors and feed them to the network as a single representation of a scene. These methods, however, do not provide detailed information about the rationale for choosing such methods and the impact of choosing the wrong concatenation method on the overall accuracy of the model.

In this paper, we study the impact of using different combinations of features from pre-trained models on the quality of video captioning model. We base our model on the state-of-the-art S2VT-encoder-decoder architecture presented in [11]. While the original S2VT model uses 2D RGB visual features extracted from VGG-19 network as the only descriptors for the input video frames, we compared the performance of the model using different combinations of rich video features that describe various modalities such as motion dynamics, events and sound. Additionally, We have added an attention-based model to improve the overall quality of the decoder. All video features are extracted using pre-trained models and saved in a combined representation as to decrease the complexity of computations at training time. The encoder network was fed the combine video frame feature vectors and used to construct corresponding representation. This representation is fed to the decoder network to construct the text.

2 Related Work

Video captioning methods rely on two basic approaches to generate video descriptions: template-based and sequencing models [13]. The first approach, template-based models, generate video descriptions using a set of predefined rules and templates that construct natural language sentences [14,15]. The visual features of the video are identified and utilized to generate semantic representations that can be encoded and utilized withing the predefined templates. While template-based models generally produce more robust video descriptions, they have limited ability to represent semantically-rich domains which incorporate large number of entities, sophisticated structures and complex relationships. The effort to construct rules and templates for such domains is prohibitively expensive [14].

The second approach is to construct sequence to sequence models that map video features representation into a chain of words. Inspired by recent advancements in computer vision and neural machine translation techniques, these models employ an encoder-decoder framework in which various sets of video features, typically extracted from pre-trained 2D or 3D convolutional neural networks (CNN), are fed to a neural network model. These sets are encoded to create a representation of the video content, and then decoded to produce a sequence of words summarizing this representation. For example, Venugopalan et al. [11], presented a long short-term memory (LSTM) model that encodes visual features from video frames to generate video descriptions. At the encoding stage, the model combines features generated from a CNN at frame level and uses them to generate a feature vector representing the video content. The resulting vector is then fed to an LSTM decoder, at the decoding stage, to generate the corresponding video caption.

Recent research efforts have focused on examining different video features to improve the accuracy of sequencing video captioning models. Yao et al. [16] focus on using a spatial temporal 3-D convolutional neural network (3-D CNN) to capture short temporal dynamics within videos. Their approach utilizes an attention mechanism that gives different weights to video frames during the decoding stage. In addition visual features, audio features have also been investigated to improve the accuracy of the sequencing models. Ramanishka et al. [8] proposed a multi-modal model approach which incorporates visual features, audio features and video topic to produce a video representations. Their model uses Mel Frequency Cepstral Coefficients (MFCCs) of the audio stream to represent the audio features. Jin et al. [17] also propose a multi-modal approach which combines different video modalities including visual, audio, and meta-data. The multi-modal approach was also employed by Long et al. [18] and Hori et al. [19], who utilize attention mechanisms to extract the most salient visual and audio features and use them to generate video descriptions.

Although the above-mentioned model have demonstrated effective methods to combine multi-modal video features, there is a lack of studies that explores which visual, audio and scene features can represent the video content better. In this paper, we focus on features extracted from pre-trained networks providing a comparative assessment of the impact of choosing these features on the overall quality of the model.

3   Approach

As mentioned earlier, the model used in this paper was based on the S2VT-encoder-decoder architecture [11]. Training the model involves the following four stages:

- The first stage involves the conversion of input video frames into high level descriptive feature vectors describing the video content. Generated feature vectors are used as input to the second phase.
- The generated feature vectors from the first stage is then projected using linear embedding onto a 512-sized feature vector. Input videos are sampled into 40 evenly-spaced frames, resulting in a concatenated vector of size (40×512).
- An encoding stage, in which the concatenated $[40 \times 512]$ vector is encoded into a latent representation describing the given video scene.
- A decoding stage, in which the representation generated from the encoder is decoded to produce a chain of words representing the video description.

In the following subsection, we describe these stages in more details providing information of how the basic architecture was extended to test different modalities of the video scene. We start, in subsection A, by discussing the details of the S2VT Model used in our experiments highlighting its underlying operation principles. Then, we discuss the type of information and selected input frame features that were fed to the S2VT model.

3.1   Basic Operation of the S2VT Model

The S2VT or Sequence to Sequence - Video to Text model [11] is a sequence to sequence encoder-decoder model. The model takes, as input, a sequence of video frames $(x1, ..., xn)$, and produces a sequence of words $(y1, ..., ym)$. Both input and output are allowed to have variable sizes. The model tries to maximize the conditional probability of producing output sentence give the input video frames, that is:

$$p(y1, ..., ym|x1, ..., xn)$$

To do so, the S2VT model relies on using an LSTM model for both the encoding and decoding stage. In our experiments, we used a stack of two LSTMs with 512 hidden units each, where the first layer produces the video representation, and the second layer produces the output word sequence.

Similar to original design, which is explained in more detail in [11], the top LSTM layer receives a sequence of input video features and encodes them to produce a latent representation of the video content (h_t). No loss is calculated in the encoding stage. When all the frames are encoded, the second LSTM is fed the beginning-of-sentence tag BOS, which triggers it to start decoding and producing the corresponding sequence of words.

During the decoding stage, the model maximizes the log-likelihood of the predicted output word given the hidden representation produced by the encoder, and given the previous words produced. That is, produced words rely on the hidden representation and previous generated words.

For a model with parameters θ, the maximization equation can be expressed as:

$$\theta^* = \arg\max_{\theta}(\sum_{t=1}^{m} \log(p(y_t|h_{n+t-1}, y_{t-1}; \theta)))$$

Optimization occurs over the entire dataset using gradient descent. During optimization, the loss is back-propagated to allow for a better latent representation (h_t). An end-of-statement tag EOS is used to indicate that the sentence is terminated.

The original S2VT model represents input Videos using 2D object recognition features obtained using a pre-trained CNN. These features were tested using Microsoft Video Description dataset (MSVD) [32], which contains 1970 youtube videos, and achieved 29.2% using the METEOR score. This paper investigates the impact of several other features, using a broader dataset as we will discuss in the upcoming sections.

3.2 Selected Multi-modal Input Video Frame Features

In our experiments, we investigate the impact of the following features extracted from pre-trained models. We examined five types of information and their impact on the quality of the S2VT model. The dimensions of these feature vectors are shown in Table 1. Extracted features are:

- 2D object recognition Features: As the name indicates, these features rely on the 2D characteristics of the video frames. They act to provide information about different objects in the scene and how the existence/ non-existence of objects impact the video description. To get these features, we used pre-trained ResNext-101(32*16d) model [20]. The ResNext models are pre-trained in weakly-supervised fashion on 940 million public images with 1.5 K hashtags matching with 1000 ImageNet1K synsets.
- 2D object recognition Intermediate Features: Intermediate features provides more abstract information about the different objects in a frame and how they are related. These features are applied in near duplicate video retrieval [25], and there are generally used to provide more comprehensive information about the objects in each frame.
- Scene Recognition Features: These features are used to present a descriptive view of each video scene. Scene information was provided using a ResNet50 network [21] trained on Places365 dataset [22].
- 3D Action Recognition Features: These features are used to provide information about the activities of the different objects detected in the movie scene. We used the kinetics-i3d model [23], which is a CNN model that is pre-trained on Kinetics dataset (contains around 700 action classes).
- Deep Sound Features: Deep Sound features are used to provide information about the different sounds in the video to use it in generating descriptions. In our experiments, we used SoundNet proposed in [24], which is used to recognize objects and scenes from sounds. The network is pre-trained using the visual features from over 2,000,000 videos. It uses a student-teacher training procedure that transfers visual knowledge from visual models such as ImageNet into sound modality. Therefore, it helps associate sounds with scene and object recognition tasks. We extracted sound features from (conv7) layer in

Table 1. Dimensions of the selected feature vectors

Feature type	Feature vector dimension
2D object recognition Features	2048
2D object recognition Intermediate Features	4096
Scene Recognition Features	2048
3D Action Recognition Features	1024
Deep Sound Features	1024
place365, I3D, ResNext101	5120
place365, I3D, ResNext101, SoundNet	6144
place365, I3D, ResNext101, SoundNet, Intermediate	10240

the 8 layer Soundnet model. Features extracted from this layer were reported in [24] to produce the best classification accuracy when used in a Multi-Modal Recognition setting.

4   Experimental Setup

4.1   Datasets

The model was trained and evaluated using the Microsoft Research Video to Text (MSR-VTT) [26]. MSR-VTT is a video description dataset that includes 10,000 video clips divided into 20 categories of video content such as music, sports, gaming, and news. Each clip is annotated with 20 natural language sentences, which provides a rich set for training and validation. Compared with the MSVD dataset using to validate the original S2VT model, the dataset includes more videos and a broader variety of video descriptions. Examples of the dataset are shown in Fig. 2. When training the model, we followed the original split in MSR-VTT, we split the data according to 65%:30%:5%, for the training, testing and validation sets respectively, which corresponds to 6,513, 2,990 and 497 clips for each set. 8,810 videos out of the 10000 clips has audio content. The audio content of these silent clips is replaced by zeros (Fig. 1).

4.2   Implementation

Feature Extraction As discussed earlier, we extract five types of video features: the 2D object recognition visual features using the ResNext101 ($32 \times 16D$) pre-trained network; 2D object recognition Intermediate Features using a pre-trained VGG-19 network; scene recognition features using a ResNet50 network trained on Places365 dataset; action recognition features using I3D pertained network; and deep sound features using the pre-trained SoundNet. These features are used, as we will discuss in the next section, to conduct several comparisons that characterize the impact of each video information type on the accuracy

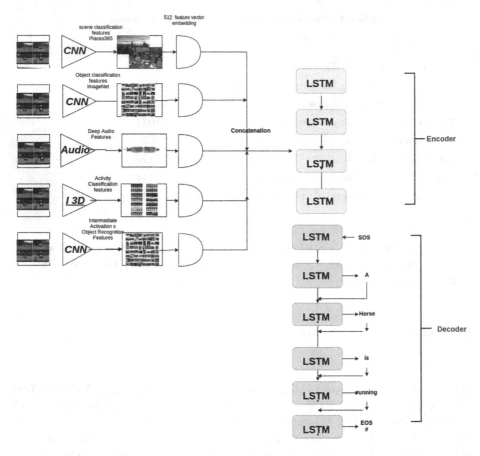

Fig. 1. The figure describes our proposed approach for the video captioning task. It extends the S2VT state of the art model and exploits a multi modal video representation to best understand the video content. The input video features are encoded using LSTM encoder, then accordingly the encoder hidden states are inputs to the decoder that generates a caption that summarizes the representation. We used a combination of the 2D object features, scene features, activity features and sound features extracted from the video as an input to the network.

of the model. The following paragraphs provide more details about the feature extraction process:

– The 2D object recognition visual features: Input videos are sampled into 40 evenly-spaced frames, if the video instance contains less than 40 frames, we pad zero frames at the end of original frames. The frames are fed to the pre-trained ResNext101 (32 × 16d) network.Features are extracted directly from the last fully connected layer (FC7) following the maxpool layer. We followed the pre-processing of inputs for ResNxt 32 × 16d. The height and width of each frame of the video were resized to 256, to then use the central crop patch

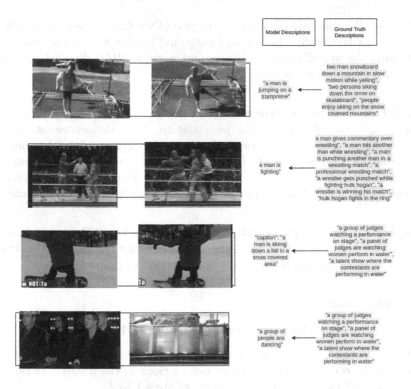

Fig. 2. Snapshots from the dataset

of 224×224 pixels of each frame. A normalisation of the pixel using mean and standard deviation (mean $= [0.485,\ 0.456,\ 0.406]$, std $= [0.229,\ 0.224,\ 0.225]$) are applied.

- 2D object recognition Intermediate Features: Similar to [25], a sampling rate of 1 frame per second was used, and features are extracted using VGG-16 from the intermediate convolutional layers. To generate a single frame descriptor, the Maximum Activation of Convolutions (MAC) function is applied on the activation of each layer before concatenating them to produce a single frame descriptor.
- Scene Recognition Features: we extracted the scene description from the 2048 dimensional feature vector from last fully connected layer after the mean-pool layer of the Resnet50.
- 3D Action Recognition Features: As discussed earlier, we utilized the pre-trained I3D model initially trained on Kinetics dataset for human action recognition. We extracted the 3D motion and activity features from the 1024 dimensional vector form fully-connected layer (FC6). The resulting extracted feature size is 1024 per video frame.
- Audio features: For extracting the associated audio features we used the state-of-the-art SoundNet architecture which represents a CNN pretrained on AudioSet. We split the videos into both audio wav files and video content.

The audio files are used as input to the SoundNet model, then the output feature vector is extracted from 24th layer(conv7) of the 8 layer Soundnet which contains the model's sound features and provides 1024 dimensional feature vector for the input audio file.

Description Sentences Processing. We started by converting all sentences to lower case and removing all the punctuation. Then the sentences were split up using blank spaces to generate the chain of words used in the modelling procedure. Both *BOS* and *EOS* tags were added to the begging and end of the annotation sentence with a max length of 28 words. To input words, we employ a 512 dimensional embedding layer that converts words to vector representations.

Optimization and Regularization Parameters. We utilized an Adam optimizer with a learning rate of 10^{-4} and batch size of 128. An LSTM's dropout percentage of 0.2 was employed as a regularization technique.

4.3 Evaluation

We used four main metrics to evaluate the quality of the sequence model:

- BLEU score [27], which is a standard technique to compare an output statement to its corresponding reference descriptions in the dataset. In our analysis, we used four n-gram BLEU scores of sizes 1 to 4.
- METEOR score [28], which, in contrast to the BLEU score that uses precision-based metrics, emphasizes recall in addition to precision when comparing generated statements to their reference descriptions.
- CIDEr [29], which is tailored for image description tasks and uses human consensus to evaluate generated description. In our analysis, we used the standard CIDEr-D version, which is available as part of the Microsoft COCO evaluation server [30].
- ROUGE-L [31], which is tailored to evaluate summarization tasks. The metric rely on assessing the number of overlapping units between the statements generated by the model and reference statements. These units include n-gram, word sequences, and word pairs.

5   Results and Analysis

We conducted several experiments using the five video features discussed earlier. The results are shown in Table 2. In the first set of experiments, we tested the impact of using each feature type when it is used separately to train the model. We found that the 2D object recognition visual features acquired using the pre-traing ResNext101 network achieved the best results compared to other feature types in BLEU1, BLEU2, BLEU3, BLUE4, METEOR, and CIDEr scores. The 3D action recognition features, acquired form the pre-trained I3D model, achieved the best score in ROUGE-L score.

Table 2. Results

FEATURES	BLEU1	BLEU2	BLEU3	BLEU4	METE-OR	CIDEr	ROUGE-L
ResNext50 (place365)	0.77132	0.61591	0.47293	0.34740	0.26163	0.38299	0.57909
ResNext101	0.79972	0.65690	0.51847	0.39288	0.27851	0.59977	0.47940
SoundNet features	0.68704	0.50683	0.37306	0.25742	0.20562	0.51485	0.13247
Intermediate features	0.76530	0.61260	0.47171	0.35078	0.26223	0.37009	0.57675
I3D (Kinetics)	0.79185	0.63785	0.49557	0.37054	0.27282	0.41092	0.59103
place365, I3D, ResNext101	0.81020	0.66874	0.53141	0.40756	0.28018	0.46732	0.60699
ResNext101, SoundNet, place365, I3D	0.80441	0.65539	0.51480	0.39106	0.28262	0.59949	0.47938
Intermediate, ResNext101, SoundNet, place365, I3D	0.80784	0.65840	0.51841	0.39274	0.27811	0.45747	0.59797

In the second set of experiments, we tried different combinations of pre-trained features. First, we tried detailed visual features only (2D object recognition visual features, scene recognition features, and 3D action recognition features). Audio features and 2D object recognition intermediate features were added in the subsequent experiments as shown in Table 2. The combination of 2D object recognition visual features, scene recognition features, and 3D action recognition features achieved the best results in BLEU1, BLEU2, BLEU3, BLUE4, METEOR, and ROUGE-L scores. Combining audio features with these features results in a slight improvement in CIDEr score. The results of using a combination of features exceed the performance of using single features in the majority of the scores.

6   Conclusion

The implementation of efficient multi-modal video involves the use of several features to represent objects, actions, scenes and sounds. These features help the model extract the most pertinent information related to the scene and use this information to generate appropriate text. In this paper, we focus on the characterization of the impact of using features from pre-trained model to implement video captioning sequence-to-sequence models. Our approach is based on the Sequence to Sequence - Video to Text (S2VT) model which uses an encoder-decoder structure to generate video descriptions.

In our experiments, we used five types of features: 2D object recognition features, extracted from a ResNext network; scene recognition features, extracted a ResNet50 network that is trained on places 365 dataset; 3D action recognition features, extracted from kinetics-i3d model, audio features, extracted using soundnet; and object recognition Intermediate Features, which provides more abstract information about the different objects in the frame and their relations.

We conducted two sets of experiments, in which we examined the impact of each feature type separated form the others and the impact of using different combinations of features.

We used four main metrics for comparison: BLEU, METEOR, SIDEr, and ROUFE-L. Our results, when using a single feature type as an input to the mode, indicate that the use of 2D object recognition features achieved best score among the other feature types. When using a combination of features, we found that 2D object recognition visual features, scene recognition features, and 3D action recognition features are the most relevant. The performance of the model when using a multi-modal approach exceeds the performance when using a single feature type for the majority of cases.

Our work can be extended in many ways. First, we would like to investigate different concatenation techniques that reduces the size of the multi-modal input data, and examine the impact of using these techniques on improving the overall efficiency of the model. Second, we would like to increase the capacity of our model, by adding more hidden nodes to the LSTM network and better regularization techniques, and examine the impact of such increase on the efficiency and accuracy of the model. Finally, we would like to investigate better modelling techniques that allow describing, summarizing and linking the salient moments in related videos.

References

1. Gao, L., Guo, Z., Zhang, H., Xu, X., Shen, H.T.: Video captioning with attention-based LSTM and semantic consistency. IEEE Trans. Multimed. **19**, 2045–2055 (2017)
2. Yu, H., Wang, J., Huang, Z., Yang, Y., Xu, W.: Video paragraph captioning using hierarchical recurrent neural networks. In: Proceedings of the IEEE Conference on Computer Vision and Pattern Recognition, pp. 4584–4593 (2016)
3. Pan, P., Xu, Z., Yang, Y., Wu, F., Zhuang, Y.: Hierarchical recurrent neural encoder for video representation with application to captioning. In: Proceedings of the IEEE Conference on Computer Vision and Pattern Recognition, pp. 1029–1038 (2016)
4. Song, J., Guo, Y., Gao, L., Li, X., Hanjalic, A., Shen, H.T.: From deterministic to generative: multimodal stochastic RNNs for video captioning. IEEE Trans. Neural Networks Learn. Syst. **30**, 3047–3058 (2019). https://doi.org/10.1109/TNNLS.2018.2851077
5. Venugopalan, S., Xu, H., Donahue, J., Rohrbach, M., Mooney, R., Saenko, K.: Translating videos to natural language using deep recurrent neural networks. arXiv preprint arXiv:1412.4729 (2014)
6. Vinyals, O., Toshev, A., Bengio, S., Erhan, D.: Show and tell: A neural image caption generator. In: Proceedings of the IEEE Conference on Computer Vision and Pattern Recognition, pp. 3156–3164 (2015)
7. Yu, Y., Ko, H., Choi, J., Kim, G.: Video captioning and retrieval models with semantic attention. arXiv preprint arXiv:1610.02947 (2016)
8. Ramanishka, V., et al.: Multimodal video description. In: Proceedings of the 24th ACM International Conference on Multimedia, pp. 1092–1096 (2016)

9. Shetty, R., Laaksonen, J.: Frame-and segment-level features and candidate pool evaluation for video caption generation. In: Proceedings of the 24th ACM International Conference on Multimedia, pp. 1073–1076 (2016)
10. Huang, F., Zhang, X., Li, Z., Mei, T., He, Y., Zhao, Z.: Learning social image embedding with deep multimodal attention networks. In: Proceedings of the on Thematic Workshops of ACM Multimedia 2017, pp. 460–468 (2017)
11. Venugopalan, S., Rohrbach, M., Donahue, J., Mooney, R., Darrell, T., Saenko, K.: Sequence to sequence-video to text. In: Proceedings of the IEEE International Conference on Computer Vision, pp. 4534–4542 (2015)
12. Karpathy, A., Fei-Fei, L.: Deep visual-semantic alignments for generating image descriptions. In: Proceedings of the IEEE Conference on Computer Vision and Pattern Recognition, pp. 3128–3137 (2015)
13. Pan, Y., Yao, T., Li, H., Mei, T.: Video captioning with transferred semantic attributes. In: Proceedings of the IEEE Conference on Computer Vision and Pattern Recognition, pp. 6504–6512 (2017)
14. Rohrbach, M., Qiu, W., Titov, I., Thater, S., Pinkal, M., Schiele, B.: Translating video content to natural language descriptions. In: Proceedings of the IEEE International Conference on Computer Vision, pp. 433–440 (2013)
15. Kulkarni, G., Premraj, V., Ordonez, V., Dhar, S., Li, S., Choi, Y., Berg, A.C., Berg, T.L.: BabyTalk: understanding and generating simple image descriptions. IEEE Trans. Pattern Anal. Mach. Intell. **35**, 2891–2903 (2013). https://doi.org/10.1109/TPAMI.2012.162
16. Yao, L., Torabi, A., Cho, K., Ballas, N., Pal, C., Larochelle, H., Courville, A.: Describing videos by exploiting temporal structure. In: Proceedings of the IEEE International Conference on Computer Vision, pp. 4507–4515 (2015)
17. Jin, Q., Chen, J., Chen, S., Xiong, Y., Hauptmann, A.: Describing videos using multi-modal fusion. In: Proceedings of the 24th ACM International Conference on Multimedia, pp. 1087–1091 (2016)
18. Long, X., Gan, C., de Melo, G.: Video captioning with multi-faceted attention. Trans. Assoc. Comput. Linguist. **6**, 173–184 (2018). https://doi.org/10.1162/tacl_a_0001
19. Hori, C., et al.: Attention-Based Multimodal Fusion for Video Description. In: Proceedings of the IEEE International Conference on Computer Vision, pp. 4193–4202 (2017)
20. Xie, S., Girshick, R., Dollár, P., Tu, Z., He, K.: Aggregated residual transformations for deep neural networks. In: Proceedings of the IEEE Conference on Computer Vision and Pattern Recognition, pp. 1492–1500 (2017)
21. He, K., Zhang, X., Ren, S., Sun, J.: Deep residual learning for image recognition. In: Proceedings of the IEEE Conference on Computer Vision and Pattern Recognition, pp. 770–778 (2016)
22. Bolei Zhou, Lapedriza, A., Khosla, A., Oliva, A., Torralba, A.: Places: a 10 million image database for scene recognition. IEEE. Trans. Pattern Anal. Mach. Intell. **40**, 1452–1464 (2018).https://doi.org/10.1109/TPAMI.2017.2723009
23. Carreira, J., Zisserman, A.: Quo vadis, action recognition? a new model and the kinetics dataset. In: Proceedings of the IEEE Conference on Computer Vision and Pattern Recognition, pp. 6299–6308 (2017)
24. Aytar, Y., Vondrick, C., Torralba, A.: SoundNet: learning sound representations from unlabeled video. In: Advances in Neural Information Processing Systems, pp. 892–900 (2016)

25. Kordopatis-Zilos, G., Papadopoulos, S., Patras, I., Kompatsiaris, Y.: Near-Duplicate Video Retrieval by Aggregating Intermediate CNN Layers. In: Amsaleg, L., Guðmundsson, G.Þ., Gurrin, C., Jónsson, B.Þ., Satoh, S. (eds.) MMM 2017. LNCS, vol. 10132, pp. 251–263. Springer, Cham (2017). https://doi.org/10.1007/978-3-319-51811-4_21

26. PA Xu, J., Mei, T., Yao, T., Rui, Y.: MSR-VTT: a large video description dataset for bridging video and language. In: Proceedings of the IEEE Conference on Computer Vision and Pattern Recognition, pp. 5288–5296 (2016)

27. Papineni, K., Roukos, S., Ward, T., Zhu, W.-J.: BLEU: a method for automatic evaluation of machine translation. In: Proceedings of the 40th Annual Meeting on Association for Computational Linguistics, pp. 311–318. Association for Computational Linguistics (2002)

28. Lavie, A., Denkowski, M.J.: The METEOR metric for automatic evaluation of machine translation. Mach. Transl. **23**, 105–115 (2009). https://doi.org/10.1007/s10590-009-9059-4

29. Vedantam, R., Zitnick, C.L., Parikh, D.: CIDEr: consensus-based image description evaluation. In: Proceedings of the IEEE Conference on Computer Vision and Pattern Recognition, pp. 4566–4575 (2015)

30. Chen, X., et al.: Microsoft coco captions: data collection and evaluation server. arXiv preprint arXiv:1504.00325 (2015)

31. Lin, C.-Y.: ROUGE: a package for automatic evaluation of summaries. In: Proceedings of Workshop on Text Summarization Branches Out, Post Conference Workshop of ACL, pp. 74–81 (2004)

32. Chen, D.L., Dolan, W.B.: Collecting highly parallel data for paraphrase evaluation. In: Proceedings of the 49th Annual Meeting of the Association for Computational Linguistics: Human Language Technologies-Volume 1, pp. 190–200. Association for Computational Linguistics (2011)

Prediction of Subsequent Memory Effects Using Convolutional Neural Network

Jenifer Kalafatovich[1] , Minji Lee[1] , and Seong-Whan Lee[1,2(✉)]

[1] Department of Brain and Cognitive Engineering, Korea University, Seoul, Korea
sw.lee@korea.ac.kr
[2] Department of Artificial Intelligence, Korea University, Seoul, Korea

Abstract. Differences in brain activity have been associated with behavioral performance in memory tasks; in order to understand memory processes, previous studies have explored them and attempted to predict when items are later remembered or forgotten. However, reported prediction accuracies are low. The aim of this research is to predict subsequent memory effects using a convolutional neural network. We additionally compare different methods of feature extraction to understand relevant information related to memory processes during pre and on-going stimulus intervals. Subjects performed a declarative memory task while electroencephalogram signals were recorded from their scalp. The signals were epoched regarding stimulus onset into pre and on-going stimulus and used for prediction evaluation. A high prediction accuracy was obtained when using convolutional neural networks (pre-stimulus: 71.64% and on-going stimulus: 70.50%). This finding showed that it is possible to predict successful remember items on a memory task using a convolutional neural network, with a higher accuracy than conventional methods.

Keywords: Declarative memory · Convolutional neural network · Prediction · Subsequent memory effects · Electroencephalogram

1 Introduction

Storage of new information into memory is an essential process in people's daily life [1]. Memory mechanisms allow us to retrieval storage information when it is required. However, memory abilities can be altered by several reasons such as aging and brain lesions. Many studies have as a goal not only understand this cognitive process but also improve it. Most of these studies involve the use of transcranial direct current stimulation (tDCS), transcranial magnetic stimulation (TMS) and neurofeedback [2, 3]. In specific, improvement of memory abilities can increase learning efficiency and attenuate the effects of memory disorders (dementia, mild cognitive disease, and Parkinson's disease). Additionally, some studies explored delayed retrieval (around 30 min between study and retrieval task in which subjects are asked to perform a distraction task) on memory task due to its importance in the formation of long-term memories [4].

Brain activity has been analyzed while trying to understand memory processes with respect to pre and on-going stimulus neural fluctuations [5–7]. Many studies reported

© Springer Nature Switzerland AG 2020
Y. Lu et al. (Eds.): ICPRAI 2020, LNCS 12068, pp. 251–263, 2020.
https://doi.org/10.1007/978-3-030-59830-3_22

significant changes in frequency bands as well as in event-related potential (ERP). In spectral analysis, previous researches have reported increased activity in theta frequency band during pre-stimulus when compared later remember and forgotten items at medial temporal sites [8] and a decrease in power over alpha and low beta bands of EEG signals during on-going stimulus [5]. Salari and Rose [9] reported an increase of theta and low beta activity during on-going stimulus on frontal and temporal electrodes, which was used in a brain-computer interface (BCI). Stimulus presentation was determined by the presence of high or low neural oscillations, as a result, the stimulus presented during the increase of beta power was more likely to be remembered. However, no difference in performance was associated with increased theta activity state. Scholz [1] attempted to differentiate effects of brain oscillation between memory formation process and attentional states. Results reported an increase of theta activity over frontal midlines areas and an increased over low beta prior stimulus presentation associated with memory formation. ERP analysis in memory tasks reported a difference between 400–800 ms post-stimulus onset [6] and a frontal negative activity around 250 ms prior stimulus related to later remembered words [7].

Memory studies referred to subsequent memory effects as the changes in activation of a particular brain region during memory encoding phase for trials that were later successfully remembered [10]. Prediction of successful subsequent memory effects has been attempted before in multiple paradigms (episodic memory and working memory) [11, 12]. These studies involved the use of functional magnetic resonance imaging (fMRI) [13], intracranial electroencephalogram (ECoG) [11], and electroencephalogram (EEG) [12]. From these studies, high prediction accuracy was achieved using ECoG on a working memory task, however, its use is not feasible in daily life applications. On the other hand, EEG is cheap and easy to use when acquiring brain information, due to that it has been widely used in different studies, which make used of different EEG paradigms to interpret brain signals (steady-state visual-evoked potentials [14, 15], ERPs [16–19], modulation of sensorimotor rhythms through motor imagery [20–22]). However, EEG's disadvantage is that the signal to noise ratio is high, which results, in this case, in low prediction accuracy [23]. Thus, better methods for prediction of subsequent memory effects using EEG signals need to be explored.

An fMRI study had reported an approximate accuracy of 67.0% when predicting later remember items during phonograms encoding tasks for 19 subjects [24]. This study used multivoxel pattern analysis focusing on the medial temporal lobe to find clusters that contained the most relevant features. In Arora et al. [11], prediction on working memory was explored using ECoG, task consisted on the presentation of 12 items during 1600 ms each separated by a black screen of 300–500 ms, after each list subjects performed a math distractor and the free recall period started, where subjects had to recalled as much word hey could during 30 s. Wavelet filter was used to extracted power from 2 to 100 Hz, later logistic regression, support vector machine (SVM) and a deep learning model were used as classification methods. An accuracy of 72.2% was obtained using a recurrent neural network, which outperformed other used methods, this high accuracy can be due to the low noise to signal ratio characteristic of intracranial signals. Noh et al. [12] used EEG signals to predict subsequent memory effects on an episodic memory task. In order to extract spectral features, common spatial patterns (CSP) algorithm was used over signals during pre-stimulus and after stimulus onset and classification was performed using SVM. Linear discriminant analysis (LDA) was also applied over on-going stimulus signals to extract temporal features. Probabilities of the three different classifiers were combined and an accuracy of 59.6% was reported.

This study explores the possibility to predict whether a shown stimulus during the encoding phase of a declarative memory task is going to be successfully recalled or not when tested during the decoding phase. If a certain stimulus is delivered when a likely to remember brain state is present, then performance in memory task can be improved [25]. In order to accomplish that, high prediction accuracy is needed. Deep learning model has been applied in many studies such as image classification, voice recognition, and others, having successful results [14]. Its application to brain signals is been increasing recently; models like CNN can extract not only time but also spatial and spectral information from EEG signals, resulting in a better performance than traditional machine learning algorithms [26, 27]. We proposed a convolutional neural network (CNN) for predictive performance. In addition, we performed different analyses to extract important features related to the prediction of subsequent memory effect and explore diverse classification methods. Temporal analysis was performed using ERP as feature extraction, and spatial analysis was performed using CSP. As for classification methods, LDA, SVM, and random forest (RF) were used. We compared these methods in order to find significant differences in prediction performance between classifiers and feature extraction methods. Previous studies related to the prediction of successful subsequent memory effects applied machine learning algorithms, however, most of them have not applied deep learning models. Our findings lead to a better prediction of later remembers items and open the possibility to improve memory performance.

2   Methods

2.1   Data Description

EEG signals from seven right-handed healthy subjects (3 females, 21–31 years old) with normal or correct-to-normal vision and no history of neurological disease were used for the analysis. All the experiments were conducted according to the principles described in the Declaration of Helsinki. This study was reviewed and approved by the Institutional Review Board at Korea University (KUIRB-2019-0269-01).

2.2   Experimental Paradigm

Experimental paradigm was implemented using Psychophysics Toolbox, and all shown words were randomly selected from a pool of 3,000 most commonly used English nouns according to Oxford University. Data were recorded during encoding and decoding phases of a declarative memory task. During the encoding phase, subjects were shown 150 English nouns divided into five lists of fifty nouns each, there was a black screen interval of 5 s between lists. They were shown a fixation cross for 1 s, followed by the presentation of a word during 2 s, after which they had to select whether the presented word was an abstract or concrete noun, for which they were given 2 s.

254   J. Kalafatovich et al.

After the encoding phase, subjects had to perform an arithmetical distraction task for 20 min in order to avoid rehearsal of words [8] (subjects were asked to count backward from 1,000 to zero in steps of seven). During the decoding phase, all previous words presented during the encoding phase plus 150 new words were presented in random order. A fixation cross was presented for 1 s followed by the presentation of a word during 2 s; after which subjects have to choose whether the presented word was an old or new word using a confidence scales ranging from 1 (certain new) to 4 (certain old). There was a black screen interval of 1 s between trials, free time was given for the recognition phase, thus the total duration of the decoding phase varied (27 min as average).

2.3 Data Acquisition and Pre-processing

EEG signals were recorded using Brain Vision/Recorder (BrainProduct GmbH, Germany) from 62 Ag/AgCl electrodes. The electrode placements followed the international 10–20 system with the sampling rate of 1,000 Hz. FCz channel was used as a reference electrode, and FPz channel was used for the ground electrode. All impedances were maintained below 10 Hz before and during signal measurement.

EEG data were down-sampled to 250 Hz and band-pass filtered from 0.5 to 40 Hz using a fifth-order Butterworth filter, this was done to reduce artifacts due to muscle or head movement and environment noise. Trials were separated into two condition remembered or forgotten trials. Each trial was epoched with reference of stimulus onset into pre-stimulus and on-going stimulus (−1,000 and 1,000 ms).

2.4 Proposed Convolutional Neural Network

We proposed an architecture based on CNN to predict the subsequent memory effects. Figure 1 shows the shallow CNN architecture, raw signals were filtered (band-pass: 0.5 to 40 Hz) and result signal with the size of 250 × 62 (time points × number of electrodes) was the input to the model. A kernel of 30 × 62 was used to segment temporal signal stride of 25. Dropout randomly sets some inputs for a layer to zero in each training update [28]

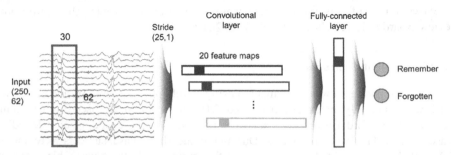

Fig. 1. Proposed CNN architecture. Input (tensor of 250 × 62) is transformed progressively until final output (remember or forgotten trial)

which prevents overfitting; we drop out the input to the convolutional layer with a rate set at 0.25. Batch normalization was applied at the output of the convolutional layer to keep inputs of layers closer to a normal distribution during training [29]. Rectifier linear unit (ReLU) was used as an activation function. This layer generates 20 feature maps. The final layer is a fully connected layer with softmax activation. Loss is minimized using cross entropy loss function and parameters are optimized via Adam method. We applied 10-fold cross validation. Data was divided into training and test set (90% and 10% of total trials respectively). The model was trained using only training set and evaluated using test set, this was done until each was used as test set. Accuracies were averaged and later reported in the result section. For evaluating the model precision, recall and F1-measure were calculated using the following equations.

$$Accuracy = \frac{True\ Positive + True\ Negative}{Total}$$

$$Precision = \frac{True\ Positive}{True\ Positive + False\ Positive}$$

$$Recall = \frac{True\ Positive}{True\ Positive + False\ Negative}$$

$$F1-measure = \frac{2 \times Precision \times Recall}{Precision + Recall}$$

2.5 Comparison of Conventional Methods

We compared the conventional classification methods with the proposed CNN. Feature extraction was performed using ERP and CSP. ERP was calculated between −900 and −200 ms before stimulus presentation for pre-stimulus and 0–1,000 ms after stimulus onset for on-going interval evaluation. CSP is based on the decomposition of EEG signals into spatial patterns to optimally distinguish two classes [30]. Patterns were calculated over −1,000–0 ms pre-stimulus and 0–1,000 ms on-going stimulus. The significant difference in power over different frequency bands have been reported on previous memory task studies; therefore, signals were divided into four relevant frequency bands: theta (3–7 Hz), alpha (7–13 Hz), beta (13–30 Hz), and gamma (30–40 Hz) bands. CSP was applied to each frequency band. Additionally, power was calculated over the filtered signal (0.5–40 Hz) and CSP was also applied.

The classification was performed using LDA, linear-SVM (other kernel types were evaluated; however, prediction accuracies were higher when linear-SVM was used), and RF using 10-fold cross-validation for each subject, therefore data was divided into 10 groups or folds. Model was evaluated using one group as test set and the remaining as training set, this was done until each of the group was used as test set. The reported classification accuracy was calculated averaging accuracies of test sets. RF consists of many individual classification trees; each tress is given a certain weight for the final classification [31].

For the comparison between classifiers and feature extraction methods, we performed statistical analysis. Non-parametric bootstrap analysis was performed having the significance level set at $p = 0.05$.

3 Results

3.1 Prediction Using Proposed CNN

We predicted subsequent memory effects using filtered raw EEG signals. Table 1 and Table 2 show classification accuracies, precision recall and F1-measure for pre and on-going stimulus respectively, when applying CNN. For pre-stimulus the achieved average accuracy was 71.64 ± 9.73% and for on-going stimulus 70.50 ± 10.15%. Sub01 had the highest performance on pre (88.60%) and on-going stimulus (89.87%). On the other hand, Sub02 had the lowest performance of 58.94% on pre-stimulus and 54.73% on on-going stimulus. No significant differences in classification performance between pre and on-going stimulus were found.

Table 1. Classification accuracies using CNN over pre-stimulus segments

Subjects	Accuracy (%)	Precision	Recall	F1-measure
Sub01	88.60	1.00	0.89	0.94
Sub02	58.94	0.73	0.51	0.60
Sub03	78.82	0.96	0.80	0.87
Sub04	71.59	0.89	0.73	0.80
Sub05	74.41	0.95	0.76	0.85
Sub06	59.34	0.98	0.56	0.72
Sub07	69.76	0.95	0.75	0.84
Mean	**71.64 ± 9.73**	**0.92 ± 0.08**	**0.71 ± 0.12**	**0.80 ± 0.10**

Table 2. Classification accuracies using CNN over on-going stimulus segments

Subjects	Accuracy (%)	Precision	Recall	F1-measure
Sub01	89.87	0.96	0.89	0.93
Sub02	54.73	0.51	0.63	0.56
Sub03	75.29	0.94	0.79	0.86
Sub04	73.86	0.86	0.75	0.80
Sub05	69.76	0.98	0.69	0.81
Sub06	63.73	0.88	0.68	0.77
Sub07	66.27	0.95	0.67	0.79
Mean	**70.50 ± 10.75**	**0.87 ± 0.15**	**0.73 ± 0.08**	**0.79 ± 0.10**

3.2   ERP Analysis

Figure 2 shows the grand-averaged ERP waveform of Fp1 electrode for remember and forgotten trials across subjects. Prediction was performed using ERP as feature extraction with different classification methods (see Table 3). For pre-stimulus interval, an average accuracy using LDA, SVM, and RF reached $55.89 \pm 3.86\%$, $58.50 \pm 7.78\%$, and $62.46 \pm 5.29\%$, respectively. There was a significant difference in prediction performance between LDA and RF during pre-stimulus ($p = 0.011$). For on-going stimulus interval, classification accuracies of $58.98 \pm 8.40\%$, $59.61 \pm 4.67\%$, and $66.32 \pm 4.46\%$ were obtained using LDA, SVM, and RF, respectively. Statistical analysis revealed a significant difference in prediction accuracies between LDA and SVM ($p = 0.033$). Also, a significance difference between LDA and RF classifiers was found ($p = 0.009$).

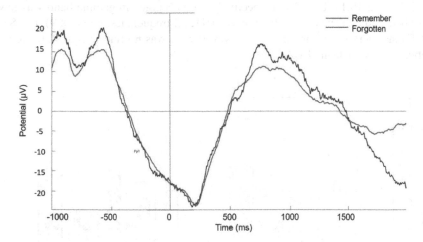

Fig. 2.   Grand-averaged ERP waveforms for remembered/forgotten of Fp1

Table 3.   Classification accuracies per subject using ERP

Subjects	Pre-stimulus (%)			On-going (%)		
	LDA	SVM	RF	LDA	SVM	RF
Sub01	59.57	63.58	58.39	72.01	66.69	73.12
Sub02	54.34	51.89	65.42	49.10	57.17	65.25
Sub03	55.50	61.65	63.08	51.30	57.79	63.28
Sub04	54.10	44.24	55.74	56.69	61.98	62.2
Sub05	53.71	58.70	69.59	67.37	60.05	64.79
Sub06	59.60	59.03	68.58	51.45	62.79	71.44
Sub07	54.39	70.40	56.36	64.96	50.79	64.17
Mean	**55.89 ± 3.86**	**58.50 ± 7.78**	**62.46 ± 5.29**	**58.98 ± 8.40**	**59.61 ± 4.67**	**66.32 ± 4.46**

3.3 CSP Analysis

EEG signals were analyzed in theta, alpha, beta, gamma bands using CSP as a feature extraction method, additionally filtered raw EEG signals were also analyzed. Figure 3 shows the significant difference between prediction accuracies of each frequency band evaluated using LDA. The highest accuracy was obtained when CSP was applied to filtered raw EEG signals for both pre-stimulus and on-going stimulus epochs (57.32 ± 8.04% and 56.71 ± 12.14%, respectively). For pre-stimulus, classification accuracies using theta, alpha, beta, gamma bands were 52.17 ± 10.57%, 54.44 ± 8.07%, 51.20 ± 11.44%, and 52.78 ± 7.66%, respectively. The classification accuracy in all frequencies was higher than in each four of the frequency bands (theta: $p = 0.017$, alpha: $p = 0.398$, beta: $p = 0.011$, gamma: $p = 0.007$). For on-going stimulus, the predictive accuracy using theta, alpha, beta, gamma bands were 52.36 ± 12.84%, 54.25 ± 11.70%, 55.32 ± 11.26%, and 48.14 ± 8.02%, respectively. The accuracy in gamma band was lower than in beta and all frequencies (beta: $p = 0.041$, all frequencies: $p = 0.009$). For SVM and RF classifiers were implemented however there was no relevance difference with the obtained results using LDA.

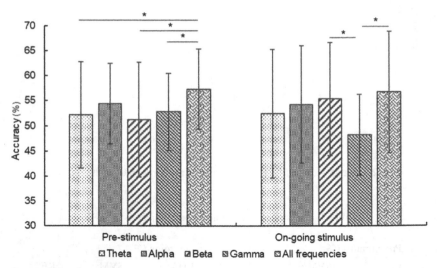

Fig. 3. Classification accuracy of different bands using CSP as feature extraction and LDA as classification method. (* p-value < 0.05, error bar: standard deviation).

Figure 4 shows CSP pattern per class during pre-stimulus and on-going stimulus for filtered raw EEG signals. CSP patterns shows the importance of channels for differentiation between two classes. Channels that carried the most relevant feature per class are located in temporal sites. In specific, differences between remember and forgotten trials was found over left temporal regions in pre-stimulus. For on-going stimulus, inactivation was observed over parietal regions in remember trials, whereas activation and inactivation were simultaneously explored over left temporal regions.

Prediction of Subsequent Memory Effects using CNN   259

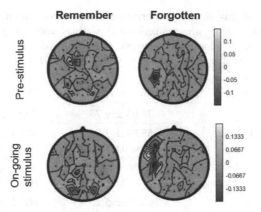

Fig. 4. CSP patterns per class. Pre-stimulus patterns (up), On-going stimulus patterns (down).

3.4 Combined Classifiers

As proposed in Noh et al. [12], we combined two different classifiers. ERP and CSP features were classified using random forest and classification probabilities were averaged to determine the final output. This method achieved $60.05 \pm 6.66\%$ during pre-stimulus and $62.42 \pm 6.30\%$ during on-going stimulus.

3.5 Classifiers Comparison

Table 4 shows classification accuracy of different pattern recognition methods and p-values when compare other classification methods to CNN prediction accuracies. All classifiers were evaluated to find significant differences. The prediction accuracies in CNN were higher than in ERP based on RF (pre-stimulus: $p = 0.031$, on-going stimulus: $p = 0.066$), CSP based on RF (pre-stimulus: $p = 0.009$, on-going stimulus: $p = 0.023$), ERP + CSP based on LDA + SVM (pre-stimulus: $p = 0.009$, on-going stimulus: $p = 0.007$), and ERP + CSP based on RF (pre-stimulus: $p = 0.017$, on-going stimulus: $p = 0.031$). For pre-stimulus, statistical analysis revealed significant differences in CNN compared to other models. For on-going stimulus, there was a significant difference in all comparisons except when comparing to ERP based on RF.

For other methods, there were significant differences in predictive accuracies. In specific, classification performance in ERP based on RF and LDA based on SVM was significantly different (pre-stimulus: $p = 0.007$, on-going stimulus: $p = 0.021$). Also, accuracy in ERP based on RF was higher than in CSP based on RF (pre-stimulus: $p = 0.015$, on-going stimulus: $p = 0.031$). Finally, significant differences of classification accuracy in ERP + CSP between LDA + SVM and RF were observed (pre-stimulus: $p = 0.007$, on-going stimulus: $p = 0.033$).

Table 4 Comparison of classification performance across classification methods using EEG signals during pre and on-going stimulus period and p-values when compared all classification methods to CNN. ([1]Convolutional neural network, [2]Event-related potentials, [3]Common spatial pattern, [4]Linear discriminant analysis, [5]Support vector machine, [6]Random forest)

Feature	Classifier	Pre-stimulus (%)	On-going stimulus (%)
Proposed CNN[1]		71.63 ± 9.73	70.50 ± 10.75
ERP[2]	LDA[4]	55.89 ± 3.86 ($p = 0.007$)	58.98 ± 8.40 ($p = 0.005$)
	SVM[5]	58.50 ± 7.78 ($p = 0.007$)	59.61 ± 4.67 ($p = 0.007$)
	RF[6]	62.46 ± 5.29 ($p = 0.031$)	66.32 ± 4.46 ($p = 0.066$)
CSP[3]	LDA	57.32 ± 8.04 ($p = 0.013$)	56.71 ± 12.14 ($p = 0.021$)
	SVM	54.57 ± 8.35 ($p = 0.003$)	51.41 ± 13.64 ($p = 0.011$)
	RF	57.40 ± 8.20 ($p = 0.009$)	57.47 ± 12.27 ($p = 0.023$)
ERP + CSP	LDA + SVM [Noh et al. 2014]	53.93 ± 5.92 ($p = 0.007$)	55.46 ± 8.26 ($p = 0.007$)
	RF	60.05 ± 6.66 ($p = 0.017$)	62.42 ± 6.30 ($p = 0.031$)

4 Discussion

Our results showed that it is possible to predict subsequent memory effect using EEG signals with higher accuracy than reported by previous studies. The highest accuracy was obtained when applying CNN (pre-stimulus: 71.64% and on-going stimulus: 70.50%). CNN can not only extract temporal but also spatial features, which can increase classification accuracy. Paller [6] and Otten [7] had studied ERP and its relationship with memory process, thus temporal analysis was performed. We obtained an accuracy of 62.46% when using EEG signals before stimulus onset and 66.32% when using EEG signals during on-going stimulus. This high accuracy compared to other conventional methods can be explained by the low noise to signal ratio characteristic of ERP.

Previous studies reported differences in frequency band signals associated with different brain regions [5, 6, 8, 12] for later remembered items. Thus, we analyzed signals for different frequency bands using CSP, however, classification accuracies were not relevant. For pre-stimulus, prediction accuracy using alpha band power was 54.44%, and for on-going stimulus, prediction accuracy using beta band power was 55.32%, these accuracies were higher when compared to other frequency band analysis. This is in line with previous studies that claim differences in alpha band and beta band power during pre and on-going stimuli. Activity in frontal and temporal areas has been associated with attentional and memory processes [5, 8], our spatial analysis showed the relevance of frontal and temporal areas, as well.

One limitation of this study is the number of trials analyzed per subject, which it is un intrinsic constrain when acquiring data of brain activity [32]. Previous studies have showed the viability to classify EEG data regardless the small number of trials used [12, 14, 32]. Additionally, instead of analyzing frequency power separately, it would be interesting to implement combined features in different frequency bands, however, this analysis increments computational time. Other studies have reported better results when analyzing combined signals (pre-stimulus and on-going stimulus). In this study, we wanted to compare important features in each of the intervals. As mentioned before, we obtained the highest accuracy when using CNN, however, it is not clear which is the most relevant information used for classification. Consequently, additional analysis is needed to explain how CNN performs classification and if the results are congruent with relevant features found when using CSP and ERP. Even though initial accuracies were not high, better results were obtained later, however more experiments are required to improve obtained accuracies including better methods for the extraction of relevant features.

In conclusion, this study explored the prediction of later remember items using different pattern recognition techniques. We decided to pursue better results and further investigate the contribution of frequency band signals to memory process using deep learning algorithms, which can integrate temporal, spatial, and spectral features in order to obtained higher prediction accuracy. Brain connectivity of remembered and forgotten words can also be used as a feature extraction method, this can also give a clue of the mechanism involve in memory process. These results could be applied improvement in memory performance, which until now has only been intended by the modulation of neural oscillations (transcranial direct current stimulation, transcranial magnetic stimulation, or neurofeedback) however results vary from study to study [2], which makes it unclear its effectiveness.

Acknowledgements. This work was partly supported by Institute for Information & Communications Technology Promotion (IITP) grant funded by the Korea government (No. 2017-0-00451, Development of BCI based Brain and Cognitive Computing Technology for Recognizing User's Intentions using Deep Learning) and Institute of Information & communications Technology Planning & Evaluation (IITP) grant funded by the Korea government (MSIT) (No. 2019-0-00079, Department of Artificial Intelligence (Korea University)).

References

1. Scholz, S., Schneider, S., Rose, M.: Differential effects of ongoing EEG beta and theta power on memory formation. PLoS ONE **12**(2), 1–18 (2017)
2. Kim, K., Ekstrom, A.D., Tandon, N.: A network approach for modulating memory processes via direct and indirect brain stimulation: toward a causal approach for the neural basis of memory. Neurobiol. Learn. Mem. **134**(1), 162–177 (2016)
3. Staufenbiel, S., Brouwer, A., Keizer, A., van Wouwe, N.: Effect of beta and gamma neurofeedback on memory and intelligence in the elderly. Biol. Psychol. **95**(1), 74–85 (2014)

4. Mameniskiene, R., Jatuzis, D., Kaubrys, G., Budrys, V.: The decay of memory between delayed and long-term recall in patients with temporal lobe epilepsy. Epilepsy Behav. **8**(1), 278–288 (2006)
5. Hanslmayr, S., Spitzer, B., Bäuml, K.-H.: Brain oscillations dissociate between semantic and nonsemantic encoding of episodic memories. Cereb. Cortex **9**(7), 1631–1640 (2009)
6. Paller, K.A., Wagner, A.D.: Observing the transformation of experience into memory. Trends Cogn. Sci. **6**(2), 93–102 (2002)
7. Otten, L.J., Quayle, A.H., Akram, S., Ditewig, T.A., Rugg, M.D.: Brain activity before an event predicts later recollection. Nat. Neurosci. **9**(4), 489–491 (2006)
8. Merkow, M.B., Burke, J.F., Stein, J.M., Kahana, M.J.: Prestimulus theta in the human hippocampus predicts subsequent recognition but not recall. Hippocampus **24**(12), 1562–1569 (2014)
9. Salari, N., Rose, M.: Dissociation of the functional relevance of different pre-stimulus oscillatory activity for memory formation. Neuroimage **125**(1), 1013–1021 (2016)
10. Rypma, B., Esposito, M.D.: A subsequent-memory effect in dorsolateral prefrontal cortex. Cognit. Brain Res. **16**(2), 162–166 (2003)
11. Arora, A., et al.: Comparison of logistic regression, support vector machines, and deep learning classifiers for predicting memory encoding success using human intracranial EEG recordings. J. Neural Eng. **15**(6), 1–15 (2018)
12. Noh, E., Herzmann, G., Curran, T., de Sa, V.R.: Using single-trial EEG to predict and analyze subsequent memory. Neuroimage **84**(1), 712–723 (2014)
13. Wagner, A.D., et al.: Building memories: remembering and forgetting of verbal experiences as predicted by brain activity. Science **281**(5380), 1188–1191 (1998)
14. Kwak, N.-S., Müller, K.R., Lee, S.-W.: A convolutional neural network for steady state visual evoked potential classification under ambulatory environment. PLoS ONE **12**(2), 1–20 (2017)
15. Won, D.-O., Hwang, H.-J., Dähne, S., Müller, K.-R., Lee, S.-W.: Effect of higher frequency on the classification of steady-state visual evoked potentials. J. Neural Eng. **13**(1), 1–11 (2015)
16. Yeom, S.-K., Fazli, S., Müller, K.R., Lee, S.-W.: An efficient ERP-based brain-computer interface using random set presentation and face familiarity. PLoS ONE **9**(11), 1–13 (2014)
17. Lee, M.-H., Williamson, J., Won, D.-O., Fazli, S., Lee, S.-W.: A high-performance spelling system based on EEG-EOG signals with visual feedback. IEEE Trans. Neural Syst. Rehabil. Eng. **26**(7), 1443–1459 (2018)
18. Kim, I.-H., Kim, J.-W., Haufe, S., Lee, S.-W.: Detection of braking intention in diverse situations during simulated driving based on EEG feature combination. J. Neural Eng. **12**(1), 1–12 (2014)
19. Chen, Y., et al.: A high-security EEG-based login system with RSVP stimuli and dry electrodes. IEEE Trans. Inf. Forensic. Secur. **11**(12), 2635–2647 (2016)
20. Suk, H.-I., Lee, S.-W.: Subject and class specific frequency bands selection for multiclass motor imagery classification. Int. J. Imaging Syst. Technol. **21**(2), 123–130 (2011)
21. Kim, J.-H., Bießmann, F., Lee, S.-W.: Decoding three-dimensional trajectory of executed and imagined arm movements from electroencephalogram signals. IEEE Trans. Neural Syst. Rehabil. Eng. **23**(5), 867–876 (2014)
22. Kam, T.-E., Suk, H.-I., Lee, S.-W.: Non-homogeneous spatial filter optimization for ElectroEncephaloGram (EEG)-based motor imagery classification. Neurocomputing **108**(2), 58–68 (2013)
23. Lee, M., et al.: Motor imagery learning across a sequence of trials in stroke patients. Restor. Neurol. Neurosci. **34**(4), 635–645 (2016)
24. Watanabe, T., Hirose, S., Wada, H., Katsura, M., Chikazoe, J., Jimura, K.: Prediction of subsequent recognition performance using brain activity in the medial temporal lobe. Neuroimage **54**(4), 3085–3092 (2011)

25. Ezzyat, Y., et al.: Direct brain stimulation modulates encoding states and memory performance in humans. Curr. Biol. **27**(9), 1251–1258 (2017)
26. Lee, M., et al.: Spatio-temporal analysis of EEG signal during consciousness using convolutional neural network. In: 6th International Conference on Brain-Computer Interface (BCI), vol. 37, no. 1, pp. 1–3 (2018)
27. Jeong, J.-H., Yu, B.-W., Lee, D.-H., Lee, S.-W.: Classification of drowsiness levels based on a deep spatio-temporal convolutional bidirectional LSTM network using electroencephalography signals. Brain Sci. **9**(12), 348 (2019)
28. Ioffe, S., Szegedy, C.: Batch normalization: Accelerating deep network training by reducing internal covariate shift. In: 32nd International Conference on Machine Learning, vol. 37, no. 1, pp. 448–456 (2015)
29. Schirrmeister, R.T., et al.: Deep learning with convolutional neural networks for EEG decoding and visualization. Hum. Brain Mapp. **38**(11), 5391–5420 (2017)
30. Ramoser, H., Gerking, J.M., Pfurtscheller, G.: Optimal spatial filtering of single trial EEG during imagined hand movements. IEEE Trans. Rehab. Eng. **8**(4), 441–446 (2000)
31. Breiman, L.: Random Forests. Mach. Learn. **45**(1), 5–32 (2001)
32. Chiarelli, A.M., Croce, P., Merla, A., Zappasodi, F.: Deep learning for hybrid of EEG-fNIRS brain-computer interface: application to motor imagery classification. J. Neural Eng. **15**(3), 1–12 (2018)

Secure Data Transmission of Smart Home Networks Based on Information Hiding

Haiyu Deng[1] , Lina Yang[1(✉)] , Xiaocui Dang[1] , Yuan Yan Tang[2] ,
and Patrick Wang[3]

[1] School of Computer, Electronics and Information, Guangxi University,
Nanning 530004, China
lnyang@gxu.edu.cn

[2] Beijing Advanced Innovation Center for Big Data and Brain Computing,
Beihang University, Beijing 100191, China

[3] Computer and Information Science, Northeastern University, Boston 02115, USA

Abstract. Smart home is an emerging form of the Internet of things
(IoT), which can enable people to master the conditions of their smart
homes remotely. However, privacy leaking in smart home network is
neglected. Therefore, it is strongly necessary to design a secure and effective scheme to ensure the privacy protection of the smart home network.
Considering the privacy leaking in the smart home network system, this
paper proposes a scheme combining encryption and information hiding
to ensure the security of data transmission. In the scheme, the sensitive data are encrypted to ciphertext before it is transmitted, and then
the ciphertext is embedded into the Least Significant Bit (LSB) of the
cover image. On the receiving terminal, the ciphertext can be extracted
from the stego image, then the ciphertext is decrypted into the sensitive
data. Comparing to the method of using encryption algorithm merely,
the scheme greatly improved the security of data transmission, making
it more suitable for the real environment.

Keywords: Smart home network · Data encryption · Information
hiding · Privacy protection

1 Introduction

With the emergence of the Internet, the worldwide data show an explosive
growth, the data of IoT [1] are expected to exceed 44zettabytes by 2020. The
development of IoT is changing People's daily life [2,3], it greatly facilitates people's communication and improves their perception of the world. As a branch

Supported by the Nature Science Foundation with No. 61862005, the Guangxi Nature
Science Foundation with No. 2017GXNSFBA198226, the Scientific Research Founda-
tion of Guangxi University with No. XGZ160483, the Higher Education Undergraduate
Teaching Reform Project of Guangxi with No. 2017JGB108, and the project with No.
DD3070051008.

© Springer Nature Switzerland AG 2020
Y. Lu et al. (Eds.): ICPRAI 2020, LNCS 12068, pp. 264–275, 2020.
https://doi.org/10.1007/978-3-030-59830-3_23

of the IoT, smart home is also developing rapidly. Privacy protection is one of the most sensitive issues for smart home users, since most of data collected by smart home sensors relating to personal privacy during transmission. Despite the fact that the security architecture of the Internet is relatively perfect, it is not suitable for smart home due to the high cost, limited computing power and other problems [4].

In the past few years, in order to conceal the user's identity, the widely used privacy protection schemes that based on the idea of intermediate agent were to use adopt onion route anonymity. The real identity of the user is often known by the proxy server, which will raise the risk for the disclosure of user privacy due to personal home devices. Mehdi Nobakht et al. [5] developed an IoT-IDM framework, it is deployed at home, monitoring the malicious network activity, providing network level protection for the smart devices. Selecting one or more sensor nodes to simulate the behavior of a real data source to confuse adversaries is the idea of the fake data source [6]. The more fake sources there are, the more privacy is protected. Unfortunately, false data sources sent frequently by sensors will increase the energy consumption of sensors and reduce the lifetime of sensors. In [7], Jianzhong Zhang applied chaos encryption to smart home data transmission system; In [8], WEI SHE et al. proposed a homomorphic encryption and blockchain integration method for privacy protection of smart home. Although the above encryption algorithm can solve the privacy protection problem of smart home to a certain extent, these encryption methods are complex and contain large amount of information, they are not suit to smart home devices due to their limited computing and storage capacity. Moreover, the security transmission of sensitive information can not be realized by relying solely on the encryption. In particular, the readability of encrypted ciphertext is poor, the ciphertext is greatly different from the original data. Therefor, it can be attracted the attention by attackers easily and can be intercepted and attacked [9].

Information hiding technology [10] is produced under those background, it can conceal the existence of sensitive information. With high robustness and concealment, information hiding technology is suitable for data-centered smart home networks. It can protect data security with low computing overhead, realizing information concealment transmission effectively, and can make up for the shortage of data concealment transmission in smart home networks while using encryption technology.

Innovation of this paper:

- Information hiding is applied to the data transmission of smart home networks, it can avoid exposing the sensitive data to the Internet directly and realize the protection of privacy.
- Data encryption and information hiding are combined to realize the double protection of data privacy in smart home. If an attacker intend to carry out a violent attack, it will need to break at least two lines of defense, which seems impossible to achieve.

2   Technical Overview

This section describes LSB information hiding algorithm and DES encryption algorithm briefly.

2.1   LSB Information Hiding Algorithm

One of classical information hiding algorithms in spatial domain steganography is LSB information hiding [11]. LSB information hiding algorithm takes advantage of the fact that HVS cannot detect subtle changes in images, and the redundant of images. The scheme can achieve the purpose of information hiding [12] by replacing the digital image LSB space with secret information bits.

LSB information hiding technology is used widely in image domain, but used rarely in smart home network. Image pixels are the basic elements of digital images, and they exist in the form of two-dimensional matrix elements in the computer [13]. The value of a pixel can be expressed as an 8-bit binary bit, with the left-most bit (the 8th bit) called the most significant bit (MSB) and the right-most bit (the 1st bit) called the least significant bit (LSB). If the 8th bit is flipped, 128 grayscale change will occur to the pixel value. If the first bit is flipped, the pixel value will undergo 1 grayscale change. As can be seen from Fig. 1, the energy of the image is mainly concentrated in the 8th bit. If the pixel value of the 8th bit is taken as the value of this pixel, the contour of the image can be seen clearly. However, 1th bit contains less image energy. If the pixel value of 1st bit is taken as the value of this pixel, the image will display as unreadable codes.

For the difference below 4 grayscale, it is impossible for the naked eye to detect the difference. When using LSB information hiding, the secret information can be randomly embedded into the LSB of the image. It will do little modification to the cover image. In other words, after finishing information hiding, the stego image is same as cover image. So, it will not attract the attention of the attacker, realizing the fact that the secret information can be transmitted safely.

2.2   DES Encryption Algorithm

DES encryption algorithm is the United States data encryption standard, which is developed by IBM company of the United States, and it is a symmetric encryption algorithm. The encryption speed of DES encryption algorithm is fast, according to the statistics of [14], it needs only about 0.000137 s when encrypting a 64-bit plaintext; What is more, DES encryption algorithm is simple, to achieve the purpose of encryption, it just needs small calculation. So it is suitable for the use of lightweight electronic units such as smart home devices. Although DES algorithm has the shortcoming that plaintext data are not easy to hide, information hiding technology can just make up for this shortcoming, so DES algorithm can be well combined with information hiding technology. DES algorithm is a grouping encryption algorithm [15], from the encryption process of

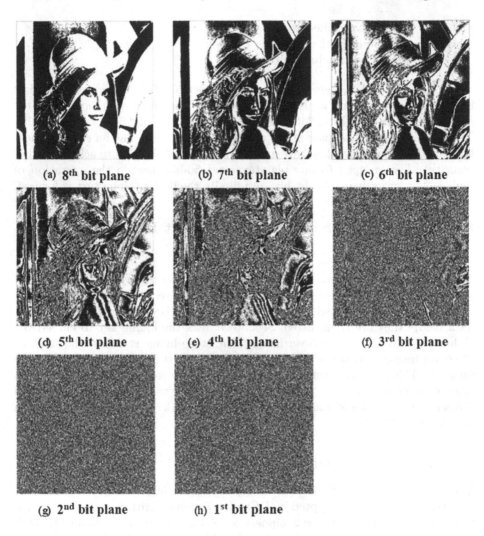

(a) 8th bit plane

(b) 7th bit plane

(c) 6th bit plane

(d) 5th bit plane

(e) 4th bit plane

(f) 3rd bit plane

(g) 2nd bit plane

(h) 1st bit plane

Fig. 1. The eight bit planes of Lena.

DES encryption algorithm in Fig. 2, it can be seen that DES algorithm is composed of data encryption and sub-key generation [16]. It divides 64 bits plaintext into two groups: the first group of 32 bits and the second group of 32 bits; these two sets of plaintext are respectively encrypted by sub-secret key for 16 round iterative encryptions by doing the calculation as (1) shown:

$$\left\{ \begin{array}{l} L_i = R_{i-1} \\ R_i = L_{i-1} \bigoplus F\left(R_{i-1}, K_i\right) \end{array} \right. (i = 1, 2 \ldots 15, 16) \tag{1}$$

In the formula:

Li,Ri—transitional ciphertext
Ki—the ith sub-secret key
F—the transformation function

Finally L_{16} and R_{16} are integrated into a set of 64-bit ciphertext. The key length of DES algorithm is also 64 bits, which is composed of 56 operation bits and 8 check bits (8th, 16th, 24th, 32th, 40th, 48th, 56th and 64th are check bits. These bits do not participate in the encryption operation. The function of these bits is to make sure each key have an odd number of "1", so as to achieve higher security). The grouped 64-bits plaintext and 56-bits key do the operation of replacement or exchange bit by bit, and finally the 64-bits ciphertext is formed.

3   The Proposed Scheme in the Smart Home Network System

This paper proposes a scheme of privacy protection for data transmission of smart home networks by combining DES encryption technology with LSB information hiding. First, the scheme uses the encryption key to encrypt the sensitive data before they are transmitted. Second, it uses the hiding key to embed the ciphertext into LSB of the cover image. The stego image is almost the same as the cover image, so that the existence of the ciphertext can not be noticed by the attacker. Third, the receiver uses the extraction key to extract ciphertext from the stego image. Fourth, it uses the decryption key to decrypt ciphertext, and the sensitive data are obtained finally. The framework of the scheme is shown in the following Fig. 3:

3.1   Encryption and Information Hiding

Data encryption and information hiding are shown in Algorithm 1. First, the sensitive data (R) are encrypted by DES encryption algorithm using the encryption key (Ke) in line 1. In line 2, ciphertext (C) is converted into binary bits C'. According to the sequence of the hiding key (Kh), binary bits C' are embedded to the cover image's LSB bit by bit, as in line 3. Finally, the stego image I' is returned and transmitted on the ordinary channel, as in line 4.

Algorithm 1. Encryption and information hiding
Input: sensitive data R, encryption key Ke, hiding key Kh, cover image I
Output: stego image I'
1. $C=$ encryption (R, Ke)
2. $C'=$ char 2 bit (C)
3. for $i = 0$ to len(C')-1
$\quad I'=$Do Embedded bit to LSB (C', I, Kh)
4. return I'

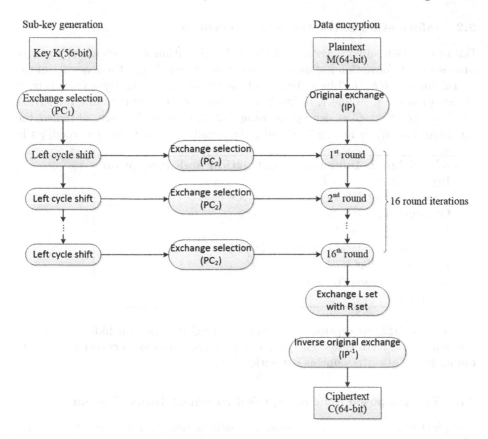

Fig. 2. DES encryption algorithm framework.

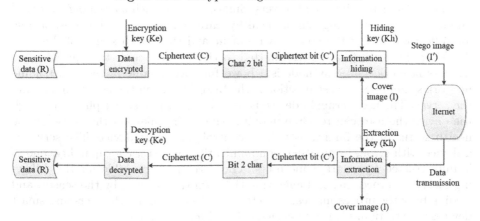

Fig. 3. The framework of the scheme.

3.2   Information Extraction and Decryption

Data extraction and decryption are shown in Algorithm 2, which are the reverse processes of information hiding and encryption. According to the sequence of the extraction key Kh, the binary bits C'' of the ciphertext are firstly extracted from the stego image I' in line 1. In line 2, the binary bits C' are turn into ciphertext C. The ciphertext C is decrypted using the decryption key Ke, obtained the sensitive data R, as in line 3. Finally, the sensitive data R are returned, as in line 4.

Algorithm 2. Information extraction and decryption
Input: stego image I',
extraction key Kh, decryption key Ke
Output: sensitive data R
1. for $i = 0$ to len (I')-1
   C'=do information extraction(I',Kh)
2. C= bit 2 char (C')
3. R= decryption (C, Ke)
4. return R

Because DES encryption algorithm and LSB information hiding algorithm are symmetric algorithms, encryption key is the same as decryption key, and hiding key is also the same as extraction key.

3.3   The Proposed Scheme Applied to Smart Home System

A typical framework of smart home network model [17,18] is shown in Fig. 4. The smart home network model is composed of four parts: smart sensors, gateway nodes, user terminals and servers. Smart sensors, such as, temperature and humidity sensor, imaging sensor, healthy care sensor, are mainly responsible for collecting data of the home environment and the health status of the family. After that, the data are transmitted to the gateway node through wired or wireless network. Gateway node is a powerful master node in the smart home network system [18], whose function is the bridge between the user terminal and the server. The user terminal device is usually the user's smart phone, through the device the user can remotely monitor the environment of the house or the health status of the family, as well as control the smart device. The server is a device with strong computing, communication and storage capability, and it' is fully trusted by smart home users. The sever is responsible for the following tasks: (a) processing and analyzing the data transmitted by the sensor, and feeding back the processing results to the smart home users. (b) users and smart devices registration. (c) key management and maintenance.

Due to the limited computing and storage capacity of the smart sensors, apart from collecting data, encrypting data is another task of smart sensors. DES encryption algorithm is simple, so it is suitable for the application of smart sensors. Gateway is a node with relatively strong computing and storage capabilities. It can store images used as information hiding carriers and perform slightly

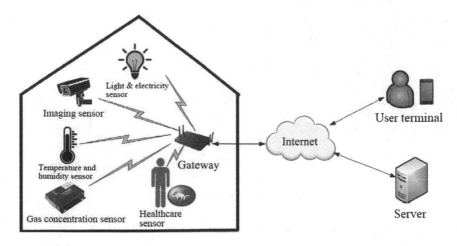

Fig. 4. A typical framework of smart home network system.

complex information hiding operations. After the sensitive data are encrypted and hidden, they are uploaded to the server for the calculation and analysis through a ordinary channel, and the transmitting process will not be discovered by malicious attackers. The server has strong computing capacity, and it is completely trusted by users. After receiving stego image, it extracts the sensitive information from the stego image and decrypts it. After that, the sensitive data are calculated and analyzed by the sever, and the final results are fed back to the user terminal.

4   Experiments and Analysis

4.1   Simulation Experiment

This simulation experiment was conducted in MATLAB R2016a. The dataset for this experiment is from the UMass Trace Repository, a data-sharing platform operated by the advanced systems software laboratory at the university of Massachusetts. The environment dataset and application dataset of Smart Home are included, part of the dataset showing as Fig. 5. Environmental dataset is collected by environmental sensors, such as indoor and outdoor humidity sensors, indoor and outdoor temperature sensors, wind speed and direction sensors and rainfall sensors. The dataset of application is the power consumption of various electrical devices. In this experiment, we assume that the dataset is sensitive to the smart home users. Assuming we use the encryption key "133457799BBCDFF3" to encrypt the dataset, we can obtain the ciphertext of the dataset, shown as Fig. 6.

The above ciphertext presented unreadable garbled state, if directly exposed to the Internet, it is easy to attract the attention of adversaries. The adversary will try to intercept the ciphertext and decrypt it. In this scheme, the

temperature	icon	humidity	visibility	summary	pressure	windSpeed	cloudCover
13.3	clear-night	0.74	9.89	Clear	1018.62	7.24	0
15.41	clear-night	0.71	10	Clear	1017.65	5.47	0.22
16.93	clear-night	0.67	9.97	Clear	1017.18	5.53	0.18
19.04	clear-night	0.57	9.97	Clear	1016.63	6.02	0.21
18.32	clear-night	0.55	9.94	Clear	1016.33	6.09	0.05
18.64	clear-night	0.54	9.94	Clear	1016.72	7.41	0.06
31.34	partly-cloudy	0.55	9.8	Partly Cloudy	1017.93	11.09	0.31
32.01	partly-cloudy	0.53	9.81	Partly Cloudy	1018.87	11.42	0.31
32.9	partly-cloudy	0.51	10	Partly Cloudy	1019.55	12.4	0.42
33.09	clear-day	0.52	9.88	Clear	1019.72	11.9	0.42
32.56	clear-day	0.54	10	Clear	1019.66	14.88	0.75
32.94	partly-cloudy	0.51	9.6	Mostly Cloudy	1020.1	10.91	0.75
28.63	snow	0.82	3.99	Snow	1023.49	4.95	1
29.81	snow	0.87	4.04	Snow	1021.46	9.83	1
32.05	rain	0.89	5.01	Light Rain	1019.54	10.51	1
32.78	rain	0.89	3.91	Light Rain	1017.75	10.01	1
33.71	rain	0.88	7.52	Light Rain	1016.71	12.14	1

(a) Dataset-environment

Date & Time	KitchenDenLights [kW]	MasterBedBathLights [kW]	DenOutdoorLights [kW]	DenOutlets [kW]
2016/1/1 0:00	0.007657778	0.021298889	0.021405556	0.000455556
2016/1/1 0:15	0.007606667	0.004708889	0.021478889	0.000486667
2016/1/1 0:30	0.007591111	0.004655556	0.020736667	0.00048
2016/1/1 0:45	0.007613333	0.004643333	0.02087	0.000495556
2016/1/1 1:00	0.007593333	0.004641111	0.02009	0.000494444
2016/1/1 1:15	0.007591111	0.00466	0.018381111	0.00049
2016/1/1 1:30	0.007614444	0.004855556	0.012665556	0.000503333
2016/1/2 10:00	0.082562222	0.055713333	0.00796	0.000195556
2016/1/2 10:15	0.082652222	0.055367778	0.007981111	0.000193333
2016/1/2 10:30	0.091895556	0.026654444	0.007972222	0.000203333
2016/1/2 10:45	0.094245556	0.029964444	0.007995556	0.000272222
2016/1/2 11:00	0.090443333	0.026102222	0.007926667	0.000231111
2016/1/2 11:15	0.086024444	0.00436	0.007978889	9.44E-05

(b) Dataset-application

Fig. 5. Part of the dataset.

(a) Dataset-environment_encrypted

(b) Dataset-application_encrypted

Fig. 6. The result of encryption.

above ciphertext is hidden into cover image to reduce the probability of sensitive information being discovered by the attacker. After receiving the ciphertext, the gateway embeds the ciphertext into the cover image. After the operation of information hiding, the stego image is not different from the cover image, so the adversary will not notice the existence of the sensitive information, as shown in Fig. 7. After the server receives the stego image, it extracts the sensitive information from the stego image and decrypts it.

Fig. 7. The result of information hiding.

4.2 Security Analysis

The security of DES algorithm is high, so far, there is not a well performance crack method. Although DES algorithm would be attacked by exhaustive search brute attack method, it will really need a high cost. For example, if an exhaustive search is used to force an attack on ciphertext that encrypted with a 56-bit key, then the exhaustive space will reach to 2^{56}. If a computer could detect one million keys every second, an attack on the above ciphertext would take nearly 2,285 years to search for the full key. Obviously, the price of this cracking method is costly, which is unrealistic.

Security analysis of information hiding: Assuming that the length of the string to be hidden is 16, one char consists of 8 bits binary, with a total of $16*8 = 128$ bits. The proposed scheme uses a discrete method to embed the 128 bits into the $512 * 512$ pixels cover image according to a random generation matrix (the key to hide the sensitive information or extract the sensitive information), the number of the possible embedding methods is $C_{512*512}^{128}$. It is almost impossible to break by brute attack, which proves that LSB information hiding method has a good security performance.

Based on the security analysis of the above two algorithms, the security strategy has the characteristics of high concealment, high difficulty in cracking and high neglect, so the security of the proposed scheme is compellent.

5 Conclusion

Base on the fact that smart home users want to protect their privacy from disclosure, but the current privacy protection of smart home network based on encryption is easy to attract the attention by attackers. Therefore, this paper proposes a scheme combining information hiding with encryption algorithm. In the scheme, the sensitive data are encrypted at first and then the ciphertext is embedded into the LSB of the cover image. After the sensitive information is embedded, the stego image is the same as cover image, so the adversary cannot easily find out the exist of sensitive information. The feasibility and security of the scheme have been verified by the simulation experiment and the security analysis. Comparing to the method of using encryption algorithm merely, the scheme do a better performance in privacy protection, making it more suitable for the smart home network.

Acknowledgment. This work is financially supported by the Nature Science Foundation with No. 61862005, the Guangxi Nature Science Foundation with No. 2017GXNSFBA198226, the Scientific Research Foundation of Guangxi University with No. XGZ160483, the Higher Education Undergraduate Teaching Reform Project of Guangxi with No. 2017JGB108, and the project with No. DD3070051008.

Special thanks to Dr Lina Yang for her careful guidance on the paper writing, and thanks to Miss Xiaocui Dang for her help in the paper writing.

References

1. Santoso, F.K., Vun, N.C.H.: Securing IoT for smart home system. In: 2015 International Symposium on Consumer Electronics (ISCE), Madrid, pp. 1–2. IEEE (2015)
2. Li, S., Xu, L.D., Zhao, S.: 5G Internet of Things: a survey. J. Ind. Inf. Integr. **10**, 99–110 (2016)
3. Yadav, P., Mittal, A., Yadav, H.: IoT: challenges and issues in Indian perspective. In: 3rd International Conference on Internet Things (IoT-SIU), New Delhi, pp. 1–15. IEEE (2018)
4. Qiu, T.: The main scheme analysis of smart home network control system. Innov. Appl. Sci. Technol. **10**(20), 80–90 (2016)
5. Nobakht, M., Sivaraman, V., Boreli, R.: A host-based intrusion detection and mitigation framework for smart home IoT using OpenFlow. In: 2016 11th International Conference on Availability, Reliability and Security (ARES), Salzburg, pp. 147–156. IEEE (2015)
6. Shao, M., Yang, Y., Zhu, S., Cao, G.: Towards statistically strong source anonymity for sensor networks. ACM Trans. Sensor Netw. **9**(3), 1–23 (2013)
7. Zhang, J.: Application research of chaos encryption in intelligent home data transmission system. M.S. thesis, Hangzhou Electronic Science and Technology University, Hangzhou, China (2015)
8. She, W., Gu, Z., Lyu, X., Liu, Q., Tian, Z., Liu, W.: Homomorphic consortium blockchain for smart home system sensitive data privacy preserving. IEEE Access **7**, 62058–62070 (2019)
9. Yang, J.: Research on security enhancement of information hiding in sensor networks. M.S. thesis, Hunan University, Changsha, China (2011)
10. Halpern, J.Y., O'Neill, K.R.: Anonymity and information hiding in multi-agent systems. J. Comput. Secur. **13**(3), 483–514 (2005)
11. Sharp, T.: An implementation of key-based digital signal steganography. In: Moskowitz, I.S. (ed.) IH 2001. LNCS, vol. 2137, pp. 13–26. Springer, Heidelberg (2001). https://doi.org/10.1007/3-540-45496-9_2
12. Luo, F.: Adaptive steganographic analysis of gray image LSB algorithm. M.S. thesis, Xidian university, Xi'an, China (2013)

13. Deng, M.: Research on high capacity reversible information hiding algorithm based on encrypted image. M.S. thesis, Southwest Jiaotong University, Chengdu, China (2018)
14. Zhu, J.: Research and implementation of QR code anti-counterfeiting algorithm based on DES improved algorithm. M.S. thesis, Shanxi University of Science and Technology, Taiyuan, China (2018)
15. Cheng, Z., Chen, C., Qiu, X.: Cache timing attack based on Flush+Reload DES algorithm. Comput. Eng. **44**(12), 163–167 (2018)
16. Wei-Xing, Q., Qin, L.I., Jin-Lian, X.U.: A combination algorithm based on DES. J. Nanjing Univ. Posts Telecommun. **31**(5), 83–86 (2011)
17. Kumar, P., Gurtov, A., Iinatti, J., Ylianttila, M., Sain, M.: Lightweight and secure session-key establishment scheme in smart home environments. IEEE Sens. J. **16**(1), 254–264 (2016)
18. Spreitzer, R., Moonsamy, V., Korak, T., Mangard, S.: Systematic classification of side-channel attacks: a case study for mobile devices. IEEE Commun. Surv. Tutor. **20**(1), 465–488 (2018)

Differences and Similarities Learning for Unsupervised Feature Selection

Xinyi Li[ID], Ning You[ID], Jun Tan[(⊠)][ID], and Ning Bi[ID]

School of Mathematic, Sun Yat-sen University, Guangzhou, China
{lixy386,youn7}@mail2.sysu.edu.cn,
{mcstj,mcsbn}@mail.sysu.edu.cn

Abstract. In this paper, a novel feature selection algorithm, named Feature Selection with Differences and Similarities (FSDS), is proposed. FSDS jointly exploits sample differences from global structure and similarities from local structure. To reduce the disturbance from noisy feature, a row-wise sparse constraint is also merged into the objective function. FSDS, then combines the underlying subspace features with original feature to construct a more reliable feature set. Furthermore, a joint version of FSDS (FSDS2) is introduced. To optimize the proposed two-step FSDS and the joint version FSDS2 we also design two efficient iterative algorithms. Experimental results on various datasets demonstrate the effectiveness of the proposed algorithms.

Keywords: Feature selection · Spectral learning · Row sparsity · Sample differences and similarities exploiting

1 Introduction

The dilemma of high-dimension data i.e. time and space pressure, has brought about great challenges to various areas, including computer vision, pattern recognition and biological science. Moreover, the information redundancy together with noise in original data makes it significantly necessary to implement feature selection and dimension reduction.

Recent years has witnessed the increase on creative feature selection algorithms proposed to address dimension dilemma. Generally, they can be classified into three groups, 1) supervised algorithm, 2) semi-supervised algorithms and 3) unsupervised algorithms. Supervised feature selection methods [1, 2] select important features by considering the correlation between features and label information. Supervised methods are comparatively efficient since the goal information is remaining in labels. However, labeling data requires excessive time and human labor. Semi-supervised feature selection methods discuss about both labeled data and unlabeled data based on some regularization frameworks, such as normalized min-cut with NMI classification loss [3] and semi-supervised SVM with manifold regularization [4]. However, atypical labeled data may fail to represent data structure and yield negative effect to feature selection procedure. In lack of sufficient label information, it is promising and sensible to develop unsupervised feature selection methods.

© Springer Nature Switzerland AG 2020
Y. Lu et al. (Eds.): ICPRAI 2020, LNCS 12068, pp. 276–292, 2020.
https://doi.org/10.1007/978-3-030-59830-3_24

In unsupervised scenarios, feature selection is achieved according to different criteria (e.g. [5, 6, 8–10, 12–14]), such as maximum variance. However, it ignores locality preserving whereas the local structure of the data space is more important than the global structure as mentioned in [6]. For this reason, many approaches utilizes a transformation matrix which can map the original space into a lower-dimensional one and simultaneously model the local geometric structure of the original space. Among these methods, some [6, 8–10] emphasize similarity preserving, which is achieved by preserving manifold structure of the original data space. Others [5] focuses on exploiting features that involves the most discriminative information between samples. Nevertheless, taking either sample similarities or differences into account may not be comprehensive enough and thus this two procedures will be joined in the proposed method. Additionally, it is conducive that feature selection should be well-directed and more label information should be assimilated into the procedure of feature selection, revealing the goal of final classification or clustering. For example, reference [5, 10, 12–14] employ pseudo labels to select effective features. To strengthen the effect of label information, not only pseudo labels are applied in this paper, but k-means algorithm will also be utilized since the parameter k is related to experiment target to some extent. What's more, most references [8–10, 12–14] unify subspace learning technique into feature selection. There is no reason to assume that separating those two procedures must gain worse experimental performance. In this paper, interesting experiments are conducted to explore the two situations.

The main contribution of this study relies on four aspects:

- We propose a new algorithm named Feature Selection with Differences and Similarities (FSDS) by simultaneously exploiting the sample differences from the global structure and the sample similarities from the local structure.
- Our method (FSDS) also considers selected original space feature and subspace features independently.
- We introduce a unifying framework of a "two-step" FSDS, named as FSDS2, in which the information of subspace features and original features fuses together, compared to FSDS.
- Corresponding iterative optimization algorithms of FSDS and FSDS2 are also proposed respectively in our work.

The remaining of this paper is arranged as follows. We elaborate our proposed formulation FSDS in Sect. 2. Then optimization algorithm is presented in Sect. 3 followed with discussions about algorithm FSDS2 introduced in Sect. 4. Extensive experiments are conducted and analyzed in Sect. 5. Section 6 illustrates the conclusion of our work.

2 The Objective Function

2.1 Preliminary

In this section, we first summarize some notations used throughout this paper. We use bold uppercase characters to denote matrices, bold lowercase characters to denote vectors. For an arbitrary matrix A, a_i means the i-th row vector of A, a^j means the j-th column vector of A, A_{ij} denotes the *(i, j)*-th entry of A, $\|A\|_F$ is Frobenius norm of A and Tr[A] is the trace of A if A is square. The -norm is defined as

$$\|A\|_{2,1} = \sum_{i=1}^{r} \sqrt{\sum_{j=1}^{t} A_{ij}^2} \tag{1}$$

Assume that we have n samples $\chi = \{x_i\}_{i=1}^n$. Let $X = [x_1, \ldots, x_n] \in \mathcal{R}^{d \times n}$ denote the data matrix, in which $x_i \in \mathcal{R}^{d \times n}$ represents the *i*-th sample. Suppose these n samples are sampled from c classes and there are n_i samples in the *i*-th class. In this paper, I is identity matrix, $1_m \in \mathcal{R}^m$ is a column vector with all elements being 1 and H_n is used for data centralizations, as in Eq. (3), $\tilde{X}$ is the data matrix after being centered.

$$H_n = I - \frac{1}{n}1_n 1_n^T \in \mathfrak{R}^{n \times n} \tag{2}$$

$$\tilde{X} = X H_n \tag{3}$$

We define $Y = [y_1, \ldots, y_n]^T \in \{0, 1\}^{n \times c}$ to be the label matrix, which means that the *j*-th element of y_i is 1 if sample x_i belongs to the *j*-th class, and 0 otherwise. The scaled label matrix G is defined as

$$G = [G_1, \ldots, G_n]^T = Y(Y^T Y)^{-1/2} \in \mathfrak{R}^{n \times c} \tag{4}$$

It turns out that

$$G^T G = (Y^T Y)^{-1/2} Y^T Y (Y^T Y)^{-1/2} = I_c \tag{5}$$

The intraclass scatter matrix S_e and between class scatter matrix S_b are defined as follows.

$$S_e = \sum_{i=1}^{c} \sum_{x_j \in class i} (x_j - \mu_i)(x_j - \mu_i)^T = \tilde{X}(I_n - GG^T)\tilde{X}^T \tag{6}$$

$$S_b = \sum_{i=1}^{c} n_i(\mu_i - \mu)(\mu_i - \mu)^T = \tilde{X} GG^T \tilde{X}^T \tag{7}$$

where μ is the mean of all samples, μ_i is the mean of samples from the *j*-th class.

2.2   The Proposed Framework

Many algorithms [5, 8–10, 12, 14] complete the task of feature selection by comparing the l_2 norm of each row in transformation matrix $W \in \mathcal{R}^{d \times c}$, which can map the d-dimensional original space to the c-dimensional subspace. And the matrix W is usually gained through optimizing different objective functions. Others [6] may use the graph Laplacian to learn some abstract subspace features, which aim to characterizing original high-dimensional data structure. The abstract subspace features representing compressed information from new low-dimensional space, is similar to the "viewpoints".

And these "viewpoints" need to be supported by exact "evidence", that is, some concrete selected features in the original space. Inspired by this, we propose to combine the exact selected features with the subspace features to construct a new interpretive feature set. Our framework is formulated as follows

- feature selection

$$\min_{W} \alpha D(W) + S(W) + \gamma \|W\|_{2,1}$$
$$s.t. W^T W = I \tag{8}$$

For efficient classification and clustering, both similarities among samples in the same classes and differences of samples from different classes are important. Here, D(W) is a clustering criterion of sample differences between different classes, S(W) is another clustering criterion to evaluate sample similarities in the same classes and the $l_{2,1}$-norm of transformation matrix W aims to make it sparse and suitable for feature selection. The orthogonal constraint is imposed to avoid arbitrary scaling and trivial solution of all zeros.

- Subspace feature learning

In this step, via Laplacian Eigenmap algorithm we learn subspace features $F = \left[F^1, \ldots, F^c\right] \in \mathcal{R}^{n \times c}$.

- Feature combination

Finally, we combine the selected features with the subspace features.

2.3 Exploiting Differences from Global Structure

A satisfactory classifier ought to amplify the difference between different classes and as a result we use the trace of global between class scatter matrix S_b to evaluate the difference. Moreover, it is appropriate to limit the intraclass deviation, which can be viewed as error to some extent. Analogously, the trace of intraclass scatter matrix S_e is used to evaluate intraclass deviation.

$$\max tr[S_b] = \max tr[\tilde{X} G G^T \tilde{X}^T]$$
$$\min tr[S_e] = \min tr[\tilde{X} (I_n - G G^T) \tilde{X}^T] \tag{9}$$
$$\Leftrightarrow \min tr(-\tilde{X} G G^T \tilde{X}^T (\tilde{X} \tilde{X}^T + \lambda I)^{-1})$$

where λ is a parameter and λI is added to ensure the term $\left(\tilde{X} \tilde{X}^T + \lambda I\right)$ invertible. Since the proposed algorithm is an unsupervised one, G defined in (4) is unknown. We assume there is a linear classifier $W \in \mathcal{R}^{d \times r}$ to transform the original features matrix to the corresponding label matrix, that is, $G = X^T W$. It's worth noting that our framework simultaneously utilizes linear features (feature selected) and nonlinear features (subspace feature), which may improve interpretability. Besides, the orthogonal constraint is added to prevent arbitrary scaling and trivial solution, which can also avoid redundancy in learnt transformation matrix W. In this way, the objective function $\alpha D(W)$ in this subsection can be summarized as

$$\min_{W^T W = I} tr[W^T M^{(1)} W], M^{(1)} = -\alpha X \tilde{X}^T (\tilde{X} \tilde{X}^T + \lambda I) \dot{X} X X^T \tag{10}$$

2.4 Exploiting Similarities from Local Perspective

To exploit the similarities among the same classes, we first construct a local set $N_k(centre_i)_{i=1}^c$ for each class center, which can also preserve local structure. Additionally, the k-means algorithm is performed for c center points of c classes. $X_i = [x_{i1}, \ldots, x_{ik}]$ denotes the local data matrix for i-th class center $G_{(i)} = [G_{i1}, \ldots, G_{ik}]^T \in \mathcal{R}^{k \times c}$ represents as the corresponding scaled label matrix.

Similar to (7), the between class scatter matrix $S_b^{(i)}$ for $N_k(centre_i)$ is defined as

$$\begin{aligned}
S_b^{(i)} &= \sum_{j=1}^{c} n_j^{(i)} \left(\mu_j^{(i)} - \mu^{(i)} \right) \left(\mu_j^{(i)} - \mu^{(i)} \right)^T \\
&= \tilde{X}_i G_{(i)} G_{(i)}^T \tilde{X}_i^T
\end{aligned} \tag{11}$$

where $\tilde{X}_i = X_i H_k$. $\mu^{(i)}$ represents the mean of samples in $N_k(centre_i)$, $\mu_j^{(i)}$ represents the mean of samples in the j-th class, $N_k(centre_i)$. If k is a set small enough, it is appropriate to assume that all samples in $N_k(centre_i)$ belong to the same class. Therefore, $S_b^{(i)}$ is close zero. We then define the selection matrix $S_i \in \{0, 1\}^{n \times k}$, each column of which only contains one element being 1 and all others being 0. The first column of S_i is used to record the nearest sample of center i, while the last column is for the farthest away, i.e., $(S_i)_{hg} = 1$ if and only if the h-th sample is the g-th nearest sample of the i-th center. Note that $XS_i = X_i$, $G_{(i)} = S_i^T G = S_i^T X^T W$. Then, we define each term $S_i(W)$ of the whole objective function S(W)

$$\begin{aligned}
S_i(W) &= tr[S_b^{(i)}] + tr[G_{(i)}^T H_k G_{(i)}] \\
&= tr[\tilde{X}_i G_{(i)} G_{(i)}^T \tilde{X}_i^T + G_{(i)}^T H_k G_{(i)}] \\
&= tr[G_{(i)}^T \tilde{X}_i^T \tilde{X}_i G_{(i)} + G_{(i)}^T H_k G_{(i)}] \\
&= tr[G^T S_i \tilde{X}_i^T \tilde{X}_i S_i^T G + G^T S_i H_k S_i^T G] \\
&= tr[W^T X S_i \tilde{X}_i^T \tilde{X}_i S_i^T X^T W + W^T X S_i H_k S_i^T X^T W]
\end{aligned} \tag{12}$$

Because the learned W and G ought to be suitable for all c classes, we add an $G_{(i)}^T H_k G_{(i)}$ to each $S_i(W)$ to avoid overfitting. Finally, we conclude the objective function S(W) in this section.

$$\min_{W^T W = I} tr[W^T M^{(2)} W], M^{(2)} = \sum_{i=1}^{c} X S_i (\tilde{X}_i^T \tilde{X}_i + H_k) S_i^T X^T \tag{13}$$

2.5 Sparseness Control

After adding $l_{2,1}$-norm of the transformation matrix W, the objective function for feature selection (8) can be rewritten as

$$
\min_{W^T W = I} tr\left(W^T M^{(3)} W\right) + \gamma \|W\|_{2,1}
$$
$$
M^{(3)} = M^{(1)} + M^{(2)}
$$
$$
= \sum_{i=1}^{c} X S_i \left(\tilde{X}_i^T \tilde{X}_i + H_k\right) S_i^T X^T - \alpha X \tilde{X}^T \left(\tilde{X} \tilde{X}^T + \lambda I_d\right)^{-1} \tilde{X} X^T
$$

(14)

2.6 Subspace Feature Learning and Feature Combination

Construct the affinity graph W [6]:

$$
W_{ij} = \begin{cases} \exp\left(-\dfrac{\|x_i - x_j\|^2}{\sigma^2}\right), & x_i \in N_k(x_j) \ or \ x_j \in N_k(x_i) \\ 0, & otherwise \end{cases}
$$

(15)

where $N_k(x)$ is the set of k-nearest neighbors of x. D is defined as a $D_{ii} = \sum_{j=1}^{n} W_{ij}$ diagonal matrix whose entries are column(or row) sums of W, that is . We can gain the graph Laplacian matrix $L = D - W$. The subspace feature can be computed by solving the following generalized eigen-problem:

$$
LF = \lambda DF
$$

(16)

$\left[F^1, \ldots, F^c\right] \in \mathcal{R}^{n \times c}$, F^i 's are the eigenvectors of (16) with the smallest eigenvalues, which are also the solution of $min_{F^T DF = I} tr\left[F^T LF\right]$.

Finally, we combine the selected features X* with the subspace features F to form a new feature set [X* F].

3 Optimization of FSDS Algorithm

3.1 Iterative Algorithm

The optimization problem in (14) involves orthogonal constraint and the $l_{2,1}$-norm, which is not smooth, both increasing the difficulties of optimization. In this subsection, we present an iterative algorithm to solve the optimization problem of FSDS. Similar to [5], Lagrange multiplier is applied and L(W) is defined considering the orthogonal constraint in (14) as well.

$$
\begin{aligned}
L(W) &= tr[W^T M^{(3)} W] + \gamma \|W\|_{2,1} + \lambda(W^T W - I_c) \\
&= tr[W^T M^{(3)} W] + \gamma tr[W^T U W] + \lambda(W^T W - I_c) \\
&= tr[W^T (M^{(3)} + \gamma U) W] + \lambda(W^T W - I_c)
\end{aligned}
$$

(17)

where U is a diagonal matrix with $U_{ii} = \frac{1}{2\|w_i\|_2}$. Setting $\frac{\partial L(W)}{\partial W} = 0$, we have

$$
\begin{aligned}
\frac{\partial L(W)}{\partial W} &= 2(M^{(3)} + \gamma U)W + 2\lambda W = 0 \\
\Rightarrow (M^{(3)} + \gamma U)W &= \lambda W = \lambda^* W
\end{aligned}
\tag{18}
$$

Therefore, we can update W when we fix U through solving the eigen-problem (18). Note that $W = [w^1, \ldots, w^c]$, w^i are the eigenvectors with the smallest eigenvalues. In practice, $\|w_i\|_2$ could be close to zero. For this case, we can regularize $2U_{ii} = (w_i^T w_i + \varepsilon)^{-1/2}$, where ε is a small constant to prevent complex solution. When ε is close to zero, $2U_{ii}$ approximates $(w_i^T w_i)^{-1/2}$.

We elaborate the details of FSDS algorithm in Algorithm 1 as follow.

Algorithm 1: The FSDS algorithm

1 $M^{(1)} = -\alpha X \tilde{X}^T (\tilde{X}\tilde{X}^T + \lambda I)\tilde{X}X^T$;

2 perform k-means 20 times to find the c class centre;

3 $M^{(2)} = \sum_i^c XS_i(\tilde{X}_i^T \tilde{X}_i + H_k)S_i^T X^T$;

4 $M^{(3)} = M^{(1)} + M^{(2)}$;

5 **Set** t=0 and initialize $U_0 \in \Re^{d \times d}$ as an identity matrix;

6 **repeat**

7 　　$P_t = M^{(3)} + \gamma U_t$;

8 　　$W_t = [w^1, \ldots, w^c]$,where $w^1, \ldots, w^c$ are the eigenvector of P_t with smallest c eigenvalues

　　　　(c,the class number);

9 　　Update matrix U_{t+1} as $U_{t+1} = \begin{bmatrix} \frac{1}{2}\|w_t^1\|_2 & & \\ & \cdots & \\ & & \frac{1}{2}\|w_t^d\|_2 \end{bmatrix}$

10 　　t=t+1;

11 **until** *convergence*

12 Sort all d features according to $\|w_t^i\|_2$ in descending order and select the top r ranked features. The selected feature matrix can be marked as $X^* \in R^{n \times r}$

13 Construct the k-nearest neighbor graph and calculate the graph Laplacian matrix L

14 Gain subspace feature $F=[F^1, \ldots, F^c] \in R^{n \times c}$ by solving the generalized eigen-problem $LF = \lambda DF$. Then combine W and F to form a new feature set $W = [X^* \ F] \in R^{n \times (r+c)}$

3.2 Convergence Analysis

Next, we prove the convergence of proposed iterative procedure in Algorithm 1.

Theorem 1. The iterative approach in Algorithm 1 (line 6 to line 11) makes the objective function (14) monotonically decrease in each iteration.

Proof. According to Eq. (17), Eq. (18), we can see that

$$
L(W_t, U_{t-1}) \leq L(W_{t-1}, U_{t-1})
\tag{19}
$$

That is to say,

$$tr[W_t^T(M^{(3)} + \gamma U_{t-1})W_t] \leq tr[W_{t-1}^T(M^{(3)} + \gamma U_{t-1})W_{t-1}]$$

$$\Leftrightarrow tr[W_t^T M^{(3)} W_t] + \gamma \sum_{i=1}^{d} \frac{\|w_i^t\|_2^2}{2\|w_i^{t-1}\|_2} \leq tr[W_{t-1}^T M^{(3)} W_{t-1}] + \gamma \sum_{i=1}^{d} \frac{\|w_i^{t-1}\|_2^2}{2\|w_i^{t-1}\|_2}$$

$$\Leftrightarrow tr[W_t^T M^{(3)} W_t] + \gamma \|W_t\|_{2,1} - \gamma \left(\|W_t\|_{2,1} - \sum_{i=1}^{d} \frac{\|w_i^t\|_2^2}{2\|w_i^{t-1}\|_2} \right) \qquad (20)$$

$$\leq tr[W_{t-1}^T M^{(3)} W_{t-1}] + \gamma \|W_{t-1}\|_{2,1} - \gamma \left(\|W_{t-1}\|_{2,1} - \sum_{i=1}^{d} \frac{\|w_i^{t-1}\|_2^2}{2\|w_i^{t-1}\|_2} \right)$$

Note that the parameter γ is non-negative. According to Lemmas in (Ref. [2]),

$$\sqrt{a} - \frac{a}{2\sqrt{b}} \leq \sqrt{b} - \frac{b}{2\sqrt{b}}, \quad \|W_t\|_{2,1} - \sum_{i=1}^{d} \frac{\|w_i^t\|_2^2}{2\|w_i^{t-1}\|_2} \leq \|W_{t-1}\|_{2,1} - \sum_{i=1}^{d} \frac{\|w_i^{t-1}\|_2^2}{2\|w_i^{t-1}\|_2}$$

Thus, we can prove that

$$tr[W_t^T M^{(3)} W_t] + \gamma \|W_t\|_{2,1} \leq tr[W_{t-1}^T M^{(3)} W_{t-1}] + \gamma \|W_{t-1}\|_{2,1} \qquad (21)$$

This is to say,

$$L(W_t, U_t) \leq L(W_{t-1}, U_{t-1}) \qquad (22)$$

which indicates that the objective function value monotonically decreases in each iteration.

4 Discussion

The FSDS is a "two-step" strategy that separates the procedure of subspace feature learning and feature selection in the original space. Inspired by [9], we will discuss about the unified framework of FSDS.

The FSDS algorithm can be regarded as solving the following problems in sequence:

- $\arg \min_{W^T W = I} tr(W^T M^{(3)} W) + \gamma \|W\|_{2,1}^T$
- $\arg \min_{F^T F = I} tr(F^T L F)$
- $[X^* F]$

The connection of the first two steps can be realized by adding a loss function $\|X^T W - F\|_F^2$. Then the objective function will be formulated as

$$\min_{F,W} tr[F^T L F] + \alpha \|X^T W - F\|_F^2 + \beta tr(W^T M^{(3)} W) + \gamma \|W\|_{2,1}$$

$$s.t. F^T F = I, F_{ij} \geq 0 \qquad (23)$$

Here, we substitute the orthogonal constraint on W with the one on F, which is analogous. It's worth noting that we should regard F as pseudo subspace label but not subspace feature, which may contain more accurate and concise information. Moreover, the nonnegative constraints on F will force only one element in each row of F greater than zero. In this way, F is more like the label.

As for the problem of optimization, let us define

$$L(W, F) = tr[F^T L F] + \alpha \|X^T W - F\|_F^2 + \beta tr[W^T M^{(3)} W] + \gamma \|W\|_{2,1}$$
$$+ \lambda^{(2)} \|F^T F - I\|_F^2 s.t. F_{ij} \geq 0, \tag{24}$$

When we fix W and upgrade F. We have

$$\frac{\partial L(F,W)}{\partial W} = 0 = 2\alpha X (X^T W - F) + 2\gamma U W + 2\beta M^{(3)} W$$
$$\Rightarrow W = \alpha(\alpha X X^T + \gamma U + \beta M^{(3)})^{-1} X F \tag{25}$$

where U has been discussed above (17). Substituting W by (25), the objective function (24) can be rewritten as

$$min tr[F^T M^{(4)} F] + \lambda^{(2)} \|F^T F - I\|_F^2 \ s.t. F_{ij} \geq 0,$$
$$M^{(4)} = L + \alpha \left(I_n - \alpha X^T \left(\alpha X X^T + \gamma U + \beta M^{(3)} \right)^{-1} X \right) \tag{26}$$

Next, we apply multiplicative updating rules to update F according to [9, 10, 12, 14]. Matrix $\phi = (\phi_{ij})$ is introduced to be Lagrange multiplier for the nonnegative constraint on F. Then we get

$$tr[F^T M^{(4)} F] + \lambda^{(2)} \|F^T F - I\|_F^2 + tr[\varphi F^T] \tag{27}$$

Setting its derivative with respect to F to 0 and using the KKT constraints $\phi_{ij} F_{ij} = 0$, we obtain the updating rules:

$$F_{ij} \leftarrow F_{ij} \frac{\left(\lambda^{(2)} F\right)_{ij}}{\left(M^{(4)} F + \lambda F F^T F\right)_{ij}} \tag{28}$$

Then, we normalize F as $\left(F^T F\right)_{ii} = 1, i = 1, \ldots, c$. The detailed algorithm is present below, and the name for the unified approach is FSDS2.

Algorithm 2: The FSDS2 algorithm

---

1 $M^{(1)} = -\alpha X \tilde{X}^T (\tilde{X}\tilde{X}^T + \lambda I)\tilde{X}X^T$;

2 perform k-means 20 times to find the c class centre;

3 $M^{(2)} = \sum_{i}^{c} XS_i(\tilde{X}_i^T \tilde{X}_i + H_k)S_i^T X^T$;

4 $M^{(3)} = M^{(1)} + M^{(2)}$;

5 **Set** t=0 and initialize $U_0 \in \Re^{d \times d}$ as an identity matrix and initialize $F_t \in \Re^{m \times c}$

6 **repeat**

7      $M^{(4)^{t+1}} = L + \alpha(I_n - \alpha X^T(\alpha XX^T + \gamma U^t + \beta M^{(3)})^{-1}X)$

8      $F_{ij} \leftarrow F_{ij} \dfrac{(\lambda^{(2)}F)_{ij}}{(M^{(4)}F + \lambda FF^TF)_{ij}}$

9      $W^{t+1} = \alpha(\alpha XX^T + \gamma U^t + \beta M^{(3)})^{-1}XF^t$

10   Update the diagonal matrix U_{t+1} as $U_{t+1} = \begin{bmatrix} 1/2\left\|w_{t+1}^1\right\|_2 & & \\ & \cdots & \\ & & 1/2\left\|w_{t+1}^d\right\|_2 \end{bmatrix}$

11   t=t+1;

12 **until** *convergence*

13 Sort all d features according to $\|w_t^i\|_2$ in descending order and select the top r ranked features

---

5 Experiments

In this section, we conduct several experiments to evaluate the performance of FSDS and FSDS2. We test the performance of the proposed algorithms in terms of clustering results.

5.1 Datasets

The experiments are conducted on 5 public datasets, including 3 face image datasets, Yaleface, JAFFE [18] and UMIST [17], one hand-written digit image dataset, USPS [17] and one object image dataset, coil20 [19]. Table 1 summarizes the details of these datasets mentioned above.

Table 1. Dataset description

Dataset	# of Samples	# of Features	# of Class
Yaleface	165	1024	15
JAFFE	213	676	10
UMIST	575	644	20
USPS	400	256	10
coil20	1440	1024	20

5.2 Experiment Settings

We compare the performance of FSDS and FSDS2 with the following unsupervised feature selection algorithm.

- **Baseline:** All original features are adopted.
- **MaxVar:** Features with maximum variance are selected.
- **LPP [7]:** Subspace features gain by Laplacian Eigenmap algorithm.
- **LS [6]:** Features consistent with Gaussian Laplacian matrix are selected.
- **MCFS [8]:** Features are selected through spectral analysis and sparse regression problem.
- **UDFS [5]:** Features are selected based on local discriminative feature analysis and $l_{2,1}$-norm minimization.
- **NDFS [10]:** Features are selected by nonnegative spectral analysis and linear regression with $l_{2,1}$-norm regularization.
- **CGSSL [12]:** Features are selected by also considering the original features when predicting the pseudo label based on NDFS.
- **FSDS:** Features are selected by the proposed method Feature Selection with Differences and Similarities (Algorithm 1).
- **FSDS2:** Features are selected by the joint framework of FSDS (Algorithm 2).

For all algorithm except Baseline, MaxVar, we set k, the size of neighborhoods, to be 5 for all the datasets. For NDFS, CGSSL, FSDS2, we fix $\lambda^{(2)} = 10^8$ to guarantee the orthogonality of F and fix $\lambda = 10$ to ensure the term $\left(\tilde{X}\tilde{X}^T + \lambda I \right)$ invertible. To fairly compare the performance of all algorithms, we tune the parameters for all approaches by a "grid search" strategy from $\{10^{-6}, 10^{-4}, \ldots, 10^6\}$. The numbers of selected features are set as $\{50, 100, 150, 200, 250, 300\}$ for all of the datasets except for USPS, whose total feature number is 256. We set the selected feature number of USPS as $\{50, 80, 110, 140, 170, 200\}$. We report the best results of all these 10 algorithms with different parameters.

After feature selection through different algorithm, k-means clustering is conducted based on the selected features. Besides, k-means is also performed in FSDS and FSDS2 feature selection procedure. Since the results of the k-means is greatly affected by its initialization, we repeat it 20 times with random initialization every time we use it. We report the average results with standard deviation (std).

We apply two evaluation metrics, Accuracy (ACC) and Normalized Mutual Information (NMI), to evaluate the clustering performance. The larger ACC and NMI are, the better the performance is. ACC is defined as follow.

$$ACC = \frac{\sum\limits_{i=1}^{n} \delta(p_i, map(q_i))}{n} \tag{29}$$

where $\delta(a, b) = 1, if\ a = b, \delta(a, b) = 0\ otherwise$, and q_i is the clustering label and p_i is the ground-truth label of x_i.*map* is a kind of permutation function that maps each cluster index to the best ground-truth labels using the Kuhn-Munkres algorithm. Mutual Information (MI) and Normalized Mutual Information (NMI) are defined as follow

$$MI(P, Q) = \sum_{c_i \in P} \sum_{c_j' \in Q} p(c_i, c_j') \log_2 \frac{p(c_i, c_j')}{p(c_i)p(c_j')}$$

$$NMI(P, Q) = \frac{MI(P, Q)}{\max(H(P), H(q))}$$

(30)

where p(c) is the probability that a sample randomly selected from the dataset belongs to class c. P represents clustering label set whereas Q represents groud truth label set. And H(c) is the entropy of c. $H(c) = -\sum_{c_i \in c} p(c_i) \log_2 p(c_i)$

5.3 Experimental Results

We summarize the clustering results of different methods on the 5 datasets in Table 2.

Through observing this table, we can gain the following conclusion. First, it is necessary to apply feature selection not only because feature selection can increase efficiency and save experiment time, but also because it can reduce useless noise in datasets and improve clustering performance, which can be supported by the fact that most feature selection algorithm outperform Baseline. Second, it places great importance to consider local structure in feature selection, which is consistent with the fact that other algorithms get better performance than MaxVar, the only one algorithm without preserving local structure. Third, the proposed Algorithm FSDS and FSDS2 achieve best performance in most datasets by learning the linear classifier W that focuses on both sample differences and sample similarities. Other algorithms just take either of sample differences or similarities into account. For example, LS, MCFS, NDFS, CGSSL preserve the manifold structure by mapping the neighboring data point close to each other in the mapping subspace, which is a emphasis on sample similarities. And UDFS learn the classifier W that can recognize the sample difference. It may be more comprehensive to consider both of differences and similarities. Hence, FSDS and FSDS2 are proved to be capable of more efficient feature selection.

Moreover, FSDS and FSDS2 achieve similar ACC and NMI, which may be owing to the following reasons. We can easily find that the information of subspace features and the information of original features remain independent and complete after FSDS algorithm. For FSDS2, according to (23), the information of subspace features and original features fuses together and they affect each other. Both these two solutions have their advantages, thus resulting in the similar results (Table 3).

5.4 Convergence Study

In the previous section, we have proved the convergence of the proposed alternative optimization algorithm for FSDS and it is analogy for FSDS2. Our algorithms converge within 50 iterations for all the data sets, which indicates that the proposed update algorithms are effective and converge fast, especially the one for FSDS2. Moreover, the results of convergence are shown in Fig. 1.

Table 2. Clustering results of different feature selection algorithm

Dataset	Baseline	MaxVar	LS	LPP	MCFS	UDFS	NDFS	CGSSL	FSDS	FSDS2
ACC ± std(%)										
Yale	46.7 ± 3.3	40.6 ± 2.9	50.5 ± 3.5	45.5 ± 3.2	50.3 ± 5.1	49.1 ± 3.8	45.5 ± 3.0	–	**50.9 ± 4.4**	**50.9 ± 3.7**
JAFFE	72.5 ± 9.2	67.3 ± 5.8	74.0 ± 7.6	69.7 ± 2.4	78.8 ± 9.1	76.7 ± 7.1	81.2 ± 8.1	82.3 ± 7.5	95.8 ± 8.0	**95.3 ± 9.1**
USPS	62.6 ± 5.3	63.8 ± 4.3	64.9 ± 5.1	64.0 ± 1.8	64.4 ± 3.1	66.2 ± 4.7	67.3 ± 4.7	68.3 ± 4.5	**78.8 ± 5.8**	**78.8 ± 5.5**
coil20	59.0 ± 5.7	58.4 ± 4.0	57.3 ± 3.0	50.8 ± 5.0	61.7 ± 4.3	62.9 ± 2.6	63.8 ± 2.8	–	72.6 ± 4.7	**72.9 ± 4.3**
UMIST	41.8 ± 2.7	45.8 ± 2.8	45.9 ± 2.9	56.5 ± 2.4	46.3 ± 3.6	48.6 ± 3.7	51.3 ± 3.9	53.4 ± 3.1	58.8 ± 4.2	**60.5 ± 3.8**
NMI ± std(%)										
Yale	48.3 ± 2.8	46.6 ± 2.0	54.3 ± 2.6	50.6 ± 2.9	**57.7 ± 3.6**	53.9 ± 3.4	49.4 ± 2.7	–	55.2 ± 3.4	57.0 ± 3.1
JAFFE	80.0 ± 5.7	70.3 ± 4.2	79.4 ± 7.0	80.8 ± 3.3	83.4 ± 5.0	82.3 ± 6.5	86.3 ± 7.1	87.5 ± 5.1	**93.8 ± 5.5**	93.5 ± 6.9
USPS	56.9 ± 3.1	58.1 ± 2.7	58.7 ± 3.0	62.4 ± 1.5	59.3 ± 2.9	60.1 ± 4.3	61.3 ± 2.5	62.0 ± 2.8	**68.1 ± 3.4**	**68.1 ± 3.3**
coil20	72.9 ± 2.8	70.5 ± 0.9	70.4 ± 1.1	61.9 ± 3.4	74.7 ± 2.3	75.9 ± 1.1	77.1 ± 1.8	–	**80.6 ± 2.1**	30.2 ± 3.1
UMIST	62.3 ± 2.3	63.5 ± 1.5	63.9 ± 1.8	**74.2 ± 3.0**	66.7 ± 1.9	67.3 ± 3.0	69.7 ± 2.3	70.9 ± 2.2	73.3 ± 2.8	73.1 ± 2.7

Table 3. Clustering Results of FSDS with subspace feature learning and without

Dataset	ACC ± std(%)	
	With subspace feature	Without subspace feature
Yale	50.9 ± 4.4	47.3 ± 4.3
JAFFE	95.8 ± 8.0	92.5 ± 7.8
USPS	78.8 ± 5.8	76.8 ± 4.3
coil20	72.6 ± 4.7	65.5 ± 4.0
UMIST	58.8 ± 4.2	55.5 ± 3.8

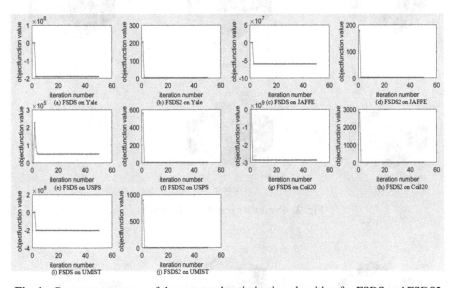

Fig. 1. Convergence curve of the proposed optimization algorithm for FSDS and FSDS2

5.5 Sensitivity Analysis

For FSDS method, there are three parameters α, γ, f (*featurenumber*) in need of setting in advance. And FSDS2 algorithm requires five parameters to be tuned, i.e. $\alpha, \gamma, \beta, \alpha^s$ (*parameterin* $M^{(3)}$), f (*featurenumber*). For the results reported in the first subsection, we do not discuss about number of selected feature. Thus, we will investigate the sensitivity of α, γ for FSDS. As for FSDS2, we observe that the parameters γ, β have more effect on the performance than others in our experiment, which will be studied in the following.

All the parameters mentioned above are tuned from $\{10^{-6}, 10^{-4}, \ldots, 10^6\}$. The results over the five data sets are shown in Fig. 2 and Fig. 3. In conclusion, there seems no analogous rule to determine the optimal parameters on all condition due to the varying performance with respect to different data sets. In terms of FSDS, firstly, the parameter γ

controls the row sparsity of the feature selection matrix W. when γ is set to be too small, noisy or redundant features can't be spotted and removed. Instead, not only useless features but also informative features will be discarded if γ is too large. These are consistent with the observation in Fig. 2 that extremely large or extremely small γ has negative influence on the performance. Secondly, the α parameter balances the importance between similarity preserving and difference exploiting. It can be concluded from Fig. 3 that we always attain the best performance while α is in the middle interval, demonstrating that similarity preserving and difference exploiting are of equal importance.

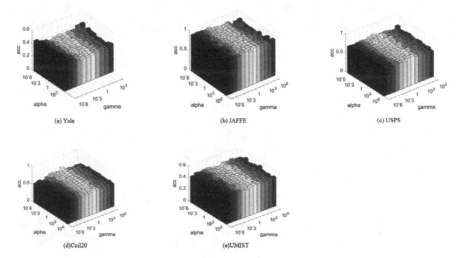

Fig. 2. Parameter sensitivity for FSDS

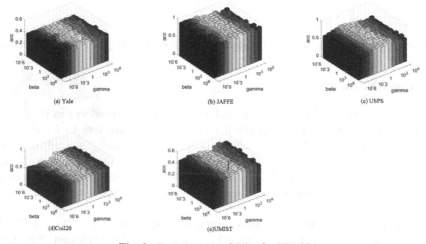

Fig. 3. Parameter sensitivity for FSDS2

With regard to FSDS2, on one hand, the analysis for parameters γ similar to the one mentioned above. On the other hand, the parameter β is the trade-off parameter between feature selection according to (23) and subspace feature learning. It is observed from Fig. 3 that parameter β, neither too small nor too large, frequently accompanies optimal results. This indicates that the information in the original feature space is as essential as the information in abstract subspace.

Next, we evaluate the performance of both proposed methods and compared methods with different value of f (*feature number*). The experimental results are present in Fig. 4, from which we can observe the following conclusions. First, the performance is comparatively sensitive to the number of selected features. Second, the proposed algorithms FSDS and FSDS2 can always reach the best results with some small values of selected features. It can be verified that the proposed algorithms not only comprehensively exploit the differences and similarities among the samples, but they also synthetically investigate the original feature space and the underlying manifold space, thus resulting in efficient feature selection.

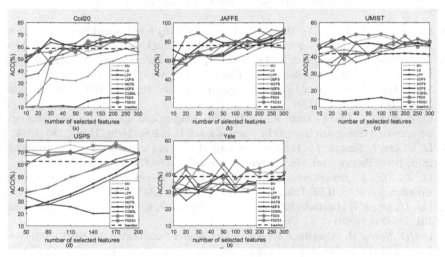

Fig. 4. The clustering performance in terms of ACC (%) with respect to the number of selected features on all the five datasets

6 Conclusion

In this paper, we propose two feature selection approaches FSDS and FSDS2, which combine the feature selection procedure and spectral subspace feature learning. Utilizing both original feature and subspace feature, the subsequent clustering methods gains better performance. Besides, our methods also jointly exploit sample differences and sample similarities. Extensive experiments on different datasets have validated the effectiveness of the proposed methods.

References

1. Wolf, L., Shashua, A.: Feature selection for unsupervised and supervised inference: the emergence of sparsity in a weightbased approach. J. Mach. Learn. Res. **6**, 1855–1887 (2005)
2. Nie, F., Huang, H., Cai, X., Ding, C.: Efficient and robust feature selection via joint $l_{2,1}$-norms minimization. In: Proceedings of the Advances in NIPS (2010)
3. Zhao, Z., Liu, H.: Semi-supervised feature selection via spectral analysis. In: Proceedings of the SDM (2007)
4. Xu, Z., King, I., Lyu, R.T., Jin, R.: Discriminative semi-supervised feature selection via manifold regularization. IEEE Trans. Neural Netw. **21**(7), 1033–1047 (2010)
5. Yang, Y., Shen, H.T., Ma, Z., Huang, Z., Zhou, X.: $l_{2,1}$-norm regularized discriminative feature selection for unsupervised learning. In: Proceedings of the 22nd IJCAI (2011)
6. He, X., Cai, D., Niyogi, P.: Laplacian score for feature selection. In: Proceedings of the Advances in NIPS (2005)
7. He, X., Niyogi, P.: Locality preserving projections. In: Proceeding of the International Conference on Neural Information Processing System (2003)
8. Cai, D., Zhang, C., He, X.: Unsupervised feature selection for multi-cluster data. In: Proceedings of the ACM SIGKDD International Conference on KDD, Washington, DC, USA (2010)
9. Hou, C., Nie, F., Li, X., Yi, D., Wu, Y.: Joint embedding learning and sparse regression: a framework for unsupervised feature selection. IEEE Trans. Cybern. **44**(6), 793–804 (2014)
10. Li, Z., Yang, Y., Liu, J., Zhou, X., Lu, H.: Unsupervised feature selection using nonnegative spectral analysis. In: Proceedings of the Conference on AAAI (2012)
11. Shi, J., Malik, J.: Normalized cuts and image segmentation. IEEE Trans. Pattern Anal. Mach. Intell. **22**(8), 888–905 (2000)
12. Li, Z., Yang, Y., Liu, J., Zhou, X., Lu, H.: Clustering-guided sparse structural learning for unsupervised feature selection. IEEE Trans. Knowl. Data Eng. **26**(9), 2138–2150 (2014)
13. Li, Z., Liu, J., Tang, J., Lu, H.: Robust structured subspace learning for data representation. IEEE Trans. Pattern Anal. Mach. Intell. **37**(10), 2085–2098 (2015)
14. Li, Z., Tan, J.: Unsupervised feature selection via nonnegative spectral analysis and redundancy control. IEEE Trans. Image Process. **24**(12), 5343–5355 (2015)
15. Lee, D., Seung, H.: Learning the parts of objects by nonnegative matrix factorization. Nature **401**, 788–791 (1999)
16. Lee, D., Seung, H.: Algorithms for nonnegative matrix factorization. In: NIPS (2001)
17. Data for MATLAB Hackers. http://cs.nyu.edu/~roweis/data.html. Accessed 16 Oct 2012
18. Lyons, M.J., Budynek, J., Akamatsu, S.: Automatic classification of single facial images. IEEE Trans. Pattern Anal. Mach. Intell. **21**(12), 1357–1362 (1999)
19. Nene, S.A., Nayar, S.K., Murase, H.: Columbia object image library (COIL-20), Dept. Computation. Sci., Columbia Univ., New York, NY, USA, Techical report CUCS-005-96 (1996)
20. Luo, M., Chang, X., Nie, L., Yang, Y., Hauptmann, A.G., Zheng, Q.: An adaptive semisupervised feature analysis for video semantic recognition. IEEE Trans. Cybern. **48**(2), 648–660 (2018)
21. Zhao, Z., He, X., Cai, D., Zhang, L., Ng, W., Zhuang, Y.: Graph regularized feature selection with data reconstruction. IEEE Trans. Knowl. Data Eng. **28**(3), 689–700 (2016)
22. Cai, Z., Zhu, W.: Feature selection for multi-label classification using neighborhood preservation. IEEE/CAA J. Autom. Sinica **5**(1), 320–330 (2018)
23. Zhou, N., Xu, Y., Cheng, H., Yuan, Z., Chen, B.: Maximum correntropy criterion-based sparse subspace learning for unsupervised feature selection. IEEE Trans. Circ. Syst. Video Technol. **29**(2), 404–417 (2019)

Manifold-Based Classifier Ensembles

Vitaliy Tayanov[ID], Adam Krzyzak$^{(\boxtimes)}$[ID], and Ching Y. Suen[ID]

Department of Computer Science and Software Engineering, Concordia University,
Montreal, QC, Canada
{vtayanov,krzyzak,suen}@encs.concordia.ca

Abstract. In this paper, we briefly present classifier ensembles making use of nonlinear manifolds. Riemannian manifolds have been created using classifier interactions which are presented as symmetric and positive-definite (SPD) matrices. Grassmann manifolds as some particular case of Riemannian manifolds are constructed using decision profiles. Experimental routine shows advantages of Riemannian geometry and nonlinear manifolds for classifier ensemble learning.

Keywords: Riemannian and Grassmann manifolds · Classifier ensembles · Random Forests

1 Introduction

There are two principal approaches to enhance classification accuracy: use more advanced and complex classifiers and architectures (classifier ensembles [17] and deep learning architectures [3,15] or use data transformation (kernel-based algorithms [14] and nonlinear manifolds [6]). However, it is hard sometimes to distinguish between two categories because classifier stacking [17] performs data transformation and there are multiple data transformations when learning optimal features in deep architectures. Recently, in [18], it was proposed to use cascades of classifier ensembles such as cascades of Random Forests (RF) as an alternative to deep architectures. Again, in this situation, RF approach can be classified either as data transformation or using advanced complex architectures. Apparently architectures may be created that use all the aforementioned techniques. Focusing principally on classifier ensemble learning using geometric transformations we involve all four techniques to design different algorithms and we compare them experimentally.

Contributions. While the main idea of the proposed approach is to use Riemannian geometry to learn from classifier ensembles, we implemented together four different techniques to enhance classification accuracy. Riemannian manifolds of SPD matrices Sym_+^d were created using classifier interactions. The outputs of individual classifiers (predictions) from a classifier ensemble are taken to

This research was supported by the Natural Sciences and Engineering Research Council of Canada.

build Sym_+^d matrices. On one hand, the interactions of classifiers in ensemble induce the nonlinear space and on the other hand interactions between classifiers produce new data which generally are more informative than initial ensemble predictions. Linearisation of Sym_+^d matrices allows using geometrical properties of Riemannian manifold. Vectorizing both matrices (linearised and non-linearised) allows obtaining new data in Euclidean space. After that, any classifier can be used in the same way as in the Euclidean space. Deep learning architectures can be applied directly to tensors created using Sym_+^d matrices both linearised and non-linearised. Cascades of RFs can also be applied to vectorised versions of Sym_+^d matrices. Using kernel Support Vector Machines (SVM) allows using kernel-based approaches together with the Riemannian geometry to learn from classifier predictions. As seen we can apply both data transformation and advanced classification architectures to boost classification accuracy. To build Grassmann manifolds we use decision profiles that are decomposed using Singular Value Decomposition (SVD) to create orthonormal subspaces of Euclidean space. To compute distances in these spaces appropriate metrics are applied.

1.1   Motivations About Classifier Selection for an Ensemble

To build a good ensemble according to [9] it is important to have accurate individual classifiers. On the other hand it is important to have diverse classifiers according to kappa-error diagrams which allow to have an error as a function of classifier dissimilarity [9]. It is mentioned in [13] that if we have n independent binary classifiers each characterized by an error ϵ then the classification error of the entire ensemble can be written formally as follows [17]:

$$\epsilon_{ensemble}(n) = \sum_{i=\lfloor \frac{n}{2} \rfloor+1}^{n} \binom{n}{i}\epsilon^i(1-\epsilon)^{n-i}. \tag{1}$$

Putting $n \to \infty$ and $\epsilon < 0.5$ we can see that

$$\epsilon_{ensemble}(\infty) = \lim_{n\to\infty} \sum_{i=\lfloor \frac{n}{2} \rfloor+1}^{n} \binom{n}{i}\epsilon^i(1-\epsilon)^{n-i} = 0. \tag{2}$$

1.2   Nonlinear Manifolds and Their Applications

In some situations, our data to be classified can lie in a space other than Euclidean. This means that we are in a nonlinear space or nonlinear manifold. We assume for the next that we deal with smooth nonlinear manifolds. The principal property of smooth nonlinear manifolds is that they are homeomorphic to Euclidean space. This allows measuring nonlinear manifolds using vector norms that can be applied in the Euclidean space after manifold flattening using appropriate mappings.

1.3   Riemannian Manifolds

A Riemannian manifold of SPD matrices have a geometric representation of a convex open cone (Fig. 1):

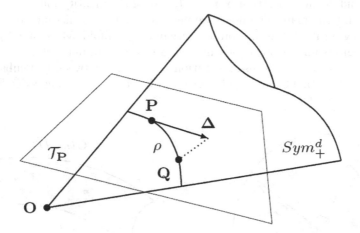

Fig. 1. Geometric visualization of Riemannian manifold of SPD matrices

A pair of Gauss maps (logarithmic and exponential maps) for the affine invariant metric can be written as follows [1,6,16]

$$\exp_{\mathbf{P}}(\mathbf{\Delta}) = \mathbf{P}^{\frac{1}{2}} \exp\left(\mathbf{P}^{-\frac{1}{2}} \mathbf{\Delta} \mathbf{P}^{-\frac{1}{2}}\right) \mathbf{P}^{\frac{1}{2}} = \mathbf{Q}$$
$$\log_{\mathbf{P}}(\mathbf{Q}) = \mathbf{P}^{\frac{1}{2}} \log\left(\mathbf{P}^{-\frac{1}{2}} \mathbf{Q} \mathbf{P}^{-\frac{1}{2}}\right) \mathbf{P}^{\frac{1}{2}} \tag{3}$$

and for the log Euclidean metric [5]

$$\exp_{\mathbf{P}}(\mathbf{\Delta}) = \exp(\log(\mathbf{P}) + \mathbf{\Delta}) = \mathbf{Q}$$
$$\log_{\mathbf{P}}(\mathbf{Q}) = \log(\mathbf{Q}) - \log(\mathbf{P}) = \mathbf{\Delta} \tag{4}$$

Geodesic on Sym_+^d in case of affine invariant metric can be computed as

$$d(\mathbf{P}, \mathbf{Q}) = \sqrt{tr[\log^2\left(\mathbf{P}^{-\frac{1}{2}} \mathbf{Q} \mathbf{P}^{-\frac{1}{2}}\right)]} \tag{5}$$

and for the log Euclidean metric as

$$d(\mathbf{P}, \mathbf{Q}) = ||\log(\mathbf{Q}) - \log(\mathbf{P})|| \tag{6}$$

1.4   Grassmann Manifolds

The Grassmann manifold $\mathcal{G}_D^m$ is the set of all D-dimensional linear subspaces of the $\mathbb{R}^m$ and is $D(m - D)$ dimensional compact Riemannian manifold [4]. So every point on $\mathcal{G}_D^m$ is $span(\mathbf{Y})$, where $\mathbf{Y}$ is an orthonormal basis matrix with D columns and m rows such that $\mathbf{Y}\mathbf{Y}^T = \mathbf{I}$. Size of the identity matrix $\mathbf{I}$ is $D \times D$. To learn on Grassmann manifolds we need to compute a distance between two points lying on it. A distance on a Grassmann manifold $\mathcal{M}$ is called geodesic. Below we give some distance measures on a Grassmann manifold.

It is very convenient to use a notation of an angle between two subspaces Y_1 and Y_2 (see Fig. 2). The principal angles can be computed using SVD. Then the geodesic is given by

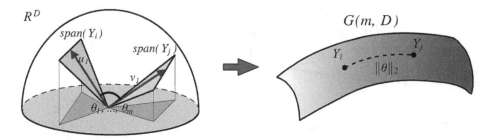

Fig. 2. Principal angles between two subspaces and how to compute distances on a Grassmann manifold

$$d_A(\mathbf{Y}_1, \mathbf{Y}_2) = \Big(\sum_i \theta_i^2\Big)^{1/2} \tag{7}$$

Using the aforementioned notation the projection metric is expressed as follows:

$$d_P(\mathbf{Y}_1, \mathbf{Y}_2) = \Big(\sum_{i=1}^m \sin^2 \theta_i\Big)^{1/2}, \tag{8}$$

the Binet-Cauchy metric by

$$d_{BC}(\mathbf{Y}_1, \mathbf{Y}_2) = \Big(1 - \prod_i \cos^2 \theta_i\Big)^{1/2}, \tag{9}$$

and the Procrustes metric by

$$d_{CF}(\mathbf{Y}_1, \mathbf{Y}_2) = 2\Big(\sum_{i=1}^m \sin^2(\theta_i/2)\Big)^{1/2}. \tag{10}$$

2 Meta-learning of Classifier Ensembles on Riemannian and Grassmann Manifolds

2.1 Homotopy of Transition Between Different Spaces Using Classifier Predictions

To learn from classifier ensemble predictions one needs to accomplish the following mapping: $f : X \rightarrow Z$ and $g : Z \rightarrow Y$. In terms of homotopy this yields $H : X \times \{0, 1\} \rightarrow Y$, s.t. $H(x, 0) = f(x), H(x, 1) = g(f(x)) = (g \circ f)(x)$.

Here X is initial, Z is transformed data obtained as classifier outputs or predictions and Y is a set of labels. Hence a total sequence of mappings or a homotopy is $X \rightarrow Z \rightarrow Y$. In general, we need to learn both $X \rightarrow Z$ and $Z \rightarrow Y$. To skip $X \rightarrow Z$ learning one can use some classifier generator which generates sufficiently independent classifiers. Notwithstanding, in general one needs classifier selection procedure using for instance genetic algorithms (GA) [8]. Classifier selection can also be done using kappa-error diagrams [9] which shows the relationship between the error of the ensemble of two classifiers and the diversity between them. Learning from classifier predictions allows focusing on classifier diversity keeping satisfactory accuracy of individual classifiers. Because of homotopy properties assuming two directional mappings, it is possible to generate class samples having their labels: $Y \rightarrow X$, classifier predictions from class labels: $Y \rightarrow Z$ or initial class instances from classifier predictions. $Z \rightarrow X$.

It is commonly known that any vector can be interpreted as a point in some Euclidean space $\mathbb{R}^n$. However the point can lie on some manifold as well. In contradiction to Euclidean linear vector space a manifold is nonlinear space. Then how do we know where we are? In this case the only indicator is the way how one measures a distance between two points. Riemannian manifolds are equipped with a measure g which is $(0, 2)$ positive definite tensor. Geodesic (the shortest path between two points on nonlinear manifold) can be computed using this measure.

Let us denote $f : \mathbb{R}^n \rightarrow \mathbb{R}^m$ ($X \rightarrow Z$), $\pi_1 : \mathbb{R}^n \rightarrow \mathbb{S}^k$ ($X \rightarrow \mathcal{M}_x$), $\pi_2 : \mathbb{R}^m \rightarrow \mathbb{S}^\ell$ ($Z \rightarrow \mathcal{M}_z$) and $h : \mathbb{S}^k \rightarrow \mathbb{S}^\ell (\mathcal{M}_x \rightarrow \mathcal{M}_z)$. We assume that all mappings are bidirectional, i.e. f^{-1}, h^{-1}, π_1^{-1} and π_2^{-1} exist. Including a point to Riemannian manifold means also a measure mapping $d \rightarrow g$, where d is a distance in Euclidean space and g is a geodesic.

Suppose that we want to accomplish the following mapping $X \rightarrow \mathcal{M}_z$ ($\mathbb{R}^n \rightarrow \mathbb{S}^\ell$). This can be done in two different ways: $X \rightarrow Z$ and $Z \rightarrow \mathcal{M}_z$ or $X \rightarrow \mathcal{M}_x$ and $\mathcal{M}_x \rightarrow \mathcal{M}_z$ (see Fig. 3). The homotopy for the first path can be written as $\pi_2(f(X)) = (\pi_2 \circ f)(X)$. For the second path we have $h(\pi_1(X)) = (h \circ \pi_1)(X)$. Because both mappings end up at the same point we can write $h \circ \pi_1 = \pi_2 \circ f$. Then mapping h ($\mathcal{M}_x \rightarrow \mathcal{M}_z$) can be defined through other mappings as $h = (\pi_2 \circ f) \circ \pi_1^{-1}$. This means that mapping manifold-to-manifold can be learned via three other mappings. This can be done for instance using generative models

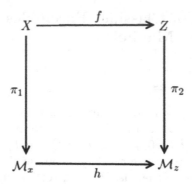

Fig. 3. Homotopy diagram of data transformation using Riemannian manifolds

of neural networks [7,11,12]. It should be noticed that such kind of learning has a local character because Riemannian manifolds are locally isomorphic to the Euclidean space and two gauss maps (logarithmic and exponential) are local.

In case of Grassmann manifolds (build using classifier predictions in form of decision templates) one has only one way of mappings $X \to Z \to \mathcal{M}_z(f \circ \pi)$. This can be done by two consecutive mappings $f : X \to Z(\mathbb{R}^n \to \mathbb{R}^m)$ and $\pi : Z \to \mathcal{M}_z(\mathbb{R}^m \to \mathbb{S}^\ell)$. In this case we end up only with one manifold $(\mathcal{M}_z)$. It is necessary to recall that Grassmann manifolds are particular case of Riemannian manifolds and can be analyzed using Riemannian geometry (geometry of a generalized sphere). For the classification problems we are going to use only $X \to Z \to \mathcal{M}_z(f \circ \pi_2)$ chain of mappings. Classical way of applying manifolds is $X \to \mathcal{M}_x(\pi_1)$ for Riemannian and Grassmann manifolds. However to understand better manifolds structure and what is the beneficial difference between manifold obtained using initial features X: $X \to \mathcal{M}_x(\pi_1)$ and transformed features Z: $Z \to \mathcal{M}_z(\pi_2)$ (Fig. 4).

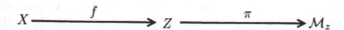

Fig. 4. Homotopy diagram of data transformation using Grassmann manifolds

2.2   Meta-learning of Classifier Ensembles on Riemannian Manifolds

Classical meta-learning in classifier ensembles is done using classifier stacking procedure. The homotopy chain can be presented as $X \to Z \to \mathcal{Y}$, where X is the initial vector of features, Z is the vector of ensemble predictions and $\mathcal{Y}$ is a class label.

Another meta-learning representation can be made using classifier predictions pairwise matrices (CPPM). Then the meta-classifier works with the new data presented as classifier interactions. Advantages of using pairwise interactions are shown in [8]. However in this work authors use deterministic rules to combine data from matrices of pair-wise interactions and we do not learn from classifier predictions. So if we have L classes then we have a vector of prediction probabilities for each classifier that the given pattern belongs to some of L classes. These predictions are conditional probabilities $p(y = c_\ell | X)$, where c_ℓ is the ℓ-th class, $\ell = 1, ..., L$. Let us assume that we have T classifiers in an ensemble. Then we compose a tensor $\mathbf{T}$ of size $T \times T \times L$, where for each class $C_\ell, \ell = 1, ..., L$ we have a CPPM $\mathbf{A}^\ell(x)$ $T \times T$ with elements $a_{ij}^\ell, \{i, j\} = 1, ..., T$:

$$a_{ij}^\ell(x) = p_i(y = c_\ell | X)p_j(y = c_\ell | X) = h_i^\ell(x)h_j^\ell(x), i \neq j;$$
$$a_{ij}^\ell = p_i(y = c_k | X) = h_i(x), i = j, \tag{11}$$

where $h_i^\ell(x) = p_i(y = c_\ell | X)$. Using matrix form we can write $A^\ell(x)$ as

$$\mathbf{A}^\ell(x) = \begin{bmatrix} h_1^\ell(x) & \dots & h_1^\ell(x)h_j^\ell(x) & \dots & h_1^\ell(x)h_T^\ell(x) \\ \vdots & \ddots & \vdots & \ddots & \vdots \\ h_1^\ell(x)h_i^\ell(x) & \dots & h_i^\ell(x) & \dots & h_i^\ell(x)h_T^\ell(x) \\ \vdots & \ddots & \vdots & \ddots & \vdots \\ h_1^\ell(x)h_T^\ell(x) & \dots & h_T^\ell(x)h_j^\ell(x) & \dots & h_T^\ell(x) \end{bmatrix} \tag{12}$$

Thus, we can extract the information presented in these CPPMs which can be used for the final classification. Instead of building every separate CPPM for every class label ℓ [8] we stack all of them in a tensor $\mathbf{T}$. Then this tensor is used as an input to CNN for learning.

To apply other classifiers (such as Support Vector Machines (SVM), k nearest neighbors, multi-layer perceptron, etc.) in Riemannian geometry of SPD matrices one needs to vectorise CPPMs.

2.3 Meta-learning of Classifier Ensembles on Grassmann Manifolds

Let us consider a decision template (DT) originally proposed in [10]. For a probe x predictions from classifiers in an ensemble can be written as decision profile (DP) in the form of the matrix

$$DP(x) = \begin{bmatrix} h_1^1(x) & \dots & h_1^j(x) & \dots & h_1^\ell(x) \\ \vdots & \ddots & \vdots & \ddots & \vdots \\ h_i^1(x) & \dots & h_i^j(x) & \dots & h_i^\ell(x) \\ \vdots & \ddots & \vdots & \ddots & \vdots \\ h_T^1(x) & \dots & h_T^j(x) & \dots & h_T^\ell(x) \end{bmatrix} \tag{13}$$

Instead of using $DP(x)$ to compute a DT

$$DT_k = \frac{1}{m_k} \sum_{i:y_i=c_k} DP(x_i), k = 1, ..., \ell, \tag{14}$$

where m_k is the number of instances belonging to class c_k, one can use them for learning. Since a number of columns represents a number of classes and assuming that $L < T$ one can consider a $DP(x)$ to compute an orthonormal basis using SVD. This is presented as an orthonormal matrix where the number of columns represents a size of subspace $\mathbb{R}^L$ of Euclidean space $\mathbb{R}^T$. To compute dissimilarity between different $DP(x)$ we can use one of the metrics considered before. The classical way of using decision profiles and decision templates assumes to apply some similarity measure and k-NN-based algorithm to compute the proximity between decision profile initiated by instance x and decision template DT_k.

3   Experiments

For the experimental load we used 11 datasets from UCI repository [2] (see Table 1). As can be seen, these datasets have different sizes ranging from 208 to 19200 instances and represent binary classification problems as well as multi-class ones. Selected datasets have different classification difficulty and equal splits between training and testing datasets have been applied to make experiments more challenging [8].

Table 1. Summary of characteristics of used data sets

Dataset	Size	Features	Classes	Tr, %	Ts, %
Balance	625	4	3	50	50
Bupa	345	6	2	50	50
Gamma	19200	10	2	50	50
German	1000	24	2	50	50
Heart	270	13	2	50	50
Mfeat-mor	2000	6	10	50	50
Mfeat-zer	2000	47	10	50	50
Pima	768	8	2	50	50
Segment	2310	19	7	50	50
Sonar	208	60	2	50	50
Spambase	4601	57	2	50	50

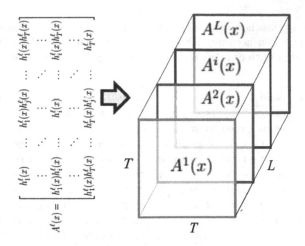

Fig. 5. Tensor formation

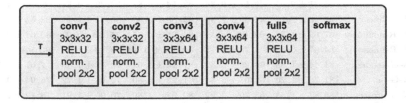

Fig. 6. CNN for prediction tensor learning

As seen from both tables using the Riemannian geometry is advantageous. However, the nonlinearity of the space is better explored for lower depths in trees when building forest-based ensembles. This means that trees with lower depth induce Riemannian manifolds with higher nonlinearity. Overall the best accuracy is obtained on Riemannian manifolds of SPD matrices with CNNs used as the learning algorithm. The Riemannian geometry of SPD matrices allows exploiting both interactions between classifiers in an ensemble and geometry of nonlinear manifolds. Grassmann manifolds perform above average for most of the datasets. Together with Riemannian manifolds of SPD matrices, they perform better than the Euclidean geometry when trees with lower depth are used (Figs. 5, 6 and Tables 2, 3, 4, 5).

Table 2. Learning classifier interactions (advanced experiments): means and standard deviations of prediction accuracy (shown in %) for each method on the benchmark data sets. A number of trees in an ETs classifier is equal to 100 and its depth is equal to 2.

Method/Dataset	Balance	Bupa	Gamma	German	Heart
ETs	$83,11 \pm 2,28$	$58,84 \pm 2,65$	$71,08 \pm 0,39$	$70,54 \pm 1,65$	$79,02 \pm 2,61$
SVM	$\mathbf{89,29 \pm 0,96}$	$59,07 \pm 1,75$	$65,45 \pm 0,40$	$70,10 \pm 1,16$	$79,02 \pm 2,61$
SVM-stacking	$85,51 \pm 1,66$	$58,26 \pm 2,01$	$80,79 \pm 0,89$	$71,98 \pm 2,20$	$79,02 \pm 2,61$
SVM-R-nonlinear	$85,77 \pm 1,76$	$58,20 \pm 1,97$	$81,01 \pm 0,87$	$73,50 \pm 1,80$	$79,02 \pm 2,61$
SVM-R-linear	$66,35 \pm 6,75$	$58,31 \pm 1,95$	$77,07 \pm 1,62$	$70,00 \pm 1,30$	$79,02 \pm 2,61$
kNN	$80,16 \pm 1,80$	$63,90 \pm 2,27$	$79,58 \pm 0,35$	$67,20 \pm 0,68$	$74,66 \pm 2,56$
kNN-stacking	$79,52 \pm 1,39$	$64,30 \pm 4,97$	$81,27 \pm 0,70$	$71,74 \pm 1,68$	$77,97 \pm 2,58$
kNN-R-nonlinear	$79,78 \pm 1,33$	$64,53 \pm 4,97$	$81,30 \pm 0,67$	$71,94 \pm 1,69$	$78,57 \pm 2,76$
kNN-R-linear	$80,19 \pm 1,40$	$63,08 \pm 2,46$	$81,23 \pm 0,72$	$71,12 \pm 1,63$	$75,49 \pm 2,74$
MLP	$47,47 \pm 2,61$	$44,83 \pm 8,44$	$83,10 \pm 0,61$	$72,24 \pm 1,47$	$45,49 \pm 23,88$
MLP-stacking	$46,57 \pm 1,67$	$44,30 \pm 6,33$	$83,66 \pm 0,47$	$73,72 \pm 1,46$	$22,18 \pm 3,85$
MLP-R-nonlinear	$46,31 \pm 2,05$	$46,57 \pm 9,34$	$83,54 \pm 0,81$	$73,32 \pm 1,58$	$27,14 \pm 18,22$
MLP-R-linear	$45,93 \pm 2,12$	$44,88 \pm 7,46$	$83,58 \pm 0,88$	$56,24 \pm 9,84$	$23,53 \pm 4,68$
CNN-R-nonlinear	$87,69 \pm 1,24$	$\mathbf{67,09 \pm 2,86}$	$84,05 \pm 0,75$	$\mathbf{75,60 \pm 0,88}$	$80,68 \pm 2,99$
CNN-R-linear	$84,97 \pm 0,66$	$63,78 \pm 3,13$	$\mathbf{84,17 \pm 0,44}$	$73,72 \pm 1,39$	$\mathbf{80,98 \pm 2,10}$
kNN-G-geodesic	$81,44 \pm 1,33$	$63,08 \pm 3,82$	$81,16 \pm 1,01$	$70,14 \pm 1,49$	$77,67 \pm 1,75$
kNN-G-bc	$81,76 \pm 1,20$	$62,67 \pm 3,84$	$81,23 \pm 0,80$	$70,64 \pm 1,72$	$75,94 \pm 4,15$
kNN-G-projection	$80,83 \pm 1,33$	$63,20 \pm 3,92$	$80,97 \pm 1,59$	$70,90 \pm 1,91$	$78,12 \pm 2,59$

Table 3. Learning classifier interactions (advanced experiments): means and standard deviations of prediction accuracy (shown in %) for each method on the benchmark data sets. A number of trees in an ETs classifier is equal to 100 and its depth is equal to 2.

Method/Dataset	Mfeat-mor	Mfeat-zer	Pima	Segment	Sonar	Spambase
ETs	$65,49 \pm 2,42$	$61,01 \pm 3,26$	$66,43 \pm 1,48$	$76,30 \pm 6,07$	$75,10 \pm 2,25$	$66,10 \pm 1,84$
SVM	$19,73 \pm 1,77$	$9,47 \pm 2,20$	$65,76 \pm 0,75$	$41,43 \pm 3,13$	$54,04 \pm 2,85$	$80,92 \pm 0,45$
SVM-stacking	$57,98 \pm 1,84$	$38,84 \pm 3,95$	$75,78 \pm 1,62$	$71,90 \pm 5,64$	$75,87 \pm 2,96$	$89,88 \pm 0,64$
SVM-R-nonlinear	$28,78 \pm 5,64$	$14,66 \pm 2,21$	$76,22 \pm 1,54$	$65,00 \pm 6,99$	$75,38 \pm 2,48$	$90,40 \pm 0,61$
SVM-R-linear	$69,18 \pm 0,79$	$70,51 \pm 1,27$	$65,70 \pm 0,74$	$84,21 \pm 3,94$	$64,81 \pm 4,35$	$79,25 \pm 3,70$
kNN	$39,02 \pm 1,20$	$\mathbf{80,82 \pm 0,88}$	$71,06 \pm 1,81$	$91,72 \pm 0,84$	$75,87 \pm 4,22$	$78,77 \pm 0,71$
kNN-stacking	$67,18 \pm 1,55$	$71,76 \pm 0,79$	$71,67 \pm 1,98$	$82,57 \pm 5,64$	$78,75 \pm 2,67$	$86,55 \pm 0,80$
kNN-R-nonlinear	$67,07 \pm 1,41$	$70,33 \pm 0,91$	$71,54 \pm 1,83$	$82,35 \pm 5,35$	$\mathbf{78,85 \pm 2,72}$	$87,48 \pm 1,66$
kNN-R-linear	$66,69 \pm 1,14$	$70,88 \pm 1,54$	$72,11 \pm 1,91$	$88,54 \pm 1,95$	$73,75 \pm 4,53$	$90,87 \pm 0,86$
MLP	$17,31 \pm 4,50$	$10,59 \pm 1,17$	$73,62 \pm 2,40$	$15,74 \pm 4,36$	$47,98 \pm 3,11$	$77,59 \pm 5,29$
MLP-stacking	$11,42 \pm 2,84$	$10,55 \pm 0,60$	$73,44 \pm 3,88$	$14,64 \pm 0,40$	$48,75 \pm 2,36$	$92,08 \pm 0,56$
MLP-R-nonlinear	$10,69 \pm 1,48$	$10,43 \pm 1,12$	$72,92 \pm 5,07$	$18,62 \pm 5,07$	$53,56 \pm 10,19$	$91,38 \pm 0,58$
MLP-R-linear	$10,70 \pm 1,02$	$10,55 \pm 0,87$	$72,99 \pm 3,20$	$20,31 \pm 6,56$	$46,92 \pm 1,96$	$91,80 \pm 0,83$
CNN-R-nonlinear	$71,79 \pm 1,33$	$74,20 \pm 1,25$	$\mathbf{77,32 \pm 1,20}$	$86,83 \pm 5,70$	$78,17 \pm 3,13$	$92,45 \pm 0,68$
CNN-R-linear	$\mathbf{72,98 \pm 0,75}$	$76,45 \pm 1,22$	$75,89 \pm 1,12$	$90,21 \pm 1,53$	$74,81 \pm 3,24$	$\mathbf{92,60 \pm 0,81}$
kNN-G-geodesic	$67,83 \pm 1,51$	$75,45 \pm 1,47$	$72,06 \pm 1,55$	$94,36 \pm 0,88$	$76,15 \pm 3,03$	$88,20 \pm 1,09$
kNN-G-bc	$68,24 \pm 1,39$	$74,44 \pm 1,48$	$71,93 \pm 0,96$	$93,86 \pm 1,77$	$74,52 \pm 3,85$	$86,63 \pm 1,64$
kNN-G-projection	$69,00 \pm 1,46$	$75,85 \pm 1,21$	$72,14 \pm 1,46$	$\mathbf{94,53 \pm 1,36}$	$73,56 \pm 3,20$	$86,80 \pm 1,06$

Table 4. Learning classifier interactions (advanced experiments): means and standard deviations of prediction accuracy (shown in %) for each method on the benchmark data sets. A number of trees in an ETs classifier is equal to 100 and its depth is equal to 5.

Method/Dataset	Balance	Bupa	Gamma	German	Heart
ETs	$86, 89 \pm 2, 08$	$63, 60 \pm 4, 44$	$78, 68 \pm 0, 50$	$72, 34 \pm 1, 44$	$79, 02 \pm 2, 80$
SVM	$\mathbf{89, 29 \pm 0, 96}$	$59, 07 \pm 1, 75$	$65, 45 \pm 0, 40$	$70, 10 \pm 1, 16$	$79, 02 \pm 2, 61$
SVM-stacking	$87, 12 \pm 1, 65$	$68, 55 \pm 3, 77$	$84, 60 \pm 0, 53$	$74, 70 \pm 1, 55$	$81, 73 \pm 2, 33$
SVM-R-nonlinear	$86, 70 \pm 1, 99$	$69, 19 \pm 3, 86$	$79, 32 \pm 0, 71$	$\underline{75, 08 \pm 1, 48}$	$81, 80 \pm 1, 98$
SVM-R-linear	$86, 70 \pm 1, 99$	$59, 42 \pm 3, 84$	$84, 31 \pm 0, 48$	$71, 22 \pm 1, 41$	$79, 55 \pm 3, 04$
kNN	$80, 16 \pm 1, 80$	$63, 90 \pm 2, 27$	$79, 58 \pm 0, 35$	$67, 20 \pm 0, 68$	$74, 66 \pm 2, 56$
kNN-stacking	$81, 06 \pm 1, 11$	$67, 97 \pm 3, 27$	$83, 74 \pm 0, 41$	$72, 60 \pm 1, 11$	$80, 08 \pm 2, 50$
kNN-R-nonlinear	$81, 19 \pm 1, 54$	$68, 72 \pm 3, 09$	$83, 66 \pm 0, 32$	$72, 60 \pm 1, 58$	$81, 50 \pm 2, 43$
kNN-R-linear	$79, 10 \pm 1, 91$	$64, 13 \pm 3, 45$	$83, 40 \pm 0, 46$	$69, 62 \pm 1, 52$	$75, 04 \pm 2, 12$
MLP	$47, 47 \pm 2, 61$	$44, 83 \pm 8, 44$	$83, 10 \pm 0, 61$	$72, 24 \pm 1, 47$	$45, 49 \pm 23, 88$
MLP-stacking	$46, 47 \pm 1, 84$	$44, 13 \pm 6, 42$	$85, 75 \pm 0, 41$	$75, 00 \pm 0, 93$	$21, 20 \pm 2, 77$
MLP-R-nonlinear	$46, 76 \pm 1, 66$	$43, 49 \pm 4, 22$	$85, 40 \pm 0, 96$	$\mathbf{75, 84 \pm 1, 30}$	$29, 92 \pm 19, 11$
MLP-R-linear	$46, 67 \pm 1, 84$	$42, 27 \pm 2, 60$	$85, 58 \pm 0, 27$	$72, 12 \pm 1, 98$	$28, 27 \pm 14, 90$
CNN-R-nonlinear	$\underline{87, 98 \pm 2, 31}$	$\mathbf{71, 86 \pm 3, 01}$	$\mathbf{85, 76 \pm 0, 41}$	$74, 74 \pm 1, 64$	$\mathbf{82, 93 \pm 2, 65}$
CNN-R-linear	$87, 44 \pm 1, 95$	$66, 51 \pm 2, 62$	$85, 52 \pm 0, 26$	$73, 94 \pm 1, 08$	$\underline{81, 88 \pm 2, 39}$
kNN-G-geodesic	$79, 48 \pm 2, 04$	$64, 13 \pm 2, 65$	$83, 18 \pm 0, 61$	$70, 88 \pm 1, 21$	$75, 34 \pm 3, 65$
kNN-G-bc	$79, 17 \pm 1, 71$	$63, 55 \pm 2, 11$	$83, 76 \pm 0, 96$	$70, 54 \pm 1, 23$	$74, 74 \pm 3, 23$
kNN-G-projection	$80, 03 \pm 1, 91$	$64, 13 \pm 2, 01$	$83, 27 \pm 0, 97$	$70, 84 \pm 1, 32$	$74, 29 \pm 2, 85$

Table 5. Learning classifier interactions (advanced experiments): means and standard deviations of prediction accuracy (shown in %) for each method on the benchmark data sets. A number of trees in an ETs classifier is equal to 100 and its depth is equal to 5.

Method/Dataset	Mfeat-mor	Mfeat-zer	Pima	Segment	Sonar	Spambase
ETs	$69, 35 \pm 0, 76$	$74, 78 \pm 1, 66$	$74, 14 \pm 1, 41$	$90, 37 \pm 1, 54$	$79, 42 \pm 2, 62$	$82, 43 \pm 1, 33$
SVM	$19, 73 \pm 1, 77$	$9, 47 \pm 2, 20$	$65, 76 \pm 0, 75$	$41, 43 \pm 3, 13$	$54, 04 \pm 2, 85$	$80, 92 \pm 0, 45$
SVM-stacking	$69, 58 \pm 0, 60$	$73, 32 \pm 0, 80$	$76, 28 \pm 1, 84$	$92, 00 \pm 1, 27$	$\underline{82, 12 \pm 1, 78}$	$92, 37 \pm 0, 45$
SVM-R-nonlinear	$65, 79 \pm 2, 33$	$39, 13 \pm 3, 91$	$76, 30 \pm 1, 58$	$90, 20 \pm 1, 22$	$82, 12 \pm 1, 98$	$92, 46 \pm 0, 50$
SVM-R-linear	$69, 72 \pm 0, 62$	$78, 19 \pm 0, 48$	$70, 44 \pm 1, 23$	$93, 14 \pm 1, 30$	$78, 08 \pm 2, 71$	$85, 94 \pm 1, 27$
kNN	$39, 02 \pm 1, 20$	$\mathbf{80, 82 \pm 0, 88}$	$71, 06 \pm 1, 81$	$91, 72 \pm 0, 84$	$75, 87 \pm 4, 22$	$78, 77 \pm 0, 71$
kNN-stacking	$68, 47 \pm 1, 02$	$75, 95 \pm 1, 24$	$72, 60 \pm 0, 99$	$95, 61 \pm 0, 93$	$80, 76 \pm 2, 94$	$91, 01 \pm 1, 22$
kNN-R-nonlinear	$67, 90 \pm 1, 10$	$70, 61 \pm 1, 26$	$72, 50 \pm 1, 35$	$95, 29 \pm 1, 04$	$80, 87 \pm 2, 56$	$91, 68 \pm 0, 82$
kNN-R-linear	$67, 74 \pm 1, 00$	$75, 72 \pm 0, 74$	$68, 57 \pm 2, 50$	$95, 31 \pm 0, 97$	$79, 33 \pm 2, 99$	$91, 74 \pm 0, 56$
MLP	$17, 31 \pm 4, 50$	$10, 59 \pm 1, 17$	$73, 62 \pm 2, 40$	$15, 74 \pm 4, 36$	$47, 98 \pm 3, 11$	$77, 59 \pm 5, 29$
MLP-stacking	$10, 27 \pm 0, 87$	$10, 27 \pm 0, 68$	$75, 78 \pm 1, 38$	$14, 48 \pm 0, 47$	$55, 38 \pm 12, 56$	$93, 25 \pm 0, 59$
MLP-R-nonlinear	$10, 62 \pm 1, 10$	$11, 43 \pm 2, 47$	$\underline{76, 77 \pm 0, 98}$	$15, 83 \pm 3, 80$	$46, 73 \pm 1, 62$	$93, 21 \pm 0, 52$
MLP-R-linear	$11, 15 \pm 1, 73$	$10, 53 \pm 0, 66$	$74, 79 \pm 1, 41$	$16, 45 \pm 4, 71$	$51, 73 \pm 9, 68$	$93, 00 \pm 0, 61$
CNN-R-nonlinear	$72, 18 \pm 0, 81$	$78, 10 \pm 0, 74$	$\mathbf{77, 11 \pm 1, 40}$	$95, 08 \pm 0, 87$	$\mathbf{82, 79 \pm 1, 94}$	$\mathbf{93, 57 \pm 0, 54}$
CNN-R-linear	$\mathbf{73, 31 \pm 0, 62}$	$78, 71 \pm 1, 12$	$75, 13 \pm 1, 70$	$\mathbf{96, 76 \pm 0, 61}$	$82, 12 \pm 1, 83$	$93, 38 \pm 0, 56$
kNN-G-geodesic	$69, 28 \pm 0, 68$	$77, 31 \pm 0, 83$	$72, 11 \pm 2, 23$	$95, 75 \pm 0, 74$	$68, 17 \pm 4, 09$	$90, 80 \pm 1, 59$
kNN-G-bc	$68, 80 \pm 0, 96$	$74, 73 \pm 0, 63$	$72, 89 \pm 1, 64$	$95, 67 \pm 0, 85$	$65, 67 \pm 4, 43$	$90, 05 \pm 1, 51$
kNN-G-projection	$69, 27 \pm 0, 57$	$77, 06 \pm 1, 17$	$72, 11 \pm 1, 85$	$95, 45 \pm 0, 81$	$65, 87 \pm 4, 68$	$89, 45 \pm 1, 07$

4   Conclusions and Future Work

In this work we reported results of our research on classifier ensembles stacking in the Euclidean and Riemannian geometries. Experimental results show advantage of using the Riemannian manifolds of SPD matrices to learn different meta-classifiers such as SVMs, k-NNs, MLPs and CNNs. This is confirmed for the variety of classification data sets from UCI depository.

As part of our future research we are going to compute more characteristics of a manifold such as the Riemannian curvature tensor and Ricci flow tensors on Riemannian manifolds of SPD matrices and Grassmann manifolds. It is interesting to implement Riemannian geometry in deep learning architectures. There are some works on how to use Riemannian and Grassmann manifolds for classification and generative models.

Acknowledgements. The Authors would like to thank the anonymous referees for valuable comments which helped to improve the paper.

References

1. Dong, G., Kuang, G.: Target recognition in SAR images via classification on Riemannian manifolds. IEEE Geosci. Remote Sens. Lett. **12**(1), 199–203 (2014)
2. Frank, A.: UCI machine learning repository (2010). http://archive.ics.uci.edu/ml
3. Goodfellow, I., Bengio, Y., Courville, A.: Deep Learning. MIT Press, Cambridge (2016)
4. Huang, Z., Wang, R., Shan, S., Chen, X.: Projection metric learning on Grassmann manifold with application to video based face recognition. In: Proceedings of the IEEE Conference on Computer Vision and Pattern Recognition, pp. 140–149 (2015)
5. Huang, Z., Wang, R., Shan, S., Van Gool, L., Chen, X.: Cross Euclidean-to-Riemannian metric learning with application to face recognition from video. IEEE Trans. Pattern Anal. Mach. Intell. **40**(12), 2827–2840 (2017)
6. Jayasumana, S., Hartley, R., Salzmann, M., Li, H., Harandi, M.: Kernel methods on Riemannian manifolds with Gaussian RBF kernels. IEEE Trans. Pattern Anal. Mach. Intell. **37**(12), 2464–2477 (2015)
7. Kingma, D.P., Welling, M.: Auto-encoding variational Bayes. In: Proceedings of the 2nd International Conference on Learning Representations (2013)
8. Ko, A.H., Sabourin, R., de Souza Britto, A., Oliveira, L.: Pairwise fusion matrix for combining classifiers. Pattern Recogn. **40**(8), 2198–2210 (2007)
9. Kuncheva, L.I.: A bound on kappa-error diagrams for analysis of classifier ensembles. IEEE Trans. Knowl. Data Eng. **25**(3), 494–501 (2013)
10. Kuncheva, L.I., Bezdek, J.C., Duin, R.P.: Decision templates for multiple classifier fusion: an experimental comparison. Pattern Recogn. **34**(2), 299–314 (2001)
11. Oord, A.v.d., Kalchbrenner, N., Kavukcuoglu, K.: Pixel recurrent neural networks. In: Proceedings of the 33rd International Conference on Machine Learning, pp. 1747–1756 (2016)
12. Radford, A., Metz, L., Chintala, S.: Unsupervised representation learning with deep convolutional generative adversarial networks. In: Proceedings of the 4th International Conference on Learning Representations (2016)
13. Rokach, L.: Decision forest: twenty years of research. Inf. Fusion **27**, 111–125 (2016)

14. Schölkopf, B., Smola, A.J.: Learning with Kernels: Support Vector Machines, Regularization, Optimization, and Beyond. MIT Press, Cambridge (2001)
15. Szegedy, C., et al.: Going deeper with convolutions. In: Proceedings of the IEEE Conference on Computer Vision and Pattern Recognition, pp. 1–9 (2015)
16. Xie, X., Yu, Z.L., Lu, H., Gu, Z., Li, Y.: Motor imagery classification based on bilinear sub-manifold learning of symmetric positive-definite matrices. IEEE Trans. Neural Syst. Rehabil. Eng. **25**(6), 504–516 (2016)
17. Zhou, Z.H.: Ensemble Methods: Foundations and Algorithms. Chapman and Hall/CRC, Boca Raton, FL, USA (2012)
18. Zhou, Z.H., Feng, J.: Deep forest: towards an alternative to deep neural networks. In: Proceedings of the 26th International Joint Conference on Artificial Intelligence (IJCAI 2017), pp. 3553–3559 (2017)

PD-DARTS: Progressive Discretization Differentiable Architecture Search

Yonggang Li[iD], Yafeng Zhou[iD], Yongtao Wang[(✉)][iD], and Zhi Tang[iD]

Wangxuan Institute of Computer Technology, Peking University, Beijing, China
wyt@pku.edu.cn

Abstract. Architecture design is a crucial step for neural-network-based methods, and it requires years of experience and extensive work. Encouragingly, with recently proposed neural architecture search (NAS), the architecture design process could be automated. In particular, differentiable architecture search (DARTS) reduces the time cost of search to a couple of GPU days. However, due to the inconsistency between the architecture search and evaluation of DARTS, its performance has yet to be improved. We propose two strategies to narrow the search/evaluation gap: firstly, rectify the operation with the highest confidence; secondly, prune the operation with the lowest confidence iteratively. Experiments show that our method achieves 2.46%/2.48% (test error, Strategy 1 or 2) on CIFAR-10 and 16.48%/16.15% (test error, Strategy 1 or 2) on CIFAR-100 at a low cost of 11 or 8 (Strategy 1 or 2) GPU hours, and outperforms state-of-the-art algorithms.

Keywords: Differentiable Architecture Search · Neural Architecture Search · Convolutional neural network

1 Introduction

Designing an effective convolution neural network is important for computer vision tasks. In the early years, the networks, such as AlexNet, ResNet, and DenseNet, are usually designed by experts.

In recent years, the community focuses on the automatic neural architecture search (NAS) [7], which has achieved promising results in tasks like image classification. The early architecture search algorithms utilize reinforce learning and evaluation algorithm to search a reusable cell and stack the searched cell to compose the final network.

However, those search algorithms require thousands of GPU days to search for an effective neural network, which is very time-consuming. DARTS [3] propose to relax the discrete search space to continuous search space, therefore the architecture parameters become differentiable and search algorithm can directly adopt stochastic gradient descent (SGD), which reduces the search cost hugely.

Nevertheless, the architecture of DARTS in the search stage is hugely different from the architecture used in the evaluation stage, which hinders the performance of DARTS. In this paper, we focus on the architecture gap and propose the progressive discretization DARTS (PD-DARTS).

© Springer Nature Switzerland AG 2020
Y. Lu et al. (Eds.): ICPRAI 2020, LNCS 12068, pp. 306–311, 2020.
https://doi.org/10.1007/978-3-030-59830-3_26

2   Proposed Method

2.1   Preliminary: DARTS

DARTS, or Differentiable Architecture Search, solves the problem of extremely high computation cost in previous Neural Architecture Search (NAS) methods. It breaks down the network search problem to the search of cells and nodes within cells. The network is composed of stacked or recursively connected cells, and every cell is a directed acyclic graph of nodes. Figure 1a shows a cell with four nodes and there are three default operations between every possible pair of nodes. Mathematically, if we rank the nodes topologically, the representation of node x_j is based on all of its predecessors:

$$x^{(j)} = \sum_{i<j} o^{(i,j)} \left(x^{(i)} \right). \tag{1}$$

Nodes x_i, x_j is connected by the edge $o^{(i,j)}$. The operation behind the edge is unknown initially but will be determined as the most likely operation from a set of pre-defined operations after training. It can be represented as a softmax mixture over all possible operations:

$$\bar{o}^{(i,j)} \left(x^{(i)} \right) = \sum_{o \in \mathcal{O}} \frac{\exp(\alpha_o^{(i,j)})}{\sum_{o' \in \mathcal{O}} \exp(\alpha_{o'}^{(i,j)})} o(x^{(i)}), \tag{2}$$

where $\alpha^{(i,j)}$ is the operation weight. When the search stops, the operation with the highest weight is chosen.

The search goal of DARTS is two-fold: the system needs to find the optimal weights within the network to minimize training loss, and to eventually obtain the optimal α to minimize validation loss and guarantee the generalization ability.

Simply put, α determines the network architecture while w represents the weights of convolution filters for all possible architectures. In the training stage, they fix the network architecture and search for the best weights. In the validation stage, they fix the weights and search for the best architecture. Therefore, the objective function of DARTS is:

$$\min_{\alpha} \mathcal{L}_{val} \left(w^*(\alpha), \alpha \right),$$
$$\text{s.t. } w^*(\alpha) = \operatorname*{argmin}_{w} \mathcal{L}_{train}(w, \alpha). \tag{3}$$

2.2   PD-DARTS: Progressive Discretization DARTS

We find that there exists a large **architecture gap** between the search stage and the evaluation stage of DARTS. In the search stage of DARTS, there exist 8 candidate operations in every connection of DAG nodes. However, in the evaluation stage, DARTS only reserves the operation with the highest probability

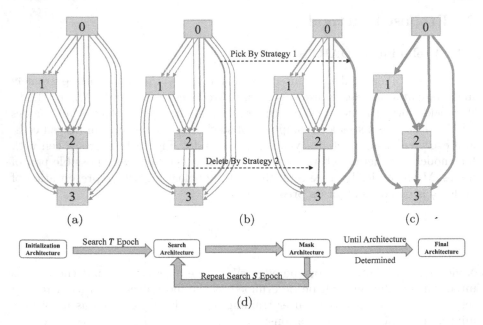

Fig. 1. The pipeline of PD-DARTS. Firstly, we randomly initialize the architecture (or cells) as in DARTS, which ensures all candidate operations have nearly equal probability. Secondly, we utilize the 1st order search algorithm of DARTS to search T epoch. Thirdly, we mask the architecture by **Strategy 1** or **Strategy 2** and continue search S epoch. This procedure is repeated until the operation in each connection is determined. Finally, we can get the final architecture from the above procedure directly.

in each connection of the final architecture and re-trains the found architecture to get the final performance. Consequently, the architecture found by the search stage of DARTS cannot guarantee the best performance in the evaluation stage.

To narrow the aforementioned gap between search and evaluation, we propose an algorithm to progressively eliminate some candidate operations in edges. Formally speaking, we add a mask parameter $M^{(i,j)} \in \mathbb{R}^m$ for each edge (i,j), where $m = \|O\|$ is the number of candidate operations. The mask parameter represents the status of whether the operation is reversed in the search stage. With the mask parameter, the probability of each operation $o \in \mathcal{O}$ in edge (i,j) can be represented as below:

$$p_o^{(i,j)} = \frac{\exp(\alpha_o^{(i,j)}) \cdot M_o^{(i,j)}}{\sum_{o' \in \mathcal{O}} \exp(\alpha_{o'}^{(i,j)}) \cdot M_{o'}^{(i,j)}}. \tag{4}$$

We first initialize each element of parameter M to one, which means each operation is reversed in the initialization stage. Then, we update the mask parameter by any of the below strategies. The pipeline of our method is shown as Fig. 1. With the proposed strategies to update M, we progressively close the gap between search and evaluation stages.

Strategy 1. Fix the operation in the edge with the highest confidence, that is:

$$M_o^{(i,j)} = 0, \quad \forall o \neq \underset{o' \in \mathcal{O}}{\operatorname{argmax}} \, p_{o'}^{(i',j')}, \quad \text{s.t. } (i,j) = \underset{(i',j')}{\operatorname{argmax}} \, \underset{o' \in \mathcal{O}}{\max} \, p_{o'}^{(i',j')}. \tag{5}$$

As shown in Eq. (5), Strategy 1 uses the maximum probability of each operation in the edge as the confidence of the edge. It is reasonable since the larger the maximum probability of the edge is, the more likely it is that the operation in the edge will not change. When we find the edge with the highest confidence, we fix its operation to the operation with the largest probability.

Strategy 2. Remove the operation with the lowest probability among all remaining operations, that is:

$$M_o^{(i,j)} = 0, \quad \text{s.t.} (i,j,o) = \underset{(i',j',o')}{\operatorname{argmin}} \, p_{o'}^{(i',j')}, \quad \text{where } M_{o'}^{(i',j')} \neq 0. \tag{6}$$

Different from Strategy 1, we narrow the gap of search space in a more fine-grained approach in Strategy 2, as shown in Eq. (6). In the search stage, we only remove the most unlikely operation in each edge, which enables other operations to stand out in further search.

3   Experiments

Dataset. We search architecture on CIFAR-10, and evaluate the found architecture on CIFAR-10 and CIFAR-100.

Implementation Details. Our method is implemented using PyTorch 0.3 and evaluated on a single NVIDIA Titan Xp GPU. We build our method upon DARTS, and follow the search space and settings as in DARTS.

In the search stage, the hyper-parameters are the same with DARTS except the initial learning rate is set to 0.1. For Strategy 1, we search 50 epochs totally, where we first search $T = 30$ epochs and then update M every $S = 2$ epochs. For Strategy 2, we search 40 epochs totally, where we first search $T = 10$ epochs and then update M every $S = 100$ iterations. Specifically, for Strategy 2, in order to keep the search procedure stable, we explicitly force the range of the number of removed operations in all edges to be less than or equal to 1, i.e. $|\max(\#OpsRemoved) - \min(\#OpsRemoved)| \leq 1$. Figure 2 shows the final architecture produced by PD-DARTS.

In the evaluation stage, we re-train the architecture for another 600 epochs with batch size 96. We use the same hyper-parameters as DARTS except that weight decay is set to 3×10^{-4} for CIFAR-10 and 5×10^{-4} for CIFAR-100. Furthermore, we set the initial channel to 32 rather than 36 to keep the params comparable with DARTS.

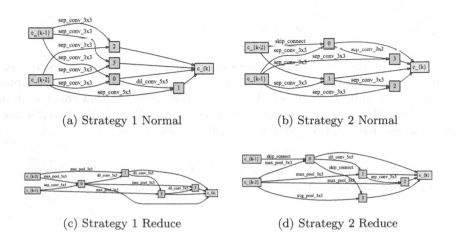

(a) Strategy 1 Normal   (b) Strategy 2 Normal

(c) Strategy 1 Reduce   (d) Strategy 2 Reduce

Fig. 2. The found architecture of PD-DARTS.

Table 1. Comparison with state-of-the-art NAS methods on CIFAR-10 and CIFAR-100. Some methods only report the best results, while others report the average results as *mean + std*.

Architecture	Search dataset	Test error. (%)		Param (M)	Search cost (GPU days)	Search method
		CIFAR-10	CIFAR-100			
NASNet-A + cutout [7]	CIFAR-10	2.65	–	3.3	1800	RL
AmoebaNet-B [5]	CIFAR-10	2.55 ± 0.05	–	2.8	3150	Evolution
ENAS [4]	CIFAR-10	2.89	–	4.6	0.5	RL
DARTS (1st order) + cutout [3]	CIFAR-10	2.94	17.76	2.9	1.5	Gradient-based
DARTS (2nd order) + cutout [3]	CIFAR-10	2.83 ± 0.06	17.54	3.4	4.0	Gradient-based
P-DARTS + cutout [2]	CIFAR-10	2.50	16.55	3.4	0.3	Gradient-based
P-DARTS + cutout [2]	CIFAR-100	2.62	15.92	3.6	0.3	Gradient-based
PC-DARTS + cutout [6]	CIFAR-10	2.57 ± 0.07	–	3.6	0.1	Gradient-based
ProxylessNAS + cutout [1]	CIFAR-10	2.08	–	5.7	4.0	Gradient-based
PD-DARTS (Strategy 1) + cutout	CIFAR-10	2.46 (2.63 ± 0.17)	16.48 (16.64 ± 0.23)	3.43	0.46	Gradient-based
PD-DARTS (Strategy 2) + cutout	CIFAR-10	2.48 (2.57 ± 0.12)	16.15 (16.19 ± 0.06)	3.20	0.33	Gradient-based

Result. We evaluate our methods on both CIFAR-10 and CIFAR-100 datasets. In order to verify the generalization ability of the found architecture, we search architecture with only CIFAR-10 dataset but train it with both CIFAR-10 and CIFAR-100 datasets. As shown in Table 1, Strategy 2 achieves better results compared with Strategy 1. Although the performance of Strategy 1 is better than Strategy 2 in CIFAR-10 dataset, the architecture found by Strategy 2 is more stable than Strategy 1. Furthermore, the proposed PD-DARTS achieves comparable results as opposed to P-DARTS and PC-DARTS, and surpasses DARTS by a large margin. Compared with P-DARTS and PC-DARTS, the architecture found by our method obtains better performance and the parameter scale is smaller than architectures found by those methods. Moreover, our method has a lower search cost: less than 0.5 GPU days.

Acknowledgments. This work is supported by the National Natural Science Foundation of China under Grant 61673029. This work is also a research achievement of Key Laboratory of Science, Technology, and Standard in Press Industry (Key Laboratory of Intelligent Press Media Technology).

References

1. Cai, H., Zhu, L., Han, S.: ProxylessNAS: direct neural architecture search on target task and hardware. In: International Conference on Learning Representations (2019)
2. Chen, X., Xie, L., Wu, J., Tian, Q.: Progressive differentiable architecture search: bridging the depth gap between search and evaluation. In: International Conference on Computer Vision (2019)
3. Liu, H., Simonyan, K., Yang, Y.: DARTS: differentiable architecture search. In: International Conference on Learning Representations (2019)
4. Pham, H., Guan, M.Y., Zoph, B., Le, Q.V., Dean, J.: Efficient neural architecture search via parameter sharing. In: International Conference on Machine Learning, pp. 4092–4101 (2018)
5. Real, E., Aggarwal, A., Huang, Y., Le, Q.V.: Regularized evolution for image classifier architecture search. In: AAAI Conference on Artificial Intelligence (2019)
6. Xu, Y., et al.: PC-DARTS: partial channel connections for memory-efficient architecture search. In: International Conference on Learning Representations (2020)
7. Zoph, B., Le, Q.V.: Neural architecture search with reinforcement learning. In: International Conference on Learning Representations (2017)

Transformed Network Based on Task-Driven

Hongjun Li[1,2,3,4](✉) ⓘ, Ze Zhou[1] ⓘ, Chaobo Li[1] ⓘ, and Qi Wang[1] ⓘ

[1] School of Information Science and Technology, Nantong University, Nantong 226019, China
lihongjun@ntu.edu.cn
[2] State Key Laboratory for Novel Software Technology,
Nanjing University, Nanjing 210023, China
[3] Nantong Research Institute for Advanced Communication Technologies, Nantong 226019,
China
[4] TONGKE School of Microelectronics, Nantong 226019, China

Abstract. In view of the fact that the current networks mostly improve the network performance and robustness by proposing multiple strategies to update the network parameters, this paper puts forward a novel task-driven transformed network idea. With optimizing network parameters, the network structure can be optimized by vertical update, horizontal update and cross update based on the flat network only containing the input layer and the output layer. A multi-branch error back-propagation mechanism is proposed to match the unfixed network on purpose, which is beneficial to explore the function of each hidden layer and flexible to train the network, ensuring optimal network structure for each update. Objective contribution evaluation indicators based on the tasks are also established to analyze the contribution of each strategy to the network performance and guide the subsequent network structure update. Extensive simulation results show that the method proposed is feasible, specially, the recognition accuracy of testing obtains fine results and the network training of speed is real-time on one-dimension and two-dimension image datasets.

Keywords: Network structure update · Multi-branch error · Contribution evaluation indicators

1 Motivation

In order to improve the performance and robustness of networks, the current networks focus on updating network parameters mostly [1, 2]. For example, transfer learning [3–5] is to transfer the knowledge model learned on other tasks into the target task, where the restructured network obtains initial parameters well, and continuously fine-tunes the network parameters to achieve satisfactory performance and robustness [6]. Since the parameters of the restructured network are optimized under a fixed structure, it's difficult to convince this structure is most suitable for the task, and the reliability of the optimal performance obtained is insufficient. Naturally, we think the network structure can also be transformed like updating the network parameters. The conventional error back-propagation mechanism [7, 8], is performed under a fixed structure, where the error

© Springer Nature Switzerland AG 2020
Y. Lu et al. (Eds.): ICPRAI 2020, LNCS 12068, pp. 312–318, 2020.
https://doi.org/10.1007/978-3-030-59830-3_27

is calculated between the network output and the target output, and then backpropagated layer by layer, which may not work on unfixed network. Therefore, it is necessary to propose an error back-propagation mechanism for matching the transformed network. In addition, various strategies are proposed to improve the network performance. For example, Rehman et al. [9] proposed to supervise a Convolutional Neural Network (CNN) classifier by introducing a disparity layer within CNN to learn the dynamic disparity-maps, obtaining promising performance and good generalization ability. To accelerate convolutional neural network, Zeng [10] removed both types of redundancies in network, in which the interspatial redundancy is converted into interkernel redundancy, and then the interkernel redundancy is removed by rank-selection and pruning methods. However, little attention are payed on exploring the contribution of strategies to the network performance, and it's hard to be aware of how to improve the performance further according to the contribution rate. Thus some objective evaluation indicators need to be proposed to analyze the contribution of various strategies.

2   Method

Aim to improve the robustness of the network to multiple tasks, the network structure is able to be updated as well as the network parameters. We take a flat structure only containing the input and output layer as the initial network. The direction of network updating can be concluded as vertical update, horizontal update and cross update, as shown in Fig. 1. Among them, the vertical update is considered as cascading more hidden layers vertically to extract multiple levels features, improving the network capability of representing samples, the horizontal update means to widen the network horizontally in each layer, trying to mine diverse features from samples through multiple feature extraction units, and the cross update refers to the selective use of vertical update or horizontal update based on network performance feedback.

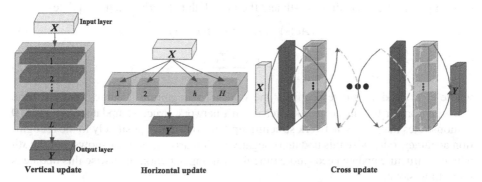

Fig. 1. Three ways of updating network structure.

In order to address the problem of calculating the approximation error of network in unfixed structure, we propose a multi-branch error back-propagation mechanism for exploring the contribution produced by each network structure update, where the error is calculated between the target output and each hidden layer that is the result of each vertical update or horizontal update. The advantage of the multi-branch error back-propagation mechanism applied to the unfixed network structure is that the multi-branch error back-propagation is performed from the beginning of the network structure update, thus which allows the performance of network obtained after each update is optimal, as shown in Fig. 2.

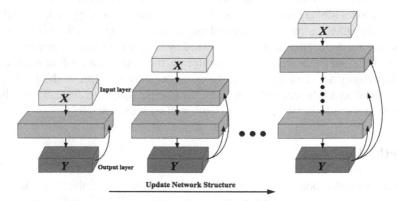

Fig. 2. Multi-branch error back-propagation.

We combine with specific tasks to put forward objective indexes to judge the contribution of structure update. For example, the recognition rate and real-time are important indicators to measure various strategies in the network for target recognition tasks, thus we propose the growth rate of accuracy and the time-consuming for training network as contribution evaluation indicators.

The growth rate of accuracy $\nabla Ac(s)$ and the growth rate of time-consuming for training network η_s respectively refers to the testing recognition accuracy and time consumption difference after the sth and the $(s-1)$th network structure update.

$$\nabla Ac(s) = Ac(s) - Ac(s-1) \tag{1}$$

$$\eta_s = \frac{|T_s - T_{s-1}|}{T_{s-1}} \tag{2}$$

where $Ac(s)$ and T_s are respectively the optimal recognition accuracy and the time consumption for training network after the sth network structure update. $\nabla Ac(s) > 0$ demonstrates that the sth network structure update contributes positively to the recognition accuracy, otherwise this update is negative. Similarly, $\eta_s > \eta_{s-1}$ shows that the sth network structure update needs more time for training network, otherwise this update is more time-saving.

3   Simulation Results

In order to demonstrate the feasibility of our proposed method, we have conducted some experiments about recognition, which are all performed on a Windows 10 PC laptop Intel Core i7-8700 3.20 GHz CPU, 8 GB RAM and MATLAB R2016a 64-bit.

3.1 Datasets

The databases used in all experiments are respectively one-dimension (1D) datasets from the UCI repository [11], such as Balance Scale (BS), Waveform Database Generator (WDG), Pen-Based Recognition of Handwritten Digits (Pen), Wall-Following Robot Navigation (Robot), Landsat Satellite (LS). And four 2D image databases, for example The ORL Database of Faces (ORL) [12] is composed of 400 grayscale images of 40 different persons with size of 32×32 pixels. There are 6 images of each person for training and the remain 160 images for testing. Extend YaleB (EYB) [13] consists of 2414 cropped grayscale images of 38 people with size of 32×32, where each person randomly selects 30 images for training and the other for testing. The images of everyone with large variations in terms of illumination conditions and expressions. NORB [14] contains 48600 images totally, each has $2 \times 32 \times 32$ pixels. Images of 50 different 3-D toy objects belonging to five distinct categories: animals, humans, airplanes, trucks and cars. Among them, 24300 images are selected for training and the others are used for testing. Fashion-mnist (Fmnist) [15] is made up of 60000 training samples and 10000 testing samples, in which each gray-level image with size of 28×28 belongs to ten categories: Ankle boot, Bag, Coat, Dress, Pullover, Sandal, Shirt, Sneaker, Trouser, T-shirtis. The details of one-dimension datasets and two-dimension databases are listed in Table 1.

Table 1. Details of datasets

Datasets	No. of samples		Attributes	Categories
	Training	Testing		
Balance	562	63	4	3
Wave1	4500	500	21	3
Pen	7494	3498	16	10
Robot	4910	546	24	4
LS	4435	2000	36	6
ORL	240	160	1024	40
EYB	1140	1274	1024	38
NORB	24300	24300	2048	5
Fmnist	60000	10000	784	10

3.2 Network Configuration and Parameter Settings

The initial network only contains the input layer and the output layer. The cross update manner is used to adjust the network structure, and the fuzzy system [16] is utilized as the feature extraction unit. The network parameters corresponding to each dataset are shown in Table 2.

Table 2. Parameter settings of each dataset

Datasets	Parameter settings	
	L	S
Balance	1	7
Wave1	3	(18, 2, 2)
Pen	3	(18, 6, 3)
Robot	4	(19, 3, 6, 4)
LS	3	(6, 4, 3)
ORL	2	(24, 30)
EYaleB	3	(30, 10, 8)
NORB	4	(32, 20, 8, 4)
Fmnist	5	(45, 10, 8, 5, 3)

Here, L and S are respectively the number of layers produced by the vertical update and the number of feature extraction unit distributed by the horizontal update. For example, the parameter settings of Pen dataset are $L = 3$ and $S = (18, 6, 3)$, and it can be exactly explained that the initial network are updated cross, in which three hidden layers are updated vertically and each hidden layer is distributed with 18, 6, 3 feature extraction unit horizontally.

3.3 Experimental Results

The network configuration and parameter settings in the previous section are adopted to verify the feasibility of proposed method on one-dimension and two-dimension datasets. Specially, the recognition accuracy of testing and network training speed are utilized as objective quantitative indicators. The experimental results are shown in Fig. 3.

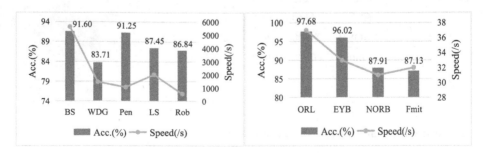

Fig. 3. Experimental results on various datasets

It can been seen from Fig. 3 that the method proposed in this paper can achieve fine recognition accuracy on one-dimension dataset and two-dimension image datasets. The reason is that the network structure update is performed according to the recognition task closely, i.e., the datasets, and each network update has a positive effect on recognizing samples. Especially, the network training speed of proposed method is real-time on all datasets, i.e., the recognition speed exceeds 30 samples per second. In addition, by comparing the size of datasets in Table 1 and networking parameter settings in Table 2, it can be found that the more complex the dataset is, the more times the network structure is updated, i.e., the more hierarchical the network structure becomes.

4 Conclusion

In view of network performance and robustness, this paper proposes a novel idea of network structure update. Based on different tasks, the network structure can be updated vertically or horizontally with the optimization of network parameters. A multi-branch error back-propagation mechanism is designed to match the unfixed network structure, exploring the effect of each hidden layer on the network performance. Some contribution evaluation indexes are proposed to measure the contribution of various strategies to network performance, and provide guidance for subsequent network updates and meeting task requirements. In addition, the experimental results on recognition task demonstrate the feasibility of proposed method in terms of recognition accuracy of testing and network training speed. In the future, we will further explore the strategies of network structure update and verify the contribution evaluation index combining with specific tasks.

Acknowledgment. This work is supported by National Natural Science Foundation of China (NO. 61871241, NO. 61976120); Ministry of education cooperation in production and education (NO. 201802302115); Educational Science Research Subject of China Transportation Education Research Association (Jiaotong Education Research 1802-118); the Science and Technology Program of Nantong (JC2018025, JC2018129); Nantong University-Nantong Joint Research Center for Intelligent Information Technology (KFKT2017B04); Nanjing University State Key Lab. for Novel Software Technology (KFKT2019B15); Postgraduate Research and Practice Innovation Program of Jiangsu Province (KYCX19_2056).

References

1. Li, H., Suen, C.Y.: Robust face recognition based on dynamic rank representation. Pattern Recogn. **60**(12), 13–24 (2016)
2. Li, H., Zhou, Z., Li, C., Zhang, S.: RCCM: reinforce cycle cascade model for image recognition. IEEE Access **8**, 15369–15376 (2020)
3. Pan, S., Yang, Q.: A survey on transfer learning. IEEE Trans. Knowl. Data Eng. **22**(10), 1345–1359 (2010)
4. Yosinski, J., Clune, J., Bengio, Y., Lipson H.: How transferable are features in deep neural networks? In: 28th Conference on Neural Information Processing Systems, NIPS, La Jolla, California, USA, pp. 3320–3328 (2014)
5. Zhao, B., Huang, B., Zhong, Y.: Transfer learning with fully pretrained deep convolution networks for land-use classification. IEEE Geosci. Remote Sens. Lett. **14**(9), 1436–1440 (2017)

6. Fang, X., Teng, S., Lai, Z., He, Z.: Robust latent subspace learning for image classification. IEEE Trans. Neural Netw. Learn. Syst. **29**(6), 2502–2515 (2018)
7. Tollenaere, T.: SuperSAB: fast adaptive back propagation with good scaling properties. Neural Network **3**(5), 561 573 (1990)
8. Zhang, L., Wu, K., Zhong, Y., Li, P.: A new sub-pixel mapping algorithm based on a BP neural network with an observation model. Neurocomputing **71**(10), 2046–2054 (2008)
9. Rehman, Y.A.U., Po, L.M., Liu, M.Y.: SLNet: stereo face liveness detection via dynamic disparity-maps and convolutional neural network. Expert Syst. Appl. **142**(35), 325–331 (2020)
10. Zeng, L., Tian, X.: Accelerating convolutional neural networks by removing interspatial and interkernel redundancies. IEEE Trans. Cybern. **50**(2), 452–464 (2020)
11. Blake, C.L., Merz, C.J.: UCI repository of machine learning databases. Preprint at http://arc hive.ics.uci.edu/ml/datasets.html
12. Samaria, F.S., Harter, A.C.: Parameterisation of a stochastic model for human face identification. In: IEEE Workshop on Application of Computer Vision, Sarasota, USA, pp. 138–142. IEEE (1994)
13. Lee, K.C., Ho, J., Kriegman, D.J.: Acquiring linear subspaces for face recognition under variable lighting. IEEE Trans. Pattern Anal. Mach. Intell. **27**(5), 684–698 (2005)
14. LeCun, Y., Huang, F.J., Bottou, L.: Learning methods for generic object recognition with invariance to pose and lighting. In: 2004 IEEE Computer Society Conference on Computer Vision and Pattern Recognition, Washing, D.C., USA, pp. II-92–II-104. IEEE (2004)
15. Han, X., Kashif, R., Roland, V.: Fashion-MNIST: a novel image dataset for benchmarking machine learning algorithms. arXiv preprint arXiv:1708.07747 (2017)
16. Takagi, T., Sugeno, M.: Fuzzy identification of systems and its applications to modeling and control. IEEE Trans. System Man Cybern. **SMC-15**(1), 387–403 (1985)

A Blockchain-Based Online Game Design Architecture for Performance Issues

Dong Wenhao[1] ⓘ, Tang Yufang[2,3] ⓘ, and Xu Yan[2,3(✉)] ⓘ

[1] Beijing University of Posts and Telecommunications, Beijing 100876, China
[2] Shandong Normal University, Jinan 250014, China
yan.soe1011@gmail.com
[3] Concordia University, Montreal H3G 1M8, Canada

Abstract. Online games are one of the ideal application scenarios of blockchain technology, but the high real-time requirements of games and the throughput speed of blockchain networks have become obstacles to the development of blockchain games. In real application scenarios, because of the inevitable fork of blockchain, it is difficult to significantly improve the processing speed of the blockchain through the block generation rate when the block size and the size of each transaction data are unchanged. Finally, it will affect the security of blockchain.In this article, the theoretical model and actual situation of blockchain technical performance are discussed by data modeling. And we propose a new blockchain game design architecture. The result show that our architecture can process data at higher speeds adn solve above problem efficiently.

Keywords: Blockchain · Design architecture · Online game

1 Introduction

Blockchain technology was first proposed by Satoshi Nakamoto [1]. It is a decentralized, non-tamperable, traceable, and trusted distributed database that is jointly maintained by multi-participants who do not trust each other. Each transaction will be consistently recorded in the nodes of the entire network after it is verified and agreed by the majority of participants on the entire network. Once the transaction is recorded, no one can tamper with it, thereby achieving credible multi-party data sharing, avoiding manual reconciliation, eliminating intermediaries, and reducing transaction delays and costs.

In recent years, with the rapid development and widespread application of blockchain technology, blockchain games have begun to appear. As a single closed scene, the game's own virtual asset circulation, community autonomy, and virtual economic gene, making it consider as one of the important application scenarios of blockchain technology. Based on the new features of blockchain technology, blockchain games will bring new design ideas and player experience into online game, and have high application value for protecting the safety of game assets and the liquidity of player assets. However, the blockchain system does not perform well in application scenarios such as games that require high latency and concurrency [2]. Therefore, we need to find a solution to enable the blockchain technology to work well in such scenarios.

© Springer Nature Switzerland AG 2020
Y. Lu et al. (Eds.): ICPRAI 2020, LNCS 12068, pp. 319–324, 2020.
https://doi.org/10.1007/978-3-030-59830-3_28

Although there are many applications of blockchain in games, the research on blockchain games is rare. Among them, Guy Barshap [3] proposed a solution to transplant a Battleships game (a traditional guessing game for two players) to the Ethereum network, and solved the trust problem between players through the smart contract of the Ethereum network. In his software architecture, there are three layers, presentation layer, logic layer and data layer. The data is stored in the data layer on the blockchain network, and the logic layer uses blockchain smart contracts for game control. Inspired by the concept of sidechain [4], Daniel Kraft [5] proposed an off-chain interaction in a game where players take turns performing operations. There is almost no need for public blockchain space during the game. Once disputes arise, the public blockchain network can be used to resolve these disputes. At the same time, Daniel Kraft proposed a mechanism that allows players to play games with blockchain miners to solve the problem of random number security in current blockchain games. Han Shuang [6] analyzed the feasibility and advantages of digital asset trading methods based on blockchain technology, and demonstrated the application of blockchain technology in virtual asset trading. At present, few researches of blockchain games focus on blockchain game security and mutual trust, and lack of consideration and research on the operation speed of blockchain games from the perspective of game design.

2 Blockchain Performance Issues

2.1 Theoretical Model of Blockchain Performance

TPS (transaction-per-second) is an important indicator used to measure the performance of the blockchain. According to Satoshi Nakamoto, the TPS is proportional to the block size and the growth rate of the main chain (blocks per second), inversely proportional to the byte size of each transaction.

$$\text{TPS} = \mathbb{V}\frac{\mathbb{S}}{\mathbb{T}} \tag{1}$$

where $\mathbb{V}$ is the growth rate of the main chain, $\mathbb{S}$ is the block size, and $\mathbb{T}$ is the storage space required for a single transaction.

Assume that there are N nodes in a blockchain system, and each node has the same computing power, so the probability of any node obtaining a block bookkeeping right is $\frac{1}{N}$. Assume that the highest block number of the current main chain is j, and the node number that obtained the accounting right is E_j. Ideally, when the blockchain network does not fork, that is, when the time of the $(j-1)^{\text{th}}$ block is transmitted from the E_{j-1} node to the E_j node meets $t_j \leq \frac{j}{v}$, then, the E_j node is the information of the $(j-1)^{\text{th}}$ block has been received at the jth block. Then, the system IO time is ignored, and all newly generated blocks are linked to the main chain. At this time, the growth rate of the main chain is equal to the block. The block output speed v of the chain system is the maximum performance of the blockchain.

$$\text{TPS} = v\frac{\mathbb{S}}{\mathbb{T}} \tag{2}$$

2.2 Blockchain Performance Model in Real Environment

Actually, when a fork causes some blocks to be linked outside the main chain, there exists $\mathbb{V} \le v$. At this time, if $t_1 > \frac{1}{v}$ and $t_2 > \frac{2}{v}$, the system generates a fork, with $\mathbb{V} = \frac{2}{v}$. By analogy, the expected value of is $\mathbb{V}$:

$$\mathbb{V} = P\left(t_1 \le \frac{k}{v}\right)v + \sum_{i=2}^{\infty}\left\{\prod_{j=1}^{i-1} P\left(t_j > \frac{i}{v}\right) P\left(t_i \le \frac{i}{v}\right)\frac{v}{i}\right\} \tag{3}$$

assuming that t_j follows a uniform distribution with a mean of α, the range of t_j is [0, 2α]. Then, the expectations of $\mathbb{V}$ is:

$$\mathbb{V} = \begin{cases} v, & \alpha \le \frac{1}{2v} \\ \frac{v}{2} + \frac{1}{4\alpha}, & \frac{1}{2v} < \alpha \le \frac{1}{v} \\ \sum_{i=1}^{\lfloor 2\alpha v \rfloor}\left\{\frac{1}{2\alpha}\prod_{j=0}^{i-1}\left(1 - \frac{i}{2\alpha v}\right)\right\} + \frac{v}{\lfloor 2\alpha v \rfloor + 1}\prod_{j=1}^{\lfloor 2\alpha v \rfloor}\left(1 - \frac{k}{2\alpha v}\right), & \alpha > \frac{1}{v} \end{cases} \tag{4}$$

where a is the average transmission time of the block in the P2P network. Taking the Bitcoin network as an example, the size of each block is 1 MB. Assume that the transmission speed of the network is 512 Kb/s, then

$$\alpha = \frac{1 \times 1024 \times 8}{512} = 16$$

Assume that the average value of T is 512 B, then a is substituted into the above formula of the classification discussion. With the help of python's matplotlib library, an image of the function can be drawn (see Fig. 1).

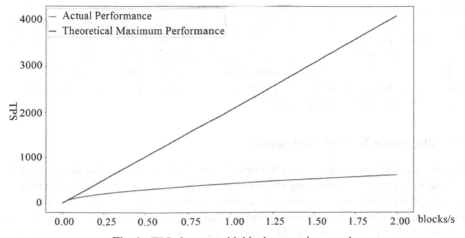

Fig. 1. TPS changes with block generation speed

In the real production environment, because the fork of the blockchain is inevitable, the line chart proves that it is difficult to significantly improve the performance through the block generation rate, when the size of the block and each transaction data are unchanged. While if we try to increase TPS by increasing the capacity of a single block, the expansion of the block capacity will increase the amount of data that nodes in the blockchain need to store. For example, BCH uses 8 M block capacity, and the amount of data will reach 400 GB in a year. It can not imagine how this will achieve sustainable development. Increasing the size of the block will increase the transmission time of the block in the network. Thus, it will lead to an increase in the probability of the block chain forking, and then affect the security of the system. In addition, if we try to reduce the space of data stored in a single transaction, equivalent to increasing the block capacity in another way, the content of the block will increase, and the speaking right (the number of recorded transactions) of the node with the block bookkeeping right will increase accordingly, and the blockchain network will move toward the center. The continuous evolution of the network has weakened the advantages of blockchain decentralization.

Obviously, no matter how to increase the block production speed, increase the block capacity or reduce the storage space required for a single transaction, the performance of the blockchain network can not be effectively improved.

3   Blockchain Online Games Design Strategy

3.1   Overall System Architecture

We propose a design strategy suitable for blockchain games, with the same response speed as traditional online game, and taking advantage of blockchain technology in game scenarios. We consider using a blockchain network in a game application with a traditional server instead of just using the blockchain network to complete all network requests. Considering the application value of blockchain technology in the game scenario, requests related to game asset changes must be implemented using the blockchain network, while requests from other subsystems of the game, such as in-game chat systems, battle systems with high real-time demand, and other systems that do not involve changes to the ownership of virtual assets, can request data from traditional servers to reduce delays in the game, while enhancing the concurrency of the game. Thus, the overall design architecture (see Fig. 2) of our system can be proposed and it distinguishes network requests in blockchain games.

3.2   Blockchain Network Architecture

For the network request to process through the blockchain network, we design a cache system to improve the processing efficiency of the system (see Fig. 3). The data request is first retrieved in the Redis cache. If the query key is hit and the value is correct, the key-value pair is directly manipulated and returned, then synchronize to each blockchain network node. If the record is not hit, data operations will be added to a message queue, waiting on the blockchain network. If multiple requests for the same data object are initiated during the waiting period, they will be merged in the message queue and re-queued to execute. After the blockchain is packaged to form a new block for the transaction, the system will return the data result and synchronize it to the Redis system.

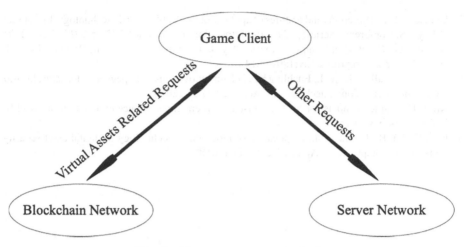

Fig. 2. Game request processing logic

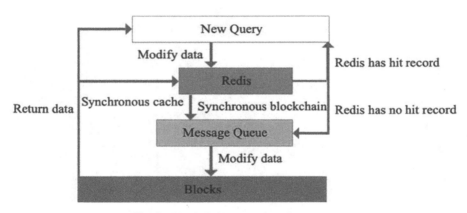

Fig. 3. Blockchain network cache system

4 Conclusion

In this paper, we analyze the problems existing in the application of blockchain technology to online game, and propose a theoretically feasible strategy to solve this problem. This solution can reduce the number of concurrent requests to the blockchain network, and optimize the execution effect of a single transaction, effectively solving the network congestion problem in game applications. In future work, we will verify and further improve the strategy through real online game.

References

1. Nakamoto, S.: Bitcoin: a peer-to-peer electronic cash system. https://bitcoin.org/bitcoin.pdf. Accessed 31 Jan 2020

2. Utakaeva, I.K.: Directions and features of application of the blockchain technology. In: Journal of Physics: Conference Series, p. 5 (2019). https://doi.org/10.1088/1742-6596/1353/1/012103
3. Barshap, G.: Crypto-Battleships or How to Play Battleships Game Over the Blockchain?, vol. 16 (2018). arXiv preprint arXiv:1807.08142
4. Back, A., Corallo, M., et al.: Enabling blockchain innovations with pegged sidechains. https://www.blockstream.com/sidechains.pdf. Accessed 31 Jan 2020
5. Kraft, D.: Game channels for trustless off-chain interactions in decentralized virtual worlds. Ledger 1, 84–98 (2016)
6. Han, S., Pu, B., Li, S., et al.: Application of block chain technology in digital asset security transaction. Comput. Syst. Appl. 27, 205–209 (2018)

Predicting US Elections with Social Media
and Neural Networks

Ellison Chan, Adam Krzyzak(✉), and Ching Y. Suen

Department of Computer Science and Software Engineering, Concordia University, Montreal
H3G 1M8, Canada
ellison.chan@gmail.com, krzyzak@cs.concordia.ca,
suen@cse.concordia.ca

Abstract. Increasingly, politicians and political parties are engaging their electors using social media. In the US Federal Election of 2016, candidates from both parties made heavy use of social media, particularly Twitter. It is then reasonable to attempt to find a correlation between popularity on Twitter, and eventual popular vote in the election. In this paper, we will focus on using the subscriber 'location' field in the profile of each candidate to estimate support in each state.

A major challenge is that the Twitter location field in a user profile is not constrained, requiring the application of machine learning techniques to cluster users according to state.

In this paper, we will train a Deep Convolutional Neural Network (CNN) to classify place names by state. Then we will apply the model to the Twitter Subscriber 'location' field of Twitter subscribers collected from each of the two candidates, Hillary Clinton (D), and Donald Trump (R). Finally, we will compare predicted popular votes in each state, to the actual results from the 2016 Presidential Election.

In addition to learning the pattern related to the state names, this additional information may help a machine learning model learn to classify locations by state.

The results from our experiments are very promising. Using a dataset containing 695,389 cities, correctly labelled with their state, we partitioned the cities into a training dataset containing 556,311 cities, a validation dataset containing 111,262, and a test dataset containing 27,816. After the trained model was applied to the test dataset. We achieved a correct prediction rate of 84.4365%, a false negative rate of 1.6106%, and a false positive rate of 1.0697%.

Applying the trained model on Twitter location data of subscribers of the two candidates, the model achieved an accuracy of 90%. The trained model was able to correctly pick the winner, by popular vote, in 45 out of the 50 states. With another US and Canadian election coming up in 2019, and 2020, it would be interesting to test the model on those as well.

Keywords: Predicting elections · Twitter data · CNN

© Springer Nature Switzerland AG 2020
Y. Lu et al. (Eds.): ICPRAI 2020, LNCS 12068, pp. 325–335, 2020.
https://doi.org/10.1007/978-3-030-59830-3_29

1   Introduction

1.1   Motivation

Increasingly, politicians and political parties are engaging their electors using social media. In the US Federal Election of 2016, candidates from both parties made heavy use of social media, particularly Twitter. It is then reasonable to research a correlation between popularity on Twitter, and eventual popular vote in the election. In this paper, we will focus on using the subscriber location information of each candidate to estimate support in each state.

For this paper, over 7.8 million subscriber profiles were gathered from Twitter for the two candidates, Hillary Clinton and Donald Trump, between 2015 and 2016.

The goal is to use the locations of subscribers to each Candidate to determine who will win the most popular votes by state.

In fact, cities and towns in the US have been named by the people who colonized the area. For instance, in the Eastern US states, the northern states have primarily been colonized by people of French descent, and many place names are French, whereas place names get increasingly English as you move south. Same patterns are also seen. as you head west, with the place names adopting more Native American languages, and progressively more Spanish as you head towards the west and south. With this technique, as well as other features, we will classify city and town names into states by using a suitably trained CNN classifier. It is felt from after having studied the analysis done by LeCun [1] that for its accuracy and shorter convergence time, especially when using GPU acceleration that CNN is an ideal algorithm for classifying the state names when representing them as sparse images.

Furthermore, the CNN should also be able to recognize letter patterns, such as the short state names along with full state names. (i.e. CA/California, AK/Alaska, ME/Maine …).

1.2   Challenges

A major challenge is that the Twitter "location" field in a user profile is not constrained, requiring the application of AI and Machine Learning algorithms to create clusters corresponding to state names.

The task at hand is to turn the chaos into order. The Twitter "location" field in a Twitter subscriber profile provides information about where the user resides. However, the text is unconstrained. To obtain information from this Twitter field requires filtering out text that has a low likelihood of being a location, then grouping the remaining locations into US States. Additionally, we need to remove locations with a low likelihood of being a US State.

Since we have over 7.8 million locations to classify, we need an algorithm suitable for such a large dataset.

One possibility is an unsupervised clustering algorithm, such as K-Means, which has an algorithmic complexity of $O(n^2)$. The challenge of using this algorithm is the time required to achieve the clustering on such a large dataset and the second challenge is that we need to apply the long clustering process to every new dataset. Research has been done to speed up this algorithm with GPU acceleration, with good success. However, this does not achieve our goal of having an algorithm that can easily be applied to input data.

Using a supervised algorithm, such as Convolutional Neural Network (CNN), will achieve the goal of training once, and applying to future data. Additionally, many modern Deep Learning frameworks such as Tensorflow, and Keras can use GPU acceleration to speed up training.

1.3 Approach

In recent years there has been a big resurgence in Neural Networks, primarily due to an increase in computing power, which enables training dense Deep Learning Neural Networks quickly.

The company Google has created a computational framework called Tensorflow, which enables GPU accelerated high-performance computations. This is especially suitable for creating and training Deep Learning Neural Networks. Weight calculations are matrix multiplications, which are well suited for parallel execution on massively parallel systems, such as the GPU graphics platform.

In this paper, our primary tool for learning the similarity function for state classification will be done using a Convolution Neural Network, with the Keras library running in the Tensorflow framework. This enables high-speed training, and inferencing.

After training a CNN to recognize an input dataset of cities labelled with their corresponding states, we will apply this CNN to predict the US state of a Twitter Subscriber, based on what they input into the 'location' field in their profile. We will then compare the predicted result to the actual election results for the 2016 US Federal Election.

We will encode the cities as bigrams, with a vocabulary of 26 uppercase and 26 lowercase letters, including 3 non-alphabet characters, giving us a total vocabulary size of 55 characters. We will then convert this bigram matrix into 55×55 RGB images. This will enable a CNN to train on the city names. In fact, once encoded as images, the images resemble what's in the MNIST database, which is a standard dataset used to train models to recognize handwritten numbers. A sample from the MNIST Database is in Fig. 1. A sample of our encoding of the city name bigrams can be found in Fig. 2. They are similar in size and sparseness of the image.

2 Data

2.1 Data Sources

The data source used to train the neural network on city names to state mapping was obtain from these sources:

Purchased Dataset:

https://www.uscitieslist.org/cart/packages/

Open Source Dataset:

https://github.com/grammakov/USA-cities-and-states

The location data consists of subscriber profiles of the two front runners in the 2016 US Presidential election, Donald Trump and Hillary Clinton. In order to gather the data,

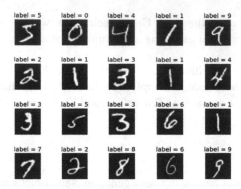

Fig. 1. MNIST database sample

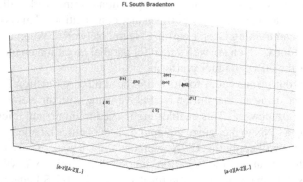

Fig. 2. Bigram frequency of South Bradenton, FL (Side)

cron jobs on an internet connected server was executed which scraped both candidate's accounts for their subscriber information.

For this thesis, over 7.8 million subscriber profiles were gathered from Twitter for the two candidates, Hillary Clinton and Donald Trump, between 2015 and 2016.

2.2 Data Encoding

As previously stated, we want to reframe the place name classification problem into an image processing problem, and to apply state-of-the-art hardware and software, actively being used in image classification research. Our data encoding is simple. The x and y-axis of our image will be one character in the bigram and each pixel in our images will represent the 24-bit frequency for that bigram. By convention, we will encode the first letter of the bigram in the x-axis and the second character in the y-axis.

2.3 Converting to Image Representation

To enable a CNN to train on our histograms, we will convert them into a two-dimensional image. The x and y-axis are indexes to each letter of our vocabulary, and the RGB values represent a 24-bit sized frequency of occurrence for the bigram. More details of the encoding process are found in Sect. 2 data. These are composite images of all the bigrams of all cities in that state.

A sample image built in this manner is shown in Fig. 3. These images represent the superposition of all bigram frequencies for all cities in the respective state. The images help to illustrate that some bigrams combinations occur more frequently with some states. Together this represents a "fingerprint" of sorts that we hope a neural network can learn the pattern to and thus be able to classify cities by their bigrams. As in fingerprint recognition, each fingerprint is composed of recognizable features. The CNN will be trained to recognize the features that each state fingerprint is composed of. Using our method, the image in Fig. 3 represents bigram frequencies of all cities in the US.

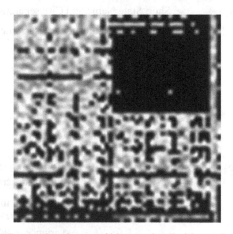

Fig. 3. Composite picture of bigrams in all cities and all states

2.4 Augmenting the Dataset

To augment the basic list of city names and states, the operations in Table 1 are applied to the original list. This will help to capture the variations that people will use in the location field of their Twitter profile. It's interesting to note that these permutations do not contain the city name alone with no state name. This is done intentionally to avoid confusing the neural network. Many states carry common city names and adding this permutation would result in adjusting weights towards one state, and then when the same city came in for another state it would undo that training resulting in the neural network oscillating back and forth, until the last entry of that city.

2.5 Splitting Data into Training, Validation, and Test Sets

The final step is to split the data into the separate components. We will keep 80% of the original data for training while splitting the remaining 20% into validation (16%), and test (4%). The details on actual sizes of each and descriptive statistics are listed in Table 2.

Table 1. Permutation operation on state names and cities

Operation	Description
[Long State Name]	Long state name alone
[City], [Long State Name]	City and Long State separated by a comma
[City] [Long State Name]	City and Long State Name separated by space
[Long State Name], [City]	Long State Name and City separated by a comma
[Long State Name] [City]	Long State Name and City separated by a space
[City], [Short State Name]	City and Short State Name separated by a comma
[City] [Short State Nam]	City and Short State name separated by a space
[Short State Name], [City]	Short State Name and City separated by a comma
[Short State Name] [City]	Short State Name and City separated by a space

3   CNN Model

Table 2. Training, validation, test data

	City entries	File name	Description statistics Numpy.describe()
Full data (100%)	695389	cities_shuffled_all.json	nobs = 52, minmax = (0, 46271), mean = 12929.98076923077, variance = 94827083.3525641, skewness = 1.3722954619398018, kurtosis = 1.9301177227759876
Training (80%)	556311	cities_shuffled_train.json	nobs = 52, minmax = (0, 37041), mean = 10342.307692307691, variance = 60726949.62895928, skewness = 1.3738867581234702, kurtosis = 1.9387584597763698
Validation (16%)	111262	cities_shuffled_valid.json	(nobs = 52, minmax = (0, 7420), mean = 2069.6923076923076, variance = 2435641.3152337857, skewness = 1.3721646821298885, kurtosis = 1.915920922857353
Test (4%)	27816	cities_shuffled_test.json	nobs = 52, minmax = (0, 1810), mean = 517.9807692307693, variance = 148579.43099547512, skewness = 1.3356616905991185, kurtosis = 1.7888861304572803

We are using five 2D convolution layers with 2 × 2 max pooling layers between each convolution layer. The Keras Neural Network Framework is being used to define our Convolutional Neural Network. Keras is a popular framework for defining, training, and generating predictions, and provides GPU acceleration using a Tensorflow Backend. The programming language of choice is Python. In out CNN model, we use 5 convolution layers. The first convolution layer is an input layer and designed for our 55 × 55, 3 color image. The first layer will pick out the smaller features of our image, and pass it on to subsequent layers, after passing the features into a pooling layer.

The last two layers of our model are fully connected classification layers. The first one is a dense input layer which takes the input of the final pooling layer as input and feeds into a dense hidden layer that will classify the features into the final 51 classes. In our CNN, we have 51 classes, one for each state. We are using the Sigmoid activation function which is more suited for multiclass. For the general description of CNN refer to [2].

3.1 Model Training

We trained our CNN model on 556311 city names from our training dataset. In addition to the model definition in Table 2, when training there are hyperparameters that needed to be set, which can affect the outcome of the trained model. These are listed in Table 3. We will discuss these parameters in greater detail and provide some discussion on how training and model accuracies are affected in the next sections.

Table 3. Training hyperparameters

Parameter	Value	Description
Batch size	32	Number of samples between gradient calculation
Optimizer	Adam	Adam optimizer, with an initial learning rate of 0.001, and decay rate of 0.001
Loss function	Binary Cross Entropy	Calculates the loss between each gradient calculation
Maximum epochs	<=75	Rather than using a fixed number of epochs, we are using early stop, if the validation loss does not change for 4 epochs

3.2 Optimizer

In neural networks, the "learning rate" determines how fast it converges or whether it will converge at all. For our training, we are using the Adam optimizer to find the optimal learning rate. It has been shown to converge quickly for CNN [3]. In Figs. 4 and 5, we have plotted the training and validation accuracies for each epoch. We can tell by this that the model has converged by epoch 4 with slight oscillation. After epoch 4 the training and validation accuracies are similar along with the training and validation losses.

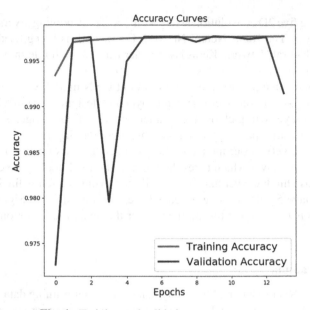

Fig. 4. Training and validation accuracy curves

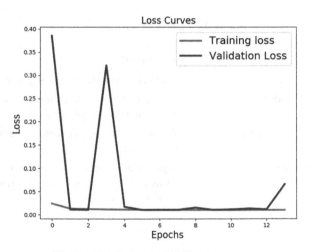

Fig. 5. Training and validation loss curves

4   Results

4.1   Cross-Validation Results

The cross-validation accuracy results are presented in Table 4. The mean Cross-Validation accuracy is 99.74% with a standard deviation of ±0.00. This is an indication that the model behaves stably. As with the full training dataset, we are only limiting cross-validation to the 51 states which are eligible to vote in the 2016 Presidential Election.

Table 4. Results of 10-fold cross-validation

Fold number	Accuracy
1	99.74%
2	99.73%
3	99.75%
4	99.74%
5	99.74%
6	99.74%
7	99.74%
8	99.74%
9	99.74%
10	99.74%

4.2 Analysis of CNN Model

We can see from Table 5 that the correct prediction rate on the test dataset is 84.4365%. The rest of the predictions are, including false positives, and false negatives, and wrong predictions. For the test dataset, we have a false positive rate, and a false negative rate of 1.0697% each and wrong predictions is at 12.2632%. This compares favourably to the training and validation dataset accuracy.

4.3 Analysis of CNN Model Applied to Election Prediction

Finally, the model predicts overall that Trump won 29 states, while Clinton was predicted to win 21, see Table 6. Recall that we did not include Alaska, as there was insufficient data for that state in our source data. This is not too far off the actual results, which is 28 states for Trump and 22 states for Clinton.

When comparing the number of states accurately attributed to each candidate by the model a 90% match rate was obtained, see Table 7.

Table 5. Prediction results

Prediction type	Test	Training	Validation	Description
False positives	1.0697%	1.5833%	1.7077%	Predictions with a probability greater than and equal to 50%, where the labelled state does not match with the predicted state
False negatives	1.6106%	2.0073%	1.1.5513%	Predictions with a probability less than 50%, where the labelled state matches the predicted state
Wrong prediction	12.2632%	11.4952%	12.0743%	Predictions with probability less than 50%, where the labelled state and predicted state do not match. (Note that this means the model assigned a state but was able to determine that there was a low probability)
Correct prediction	84.4365%	84.9142%	84.6667%	Predictions with probability greater than and equal to 50%, where labelled and predicted states match

Table 6. States won predicted vs actual

Candidate	#States won (Predicted)	#States won (Actual)
Trump	28	29
Clinton	22	21

Table 7. Prediction matches by state

Matched	45
Mismatched	5
% Matched	90%

5 Conclusions

In this paper we have successfully applied a trained machine learning model to help predict locations of Twitter subscribers of the two major candidates in the 2016 US general election, and have our predicted results match actual results. However, the model needs to be refined further to prevent breaking down, in cases that location names do not carry enough clues for classification.

References

1. LeCun, Y.: Gradient-based learning applied to document recognition. Proc. IEEE **86**(11), 2278–2324 (1998)
2. Goodfellow, I.., Bengio, Y., Courville, A.: Deep Learning. MIT Press, Cambridge (2016)
3. Xiong, F., Xiao, Y., Cao, Z., Gong, K., Fang, Z., Zhou, J.T.: Towards good practices on building effective cnn baseline model for person re-identification. In: Proceedings of the IEEE Conference on Computer Vision and Pattern Recognition, vol. abs/1807.11042 (2018)

Deep Learning

From Pixels to Random Walk Based Segments for Image Time Series Deep Classification

Mohamed Chelali[1]([✉])(iD), Camille Kurtz[1](iD), Anne Puissant[2](iD),
and Nicole Vincent[1](iD)

[1] Université de Paris, LIPADE, Paris, France
{mohamed.chelali,camille.kurtz,nicole.vincent}@u-paris.fr
[2] Université de Strasbourg, LIVE, Strasbourg, France
anne.puissant@unistra.fr

Abstract. Image time series, such as Satellite Image Time Series (SITS) or MRI functional sequences in the medical domain, carry both spatial and temporal information. In many applications such as image classification, taking into account such rich information may be crucial and discrimative during the decision making stage. However, the extraction of spatio-temporal features from image time series is difficult due to the complex representation of the data cube. In this article, we present a strategy based on Random Walk to build a novel segment-based representation of the data, passing from a $2D + t$ dimension to a $2D$ one, more easily manipulable and without loosing too much spatial information. Such new representation is then used to feed a classical Convolutional Neural Network (CNN) in order to learn spatio-temporal features with only $2D$ convolutions and to classify image time series data for a particular classification problem. The interest of this approach is highlighted on a remote sensing application for the classification and the mapping of complex agricultural crops.

Keywords: Satellite Image Time Series · Spatio-temporal features · Random Walk · Convolutional Neural Networks

1 Introduction

An image time series is an ordered set of images taken from the same scene at different dates. Such data provide rich information with the temporal evolution of the studied area. In remote sensing applications, many constellations of satellites acquire images with a high spatial, spectral and temporal resolution around the world leading to Satellite Image Time Series (SITS). For example, the Sentinel-2 sensors produce optical SITS with a revisit time of 5 days and a spatial resolution of 10–20 m.

The authors thank the French ANR for supporting this work under Grant ANR-17-CE23-0015.

© Springer Nature Switzerland AG 2020
Y. Lu et al. (Eds.): ICPRAI 2020, LNCS 12068, pp. 339–351, 2020.
https://doi.org/10.1007/978-3-030-59830-3_30

SITS help understanding environmental evolution, studying the causes of various changes, and predicting future evolution. Temporal information, integrated with spectral and spatial dimensions, enables in particular the analysis of complex patterns related to applications related to land cover mapping (e.g. agricultural zones, urban areas) or the identification of land use changes (e.g. urbanization, deforestation) and the production of accurate land-cover maps of a territory [11].

A major issue when analyzing image time series is to consider simultaneously the temporal and the spatial dimensions of the $2D + t$ data-cube. In this context, methods for SITS analysis are actually mainly based on temporal information [15] at the pixel level. But in some specific applications, this may not be sufficient to get satisfactory results. Taking both temporal and spatial aspects into account at the same time can, for example, make it easier to discriminate between different complex land cover classes (e.g. agricultural practices, urban vs. peri-urban areas). Note that here our objective is to map complex land-cover classes prone to confusions when a single date image is used.

This article focuses on the problem of spatio-temporal features extraction for the classification of image time series, using a deep learning strategy. In this context, we define a novel spatio-temporal representation of image time series that makes it possible to consider classical Convolutional Neural Network (CNN) frameworks (proposed for the analysis of $2D$ images). Our main contribution is the proposal of a transformation to represent $2D + t$ data as $2D$ images without loosing too much spatial information. It relies on the construction of sets of $(1D)$ segments using a Random Walk paradigm to decrease the spatial dimension of the data. This new data representation is then used to feed a CNN in order: (1) to learn spatio-temporal features with only $2D$ filters, involving at the same time temporal and spatial information, and (2) to classify image time series data according to a particular thematic problem.

The remainder of this article is organized as follow. Section 2 presents related works for SITS analysis. Section 3 introduces the proposed representation of the image time series for a CNN analysis. An experimental study, in the remote sensing domain, focusing on the classification of agricultural crops is described in Sect. 4. Section 5 discusses the obtained results. Finally, conclusion and perspectives will be found in Sect. 6.

2   Related Works on SITS Analysis

SITS allow the observation of the Earth surface. Such data improves our knowledge and understanding of environmental evolution and changes, which may be of different types, origins and duration. For a detailed survey, see [5].

Pioneer methods processed single images from image stacks. On each image, different measurements per pixel were considered as independent features and involved in classical machine learning-based procedures. Methods designed for bi-temporal analysis locate and study changes occurring between the two observations. These methods include image differencing [3], ratioing [13] or vector change analysis [14].

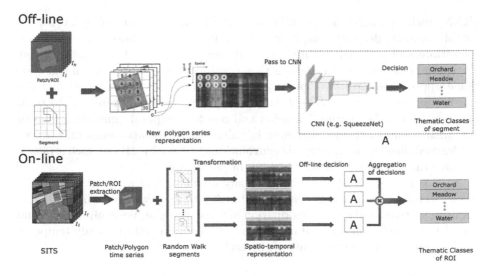

Fig. 1. Flowchart of our method for image time series deep classification based on a planar spatio-temporal data representation obtained from Random Walk based segments. (top) off-line (i.e. learning) phase and (bottom) on-line (i.e. testing) phase of the classification process.

Another family of methods are directly dedicated to the analysis of image time series. Most of them are based on multi-date classification. Among them, we find radiometric trajectory analysis [22]. These methods exploit the evolution of land cover (e.g. seasons, vegetation evolution [20]), and take into account the chronology by using dedicated time series analysis methods [2]. Every pixel is considered as time ordered (and aligned) series of measurements, and the changes of the measurements through time are analyzed to find (temporal) patterns, using statistical or symbolic approaches.

Some methods first propose a new representation of the SITS into a new space. We may cite "frequency-domain" approaches that include spectral analysis, wavelet analysis [1]. Other methods extract more discriminative "hand-crafted" features from a new enriched space [4,17,18]. Concerning the classification step, the classical approaches measure similarity between any incoming sample (that can be enriched with the "hand-crafted" features) and the training set. They assign the label of the most similar class using e.g. the Euclidean distance based on a nearest neighbor algorithm or/and the Dynamic Time Wrapping method [16].

More recently, deep learning paradigms have been considered to classify remote sensing images and generate land-cover maps. In general, Convolutional Neural Networks (CNN) are used to deal with the spatial domain of the data by applying $2D$ convolutions [8]. When dealing with image time series, convolutions can be applied in the temporal domain [15]. Another type of deep learning architecture that is designed for temporal data is Recurrent Neural Network

(RNN) such as Long-Short Term Memory (LSTM), used successfully in [10,19]. In this context, deep learning approaches outperform traditional classification algorithms such as Random Forest [12], but they do not directly take into account the spatial dimension of the data as they consider pixels in an independent way. Some approaches have been proposed to consider both the temporal and the spatial dimensions of the $2D + t$ data-cube [6]. A common strategy is to train two models, one for spatial dimension and one for temporal dimension, then to fuse their results at the decision level. In video analysis, spatio-temporal features are learned directly using deep $3D$ convolutional networks [21] but such strategy requires the learning of an huge number of parameters.

In this paper, our strategy is to classify a SITS using a classical $2D$ CNN model, thanks to a new representation of image time series embedding simultaneously temporal and spatial information of the data-cube. We compare with the use of $1D$ convolutions applied in the temporal domain [15] to classify temporal pixels (which is the current state of the art).

3   Proposed Method

The proposed method aims to classify image time series from spatio-temporal features. The underlying strategy is to use a $2D$ input classical deep neural network architecture in order to learn a spatio-temporal model from the $2D + t$ data. Figure 1 illustrates the global workflow of our system, with the traditional off-line (i.e. learning) and on-line (i.e. testing) phases of a classification process. Since our system is dedicated to classification of objects of interest (e.g. agricultural crops), the initial input data may be an image centered over a specific object, an image patch, or only the connected pixels of a region of interest (ROI), modeled as a polygon. In any case, we will use the term "image" for the input data.

We manage to consider some $2D$ elements to perform the learning phase in the off-line process. In this way we differ from other approaches considering a $1D$ structure [15] or a $3D$ one [21] as we find in the state-of-the-art methods.

We start by transforming the original $2D + t$ data into planar entity containing spatio-temporal data built from $1D$ spatial segments over time. Such a representation is then transferred as the input of a CNN to achieve a classification of the segments that are built in off-line and used on-line. The network can be trained in order to learn the labels from the spatial, as well as temporal information contained in the data.

3.1   Data Transformation

We first explain how to transform the original $2D + t$ data to less complex $2D$ representations that contain spatio-temporal data built from $1D$ spatial segments.

From Pixels to Segments. Traditional methods that only handle temporal information consider the $2D$ domain as a set/bag of pixels, $0D$ entities. The pixels are generally characterized by the temporal series of the pixel intensities. In our case, we include some spatial information, leading to the notion of segments which are $1D$ spatial entities. An input image is then replaced by a set of $1D$ segment entities, where L is the length of the segments included in the input image.

In a $1D$ segment each pixel has 2 neighbors, except for the two extreme pixels. Our transformation will then decrease the spatial information with keeping only 2 nearest neighbors.

Different strategies to define $1D$ segments in the original $2D$ space are here studied and compared in this work. For each chosen strategy, we apply the process N_p times from an input data, producing N_p different segments, in order to keep enough neighbors; The pixel orders are then chosen following these segments. In this way, the spatial representation complexity of the images is decreased, from $2D$ to $1D$ segments.

Next, segments characterized by temporal information leading to $2D$ spatio-temporal data are classified.

From Segments to $2D$ Representations. For a given series composed of N images (i.e. N temporal acquisitions), segments are first extracted. They are used for the learning of the classification model. The segment pixels are spatially represented by the pixel index within the segment. These $1D$ spatial segments will now be enriched with temporal information to build $2D$ spatio-temporal data.

With each of the N_p segments, we associate a $2D$ structure. In the abscissa X axis, is considered the index of the pixel in the segment (from the initial pixel) and in the ordinate Y axis is considered the evolution of the intensity of the pixels over the time. This leads to a novel $2D$ representation composed of N rows (N is the number of images in the SITS) in the temporal domain and L columns (L is the length of the considered segment) in the spatial domain. This image can then be interpreted as a partial spatio-temporal $2D$ representation of the $2D + t$ image time series.

When applying the transformation process to the N_p segments, we finally obtain N_p spatio-temporal $2D$ representations from the original image time series. These representations will be used as input of a learning process, the segments classes are the classes of the annotated input image they belong to.

3.2   Segment Construction Strategies

Two different strategies were considered to build segments:

- **Scanning strategy (scan).** Here we consider all the rows and columns in the input image to build $1D$ segments. The dimensions of the input image limit both the number and the lengths of possible segments. To guarantee similar lengths L for each segment, it is needed to replicate the values on too short segments (this may correspond to segments extracted from the borders of the image. In this way, the pixels are considered only twice in the new representation, and each pixel has only 4 neighbors.

– **Random Walk (RW) based segment.** A Random Walk [7] is a mathematical process based on a random iterative system. Each iteration is a step with Markovian properties.

Here, the Random Walk is used to generate a random segment in a $2D$ image space with length L, noted $RW(L)$. The first point of the segment is chosen randomly on the $2D$ image and for next point, 8 directions are possible.

Given an input image, we proceed to N_p initializations of N_p Random Walk segments. For each one a $2D$ image is then built, where the rows correspond to the pixel values of the pixels in the segment extracted from the different images of the series. The chronology is related to the line number. The middle of the on-line part of Fig. 1 illustrates the spatio-temporal representations from three different segments built from an input image.

3.3   CNN Model (Architecture)

Convolutional Neural Networks (CNN) refer to the family of deep learning algorithms. Systems are composed of two parts. The first one is designed to feature extraction, it has many neuron layers that compute the convolutions of the previous ones. The neurons of each layer are activated by non-linear functions (e.g. sigmoïde, ReLU) in order to keep the most representative features (high order features). We find also max-pooling layers between convolutional layers to reduce progressively the quantity of the inputs and the number of the parameters to be computed to define the network, and hence to also control over-fitting. The second part may be a classifier. Generally, it is a fully connected layer that provides a probability vector, on which is plugged a softmax function to predict the class label of input data.

We have chosen the SqueezeNet model [9] but any other $2D$ CNN model can be used. SqueezeNet has interesting properties, few parameters, and same accuracy level as the AlexNet model on the ImageNet dataset. The training of the model is then faster. The architecture of SqueezeNet introduces a new module called Fire composed of a squeeze layer using 1×1 convolution filters followed by expand layer that contains a mix of 1×1 and 3×3 convolution filters. Also, its classifier is based on a global average pooling over feature maps, potentially decreasing the overfitting effect. We used the PYTORCH implementation of SqueezeNet[1]. The CNN model is trained with the $2D$ spatio-temporal representations obtained from each input image time series from the training set.

3.4   Decision Making at Polygon Level

As already mentioned, our input data are polygons representing objects of interest in SITS. Each input data is associated with a set of N_p segments, N_p is consequently a parameter of the method. With each segment is associated a

---

[1] https://github.com/pytorch/vision/blob/master/torchvision/models/squeezenet.py.

$2D$ planar spatio-temporal representation. Thanks to the classifier described in Sect. 3.3, a class label is predicted for each $2D$ spatio-temporal representation (i.e. for each segment) with some probability. We proceed by taking average of the returned probabilities by the model for the N_p segments of the polygon and we affect the class label with the highest probability ensuring a unique decision per image.

4 Experimental Study

The experimental study is focused on a remote sensing application, the classification of agricultural crop fields from a SITS. The goal is to discriminate within agricultural thematic classes (e.g. traditional vs. intensive orchards). The automatic identification of these classes is a complex task since orchards are subject of many agricultural practices depending on the season and the territory management policy. In order to differentiate these two classes, spatio-temporal features carry useful information to discriminate the agricultural practices.

4.1 Material

We dispose of a SITS provided by the satellite Sentinel-2, it contains $N = 50$ optical images sensed in 2017 over the same geographical area (East of France – tile 32ULU). Figure 2 displays the temporal distribution of the images of this SITS. The images have been corrected and orthorectified by the French Theia program[2] to be radiometrically comparable. We also dispose of the cloud, shadow and saturation masks associated with each image. A pre-processing step was applied on the images with a linear interpolation on masked pixels to fill the missing values in the SITS.

For each image, only three bands are kept which are near-infrared (Nir), red (R) and green (G). The blue band (B) is considered as useless in the literature to discriminate different kinds of agricultural fields and is also sensitive to atmospheric effects. All these bands have a spatial resolution of 10 m.

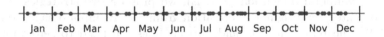

Jan Feb Mar Apr May Jun Jul Aug Sep Oct Nov Dec

Fig. 2. Distribution of the images from the SITS (2017).

The used reference data are extracted from the (freely distributed) RPG[3], which is the agricultural parcel delineations (in our context orchards). Some examples of polygons are represented in Fig. 1. These polygons have been corrected by photo-interpretation to ensure a good delimitation of the parcels.

---

[2] https://theia.cnes.fr/.
[3] http://professionnels.ign.fr/rpg.

The reference data used in our experiment are the semantic labels of these polygons (traditional or intensive orchards). These polygons are leading to a new time series of polygons, noted Polygon Image Time Series (PITS).

Table 1. Summary of the data; (first col.) Initial number of polygons per class; (two last col.) Number of spatio-temporal segments depending on the segment construction strategy.

Classes	# polygons	# Spatio-temp. rep. for *scan*	# Spatio-temp. rep. for *RW*
Int. orchards	100	3084	3000
Trad. orchards	100	3059	3000
Total	200	6143	6000

4.2 Data Preparation

First, PITS are formed, then we analyze the importance of the spatial relationship of pixels, so N_p segments are extracted from the ROI. According to the ROI sizes, we set N_p to 30 for the RW strategy. For the *scan* strategy, the number of possible $1D$ segments depends on the ROI size. The average number of segments for the *scan* strategy is 487 ± 110. In the off-line part of Fig. 1, we illustrate the transformation process of PITS with $RW(10)$. Table 1 displays the number of instances of polygons per class and the number of segments built from these data according to the segment construction strategy.

In the following, we study the impact of the length L of the segments. This enables to evaluate the impact of adding more spatial information to learn spatio-temporal features instead of considering single $0D$ pixels, as this is the case in most of the classical approaches. The used lengths L are $10, 50, 100$ and 224. The largest one depends on the maximum input size of the CNN SqueezeNet model. When building the 224×224 $2D$ image from the segments, if the segments are less than 224, we center them horizontally and the rest of columns are fixed to zero value. Table 1 indicates the actual number of segments.

For the temporal dimension (Y axis), we propose two strategies. The first one is to center the original information from the N input images vertically ($N = 50$). The remaining top and bottom lines are fixed to zero value. The second one is to fill the 224 values by applying a linear interpolation in the STIS on time information. We assume that the temporal information between two consecutive dates is monotonic and linear. The interpolation is then done by considering that we only have 224 days in the year so that one day is done with about 39 h. For the initial dates (begin of the year of 224), we affect the temporal information of the first date in the SITS. For the last dates (end of the year of the 224), we affect the last known temporal information in the SITS.

The data normalization is a linear transform based on the maximum and the minimum values of the dataset after values are limited with 2% (or 98%) percentile, as proposed in [15].

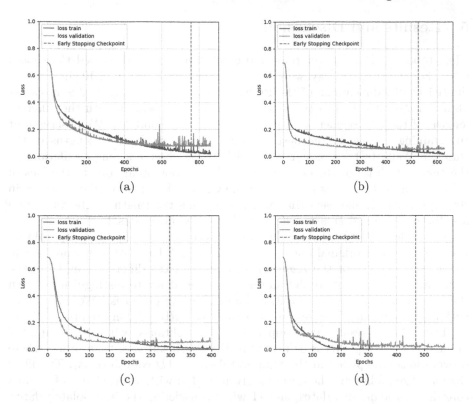

Fig. 3. Training phase loss curves, according to the length L of Random Walk paths: (a) $L = 10$; (b) $L = 50$; (c) $L = 100$ and (d) $L = 224$.

4.3 Learning and Validation Protocol

The experiments are validated using a five-fold cross validation strategy. Each time, we split the dataset into three subsets at polygon level with sizes of 60%, 20% and 20% representing respectively training, validation and test sets. The CNN model is then trained and evaluated five times at decision level. In the end, we report the average overall accuracy (OA) of the five splits and indicate the standard-deviation (STD).

The model is trained using *Adam* optimizer with a learning rate of 10^{-6} and default values of the other parameters ($\beta_1 = 0.9$, $\beta_2 = 0.999$ and $\epsilon = 10^{-8}$) with a batch size of 8. We limit the number of epochs to 2000, following an early stopping technique with a patience number of 100. The experiments are done on a laptop machine with a Nvidia GPU model GTX 1050 Ti with Max-Q Design (4 GB).

According to the limited number of polygons, the training is operated with two strategies. In the first one, the model is trained from scratch and in the second one it is initialized with weights obtained with the IMAGENET database in a classification problem and then fine-tuned with our data.

5   Results and Discussion

The proposed $2D$ spatio-temporal representations are used to feed the chosen CNN. For the *scan* strategy, we just use the segment length L of 10 since we are limited by the dimensions of the ROIs. At segment level, Fig. 3 illustrates the obtained loss curves when the model is trained from scratch with the different lengths of the RW segments, respectively 10, 50, 100 and 224. We observe that the training is done in the best conditions with the different lengths of the Random Walk. In the loss curve of $RW(10)$, we observe strong oscillations in the curve which is not the case in others. This is potentially due to the lack of information in the images provided to the CNN (a lot of zero –black– values in the input image), and each time when increasing the length L, the validation loss (orange curve) decreases leading to better learning rates.

The on-line classification results (overall accuracy) with spatio-temporal representations (with original dates) are reported in Table 2. All the scores are in the same range except the *scan(10)*. Indeed, with the different lengths of the Random Walk, we kept spatial information that allows to distinguish between the two considered classes (traditional and intensive orchards). We notice that with fine-tuning, all the scores are increased, with $RW(224)$ in the first position.

The obtained results are compared to those obtained with the TempCNN method [15][4]. TempCNN is dedicated to the classification of time series, where convolutions are applied in the temporal domain ($1D$ convolutions). The filter sizes are fixed following the criterion given in [15]: with a kernel size of 5 when considering the original dates, and 11 when considering the interpolated dates. For comparison purpose, we trained and validated the TempCNN model using the same data and validation protocol than the one used for our model. Note that the TempCNN model is proposed in [15] with different architectures (depths of the network), leading to different numbers of filters.

Table 2. Classification results (overall accuracy – OA and standard deviation – STD) obtained with our spatio-temporal representations (with original temporal information).

Lengths of the segments	From scratch		Fine-tuning	
	OA	STD	OA	STD
scan(10)	73.00	9.27	80.50	5.09
RW(10)	85.50	5.56	90.50	5.78
RW(50)	80.00	7.27	92.00	2.91
RW(100)	84.50	5.33	**94.00**	**2.00**
RW(224)	**86.00**	**5.61**	92.50	4.18

---

[4] https://github.com/charlotte-pel/temporalCNN.

Table 3. Classification results (overall accuracy – OA and standard deviation – STD) with the different architectures of TempCNN [15] (with original dates and kernel size of 5).

Nb filters	16	32	64	128	256	512	1024
OA	78.81	77.38	81.66	78.45	**85.37**	81.73	84.80
STD	6.08	6.51	4.59	4.79	**3.44**	5.75	6.48

Table 3 reports the obtained results with the TempCNN method. Best scores were obtained with 256 filters. The best obtained score with our method (when we train from scratch) is slightly better than those obtained with TempCNN. However, when we use fine-tuning, we outperform them. This highlights, for our applicative context, the benefit of considering a classical $2D$ CNN model for classifying $2D + t$ images combined with our spatio-temporal representations.

Table 4 presents the classification results when considering the spatio-temporal images with the temporal interpolation strategy. We remark that all the scores are increased compared to those with less temporal information (Table 2). This is explained by the non-regular temporal distribution of the original images. So with the linear interpolation, we make the temporal distribution regular to obtain 224 dates. $scan(10)$ is always less efficient than those that are based on RW. All the obtained scores are in the same rank with $RW(100)$ in the first position with and without fine-tuning.

Table 4. Classification results (overall accuracy – OA and standard deviation – STD) obtained with our spatio-temporal representations (with temporal interpolation).

Lengths of the segments	From scratch		Fine-tuning	
	OA	STD	OA	STD
$scan(10)$	78.50	6.44	83.00	1.87
$RW(10)$	90.00	4.18	**93.00**	**4.30**
$RW(50)$	90.50	1.87	**93.00**	**2.44**
$RW(100)$	**93.50**	**2.00**	**93.00**	**2.44**
$RW(224)$	91.50	1.22	91.00	2.00

Table 5. Obtained results (overall accuracy – OA and standard deviation – STD) with the different architectures of TempCNN [15] (with temporal interpolation and kernel size of 11).

Nb filters	16	32	64	128	256	512	1024
OA	78.96	81.40	83.96	81.86	85.93	84.23	**87.21**
STD	7.34	6.32	7.14	5.18	8.03	6.23	**8.28**

Table 5 reports scores when classifying with TempCNN [15] with more temporal data. The overall accuracy is slightly increased compared to the previous one (Table 3) and the best result is obtained with 1024 filters. The scores obtained with our method are higher (with and without fine tuning) than with TempCNN.

6   Conclusion

In this article, we present a new method to classify an image time series based on a spatio-temporal representation. This representation aims to reduce the structure of the data from $2D + t$ to $2D$ without loosing too much the spatial relationship of pixels and the temporal one. Then, these new representations images are used to feed a classical CNN to perform a classification. With the proposed representation, the applied $2D$ convolutions lead to a spatio-temporal features extraction. The trained filters have weights linked to the temporal evolution and others linked to spatial evolution, finally, the combination of both carry information on spatio-temporal evolution. By considering $2D$ convolutions on this kind of images, we can also benefit of a pre-trained model, e.g. trained on the ImageNet database on a similar classification problem. Such initialization of the weights of the CNN is less tractable for $1D$ studies as no large public dataset, at the scale of ImageNet, and pre-trained networks, are available.

References

1. Andres, L., Salas, W., Skole, D.: Fourier analysis of multi-temporal AVHRR data applied to a land cover classification. Int. J. Remote Sens. **15**(5), 1115–1121 (1994)
2. Bagnall, A., Lines, J., Bostrom, A., Large, J., Keogh, E.: The great time series classification bake off: a review and experimental evaluation of recent algorithmic advances. Data Min. Knowl. Disc. **31**(3), 606–660 (2016). https://doi.org/10.1007/s10618-016-0483-9
3. Bruzzone, L., Prieto, D.: Automatic analysis of the difference image for unsupervised change detection. IEEE Trans. Geosci. Remote Sens. **38**(3), 1171–1182 (2000)
4. Chelali, M., Kurtz, C., Puissant, A., Vincent, N.: Urban land cover analysis from satellite image time series based on temporal stability. In: JURSE, Proceedings, pp. 1–4 (2019)
5. Coppin, P., Jonckheere, I., Nackaerts, K., Muys, B., Lambin, E.: Digital change detection methods in ecosystem monitoring: a review. Int. J. Remote Sens. **25**, 1565–1596 (2004)
6. Di Mauro, N., Vergari, A., Basile, T.M.A., Ventola, F.G., Esposito, F.: End-to-end learning of deep spatio-temporal representations for satellite image time series classification. In: DC@PKDD/ECML, Proceedings, pp. 1–8 (2017)
7. Grady, L.: Multilabel random walker image segmentation using prior models. In: CVPR, Proceedings, pp. 763–770 (2005)
8. Huang, B., et al.: Large-scale semantic classification: outcome of the first year of inria aerial image labeling benchmark. In: IGARSS, Proceedings, pp. 6947–6950 (2018)

9. Iandola, F., Moskewicz, M., Ashraf, K., Han, S., Dally, W., Keutzer, K.: Squeezenet: AlexNet-level accuracy with 50x fewer parameters and <1 MB model size. Computing Research Repository abs/1602.07360 (2016)

10. Ienco, D., Gaetano, R., Dupaquier, C., Maurel, P.: Land cover classification via multitemporal spatial data by deep recurrent neural networks. IEEE Geosci. Remote Sens. Lett. **14**(10), 1685–1689 (2017)

11. Inglada, J., Vincent, A., Arias, M., Tardy, B., Morin, D., Rodes, I.: Operational high resolution land cover map production at the country scale using satellite image time series. Remote Sens. **9**(1), 95–108 (2017)

12. Ismail Fawaz, H., Forestier, G., Weber, J., Idoumghar, L., Muller, P.: Deep learning for time series classification: a review. Data Min. Knowl. Disc. **33**(4), 917–963 (2019)

13. Jensen, J.R.: Urban change detection mapping using Landsat digital data. Cartography Geogr. Inf. Sci. **8**(21), 127–147 (1981)

14. Johnson, R., Kasischke, E.: Change vector analysis: a technique for the multispectral monitoring of land cover and condition. Int. J. Remote Sens. **19**(16), 411–426 (1998)

15. Pelletier, C., Webb, G., Petitjean, F.: Temporal convolutional neural network for the classification of satellite image time series. Remote Sens. **11**(5), 523–534 (2019)

16. Petitjean, F., Inglada, J., Gançarski, P.: Satellite image time series analysis under time warping. IEEE Trans. Geosci. Remote Sens. **50**(8), 3081–3095 (2012)

17. Petitjean, F., Kurtz, C., Passat, N., Gançarski, P.: Spatio-temporal reasoning for the classification of satellite image time series. Pattern Recogn. Lett. **33**(13), 1805–1815 (2012)

18. Ravikumar, P., Devi, V.S.: Weighted feature-based classification of time series data. In: CIDM, Proceedings, pp. 222–228 (2014)

19. Russwurm, M., Korner, M.: Temporal vegetation modelling using long short-term memory networks for crop identification from medium-resolution multi-spectral satellite images. In: EarthVision@CVPR, Proceedings, pp. 1496–1504 (2017)

20. Senf, C., Leitao, P., Pflugmacher, D., Van der Linden, S., Hostert, P.: Mapping land cover in complex mediterranean landscapes using Landsat: improved classification accuracies from integrating multi-seasonal and synthetic imagery. Remote Sens. Environ. **156**, 527–536 (2015)

21. Tran, D., Bourdev, L., Fergus, R., Torresani, L., Paluri, M.: Learning spatiotemporal features with 3D convolutional networks. In: ICCV, Proceedings, pp. 4489–4497 (2015)

22. Verbesselt, J., Hyndman, R., Newnham, G., Culvenor, D.: Detecting trend and seasonal changes in satellite image time series. Remote Sens. Environ. **114**(1), 106–115 (2010)

Handwriting and Hand-Sketched Graphics Detection Using Convolutional Neural Nctworks

Song-Yang Cheng[1] , Yu-Jie Xiong[1]([✉]) , Jun-Qing Zhang[1] ,
and Yan-Chun Cao[2]

[1] School of Electronic and Electrical Engineering, Shanghai University of Engineering
Science, Shanghai 201620, People's Republic of China
xiong@sues.edu.cn
[2] Faculty of Economics and Management, School of Public Administration,
East China Normal University, Shanghai 200062, People's Republic of China

Abstract. Handwriting and hand-sketched graphics carry rich information to reveal the insights of the physical and emotional state of the writer. Before analyzing the personal traits of handwriting and hand-sketched graphics, detecting them from the image is the most immediate subproblem in handwritten document analysis and understanding. In this paper, we introduce two Convolutional Neural Networks (CNN) based methods to extract multimodal information (handwriting and hand-sketched graphics) from questionnaire documents. A Connectionist Text Proposal Network (CTPN) based method is proposed to detect handwriting. The first stage employs the VGG-16 model to generate the convolutional feature maps of the original images. Then the second stage adopts a BLSTM based detector to predict the scores of candidate zones. An instance segmentation method using the Mask Region Convolutional Neural Network (Mask-RCNN) is also proposed to solve hand-sketched graphics detection problem. The Mask-RCNN based approach has two parts: the backbone and the head. The backbone is to extract the features over the original image, and the head is to perform bounding boxes regression and mask prediction. At first, a simple Region Proposal Network (RPN) is adopted to generate the proposals of hand-sketched graphics efficiently. Then, the Region of Interest (RoI) features of the above proposals are fed into the Fast-RCNN branch and the mask branch to obtain the bounding boxes and the graphics segmentation results. The best handwriting detection performance of 200 test questionnaire images is that the precision rate is 99.5%, the recall rate is 99.2% and the F-measure score is 99.4%. The best detection performance of hand-sketched graphics is that the precision rate is 99.0%, the recall rate is 99.5% and the F-measure score is 99.3%. Experiments demonstrate that the proposed methods achieve promising results in both handwriting and hand-sketched graphics detection tasks.

Keywords: Handwriting detection · Hand-sketched graphics
detection · Convolutional neural networks

© Springer Nature Switzerland AG 2020
Y. Lu et al. (Eds.): ICPRAI 2020, LNCS 12068, pp. 352–362, 2020.
https://doi.org/10.1007/978-3-030-59830-3_31

1 Introduction

Modern science has confirmed that handwriting and hand-sketched graphics play significant roles on human communication, perception, emotional behavior and so on. With the wide and rapid development of pattern recognition and artificial intelligence, it is reasonable to discover the automatic analysis of individual characteristics based on his/her handwriting and hand-sketched graphics. In order to achieve this ambitious goal, extracting structured information from handwritten documents is the first and fundamental process.

In daily life and business, handwritten documents are everywhere. Examples of handwritten documents are purchase receipts, questionnaires and so on. Handwritten documents contain information in the forms of both text (handwriting) and vision (hand-sketched graphics). Handwriting usually has a strong correlation between the horizontal and vertical texture properties around a certain patch. The position of the handwriting can be represented by the four coordinates of the handwriting bounding box, however, the hand-sketched graphics focus on the visual features of the global structural information. The zone of the hand-sketched graphics can be marked as the segmented region of the graphics. Thus, multimodal information extraction from handwritten documents via a unified framework is not easy, besides the various printed texts and lines, the main difficulty comes from the diversity of the textural and visual cues between handwriting and hand-sketched graphics.

Convolutional Neural Networks (CNN) have demonstrated excellent capabilities in the fields of machine learning, and are widely used in many vision-based applications [1]. With millions of cameras acting as sensors around the world, there are lots of significant opportunities for images analysis of these photos to provide actionable insights. These insights will benefit a wide variety of fields, from public safety to commercial activity. In this paper, we introduce two methods based on CNN to extract handwriting and hand-sketched graphics from the questionnaire documents. An end-to-end method based on CTPN [2] is proposed and used for spotting the handwriting. The first stage employs the VGG-16 model [3] as its backbone network to generate the convolutional feature maps of the initial samples. Then the second stage adopts a Bi-directional LSTM-based encoder, and connects it with the Fully Connected layer (FC) to predict the scores and locations of handwriting zones. An instance segmentation based approach using Mask-RCNN [4] is also proposed to deal with the problem of hand-sketched graphics detection. The approach is divided into two parts: the backbone and the head. The role of the backbone part aims to extract the features over the original images through the convolution layers, and the head part is to achieve three tasks: classification, bounding box regression and mask prediction. Firstly, a simple CNN called as Region Proposal Network (RPN) is assigned to generate the region proposals of hand-sketched graphics efficiently. Secondly, the Region of Interest (RoI) features of the above region proposals are fed into the Fast-RCNN [5] branch and the mask branch to obtain the bounding boxes and graphic instances segmentation results. Experiments demonstrate that the proposed two methods achieve promising results in both handwriting and hand-sketched graphics detection tasks. The remainder of this paper is organized as follows: Sect. 2 gives a brief review of recent works on text and object detection; Sect. 3 describes details of the proposed methods; Sect. 4 provides experimental results, and Sect. 5 concludes this paper.

2   Related Work

Our work is inspired by recent researches in the field of text detection and object detection. Text and object detection have attracted great attention in document analysis and computer vision in recent years. There are two commonly used deep learning methodologies in text and object detection systems: one-stage detectors and two-stage detectors. Generally speaking, the dominant methods in text/object detection are based on a two-stage approach, and one-stage detectors are tuned for realtime applications. While, recent researches reported that two-stage detectors can be made fast simply by reducing input image resolution and the number of proposals [6]. Detection tasks are still very challenging due to the complexity of environments, flexible imagine acquisition styles and variations of text/object contents [7].

2.1   Text Detection

The earlier work on scene text detection usually focused on handcraft features, like Maximally Stable Extremal Regions (MSER) [8], Extremal Regions (ER) [9] and Stroke Width Transform (SWT) [10, 11]. These methods utilized the texture and shape of text characters, and they were workable when the images were clean [12]. Deep learning is able to extract different discriminative features for improving the performance of text detection from images with complex backgrounds. For example, Single Shot Multiboxes Detector (SSD) [13] can be used to detect multi-scale texts in scenes. Faster-RCNN achieved good results on horizontal text detection with the help of the LocNet which was based on localization module [14]. Connectionist Text Proposal Network (CTPN) used the LSTM module to realize accurate text location [2]. B. Shi et al. combined the core ideas of small candidate proposals with SSD to deal with multi-oriented problem [15]. Mask TextSpotter [16] can recognize the instance sequence inside character maps rather than only predict an object region.

2.2   Object Detection and Instance Segmentation

Object detection and instance segmentation made great progress in the past decade. Faster-RCNN [17] classified object proposals and predicted their spatial locations jointly. YOLO [18] was simple to construct and can be trained directly on full images. It pushed the state-of-the-art into real-time object detection. FPN [19] presented a clean and simple framework for building feature pyramids inside networks, and showed good performance on object detection. J. Dai et al. exploited image local coherence to provide instance-level segment candidates [20].

3 The Proposed Methods

3.1 CTPN Based Handwriting Detection

The overall framework of the proposed method is shown in Fig. 1. CTPN receives the images of different sizes, and predicts the positions of handwriting by moving the sliding-windows over the extracted feature maps. For a document image, CNN features are firstly extracted using VGG16 pre-trained model. Secondly, a sliding-window (3 × 3) moves densely over the extracted feature to create sequential features as the input for B-LSTM. At last, the output of B-LSTM is connected to the following fully connected layer that calculates text/non-text score and coordinates information of each proposal.

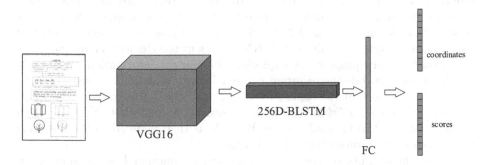

Fig. 1. The framework of CTPN based method for handwriting detection

Convolutional Feature Extraction. VGG-16 model is utilized as a feature extractor in our method. In order to share convolutional computation, a small 3 × 3 spatial window is slide over the output of VGG-16 model. The total stride and receptive field of the obtained feature map are fixed as 16 and 228 pixels by the network architecture.

Fine-Scale Proposals. Compared with general object, handwriting has strong correlation between the horizontal pixel sequences. Thus, the fine-scale text proposal is defined as a sequence of text pieces with the fixed (16-pixel) width. Each proposal contains a small patch (such as single strokes, a part of a character, or a single character) of a certain horizontal text lines. By fixing the horizontal coordinates, the prediction of vertical location will be easier and more effective. The specific implementation is as follows. Due to the structure of VVG-16 model, the receptive field of the obtained feature map is 16 × 16. Thus, the minimum horizontal interval is 16 pixels in the original image. Ten vertical anchors are created to predict coordinates of y-axis in each proposal, and the height of these anchors in the original image ranges from 11 to 273 pixels with equal ratio.

Then we get the coordinates measured by the height and y-axis center of a proposal. The relative predicted vertical coordinates are calculated:

$$v_c = (c_y - c_y^a)/h^a \quad v_h = log(h/h^a) \tag{1}$$

$$v_c^* = (c_y^* - c_y^a)/h^a \quad v_h^* = log(h^*/h^a) \tag{2}$$

where v_c, v_h and v_c^*, v_h^* are the predicted coordinates and ground truth coordinates. c_y^a and h^a are the y-axis center and height of each anchor box, it can be calculated from the original input image. c_y and h are the predicted value of y-axis, c_y^* and h^* are the ground truth value. In this way, the bounding box with size of $h \times 16$ in each proposal from original images is obtained.

Recurrent Connectionist Text Proposals. Text lines are split into a sequence in order to get more accurate location of each proposal. It has been verified that Recurrent Neural Network (RNN) can be used to encode context information for text recognition [21]. RNN gives us the chance to make accurate detection for every proposal by exploring the context information. A BLSTM which allows to encode the recurrent context in both directions is used to extend the RNN layer, so that the connectionist receipt field is able to cover the whole image width. Each LSTM has 128 nodes, resulting in a 256D RNN hidden layer in our network. The internal state of BLSTM is then mapped to the following FC to compute the predictions of each proposal.

As for the optimization of the model, the loss function L is the sum of the classification loss L_s^{cl} and the regression loss L_v^{re}:

$$L = \frac{1}{N_s} \sum_i L_s^{cl}(s_i, s_i^*) + \frac{1}{N_v} \sum_j L_v^{re}(v_j, v_j^*) \tag{3}$$

where i is the index of anchors in one batch, j is the index in a set of interested anchors for y-axis coordinates. An interested anchor represents the anchor whose Intersection-over-Union (IoU) with the ground truth text proposal is larger than 0.7. s_i is the confidence of judging an anchor as a handwriting, and s_i^* is the ground truth. v_j are the prediction y-axis coordinates while v_j^* are the ground truth coordinates.

Proposal Connection. In this part, the fine-scale proposals are connected into an integral proposal which contains all information of an area that we are interested. In this way, we can get proper results which are meaningful for human beings. As done in Ref. [2], text line construction is straightforward by connecting continuous text proposals whose text/non-text score is >0.7. At First, a paired neighbour (B_j) for a proposal is defined as B_i as $B_j \Rightarrow B_i$, when (i) B_j is the nearest horizontal distance to B_i, and (ii) this distance is less than 50 pixels, and (iii) their vertical overlap is larger than 0.7. Secondly, two proposals are grouped into a pair, if $B_j \Rightarrow B_i$ and $B_i \Rightarrow B_j$. Then a text line is constructed by sequentially connecting the pairs having a same proposal.

3.2   Mask-RCNN Based Hand-Sketched Graphics Detection

The Mask-RCNN [4] based framework for hand-sketched graphics detection is divided into four parts: Feature Pyramid Network (FPN), Region Proposal Network (RPN), Fast-Region Convolutional Neural Network (Fast-RCNN) and Fully convolutional network (FCN). The overall architecture of the proposed method is presented in Fig. 2. For a document image, CNN features are firstly extracted using FPN. During this period, the top-level features are merged with the bottom-level features by up-sampling, and each layer provides feature maps independently. Secondly, the RPN generates a lot of graphics proposals. Then, the features of the proposals are fed into the RCNN and the mask branch by applying RoIAlign [4] on the output of FPN. The RCNN branch performs classification and regression to produce horizontal bounding boxes, meanwhile, the Mask branch predicts the global graphics instance segmentation results.

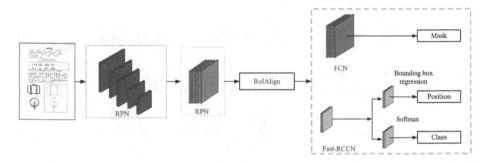

Fig. 2. The framework of Mask-RCNN based method for hand-sketched graphics detection

FPN. To demonstrate the generality of the original Mask-RCNN, multiple architectures of network such as ResNet-50, ResNet-101, ResNeXt-50 and ResNeXt-101 are used as the backbone [4]. It is well known that FPN takes the advantages of multi-scale feature spaces for accurate localization, thus, we utilize it to enhance the backbone of our approach. ResNet-101 consists of five stages, and each stage corresponds to one scales of feature map $[C_1, C_2, C_3, C_4, C_5]$. These five feature maps are used to establish the feature pyramid of FPN, and get new features respectively $[P_1, P_2, P_3, P_4, P_5]$:

$$\begin{cases} P_1 = Conv(Sum(Upsample(P_2, Conv(C_2)))) \\ P_2 = Conv(Sum(Upsample(P_3, Conv(C_3)))) \\ P_3 = Conv(Sum(Upsample(P_4, Conv(C_4)))) \\ P_4 = Conv(C_5) \\ P_5 = Downsample(P_4) \end{cases}$$

where $Conv$ represents the convolution, Sum represents the element-by-element alignment operation, $Upsample$ and $Downsample$ represent upsampling and downsampling respectively.

RPN. RPN is used to scan the above feature maps to estimate the Region of Interest (RoI) where the hand-sketched graphics may exist. The size of RoI on each stage depends on the scale of the layer of the feature pyramid. According to the Ref. [5], the RoI sizes are set to $\{32^2, 64^2, 128^2, 256^2, 512^2\}$ pixels on five stages respectively, and two aspect ratios $\{0.5, 1, 2\}$ are also adopted in each stage. The output of RPN is a series of bounding boxes with their anchors. If bounding boxes are overlapped, the non-maximum suppression (NMS) is applied to obtain the refined bounding box. After that, RoIAlign [4] is used to pool the RoIs into fixed-size feature maps. Compared to RoI Pooling, RoIAlign aligns the extracted features with the original region proposal properly, and is beneficial to the segmentation task in the mask branch. For the training of RPN, two loss functions to indicate the difference between the generated RoIs and the ground truth are utilized: L_{class}^{RPN} used to classify the generated RoIs and L_{bbx}^{RPN} used to modify the coordinates of the anchors. L_{class}^{RPN} is computed by a softmax over the outputs of a fully connected layer:

$$L_{class}^{RPN} = -\frac{1}{N_{class1}} \sum_i^{N_{class1}} log[p_i^* p_i + (1 - p_i^*)(1 - p_i)], \tag{4}$$

L_{bbx}^{RPN} is defined over the output of the i-th anchor $(x_i^o, y_i^o, h_i^o, w_i^o)$ and the corresponding ground truth of bounding box $(x_i^g, y_i^g, h_i^g, w_i^g)$:

$$L_{bbx}^{RPN} = \frac{1}{N_{reg1}} \sum_{i,j \in \{x,y,w,h\}}^{N_{reg1}} smooth_{L_1}(j_i^o - j_i^g), \tag{5}$$

in which

$$smooth_{L_1}(x) = \begin{cases} 0.5x^2 & \text{if } |x| < 1, \\ |x| - 0.5 & \text{otherwise,} \end{cases}$$

where N_{class1} indicates the number of RoIs; If the RoIs is positive sample, $p^* = 1$; otherwise, $p^* = 0$. p_i is the probability that the i-th RoI is predicted to be positive sample, and N_{reg1} is the number of corresponding anchors (background RoIs are ignored for training).

Fast-RCNN. In this part, two tasks are accomplished: box regression and classification. Thus, this part is called as box regression and classification branch. The fixed-size feature maps of RoIs are fed into fully connected layers, then the output vectors are used for the classification and bounding boxes of detected targets. Similar to the above RPN, two loss functions (L_{class}^{RCNN} and L_{bbx}^{RCNN}) are obtained by:

$$L_{class}^{RCNN} = -\frac{1}{N_{class2}} \sum_i^{N_{class2}} log[p_i^* p_i + (1 - p_i^*)(1 - p_i)], \tag{6}$$

$$L_{bbx}^{RCNN} = \frac{1}{N_{reg2}} \sum_{i,j \in \{x,y,w,h\}}^{N_{reg2}} smooth_{L_1}(j_i^o - j_i^g), \tag{7}$$

where N_{class2} indicates the number of detected targets, p_i is the probability that the i-th target is predicted to be positive sample, and N_{reg2} is the number of corresponding bounding boxes.

FCN. The difference between Mask-RCNN and Fast-RCNN is that the former not only predicts the class and bounding box, but also produces a binary mask for each detected target. As a result, this part is also called as mask branch. A mask should represent the whole spatial information of a target. Thus, FCN is the best choice that allows each layer in the mask branch to keep the spatial layout without compressing it into a feature vector that lose the spatial dimensions. The mask loss function L_{mask} is defined as:

$$L_{mask} = -\frac{1}{N} \sum_{n=1}^{N} \left[log(S(o_n))g_n + log(1 - S(o_n))(1 - g_n) \right], \tag{8}$$

where N is the area of the mask, g_n is the pixel label ($g_n \in 0, 1$), o_n is the output pixel, and $S(x) = \frac{1}{1+e^{-x}}$.

4   Experiments

To validate the effectiveness of the proposed methods, we conduct experiments on our own SUES-1000 database. **SUES-1000 database** is our own database of text and object detection. In this database, totally 1000 questionnaire images of the primary and secondary school students were collected as the handwritten samples (as shown in Fig. 3). Each student filled out one questionnaire independently and anonymously. The questionnaire contains two zones for writing

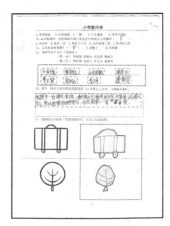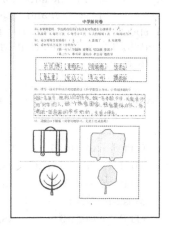

Fig. 3. Two questionnaire images of SUES-1000 database. The left one is written by a elementary school student, and the right one is written by a middle school student. The rectangles (red dashed lines) are the ground truth for the handwriting, and the polygons (red lines) are that of hand-sketched graphics. (Color figure online)

(Z_1, Z_2), and two areas for drawing (A_1, A_2). The handwriting in Z_1 is a copy of eight specified words (24 Chinese characters), and the handwriting in Z_2 is a free description about the future life (10–80 Chinese characters). The hand-sketched graphics in A_1 and A_2 are two tracings of specified illustrations (a suitcase and a tree).

During the experiment, the precision rate (P), recall (R) rate, and F-Measure are applied to evaluate the detection performance:

$$P = \frac{TP}{TP + FP}, R = \frac{TP}{TP + FN}, F = \frac{2 * P * R}{P + R},$$

where TP is the number of cases that are positive and detected positive, FP is the number of cases that are negative but detected positive, and FN is the number of cases that are positive but detected negative.

The experiments are performed under the framework of TensorFlow(1.14.0), with NVIDIA2080 for GPU acceleration, Inter (R) Core (TM) i7-9700k CPU and 16G memory. During the experiments, 800 document images were selected for training the models, and 200 document images are used to verify the stability and reliability of the trained models.

Table 1 shows the handwriting detection results of the proposed methods on SUES-1000 database. When the threshold of Intersection over Union (IOU) is set to 0.5 or 0.7, there is not much difference between the results of two methods. When the threshold of IOU is 0.9, CTPN based method has quite better performance than that of Mask-RCNN based method. It demonstrates that CTPN keeps more local structural information of handwriting than Mask-RCNN.

Table 2 shows the results of the proposed methods for hand-sketched graphics detection. Like handwriting detection task, when the threshold of Intersection over Union (IOU) is small (<0.7), both methods work very well (>99%). With the increasing of threshold, the superiority of Mask-RCNN raises quickly. When IoU = 0.9, the overall precision and recall rates of 200 testing samples are 60.5% and 60.8%, respectively. However, the performance of CTPN based method is only 45.7% and 45.6%. It means that Mask-RCNN is more suitable for the task of graphics detection.

Table 1. The handwriting detection results on SUES-1000.

	Precision	Recall	F-Measure
CTPN based@0.5	99.5%	99.2%	99.4%
Mask-RCNN baed@0.5	98.1%	99.0%	98.6%
CTPN based@0.7	95.0%	94.7%	94.9%
Mask-RCNN baed@0.7	97.1%	95.4%	95.0%
CTPN based@0.9	48.4%	48.3%	48.4%
Mask-RCNN baed@0.9	33.6%	33.9%	33.8%

CTPN based@0.5 means that CTPN based method is tested with the threshold of Intersection over Union (IoU) as 0.5.

Table 2. The hand-sketched graphics detection results on SUES-1000.

	Precision	Recall	F-Measure
CTPN based@0.5	99.5%	99.2%	99.3%
Mask-RCNN baed@0.5	99.0%	99.5%	99.2%
CTPN based@0.7	90.8%	90.6%	90.7%
Mask-RCNN baed@0.7	98.5%	99.0%	98.7%
CTPN based@0.9	45.7%	45.6%	45.6%
Mask-RCNN baed@0.9	60.5%	60.8%	60.6%

CTPN based@0.5 means that CTPN based method is tested with the threshold of Intersection over Union (IoU) as 0.5.

5   Conclusions

In this paper, we introduce two CNN based methods to extract handwriting and hand-sketched graphics from questionnaires. CTPN based method used fine-scaled proposals and vertical anchor mechanism to accurate and effective text detection. With the help of instance segmentation, Mask-RCNN based method dominate the global graphics detection. The results achieved by our proposed methods on the SUES-1000 database validate their effectiveness. The former is more workable in detecting handwriting, and the latter has better performance in the detection of hand-sketched graphics. In the future, we would like to apply the proposed methods into further handwriting analysis, such as writer identification and signature verification.

Acknowledgment. This work is sponsored by Shanghai Sailing Program (Grant No. 19YF1418400).

References

1. Gatys, L., Ecker, A., Bethge, M.: Image style transfer using convolutional neural networks. In: Proceeding of IEEE Conference on Computer Vision Pattern Recognition, pp. 2414–2423 (2016)
2. Zhi, T., Huang, W., He, T., He, P., Qiao, Y.: Detecting text in natural image with connectionist text proposal network. arXiv e-prints. arXiv:1609.03605 (2016)
3. Simonyan, K. and A. Zisserman. Very deep convolutional networks for large-scale image recognition. arXiv preprint. arXiv:1409.1556 (2014)
4. He, K., Gkioxari, G., Dollar, P., Girshick, R.B.: Mask R-CNN. In: Proceeding of IEEE International Conference on Computer Vision, pp. 2980–2988 (2017)
5. Girshick, R.B.: Fast R-CNN. In: Proceeding of IEEE International Conference on Computer Vision, pp. 1440–1448 (2015)
6. Lin, T., Goyal, P., Girshick, R., He, K., Dollár, P.: Focal loss for dense object detection. In: Proceeding of IEEE International Conference on Computer Vision, pp. 2980–2988 (2017)
7. Ye, Q., Doermann, D.: Text detection and recognition in imagery: a survey. IEEE Trans. Pattern Anal. Mach. Intell. **37**(7), 1480–1500 (2015)

8. Neumann, L., Matas, J.: A method for text localization and recognition in realworld images. In: Proceeding of Asian Conference on Computer Vision, pp. 770–783 (2010)

9. Neumann, L., Matas, J.: Real-time scene text localization and recognition. In: Proceeding of IEEE Conference on Computer Vision Pattern Recognition, pp. 3538–3545 (2012)

10. Epshtein, B., Ofek, E., Wexler, Y.: Detecting text in natural scenes with stroke width transform. In: Proceeding of IEEE Conference on Computer and Vision Pattern Recognition, pp. 2963–2970 (2010)

11. Yao, C., Bai, X., Liu, W.: A unified framework for multioriented text detection and recognition. IEEE Trans. Image Process. **23**(11), 4737–4749 (2014)

12. Shi, B., Yang, M., Wang, X., Lyu, P., Yao, C., Bai, X.: ASTER: an attentional scene text recognizer with flexible rectification. IEEE Trans. Pattern Anal. Mach. Intell. **41**(9), 2035–2048 (2019)

13. Liu, W., et al.: Single shot multibox detector. In: Proceeding of European Conference on Computer Vision, pp. 21—37 (2016)

14. Zhong, Z., Sun, L., Huo, Q.: Improved localization accuracy by LocNet for faster R-CNN based text detection in natural scene images. In: Proceeding of IEEE Conference Document and Analysis, in press

15. Shi, B., Bai, X., Belongie, S.: Detecting oriented text in natural images by linking segments. In: Proceeding of IEEE Conference on Computer Vision and Pattern Recognition, pp. 2550–2558 (2017)

16. Liao, M., Lyu, P., He, M., Yao, C., Wu, W., Bai, X.: Mask TextSpotter: an end-to-end trainable neural network for spotting text with arbitrary shapes. In: IEEE Transactions on Pattern Analysis and Machine Intelligence, in press

17. Ren, S., He, K., Girshick, R., Sun, J.: Faster R-CNN: towards real-time object detection with region proposal networks. IEEE Trans. Pattern Anal. Mach. Intell. **39**(6), 1137–1149 (2017)

18. Redmon, J., Divvala, S., Girshick, R., Farhadi, A.: You only look once: unified, real-time object detection. In: Proceeding of IEEE Conference on Computer Vision and Pattern Recognition, pp. 779–788 (2016)

19. Lin, T., Dollar, P., Girshick, R., He, K., Hariharan, B., Belongie, S.: Feature pyramid networks for object detection. In: Proceeding IEEE Conference on Computer Vision and Pattern Recognition, pp. 2117–2125 (2017)

20. Dai, J., He, K., Li, Y., Ren, S., Sun, J.: Instance-sensitive fully convolutional networks. In: Proceeding European Conference on Computer Vision, pp. 534–549 (2016)

21. He, P., Huang, W., Qiao, Y., Loy, C., Tang, X.: Reading scene text in deep convolutional sequences. In: Proceeding of AAAI Conference on Artificial Intelligence, pp. 3501–3508 (2016)

Deep Active Learning with Simulated Rationales for Text Classification

Paul Guélorget[1,2(✉)] , Bruno Grilheres[2] , and Titus Zaharia[1]

[1] Institut Polytechnique de Paris, Institut Mines-Télécom, Télécom SudParis,
Palaiseau, France
[2] Airbus Defence and Space, Élancourt, France
paul.guelorget@airbus.com

Abstract. Neural networks have become a preferred tool for text classification tasks, demonstrating state of the art performances when trained on a large set of labeled data. However, in an early active learning setup, the scarcity of the ground-truth labels available severely penalizes the generalization capability of the neural network. In order to overcome such limitations, in this paper, we introduce a new learning strategy, which consist of inserting in the early stages of the learning process some additional, local and salient knowledge, presented under the form of simulated, human like rationales. We show how such knowledge can be automatically extracted from documents by analyzing the class activation maps of a convolutional neural network. The experimental results obtained demonstrate that the exploitation of such rationales permits to significantly speed-up the learning process, with a spectacular increase of the accuracy rates, starting from a very reduced number of documents (10–20).

Keywords: Deep neural networks · Active learning · Rationales · Class activation maps · Text classification

1 Introduction

Numerous classification techniques and neural networks in particular have shown to be very effective when large labeled, ground truth data sets are available for training. However, obtaining reliable ground-truth in real-life, user-specific applications may be an extremely expensive, tedious and time-consuming task. Also called *query learning*, *active learning* [2,31] is a sub-field of machine learning designed to address the scarcity of ground-truth samples in supervised training applications, whether it is caused by the lack of reliable labels or by the progressive exploitation of a stream of unlabeled data. Such learning strategies are mandatory for various applications, including the classification of open source contents (online press, social networks, sentiment analysis, etc.) or medical imagery [15,32]. The underlying principle is the following. Over a sequence of active learning cycles, unlabeled documents are sampled or synthesized, according to criteria dictated by a sampling strategy, and presented to a human,

© Springer Nature Switzerland AG 2020
Y. Lu et al. (Eds.): ICPRAI 2020, LNCS 12068, pp. 363–379, 2020.
https://doi.org/10.1007/978-3-030-59830-3_32

expert oracle for labeling. The labeled documents are then successively added to the training set and the neural network is re-trained with these new examples. However, in an usual active classification setup, the oracle only provides a label information for a given document. Yet, as an expert of the classification task, the oracle holds a more detailed knowledge than solely the class membership.

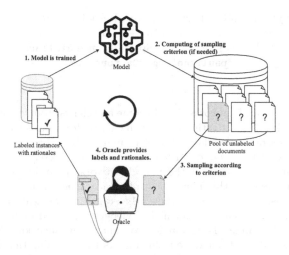

Fig. 1. An iteration of active learning with rationales in a pool-based scenario

This observation is notably exploited in [34] and [30]. In order to further benefit from the oracle's expert knowledge, authors propose to ask the oracle, in addition to a simple label, for rationales that are guiding the labeling process, for text categorization applications. In this scenario called active learning with rationales (Fig. 1), the oracle has the possibility to specify a set of words in a given document that guided the classification choice. This information is further exploited in order to speed-up and enhance, in terms of classification performances, the learning process. In this paper, we present a method to transfer local rationales to a fully-convolutional neural network dedicated to text classification, within an active learning setting. The rest of the paper is organized as follows. Related work is presented in Sect. 2. In Sect. 3, we describe our approach for training a neural text classifier with additional rich spatial knowledge and with sparse, human like rationales. Experimental results are presented and discussed in Sect. 4. Finally, Sect. 5 concludes the paper and opens perspectives of future work.

2   Related Work

Text classification tasks have been widely addressed in numerous real applications over the last few years, such as emotions, sentiments and opinions classification [1,6,14], language identification [5] and even irony detection [7].

For the exploitation of raw textual data, the classification task is addressed by two main processes: feature extraction and classification technique. Common feature extraction methods are Term Frequency-Inverse Document Frequency (tf-idf) [28], Word2Vec [25], GlobalVectors for Word Representation (GloVe) [26], FastText [16], ELMo [27], BERT [11]. Naive Bayes, Logistic Regression and Support Vector Machines are traditional classification techniques, whereas Convolutional Neural Networks (CNN) [17] and Recurrent Neural Networks (RNN) [8] have been recently increasing in popularity.

The key principle of active learning methods consists in selecting the examples which are the most appropriate for boosting the learning process. Most often a human labeler, called oracle, is involved in the learning process. His mission is to provide the correct labels for a set of sampled examples. Depending on the active learning scenario, samples may come from a static pool, from a continuous stream of unlabeled documents, or they can be even synthesized by generative models [13].

Active learning frameworks are characterized by their sampling strategy, which dictates what examples are presented to the oracle. Some common pool-based scenario sampling criteria that have shown to be efficient are uncertainty [9,19], information density [21,29], diversity [4,12], or query-by-comittee [10]. For evaluation purposes, active learning iterations are simulated by splitting the full-size labeled training set into an initial restrained training set and a larger pool of candidates for sampling. The restrained training set is then progressively enriched during the learning process with samples that satisfy the most the criterion under consideration. The performance of an active learning framework is generally evaluated by: (1) the amount of human labeling that is necessary to achieve a given accuracy score, or (2) examining the model's accuracy as a function of the size of the training set.

Within the framework of active learning with rationales, to further benefit from the underlying knowledge and understanding of the oracle, in [30,34] authors propose to ask the labeler, in addition to a simple label, for rationales that are guiding the labeling process, for text categorization applications. In this scenario, the oracle specifies which words are determinant for the classification choice. The influence of the words picked as rationales is then artificially inflated during the classification process. To this purpose, various methods can be employed, including multinomial naïve Bayes, logistic regression or support vector machines. By considering solely sets of words, such approaches do not take into account the local context of the rationales. However, the significance of a word and its relevance with respect to a given category strongly depends on the context of its appearance.

The issue of active learning with rationales has been until now poorly addressed within the context of deep convolutional neural networks (CNNs). In our work, we notably tackle this issue and propose the following contributions: (1) We first show that the structure of a fully convolutional network defines receptive fields that are naturally adapted to detecting, in an automatic manner, groups of successive words that reveal local, contextual and salient knowledge that can be used as rationale for identifying the corresponding category. (2) We show these rationales can be efficiently transferred from a teacher to a student network within the framework of an active learning process. (3) We finally derive a binarized rationale representation that is resembling with human like rationales and demonstrate its pertinence for active learning objectives. The proposed methodology is described in details in the following section.

3 Learning with Rationales

3.1 Background

This section introduces the notation used throughout the paper and defines the concepts of dense, comprehensive spatial domain knowledge and sparse spatial domain knowledge.

Active Learning Without Rationales. Let $\mathcal{L} = \{\langle x, y \rangle\}$ be a set of document-label pairs and $\mathcal{U} = \{x\}$ a set of unlabeled documents. Documents are 1-dimensional arrays of words or punctuation characters. In a C-label multi-label setup, a label y is represented as a vector in $\{0, 1\}^C$. At each active learning iteration, the underlying model $\mathcal{M}$ is trained on $\mathcal{L}$, then documents from $\mathcal{U}$ are sampled and shown to the oracle for labeling. These newly-labeled documents are removed from $\mathcal{U}$ and added to $\mathcal{L}$ with their respective labels.

Active Learning with Rationales. Let $\mathcal{L} = \{\langle x, y, R \rangle\}$ be a learning set of document-label-rationale triplet. New component R is a matrix containing expert spatial domain knowledge. As an example, let $x = \{x_t\}$ be a text document of length T (with T denoting the number of words). The rationales matrix $R = (r_{c,t})_{1 \leq c \leq C, 1 \leq t \leq T}$ stores the influence of each in-context word x_t on the label-memberships expressed by y. A high, positive $r_{c,t}$ value indicates a high positive influence of word x_t, in favor of label c. On the contrary, a low negative $r_{c,t}$ value represents a high negative influence of word x_t in the detriment of label c. A close to zero value tends to express neutrality towards the concerned label. The interaction with the user is carried out as in the previous case. A set $\mathcal{U} = \{u\}$ of unlabeled documents is supposed to be available and presented to the oracle. At each iteration, for a set of sample documents from $\mathcal{U}$, the oracle is asked to provide the corresponding label vectors as well as the rationale matrices R. Finally, the learning set $\mathcal{L}$ is updated with the new documents (which are in the same time withdrawn from $\mathcal{U}$) and the learning process is iterated on the updated set $\mathcal{L}$.

3.2 Training a Classifier with Contextual and Salient Knowledge

This section presents our main contribution, which defines how to extract and inject expert knowledge within a context of restrained ground-truth, as encountered in an active learning setup. We propose original solutions for the following issues: (1) the automatic generation of *local, contextual and salient* (LCS) knowledge that provides rationales for the learning process; (2) the transfer of such knowledge to another network in an active learning setup; (3) the pruning of the contextual knowledge in order to make it similar to human-generated rationales. Under this framework, the central concept is the one of LCS knowledge. We claim that useful rationales should be: (1) *local*: solely some sub-parts of the document are relevant for determining the corresponding category; (2) *contextual*: words become meaningful only when considered within their context; (3) *salient*: the selected words should be discriminative for establishing a given category. Let us now detail how such LCS knowledge can be extracted automatically from textual documents.

Automatic Extraction of Local, Contextual and Salient Knowledge. Obtaining comprehensive class-membership cues from human labelers is a highly tedious task, significantly more complex than a simple labeling one. For this reason, we propose an automatic approach for generating rationales, based on a preliminary teacher model that is further exploited for knowledge transfer towards a student model. The teacher neural network is trained on a large set of labeled data (different from the one considered in the user's application) in order to extract some form of dense, comprehensive domain knowledge. To this purpose, two techniques are often considered to characterize sub-documents segments. A first one consists in observing the behavior of an attention mechanism [3]. The second approach is based on the extraction of class activation maps [35]. Attention mechanisms represent a first manner to obtain spatial or temporal saliency maps with neural networks. In [3], authors address the issue of text translation and point out that attention scores produce a meaningful unsupervised alignment between input and output tokens. In [22], a self-attention mechanism is introduced. Here, high attention scores assigned to crucial words are observed when performing text classification. Also, in [33] authors show that attention is transferable. However, attention scores are flawed when considering the purposes of this paper. They can be interpreted as the saliency of a sub-part of the document relatively to a generic classification task, and not relatively to a precise label. They are ranging from zero (no interest) to one (high interest) without carrying information whereas the considered sub-parts play in favor or against class membership. For example, in [22], a self-attention mechanism is used to predict opinion polarity: the observed attention scores are high for both positively and negatively polarized words, without distinction. We would benefit from a finer characterization technique acknowledging for negative and positive contributions of words to class membership.

A different approach concerns the so-called class activation maps (CAMs), introduced in [35]. The CAMs are able to define the contribution of each sub-part of the document to each considered class. Initially used for image classification purposes, we propose to adapt this technique to textual data. The method relies on the observation that spatiality is preserved across convolutional layers, whereas it is lost in the last fully-connected layers used by some CNNs. Hence, it only concerns fully-convolutional networks where global average pooling (GAP) or global max pooling (GMP) is applied to squash the T spatial features vectors associated to the deepest feature maps $F = \{F_1, ..., F_T\} \in \mathbb{R}^{T \times K}$ into a single, global feature vector $\Gamma_g \in \mathbb{R}^K$ in which all spatiality is lost. Here, K denotes the size of the considered feature maps.

The model that we have adopted is a fully-convolutional neural network made of 3 layers of 128 kernels of size 5 followed by a global average pooling and a sigmoid classification layer. We used the pre-trained FastText word embedding [16].

Because we need to preserve the temporality across layers, we use same padding for convolutions, so that there is exactly one output layer directly corresponding to each input token. Thus, the last convolutional layer presents a number of outputs that is equal to the number of input words. In this way, the tth output of the last convolutional layer describes the tth word of the input sentence, while taking into consideration its context within the convolutional receptive field.

The CAM extraction is explained hereafter. If there are C different labels, then the softmax input is defined as:

$$S = W^T F_g + b \tag{1}$$

where $W = \{w_c^k\} \in \mathbb{R}^{K \times C}$ and $b \in \mathbb{R}^C$ are the final weights and biases. For any input example x, the class activation map for label c at location t is obtained by summing the contribution $w_c^k F_t^k$ of each scalar feature F_t^k to the final score of label c, as described in the following equation:

$$\text{CAM}(x, c, t) = \sum_k w_c^k F_t^k \tag{2}$$

The CAM address the aforementioned limitations of the attention scores to fulfill our purpose: in a text classification context, it provides a signed contribution of each word to a class membership, for every class tackled by the model. Hence, we chose to use CAMs to extract dense, comprehensive LCS knowledge from a network previously trained on a large set of documents, thus $R = \text{CAM}$. Our CAM-extracted model is illustrated in Fig. 2. Figure 3 is an example of LCS knowledge.

Knowledge Transfer of LCS Knowledge. We now formulate the assumption that we are considering under the following active learning circumstance. At the active learning bootstrap stage and at each iteration, some unlabeled documents are sampled and shown to a simulated oracle who provides (1) the actual labels L of the sampled documents and (2) the dense, comprehensive domain knowledge matrix $R \in \mathbb{R}^{C \times T}$ as described in Sect. 3.1, extracted from a learned teacher model.

At each active learning iteration, a fully-convolutional neural network is trained on the (L, R) ground-truth pairs. From a knowledge transfer perspective, we consider this model as a student model whereas the model used for

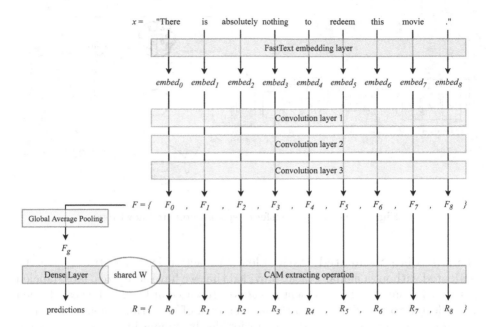

Fig. 2. Extraction of Class Activation Maps from a tokenized sentence.

label	"supplies	for	deployed	troops"
sports	0.22	0.12	-0.89	-0.91
military	0.41	0.15	0.92	1.00

Fig. 3. LCS knowledge: class specificity of words taken in their context.

the extraction of transferred knowledge R is a teacher model, as illustrated in Fig. 4. Both teacher and student network use the architecture described above, and illustrated in Fig. 2. First, the teacher neural network is trained on a large set of labeled documents. Then, the rationales matrix R of each document are extracted using the trained teacher model after what active learning iterations are carried out to train the student model. When a document x is sampled, the student model gains access to its label y and rationales matrix R.

The objective is twofold. On one hand, we want to train the student model to correctly classify the scarce available documents. On the other hand, we aim at correctly learning the class specificity for each sub-part of these documents. To achieve this purpose, we introduce a rationales loss L_R that is the Mean Squared Error (MSE) between the spatial knowledge R and the CAMs directly extracted from the student model during training. This new loss function term is added to the standard cross-entropy classification loss function. The hypothesis backing this process is that the accurate spatial knowledge embedded within an even scarce set of documents is likely to be beneficial to the generalization power of the network.

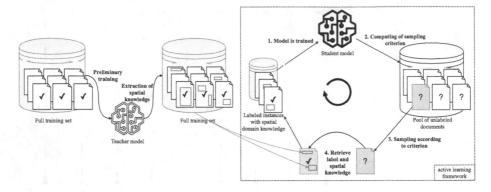

Fig. 4. Knowledge transfer of spatial domain knowledge

Human Like Simulated Rationales. An active learning strategy can be evaluated without a human oracle, provided that we have access to a large labeled dataset. The label y of a document $x \in \mathcal{U}$ is obfuscated at first, and revealed when x is selected to join $\mathcal{L}$. To avoid the tedious task of gathering human rationales, we propose to do the same with class rationales by simulating them beforehand. To this end, [30] decided of a vocabulary specific to each label. Every time a word from this vocabulary is encountered, it is considered as a simulated rationale, no matter the context of this word. Our method brings some contextualization to simulated rationales, since the same word in a different context may or may not be retained as a rationale depending on the words surrounding it.

We propose to differentiate the *dense, comprehensive* LCS knowledge (not likely to be informed by a human) from the *sparse* LCS knowledge, more likely to constitute realistic simulated rationales. We used Class Activation Mappings to extract the former. We propose to prune CAM values according to the following hypotheses: (1) a human oracle may input zero, one or more rationales per text for his labeling choice; (2) it would be tedious for a human oracle to input a score for every rationale.

Hence, we used a threshold to prune-out and discretize CAM values, so texts may keep from zero to many rationales. First, the teacher model is used to extract all CAM values for all (token, label) couples. For texts actually belonging to a certain label, only a percentage α of the highest values are kept as rationales and set to 1. In a similar manner, for negative CAM values, the same percentage α of the lowest (negative) values are kept as rationales, and set to -1. All other values are set to 0. Here, the percentage threshold α is a user-defined parameter. In our work, we have considered a variation range from 1% to 10%. Also, rationales loss L_R is a hinge loss function modified to ignore null values:

$$L_R(R, \hat{R}) = \frac{1}{C} \sum_{c=1}^{C} \frac{1}{|R'_c|} \sum_{t=1}^{T} \begin{cases} \max(0, 1 - \tanh(\hat{r}_{c,t})) & \text{if } r_{c,t} = 1 \\ \max(0, 1 + \tanh(\hat{r}_{c,t})) & \text{if } r_{c,t} = -1 \\ 0 & \text{if } r_{c,t} = 0 \end{cases} \quad (3)$$

where $R = (r_{c,t})_{1 \leq c \leq C, 1 \leq t \leq T}$ are the ground-truth rationales and $\hat{R} = (\hat{r}_{c,t})_{1 \leq c \leq C, 1 \leq t \leq T}$ are the CAM values extracted from the student network at training time, corresponding to labels c and words offsets t; and $R'_c = \{r_{c,t} \in R_c | r_{c,t} \neq 0\}$.

Figure 5 presents pruned, binarized spatial knowledge.

label	"supplies	for	deployed	troops"
sports	0	0	-1	-1
military	0	0	1	1

Fig. 5. Sparse spatial domain knowledge: label specificity of words taken in their context.

4   Experimental Results

In this section we describe the datasets and the active learning settings used for our experiments. Then, we present the simulated, sparse rationales obtained when applying the CAM-pruning method presented in Sect. 3.2. Finally, we present and comment the learning curves obtained with learning without rationales (Lw/oR), learning with dense, comprehensive rationales (LwDR) and learning with sparse, human like rationales (LwSR).

4.1   Evaluation Protocol

We used the following text classifications datasets:

- The IMDB dataset consists in 25K movie reviews [24] labeled as positive or negative.
- The WvsH dataset is a 20 Newsgroups dataset [18] reduced to Windows versus Hardware topics.

In order to compare our results with those of [30], we have used the AUC (Area Under ROC Curve) measure as evaluation metric. For LwSR, the values considered for the α parameter are of 10, 3 and 1. We have used a budget of 200 documents to feed the active learning process.

Concerning the active learning process, we have used a bootstrap of 10 random documents and we labeled 5 documents at each active learning iteration, which have been sampled according to the uncertainty criterion [20]. In this binary classification context, our uncertainty function is defined by

$$u(x) = \frac{1}{2} - |\hat{y} - \frac{1}{2}| \tag{4}$$

where $\hat{y} \in [0, 1]$ is the student model's prediction. At the beginning of each active learning iteration, the 5 most uncertain unlabeled documents are selecting using this criterion.

The Lw/oR, LwDR and LwSR scenarios have the same initial documents boot-straps and all models are initialized with the same set of random weights and biases.

4.2   Contextualized Simulated Rationales

Figures 6 and 7 illustrate Class Activation Maps extracted from the teacher model, and the results of their pruning to get sparse class rationales. First, the extracted dense, comprehensive spatial knowledge shows that the trained teacher model is able to detect what words are discriminative for establishing a given category: in these examples, the highest values are "worst" (negative review), "wonderful" (positive review), "interface" (hardware) and "windows" (Windows). It appears that the teacher model is sensitive to context, as shown by the apparent continuity of values across words. Two examples notably contains

	Spatial domain knowledge (green: positive, red: negative)
Dense, comprehensive	This is one of the worst movies I have ever seen . However , the little slave girl , Alice and Jared Harris imitating Christopher Walken is what makes this movie entertaining . [...]
Sparse (α = 10%)	This is one **of the worst movies I** have ever seen. However, the little slave girl, Alice and Jared Harris imitating Christopher Walken is what makes this movie entertaining. [...]
Sparse (α = 3%)	This is one of **the worst movies** I have ever seen. However, the little slave girl, Alice and Jared Harris imitating Christopher Walken is what makes this movie entertaining. [...]
Sparse (α = 1%)	This is one of the worst movies I have ever seen. However, the little slave girl, Alice and Jared Harris imitating Christopher Walken is what makes this movie entertaining. [...]
Dense, comprehensive	This is the definite Lars von Trier Movie , my favorite , I rank it higher than " Breaking the waves " or the latest " Dancer in the Dark "... I simply love the beauty of the pictures ... The framing is so original ; acting is wonderful , A MUST SEE .
Sparse (α = 10%)	This is **the definite Lars von Trier Movie, my favorite, I rank it** higher than "Breaking the waves" or the latest "Dancer in the Dark"... I simply love the beauty of the pictures... The framing is so original; **acting is wonderful, A MUST SEE.**
Sparse (α = 3%)	This is the definite Lars von Trier Movie, my **favorite,** I rank it higher than "Breaking the waves" or the latest "Dancer in the Dark"... I simply love the beauty of the pictures... The framing is so original; acting is **wonderful, A** MUST SEE.
Sparse (α = 1%)	This is the definite Lars von Trier Movie, my favorite, I rank it higher than "Breaking the waves" or the latest "Dancer in the Dark"... I simply love the beauty of the pictures... The framing is so original; acting is **wonderful,** A MUST SEE.

Fig. 6. Extracted spatial domain knowledge, IMDB dataset (best viewed with colors) (Color figure online)

	Spatial domain knowledge (red: Windows, blue: hardware)
Dense, comprehensive	I have a DFI Handy Scanner Model HS - 3000Plus and a little bit of software running under dos to use it . I ' d like to make more extensive use of this device (in particular , write a driver for it on unix) . So , can anyone give me a description of how to talk to this device . It connects to the system via it ' s own interface card . Any info would help , it can ' t be too difficult to talk to : -)
Sparse ($\alpha = 10\%$)	I have a **DFI Handy Scanner Model HS - 3000Plus** and a little bit of software running under dos to use it. I' d like to make more extensive use of **this device (in particular,** write a driver for it on unix). So , can anyone give me a description of how to talk to this **device. It connects** to the system **via it' s own interface card**. Any info would help, it can' t be too difficult to talk to :-)
Sparse ($\alpha = 3\%$)	I have a DFI Handy Scanner Model HS - 3000Plus and a little bit of software running under dos to use it. I' d like to make more extensive use of this device (in particular, write a driver for it on unix). So , can anyone give me a description of how to talk to this **device. It connects** to the system via it' s **own interface card**. Any info would help, it can' t be too difficult to talk to :-)
Sparse ($\alpha = 1\%$)	I have a DFI Handy Scanner Model HS - 3000Plus and a little bit of software running under dos to use it. I' d like to make more extensive use of this device (in particular, write a driver for it on unix). So , can anyone give me a description of how to talk to this device. It connects to the system via it' s own **interface card**. Any info would help, it can' t be too difficult to talk to :-)
Dense, comprehensive	I remember reading about a program that made windows icons run away from the mouse as it moved near them . Does anyone know the name of this program and the ftp location
Sparse ($\alpha = 10\%$)	I remember reading about a **program that made windows icons run away from the** mouse as it moved near them. Does anyone know the name of this **program and the ftp location**
Sparse ($\alpha = 3\%$)	I remember reading about a program **that made windows icons run away** from the mouse as it moved near them. Does anyone know the name of this program and **the ftp location**
Sparse ($\alpha = 1\%$)	I remember reading about a program that **made windows icons run** away from the mouse as it moved near them. Does anyone know the name of this program and the ftp location

Fig. 7. Extracted spatial domain knowledge, WvsH dataset (best viewed with colors) (Color figure online)

contradictory series of words. The teacher model detects positiveness when a movie reviewer admits being entertained by one of the worst movies he/she has ever seen; the teacher model detects software references in an overall hardware related message. The transfer of this very comprehensive knowledge to a student network corresponds to our LwDR scenario. The continuity of the CAM values makes it difficult to prune them to realistic, human like rationales. Pruning with $\alpha = 10\%$ tends to contain more information than a human oracle could efficiently provide with a highlighting tool, whereas pruning with $\alpha = 1\%$ is more realistic but may omit every salient word of some sentences (Fig. 6). The transfer of these sparse, simulated rationales to a student network corresponds to our LwSR scenarios.

IMDB positive reviews	IMDB negative reviews
Kusturica brilliantly examines this theme	lack of much evidence of
compellingly explores the emotional chasm	of confusion and disappointment
well-written and well	preposterously ugly and annoying girl
are remarkably futuristic today.	disrespectful boring shame that will
perfect supporting cast and great	was a crass attempt at
definitely recommend this. 9	horrible acting, lame porn
I really liked Dana Plato	was pretty crap-freaking
Everyone was great! 10	boring with its clichés and
the unmatchably billiant and ingenious	create a cheap idiotic show
Brashear deeply touches the heart	was the worst ending I
WvsH Windows messages	**WvsH hardware messages**
in the apps icon (	with 2MB of DRAM.
for Windows? Will it	. Are SCSI drives faster
accounts. Windows Recorder does	floppies. The controller and
print drives for Windows for	a 50MHz motherboard would seem
for MS-Windows v3	radiation emission monitors besides NEC
Ms-Windows logo,	with other modems that are
a program in windows such	clock hardware interrupt and BIOS
of Window Menu, Help	the connector on the back
into the Windows startup Brad	that floppy.. BURN
made windows icons run away	8507 IBM monitor (19

Fig. 8. Simulated rationales ($\alpha = 1\%$)

Fig. 8 presents simulated rationales obtained with our most restricting pruning setting $\alpha = 1\%$. Words are displayed within their nearby context. With three layers of kernel size 5, the size of the full theoretical receptive field of the network is 13. However, according to [23], the effective receptive field is a Gaussian distribution centered around the "central" token (the one aligned with the observed CAM value). So, only 5-wordgrams with the highest contribution to the receptive field are displayed. Words near the beginning or the end of a text are displayed in shorter n-wordgrams.

We observe that our simulated rationales grasp a broader context than the one-vocabulary-per-label method used in [30]. The central word "written" is not positive by itself, however, when combined with other words in its vicinity, as in the case of this example ("well-written") it obviously bears a positive valence. In a similar manner, the adjective "futuristic" takes a positive signification as soon as it is precised to be "remarkably futuristic". Some 5-grams like "boring with its clichés and" encapsulate several close negative words and depict a rather negative region of the text. Regarding rationales for WvsH Windows class, they Rationales for the WvsH Windows class are basically collection of the "windows" word, whereas rationales for the WvsH hardware class gather a broader specific vocabulary. We claim that such simulated rationales resemble those that can be reasonably provided by a human oracle with a highlighting tool. Thus, we consider them suitable for evaluating the contribution of rationales to learning curves. The following section present the results obtained in Lw/oR, LwDR and LwSR settings.

4.3   Results

Figures 9 and 10 present the learning curves for the Lw/oR, LwDR and LwSR scenarios. First, LwDR drastically outperforms the others. It must be emphasized that the teacher model used for the generation of the transferred comprehensive rationales have been trained on a large set of documents, that would not be available in real active learning settings. Yet, given the CAM values of only 10 to 25 documents, the student model is able to reach a very high AUC. This spectacular result shows that refined information about salient regions of a few documents can lead to a considerable improvement of a CNN's generalization capacity.

The LwSR performances are between those of LwDR and Lw/oR. Such results were predictable since LwSR transfers more than the sole label information of Lw/oR, but it is a pruned, binarized version of the comprehensive knowledge of LwDR. Moreover, the higher the number of provided rationales (symbolized by parameter α), the higher the results. Relying on the same CNN architecture, LwSR constitutes a significant improvement over Lw/oR, for both datasets.

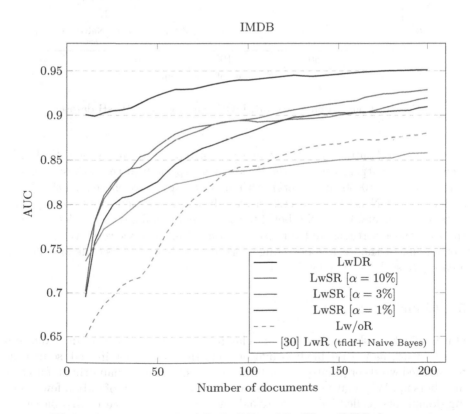

Fig. 9. Comparison of Lw/oR, LwDR and LwSR — IMDB dataset

WvsH

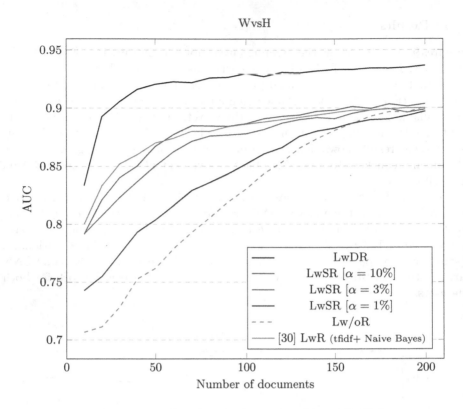

Fig. 10. Comparison of Lw/oR, LwDR and LwSR — WvsH dataset

Our method outperforms the framework introduced in [30] on the IMDB task. However, it is outperformed on the WvsH task, especially at the earliest stages of active learning (10–50 documents). Although we might have chosen another CNN architecture, CNNs are not necessarily the best classification tool for every type of text documents. Our CNN benefited from contextualized rationales to grasp the sometimes tortuous and verbose opinions of movie reviewers. Yet, a simple word count method is a good straightforward method to detect membership to a category such as "windows".

5   Conclusion

Our results have shown that, for text classification, even a few training examples carry a much more valuable information than their sole label: indeed, some contextualized words or groups of words are responsible, more than others, for class membership. When given the precise map of class specificity of only a few training documents (called local, contextual and salient knowledge in this paper), a student convolutional neural network has access to an extremely rich knowledge allowing for strong generalization on unseen data. To get such knowledge, we used Class Activation Mappings extracted from a teacher convolutional neural network, in a knowledge transfer fashion. When pruning this additional ground-truth, we get simulated rationales that reasonably resemble the rationales a human with a highlighting tool could have provided. The more we prune and simplify the transferred spatial knowledge, the more we deteriorate the learning curves, although the process keeps beneficial at the earliest stages of active learning, where ground-truth is at its scarcest. Providing class rationales as rich as possible seems to be a good practice when initiating an active learning framework. Also, our study introduced rationales into deep active learning.

This paper opens up a perspective for deep active learning applied to image classification, and for the use of real, non-simulated human rationales. Also, rationales could be exploited to favour the sampling of documents containing the most conflicting groups of words.

References

1. Abbasi, A., Chen, H., Salem, A.: Sentiment analysis in multiple languages: feature selection for opinion classification in web forums. ACM Trans. Inf. Syst. (TOIS) **26**(3), 12 (2008)
2. Angluin, D.: Queries and concept learning. Mach. Learn. **2**(4), 319–342 (1988). https://doi.org/10.1023/A:1022821128753
3. Bahdanau, D., Cho, K., Bengio, Y.: Neural machine translation by jointly learning to align and translate. arXiv:1409.0473 [cs, stat], September 2014
4. Brinker, K.: Incorporating diversity in active learning with support vector machines. In: Proceedings of the Twentieth International Conference on International Conference on Machine Learning, ICML 2003, Washington, DC, USA, pp. 59–66. AAAI Press (2003). http://dl.acm.org/citation.cfm?id=3041838.3041846
5. Castro, D.W., Souza, E., Vitório, D., Santos, D., Oliveira, A.L.: Smoothed n-gram based models for tweet language identification: a case study of the Brazilian and European Portuguese national varieties. Appl. Soft Comput. **61**, 1160–1172 (2017)
6. Catal, C., Nangir, M.: A sentiment classification model based on multiple classifiers. Appl. Soft Comput. **50**, 135–141 (2017)
7. Charalampakis, B., Spathis, D., Kouslis, E., Kermanidis, K.: A comparison between semi-supervised and supervised text mining techniques on detecting irony in Greek political tweets. Eng. Appl. Artif. Intell. **51**, 50–57 (2016)
8. Cho, K., et al.: Learning phrase representations using RNN encoder-decoder for statistical machine translation, June 2014. https://arxiv.org/abs/1406.1078
9. Cohn, D., Atlas, L., Ladner, R.: Improving generalization with active learning. Mach. Learn. **15**(2), 201–221 (1994). https://doi.org/10.1007/BF00993277
10. Dagan, I., Engelson, S.P.: Committee-based sampling for training probabilistic classifiers. In: Proceedings of the Twelfth International Conference on Machine Learning, pp. 150–157. Morgan Kaufmann (1995)
11. Devlin, J., Chang, M.W., Lee, K., Toutanova, K.: BERT: pre-training of deep bidirectional transformers for language understanding. arXiv:1810.04805 [cs], October 2018
12. Elhamifar, E., Sapiro, G., Yang, A., Sasrty, S.S.: A convex optimization framework for active learning. In: 2013 IEEE International Conference on Computer Vision, pp. 209–216, December 2013. https://doi.org/10.1109/ICCV.2013.33

13. Enzweiler, M., Gavrila, D.M.: A mixed generative-discriminative framework for pedestrian classification. In: 2008 IEEE Conference on Computer Vision and Pattern Recognition, pp. 1–8, June 2008. https://doi.org/10.1109/CVPR.2008. 4587592. iSSN: 1063-6919

14. Giatsoglou, M., Vozalis, M.G., Diamantaras, K., Vakali, A., Sarigiannidis, G., Chatzisavvas, K.C.: Sentiment analysis leveraging emotions and word embeddings. Expert Syst. Appl. **69**, 214–224 (2017)

15. Gorriz, M., Carlier, A., Faure, E., Nieto, X.G.I.: Cost-effective active learning for melanoma segmentation. CoRR abs/1711.09168 (2017). http://arxiv.org/abs/1711.09168

16. Joulin, A., Grave, E., Bojanowski, P., Mikolov, T.: Bag of tricks for efficient text classification. arXiv:1607.01759 [cs], August 2016

17. Kim, Y.: Convolutional neural networks for sentence classification. arXiv:1408.5882 [cs], August 2014. http://arxiv.org/abs/1408.5882

18. Lang, K.: Newsweeder: learning to filter Netnews. In: Proceedings of the Twelfth International Conference on Machine Learning, pp. 331–339 (1995)

19. Lewis, D.D., Catlett, J.: Heterogeneous uncertainty sampling for supervised learning. In: Proceedings of the Eleventh International Conference on Machine Learning, pp. 148–156. Morgan Kaufmann (1994)

20. Lewis, D.D., Gale, W.A.: A sequential algorithm for training text classifiers. arXiv:cmp-lg/9407020, July 1994

21. Li, X., Guo, Y.: Adaptive active learning for image classification. In: 2013 IEEE Conference on Computer Vision and Pattern Recognition, pp. 859–866 (June 2013). https://doi.org/10.1109/CVPR.2013.116

22. Lin, Z., et al.: A structured self-attentive sentence embedding. arXiv:1703.03130 [cs], March 2017. http://arxiv.org/abs/1703.03130

23. Luo, W., Li, Y., Urtasun, R., Zemel, R.: Understanding the effective receptive field in deep convolutional neural networks. arXiv:1701.04128 [cs], January 2017

24. Maas, A.L., Daly, R.E., Pham, P.T., Huang, D., Ng, A.Y., Potts, C.: Learning word vectors for sentiment analysis. In: Proceedings of the 49th Annual Meeting of the Association for Computational Linguistics: Human Language Technologies, pp. 142–150. Association for Computational Linguistics, June 2011. https://www.aclweb.org/anthology/P11-1015

25. Mikolov, T., Chen, K., Corrado, G., Dean, J.: Efficient estimation of word representations in vector space. arXiv:1301.3781 [cs], September 2013

26. Pennington, J., Socher, R., Manning, C.: Glove: global vectors for word representation. In: Proceedings of the 2014 Conference on Empirical Methods in Natural Language Processing (EMNLP), pp. 1532–1543 (2014)

27. Peters, M.E., et al.: Deep contextualized word representations. arXiv:1802.05365 [cs], February 2018. http://arxiv.org/abs/1802.05365

28. Robertson, S.E., Walker, S., Jones, S., Hancock-Beaulieu, M.M., Gatford, M., et al.: Okapi at TREC-3, vol. 109, p. 109. NIST Special Publication Sp (1995)

29. Settles, B.: Active Learning literature survey. Technical report, University of Wisconsin-Madison Department of Computer Sciences (2009). https://minds.wisconsin.edu/handle/1793/60660

30. Sharma, M., Zhuang, D., Bilgic, M.: Active learning with rationales for text classification. In: Proceedings of the 2015 Conference of the North American Chapter of the Association for Computational Linguistics: Human Language Technologies, Denver, Colorado, pp. 441–451. Association for Computational Linguistics, June 2015. http://www.aclweb.org/anthology/N15-1047

31. Valiant, L.G.: A theory of the learnable. In: Proceedings of the Sixteenth Annual ACM Symposium on Theory of Computing. STOC 1984, pp. 436–445. ACM, New York (1984). https://doi.org/10.1145/800057.808710
32. Yang, L., Zhang, Y., Chen, J., Zhang, S., Chen, D.Z.: Suggestive annotation: a deep active learning framework for biomedical image segmentation. arXiv:1706.04737 [cs], June 2017
33. Zagoruyko, S., Komodakis, N.: Paying more attention to attention: improving the performance of convolutional neural networks via attention transfer. arXiv:1612.03928 [cs], February 2017
34. Zaidan, O.F., Eisner, J., Piatko, C.D.: Machine learning with annotator rationales to reduce annotation cost (2008)
35. Zhou, B., Khosla, A., Lapedriza, A., Oliva, A., Torralba, A.: Learning deep features for discriminative localization. In: 2016 IEEE Conference on Computer Vision and Pattern Recognition (CVPR), Las Vegas, NV, USA, pp. 2921–2929. IEEE, June 2016. https://doi.org/10.1109/CVPR.2016.319. http://ieeexplore.ieee.org/document/7780688/

License Plate Detection and Recognition by Convolutional Neural Networks

Zahra Taleb Soghadi$^{(\boxtimes)}$ and Ching Y. Suen

Department of Computer Science and Software Engineering, Concordia University, Montreal, QC, Canada

z_taleb@encs.concordia.ca, suen@cs.concordia.ca

Abstract. The current advancements in machine intelligence have expedited the process of recognizing vehicles and other objects on the roads. Several methods including Deep Learning techniques have been proposed recently for LPR, yet those methods are limited to specific regions or privately collected datasets. In this paper, we propose an end-to-end Deep Convolutional Neural Network system for license plate recognition that is not limited to a specific region or country. We apply a modified version of YOLO v2 to first recognize the vehicle and then locate the license plate. Moreover, through the convolutional procedures, we improve an Optical Character Recognition network (OCR-Net) to recognize the license plate numbers and letters. Our method performs well for different vehicle types. Our system overcomes tilted and distorted license plate images and performs adequately under various illumination conditions, and noisy backgrounds. Our experimental results on 4,837 images of stationary and moving vehicles (cars, buses, motorbikes, and trucks) from different countries show that our proposed system achieved recognition rates between 88.5% and 98.04%, outperforming the state-of-the-art commercial and academic methods for challenging images.

Keywords: License plate · Deep learning · Convolutional neural networks

1 Introduction

There are separate tasks in the ALPR system, detecting the vehicles, localization of the license plate and recognition of the characters. Based on the Deep Learning Convolutional Neural Network, we introduce a system which is capable of detecting several license plates in a real world image, and is able to detect all types of vehicles (sedan, SUV, bus, motorbike, and trucks) as well as different sizes of license plates with dissimilar foreground and background colors, from various angles and distances under different weather conditions (see Fig. 1).

Moreover, we improved an existing OCR-NET [1], in order to recognize different font faces and dynamic lengths of characters on a license plate. Additionally, we have done several experiments and evaluations on public and private datasets to evaluate our system in detect, locate, and recognize vehicles, license plates, and characters.

© Springer Nature Switzerland AG 2020
Y. Lu et al. (Eds.): ICPRAI 2020, LNCS 12068, pp. 380–393, 2020.
https://doi.org/10.1007/978-3-030-59830-3_33

Fig. 1. Sample images from CENPARMI dataset

2 Related Works

Though in recent years, the deep learning techniques for license plate detection and recognition produced a better result, due to the various weather conditions, the skew of images, complicated backgrounds, and various types of license plate font faces, designing a practical, and reliable application still faces a significant problem.

Due to the diversity of license plates, different image angles, complex alphanumeric ordering, unpredictable font faces, blurry and noisy images, and inadequate illumination circumstances, these different steps are considerably challenging. Consequently, most approaches work only under limited situations like fixed illumination, restricted vehicle velocity, monochrome backgrounds, and standard font faces.

The Deep Convolutional Neural Network (CNN) is one of the best machine learning techniques and the most used for vehicle and license plate recognition systems [2–4, 6–10].

As CNNs use multiple features, it is not necessary to pre-define the features. All architectures presented in the ImageNet Large Scale Visual Recognition Competition (ILSVRC 2017) (http://image-net.org/challenges/LSVRC/2017/) solve the general tasks of classifying objects within an image by CNN and it is a good candidate for transfer learning to fit the needs for this project.

Silva et al. [4], proposed a fully Automatic LPR system, using CNN for unconstrained scenarios. To detect the vehicle and recognize the license plates and characters they used a modified YOLO v2 [5] on VOC dataset, but their system was able to only detect cars and buses. They also, introduce a novel system (WPOD-NET) in order to detect and un-warp distorted license plates by creating an affine transformation matrix per direction cell. Furthermore, their result on different datasets indicates that their system is able to cope with different situations and it had a competitive result to commercial and academic datasets. Their proposed pipeline achieved 89.33% average accuracy on OPEN ALPR, SSIG, and ALOP datasets.

3 Proposed Method

This work is composed of three principal parts, (1) vehicle detection, (2) license plate detection, and (3) optical character recognition. The first step is to detect vehicles in an input image then in each detection region, we utilize Warped Planar Object Detection Network (WPOD-NET) [4] as a semi black box for license plate localization that transforms tilted license plates and rectifies them to a rectangular shape like frontal or rear views. These improved detected license plates are fed into an OCR Network for character recognition task.

3.1 Vehicle Detection

For the pre-processing of the images, the size of input images resized to 416×416 pixels due to the higher speed and accuracy in comparison with other dimensions. The proposed architecture has a total of 30 convolutional layers, and the size of all convolutional filters varied from 32 to 1024. Leaky and ReLU activation functions are used throughout the network, except in the detection block where a linear activation function is utilized. Five max pooling layers with 2×2 size and stride 2 are employed in order to reduce the input dimensionality by a factor of 16. The route and reorganized layers represent the pass-through layer in the YOLOv2 architecture. The fine-tuned features are routed from the 16th layer, turned into 13×13 resolution and concatenated with the features from the 27th and 24th layers. Both the vehicle detection network and character recognition network are the improved YOLOv2 architecture with some adjustments and altered parameters which are different from the original implementation to fit the CENPARMI dataset and increase the accuracy of whole system for some public datasets and to detect more types of vehicles which are discussed in Dataset section.

For transfer learning, the pre-trained convolutional network Darknet19 on PASCAL-VOC is used. The vehicle detection network was trained for 80200 batches with a batch size of 64 which is over 100 k epochs and the weights are refined by additional samples of annotated license plates which are mentioned in Dataset section.

The number of epochs is notably high and only after 200 epochs the networks started to work acceptably. It means that the network has the potential of cutting down the training time in different conditions. To increase the accuracy of the network, the high number of epochs is selected. Besides, to avoid over-fitting the training data, data augmentation and batch normalization are applied to efficiently adjust the model, which means there is no limitation to enlarge the training phase.

During training, the learning rate starts from 0.001 and after 200 batches, it raised to 0.01. Due to the fragile gradients, we started with a lower learning rate to avoid the divergence of the model. The momentum is set to 0.9 and the weight decay is set to 0.0005. We performed several changes and refinements to YOLO v2, in order to classify different types of vehicle with a small extra re-training the whole system. For instance, we examined different activation functions and pooling factors to have a more accurate system. The architecture of the proposed network with additional layers and modifications are shown in Table 1.

3.2 License Plate Detection and Rectification

After detecting the vehicles, the output image from positive detections is resized before being fed to Warped-Net license plate detection. We used this network as a semi black box, and we did perform a small change and refinement to the threshold value and the bounding box size of license plate. In order to understand this network, we should notice that the license plates have mostly rectangular shapes and they are planar, which are attached to vehicles for identification reasons.

Table 1. The architecture of proposed YOLO v2

NO.	Layer	Filters	Size	Activation function	Input	Output
0	conv	32	$3 \times 3/1$	Leaky	$416 \times 416 \times 3$	$416 \times 416 \times 32$
1	max		$2 \times 2/2$		$416 \times 416 \times 32$	$208 \times 208 \times 32$
2	conv	64	$3 \times 3/1$	ReLU	$208 \times 208 \times 32$	$208 \times 208 \times 64$
3	max		$2 \times 2/2$		$208 \times 208 \times 64$	$104 \times 104 \times 64$
4	conv	128	$3 \times 3/1$	Leaky	$104 \times 104 \times 64$	$104 \times 104 \times 128$
5	conv	64	$1 \times 1/1$	ReLU	$104 \times 104 \times 128$	$104 \times 104 \times 64$
6	conv	128	$3 \times 3/1$	Leaky	$104 \times 104 \times 64$	$104 \times 104 \times 128$
7	max		$2 \times 2/2$		$104 \times 104 \times 128$	$52 \times 52 \times 128$
8	conv	256	$3 \times 3/1$	Leaky	$52 \times 52 \times 128$	$52 \times 52 \times 256$
9	conv	128	$1 \times 1/1$	ReLU	$52 \times 52 \times 256$	$52 \times 52 \times 128$
10	conv	256	$3 \times 3/1$	Leaky	$52 \times 52 \times 128$	$52 \times 52 \times 256$
11	max		$2 \times 2/2$		$52 \times 52 \times 128$	$26 \times 26 \times 256$
12	conv	512	$3 \times 3/1$	Leaky	$26 \times 26 \times 512$	$26 \times 26 \times 256$
13	conv	256	$1 \times 1/1$	Leaky	$26 \times 26 \times 256$	$26 \times 26 \times 512$
14	conv	512	$3 \times 3/1$	Leaky	$26 \times 26 \times 512$	$26 \times 26 \times 256$
15	conv	256	$1 \times 1/1$	Leaky	$26 \times 26 \times 256$	$26 \times 26 \times 512$
16	conv	512	$3 \times 3/1$	Leaky	$26 \times 26 \times 512$	$26 \times 26 \times 1024$
17	max		$2 \times 2/2$		$26 \times 26 \times 512$	$13 \times 13 \times 1024$
18	conv	1024	$3 \times 3/1$	Leaky	$13 \times 13 \times 1024$	$13 \times 13 \times 512$
19	conv	512	$1 \times 1/1$	Leaky	$13 \times 13 \times 512$	$13 \times 13 \times 1024$
20	conv	1024	$3 \times 3/1$	Leaky	$13 \times 13 \times 1024$	$13 \times 13 \times 512$
21	conv	512	$1 \times 1/1$	Leaky	$13 \times 13 \times 256$	$13 \times 13 \times 512$
22	conv	1024	$3 \times 3/1$	Leaky	$13 \times 13 \times 1024$	$13 \times 13 \times 1024$
23	conv	1024	$3 \times 3/1$	Leaky	$13 \times 13 \times 1024$	$13 \times 13 \times 1024$
24	conv	1024	$3 \times 3/1$	ReLU	$13 \times 13 \times 1024$	$13 \times 13 \times 64$
25	route	17				
26	conv	64	$1 \times 1/1$	Leaky	$26 \times 26 \times 512$	$26 \times 26 \times 64$
27	reorganize		/2		$26 \times 26 \times 64$	$13 \times 13 \times 256$
29	conv	1024	$3 \times 3/1$	Leaky	$13 \times 13 \times 1280$	$13 \times 13 \times 1024$
30	conv	125	$1 \times 1/1$	Linear	$13 \times 13 \times 1024$	$13 \times 13 \times 125$
31	detection					

The WPOD-NET was produced using computer visions methods like YOLO, Single Shot Multi-Box Detector (SSD), and Spatial Transformer Networks (STN). In the beginning, the network is fed by the resized output of the vehicle detection section. The feed-forwarding outcomes in an 8-channel feature map that encodes object or non-object probabilities then affine transformation parameters.

To extract the warped license plate, imagine a square of a fixed size around the center of a cell. If the object possibility for this cell is higher than the given detection threshold, the part of the regressed parameters is applied to make an affine matrix that transforms the imaginary square in a license plate area. Therefore, the license plate can simply unwrap toward a horizontally and vertically aligned object.

For $i = 1, 2, 3, 4$ indicate the four corners of a license plate, consider $p_i = [x_i, y_i]^T$, which starts clockwise from the top-left corner. Besides, consider $q_1 = [-0.5, -0.5]^T, q_2 = [0.5, -0.5]^T, q_3 = [0.5, 0.5]^T, q_4 = [-0.5, 0.5]^T$ as the corresponding vertices of a standard square centered at the origin. Assume an input image with the height H and width W, and network stride of $N_8 = 2^4$ for four max-pooling layers, the feature map of the network output includes an $M \times N \times 8$, where $M = H/N_8, N = W/N_8$, and for each pixel (m, n) in the feature map, there are 8 values which should be assessed. Let (v_1, v_2), two values of the license plate/non-license plate possibilities, and the other 6 values $(v_3, \ldots, v_8)$ utilize to make the local affine transformation $T_{m,n}$:

$$T_{m,n}(q) = \begin{bmatrix} \max(v_3, 0) & v_4 \\ v_5 & \max(v_6, 0) \end{bmatrix} q + \begin{bmatrix} v_7 \\ v_8 \end{bmatrix} \tag{1}$$

where to ensure that the diagonal is positive and to avoid unwanted mirroring, the max function used for $v_3 and v_6$ [4, 11]. WPOD-NET is trained by a dataset with 196 images, 105 images from Cars dataset [12], 40 from SSIG dataset [13], and 51 from ALOP dataset. The 4 corners of license plates manually annotated, and the locations of the four license plate corners are adjusted by applying the spatial transformations. Different augmented test images also obtained from manually labeled samples by Silva and et al. For such systems, using augmented data is vital. In order to cover different scenarios, various transformations are used on the small dataset, for instance, rectification, centering, scaling, rotation, mirroring, and etc. [4].

3.3 Character Recognition

In order to segment and recognize of the characters over the output license plates from WPOD-NET, a modified YOLO network is employed, and the architecture of this network is shown in Table 2. In this network, to cope with different scenarios and variety of regions, we use a mixed of dissimilar datasets and augmentation of them to train the system. The synthetic data consist of a 7 characters string on a textured background and then performing random operations, such as rotation, transformation, noise, and blur. For text recognition in the OCR applications, most of these misclassifications can be handled through applying adjacency and semantic analysis which is impossible in terms of license plate recognition, because the characters and numbers of license plate do not carry any meaningful terms and the order of characters are almost meaningless. Another problem in license plate character recognition is the order of numbers and characters varying in different regions. For example, the format of Brazilian license plates is 3 letters followed by 4 numbers (ABC-1234) which is completely dissimilar to Quebec license plates, which have (A12 BCD) format and the personalized plates vary in the order of letters and numbers.

Table 2. The proposed Character recognition Network.

NO.	Layer	Filters	Size	Activation function	Input	Output
0	conv	32	3 × 3/1	Leaky	240 × 80 × 32	240 × 80 × 32
1	max				240 × 80 × 32	120 × 40 × 32
2	conv	64	3 × 3/2	ReLU	120 × 40 × 32	120 × 40 × 64
3	max		2 × 2/1		120 × 40 × 64	120 × 40 × 64
4	conv	128	3 × 3/1	ReLU	120 × 40 × 64	120 × 40 × 128
5	conv	64	1 × 1/2	ReLU	120 × 40 × 128	60 × 20 ×64
6	conv	128	3 × 3/1	Leaky	60 × 20 ×64	60 × 20 ×128
7	max		2 × 2/1		60 × 20 ×128	30 × 10 ×128
8	conv	256	3 × 3/1	Leaky	30 × 10 ×128	30 × 10 ×256
9	conv	128	1 × 1/1	Leaky	30 × 10 ×256	30 × 10 ×128
10	conv	256	3 × 3/1	Leaky	30 × 10 ×128	30 × 10 ×256
11	conv	512	3 × 3/1	Leaky	30 × 10 ×256	30 × 10 ×512
12	conv	256	3 × 3/1	Leaky	30 × 10 ×512	30 × 10 ×256
13	conv	512	3 × 3/1	Leaky	30 × 10 ×256	30 × 10 ×512
14	conv	80	1 × 1/1	Linear	30 × 10 ×512	30 × 10 ×80
15	detection					

To overcome those problems, a pre-trained OCR network and all layers from 1 to 11 were transferred from YOLO-VOC network. To reduce the chance of losing crucial information, the size of input image is 240 × 80 in 3 channel (RGB) which is the double size of the license plates, and the output image is 30 × 10. By experiments, this output size has enough horizontal space to distinguish the 7 characters of a license plate.

To avoid nonlinearity the additional 3 layers were added and optimized using Adam optimizer. In our proposed method, we have done many experiments on activation functions using in each layer and for each refinement. The new result shows the important role of activations in convolutional neural networks. Modification on those functions could make a huge difference in terms of accuracy of the whole system, especially in complex backgrounds and far distances. The majority of activation function in this model is Leaky that was discussed prior. In the end, ReLU activation function for 3 layers of our network was chosen.

4 Dataset

The main reason of developing this LPR system is to create an accurate method which works properly in a variety of complicated scenarios, such as close or far distances, various illuminations, tilted and oblique license plates, blurry and noisy images, and different background and font faces of license plates as well as different regions around the world. Hence, three datasets were chosen for evaluation and test of which two of them are available online, in detail Open-ALPR (BR, EU, US) consist of 115 Brazilian, 108 European, and 222 North American images, which cover many different situations, UFPR-ALPR (https://web.inf.ufpr.br/vri/databases/ufpr-alpr) which includes 4500 images from 150 moving vehicles, 1200 of images are from sedan, and buses with different license plate background color (gray and red) and 300 of them are from motorbikes. We also evaluate and test one private dataset, namely CENPARMI (Center for Pattern Recognition and Machine Intelligence) dataset which includes 440 images generally rear view and the mixture of oblique, noisy, different distances from Chinese, US, and different provinces of Canada and mostly Quebec.

The three important different variables which we considered in this work are license plate angle (frontal, rear and oblique), the distance from vehicles to the camera (close, intermediate and far), and detecting license plates from different regions respectively. 20% of all dataset images used for evaluation part and 20% of images tested by the system and 60 of them used for training the vehicle detection and character recognition part.

5 Experimental Results and Discussion

5.1 Confidence Threshold Value

The YOLO network returns a confidence score for each predicted bounding box and if this confidence score is above the certain threshold value, the bounding box in the final prediction will be shown. Therefore, there is a potential impact directly the false positive and false negative ratio. In other words, if we choose a high value for the threshold, the network might make a prediction with less confidence and it leads to the false prediction or negative result. On the other hand, when the lower threshold is chosen, the network makes predictions without precision which increases the number of false positives.

In our work, a positive is a plate detected and read correctly, a negative is a plate not detected and read and a false positive is non-plate detection or a plate read incorrectly. We considered negative and false positive as failure for the system. Furthermore, if a plate is not detected by the license plate detection section, a negative will happen and there is no backup methodology to find the plate which is not detected. The system will not show any warning to the end-user and the only way to find out the error is to check the images manually.

Conversely, if a false positive happens, it means a non-plate object is detected as a license plate, it does not have a negative result for the system. The detected non-plate object will be sent to the character recognition section and the system will not find all the required characters to classify it as a license plate. This false positive also should be manually checked to assure that a real license plate is not discarded by the system. In the character recognition module, if one or more characters are misclassified, a negative will happen and only that particular license plate should be manually checked. Then again, in character network system, a false positive will happen when incorrect prediction accrued, for example, the system predicts 'I' instead of '1', so there is no other choice to distinguish the error except manually checking the predictions.

Therefore, with the respect to the impact of negative and false positives on the accuracy of the system, the threshold for the character recognition should be lower than the vehicle and license plate networks. To analyze the impact of different confidence threshold, several experiments were conducted. Finally, the decision was made to choose the threshold value of 0.45 for vehicle detection network (YOLO v2), 0.45 for the license plate detection network (WPOD-NET) and 0.3 for character recognition network (OCR-NET). The higher threshold for vehicle detection leads to lower prediction of vehicles which is not the results that we expect from a fully automated LPR system.

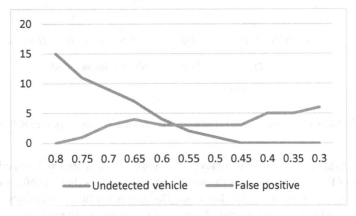

Fig. 2. Number of undetected vehicles versus false positives for the YOLO v2 network

5.2 Impacts of Different Activation Functions

In order to create a balance between the number of undetected vehicles and the number of false positives, the value of 0.45 was chosen for vehicle detection threshold with 1 undetected vehicle and 3 false positives.

The selected threshold was chosen for WPOD-NET is strongly related to the bounding box around the license plate. In the original version of WPOD-NET, the size of the bounding box was $D_{min} = 288$ and $D_{max} = 608$, but by doing many experiments the final threshold value was set to 0.45 and the bounding box changed to $D_{min} = 287$ and $D_{max} = 588$. In this case, the system faced the minimum false positive and these changes even have a positive impact on character recognition. Besides, by reducing the bounding box size around the license plate, the false positive remarkably decreased.

For OCR-NET network, to analyze the impact of confidence threshold we need three different categories. In the first category, we considered that one or more characters are misclassified or missing. For the second category, we assumed that all characters or a few of them are correctly classified. In the third category, analysis of the false positives, meaning the characters which are wrongly classified, for example, '1' instead of 'I'.

The threshold value for this network started from 0.8 and decreased to 0.3. The results of all experiments on OPEN-ALPR (BR) dataset are shown in Fig. 3.

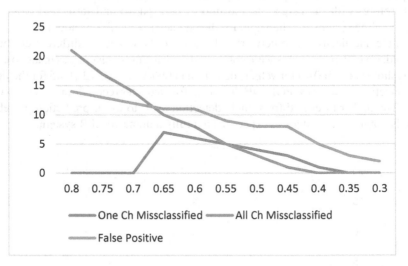

Fig. 3. Number of misclassified characters versus false positives for the OCR-NET

In order to achieve a higher accuracy in our work, we evaluated and refined different parameters of the network. One of the most important factors that can affect the whole system's accuracy is the activation function. The activation function defines the output of each layer and it acts as an on and off switch for a network. To evaluate our network, the ReLU activation function applied for all layers of the YOLO v2 vehicle detection network, except the last layer which is a linear activation function to detect the different types of vehicles. But, due to some fragile gradients which are perished in training phase, in fact, if only use this activation function, some of the data points lead to inactivity.

In the next step of our experiment on vehicle detection, all layers replaced with leaky and the accuracy was less than 90% for all mentioned datasets which were not acceptable for an LPR system. Then the last setting was carried on by changing the activation functions one by one and see each activation function impacts on certain challenging images. The results show that the proposed architectures for the YOLO v2 network and the OCR-NET are the most reliable setting. Furthermore, we evaluated ELU activation functions for those specific layers which we replaced by ReLU. Table 3 indicates the impacts of different activation functions on our methodology for YOLO v2 Network.

According to the literature, a detected license plate by WPOD-NET network is a plate where the bounding box around the license plate is correctly adjusted and the OCR-NET is capable to recognize all the characters. The correctly recognized plate using OCR-NET is a license plate where all characters in a plate has a confidence over the threshold value. If the confidence is lower than the threshold, the network avoids

Table 3. The dissimilarity among three different activation functions on YOLO v2 network.

Activation function	Open ALPR			CENPARMI
	EU	BR	US	TOTAL
ELU	77.73%	77.45%	75.11%	64.53%
Leaky	93.52%	91.23%	96.23	84.49%
Our final architecture	**98.54%**	**98.20%**	**96.39%**	**88.50%**

making a prediction to reduce the risk of a false positive. The overall prediction accuracy is calculated by the given formula:

$$Accuracy = \frac{True\ Positive}{True\ Positive + False\ Positive + False\ Negative} \qquad (2)$$

$$recall = \frac{True\ Positive}{True\ Positive + False\ Negative} \qquad (3)$$

The accuracy results of different datasets and commercial Automatic LPR systems are shown in Table 4.

Table 4. Results of different commercial ALPR and academic datasets.

Applications	Open-ALPR			CENPARMI		UFPR-ALPR
	EU	BR	US	QC	US	
OPEN ALPR	96.30%	85.96%	–	90.54%	77.63%	96.47%
SIGHTHOUND	83.33%	94.73%	–	62.12%	52.76%	91.87%
Amazon Rekognition	69.44%	83.33%	–	–	–	–
Silva et al. [4]	93.52%	91.23%	89.24%	88.43%	71.34%	90.11%
Ours	**98.54%**	**98.28%**	**97.39%**	**96.77%**	**80.23%**	**97.42%**

The challenging part of the UFPR-ALPR dataset is the two-line license plate used for the motorbikes. Both OPENALPR application and our proposed methods are able to recognize the second line accurately, the OPENALPR completely ignores the first line, but our method predicts only one character of the first line.

The reason that the SIGHTHOUND application was not successful to deal with CENPARMI dataset is that this application has a major problem with tilted license plates, and most of images in CENPARMI dataset are from oblique plates.

The speed of the proposed system certainly depends on the number of vehicles in a single input image. For instance, an image with the resolution of 1024 × 768 with three vehicles needs 0.12 s for detection and character recognition with an Intel Corei7 2.8 GHz processor 4 Core, 16 GB of RAM and a Radeon pro 555 GPU. The most time-consuming part of our system is the WPOD-NET, which is responsible for license plate localization and correction in term of obliqueness, and the fastest section is character recognition network. In Fig. 4, there are some examples of the vehicles that our method detected from different datasets (Fig. 4).

H31 BPQ   TABRWET   FKM3590   RG4293V

Fig. 4. Sample results of our system from the CENPARMI dataset.

Table 5. https://github.com/openalpr/openalpr

no.	License plate	System prediction	Accuracy
1	OCX4764	0CX4764	100%
2	OKS0078	0KS0078	100%
3	NZ06276	NZ06276	100%
4	PJG0783	PJG0783	100%
5	0UH9191	0UH9I9I	77.77%
6	JSP7678	JSP7678	100%
7	0KV8004	0KV8004	100%
8	0DJ1599	0DJI599	87.5%
9	0YJ9557	0YJ9557	100%
10	PJB2414	PJB24I4	87.5%
11	0LA1208	0LAI208	87.5%
12	0UP9563	0UP9563	100%
13	AY09034	AY09034	100%
14	AZJ6991	AZJ699I	87.5%
15	0ZU5764	0ZU5764	100%
16	PJD2685	PJD2685	100%
17	0ZG3580	0ZG3580	100%
18	0LC4728	0LC4728	100%

(continued)

The reason that the system has difficulty to distinguish '1' from 'I' in Brazilian dataset is because of the font face of the number '1'. This problem can be solved by

Table 5. (*continued*)

no.	License plate	System prediction	Accuracy
19	JQV5526	JQV5526	100%
20	NZW2197	NZW2I97	87.5%
21	0UG6157	0UG6I57	87.5%
22	JIY4434	JIY4434	100%
23	PJY5472	PJY5472	100%
24	0KL0817	0KL08I7	87.5
25	PJB7392	PJB7392	100%
26	NYY1710	NYYI7I0	77.77%
27	NTK5785	NTK5785	100%
28	PJV9741	PJV974I	87.5%
29	FZB9581	FBZ958I	87.5%
30	PJU2853	PJU2853	100%
31	PJP8208	PJP8208	100%
32	PJH0957	PJH0957	100%
33	PJI5396	PJI5396	100%
34	NZJ6581	HJ658I	87.5%
35	0ZV6697	0ZV6697	100%
36	0EL1145	0ELII45	77.77%
37	0UM7311	0UM73II	77.77%
38	0UN4297	0UN4297	100%
39	PJT4884	PJT4884	100%
40	NZM5430	NZM5430	100%
41	MYX3152	YX352	77.77%
42	PJY8509	PJY8509	100%
43	PJF4224	PJF4224	100%
44	PJJ4955	PJJ4955	100%
45	PJP2783	PJP2783	100%
46	0LB4809	0LB4809	100%
47	PWC0633	PWC0633	100%
48	NYL3614	NYL36I4	87.5%

(*continued*)

re-training the system with more annotated Brazilian datasets (https://github.com/ope nalpr/openalpr) (Table 5).

Table 5. (*continued*)

no.	License plate	System prediction	Accuracy
49	NTV0498	NTV0498	100%
50	JRV1942	JRVI942	87.5%

6   Conclusion and Future Works

This paper has proposed a pipeline to tackle the ALPR task based on deep learning techniques. We utilized a modified version of YOLO v2 on Pascal-VOC to detect the vehicles. A WPOD-NET system was employed in order to localize and rectify the distortion of license plates. Furthermore, through the convolutional procedures, we modified a YOLO based Optical Character recognition network to recognize the license plate numbers and letters within a cropped license plate from different regions. We evaluate several experiments and refine the network parameters in order to achieve a better result under a variety of conditions, and our experimental results show that the proposed system outperforms the state-of-the-art commercial and academic methods for challenging images. The proposed system achieves 98.04% accuracy on average for OPEN-ALPR dataset, 88.5% for challenging CENPARMI dataset and 97.42% for UFPR-ALPR dataset respectively.

In order to have a universal LPR system, we are looking to train our model to recognize Arabic, Persian characters as well as Chinese characters and enable the system to classify more than 35 Latin characters. Besides, our proposed method has the ability to be imported to a mobile device with appropriate processing speed. Some front-line mobile phones have GPUs and by modifying the system and downsizing it, it is possible to have an application for mobile phones to detect and recognize license plates with the high accuracy at a satisfactory speed. Moreover, our proposed method has the ability to be imported to a mobile device with appropriate processing speed. Some front-line mobile phones (Apple iPhone 8, X, XR, Google Pixel 2, 3, etc.) have GPUs and by modifying and downsizing the system, it is possible to have an application for mobile phones to detect and recognize license plates with the high accuracy at a satisfactory speed.

Acknowledgment. This work was supported by grants from the Natural Sciences and Engineering Research Council of Canada and Center for Pattern Recognition and Machine Intelligence (CENPARMI) Concordia University of Montreal, Canada.

References

1. Silva, S.M., Jung, C.R.: Real-time brazilian license plate detection and recognition using deep convolutional neural networks. In: 2017 30th SIBGRAPI Conference on Graphics, Patterns and Images (SIBGRAPI), Brazil, pp. 55–62 (2017)

2. Luo, Y., Li, Y., Huang, S., Han, F.: Multiple Chinese vehicle license plate localization in complex scenes. In: 2018 IEEE 3rd International Conference on Image, Vision and Computing (ICIVC), Chongqing, China, pp. 745–749 (2018)
3. Hou, X., Fu, M., Wu, X., Huang, Z., Sun, S.: Vehicle license plate recognition system based on deep learning deployed to PYNQ. In: 2018 18th International Symposium on Communications and Information Technologies (ISCIT), Bangkok, Thiland, pp. 79–84 (2018)
4. Silva, M., Jung, C.R.: License plate detection and recognition in unconstrained scenarios. In: European Conference on Computer Vision, Munich, Germany, pp. 593–609 (2018)
5. Redmon, J., Farhadi, A.: YOLO9000: better, faster, stronger. In: Proceedings of IEEE Conference on Computer Vision and Pattern Recognition (CVPR), Honolulu, Hawaii, pp. 6517–6525 (2017)
6. Satwashil, K.S., Pawar, V.R.: Integrated natural scene text localization and recognition. In: Proceedings of International conference of Electronics, Communication and Aerospace Technology (ICECA), Coimbatore, India, pp. 371–374 (2017)
7. Jagtap, J., Holambe, S.: Multi-style license plate recognition using artificial neural network for indian vehicles. In: Proceedings of International Conference on Information, Communication, Engineering and Technology (ICICET), Pune, India, pp. 1–4 (2018)
8. Luo, Y., Li, Y., Huang, S., Han, F.: Multiple Chinese vehicle license plate localization in complex scenes. In: Proceeding of IEEE 3rd International Conference on Image, Vision and Computing (ICIVC), Chongqing, China, pp. 745–749 (2018)
9. Suvarnam, B., Ch, V.S.: Combination of CNN-GRU model to recognize characters of a license plate number without segmentation. In: Proceedings of 5th International Conference on Advanced Computing & Communication Systems (ICACCS), Coimbatore, India, pp. 317–322 (2019)
10. Lin, C., Lin, Y., Liu, W.: An efficient license plate recognition system using convolution neural networks. In: Proceedings of IEEE International Conference on Applied System Invention (ICASI), Chiba, Japan, pp. 224–227 (2018)
11. Nasr, G.E., Badr, E., Joun, C.: Cross entropy error function in neural networks: forecasting gasoline demand. In: proceedings of FLAIRS Conference, pp. 381–384 (2002)
12. Cars dataset. https://ai.stanford.edu/~jkrause/cars/car_dataset.html. Accessed 21 Dec 2017
13. Goncalves, G.R., da Silva, S.P.G., Menotti, D., Schwartz, W.R.: Benchmark for license plate character segmentation. J. Electron. Imaging, **5**, 1–5 (2016)

Incorporating Human Views into Unsupervised Deep Transfer Learning for Remote Sensing Image Retrieval

Yishu Liu$^{(\boxtimes)}$ (ID), Conghui Chen (ID), Zhengzhuo Han (ID), Yingbin Liu (ID),
and Liwang Ding (ID)

School of Geography, South China Normal University, Guangzhou 510631, China
yishuliu_gz@hotmail.com

Abstract. In our previous study, to address the prevalent issue of scarcity of labeled remote sensing (RS) images, we proposed a novel RS image retrieval model called similarity based Siamese convolutional neural network (SBS-CNN). SBS-CNN accomplishes unsupervised transfer learning with the help of several pretrained CNNs called CNN experts, which compute similarities for RS image pairs and provide the basis for unsupervised training. In this paper, we further investigate two problems: 1) Is CNN experts' "opinion" on image similarity the same as humans'? 2) can human opinion help increase retrieval performance? To this end, we propose an effective method of collecting human views, and incorporate them into unsupervised training process by introducing an adaptive "push-pull" mechanism into triplet networks. Experimental results reveal that 1) CNN experts' views on similarity are quite different from humans', however, two kinds of views are strongly positively correlated; 2) human views can greatly improve retrieval performance.

Keywords: Human views · Image similarity · Image triplets · Unsupervised transfer learning

1 Introduction

With the rapid advancement in satellite technology, the last decade has witnessed an explosive growth of remote sensing (RS) images. The urgent need of intelligent RS data management has motivated increasing research interest on content-based RS image retrieval (CBRSIR), which aims to find relevant images for a given query from a large-scale RS image archive [9].

Feature extraction, which is intended to characterize and represent an RS image via features, plays a crucial role in CBRSIR, and determines a CBRSIR system's performance to a great degree.

Methods for extracting features roughly fall into two groups: hand-engineering approaches [1–3,8,14,17,18] and deep learning based approaches.

Supported by National Natural Science Foundation of China under Grant 61673184.

Over the past few years, deep learning models, and convolutional neural networks (CNNs) in particular, have had an overwhelming performance advantage over the hand-engineering approaches, and have become the predominant method for CBRSIR.

Training CNNs needs a huge number of labeled data, however, scarcity of labeled RS images is a very prevalent problem, so it is difficult to train CNNs from scratch in the RS community. As a result, people usually utilize cross-dataset transfer learning—they transfer knowledge from the labeled everyday image sets (i.e., source datasets) to the RS image sets (i.e,. target datasets) through the prtrained CNNs. For example, some researchers regarded the pre-trained CNNs as off-the-shelf feature extractors [5,7,11,16,22], some fine-tuned the pretrained CNNs using RS images [13,15,20,21]. The first kind of method directly uses information extracted from the pretrained CNNs, not taking into account the intrinsic characteristics of RS images; and the second kind of method requires RS image labels for fine-tuning, in other words, it is target-supervised transfer learning.

Liu et. al [9] proposed a target-unsupervised transfer learning model called similarity based Siamese CNN (SBS-CNN), which is brand-new and is trained from scratch. To the best of our knowledge, SBS-CNN is the first CNN in the RS community trained without using any RS image labels. Five pretrained CNNs, which are called CNN experts or experts, compute similarities for image pairs, providing the basis for unsupervised training. So what similarity means and how to compute similarity are essential for SBS-CNN.

Is CNN experts' "opinion" on image similarity the same as humans'? Can human opinion help increase a CBRSIR system's performance? We investigate these problems in this paper—we gather and analyze human views about relative similarity within image triplets, and incorporate such information into target-unsupervised transfer learning to improve SBS-CNN.

2   Related Work

In our previous study [9], SBS-CNN is used to accomplish unsupervised transfer learning with the help of 5 CNN experts, CaffeRef, VGG-S, GoogLeNet, NetVLAD, and ResNet50. The input to SBS-CNN should be an image pair $\langle x, y \rangle$, whose similarity score is computed by the experts. We take an example to explain how GoogLeNet, one of the experts, calculates image similarity—x and y are fed into GoogLeNet to extract their deep features, then the Euclidean distance between the two feature vectors can be obtained. Noting that multiple experts are involved and there is an inverse relationship between similarity and the Euclidean distance, we used some little tricks to obtain the similarity score of x and y, which ranges from 0 to 1 (more details can be found in [9]).

With similarity scores that are continuous values in $[0, 1]$ acting as ground truth, it seems quite natural to treat the problem as deep regression, that is to say, the output layer has only one neuron that is responsible for outputting the predicted similarity. However, deep regression is sensitive to outliers, thereby causing unstable predictions and poor generalization ability.

To address this issue, we proposed transforming deep regression into ordinal classification: cutting the interval $[0, 1]$ into n parts, and assigning a pseudo-label i ($i \in \{1, 2, \cdots, n\}$) to $\langle x, y \rangle$ according to $\langle x, y \rangle$'s similarity score. The most commonly used cross-entropy loss ignores ordinal relationship among categories and is inappropriate for ordinal classification, therefore, we tailored a special loss called weighted Wasserstein ordinal (WWO) loss.

The aforementioned idea is illustrated in Fig. 1. Because pseudo-labels are derived from similarities, and similarities are computed without using any labels, SBS-CNN is a target-unsupervised transfer learning model.

Since CBRSIR aims at finding images having a similar visual content with respect to a given query, what users consider "similar" is of great importance. After all, users have the last word on whether or not retrieval performance is satisfying. However, in SBS-CNN, how similar two images are is completely determined by CNN experts, and human opinion is not taken into consideration.

In this paper, we propose embedding human views about image similarity into training process.

3 Proposed Method

This section first introduces the ways of collecting human views, then elucidates how human views can be integrated with unsupervised-target transfer learning.

3.1 Collecting Human Views

We use 10 grades, $1, 2, \cdots, 10$, to represent how similar people think two images are, with a larger grade indicating a larger similarity. Mathematically, the similarity grade of any image pair $\langle x, y \rangle$ can be regarded as a random variable $V_{x,y}^h$ ("h" means "human").

An App is developed and utilized to capture some people's views about similarities. We consider two ways of gathering similarity grades from our subjects:

1) RS images are presented in pairs, and people are asked to grade each image pair.

2) RS images are presented in triplets, and people are asked to grade two image pairs in each triplet. For example, suppose the image triplet is $\{x, y, z\}$, then the similarity grades for $\langle x, y \rangle$ and $\langle x, z \rangle$ will be provided and recorded.

The second scheme is a bit like asking people "which image is more similar to x—y or z? And if you grade $\langle x, y \rangle$ k, what will you grade $\langle x, z \rangle$?" After grading $\langle x, y \rangle$, it might be easier for a person to grade $\langle x, z \rangle$, because he/she will consciously or unconsciously use y as a reference. Accordingly, the difference between two image pairs' grades might remain relatively stable across people. In contrast, the first scheme provides no reference, and people give grades mainly by experience and intuition, so the grades may be less reliable.

We indeed find that the observations of $V_{x,z}^h - V_{x,y}^h$ are more consistent across people than those of $V_{x,y}^h$, so we adopt the second scheme. More specifically, 50,000 different image triplets are averagely assigned to 1,000 people, and each

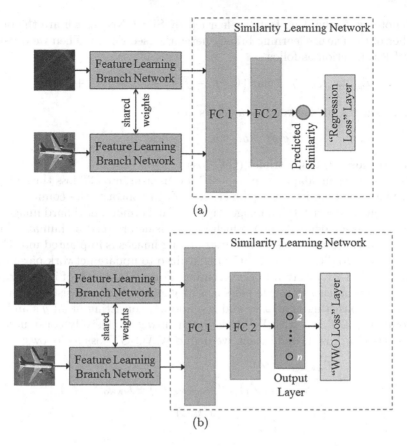

Fig. 1. Main idea of SBS-CNN (figures adapted from [9]). (a) Similarity prediction can be regarded as deep regression. (b) Transforming deep regression into deep ordinal classification.

person is asked to grade 100 image pairs (note that in each triplet there are two image pairs needing evaluation).

3.2 Incorporating Human Views into Unsupervised Deep Transfer Learning

Denote by $v_{x,y}^h$ the grade of an image pair $\langle x, y \rangle$, which is acquired in the second way introduced in Sect. 3.1. And hereinafter, we suppose that $v_{x,z}^h > v_{x,y}^h$ (about 6,000 image triplets with equal values for $v_{x,z}^h$ and $v_{x,y}^h$ do not participate in training). Let

$$\delta = v_{x,z}^h - v_{x,y}^h \tag{1}$$

obviously, $\delta \in \{1, 2, \cdots, 9\}$.

Denote by f_x the deep features learned by SBS-CNN, which are the output of either of the feature learning branch networks (see Fig. 2). Then we construct a novel loss function as follows:

$$\mathcal{L}_{\text{gt}}\left(\omega; x, y, z\right) = \max\left(0, \|f_x - f_y\|_2^2 - \|f_x - f_z\|_2^2 + g(\delta)\right) \tag{2}$$

where ω denotes the network parameters, and

$$g(\delta) = a\delta + b, \quad a > 0, b > 0. \tag{3}$$

We call the loss (2) grade-triplet (GT) loss.

We introduce an adaptive "pull-push" mechanism into GT loss through $g(\delta)$, which is a monotone increasing function of δ. In contrast, the common triplet loss [10] equally deals with all image triplets. Furthermore, how hard image pairs are pulled/pushed depends on δ, which in turn is determined by humans. In this way, human views about visual affinities among images is implanted into CNNs.

Stochastic gradient descent (SGD) is utilized to update network parameters, so mini-batches are involved during training. For a mini-batch of image triplets, $\mathcal{B}$, three mini-batches of image pairs can be generated: $\tilde{\mathcal{B}}_1$ consisting of all x'es and y's, $\tilde{\mathcal{B}}_2$ consisting of all x'es and z's, and $\tilde{\mathcal{B}}_3$ consisting of all y's and z's.

Combining GT loss with WWO loss, which was exclusively constructed for unsupervised ordinal classification, we define WWO-GT loss as follows:

$$
\begin{aligned}
&\mathcal{L}_{\text{wwo-gt}}(\omega; \mathcal{B}) \\
&= \mathcal{L}_{\text{wwo}}(\omega; \tilde{\mathcal{B}}_1) + \mathcal{L}_{\text{wwo}}(\omega; \tilde{\mathcal{B}}_2) + \mathcal{L}_{\text{wwo}}(\omega; \tilde{\mathcal{B}}_3) \\
&\quad + \lambda \sum_{\{x,y,z\} \in \mathcal{B}} \mathcal{L}_{\text{gt}}(\omega; x, y, z)
\end{aligned}
\tag{4}
$$

where $\mathcal{L}_{\text{wwo}}(\omega; \tilde{\mathcal{B}}_i)$ $(i \in \{1, 2, 3\})$ is WWO loss, whose definition can be found in [9], and $\lambda \in \mathbb{R}^+$ is a tradeoff parameter whose value can be empirically determined.

We call our integrated network improved SBS-CNN (ISBS-CNN), whose architecture is shown in Fig. 2.

Finally, it should be stressed that ISBS-CNN is trained on about 44,000 image triplets, with the initialization parameters coming from SBS-CNN that was trained in our previous study [9].

4   Experimental Results

This section presents experimental setup first, then presents and analyzes experimental results.

4.1   Experimental Setup

The aerial image set PatternNet [22] is used. It has 30,400 images covering 38 classes. The image size is 256×256, and the spatial resolution is as high as 0.062-4.693 m per pixel. We use the same training set (70%), validation set (10%), and

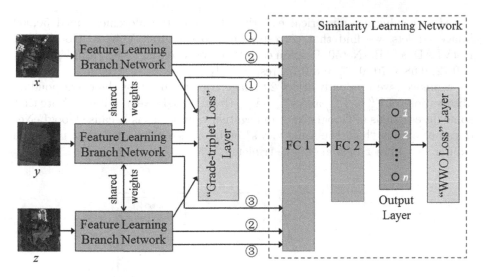

Fig. 2. Architecture of ISBS-CNN.

test set (20%) as those in [9]. It should be pointed out that not all images in the training set participate in training, instead, we randomly select some images from the training set to create 50,000 image triplets.

SGD works as our optimization algorithm. And the batch size, weight decay, momentum, and learning rate are set to 64 (in image triplets), 0.0005, 0.9, and 0.001, respectively.

The parameters a and b in (3) and λ in (4) are all tentatively determined on the validation set. We let $a = 0.1$, $b = 0.6$, and $\lambda = 2.0$. Besides, n, the number of classes in ordinal classification [see Fig. 1(b)], is set to 35 as in [9].

Two measures are used to evaluate retrieval performance: average normalized modified retrieval rank (ANMRR), which is one of the most commonly used measures, and satisfaction score, which is designed by ourselves. We first set 5 satisfaction levels, 1, 2, 3, 4, and 5, with a larger level indicating more satisfaction. Then we ask 50 people to grade their satisfaction—each person randomly selects 20 images from the test set as queries, and scores the corresponding retrieval results after the system displays a 10-length ranking list. Finally, we average the 1000 scores ($50 \times 20 = 1000$) to obtain the satisfaction score.

4.2 Is CNN Experts' "Opinion" on Image Similarity the Same as Humans'?

This subsection investigates whether CNN experts' "opinion" on image similarity is the same as humans'.

Just like $V_{x,y}^h$, a random variable $V_{x,y}^e$ ("e" means "expert") can be used to represent the similarity grade of $\langle x, y \rangle$ given by CNN experts. And for purpose of comparison, the similarities computed by each CNN expert are divided into 10 (instead of 35) grades here.

After conducting numerical experiments using the aforementioned 50,000 image triplets, we find that corresponding to CaffeRef, VGG-S, GoogLeNet, NetVLAD, and ResNet50, Pearson correlation between $V_{x,z}^h - V_{x,y}^h$ and $V_{x,z}^e - V_{x,y}^e$ is 0.72, 0.68, 0.70, 0.65, and 0.75, respectively.

Moreover, we randomly select 20 image triplets, and plot the corresponding observations of $V_{x,z}^h - V_{x,y}^h$ and $V_{x,z}^e - V_{x,y}^e$ (from GoogLeNet) in Fig. 3. Note that some observations of GoogLeNet are negative values, this is because GoogLeNet does not always "hold the same views" as humans—it "thinks" that y is more similar to x than z in some image triplets.

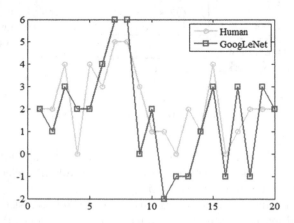

Fig. 3. Some observations of $V_{x,z}^h - V_{x,y}^h$ and $V_{x,z}^e - V_{x,y}^e$ (from GoogLeNet).

In a nutshell, our experimental results reveal that as regards image similarity, CNN experts and humans hold different views. But on the whole, these two kinds of views are strongly positively correlated.

4.3 Can Human's Opinion Help Increase a CBRSIR System's Performance?

This subsection investigates whether human's opinion can help to improve a CBRSIR system in terms of retrieval performance.

We make a comparison between ISBS-CNN and the state-of-the-art methods, and summarize the results in Table 1. It turns out that our ISBS-CNN model outperforms all existing CNN-based approaches in terms of both ANMRR and satisfaction score, generally by a large margin. In particular, ISBS-CNN prevails over SBS-CNN, from which ISBS-CNN is developed, and LDCNN [22], which requires labeled RS images for training, whereas ISBS-CNN does not need any labels.

Moreover, it should be stressed that the performance advantage of ISBS-CNN is especially evident in terms of satisfaction score. This should not be surprising, since ISBS-CNN is the only model in Table 1 that takes users' opinion into account.

So we reach the conclusion that human's opinion can indeed enhance retrieval performance.

Table 1. Performance comparison with state-of-the-art methods. A smaller value of ANMRR indicates a better performance, and the opposite is true for satisfaction score.

Method	ANMRR	Satisfaction score
VGG-F [22]	0.2995	2.39
VGG-S [22]	0.2961	2.21
GoogLeNet-m [5]	0.2784	2.43
ResNet50 [22]	0.2584	2.51
DDAM [4,6]	0.2487	2.59
LDCNN [22]	0.2416	2.67
UDAM [19]	0.2402	2.61
DAN [12]	0.2208	2.81
SBS-CNN [9]	0.2185	2.75
ISBS-CNN [this work]	**0.2073**	**3.28**

5 Conclusions

CBRSIR is an increasingly important research topic in the RS community. We previously proposed an unsupervised CBRSIR model, in which several pretrained CNNs compute similarities for RS image pairs and create pseudo-labels, enabling CNNs to be trained in an unsupervised way. In this paper, we manage to answer two questions: 1) Is pretrained CNNs' "opinion" on image similarity the same as humans'? 2) can human opinion help increase retrieval performance? We explore different approaches to gathering human views, which are then embedded into unsupervised transfer learning. We conduct experiments over the publicly available aerial image set PatternNet, and it turns out that 1) pretrained CNNs and humans hold different views on similarity, nevertheless, two kinds of views are strongly correlated; 2) human views can significantly enhance a CBRSIR system in terms of retrieval performance.

References

1. Aptoula, E.: Remote sensing image retrieval with global morphological texture descriptors. IEEE Trans. Geosci. Remote Sens. 52(5), 3023–3034 (2014)
2. Dai, O.E., Demir, B., Sankur, B., Bruzzone, L.: A novel system for content-based retrieval of single and multi-label high-dimensional remote sensing images. IEEE J. Sel. Topics Appl. Earth Observ. Remote Sens. 11(7), 2473–2490 (2018)

3. Demir, B., Bruzzone, L.: Hashing-based scalable remote sensing image search and retrieval in large archives. IEEE Trans. Geosci. Remote Sens. **54**(2), 892–904 (2016)
4. Ganin, Y., Lempitsky, V.: Unsupervised domain adaptation by backpropagation. In: Proceedings of International Conference Machine Learning, Lille, France, pp. 1180–1189, July 2015
5. Ge, Y., Jiang, S., Xu, Q., Jiang, C., Ye, F.: Exploiting representations from pre-trained convolutional neural networks for high-resolution remote sensing image retrieval. Multimedia Tools Appl. **77**(13), 17489–17515 (2017). https://doi.org/10.1007/s11042-017-5314-5
6. Li, A., Lu, Z., Wang, L., Xiang, T., Wen, J.R.: Zero-shot scene classification for high spatial resolution remote sensing images. IEEE Trans. Geosci. Remote Sens. **55**(7), 4157–4167 (2017)
7. Li, P., Ren, P., Zhang, X., Wang, Q., Zhu, X., Wang, L.: Region-wise deep feature representation for remote sensing images. Remote Sens. **10**(6), 871 (2018)
8. Li, Y., Zhang, Y., Tao, C., Zhu, H.: Content-based high-resolution remote sensing image retrieval via unsupervised feature learning and collaborative affinity metric fusion. Remote Sens. **8**(9), 709 (2016)
9. Liu, Y., Ding, L., Chen, C., Liu, Y.: Similarity-based unsupervised deep transfer learning for remote sensing image retrieval. IEEE Trans. Geosci. Remote Sens. (2020). https://doi.org/10.1109/TGRS.2020.2984703
10. Liu, Y., Huang, C.: Scene classification via triplet networks. IEEE J. Sel. Topics Appl. Earth Observ. Remote Sens. **11**(1), 220–237 (2018)
11. Napoletano, P.: Visual descriptors for content-based retrieval of remote sensing images. Int. J. Remote Sens. **39**(5), 1343–1376 (2018)
12. Othman, E., Bazi, Y., Melgani, F., Alhichri, H., Alajlan, N., Zuair, M.: Domain adaptation network for cross-scene classification. IEEE Trans. Geosci. Remote Sens. **55**(8), 4441–4456 (2017)
13. Shao, Z., Yang, K., Zhou, W.: Performance evaluation of single-label and multi-label remote sensing image retrieval using a dense labeling dataset. Remote Sens. **10**(6), 964 (2018)
14. Wang, Y., et al.: A three-layered graph-based learning approach for remote sensing image retrieval. IEEE Trans. Geosci. Remote Sens. **54**(10), 6020–6034 (2016)
15. Xia, G.S., Tong, X.Y., Hu, F., Zhong, Y., Datcu, M., Zhang, L.: Exploiting deep features for remote sensing image retrieval: a systematic investigation. CoRR abs/1409.1556, July 2017. http://arxiv.org/abs/1707.07321
16. Xiao, Z., Long, Y., Li, D., Wei, C., Tang, G., Liu, J.: High-resolution remote sensing image retrieval based on CNNs from a dimensional perspective. Remote Sens. **9**(7), 725 (2017)
17. Yang, J., Liu, J., Dai, Q.: An improved bag-of-words framework for remote sensing image retrieval in large-scale image databases. Int. J. Digit. Earth **8**(4), 273–292 (2015)
18. Yang, Y., Newsam, S.: Geographic image retrieval using local invariant features. IEEE Trans. Geosci. Remote Sens. **51**(2), 818–832 (2013)
19. Ye, F., Luo, W., Dong, M., He, H., Min, W.: SAR image retrieval based on unsupervised domain adaptation and clustering. IEEE Geosci. Remote Sens. Lett. **16**(9), 1482–1486 (2019)
20. Ye, F., Xiao, H., Zhao, X., Dong, M., Luo, W., Min, W.: Remote sensing image retrieval using convolutional neural network features and weighted distance. IEEE Geosci. Remote Sens. Lett. **15**(10), 1535–1539 (2018)

21. Zhou, W., Newsam, S., Li, C., Shao, Z.: Learning low dimensional convolutional neural networks for high-resolution remote sensing image retrieval. Remote Sens. **9**(5), 489 (2017)
22. Zhou, W., Newsam, S., Li, C., Shao, Z.: PatternNet: a benchmark dataset for performance evaluation of remote sensing image retrieval. ISPRS J. Photogramm. Remote Sens. **145**, 197–209 (2018)

Overview: Research Progress on Pest and Disease Identification

Mengyao Huo$^{(\boxtimes)}$ and Jun Tan

School of Mathematics, Sun Yat-Sen University, Guangzhou, China
huomy3@mail2.sysu.edu.cn

Abstract. In recent years, the identification of pests and diseases has become a hot topic. More and more researchers are dedicated to the detection and identification of pests and diseases to achieve precision agriculture. Automatic detection of the number of pests on crops in the area has become an important means to optimize agricultural resources. With the development of modern digital technology, image processing technology has also developed rapidly, opening a new way for the identification of harmful organisms. During the agricultural planting process, timely and accurately analyze crop pests and diseases in order to make quick and accurate responses, spray pesticides accurately on the affected area, ensure the efficient use of pesticides, and achieve high yields. This article will introduce the research progress of pest identification in the second part, including disease and pest identification, pest number and position detection, existing dataset. In the third part, this article will introduce some of the methods used in previous articles. Summarize in the fourth part.

Keywords: Pest and disease identification · Deep learning · Convolutional neural networks

1 Introduction

Pests and diseases often have adverse effects on industries such as agriculture, forestry, and animal husbandry.

Plant diseases refer to phenomena such as withering, rot, spots. During planting, plants are affected by harmful organisms or adverse environmental conditions, and their normal metabolism is disturbed. As a result, their physiological functions and tissue structure have undergone a series of changes and destructions, and they have abnormal appearances.

Pests are animals that endanger plant health, mainly insects, mites, snails and rodents. Beneficial insects should be protected, reproduced and used. Plant pests have a wide range of distribution, fast reproduction and large numbers, which will directly cause serious losses to crops, so harmful insects should be controlled earlier. Therefore, understanding insects and identifying them is important for controlling pests, protecting plant health, and achieving high quality and yield.

© Springer Nature Switzerland AG 2020
Y. Lu et al. (Eds.): ICPRAI 2020, LNCS 12068, pp. 404–415, 2020.
https://doi.org/10.1007/978-3-030-59830-3_35

2 Related Work

2.1 Disease and Pest Identification

Pest and disease identification is of great significance to the realization of precision agriculture. We know that machine learning and deep learning have achieved good results in cat and dog recognition and so on. Therefore, more and more scholars have also applied machine learning and deep learning methods to pest and disease identification, great progress has been made in pest and disease identification.

In 2010, the method proposed by Al-Bashish et al. [1] attempted to identify five different plant diseases: Early scorch, Cottony mold, ashen mold, late scorch, tiny whiteness. After the noise reduction image in the pre-processing stage, the K-means clustering algorithm was used to divide the images into four clusters. For each cluster, many color and texture features were extracted by the HSI color space co-occurrence method, and these features were imported into an MLP neural network with 10 hidden layers and finally classified.

In 2014, Rothe et al. [2] proposed a method to automatically classify cotton leaf diseases by extracting leaf symptoms from digital images. Otsu's segmentation method is used for extracting color and shape features. Support vector machine (SVM) classified the extracted features. The proposed system will help identification of three types of leaf diseases namely Bacterial Blight, Myrothecium and Alternaria. If the image fails, the efficiency of the system may be affected. Furthermore, this work can be extended to develop hybrid algorithms such as genetic algorithms and neural networks to improve the accuracy of classification.

In 2016, Sladojevic et al. [3] used a deep convolutional network to develop a new method for identifying plant diseases based on leaf image classification. The model developed in this paper can identify 13 different types of plant diseases from healthy leaves, and can distinguish plant leaves from the surrounding environment. The paper not only collected the images and created the database, but also comprehensively described all the basic steps required to the model. For individual class tests, the experimental results achieved precision between 91% and 98%. The final overall accuracy of the trained model was 96.3%.

In 2017, Wang et al. [4] used apple black rot images in the PlantVillage dataset to classify apple leaves into four types: healthy stage, early stage, middle stage, and end stage. Without complex image pre-processing, training and testing can be performed after cropping and normalization. Based on a small number of training samples, small convolutional neural networks of different depths were trained, and four latest depth models were fine-tuned: VGG16, VGG19, Inception-v3 and ResNet50. The results show that fine-tuning the pre-trained depth model can significantly improve the performance of a small amount of data. The fine-tuned VGG16 model has the best performance, with an accuracy of 90.4% on the test set.

In 2017, Ramcharan et al. [5] applied transfer learning to train a deep convolutional neural network using a cassava disease image dataset taken in Tanzania to identify three diseases and two types of pest damage (or lack of pests). The highest precision is achieved when using Inception v3 model, including brown leaf spot (BLS) 98%, red mite disease (RMD) 96%, green mite disease (GMD) 95%, cassava brown streak disease (CBSD) 98%

and cassava mosaic disease (CMD) 96%. For data not used during training, the overall accuracy of the best model is 93%. The convolutional neural network model developed in this study has been applied in the Android system, which is of great significance for miniaturizing the detection equipment.

In 2018, Nanni et al. [6] proposed an automatic classifier based on fusion of saliency methods and convolutional neural networks. The saliency method is a well-known image processing algorithm that highlights the most relevant pixels in an image. This paper uses three different saliency methods as image preprocessing, and trains four different convolutional neural networks for each saliency method. Experiments have found that the best independent network is ShuffleNet without preprocessing.

In 2018, Alfarisy et al. [7] based on the fact that diseases and pests in Indonesia pose a threat to rice production, the paper created a rice image dataset to simultaneously detect rice pests and diseases, in order to solve the problem of accurate detection of rice pests and diseases. The paper used a search engine to collect 4,511 images from four languages. The dataset can identify 13 types of rice pests and diseases, of which 9 are rice pests and 4 are rice diseases. This paper used deep learning to classify rice field pests and diseases in Indonesia. The Caffe framework was selected for processing, and the pre-trained CaffeNet model was used. The experimental results of the model reached accuracy 87%.

In 2018, Wang et al. [8] proposed a pest detection and identification system based on transfer learning, which provides a basis for controlling pests and diseases and accurately spraying pesticides. The method is capable of training and testing ten pests of tea trees with an accuracy rate of 93.84%, which is superior to most human experts and traditional neural networks. Therefore, it provides reliable technical support for precision agriculture. In order to verify the wide adaptability of the model, two types of weeds were used to train and test the model, and the accuracy rate reached 98.92%. However, this paper has not yet developed a technique for actually processing images captured by different devices and people, under different angles, illumination and different environmental conditions.

In 2019, Liu et al. [9] proposed a region-based end-to-end method, called PestNet, for large-scale multi-class pest detection and classification based on deep learning. PestNet contains three main parts: Channel-Spatial Attention (CSA), Region Proposal Network (RPN), Position-Sensitive Score Map (PSSM). CSA is first integrated into the convolutional neural network (CNN) backbone for feature extraction and enhancement. Next, use the RPN to obtain pest locations. PSSM is used to replace the fully connected (FC) layer for pest classification and bounding box regression. In addition, this paper applied context RoI as context information for pest characteristics to improve detection accuracy. They evaluated PestNet on a newly collected large-scale pest image dataset, and the experimental results show that the proposed PestNet performs well in the detection of multiple types of pests, with an average average accuracy (mAP) of 75.46%.

In 2019, Mishra et al. [10] believed that lighting, complex backgrounds, and pests that are partially visible or in different directions can lead to misclassification, resulting in huge yield losses. In order to reduce the misclassification of insects due to partial visibility or different directions, and to mitigate the misclassification caused by interference of light, complex background and so on. Taking crop rice as an example, the article proposes a pest identification system that can classify pests from the field, regardless of their orientation. Sensors such as camera (mobile) or rod-based thermal sensors can be used to provide images for morphological edge detection and bone processing. The pre-trained classifier identifies insects as friendly or harmful, and this paper used CNN as a classifier to improve accuracy. Compared to other popular classifiers, such as SVM, Naive Bayes, etc., it produced better and faster results (Table 1).

Table 1. Overall : pest and disease types

Article	Type	Description	Method
Al-Bashish et al. [1]	5 diseases	Early scorch, Cottony mold, ashen mold, late scorch, tiny whiteness	MLP
Rothe et al. [2]	3 diseases	Bacterial Blight, Myrothecium, Alternaria.	SVM
Sladojevic et al. [3]	13 diseases	Pear, cherry, and peach: porosity Peach: powdery mildew Peach: Taphrina deformans Apple, pear: Erwinia amylovora Apple, pear: Venturia Apple: powdery mildew Apple: Rust Pair: Gymnosporangium sabinae Pair: gray leaf spot Grapevine: wilt Grapevine: mites Grapevine: powdery mildew Grapevine: downy mildew	CNN
Wang et al. [4]	1 disease (4 levels)	apple black rot(healthy stage, early stage, middle stage, and end stage)	VGG16, VGG19, Inception-v3, ResNet50
Ramcharan et al. [5]	3 diseases 2 pests damage	Disease: brown leaf spot, cassava brown streak disease, cassava mosaic disease Pest: red mite damage, green mite damage	Transfer learning
Nanni et al. [6]	10 pests	Locusta migratoria, Parasa lepida, Gypsy moth larva, Empoasca flavescens, Spodoptera exigua, Chrysocus chinensis, Laspeyresia pomonella larva, Spodoptera exigua larva, Atractomorpha sinensis, Laspeyresia pomonella	AlexNet, GoogLeNet, ShuffleNet, MobileNetv2

(continued)

Table 1. (*continued*)

Article	Type	Description	Method
Alfarisy et al. [7]	4 diseases 9 pests	Disease: Pycularia oryzae leaf, Pycularia oryzae neck panicle, Tungro leaf, Xanthomonas oryzae Pest: Leptocorisa acuta, Locusta migratoria, Nephotettix virescens, Nilaparvata lugens, Pomacea canaliculata egg, Pomacea canaliculata adult, Sogatella furcifera, Stemborer adult, Stemborer larva	CaffeNet
Wang et al. [8]	10 pests	Locusta migratoria, Parasa lepida, Gypsy moth larva, Empoasca flavescens, Spodoptera exigua, Chrysochus chinensis, Laspeyresia pomonella larva, Spodoptera exigua larva, Atractomorpha sinensis, Laspeyresia pomonella	Transfer learning
Liu et al. [9]	15 pests	Cnaphalocrocis medinalis, Mythimna separata, Helicoverpa armigera, Ostrinia furnacalis, Proxenus lepigone, Spodoptera litura, Spodoptera exigua, Sesamia inferens, Agrotis ipsilon, Mamestra brassicae, Hadula trifolii, Holotrichia parallela, Anomala corpulenta, Gryllotalpa orientalis, Agriotes subrittatus	CSA + RPN + PSSM
Mishra et al. [10]	10 pests	Chilo spp, Mamestra brassicae, Euscyrtus concinnus, Gryllotalpa orientalis, Nephotettix malayanus, Brevennia reh, Nilaparvata lugens, Leptocoris oratorius F, Stenchaetothrips, Orseolia oryzae	CNN, SVM, Naive Bayes

2.2 Pest Number and Position Detection

The number of pests is the basis of the severity. The detection of the number and location of the pests lays the foundation for the efficient use of pesticides and is of great significance for the realization of precision agriculture.

In 2016, Liu et al. [11] demonstrated the effectiveness of using saliency-based methods to locate pest objects in natural images. The pipeline for the location and classification of pest insects was first implemented. First, they use a global contrast-based approach to calculate a salient map for locating pest objects. Then, the bounding square containing the target is extracted and adjusted to a fixed size, and then used to build a large standard database called "pest ID." This database is used to learn local image features for classification by DCNN. The DCNN optimized the key parameters and achieved a mean Accuracy Precision of 0.951.

In 2016, Ding et al. [12] proposed a sliding window-based detection pipeline for automatically monitoring the location and number of pests in trap images. They apply convolutional neural networks to image patches at different locations to determine the likelihood of containing a particular pest type. Then, according to the position of the image block and the associated confidence, the image block is filtered by non-maximum suppression and threshold filtering to produce the final detection result, and the effectiveness of the method is proved.

In 2018, Zhong et al. [13] proposed a new vision-based flight insect statistics and recognition system. In this system, a yellow viscous trap is installed in the surveillance area to trap the flying insects, and a camera is set up to collect real-time images. The YOLO deep learning network was used to detect and roughly count flying insects, and SVM was used to classify flying insects. The system was implemented on Raspberry PI. To assess performance, six insects were used to verify the effectiveness of the system. The experimental results show that on the Raspberry PI system, the average counting accuracy is 92.50%, the average classification accuracy is 90.18%, and the detection and recognition period is about 5 min. Different from the traditional image processing-based pest detection system, the system can monitor the spread of pests in real time, predict the occurrence probability and development trend of pests and diseases, and has important significance in the agricultural field.

In 2019, Li et al. [14] proposed an improved deep learning method to locate and count agricultural pests in the image to solve the problem of pests destroying crops. The article integrated the CNN(convolutional neural network) of ZF (Zeiler and Fergus model) and a RPN (Regional Proposal Network) with NMS (non-maximum suppression) to eliminate overlap detection. First of all, using the convolutional layer in ZF Net (without the average pool layer and the fc layer) to compute the feature map of the image, the original pixel information can be better preserved by the smaller convolution kernel. Then, several key parameters of the process were optimized, including output size, score threshold, the threshold value like the NMS. Their method achieves an accuracy of 0.93, the missed detection rate is 0.10, the mean Accuracy Precision of 0.885, which is more practical than AlexNet and ResNet. Moreover, the author puts forward 4 points worth improving in the future: (1) Enhance the training data set by changing the background of the training object; (2) More model layers can be added to help the model learn more features from the image; (3) Improve the training model by training the model trained on the image at different scales of large data sets; (4) Identify different growth processes of wheat bran and assess the hazard level (Table 2).

Table 2. Overall: pest and disease types

Article	Type	Description	Method
Liu et al. [11]	12 diseases	Cnaphalocrocis medinalis, Chilo suppressalis, Parnara Guttata, Nilaparvata lugens, Nephotettix cincticeps, Diamondback moth, Scirpophaga incertulas, Oxya chinensis, Naranga aenescens, Ostrinia rnacalis, Sogatella furcifera, Cletus punctiger	DCNN
Zhong et al. [13]	6 insects	Bee, Fly, Mosquito, Moth, Chafer, Fruit fly	YOLO, SVM

2.3 Existing Dataset

This paper finds several latest datasets for everyone to conduct experiments and explore better methods of pest and disease classification and pest quantity and location detection, providing a basis for precision agriculture.

In 2019, Wu et al. [15] addresses the problem of database related to pest identification: (1) There are only a few samples or pest species; (2) Most of the existing pest images in the public data sets are collected in a controlled laboratory environment. The requirements for pest identification in the actual environment are not well met. (3) Different pest species may be very similar in appearance but may have different forms of the same pest, including egg, larva, pupa and adult, that is, the difference between classes was significant and the similarity between species was large. As a result, they collected a large-scale dataset called IP102 for pest identification, which included 102 categories and 75,000 images. Compared to previous datasets, IP102 is consistent with several characteristics of pest distribution in real environments. At the same time, the article also evaluated some of the latest identification methods on the dataset. The results show that the current manual feature method and deep feature method still can not deal with pest identification well. Due to the variety of datasets, this paper will not show it. The dataset content is roughly shown in Table 3.

In 2019, Chen et al. [16] built the Agricultural Pest Research Library (IDADP) by collecting and integrating data. The dataset has 15 disease images of three field crops of rice, wheat and corn, each of which corresponds to a folder. Among them, there are 6 folders for rice diseases, 5 folders for wheat diseases, and 4 folders for corn diseases. Each folder contains the original JPG file of the disease image, and a text file describing the basic information and prevention methods of the disease. The dataset has a total of 17,624 high-quality image data. The dataset content is roughly shown in Table 4.

3   Methodology

3.1   KNN

K-nearest Neighbor (KNN) classification algorithm is a relatively mature method in theory and one of the simplest machine learning algorithms. The idea of this method is: if most of the k samples that are most similar in the feature space (that is, the most

Table 3. Dataset: IP102

Super-Class	Class	Train	Val	Test
Rice	14	5043	843	2531
Corn	13	8404	1399	4212
Wheat	9	2048	340	1030
Beet	8	2649	441	1330
Alfalfa	13	6230	1037	3123
Vitis	16	10525	1752	5274
Citrus	19	4356	725	2192
Mango	10	5840	971	2927
IP102	102	45095	7508	22619

Table 4. Dataset: IDADP

Crop	Disease
Rice	Rice bacterial leaf blight; Rice false smut; Rice blast; Rice brown spot; Rice sheath blight; Rice bacterial leaf streak
Wheat	Wheat powdery mildew; Wheat head blight; Wheat spindle streak mosaic; Wheat speckled snow mold; Wheat leaf rust
Corn	Corn northern leaf blight; Corn southern rust; Corn rust; Corn southern leaf blight

adjacent samples in the feature space) belong to a certain category, then the sample also belongs to this category. In KNN algorithm, the selected neighbors are all objects that have been correctly classified. This method determines the category of the subsamples according to the category of the most adjacent samples or several samples.

3.2 Support Vector Machine

Support Vector Machine (SVM) is a generalized linear classifier that classifies data according to supervised learning, and its decision boundary is the maximum-margin hyperplane of learning samples. SVM uses hinge loss function to calculate empirical risk and adds regularization terms into the solution system to optimize structural risk. It is a classifier with sparsity and robustness. Support vector machines can perform non-linear classification through kernel methods, which is one of the common kernel learning methods.SVM was proposed in 1964 and developed rapidly after the 1990s and derived a series of improved and extended algorithms, which have been applied in pattern recognition problems such as portrait recognition and text classification.

3.3 Naive Bayes

Naive Bayesian Classification (NBC) is a method based on Bayes' theorem and assuming that feature conditions are independent of each other. First, through the given training set, the independence between feature words is used as a premise, learning from input to the joint probability distribution of the output. Based on the learned model, input X finds the output Y that makes the posterior probability the largest.

3.4 Convolutional Neural Networks

The Multi-Layer Perceptron (MLP) is also called the Artificial Neural Network (ANN). In addition to the input and output layers, there can be multiple hidden layers in between. The simplest MLP requires a layer of hidden layers, the input layer, the hidden layer, and the output layer. Artificial neural network (ANN) is a general-purpose classifier and has applications in many fields. Compared with other classifiers, the advantage of ANN is that it can extract all features from the object by itself, that is, no additional feature extraction is needed. The most popular version of ANN for image processing is called Convolutional Neural Network (CNN). CNN is a deep feed forward neural network that helps visualize images by using multilayer perceptrons.

AlexNet. AlexNet [17] was designed by 2012 ImageNet competition winner Hinton and his student Alex Krizhevsky. AlexNet includes several newer technologies, and for the first time successfully applied tricks in the CNN such as ReLU, Dropout and LRN. At the same time, AlexNet also uses the GPU for computing acceleration. AlexNet promoted LeNet and applied the basic principles of CNN to a very wide and wide network.

VGG. The VGG [18] model is the second place in the 2014 ILSVRC competition, the first being GoogLeNet. But the VGG model performs better than GoogLeNet in multiple transfer learning tasks. Moreover, extracting CNN features from images, the VGG model is the preferred algorithm. Its disadvantage is that the parameter quantity is as much as 140 M, which requires more storage space. But this model is very valuable.

ResNet. ResNet [19] (Residual Neural Network) was proposed by Kaiming He and others at Microsoft Research. The 152-layer neural network was successfully trained by using the ResNet Unit and won the ILSVRC2015 competition. The amount of parameters is lower than VGGNet, and the effect is very prominent. The structure of ResNet can speed up the training of neural network, and the accuracy of the model has been greatly improved.

GoogLeNet. GoogLeNet [20] was the first place in the 2014 ILSVRC competition. GoogLeNet made a bolder network structure attempt, although the depth is only 22 layers, but the size is much smaller than AlexNet and VGG. GoogleNet parameters are 5 million, AlexNet parameters are 12 times than that of GoogleNet, and VGGNet parameters are 3 times than that of AlexNet, so GoogleNet is a good choice when memory or computing resources are limited.

ShuffleNet. ShuffleNet [21] is a highly efficient CNN architecture, which is specially applied to mobile devices with limited computing power. It uses two operations: pointwise group convolution and channel shuffle. Compared with existing advanced models, it greatly reduces the amount of calculation with similar accuracy. ShuffleNet shows superior performance on ImageNet and MS COCO over other advanced models.

MobileNetv2. MobileNetv2 [22] architecture is based on an inverted residual structure. The original residual structure has three convolutions, two point-by-point convolution channels have more channels. The inverted residual structure has exactly the opposite. The paper finds that it is effective to remove the non-linear transformation in the main branch, which can maintain the expressiveness of the model. The paper conducts comparative experiments on ImageNet, COCO, VOC, and verifies the effectiveness of the architecture.

3.5 Transfer Learning

Transfer learning is a machine learning method that refers to a pre-trained model being reused in another task. Transfer learning is related to multitasking learning and concept drifting. It is not a specialized machine learning domain. Transfer learning is very popular in certain deep learning problems, such as when there are a large number of resources required to train a depth model or a large number of data sets used to pre-train the model.

4 Conclusion

Based on the above literature, this paper finds that most scholars' research is based on smaller data sets, which contain few samples or pest species, and most of the pest images are obtained under controlled conditions. The requirements for pest and disease identification in the actual environment cannot be well met. The above will lead to the actual applicability of the model is not high. Moreover, the current literature only unilaterally studies the identification of pests and diseases, the location and enumeration of pests, or the judgment of the extent of disease. Lack of relevant research to identify pests and diseases, locate and count pests, and determine the extent of disease in the mean time, so as to achieve precision agriculture.

References

1. Al Bashish, D., Braik, M., Bani-Ahmad, S.: A framework for detection and classification of plant leaf and stem diseases. In: 2010 International Conference on Signal and Image Processing, pp. 113–118. IEEE (2010)
2. Rothe, P., Kshirsagar, D.R.: Svm-based classifier system for recognition of cotton leaf diseases. Int. J. Emerg. Technol. Comput. Appl. Sci. **7**(4), 427–432 (2014)
3. Sladojevic, S., Arsenovic, M., Anderla, A., Culibrk, D., Stefanovic, D.: Deep neural networks based recognition of plant diseases by leaf image classification. Comput. Intell. Neurosci. **2016** (2016)
4. Wang, G., Sun, Y., Wang, J.: Automatic image-based plant disease severity estimation using deep learning. Comput. Intell. Neurosci. **2017** (2017)
5. Ramcharan, A., Baranowski, K., McCloskey, P., Ahmed, B., Legg, J., Hughes, D.P.: Deep learning for image-based cassava disease detection. Front. Plant Sci. **8**, 1852 (2017)
6. Deng, L., Wang, Y., Han, Z., Yu, R.: Research on insect pest image detection and recognition based on bio-inspired methods. Biosyst. Eng. **169**, 139–148 (2018)
7. Alfarisy, A.A., Chen, Q., Guo, M.: Deep learning based classification for paddy pests & diseases recognition. In: Proceedings of 2018 International Conference on Mathematics and Artificial Intelligence, pp. 21–25 (2018)
8. Dawei, W., Limiao, D., Jiangong, N., Jiyue, G., Hongfei, Z., Zhongzhi, H.: Recognition pest by image-based transfer learning. J. Sci. Food Agric. **99**(10), 4524–4531 (2019)
9. Liu, L., Wang, R., Xie, C., Yang, P., Wang, F., Sudirman, S., Liu, W.: Pestnet: an end-to-end deep learning approach for large-scale multi-class pest detection and classification. IEEE Access **7**, 45301–45312 (2019)
10. Mishra, M., Singh, P.K., Brahmachari, A., Debnath, N.C., Choudhury, P.: A robust pest identification system using morphological analysis in neural networks. Periodicals Eng. Nat. Sci. **7**(1), 483–495 (2019)
11. Liu, Z., Gao, J., Yang, G., Zhang, H., He, Y.: Localization and classification of paddy field pests using a saliency map and deep convolutional neural network. Sci. Rep. **6**, 20410 (2016)
12. Ding, W., Taylor, G.: Automatic moth detection from trap images for pest management. Comput. Electron. Agric. **123**, 17–28 (2016)
13. Zhong, Y., Gao, J., Lei, Q., Zhou, Y.: A vision-based counting and recognition system for flying insects in intelligent agriculture. Sensors **18**(5), 1489 (2018)
14. Li, W., Chen, P., Wang, B., Xie, C.: Automatic localization and count of agricultural crop pests based on an improved deep learning pipeline. Sci. Rep. **9**(1), 1–11 (2019)
15. Wu, X., Zhan, C., Lai, Y.K., Cheng, M.M., Yang, J.: Ip102: a large-scale bench- mark dataset for insect pest recognition. In: Proceedings of the IEEE Conference on Computer Vision and Pattern Recognition, pp. 8787–8796 (2019)
16. Chen, L., Yuan, Y.: An image dataset for field crop disease identification. China Scientific Data (2019)
17. Krizhevsky, A., Sutskever, I., Hinton, G.E.: Imagenet classification with deep convolutional neural networks. In: Advances in Neural Information Processing Systems, pp. 1097–1105 (2012)
18. Simonyan, K., Zisserman, A.: Very deep convolutional networks for large-scale image recognition (2014). arXiv preprint arXiv:1409.1556
19. He, K., Zhang, X., Ren, S., Sun, J.: Deep residual learning for image recognition. In: Proceedings of the IEEE Conference on Computer Vision and Pattern Recognition, pp. 770–778 (2016)
20. Szegedy, C., et al.: Going deeper with convolutions. In: Proceedings of the IEEE Conference on Computer Vision and Pattern Recognition, pp. 1–9 (2015)

21. Zhang, X., Zhou, X., Lin, M., Sun, J.: Shufflenet: an extremely efficient convolutional neural network for mobile devices. In: Proceedings of the IEEE Conference on Computer Vision and Pattern Recognition, pp. 6848–6856 (2018)
22. Sandler, M., Howard, A., Zhu, M., Zhmoginov, A., Chen, L.C.: Mobilenetv2: inverted residuals and linear bottlenecks. In: Proceedings of the IEEE Conference on Computer Vision and Pattern Recognition, pp. 4510–4520 (2018)

Advanced Data Preprocessing for Detecting Cybercrime in Text-Based Online Interactions

John Sekeres[1]([⊠]) [ID], Olga Ormandjieva[1] [ID], Ching Suen[1] [ID], and Joanie Hamel[2] [ID]

[1] Concordia University, Montréal, Québec, Canada
j_seke@encs.concordia.ca
[2] Sûreté du Québec, Boucherville, Québec, Canada

Abstract. Social media provides a powerful platform for individuals to communicate globally. This capability has many benefits, but it can also be used by malevolent individuals, i.e. predators. Anonymity exacerbates the problem. The motivation of our work is to help protect our children from this potentially hostile environment, without excluding them from utilizing its benefits.

In our research, we aim to develop an online sexual predator identification system, designed to detect cybercrime related to child grooming. We will use AI techniques to analyze chat interactions available from different social networks. However, before any meaningful analysis can be carried out, chats must be preprocessed into a consistent and suitable format. This task poses challenges in itself. In this paper we show how different and diverse chat formats can be automatically normalized into a consistent text-based format that can be subsequently used for analysis.

Keywords: Online predator identification · Chat logs · Deep learning

1 Introduction

Chats are basically short text-based conversations between 2 or more people. The dialog often takes a form similar to a spoken conversation [1]. There are three basic parts to every chat line;

- Date and time the chat line was authored
- Author identification
- Dialog text

Online chats can come from many diverse places on the Internet, for example; instant messaging, social media, chat rooms, Internet forums, computer storage (e.g. law enforcement forensics), etc…

Our goal is to develop software that assists law enforcement in identifying child grooming related cybercrime in text-based online chats. Part of our research is to build and train an AI engine that is able to analyze chats and to report suspicious interactions.

© Springer Nature Switzerland AG 2020
Y. Lu et al. (Eds.): ICPRAI 2020, LNCS 12068, pp. 416–424, 2020.
https://doi.org/10.1007/978-3-030-59830-3_36

However, before that can be done the date/time, author and chat dialog described above need to be extracted and preprocessed from the original text [2, 3].

Normalization is the process of parsing a raw chat for the date/time, author and dialog, and, storing the data in a standardized format for use later on as needed.

The research presented in this paper highlights the issues and methods associated with taking raw chats and normalizing them in a generalized and automated/semi-automated way.

2   Chat Normalization

Chat normalization refers to the process of converting the format and structure of a raw chat into a standardized text-based format. The intent is that an application can later extract with confidence and ease the different elements of the chat when needed.

Below we outline the three steps of the chat normalization process proposed in this paper:

1. The first step is to convert all non-text-based chats (e.g. PDF, MS Word, ...) into text form. There are several software utilities that can help with the conversion from various formats into text, for example: Pandoc [4]. This paper assumes that the chats have been converted into text, and when we refer to the format of a chat, we mean the format of the text tokens within a chat, and not the binary representation of the chat file.
2. The next step is to parse the chat, separating the header (date/time, author) from the chat dialog. This process presents challenges as chats come in many different formats. For example, chats acquired from one site might have conversation lines stored as '*YYYY/MM/DD -< author > : dialog...*' and at another site as '*on MM-DD-YY at HH:MM:SS, < author > wrote: dialog...*'. Furthermore, each part of the conversation can span 1 to many lines, with no guarantee where each element begins or ends. Also, chats can either be stored in a structured (e.g. XML, JSON, HTML, ...) or unstructured (i.e. no structure information) format. Issues with parsing are the main topic of this paper, and are discussed in detail in the subsequent sections.
3. The final step to normalization is to store the parsed-out chat elements into a file format that be easily read by an application later on.

2.1   Normalizing Structured Chats

Structured chats refer to chats that include meta-data or structure information about the chat embedded in the text. For example, XML, JSON, CSV (with headers), HTML, ...

It is not possible to build a fully generalized automated normalizer for structured chats unless something is known about the structure. The complication is that the information we're interested in (i.e. author, date-time, chat dialog) is interspersed with other data/meta-data that isn't needed.

In this paper we propose two basic approaches to normalizing structured chats:

1. Convert the structured chat into an unstructured chat (we'll show later that dealing with unstructured chats offer better results).
2. Supply some information about the structure of the chat (e.g. a template) that enables parsing out the different elements.

Each point is discussed in the following sections.

Converting Structured Chats into Unstructured Chats. Better results can be obtained from the normalization of unstructured chats; therefore, one of the goals when dealing with structured chats is to first convert them into an unstructured format, and then process them as unstructured chats.

Depending on the type and complexity of the structure, conversion can generally be done using an off-the-shelf utility program. For example, Pandoc [4] supports the conversion of HTML, JSON, LaTeX,… to plain text. This method is not always feasible as the chat may contain data that is not part of the chat, for example a HTML page may contain advertisements. If this is the case then it might be better to consider a template-based normalization of the structured chat, as described in the next section.

If neither method produces the desired result, then user intervention is required to extract the chat data either manually or programmatically. The following simple example illustrates how a CSV[1] file with a header can be converted into an unstructured chat simply by removing the header, and cleaning out unnecessary delimiters (Fig. 1).

```
Date,Time,Author,ChatLine
10/10/2019,10:44,Mark,"Henri, can I ask you a question?"
10/10/2019,10:45,Henri,sure
10/10/2019,10:46,Mark,how…
```

Fig. 1. Raw chat stored as a CSV file

Removing the header line, and all unnecessary delimiters, the file can be processed as an unstructured chat (Fig. 2):

```
10/10/2019 10:44,Mark: Henri, can I ask you a question?
10/10/2019 10:45,Henri: sure
10/10/2019 10:46,Mark: how…
```

Fig. 2. Chat converted into an unstructured chat

Template Based Normalization of Structured Chats. Template based normalization of structured chats involves defining a template which describes where in the text the chat data resides. In most cases the three basic elements of the chat, i.e. date/time, author

---

[1] Comma Separated Values (CSV).

and dialog can be extracted separately from the text, thus allowing for the text to be normalized directly. If the three elements are not separated (e.g. author and dialog on the same line) then the text can be converted to an unstructured chat, and treated as such.

The complexity of this technique is that each type of format (e.g. XML, JSON, HTML, ...) requires a unique implementation, and it requires a user to define a template for each chat.

What makes this technique feasible however is that in most cases structured chats will come in one of a few standardized formats, namely; HTML, Markdown, XML or JSON. Also, chats acquired from a particular site will contain the same format, thus allowing one template specification to apply to many chats.

Example: Template Based Normalization of XML Chats. In XML the data will be stored in elements or as attributes [5]. Assuming that we're dealing with elements, then if given the element name and path we can extract the needed chat data (date/time, author, dialog). Note that in the instance where there is more than one chat conversation in an XML file, then a conversation element name/path can be specified, and the data normalized accordingly.

The following is an example of a template specification that would provide the information we need to extract the chat data from the example XML provided below (Figs. 3 and 4):

```
XML-Specification:
conversation=conversations/conversation#id
date_time=conversations/conversation/message/time
author=conversations/conversation/message/author
dialog=conversations/conversation/message/text
```

Fig. 3. Example XML normalization specification

2.2 Normalizing Unstructured Chats

Unstructured chat types refer to chats that do not contain meta-data or information about the chat structure within the chat. Good general automatic normalization results can be obtained? from unstructured chats. If an unstructured chat cannot be normalized automatically then a template specification can be used to define the structure.

We define an unstructured chat as:

- Everything in the chat is either part of a header or part of the dialog (no meta-data or structure info)
- The header (date-time and/or author) precedes the chat-line dialog

An attempt can be made to normalize an unstructured chat automatically; however, if that fails then a template-based specification outlining the format of the chat header can be used. The results of templated normalization are very good. The following 2 sections describe each method.

```
<conversations>
  <conversation id="d55846cb89c0fbf5e177b0b0d499023e">
    <message line="1">
      <author>85a7e70c36dcd431d62b528ead109ed5</author>
      <time>10:36</time>
      <text>hi</text>
    </message>
  </conversation>
  <conversation id="85f0abac6ef5a2a23814a2ced73b5fb7">
    <message line="1">
      <author>bb8b358a10488f1ce25cdeb1df4a842a</author>
      <time>14:11</time>
      <text>hello there</text>
    </message>
    <message line="2">
      <author>2ded7a428b8b4536d49393c352fe1d1c</author>
      <time>14:11</time>
      <text>hey</text>
    </message>
  </conversation>
</conversations>
```

Fig. 4. Example XML taken from the PAN2012 [6] chat dataset

Template Based Normalization of Unstructured Chats. Template based normalization of unstructured chats involves defining a template which describes the format of the chat header. As the system parses each token, it is compared against the template. If each token matches the template (which has also been tokenized) then those series of tokens become the header. The remaining text after the header becomes the chat-line dialog until the next header is uncovered.

For example, we can define the following templates 1, 2 and 3 (see Fig. 5):

```
1. NORMALIZE-HEADER-TEMPLATE: on #LONGMONTH #DAY, #YEAR #HOUR:#MINUTE #AMPM, #AUTHOR wrote:
2. NORMALIZE-HEADER-TEMPLATE: #NL #DAY/#MONTH/#YEAR #HOUR:#MINUTE #NL #AUTHOR #NL
3. NORMALIZE-HEADER-TEMPLATE: [#HOUR:#MINUTE:#SECONDS] #AUTHOR:

where:
#LONGMONTH = {January, February,..., December}
#DAY = 1-31
#MONTH = 1-12
#YEAR = 1990-3000
#HOUR = 0-59
#MINUTE = 0-59
#SECONDS = 0-59
#NL = new line
#AUTHOR = the author of the current line (the exact value is unknown at this stage,
          but tokens are concatenated until the terminating delimiter is detected)
```

Fig. 5. Example unstructured chat normalization templates

Figure 6 depicts a normalization example using template 1 (as specified in Fig. 5): A normalization example using template 2 is shown in Fig. 7:

Unstructured chat

On April 17, 2013 6:35:58 AM, Rita B wrote:

Hello, how are you?

On April 18, 2013 8:37:02 AM, Fred S wrote: good and you

...

Normalized

04/17/2013 6:35:58, Rita B: Hello, how are you?

04/18/2013 8:37:02, Fred S: good and you

...

Fig. 6. Sample of a normalized chat using template 1 (see Fig. 5)

Unstructured chat

03/02/2013 09:58

Kim

Hi niko

24/02/2013 09:48

Niko

hi Kim

...

Normalized

02/03/2013 09:58:00, Kim: Hello, how are you?

02/05/2013 09:48:00, Niko: good and you

...

Fig. 7. Sample of a normalized chat using template 2 (see Fig. 5)

Template based normalization achieves very good results; however, it does require a user to specify a template. This can be mitigated by having a strong automatic template builder as described in the next section. Fortunately, the format of a chat remains rather consistent depending on where the chat was acquired, so that one template can serve to normalize multiple raw chats. If for a given dataset of unstructured raw chats, there are multiple formats, then for each chat multiple templates can be applied until the correct template is found.

Generalized Automatic Normalization of Unstructured Chats. Automatic normalization can be achieved if the format of the chat header (i.e. date-time and/or author) can be deciphered from the unstructured chat. If the format can be identified then one can build a chat header template and invoke the services of templated normalization as described in the previous section to normalize the chat.

We'll start by making the following *assumptions* about the chat header format:

- The header starts on a new line
- The date-time is optional however if one is supplied then it should precede the author

- A non-alphanumeric character delimits the author from the dialog
- The authors' name is no longer than N words (e.g. N = 2) and must contain at least one alphabetic character

We'll use the following chat as an example as we walk through the process. We'll assume the author is composed of a maximum of 2 words (Fig. 8).

```
1. at 02:55, Mark Dee X:
2. what did you do today?
3. at 02:55, Tina: did some bird watching
4. and, went to the gym
5. at 02:55, Mark Dee X : you danse?
6. at 02:55, Tina: nope :-)
```

Fig. 8. Example unstructured chat

The first step is to find the author and delimiter that separates the header from the dialog as follows;

- Parse the chat into tokens (words, numbers, delimiters)
- For every delimiter back scan each token until:

  - a number or delimiter is encountered
  - beginning of the line
  - if neither condition is met before N tokens then exclude this as a possibility

- Keep a count of the delimited pre-text set of words found. in our example we have (Table 1):

Table 1. Count of authors and delimiters found in the example chat

Author/Delimiter	Count
Tina:	2
and,	1
you danse?	1
nope:	1

Assuming the pre-text set of words with the highest count is our author and header delimiter, we build a header template as follows:

- Scan the chat for an occurrence of the assumed author and delimiter
- Start building the header template by parsing each token from the beginning of the line to the 1st token of the author

  - if the token is a word (except words associated with dates, e.g. names of the month) or delimiter then include it as is in the template

– if the token is a number check if it is a date or time.

- If it is then start building the date/time template
- if it's not, include the token as is in the template

- Append #AUTHOR and the header delimiter ':' to the template

The deciphered template for our example chat is:

at #HOUR:#MINUTE, #AUTHOR:

Repeat for more occurrences of the assumed author and delimiter in the chat and compare the generated template. if more than 2 templates match then we can be relatively confident that we've found the right template.

Finally, run the chat and template through the template based normalization and verify the results. If it fails then you may wish to try the next assumed author and delimiter in the list above (e.g. 'and,') and repeat the process.

3 Conclusions and Future Work

Normalization is the process of converting a chat in its raw unknown format into a standardized format. It is usually a necessary step before any processing or analysis can be done on a corpus. In this paper we discuss the challenges with normalization, and proposed a series of methodologies in which raw chats can be normalized either automatically or semi-automatically.

Good automatic normalization results can be obtained from unstructured raw chats using templates. In the case where a template is not specified, one can sometimes be generated automatically.

Structured chats are more difficult to work with because the raw chat includes meta-data in it that does not directly relate to any of the 3 basic elements of a chat we're interested in (i.e. date/time, author and dialog). If a structured chat cannot be normalized via a template then external intervention is required to clean up the raw chat.

Our goal is to offer law enforcement a software tool they can use to detect cybercrime on a daily basis from thousands of chats coming from a multitude of different places. In order for a software solution to be useful it needs to be able to successfully automate the normalization process as much as possible. In an effort to achieve this future work needs to be done on dealing with structured chats.

Acknowledgements. We would like to thank the law enforcement officers at the Sûreté du Québec for their help, patience and valuable advice in helping us move forward with our research.

References

1. Wikipedia, "Online chat". https://en.wikipedia.org/wiki/Online_chat
2. Dan, L.: Identifying cyber predators by using sentiment analysis and recurrent neural networks. M. Comp.Sc. diss., Concordia University (2018)
3. Ebrahimi, M.: Automatic identification of online predators in chat logs by anomaly detection and deep learning. M.Comp.Sc. diss., Concordia University (2016)
4. Pandoc, "a universal document converter". https://pandoc.org/
5. Wikipedia, "Extensible Markup Language (XML)". https://en.wikipedia.org/wiki/XML
6. "PAN-2012 Competition - Sexual Predator Identification". https://pan.webis.de/clef12/pan12-web/author-identification.html

A Real-Time License Plate Detection Method Using a Deep Learning Approach

Saeed Khazaee[1]([✉]) [ID], Ali Tourani[2] [ID], Sajjad Soroori[2] [ID], Asadollah Shahbahrami[2], and Ching Y. Suen[1] [ID]

[1] Centre for Pattern Recognition and Machine Intelligence, Concordia University Montreal, Montreal, Canada
s_khaza@encs.concordia.ca
[2] Department of Computer Engineering, University of Guilan, Rasht, Iran

Abstract. In vision-driven Intelligent Transportation Systems (ITS) where cameras play a vital role, accurate detection and re-identification of vehicles are fundamental demands. Hence, recent approaches have employed a wide range of algorithms to provide the best possible accuracy. These methods commonly generate a vehicle detection model based on its visual appearance features such as license-plate, headlights or some other distinguishable specifications. Among different object detection approaches, Deep Neural Networks (DNNs) have the advantage of magnificent detection accuracy in case a huge amount of training data is provided. In this paper, a robust approach for license-plate detection based on YOLO v.3 is proposed which takes advantage of high detection accuracy and real-time performance. The mentioned approach can detect the license-plate location of vehicles as a general representation of vehicle presence in images. To train the model, a dataset of vehicle images with Iranian license-plates has been generated by the authors and augmented to provide a wider range of data for test and train purposes. It should be mentioned that the proposed method can detect the license-plate area as an indicator of vehicle presence with no Optical Character Recognition (OCR) algorithm to distinguish characters inside the license-plate. Experimental results have shown the high performance of the system with precision 0.979 and recall 0.972.

Keywords: Deep learning · Intelligent Transportation Systems · Automatic number-plate detection · Image processing

1 Introduction

Nowadays, transportation plays a crucial role in human daily lives and demands for fast, secure and easy transport, especially among the metropolitan population is dramatically increasing. Since transportation has always faced with critical challenges like traffic occlusions and the limitations of land resources, the advent of Intelligent Transportation Systems (ITS) is known as an increasingly vital solution to tackle these issues in recent years [1]. ITS collect data from the surrounding environment by the means of various

© Springer Nature Switzerland AG 2020
Y. Lu et al. (Eds.): ICPRAI 2020, LNCS 12068, pp. 425–438, 2020.
https://doi.org/10.1007/978-3-030-59830-3_37

devices/sensors and offers beneficial consequences using a combination of modern technologies [2]. Some examples of these devices are roadway cameras, radar sensors, and inductive loop detectors. Among the wide range of technologies used in ITS, vision-driven applications are known as durable devices with the benefits such as low rate of drawback, appropriate maintenance costs, and high expandability. Practically, roadway cameras -as the most common vision-driven devices- can deliver visual information for later processing [3]. It should be emphasized that the process of vehicle detection using visual appearance features is a principal step in almost any video-based ITS applications [4]. This process is generally performed using computer vision and image processing techniques where the main goal is to recognize some indicator features such as license-plate or headlights, to represent vehicle presence in images or video frames [4]. In this regard, there are some critical challenges in the correct detection of vehicles including diverse illumination conditions, various vehicle shapes, output image/video quality, and so on [4].

Recent studies have shown that deep learning approaches provide impressive classification results compared to traditional machine vision techniques [5]. It has to be mentioned that the reduction of false-positive outcomes is an indispensable metric to evaluate a system in most real-world classification problems. In this way, deep learning has shown a remarkable potential to reduce misclassifications and improve the performance of the system [6]. Nevertheless, utilizing Deep Neural Networks (DNNs) requires abundant processing units due to their computationally intensive nature. In this regard, employing modern Graphics Processing Units (GPUs) due to their parallel architecture can solve this problem to a large extent.

In this paper, a method to detect Iranian license-plates is presented which works as a representation of vehicle presence in an image that has the advantages of high accuracy and real-time performance. This system utilizes You Only Look Once version 3 (YOLO v.3) as a robust deep learning algorithm to provide high accuracy results along with almost real-time performance. The system has been tested on real condition data with different resolution and illumination conditions that resulted in more than 98% accuracy in the correct localization of vehicles' license plates. The main contributions of the paper are considered as below:

- Fine-tuning of YOLO v.3 algorithm to provide a high accuracy system for License Plate Detection (LPD)
- A well-designed LPD system for real condition circumstances with the advantages of high accuracy and almost real-time response
- Generating a dataset of vehicles with Iranian license-plates in various environmental conditions
- Data augmentation to provide a wider range of data variety for good model training

1.1 Preliminary Concepts

In recent years, employing DNNs has gained much popularity in classification fields due to their superb accuracy. There are two key notions in deep learning approaches that have to be considered: 1) they need a huge amount of data for training and model generation to deliver remarkable accuracy, 2) they require high processing resources due to their computationally intensive nature [6]. By providing solutions to these two demands of deep learning, they can be considered as great alternatives for common machine learning and classification approaches like Support Vector Machine [7]. Particularly, in vision-based machine learning approaches, DNNs have the power to tackle classic image processing and object detection challenges [8].

Generally, DNNs consist of three main layers including an input layer, an output layer, and several (at least two) hidden layers in between. In classification and object detection approaches, the hidden layer provides a mapping structure from the input layer to the output layer which is called the trained model. In this regard, some common types of DNN-based approaches for object detection purposes are classified as below:

- Convolutional Neural Networks (CNNs): CNNs are known to be the most common approaches of DNNs in machine vision applications which provide the ability to distinguish and extract existing relations among the features of a given data and capture the temporal and spatial dependencies by the means of convolution filters [9]. In these networks, the input data (vehicle images in our special case) is divided into small portions and each share is fed to the bottommost level. Passing the image through hidden layers triggers some functions to change neurons' weights that lead to feature extraction. These layers include the convolution layer or the kernel to extract high-level features of an image, pooling layer to decrease the number of processing operations, and fully-connected layer to learn non-linear features [10]. The process of updating the weights of the neurons in the network is performed by backpropagation that refers to the loss function existed in the final fully-connected layer. Finally, the output of this network is a set of candidate regions that may contain various detected objects inside a given image/video frame.

- Region-based Convolutional Neural Networks (R-CNNs): in order to limit the number of candidate regions detected by CNNs, R-CNN [12] approach employs a Selective Search [11] algorithm that results in lower computational cost while maintaining the accuracy. Selective Search is a greedy algorithm that recursively merges similar neighboring regions into larger covering ones. These regions -called region proposals- are then fed into a CNN to extract the existing features and finally, a Support Vector Machine (SVM) produces the final classification results by checking if the detected features lay inside the region proposals or not [12].

- Fast Region-based Convolutional Neural Networks (Fast R-CNNs): common R-CNN approaches are still computationally intensive and require a huge amount of time for classification and training processes. By feeding the input vector to the CNN, a convolutional feature map is provided which is an alternative step for direct feeding of the region proposals. This reordering of the steps results in a faster training process due to employing the convolution function only once for each image [13]. Besides, a Region of Interest (RoI) in pooling layer to generate fixed-size shapes as the inputs of a fully connected layer is obtained from the feature map. The output of this method includes a SoftMax probability layer to predict the final related class and the offset values for the detected objects' bounding boxes.

- Faster Region-based Convolutional Neural Networks (Faster R-CNNs): to provide automatic learning of region proposals alongside with bypassing Selective Search algorithm due to its computationally intensive nature, another approach which is known as Faster R-CNNs have been developed. These networks are similar to Fast R-CNNs but they employ a separate convolutional network called Region Proposal Network (RPN) instead of Selective Search algorithm to predict the region proposals [14]. Finally, region proposals are fed to the RoI pooling layer of the Fast R-CNN algorithm to classify objects.

- You Only Look Once (YOLO): different from the mentioned methods where a portion of the image is considered to detect possible objects, YOLO algorithm looks at the whole image and separates the regions with a high probability of containing an object [15]. Additionally, YOLO provides real-time object detection performance which is a great advantage comparing to other methods. The process of predicting the location of objects inside an image is done by a single CNN. In this regard, the input image is split into equal-sized rectangular grids with several bounding boxes. The CNN provides a probability of relating to the existing classes and an offset value for each bounding box. Consequently, if a bounding box has the probability higher than a predefined threshold, it is considered as a region containing the detected objects. YOLOv2 is the second version of YOLO with the advantages of higher detection accuracy and faster performance. This approach is a real-time object detection method that can detect over 9000 different objects [16]. In YOLOv2, learning and classification steps are done by COCO image dataset and ImageNet samples, respectively. Recently, a newer version of YOLO has been introduced that is so-called YOLO v.3. This implementation utilizes independent logistic classifiers instead of SoftMax algorithm besides an improved feature extractor which is a 53-layer Darknet for better probability calculation [17]. Another improvement in YOLO v.3 compare to v.2 is employing skip connections along with residual blocks.

1.2 Related Works

Employing deep learning approaches for vehicle detection has been attracted huge attention in recent years. The main methods, architectures, and models generated by most studies have been designed based on the vehicles' visual features. For instance, a group of researchers proposed a high accuracy vehicle detection approach using deep neural networks for still images and implemented their method on an embedded system to verify the capability of real-time performance and low processing time [18]. Huval et al. presented a CNN architecture for vehicle detection in autonomous driving by using an efficient mask detector alongside with a bounding box regression process [19]. Wang et al. introduced a vehicle detection/tracking approach to detect vehicles based on their tail-light signals [20]. They trained their method on "Brake Lights Patterns" dataset and tested the final model on real on-road videos which resulted in robust classification results. Other researchers proposed a simplified version of Fast R-CNN algorithm that can detect vehicles in an image through their visual appearance features with much less computational cost [21]. Zhang et al. introduced a new vehicle detection method based on a modified version of Faster R-CNN that provides the advantages of high detection and classification precision for various-sized objects [22]. Similarly, Wang et al. improved the Faster R-CNN algorithm by utilizing a ResNet architecture [23]. Some other researchers focused on the process of license-plate detection as an appropriate indicator of vehicle presence in images. For instance, Kim et al. proposed a two-step LPD system based on Faster R-CNN that firstly detects vehicles inside the input image and then, searches for the location of its license-plate [24]. Similarly, Selmi et al. introduced a general method to detect various types of license-plates based on deep learning [25]. Their method utilizes R-CNN approach for all phases of ALPR, i.e. license-plate detection, character segmentation and Optical Character Recognition (OCR). Some other researchers proposed an ALPR system for Brazilian license-plates based on CNN architectures with the advantage of correct detection of license-plates and their characters with the precision rate of 93% [26]. Abdullah et al. proposed a YOLO-based deep network for Bangla ALPR system that localizes the license-plates and recognizes the digits inside them [27]. They have achieved 92.7% accuracy in recognizing the Bangla characters. A similar approach to detect Thai characters in license-plates after correct detection of the license-plate area is introduced in [28]. The authors of this paper have proposed various techniques in the field of LPD and vehicle detection. For instance, a vehicle detection method based on deep learning approaches has been introduced that provides robust detection results and almost real-time performance [29]. In the mentioned method, a transformed version of a pre-trained ResNet-50 residual network to a Faster R-CNN was utilized to build a vehicle detection model. The main drawback of the system was that the classification and bounding box regression is made at the same time in Faster R-CNN, which makes the execution time slower.

The rest of the paper is organized as follows: In Sect. 2, the main structure of the proposed method is presented in detail. Section 3 discusses the experimental results in real condition circumstances and finally, Sect. 4 concludes the paper.

2   Proposed Method

Figure 1 shows the overall process of the proposed license-plate detection method. Accordingly, a set of images is fed to YOLO v.3 to train the network and generate an LPD model. It has to be mentioned that the ground-truth bounding boxes that contain the exact locations of the license-plates in images should be specified to be employed in the classification step. To achieve better training results in the model generation, input images should be pre-processed to remove any existing noises. This process is done by employing low pass filters on each input image. For instance, to handle high illumination conditions, histogram equalization can provide better contrast with the higher probability of distinguishing a vehicle in a given image. Finally, the trained model can be utilized as a real-time LPD system.

In this regard, YOLO algorithm turns the vehicle image into an $N \times N$ grid network, where N is calculated based on the input image and other network parameters. Hereby, a reduction factor is a major network parameter that is utilized to break the input image into equal grids. For instance, if the dimensions of the input image are 416×416, considering reduction factor 32, the value of N becomes 13 and accordingly, the dimensions of the network become 13×13. Each cell of the network is able to detect A objects. Where A is the number of anchor boxes in the network. As a default value, A is equal to 9 in YOLO v.3 algorithm. It should be noted that the correct tuning of anchor boxes can lead to great output results. Consequently, we have modified these parameters based on the

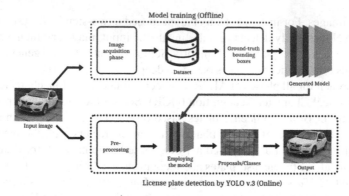

Fig. 1. Proposed method for license-plate detection.

dimensions of a standard Iranian license-plate, which has an aspect ratio of 4.298:1 (e.g. 735 × 171).

According to Fig. 2, if the centroid C of an object falls into a pre-defined grid cell, only that grid cell has the responsibility to detect the object. Each anchor box has six separate parameters as below:

- Bounding box horizontal location (bx)
- Bounding box vertical location (by)
- Bounding box width (bw)
- Bounding box height (bh)
- Box Confidence score (BC) or the objectness factor: an important parameter for classification of the detected object in the image. YOLO considers this parameter a number between 0 and 1.
- A class identifier (CL) for each class (in this special case, there is only one class for license-plates)

In this regard, BC can be calculated from the below equation:

$$BC = Pr \times IoU \tag{1}$$

Where Pr is the probability of object existence in a box and IoU is the Intersection over Union. To calculate IoU, if the ground-truth bounding box which contains the correct location of the license plate in the image has some intersections with the bounding box generated by the method, the value of IoU is obtained from Eq. (2) as below:

$$IoU = \frac{intersection\,area}{ground - truth + predicted\,area} \tag{2}$$

In the detection phase of the proposed method, each cell may find several bounding boxes, but the ones with IoU > 0.7 are acceptable. This technique is called non-maximum suppression which is a part of YOLO algorithm. Accordingly, Fig. 2 presents the overall architecture of the proposed method in brief.

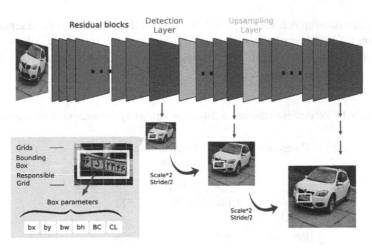

Fig. 2. The architecture of the proposed deep learning model.

3   Experimental Results

The main objectives of announcing the proposed method include: 1) designing a high-accuracy LPD system, 2) providing real-time performance, and 3) providing the ability to be employed in real-world circumstances. As a response to these demands, real condition data has been fed to the implemented version of the proposed method and two separate experiments have been performed for evaluation of accuracy and performance factors. Sub-sections A-D describe the evaluation process of the proposed method in detail.

3.1   Dataset

The accuracy of any deep learning approaches for classification purposes is highly dependent on the training data. Since we could not find any standard Iranian license-plate dataset available, we had to acquire real condition data and generate a dataset for train/test purposes. The mentioned dataset has been generated by the authors of this paper with the contribution of University of Guilan Business Incubator (UGBI) and Guilan Science and Technology Park (GSTP). For this reason, a camera was installed at the entrance of a parking lot and captured images of vehicles under the supervision of a surveyor. The main reason for generating this dataset is to provide a wide range of real condition data for the system acquired during various illumination and weather conditions. The camera was installed 1.7 meters above the ground level the resolution of the output images is 1100×624. Other properties of the mentioned dataset are presented in Table 1. As can be seen, gathered vehicle images have been classified into front and rear-view classes, each comprised of numerous samples captured in normal and challenging scenarios. Normal vehicle images have suitable illumination, acceptable contrast, and highly visible features, where challenging images may have been captured in low illumination or rainy weather, or contain some kind of obstacles in front of the license plate. Accordingly, normal samples are generally suitable for training, where challenging ones can be employed in testing the final model for experiments. Figure 2 shows some sample

432   S. Khazaee et al.

images of this dataset. Additionally, we have divided the dataset into three separate portions where 60% of the samples were used for training, 20% for the validation, and 20% for the test process. In this regard, the training data portion comprised of 80% of normal and 20% of challenging images (Fig. 3).

Table 1. Properties of the image dataset generated by the authors of this paper.

Categories	#data	Weather/illumination scenarios	
		Normal	Challenging
Front-view vehicles	2,148	1,635	513
Rear-view vehicles	969	741	228
Total	**3,117**	**2,376**	**741**

Fig. 3. Sample images of the generated dataset.

Obviously, the number of collected data is not enough for a deep learning-based method, where the final generated model is extremely dependent on data variety. To provide a wider range of data and enlarge the training samples, we have utilized some common data augmentation techniques including horizontal flipping, scaling, and crop. These processes apply minor changes to the data besides keeping its extractable visual features. Table 2 shows the result of augmentation techniques with their enlargement coefficient.

In the next step, we have utilized a manual image annotation application known as "VGG Image Annotator (VIA)" developed by Visual Geometry Group (VGG) of the University of Oxford [30]. Using VIA, a rectangular region shape drew on the license-plate location of each vehicle image along with a "ClassName: LP" value for identification. This process has been done on the whole dataset samples and a coco output was provided by VIA.

Table 2. The result of utilizing data augmentation.

Categories	Augmentation method	#augmented data
Front-view vehicles	Horizontal Flip (1x)	2,148
	Scale (3x)	6,444
	Crop (2x)	4,296
Rear-view vehicles	Horizontal Flip (1x)	969
	Scale (3x)	2,907
	Crop (2x)	1,938
Total number of augmented data		**18,702**
Total number of data (real + augmented)		**21,819**

In addition, Fig. 4 shows some sample augmented data generated by the augmentation methods in Table 2.

Fig. 4. Sample images of the generated dataset.

Finally, the coco output file has been converted into YOLO format to be able to comprehensible for the proposed method implementation.

3.2 Evaluation Criteria

In order to evaluate the system, several metrics have been considered. In this regard, if the system can detect the correct location of a vehicle's license plate in the input image, it is considered as a correctly classified sample. The number of vehicle images in which the license-plate areas are correctly detected by the proposed method is known as True-Positive (TP). On the other hand, False-Negative (FN) indicates the total number of images in which the existing license-plate areas have not been recognized. False-Positive (FP) refers to the number of images in which an object has wrongly classified as a license-plate. Consequently, precision and recall factors can be calculated by $TP/(TP + FP)$ and $TP/(TP + FN)$ fractions.

3.3 Experimental Setup and Results

In this section, we provide the results related to our experiment to evaluate the system. First, we trained the network by 20 epochs with a learning rate of 0.002. As can be seen in Fig. 5, the detected license-plates of six cars in the input image are selected by green rectangles. Table 3 shows the performance of the proposed method in terms of TP, FN, FP, precision, and recall on the test-set.

Fig. 5. Sample outputs of the proposed deep learning model.

Table 3. Performance of the proposed method.

Categories	#data (real and augmented)	TP	FN	FP	Precision	Recall
Front-view vehicles	3,759	3,608	105	46	0.987	0.972
Rear-view vehicles	1,695	1,590	42	63	0.962	0.974
Total	**5,454**	**5,198**	**147**	**109**	**0.979**	**0.972**

According to Table 3, precision and recall values prove that the proposed method can detect license-plate locations in real condition image with magnificent accuracy. The mentioned approach is able to locate vehicles' license-plates in both front and rear-view images of vehicles captured in various illumination and weather conditions. In any case, there are a few numbers of FN and FP samples where the system could not correctly classify the license-plate. We observed that all of the FN and FP samples are front/rear view images captured in the challenging circumstances. Figure 6 illustrates six samples where the license-plate has not been classified correctly by the system.

We calculated the time elapsed from providing the input image until drawing the bounding box on the image. In this regard, Fig. 7 shows the range of time to detect

Fig. 6. Challenging samples of the dataset that caused misclassification.

license-plates in dataset samples. To compare the effect of employing YOLO v.3 to provide detection results, we have done the same experiment with the first version of YOLO. It should be noted that the number of samples in Fig. 7 is the sum of TP and FP samples where a license-plate has been detected, regardless of correct or wrong classification results. Experimental results showed that this process takes between 20 to 60 ms, on the challenging situations in the license-plate area. Additionally, the proposed method produces detection output in an average of 38.79 ms, while the YOLO implementation provides the same results in 85.48 ms which is 2.2 times slower than YOLO v.3 implementation.

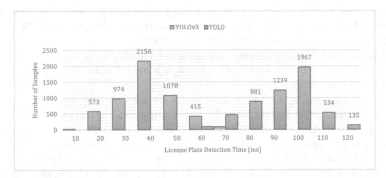

Fig. 7. Elapsed time to detect vehicle license plates on the generated dataset.

In summary, the proposed approach is a real-time method and has a high accuracy to detect Iranian license-plates in still images. For future work, the authors intend to implement the mentioned approach on real roadway camera video frames. A K-Fold Cross Validation process may also be necessary to achieve a better approximation of the performance of the system. On the other hand, we want to generate a larger dataset of Iranian license-plates in various environmental conditions, camera height and angle for train/test purposes.

4 Conclusion

Utilizing machine vision techniques combined with deep learning approaches have attracted huge attention due to its high accuracy and acceptable performance. This paper presented a license-plate detection application as a representation of vehicle presence in an image that works on real condition data with various weather and illumination conditions. The proposed method is based on YOLO v.3 deep learning architecture that provides real-time processing and high accuracy results. Since the process of training needs a large dataset including all possible angles, sizes, and location of license-plate in the image, we generated a dataset of vehicle images with Iranian license-plates. Experimental results showed that this method brings about robust detection results in distinguishing the correct location of license-plates in the images. In this regard, the precision and recall values of the system were 0.979 and 0.972, respectively.

References

1. Zhu, F., Li, Z., Chen, S., Xiong, G.: Parallel Transportation Management and Control System and Its Applications in Building Smart Cities. IEEE Trans. Intell. Transp. Syst. **17**, 1576–1585 (2016). https://doi.org/10.1109/TITS.2015.2506156
2. Zhang, J., Wang, F.Y., Wang, K., et al.: Data-driven intelligent transportation systems: a survey. IEEE Trans. Intell. Transp. Syst. **12**, 1624–1639 (2011). https://doi.org/10.1109/TITS.2011.2158001
3. Bommes, M., Fazekas, A., Volkenhoff, T., Oeser, M.: Video based intelligent transportation systems - state of the art and future development. Transp. Res. Procedia **14**, 4495–4504 (2016). https://doi.org/10.1016/j.trpro.2016.05.372
4. Tian, B., Yao, Q., Gu, Y., et al.: Video processing techniques for traffic flow monitoring: a survey. IEEE Conference on Intelligent Transportation Systems Proceedings, ITSC, pp. 1103–1108 (2011). https://doi.org/10.1109/ITSC.2011.6083125
5. O'Mahony, N., et al.: Deep learning vs. traditional computer vision. In: Arai, K., Kapoor, S. (eds.) CVC 2019. AISC, vol. 943, pp. 128–144. Springer, Cham (2020). https://doi.org/10.1007/978-3-030-17795-9_10
6. Pouyanfar, S., Sadiq, S., Yan, Y., et al.: A survey on deep learning: algorithms, techniques, and applications. ACM Comput. Surv. **51** (2018). https://doi.org/10.1145/3234150
7. Deng, L.: A tutorial survey of architectures, algorithms, and applications for deep learning. APSIPA Trans. Signal Inf. Process. **3** (2014). https://doi.org/10.1017/atsip.2013.9
8. Hatt, M., Parmar, C., Qi, J., El Naqa, I.: Machine (deep) learning methods for image processing and radiomics. IEEE Trans. Radiat. Plasma Med. Sci. **3**, 104–108 (2019). https://doi.org/10.1109/trpms.2019.2899538
9. Zeiler, M.D., Fergus, R.: Visualizing and Understanding Convolutional Neural Networks, pp. 1–9 (2012)
10. Aloysius, N., Geetha, M.: A review on deep convolutional neural networks. In: Proceedings of the 2017 IEEE International Conference on Communication and Signal Processing ICCSP 2017, January 2018, pp. 588–592 (2018). https://doi.org/10.1109/ICCSP.2017.8286426
11. Uijlings, J.R.R., Van De Sande, K.E.A., Gevers, T., Smeulders, A.W.M.: Selective search for object recognition. Int. J. Comput. Vis. **104**, 154–171 (2013). https://doi.org/10.1007/s11263-013-0620-5

12. Girshick, R., Donahue, J., Darrell, T., Malik, J.: Rich feature hierarchies for accurate object detection and semantic segmentation. In: Proceedings of the IEEE Computer Society Conference on Computer Vision and Pattern Recognition, pp. 580–587 (2014). https://doi.org/10.1109/CVPR.2014.81
13. Girshick, R.: Fast R-CNN. In: Proceedings of the IEEE International Conference on Computer Vision, pp. 1440–1448 (2015)
14. Ren, S., He, K., Girshick, R., Sun, J.: Faster R-CNN: towards real-time object detection with region proposal networks. IEEE Trans. Pattern Anal. Mach. Intell. **39**, 1137–1149 (2017). https://doi.org/10.1109/TPAMI.2016.2577031
15. Redmon, J., Divvala, S., Girshick, R., Farhadi, A.: You only look once: unified, real-time object detection. In: Proceedings of the IEEE Computer Society Conference on Computer Vision and Pattern Recognition, December 2016, pp. 779–788 (2016). https://doi.org/10.1109/CVPR.2016.91
16. Redmon, J., Farhadi, A.: YOLO9000: better, faster, stronger. In: Proceedings of the 30th IEEE Conference on Computer Vision and Pattern Recognition, CVPR 2017, January 2017, pp. 6517–6525 (2017). https://doi.org/10.1109/CVPR.2017.690
17. Redmon, J., Farhadi, A.: YOLOv3: an incremental improvement (2018). https://arxiv.org/abs/1804.02767
18. Shin, J.S., Kim, U.T., Lee, D.K., et al.: Real-time vehicle detection using deep learning scheme on embedded system. In: International Conference on Ubiquitous and Future Networks, ICUFN, Milan, pp 272–274 (2017)
19. Huval, B., Wang, T., Tandon, S., et al.: An empirical evaluation of deep learning on highway driving (2015). http://arxiv.org/abs/1504.01716
20. Wang, J.G., Zhou, L., Pan, Y., et al.: Appearance-based brake-lights recognition using deep learning and vehicle detection. In: IEEE Intelligent Vehicles Symposium Proceedings, August 2016, pp. 815–820 (2016). https://doi.org/10.1109/IVS.2016.7535481
21. Hsu, S.C., Huang, C.L., Chuang, C.H.: Vehicle detection using simplified fast R-CNN. In: 2018 International Workshop on Advanced Image Technology, IWAIT 2018, pp. 1–3 (2018). https://doi.org/10.1109/IWAIT.2018.8369767
22. Zhang, Q., Wan, C., Han, W.: A modified faster region-based convolutional neural network approach for improved vehicle detection performance. Multimedia Tools Appl. **78**(20), 29431–29446 (2018). https://doi.org/10.1007/s11042-018-6769-8
23. Wang, L., Liao, J., Xu, C.: Vehicle detection based on drone images with the improved faster R-CNN. In: ACM International Conference Proceeding Series Part F 148150, pp. 466–471 (2019). https://doi.org/10.1145/3318299.3318383
24. Kim, S.G., Jeon, H.G., Koo, H.I.: Deep-learning-based license plate detection method using vehicle region extraction. Electron. Lett. **53**, 1034–1036 (2017). https://doi.org/10.1049/el.2017.1373
25. Selmi, Z., Ben Halima, M., Alimi, A.M.: Deep learning system for automatic license plate detection and recognition. In: Proceedings of the International Conference on Document Analysis and Recognition, ICDAR, vol. 1, pp. 1132–1138 (2018). https://doi.org/10.1109/ICDAR.2017.187
26. Silva, S.M., Jung, C.R.: Real-time Brazilian license plate detection and recognition using deep convolutional neural networks. In: Proceedings of the 30th Conference on Graphics, Patterns and Images, SIBGRAPI 2017, pp. 55–62 (2017). https://doi.org/10.1109/SIBGRAPI.2017.14
27. Abdullah, S., Mahedi Hasan, M., Muhammad Saiful Islam, S.: YOLO-based three-stage network for Bangla license plate recognition in Dhaka metropolitan city. In: 2018 International Conference on Bangla Speech and Language Processing ICBSLP (2018). https://doi.org/10.1109/ICBSLP.2018.8554668
28. Puarungroj, W., Boonsirisumpun, N.: Thai license plate recognition based on deep learning. Procedia Comput. Sci. **135**, 214–221 (2018). https://doi.org/10.1016/j.procs.2018.08.168

29. Tourani, A., Soroori, S., Shahbahrami, A., et al.: A robust vehicle detection approach based on faster R-CNN algorithm. In: 4th International Conference on Pattern Recognition and Image Analysis IPRIA 2019, pp. 119–123 (2019). https://doi.org/10.1109/PRIA.2019.8785988
30. Dutta, A., Zisserman, A.: The VIA annotation software for images, audio and video. In: MM 2019 – Proceedings of the 27th ACM International Conference on Multimedia, pp. 2276–2279 (2019). https://doi.org/10.1145/3343031.3350535

Computer Vision and Image Processing

Facial Beauty Prediction Using Transfer and Multi-task Learning Techniques

Elham Vahdati$^{(\boxtimes)}$ (iD) and Ching Y. Suen$^{(\boxtimes)}$ (iD)

CENPARMI, Department of Computer Science and Software Engineering,
Concordia University, Montreal, QC H3G 1M8, Canada
g_vahdat@encs.concordia.ca, suen@cse.concordia.ca

Abstract. The objective of facial beauty prediction, which is a significant yet challenging problem in the domains of computer vision and machine learning, is to develop a human-like model that automatically evaluates facial attractiveness. Using deep learning methods to enhance facial beauty prediction is a promising and important area. This study provides a new framework for simultaneous facial attractiveness assessment, gender recognition as well as ethnicity identification using deep Convolutional Neural Networks (CNNs). Specifically, a deep residual network originally trained on massive face datasets is utilized which is capable of learning high-level and robust features from face images. Furthermore, a multi-task learning algorithm that operates on the effective features, exploits the synergy among the tasks. Said differently, a multi-task learning scheme is employed by our model to learn optimal shared features for these correlated tasks in an end-to-end manner. Interestingly, prediction correlation of 0.94 is achieved by our method for the SCUT-FBP5500 benchmark dataset (spanning 5500 facial images), which would certainly support the efficacy of our proposed model. This would also indicate significant improvement in accuracy over the other state-of-the-art methods.

Keywords: Facial attractiveness prediction · Multi-task learning · VGGFace2 · Deep Convolutional Neural Networks

1 Introduction

Facial attractiveness plays a pivotal role in human's social life, and has attracted interests of researchers from diverse fields such as computer science and medicine [1–3]. Moreover, in everyday life many attractiveness assessments would definitely be based on sexual dimorphic properties. This would mean that gender aspects are taken into account for attractiveness judgements. In other words, sexual dimorphism, which is the difference in facial features owing to sex, can impact attractiveness judgements. These sexually dimorphic features will be more evident at the age of puberty for males and females. For instance, large jawbones, thin lips, small eyes, and thick eyebrows are usually characteristics of attractive males, whereas females with beautiful faces usually have narrow jaw, full lips, wide eyes, and thin eyebrows [2]. Recently, automatic analysis

© Springer Nature Switzerland AG 2020
Y. Lu et al. (Eds.): ICPRAI 2020, LNCS 12068, pp. 441–452, 2020.
https://doi.org/10.1007/978-3-030-59830-3_38

of facial beauty has become an emerging research topic, and has led to many studies, which are of great interest in computer vision and machine learning communities. Interestingly, recent advances in the study on computer-based facial beauty assessment are derived from the rapid development of machine learning techniques. Automatic human-like facial attractiveness prediction has many potential applications, such as facial image beautification, aesthetic plastic surgery planning, cosmetic advertising as well as recommendation systems in social networks [1, 3]. Despite considerable progress achieved in face attractiveness computation research, there are some outstanding challenges. The lack of discriminative face representation is considered as one of the key challenges in this facial analysis task. The face representations proposed in the previous studies can be mainly divided into three categories, namely feature-based, holistic as well as hybrid representation [2], among which feature-based approach has grown in popularity. More specifically, geometric features have been extensively investigated in face attractiveness research studies. However, these features would certainly suffer from the insufficient facial representation ability related to attractiveness. In other words, the handcrafted features do not possess the ability to well represent facial aesthetic characteristics, which would definitely impose restrictions on the attractiveness prediction performance, to some extent. Interestingly, the rapid development of deep learning techniques, especially Convolutional Neural Networks (CNNs), enables researchers to build end-to-end computational models which are capable of learning robust and high-level facial representations [4–10]. Another challenge in facial attractiveness computation task is that training data are not extensive, and researchers would be provided with insufficient ground truth. Interestingly, transfer learning using pre-trained networks would certainly provide an effective solution to address the data insufficiency problem.

In an attempt to diminish the difference between the source task and target task (our task) in transfer learning scheme, a very deep architecture (ResNet with 50 layers [11]) first trained on MS-Celeb-1 M dataset with 10 M face images [12] and then fine-tuned on VGGFace2 dataset including 3.31 M face images [13], hereinafter "VGGFace2_ft (ResNet)" is utilized in this study. Our model is an end-to-end framework where both features and beauty predictor are automatically learned.

More recently, Multi-Task Learning (MTL) using CNNs has also been explored by researchers [14–17]. Multi-task learning indicates the ability in learning multiple related tasks in parallel using a shared representation. It is noteworthy that MTL-based methods can predict facial attributes jointly by considering the correlation between these attributes. Inspired by the success of CNNs on many facial analysis problems, an end-to-end attractiveness learning framework is presented in this study. Besides this, motivated by the human aesthetic perception mechanism where the beauty assessment standards differ for males and females, a multi-task learning based facial analysis is proposed to jointly perform gender recognition and facial beauty prediction tasks. It is worth mentioning that multi-task learning scheme would enable us to extend our proposed model to another facial attribute recognition task, namely face ethnicity. Specifically, this model utilizes a multi-task learning scheme to learn optimal shared features for these correlated tasks.

The key contributions of this paper are listed as follows:

To the best of our knowledge, this is the first attempt to employ a very deep residual network pre-trained on massive face datasets, spanning about 13 M face images, in an end-to-end manner for facial beauty prediction, which is capable of automatic learning of high-level facial representations. Interestingly, both features and beauty predictor are learned by this model. Said differently, the hierarchical structure of this deep model would definitely enable us to construct an end-to-end face analysis system where both facial representation as well as the predictors of facial attributes are automatically learnt.

Furthermore, this is the first work where both transfer and multi-task learning techniques are leveraged to investigate how beauty score prediction as a regression problem, gender recognition as well as ethnicity identification tasks, as classification problems, are combined together to enhance the accuracy of attractiveness prediction.

Notably, our model is not peculiar to only these related tasks, but also we can extend our proposed model to other facial attribute recognition tasks such as facial expression, trustworthiness and dominance which would certainly have a significant influence on the attractiveness of a face.

The remainder of the paper is organized as follows. Section 2 presents a review of related work. The proposed deep model is introduced in Sect. 3, including the network architecture, transfer learning as well as multi-task learning scheme. The experimental results are analyzed and discussed in Sect. 4. Finally, Sect. 5 concludes the paper with a brief summary and discussion.

2 Related Work

2.1 Facial Beauty Prediction

In order to represent a face, feature-based approach which has been extensively explored in previous facial attractiveness research studies includes a wide range of features, namely geometric features [6, 18–20], texture [6, 18, 19] as well as color [18, 19]. Moreover, holistic descriptors such as Eigen faces have been employed by some researchers [18]. To construct a face attractiveness predictor, the extracted features have been fed into traditional machine learning algorithms.

Recently, the state-of-the-art deep learning techniques, especially CNNs, have considerably boosted the performance of facial beauty predictors owing to the fact that high-level and discriminative features learned from these techniques are of great benefit for facial attractiveness computation task [4–10, 15, 21]. A CNN model with six convolutional layers was employed by Xie et al. [6] to analyze female facial beauty. The results reported for the six-layer CNN outperformed those for hand-crafted features, namely geometric as well as Gabor features. Xu et al. [7] indicated that superior performance can be achieved by cascaded fine-tuning methodology where several face input channels, namely face image, detail layer image as well as lighting layer image are fed into a six-layer CNN model. Furthermore, Gao et al. [15] constructed a CNN model with six convolutional layers where facial beauty assessment and landmark localization can be conducted simultaneously.

Fan et al. [5] employed a very deep convolutional residual network (ResNet [11]) pre-trained on ImageNet dataset [22] to evaluate face attractiveness. They achieved substantial increase in the performance using label distribution learning (LDL) and feature fusion. Similarly, the work of Liang et al. [8] utilized three CNN models pre-trained on the ImageNet dataset (i.e. AlexNet [23], ResNet-18 [11] and ResNeXt-50 [24]). Likewise, ResNet-18 architecture pre-trained on ImageNet has been utilized in the work of [9] where facial beauty prediction was formulated as both regression and classification problems. Very recently, VGGFace2_ft (ResNet) [13] has been used merely as a feature extractor in the work of [21]. This means the feedforward phase of the network was performed only once to extract facial features. Then, the extracted features were passed to a stacking ensemble model to predict the attractiveness of female facial images.

2.2 Multi-task Face Analysis

The efficacy of Multi-Task Learning (MTL) scheme has been verified on many facial analysis problems [14–17, 25, 26]. The aim of multi-task learning is to simultaneously learn multiple related tasks by utilizing the shared information among them. This would improve the performance of each task. Zhu et al. [14] employed a lightweight CNN with four convolutional layers, which simultaneously performs age and gender recognition tasks. A deep multi-task learning framework called HyperFace was proposed in the work of [16] for joint face detection, landmark localization, pose estimation as well as gender recognition. Gao et al. [15] attempted to build a MTL-based CNN where facial geometric landmark features are encoded and simultaneously employed for facial attractiveness prediction. Furthermore, facial attributes classification using multi-task learning has been explored by some researchers [25, 26].

3   Deep Convolutional Neural Networks

Convolutional Neural Networks (CNNs) have grown in popularity in recent years owing to the wide range of their successful applications.

3.1 Network Architecture and Transfer Learning

Inspired by the success of CNNs on facial analysis tasks such as face detection, face recognition and verification, this study exploits a very deep architecture with 50 layers (50-layer ResNet) which is capable of extracting more discriminative and more-aesthetics-aware facial features. Residual Network (ResNet) developed by He et al. [11] was the winner of ILSVRC 2015 classification task. Interestingly, deep residual networks have achieved superb performance on many challenging image recognition, localization as well as detection tasks, e.g. ImageNet and COCO object detection [11]. ResNets introduce shortcut connections, which make it possible to train very deep networks. In residual networks, gradients can be propagated directly from the loss layer to previous layers by skipping intermediate weight layers, which would have the potential to trigger vanishing or deterioration of the gradient signal. It is worth mentioning that a large amount of labeled data is required to achieve satisfying performance.

However, it is a time consuming and laborious task to collect a large number of anno-tated facial images. Consequently, transfer learning using pre-trained networks would certainly provide an effective solution to address insufficient training data. The main objective of transfer learning is to retain knowledge learnt from past data and employ it effectively for the target task. It is worthwhile leveraging deep architectures pre-trained on large datasets of similar tasks (e.g. face recognition) to diminish the difference between source task and target task in transfer learning methodology. Interestingly, weights from face recognition task possess the ability to encode features peculiar to the face which are of great benefit for our facial attributes analysis task.

As a result, a very deep residual network pre-trained on large-sized face datasets (i.e. MS-Celeb-1 M [12] as well as VGGFace2 [13] datasets with 10 M and 3.31 M face images, respectively), is employed to build an end-to-end facial beauty prediction system in which both face representation and the predictor are learnt simultaneously from input facial images. It is notable that we enlarge our training set using some augmentation techniques such as random rotation and horizontal flipping to mitigate overfitting. Afterwards, we fine-tune the ResNet-50 using the augmented training data. In other words, the pre-trained ResNet is utilized as an initialization in our task. This is followed by fine-tuning the weights of the pre-trained network which can be accomplished by continuing backpropagation.

3.2 Multi-task Learning Scheme

In order to simultaneously perform two correlated tasks, namely beauty score assessment as well as gender recognition, this study exploits the multi-task learning scheme by regarding these two tasks as a regression and classification problem, respectively. Moreover, we extend our model to another facial attribute classification (face ethnicity). In fact, our model employs a deep architecture and jointly optimizes these correlated facial attribute analysis tasks by learning a shared facial representation for both regression and classification problems. Owing to the fact that our multi-task CNN has a single-input and multi-output structure, several loss functions are utilized to optimize the model parameters. As mentioned before, facial beauty prediction, indicating to which extent a face is attractive, is formulated as a regression problem. As a result, Mean Squared Error (MSE) loss function is used for training the beauty score prediction task as in [9].

$$loss_{MSE} = \frac{1}{N} \sum_{i=1}^{N} \left(y_i - \hat{y}_i \right)^2 \tag{1}$$

Where y_i is the ground-truth beauty score of ith face image, and $\hat{y}_i$ is the predicted attractiveness score.

Furthermore, cross entropy loss function is employed for training two other tasks (i.e. gender recognition and ethnicity identification). The cross entropy loss between the output probability vector and the actual class vector, is represented as follows:

$$loss_{CE} = -\frac{1}{N} \sum_{i=1}^{N} \left[y_i \log(p_i) + (1 - y_i) \log(1 - p_i) \right] \tag{2}$$

where N is the batch size, y_i is the ground truth label and p_i is the probability produced by the network. For gender recognition task, if the gender is male, then $y = 0$. Otherwise $y = 1$ for i^{th} facial image. For ethnicity identification task, $y = 0$ and $y = 1$ are the ground truth labels for Asian and Caucasian faces, respectively. It is noteworthy that a two-dimensional probability vector obtained from the network.

Suppose we have T tasks (T = 3 in this work) to be completed. Therefore, the total loss is computed as the weighted sum of these three individual losses as shown below.

$$loss_{total} = \sum_{t=1}^{T} \lambda_t \, loss_t \tag{3}$$

The weight parameter λ is determined based on the importance of the task in the total loss. We set $\lambda_1 = 0.2, \lambda_2 = 0.8$ for joint gender recognition and beauty score prediction tasks, respectively. Furthermore, we selected $\lambda_1 = 0.2, \lambda_2 = 0.2, \lambda_3 = 0.6$ to simultaneously perform gender recognition, ethnicity identification as well as attractiveness assessment, respectively. It is notable that higher weights are assigned to attractiveness computation task since it is regarded as our original task.

As illustrated in Fig. 1, the proposed model which has a 1-input and 3-output structure, takes RGB facial images (224 × 224 × 3) as input while it outputs the prediction results of three facial attributes in three branches. The proposed framework is composed of two main steps, namely shared feature extraction as well as multi-task estimation (see Fig. 1). It is notable that FC denotes fully connected layer. Furthermore, M/F indicates male/female whereas A/C denotes Asian/Caucasian.

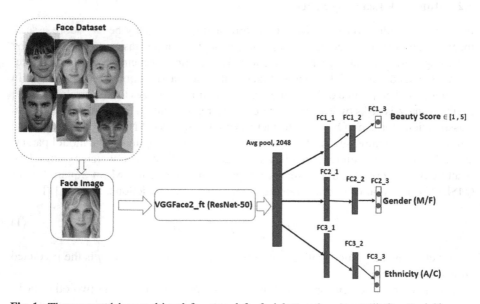

Fig. 1. The proposed deep multi-task framework for facial attractiveness prediction. Facial images were obtained from the SCUT-FBP5500 dataset [8].

As mentioned before, we employ ResNet-50 architecture where the last layer of ResNet-50 [11] is replaced by three branches which is followed by three fully connected layers. Specifically, built upon the global average pooling layer (D = 2048), the output of our model is composed of three prediction results for beauty, gender and ethnicity, respectively. In each branch, the first two fully connected layers include 256 and 64 neurons, respectively. However, the last fully connected layers contain one, two and two neurons for beauty, gender and ethnicity, respectively. It is notable that the dropout technique (with a probability of 0.5) [27] is applied on these fully-connected layers.

4   Experimental Results

4.1   Experimental Setup

The experiments are conducted on the deep learning platform of PyTorch [28]. We initialize the network weights with a public pre-trained CNN model for face recognition task (i.e. a ResNet-50 model trained on MS-Celeb-1 M and then fine-tuned on VGGFace2 dataset) [13]. Furthermore, we use the Stochastic Gradient Descent (SGD) method with the back propagation algorithm to train the network for 100 epochs. Therefore, the entire attractiveness learning framework is end-to-end. Moreover, we fix the batch size to 16. The initial learning rate is 0.001 which is divided by 10 per 50 epochs. In this study, the SCUT-FBP5500 dataset [8] which provides 5500 facial images of different gender and races (Asian and Caucasian), is employed to assess the performance of our proposed method. Figure 2 illustrates some face images of the SCUT-FBP5500 dataset [8]. Each face image was rated by 60 human raters on a 5-point scale, where 1 means most unattractive and 5 means extremely attractive. The ground-truth score is the average of the 60 ratings.

Fig. 2. Examples of facial images from the SCUT-FBP5500 dataset [8].

To achieve more robust results, we conduct the experiments using 5-fold cross validation technique as in [8], where 80% of the facial images are selected as the training set, and the remaining 20% of face images as the testing set. To assess the efficacy of the automatic rater, several prediction metrics, i.e. Pearson Correlation (PC) [3], Root Mean Squared Error (RMSE) and Mean Absolute Error (MAE), are leveraged in this study. It is notable that more accurate predictions would lead to higher values of Pearson correlation as well as smaller errors. The formulas of the three prediction metrics are given below.

$$r = \frac{\sum_{i=1}^{n}(x_l - \bar{x})(y_l - \bar{y})}{\left[\sum_{i=1}^{n}(x_i - \bar{x})^2 \sum_{i=1}^{n}(y_i - \bar{y})^2\right]^{\frac{1}{2}}} \tag{4}$$

$$MAE = \frac{1}{n}\sum_{i=1}^{n}|y_i - x_i| \tag{5}$$

$$RMSE = \sqrt{\frac{1}{n}\sum_{i=1}^{n}(y_i - x_i)^2} \tag{6}$$

The Pearson Correlation of $(x_1, x_2, \ldots, x_n)$ and $(y_1, y_2, \ldots, y_n)$ is computed using (4), where $x_1, x_2, \ldots, x_n$ are the ground-truth scores and $y_1, y_2, \ldots, y_n$ are the predicted scores by our model. Furthermore, $\bar{x}$ and $\bar{y}$ denote the mean values of $(x_1, x_2, \ldots, x_n)$ and $(y_1, y_2, \ldots, y_n)$, respectively. It is noteworthy that "r" ranges from -1 to 1, where the strongest positive linear correlation is indicated by "1".

4.2 Experiments on the SCUT-FBP5500 Benchmark Dataset

Information in Table 1 would enable us to obtain deeper insights into performance differences between single-task and multi-task models. The results of 5-fold cross validation for facial attractiveness prediction task in terms of PC, MAE and RMSE are indicated in Table 1. The results support the effectiveness of the employed multi-task schemes for facial attractiveness prediction. Said differently, multi-task learning would enable our model to generalize better on the original task i.e. beauty prediction task. The attractiveness computation, is first formulated as a regression problem, and the obtained average Pearson Correlation of 0.9297 justifies that predicted beauty scores by the single-task deep model are highly correlated with the ground truth from human raters. Furthermore, a higher accuracy is achieved by our deep model that jointly perform attractiveness assessment and gender recognition (Pearson Correlation of 0.9338). Interestingly, the highest correlation of 0.9372, minimum MAE of 0.1833 and RMSE of 0.2424 reported by the deep multi-task model which predicts three facial attributes (i.e. beauty, gender and ethnicity), reinforce the advantage of learning optimal shared features for these related tasks.

4.3 Comparison with the State of the Art on the SCUT-FBP5500 Benchmark Dataset

To verify the robustness of our method, we compare our proposed method with recent studies on the same SCUT-FBP5500 dataset. A summary of facial attractiveness prediction methods as well as their accuracy in existing works in terms of PC, MAE and RMSE are presented in Table 2.

To the best of our knowledge, there are only two studies [8, 10] that have reported performance on the benchmark SCUT-FBP5500 dataset for attractiveness score prediction. Geometric and Gabor features as well as traditional machine learning techniques have been utilized by Liang et al. [8] (the first four rows). Moreover, they employed three deep

Table 1. Comparison between the results of single-task and multi-task models using 5-fold cross validation.

PC	1	2	3	4	5	Average
Beauty	0.9273	0.9299	0.9331	0.9295	0.9287	0.9297
Beauty, Gender	0.9297	0.9333	0.9330	0.9358	0.9371	0.9338
Beauty, Gender, Ethnicity	**0.9327**	**0.9360**	**0.9379**	**0.9403**	**0.9390**	**0.9372**
MAE	1	2	3	4	5	Average
Beauty	0.1974	0.1917	0.1944	0.1965	0.1941	0.1948
Beauty, Gender	0.1951	0.1879	0.1944	0.1865	0.1847	0.1897
Beauty, Gender, Ethnicity	**0.1873**	**0.1828**	**0.1812**	**0.1821**	**0.1829**	**0.1833**
RMSE	1	2	3	4	5	Average
Beauty	0.2570	0.2556	0.2549	0.2578	0.2554	0.2561
Beauty, Gender	0.2603	0.2477	0.2605	0.2442	0.2421	0.2510
Beauty, Gender, Ethnicity	**0.2481**	**0.2441**	**0.2448**	**0.2363**	**0.2387**	**0.2424**

Table 2. Performance comparison with the related state-of-the-art works on the SCUT-FBP5500 dataset.

Method	PC	MAE	RMSE
Geometric Features + LR [8]	0.5948	0.4289	0.5531
Geometric Features + GR [8]	0.6738	0.3914	0.5085
Geometric Features + SVR [8]	0.6668	0.3898	0.5132
Gabor + SVR [8]	0.8065	0.3976	0.5126
AlexNet [8]	0.8634	0.2651	0.3481
ResNet-18 [8]	0.8900	0.2419	0.3166
ResNeXt-50 [8]	0.8997	0.2291	0.3017
MobileNetV2 [10]	0.926	0.202	0.266
Ours	**0.9372**	**0.1833**	**0.2424**

networks pre-trained on the ImageNet dataset [22], namely AlexNet [23], ResNet-18 [11] and ResNext-50 [24]. Furthermore, MobileNetV2 network [29] has been utilized in the work of [10]. Compared to the existing works, our work achieves a higher accuracy. There are two key factors which are involved in this superb performance. In fact, our method gains benefit from transfer learning methodology as well as multi-task learning scheme. Good initialization derived from a very deep architecture pre-trained on massive face datasets as well as leveraging the optimal shared information among related tasks, would definitely make our proposed method successful.

5   Conclusion

To sum up, this paper presents a deep multi-task learning method where a very deep CNN pre-trained on massive face datasets is leveraged to acquire high-level and robust features of face images, in an end-to-end manner. Furthermore, we investigate how beauty score prediction which is considered as a regression problem can be optimized by other facial attributes classification problems, namely gender and ethnicity prediction. The experimental results on the SCUT-FBP5500 benchmark dataset clearly demonstrate that the proposed method achieves better performance than the state-of-the-art methods. Thanks to the proposed multi-task learning scheme, our model has the ability to be easily extended to other facial attributes which exert a significant influence on attractiveness of a face such as trustworthiness.

Acknowledgment. This research was supported by a research grant from NSERC, the Natural Sciences and Engineering Research Council of Canada.

References

1. Laurentini, A., Bottino, A.: Computer analysis of face beauty: a survey. Comput. Vis. Image Underst. **125**, 184–199 (2014)
2. Liu, S., Fan, Y.-Y., Samal, A., Guo, Z.: Advances in computational facial attractiveness methods. Multimedia Tools Appl. **75**(23), 16633–16663 (2016). https://doi.org/10.1007/s11042-016-3830-3
3. Zhang, D., Chen, F., Xu, Y.: Computer Models for Facial Beauty Analysis. Springer, Cham (2016). https://doi.org/10.1007/978-3-319-32598-9_14
4. Lin, L., Liang, L., Jin, L.: R2-ResNeXt: a ResNeXt-based regression model with relative ranking for facial beauty prediction. In: 2018 24th International Conference on Pattern Recognition (ICPR), pp. 85–90. IEEE, Beijing (2018)
5. Fan, Y.-Y., Liu, S., Li, B., Guo, Z., Samal, A., Wan, J.: Label distribution-based facial attractiveness computation by deep residual learning. IEEE Trans. Multimed. **20**(8), 2196–2208 (2018)
6. Xie, D., Liang, L., Jin, L., Xu, J., Li, M.: SCUT-FBP: a benchmark dataset for facial beauty perception. In: 2015 IEEE International Conference on Systems, Man, and Cybernetics, pp. 1821–1826. IEEE, Kowloon (2015)
7. Xu, J., Jin, L., Liang, L., Feng, Z., Xie, D.: A new humanlike facial attractiveness predictor with cascaded fine-tuning deep learning model (2015). arXiv preprint arXiv:1511.02465

8. Liang, L., Lin, L., Jin, L., Xie, D., Li, M.: SCUT-FBP5500: a diverse benchmark dataset for multi-paradigm facial beauty prediction. In: 2018 24th International Conference on Pattern Recognition (ICPR), pp. 1598–1603. IEEE, Beijing (2018)
9. Xu, L., Xiang, J., Yuan, X.: CRNet: classification and regression neural network for facial beauty prediction. In: Hong, R., Cheng, W.-H., Yamasaki, T., Wang, M., Ngo, C.-W. (eds.) PCM 2018. LNCS, vol. 11166, pp. 661–671. Springer, Cham (2018). https://doi.org/10.1007/978-3-030-00764-5_61
10. Shi, S., Gao, F., Meng, X., Xu, X., Zhu, J.: Improving facial attractiveness prediction via co-attention learning. In: IEEE International Conference on Acoustics, Speech and Signal Processing (ICASSP), pp. 4045–4049. IEEE, Brighton (2019)
11. He, K., Zhang, X., Ren, S., Sun, J.: Deep residual learning for image recognition. In: 2016 IEEE Conference on Computer Vision and Pattern Recognition (CVPR), pp. 770–778. IEEE, Las Vegas (2016)
12. Guo, Y., Zhang, L., Hu, Y., He, X., Gao. J.: Ms-celeb-1 m: a dataset and benchmark for large-scale face recognition (2016). arXiv preprint arXiv:1607.08221
13. Cao, Q., Shen, L., Xie, W., Parkhi, O.M., Zisserman, A.: VGGFace2: a dataset for recognising faces across pose and age. In: 2018 13th IEEE International Conference on Automatic Face & Gesture Recognition (FG 2018), pp. 67–74. IEEE, Xi'an (2018)
14. Zhu, L., Wang, K., Lin, L., Zhang, L.: Learning a lightweight deep convolutional network for joint age and gender recognition. In: 23rd International Conference on Pattern Recognition (ICPR), pp. 3282–3287. IEEE, Cancún Center (2016)
15. Gao, L., Li, W., Huang, Z., Huang, D., Wang, Y.: Automatic facial attractiveness prediction by deep multi-task learning. In: IEEE 24th International Conference on Pattern Recognition (ICPR), pp. 3592–3597. IEEE, Beijing (2018)
16. Ranjan, R., Patel, V.M., Chellappa, R.: HyperFace: a deep multi-task learning framework for face detection, landmark localization, pose estimation, and gender recognition. IEEE Trans. Pattern Anal. Mach. Intell. **41**(1), 121–135 (2019)
17. Zhang, Z., Luo, P., Loy, C. C., Tang, X.: Facial landmark detection by deep multitask learning. In: 13th European Conference Computer Vision (ECCV 2014), Zurich, Switzerland, pp. 94–108 (2014)
18. Eisenthal, Y., Dror, G., Ruppin, E.: Facial attractiveness: beauty and the machine. Neural Comput. **18**(1), 119–142 (2006)
19. Kagian, A., Dror, G., Leyvand, T., Meilijson, I., Cohen-Or, D., Ruppin, E.: A machine learning predictor of facial attractiveness revealing human-like psychophysical biases. Vis. Res. **48**, 235–243 (2008)
20. Vahdati, E., Suen, C.Y.: A novel female facial beauty predictor. In: International Conference on Pattern Recognition and Artificial Intelligence (ICPRAI), Montreal, pp. 378–382 (2018)
21. Vahdati, E., Suen, C.Y.: Female facial beauty analysis using transfer learning and stacking ensemble model. In: Karray, F., Campilho, A., Yu, A. (eds.) ICIAR 2019. LNCS, vol. 11663, pp. 255–268. Springer, Cham (2019). https://doi.org/10.1007/978-3-030-27272-2_22
22. Deng, J., Dong, W., Socher, R., Li, L.-J., Li, K., Fei-Fei, L.: Imagenet: a large-scale hierarchical image database. In: 2009 IEEE Conference on Computer Vision and Pattern Recognition, pp. 248–255. IEEE, Miami (2009)
23. Krizhevsky, A., Sutskever, I., Hinton, G.E.: Imagenet classification with deep convolutional neural networks. In: Advances in Neural Information Processing Systems (NIPS), pp. 1097–1105 (2012)
24. Xie, S., Girshick, R., Doll'ar, P., Tu, Z., He, K.: Aggregated residual transformations for deep neural networks. In: IEEE International Conference Computer Vision and Pattern Recognition, Honolulu, HI, USA, pp. 5987–5995 (2017)

25. Ehrlich, M., Shields, T.J., Almaev, T., Amer, M.R.: Facial attributes classification using multi-task representation learning: In: IEEE International Conference Computer Vision Pattern Recognition Workshops (CVPRW), Las Vegas, NV, USA, pp. 752–760 (2016)
26. Hand, E.M., Chellappa, R.: Attributes for improved attributes: a multi-task network utilizing implicit and explicit relationships for facial attribute classification. In: Thirty-First AAAI Conference Artificial Intelligence (AAAI-17), San Francisco, California, USA, pp. 4068–4074 (2017)
27. Srivastava, N., Hinton, G., Krizhevsky, A., Sutskever, I., Salakhutdinov, R.: Dropout: a simple way to prevent neural networks from overfitting. J. Mach. Learn. Res. **15**(1), 1929–1958 (2014)
28. Paszke, A., et al.: Automatic differentiation in PyTorch. In: 31st Conference Neural Inform. Processing System (NIPS 2017), Long Beach, CA, USA (2017)
29. Sandler, M., Howard, A., Zhu, M., Zhmoginov, A., Chen, L. C.: MobileNetV2: Inverted residuals and linear bottlenecks (2018). arXiv preprint arXiv:1801.04381v3

CycleGAN-Based Image Translation for Near-Infrared Camera-Trap Image Recognition

Renwu Gao$^{(\boxtimes)}$, Siting Zheng , Jia He , and Linlin Shen

Shenzhen University, Shenzhen, China
{re.gao,llshen}@szu.edu.cn, m13250714546@163.com, hejia@mail.szu.edu.cn

Abstract. Due to its invisibility, NIR (Near-infrared) flash has been widely used to capture the images of wild animals in the night. Although the animals can be captured without notice, the gray NIR images are short of color and texture information and thus is difficult to analyze, for both human and machine. In this paper, we propose to use Cycle-GAN (Generative Adversarial Networks) to translate NIR image to the incandescent domain for visual quality enhancement. Example translations show that both color and texture can be well recovered by the proposed CycleGAN model. The recognition performance of a SSD based detector on the translated incandescent images is also significantly better than that on the original NIR images. Taking Wildebeest and Zebra for example, an increase of 16% and 8% in recognition accuracy has been observed.

Keywords: Generative adversarial networks · Deep learning · Image translation

Capturing the locations of wild animals within a monitored area is very important for ecologists and wild animals protection volunteers to understand the complexity of natural ecosystems, study the survival and migration of species, save wild animals in danger and protect the ecological environment. Cameras as the image capture devices naturally have been introduced for wild animal location capturing. In the field of ecology, a number of cameras are deployed at appropriate positions with intervals in the area to be monitored. To avoid the interference of human beings to the ecosystem (especially to wild animals), trigger mechanisms, such as motion sensors, infrared detector or other light beams are introduced. These image-capturing devices are known as "Camera-trap". With the help of Camera-trap, ecologists and wild animal protection volunteers can study and monitor the living conditions and behavioral characteristics and populations of wild animals by analyzing the captured images.

Camera-trap are designed to work at 24 h, from day light to night. However, at night, there is not enough light for camera to capture images with good quality. To solve this problem, incandescent flash and near-infrared (NIR) flash

© Springer Nature Switzerland AG 2020
Y. Lu et al. (Eds.): ICPRAI 2020, LNCS 12068, pp. 453–464, 2020.
https://doi.org/10.1007/978-3-030-59830-3_39

are introduced. These equipment can emit either visible or NIR illuminations when it is dark in the night. Taking the current largest camera-trap project, Snapshot Serengeti [15] for example, it consists a lot of images captured with SG565 incandescent flash camera and DLC Convert 11 NIR flash camera. While some cameras were equipped with both incandescent and NIR flash, others are equipped with NIR flash only. Figure 1 shows the example images captured with natural light, incandescent flash and NIR flash.

Although incandescent flash and NIR flash can help capture animals during the night, each of them also brings new problems. During the night, the visible light emitted by incandescent flash might scare the wild animal and keep them away from the camera. While the NIR light emitted by NIR flash in invisible to the animals, the gray NIR images contains much less textures and thus have less information than images taken under natural light or incandescent light. As shown in Fig. 1, no color information is available in NIR images and thus it is difficult for both human and computer to recognize the wild animals.

Fig. 1. Images captured with different lights. (a) natural light, (b) incandescent flash, (c) near-infrared flash.

In this paper, we focus on the NIR image recognition problem. For each category of animals, a CycleGAN [5] based model was trained for translation between NIR and incandescent domains. The trained CycleGANs are then applied to translate the NIR images to the incandescent domain to increase the visual

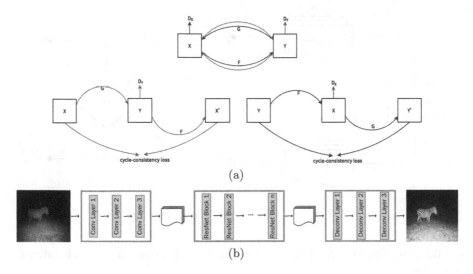

Fig. 2. CycleGAN based image translation. (a) The generators and discriminators. (b) The network of Generator.

quality of the NIR camera-trap images. The SSD based animal detector achieved much better accuracy on the translated NIR images than that on the original ones.

1   Related Works

Computer vision was firstly introduced into ecology field to classify microorganism and zooplankton communities. Later, Ramanan et al. [12] proposed to utilize geometric and texture features for the recognition of giraffe in video. Burghardt et al. [2] introduced Adaboost classifier with Haar-like features to detect and track lions. Swinnen et al. [16] utilized the low-level pixel differences between frames for the detection of wild animals in camera-trap records. Yu et al. [19] employed improved sparse coding spatial pyramid machine (ScSPM) for global and local feature extraction, and used linear support vector machine algorithm to classify images.

In recent years, deep convolutional neural networks (CNN) [8] have been introduced into ecology field as well. Chen et al. [3] constructed a network with 3 convolutional layers and 3 pooling layers to classify wild animals. Gomez et al. [4,18] investigated AlexNet [7], VGG [14], GoogLeNet [17] and ResNet [6] for wild animal recognition. Norouzzadeh et al. [10] surveyed 9 CNN models for identifying, counting and describing wild animals.

The above-mentioned methods do not distinguish NIR images from incandescent and natural light images. NIR images have less features such as color and texture, and are more difficult to classify. We focus on the NIR image recognition problem in this paper. We translate images from NIR domain into incandescent

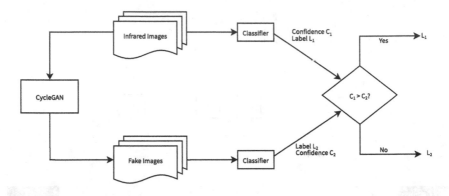

Fig. 3. The proposed NIR image recognition workflow.

domain using CycleGAN, then feed the fake incandescent images to classifier for recognition.

This paper is organized as follows. Since we utilize Snapshot Serengeti dataset for our experiments, we give brief introduction of them in Sect. 2. In Sect. 3, our proposal for NIR image translation and recognition are introduced, and the experimental results are shown in Sect. 4. Finally, Sect. 5 concludes our work.

2   Snapshot Serengeti Dataset

In 2010, Swanson et al. [15] started Snapshot Serengeti project with the purpose of evaluating spatial and temporal inter-species dynamics in Serengeti National Park, Tanzania. 225 camera traps have been deployed across $1,125\,\mathrm{km}^2$ to capture images. By 2013, total $322,653$ images that contain wild animals have been produced, and 48 different species are identified. Such a large set of images is called Snapshot Serengeti dataset.

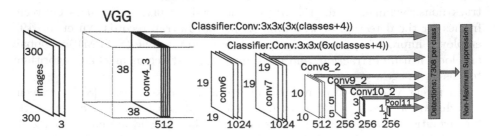

Fig. 4. SSD network.

As shown in Fig. 1, there are three types of images in Snapshot Serengeti dataset: natural light images, incandescent flash images and NIR flash images. Natural light images are captured in the daytime, while incandescent flash images and NIR flash images are captured in the night.

3 CycleGAN Based NIR Image Translation and Recognition

3.1 CycleGAN

In 2014, Ian Goodfellow et al. proposed generative adversarial networks (GAN) [5]. GAN consists of two subnets: a generator G that captures the distribution of the input, and a discriminator D that distinguishes whether the input is a real image or a fake image generated by G.

Lots of varieties have been proposed after then. Radford et al. introduced deep convolutional neural networks to construct generator and discriminator and proposed DCGAN [11]. Miraz et al. added extra information (usually label) to guide the generation and proposed conditional GAN [9]. Arjovsky et al. replaced the cross entropy with Wasserstein distance as loss and proposed wGAN [1].

In 2017, Zhu et al. introduced additional inverse mapping to prevent the generated images from mode collapse (e.g., different inputs generate similar images) [20]. These two maps constructed a cycle (therefore "CycleGAN"). Figure 2(a) shows the structure of CycleGAN. It contains two generative networks, G that maps input from domain X to domain Y, and F mapping input from domain Y to domain X. Generative network consists of three parts: encoder, translator and decoder, as shown in Fig. 2(b). The encoder encodes input to the feature space, and the translator translates features from one domain into the other while decoder decodes the translated features to image.

3.2 NIR Image Translation

There are three types of images in Snapshot Serengeti dataset: the natural light images captured during the day, and incandescent flash images and NIR flash images captured during the night. As the NIR and incandescent images are both captured at night, they have more commons. For example, they share the similar dark sky. Translating images from infrared domain into natural light domain may bring noises for the dark sky, and it is better to keep the them as they are. Therefore, we translate images from infrared domain into incandescent domain.

We employ CycleGAN for the translation. For each species, a CycleGAN is trained with images containing that species from both infrared and incandescent domain. Then fake incandescent images can be generated by feeding NIR images to the generator of CycleGAN.

3.3 SSD Based Wild Animal Recognition

A SSD based wild animal detector was used in this paper to recognize the wild animals captured in the NIR camera-trap images. Figure 4 shows the structure of SSD network for object detection. It utilizes VGG (without fully connected layer) for feature extraction and detects objects in multiply scale feature maps

Fig. 5. Fake images translated by Wildebeest CycleGAN. Columns from left to right are the input images, fake incandescent images and example incandescent images captured under incandescent flash.

by convolution operation. Detections of each scale are filtered by non-maximum suppression.

The SSD object detector is trained with Gold Standard Snapshot Serengeti Bounding Boxes Coordinates (GSBBC) [13] dataset, which is a subset of Snap-

Fig. 6. Fake images translated by Zebra CycleGAN. Columns from left to right are the input images, fake incandescent images and example incandescent images captured under incandescent flash.

shot Serengeti dataset. Image size was set as 300×300 and batch size was set as 32. SGD optimizer with momentum of 0.9 and learning rate of 0.001 was used to train SSD.

(a)

(b)

Fig. 7. Fake images translated by different CycleGAN models. (a) Fake Wildebeest images translated by Zebra CycleGAN model. (b) Fake Zebra images translated by Wildebeest CycleGAN.

After translating NIR images into incandescent domain, they are fed to SSD for recognition. To obtain better performance, both the NIR and the translated incandescent images can be fed to classifier at the same time. Classification with higher confidence is considered to be the final result. Refer to Fig. 3 for details about the fusion process.

Table 1. Number of images for training each translation model.

CycleGAN model	NIR	Incandescent
Wildebeest	2172	9142
Zebra	4799	11279

4   Experimental Results

4.1   Translation Results

We use Wildebeest and Zebra as the animals for testing in this paper. Two CycleGAN models, i.e., Wildebeest CycleGAN and Zebra CycleGAN, are trained independently using their NIR and incandescent images captured in the night. Table 1 lists the number of images used to train each of the CycleGAN models.

The size of input images was set as 256 × 256. Adam optimizer with initial learning rate of 0.0002 was used to train the CycleGAN. To speed up the train procedure, learning rate is decayed per each 50 epochs.

Figure 5 and Fig. 6 show the qualitative results of each CycleGAN translating images from infrared domain into incandescent domain. Columns from left to right are: input images, fake incandescent images and examples of real incandescent images (i.e., images taken with incandescent flash). Comparing the fake incandescent images with the original incandescent images we can conclude that, the NIR images have been colorized properly. Also the texture of the wild animals are correctly translated.

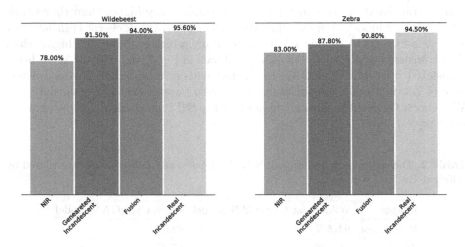

Fig. 8. The recognition accuracy of Wildebeest (left) and Zebra (right) for NIR, generated incandescent, fusion of NIR and generated incandescent, and real incandescent images.

Interesting results can be obtained if the NIR Wildebeest images were translated using the Zebra CycleGAN model, or vice versa. Figure 7(a) shows the example Wildebeest images translated with Zebra CycleGAN model. One can observe that the body of Wildebeest has been covered with a number of stripes, which is a distinct characteristic of Zebra. On the other hand, while the Zebra translated by Wildebeest CycleGAN (Fig. 7(b)) still preserves the stripes, the color of them have been translated from black and white to that of the Wildebeest.

4.2   Recognition Results

We now start to test the performance of SSD on the images translated by our CycleGAN models. We first tested the performance of SSD using 200 NIR Wildebeest and 500 NIR Zebra images, and then compared their performances with the same number of fake incandescent images translated by CycleGAN models.

As shown in Fig. 8, the recognition performances for Wildebeest and Zebra have been improved from 78.0% to 91.5%, and from 83.0% to 87.8%, respectively. By fusing the predictions from both NIR and the translated incandescent images, the accuracy can be further increased to 94.0% and 90.8%.

We also show the performance of SSD on 1000 real incandescent images for both Wildebeest and Zebra in Fig. 8. It seems that the accuracy of real incandescent images are still higher. However, the accuracy (94.0%) of our approaches for NIR Zebra images has now been very close to that (95.6%) of real incandescent images.

In addition, we tested the performance on Wildebeest images translated by Zebra CycleGAN, and vice versa. As listed in Table 2, the accuracies of NIR images translated by their own models are significantly higher than that translated by other models. For example, the NIR images of Wildebeest translated by Wildebeest CycleGAN achieved 91.5% accuracy, which is about 50% higher than that translated by Zebra CycleGAN. As shown in Fig. 7, the body of Wildebeest translated by Zebra CycleGAN is painted with stripes, which is a characteristic of Zebra. Similar circumstance might also happen for Zebra translated by Wildebeest CycleGAN. This might confuse the SSD and result in low recognition accuracy.

Table 2. The recognition accuracy of NIR Wildebeest and Zebra images translated by different CycleGAN models.

Images	Wildebeest CycleGAN model	Zebra CycleGAN model
Wildebeest	**91.5%**	42.0%
Zebra	21.80%	**87.8%**

5 Conclusion

In this paper, we proposed to translate NIR image into incandescent domain for wild animal recognition. Firstly, CycleGAN was used for the translation. Our tests on image translation (Fig. 5, Fig. 6 and Fig. 7) show that, CycleGAN learned the characteristics (e.g., stripes for Zebra) of different wild animals, and their mappings between infrared and incandescent domain. When translating NIR images, CycleGAN recovered colors and textures and generated more informative images. Then SSD trained by natural light images was used for the image recognition. Our experiments showed that, accuracy of SSD on the translated images is much higher than that of NIR images. And the performance could be further improved by fusing the predictions from both NIR and the translated incandescent images.

Translating NIR images to natural light domain is more difficult as discussed in Sect. 3.2. As detectors are generally trained with natural light images (such as SSD in our experiment), we can expect better recognition performance if NIR images are translated into natural light domain. Therefore, translating NIR images to natural light images will be one of our future works.

References

1. Arjovsky, M., Bottou, L.: Towards Principled Methods for Training Generative Adversarial Networks, January 2017. http://arxiv.org/abs/1701.04862
2. Burghardt, T., Calic, J., Thomas, B.T.: Tracking animals in wildlife videos using face detection. In: EWIMT (2004)
3. Chen, G., Han, T.X., He, Z., Kays, R., Forrester, T.: Deep convolutional neural network based species recognition for wild animal monitoring. In: 2014 IEEE International Conference on Image Processing (ICIP), pp. 858–862, October 2014. https://doi.org/10.1109/ICIP.2014.7025172
4. Gomez, A., Diez, G., Salazar, A., Diaz, A.: Animal identification in low quality camera-trap images using very deep convolutional neural networks and confidence thresholds. In: Bebis, G., et al. (eds.) ISVC 2016. LNCS, vol. 10072, pp. 747–756. Springer, Cham (2016). https://doi.org/10.1007/978-3-319-50835-1_67
5. Goodfellow, I., et al.: Generative adversarial nets. In: Advances in Neural Information Processing Systems 27 (2014)
6. He, K., Zhang, X., Ren, S., Sun, J.: Deep residual learning for image recognition, pp. 770–778, June 2016. https://doi.org/10.1109/CVPR.2016.90
7. Krizhevsky, A., Sutskever, I., Hinton, G.E.: ImageNet classification with deep convolutional neural networks. In: Proceedings of the 25th International Conference on Neural Information Processing Systems - Volume 1, NIPS 2012, pp. 1097–1105. Curran Associates Inc., USA (2012). http://dl.acm.org/citation.cfm?id=2999134.2999257
8. LeCun, Y., Bottou, L., Bengio, Y., Patrick, H.: Gradient-based learning applied to document recognition. Proc. IEEE **45** (1998). https://doi.org/10.1007/BF00774006
9. Mirza, M., Osindero, S.: Conditional Generative Adversarial Nets. arXiv:1411.1784 (2014)
10. Norouzzadeh, M.S., et al.: Automatically identifying, counting, and describing wild animals in camera-trap images with deep learning. Proc. Natl. Acad. Sci. **115**(25), E5716–E5725 (2018). https://doi.org/10.1073/pnas.1719367115. https://www.pnas.org/content/115/25/E5716
11. Radford, A., Metz, L., Chintala, S.: Unsupervised representation learning with deep convolutional GANs. In: International Conference on Learning Representations (2016). https://doi.org/10.1051/0004-6361/201527329
12. Ramanan, D., Forsyth, D.A., Barnard, K.: Detecting, localizing and recovering kinematics of textured animals. In: Proceedings of the 2005 IEEE Computer Society Conference on Computer Vision and Pattern Recognition (CVPR 2005) - Volume 2 - Volume 02, CVPR 2005, pp. 635–642. IEEE Computer Society, Washington, D.C. (2005). https://doi.org/10.1109/CVPR.2005.126
13. Schneider, S., Taylor, G.W., Kremer, S.: Deep learning object detection methods for ecological camera trap data, pp. 321–328, May 2018. https://doi.org/10.1109/CRV.2018.00052
14. Simonyan, K., Zisserman, A.: Very deep convolutional networks for large-scale image recognition. arXiv:1409.1556, September 2014

15. Swanson, A., Kosmala, M., Lintott, C., Simpson, R., Smith, A., Packer, C.: Snapshot Serengeti, high-frequency annotated camera trap images of 40 mammalian species in an African savanna. Sci. Data **2**, 150026 EP (2015). https://doi.org/10.1038/sdata.2015.26, Data Descriptor

16. Swinnen, K., Reijniers, J., Breno, M., Leirs, H.: A novel method to reduce time investment when processing videos from camera trap studies. PLoS ONE **9**, e98881 (2014). https://doi.org/10.1371/journal.pone.0098881

17. Szegedy, C., et al.: Going deeper with convolutions. In: 2015 IEEE Conference on Computer Vision and Pattern Recognition (CVPR), pp. 1–9, June 2015. https://doi.org/10.1109/CVPR.2015.7298594

18. Villa, A.G., Salazar, A., Vargas, F.: Towards automatic wild animal monitoring: identification of animal species in camera-trap images using very deep convolutional neural networks. Ecol. Inform. **41**, 24–32 (2017). https://doi.org/10.1016/j.ecoinf.2017.07.004. https://www.sciencedirect.com/science/article/pii/S1574954116302047

19. Yu, X., Wang, J., Kays, R., Jansen, P.A., Wang, T., Huang, T.: Automated identification of animal species in camera trap images. EURASIP J. Image Video Process. **2013**(1), 1–10 (2013). https://doi.org/10.1186/1687-5281-2013-52

20. Zhu, J.Y., Park, T., Isola, P., Efros, A.A.: Unpaired image-to-image translation using cycle-consistent adversarial networks. In: Proceedings of the IEEE International Conference on Computer Vision, October, pp. 2242–2251 (2017). https://doi.org/10.1109/ICCV.2017.244

Object Detection Based on Sparse Representation of Foreground

Zhenyue Zhu⊙, Shujing Lyu⁽⊠⁾ ⊙, Xiao Tu⊙, and Yue Lu⊙

Shanghai Key Laboratory of Multidimensional Information Processing, Department of Computer Science and Technology, East China Normal University, Shanghai 200062, China
51184506055@stu.cs.ecnu.edu.cn, {sjlv,ylu}@cs.ecnu.edu.cn

Abstract. Objects detection can be regard as the segmentation of foreground from background. In this paper, we propose a foreground segmentation method based on sparse representation of direction features for threat object detection in X-ray images. The threat objects are supposed as foreground and all other contents in the images are background. We extract the direction features to make up foreground dictionary firstly. Then we search the foreground area in the test image through sparse representation of their direction features by foreground dictionary. The experimental results show that this proposed method is robust to the X-ray images with different backgrounds.

Keywords: X-ray image · Object detection · Sparse representation

1 Introduction

X-ray baggage screening is widely used to maintain transport security. The in- spectors implement the security check task by watching the images of varying baggage contents within limited time. Considering the huge number of baggage to be inspected, it is desirable to fulfil this inspection automatically, which can not only reduce the workload of human inspectors but also increase the inspection speed and operator alertness consequently. In recent years, more and more scientists pay attention to automatic image recognition for security inspection. Kundegorski et al. [1] and Baştan et al. [2] generated a visual codebook to classify the image features and then detected threat objects with the use of Support Vector Machine (SVM). Mery et al. [3] proposed a model to classify the query images by voting the selected patches which had been classified by Sparse Representation Classification (SRC) [4, 5] with an adapted dictionary. Meanwhile, some methods based on Neural Networks also proposed for prohibited items detection [6–9]. Akçay et al. [10] used a pre-trained Convolutional Neural Networks (CNN) with transfer learning to classify the X-ray images.

Mery et al. [11] proposed an approach that could separate the foreground from background of an X-ray image. Its experimental results demonstrated a very good recognition performance, outperforming some alternative state-of-the-art techniques [12–26]. In Mery's work, a testing image could be modeled by the sum of the contributions of the foreground and the background, so the detection of the threat objects could be achieved

Y. Lu et al. (Eds.): ICPRAI 2020, LNCS 12068, pp. 465–473, 2020.
https://doi.org/10.1007/978-3-030-59830-3_40

by analyzing the contribution of foreground and background. Foreground dictionary was obtained from patches of the foreground images which only presented the threat objects to be detected. Whereas all the other contents in the baggage images constituted the background images and the background dictionary was obtained from patches of them. The linear combination of these two dictionaries was built for reconstructing the testing image. Usually, the testing image was assigned to the class with the minimal reconstruction error.

The model presented in [11] depends on not only the established foreground dictionary, but also the established background dictionary. So, it performs well on the baggage images which are included in the training dataset as background images. When a new type of baggage appears, the performance of the algorithm will be significantly degraded.

In order to overcome this limitation, we reconstruct the testing image only using the sparse representation by foreground dictionary. In addition, we use direction features instead of pixel features as the atoms of the foreground dictionary to provide multiple possibility for the gradient values. Experiments on the public GDXray dataset [27] illustrates that the proposed algorithm achieves a robust performance on X-ray images with different backgrounds.

2   Proposal Method

The key point of our strategy is to segment the foreground from background though sparse representation of direction features by foreground dictionary.

2.1   Formation of the Foreground Dictionary

The foreground images are consisted of clean threat objects. We use the well-known methodology of sliding window to extract the patches firstly. Each patch z_f with size $w \times w$ pixels of the foreground image is then converted into a direction feature y_f. Considering the good performance of the SURF descriptor [26], our proposed direction feature is based on similar properties that use relative strengths and orientations of gradients to reduce the effect of direction changes. We calculate the Haar wavelet responses in x and y direction on each patch. The side length of the wavelets is 4 s. The dominant orientation is estimated by calculating the sum of all responses within a sliding orientation window covering an angle of $\frac{\pi}{4}$. Therefore, each patch is split up regularly into smaller 4×4 square sub-regions. For each sub-region, we compute a few simple features at 5×5 regularly spaced sample points of size s $\times$ s. For reasons of simplicity, we call dx the Haar wavelet response in horizontal direction and dy the Haar wavelet responses in vertical direction (filter size 2 s). "Horizontal" and "Vertical" here is defined in relation to the dominant orientation. Then, the wavelet responses dx and dy are summed up over each sub-region and we obtain the sum of the absolute values of the responses, $|dx|$ and $|dy|$ according to the sign of dx and dy, respectively. Consequently, each sub-region has an eight-dimensional feature vector v for its underlying intensity structure. $v = (\sum dx\,dy < 0, \sum |dx|\,dy < 0, \sum dx\,dy > 0, \sum |dx|\,dy > 0, \sum dy\,dx < 0, \sum |dy|\,dx < 0, \sum dy\,dx > 0, \sum |dy|\,dx > 0)$. The extraction process of direction feature is exampled in Fig. 1. After converting all the n patches into feature vectors, they will combine into a foreground dictionary D of size 128 × n matric elements, as shown in Fig. 2. To deal with different contrast condition, we stack the columns of the foreground dictionary normalized to unit length.

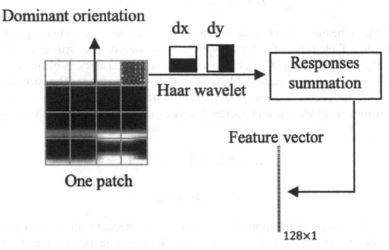

Fig. 1. Direction feature extraction.

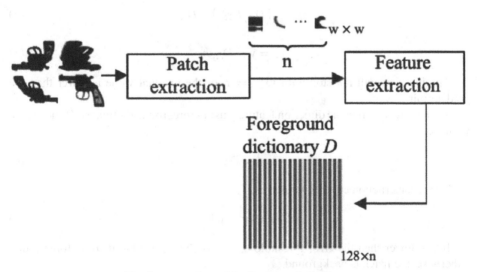

Fig. 2. Formation of the foreground dictionary.

2.2 Detection Model Design

The object detection process is detailed in Fig. 3. A patch z with $w \times w$ pixels (the same size of the patches used in the foreground dictionary) is sledded over the testing image in both horizontal and vertical directions. For each patch z, we first compute its direction feature y, and then we calculate the sparse representation of y using foreground dictionary D:

$$\|y - D_x\| \rightarrow \min \text{ subject to } \|X_0\| \leq T \tag{1}$$

where $\|x\|_0$ is the number of non-zero elements of x. Thus, in the sparse representation, no more than T elements of x are allowed to be non-zero. The Orthogonal Matching Pursuit algorithm (OMP) [28] is chosen to compute representations of function switch respect to nonorthogonal and possibly over complete dictionaries. The specific process of OMP is as follow. After initializing, we should find the atom D_{j_k} in dictionary D that has the strongest correlation with the residual vector r_{k-1} and record the location of this atom:

$$j_k = argmax\left(\left|D_{j_k}^{'} r_{k-1}\right|\right) \tag{2}$$

$$\Omega_k = \Omega_{k-1} \cup j_k \tag{3}$$

where k is the number of iterations, j is the location of selected atom in dictionary D and Ω is a collection of locations of selected atoms. We compute the sparse coefficients x_k to update the residual vector:

$$x_k = y\left(D_{\Omega_k}^{'} D_{\Omega_k}\right)^{-1} D_{\Omega_k} \tag{4}$$

$$r_k = y - D_{\Omega_k} x_k \tag{5}$$

The iteration will not stop until $\|x\|_0 = T$ or the residual value is lower than the predetermined value.

Then we reconstruct the direction feature y using foreground dictionary D and sparse vector x:

$$y = Dx \tag{6}$$

The reconstruction error is defined as:

$$e = \|y - y'\| \tag{7}$$

If e is lower than the error threshold c, the patch can be classified as foreground. Otherwise, the patch is background.

If the reconstruction error of the foreground information is low enough, then our model will put all the patches which classified as foreground into a null matrix. Z_f which has the same size of the testing image. The storage locations are the same as the locations where the patches are extracted. This candidate image Z_f shows all the potential regions of the gun detection. Now we convert Z_f into a binary image R and frame out the largest connected areas as the detected region. The proposed algorithm is summarized in Algorithm 1.

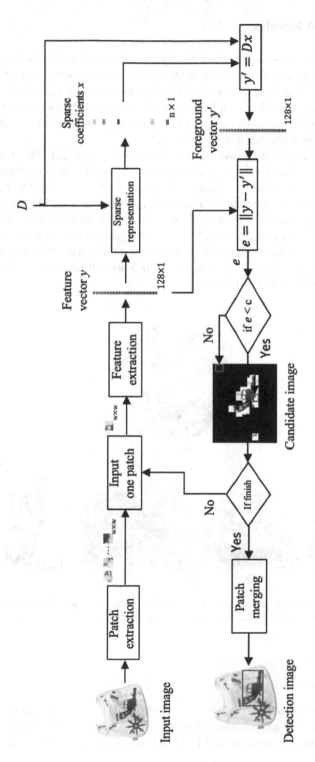

Fig. 3. Object detection in a gun case.

3   Experimental Results

For simplicity, these methods are represented by the abbreviation. The method in [11] named FBwP which means this method employs sparse representation of pixel features by foreground and background dictionaries. Our method named FwD which means it uses sparse representation of direction features by foreground dictionary. Moreover, we use pixel features rather than direction features in our algorithm in order to show the contribution of direction features and name it FwP.

The images are separated to two classes: object and non-object. The number of object X-ray images of the training sets are 200, and the number of object/non-object X-ray images of the sets of validation and testing are: 50/300, 200/1200 respectively. Furthermore, the testing images have 5 types of baggage which are shown in Fig. 4 to compare the performance of FBwP, FwP and FwD. They all come from the public GDXray dataset [27]. Figure 4(a) gives the type of baggage which is used to build the background dictionary in the method FBwP. We name it GD1. The other four types of baggage are also employed on our experiments, which are exhibited in Fig. 4(b). They are named as GD2, GD3, GD4 and GD5.

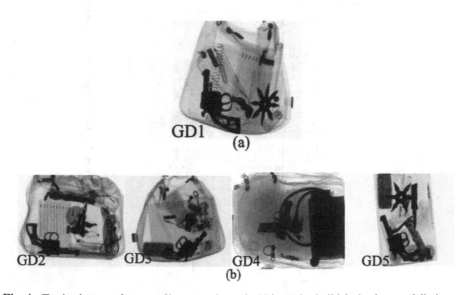

Fig. 4. Testing images: the type of baggage shown in (a) is used to build the background dictionary in the method of FBwP, while types of baggage shown in (b) are not.

3.1   Implementation

The experiments are carried out on a Windows 10 Pro ($\times$64) with a 3.2 GHz Quad-Core Intel processor and 4 GB memory. The X-ray images are obtained from GDXray dataset [27]. The experiments are implemented in MATLAB. The size of the input image is about 100 KB of pixels. To establish the foreground dictionary, we use patches with size 120 × 120 pixels. The vector y has 128 elements, and about 16,000 patches are extracted. The sparse value (parameter T) is set to 10. The error threshold (parameter c) is set to 1.2. The area threshold (parameter t) is set to one-fiftieth of the size of the input images.

3.2 Evaluation

In order to compare our method with [11], we employ f1-measure ($\beta = 1$) and accuracy as the evaluation protocol for the detection of guns in X-ray images. Table 1 describes the comparison of our method FwD and FBwP. When they are tested on GD1, the performance of FBwP is better than our method. Its f1-measure is 0.95 and the accuracy is 98.6%, which are all higher than that of FwD. When they are tested on other types of baggage, the performance of FBwP declines obviously. The average f1-measure of FBwP is only 0.59, and the accuracy of it is all no high than 84.7%. Compared with it, our method FwD obtains stable results, its f1-measure is all above 0.78 with more than 94.3% accuracy simultaneously. We also compare our method to FwP to illustrate the contribution of direction features used in our work. The experimental results are displayed in Table 2. From this table, we can see that the average performance of FwD is superior to that of FwP on the testing dataset.

Table 1. The experimental results of FBwP and FwD on GD1-5.

		GD	GD2	GD3	GD4	GD5	Average
FBwP	f1-measure	0.95	0.58	0.57	0.33	0.51	0.59
	accuracy	98.6%	82.1%	81.6%	77.1%	84.2%	84.7%
FwD	f1-measure	0.90	0.82	0.92	0.78	0.95	0.88
	Accuracy	97.1%	94.6%	97.9%	94.3%	98.7%	96.5%

Table 2. The experimental results of FwP and FwD on GD1-5.

FwP		FwD	
F1-measure	Accuracy	F1-measure	Accuracy
0.73	92.3%	0.88	96.5%

4 Conclusions

This work implements object detection based on segmenting the foreground from background using sparse representation of direction features by foreground dictionary. It is experimented on the threat object detection of X-ray images and the results demonstrate it is a robust method on different backgrounds. Future work we will focus on solving the problem of scale invariance of threat objects and employing our algorithm on a broader dataset.

References

1. Kundegorski, M.E., Akcay, S., Devereux, M.: On using feature descriptors as visual words for object detection within x-ray baggage security screening. In: Imaging for Crime Detection and Prevention, IET, pp. 1–6 (2016)
2. Baştan, M., Yousefi, M.R., Breuel, Thomas M.: Visual words on baggage X-ray images. In: Real, P., Diaz-Pernil, D., Molina-Abril, H., Berciano, A., Kropatsch, W. (eds.) CAIP 2011. LNCS, vol. 6854, pp. 360–368. Springer, Heidelberg (2011). https://doi.org/10.1007/978-3-642-23672-3_44
3. Mery, D., Svec, E., Arias, M.: Object recognition in baggage inspection using adaptive sparse representations of X-ray images. In: Bräunl, T., McCane, B., Rivera, M., Yu, X. (eds.) PSIVT 2015. LNCS, vol. 9431, pp. 709–720. Springer, Cham (2016). https://doi.org/10.1007/978-3-319-29451-3_56
4. Wright, J., Yang, A.Y., Ganesh, A.: Robust face recognition via sparse representation. IEEE Trans. Pattern Anal. Mach. Intell. 31, 210–227 (2009)
5. Mery, D., Svec, E., Arias, M.: Object recognition in X-ray testing using adaptive sparse representations. J. Nondestr. Eval. 35, 45 (2016). https://doi.org/10.1007/s10921-016-0362-8
6. Akcay, S., Kundegorski, M.E., Willcocks, C.G., Breckon, T.P.: Using deep convolutional neural network architectures for object classification and detection within x-ray baggage security imagery. IEEE Trans. Inf. Forensics Secur. 13(9), 2203–2215 (2018)
7. Jaccard, N., Rogers, T.W., Morton, E.J.: Detection of concealed cars in complex cargo X-ray imagery using deep learning. J. X-ray Sci. Technol. 25, 323–339 (2017)
8. Akçay, S., Kundegorski, M.E., Devereux, M.: Transfer learning using convolutional neural networks for object classification within X-ray baggage security imagery. In: International Conference on Image Processing, pp. 1057–1061. IEEE (2016)
9. Akcay, S., Kundegorski, M., Devereux, M., Breckon, T.: Transfer learning using convolutional neural networks for object classification within x-ray baggage security imagery. In: Proceedings of International Conference on Image Processing, pp. 1057–1061 (2016)
10. Griffin, L.D., Caldwell, M., Andrews, J.T.A., Bohler, H.: 'Unexpected item in the bagging area': anomaly detection in x-ray security images. IEEE Trans. Inf. Forensics Secur. 14(6), 1539–1553 (2019)
11. Miao, C., et al.: Sixray: A large-scale security inspection x-ray benchmark for prohibited item discovery in overlapping images. In: The IEEE Conference on Computer Vision and Pattern Recognition (CVPR) (2019)
12. Mery D., Katsaggelos A.K.: A Logarithmic X-Ray Imaging Model for Baggage Inspection: Simulation and Object Detection. In: Computer Vision and Pattern Recognition Workshops, IEEE, pp. 57–65 (2017)
13. Mery, D., Svec, E., Arias, M.: Modern computer vision techniques for x-ray testing in baggage inspection. IEEE Trans. Syst. Man Cybern. Syst. 47, 682–692 (2017)
14. Riffo, V., Mery, D.: Automated detection of threat objects using adapted implicit shape model. IEEE Trans. Syst. Man Cybern. Syst. 46, 472–482 (2016)
15. Mery, D., Arteta, C.: Automatic defect recognition in X-ray testing using computer vision. In: Applications of Computer Vision, pp. 1026–1035. IEEE (2017)
16. Megherbi, N., Han, J., Breckon, T.P.: A comparison of classification approaches for threat detection in CT based baggage screening. In: International Conference on Image Processing, pp. 3109–3112. IEEE (2012)

17. Turcsany, D., Mouton, A., Breckon, T.P.: Improving feature-based object recognition for X-ray baggage security screening using primed visual words. In: International Conference on Industrial Technology, pp. 1140–1145. IEEE (2013)
18. Riffo, V., Mery, D.: Automated detection of threat objects using adapted implicit shape model. IEEE Trans. Syst. Man Cybern. Syst. **46**(4), 472–482 (2016)
19. Uroukov, I., Speller, R.: A preliminary approach to intelligent Xray imaging for baggage inspection at airports. Signal Process. Res. **4**, 1 (2015)
20. Turcsany, D., Mouton, A., Breckon, T.P.: Improving feature-based object recognition for X-ray baggage security screening using primed visual words. In: IEEE International Conference on Industrial Technology (ICIT 2013), pp. 1140–1145 (2013)
21. Zhang, N., Zhu, J.: A study of X-ray machine image local semantic features extraction model based on bag-of-words for airport security. Int. J. Smart Sens. Intell. Syst. **8**(1), 45–64 (2015)
22. Flitton, G., Mouton, A., Breckon, T.P.: Object classification in 3D baggage security computed tomography imagery using visual codebooks. Pattern Recogn. **48**(8), 2489–2499 (2015)
23. Mouton, A., Breckon, T.P.: Materials-based 3D segmentation of unknown objects from dual-energy computed tomography imagery in baggage security screening. Pattern Recogn. **48**(6), 1961–1978 (2015)
24. Turcsany, D., Mouton, A., Breckon, T.: Improving feature based object recognition for x-ray baggage security screening using primed visual words. In: Proceedings of International Conference on Industrial Technology, pp. 1140–1145 (2013)
25. Sivic, J., Zisserman, A.: Video google: a text retrieval approach to object matching in videos. In: Proceedings International Conference Computer Vision, pp. 1470–1477 (2003)
26. Bay, H., Ess, A., Tuytelaars, T., Van Gool, L.: Speeded-up robust features (SURF). Comput. Vis. Image Underst. **110**(3), 346–359 (2008)
27. Mery, D., Riffo, V., Zscherpel, U.: GDXray: the database of X-ray images for nondestructive testing. J. Nondestr. Eval. **34**, 42 (2015). https://doi.org/10.1007/s10921-015-0315-7
28. Pati, Y.C., Rezaiifar, R., Krishnaprasad, P.S.: Orthogonal matching pursuit: recursive function approximation with applications to wavelet decomposition. In: Signals, Systems and Computers, pp. 40–44. IEEE (1993)

An Application and Integration of Machine Learning Approach on a Real IoT Agricultural Scenario

Donato Impedovo[1(✉)] , Fabrizio Balducci[1] , Giulio D'Amato[2] ,
Michela Del Prete[2] , Erminio Riezzo[2] , Lucia Sarcinella[1] ,
Mariagiorgia AgneseTandoi[1] , and Giuseppe Pirlo[1]

[1] Department of Computer Science, University of Bari Aldo Moro, Bari 70125, Italy
donato.impedovo@uniba.it
[2] Sysman Progetti and Servizi S.R.L., Bari 70125, Italy

Abstract. The Internet of Things (IoT) paradigm applied to the agriculture field provides a huge amount of data allowing the employment of Artificial Intelligence for multiple tasks. In this work, solar radiation prediction is considered. To the aim, Multi-Layer Perceptron is adopted considering a complete real complex use case and real-time working conditions. More specifically the forecasting system is integrated considering three different time forecasting horizons and, given different sites, needs and data availability, multiple input features configurations have been considered. The described work allows companies to innovate and optimize their industrial business.

Keywords: Solar irradiation forecasting · IoT · Water management · Intelligent systems · AI applications

1 Introduction

Solar index refers to the electromagnetic radiation emitted by the sun. More specifically irradiance is the sun's radiant power per unit area usually measured in terms of kWm^{-2} or in $MJm^{-2}day^{-1}$, whereas the solar irradiation is the sun's incident radiant energy on a surface of unit area and it is measured in terms of $kWhm^{-2}$ or in MJm^{-2}. It is typically expressed on an average daily basis for a given month. However, it must be argued that the two terms have been frequently mixed along literature [1, 2].

Solar radiation emitted by the sun is modified by the atmosphere (dust, clouds, etc.) as well as by land topography and surfaces. At the ground it is received as a mixture of three components: direct (Direct Normal Irradiance - DNI), diffuse, and reflected. The mixture is called Global Horizontal Irradiance (GHI), more specifically it is the sum of the direct, diffuse, and reflected components (also known as Diffuse Horizontal Irradiance - DHI) [3]. The power absorbed by the earth is proportional to the incident angle of DNI with the surface at a given moment: as this angle get closer to $90°$, more power will be absorbed. So that it is reasonable to expect higher values of solar radiation during peak sun hours.

© Springer Nature Switzerland AG 2020
Y. Lu et al. (Eds.): ICPRAI 2020, LNCS 12068, pp. 474–483, 2020.
https://doi.org/10.1007/978-3-030-59830-3_41

The solar index is one of the most challenging weather parameters to be predicted, mostly due to solar energy production forecasting and its balance with consumptions and energy routing [2]. However, energy production is not the only field in which solar radiation forecasting is important. Agricultural and farming industries are also very interested in this parameter [4]. In fact, in these businesses, the solar index supports many business-critical decisions, such as the crops irrigation and the fields fertilization scheduling within current sensors-based and data-driven approaches. The solar radiation is fundamental for the irrigation process to quantify the amount of water needed by a specific crop: its value is used to compute the Reference Evapotranspiration (ETo) through the Penman-Monteith equation, a parameter that represents a loss of water that is proportional to the actual crop evapotranspiration. The day-by-day calculation of the ETo enables agronomists to match the requirements of actively growing crops [5]. Effective irrigation scheduling decision processes typically require an estimate of the water available in the soil, the water inside the actively growing crop and the loss of water due to evapotranspiration, in such a way that a convenient water balance is maintained for a given irrigation strategy. Typically, irrigations are scheduled when rain events are unlikely and soil water is insufficient to the extent that it can no more be easily extracted by roots of the crop to meet its requirements. In this stress condition crops might still appear healthy for a while, but (depending by the specific one) their yield can be compromised.

The solar index is generally calculated, in this field, by numerical methods able to take into account only a small part of the amount of time series and a reduced set of its or related peculiar characteristics. In many practical applications a single historic value is considered thus bring the plants in water stress conditions or obtaining an overestimation of the amount of water needed. The water management process results in increased complexity being also inefficient. This work shows that it is possible to build an easy to use system placed within a modern IoT agricultural scenario to predict solar radiation. Although the innovation of the specific solar radiation predictor, from a scientific point of view, is limited, the work describes a real use case and the integration of the approach within a complex and complete IoT system, so that it can be useful for other similar real applications. In this way water stress and supply can be planned more efficiently, even considering the optimization of the production of wastewater. The system has been applied to a real scenario adopting real working conditions.

The paper is organized as follow. Section 2 briefly describes related works, Sect. 3 describes methods and the specific IoT dataset adopted in this work. Experimental results are in Sect. 4, while Sect. 5 concludes the paper.

2 Related Work

Solar radiation and irradiance forecasting can be obtained with the following techniques [21]:

- Regressive methods, which take advantage of the correlated nature of irradiance observations;
- Numerical Weather Prediction (NWP), which rely on physical laws of motion and thermodynamics;

- Artificial Intelligence models exploiting classifiers, numeric gradients and loss functions in a supervised and unsupervised way.

The forecasting tasks deal with generally large datasets, so that Artificial Intelligence based methods have been proved to be successful in solar radiation forecasting also outperforming Numerical Weather Prediction (NWP) methods [6] and empirical models [7].

The advantage of Machine Learning (ML) approaches rely in the fact that the prediction of these kind of parameters deals with time series models: past patterns can be used to predict future behaviors. There is a mapping between a finite window of the time series and the next state of the observed phenomenon as in (1):

$$y_t = f\left(y_{t-1}, \; y_{t-1-1}, \ldots, y_{t-1-n+1}\right) \tag{1}$$

where 1 is called lag time, n is the number of past values taken into consideration. Artificial Neural Networks are a convenient approach for forecasting data obtained from an IoT scenario [8], and in recent research it has been showed to perform very well to predict solar radiation [9], however those approaches can be unstable [2].

Data-driven approaches are, in general, opposed to model-driven approaches [10], temporal data (time series) can be used to learn dependencies between past and future [11].

Ren et al. [12] proposed also to consider spatial correlation for forecasting aims. Voyant et al. [13] proposed effective models for solar radiation forecasting, and conventional ensemble methods can be identified [14].

An application similar to the one proposed here, concerns the use of IoT-based systems for estimating groundnut irrigation volumes and optimal use of fertilizers in respect of soil characteristics [15]. Similarly, climatic parameters monitoring can be used for leafs and plants diseases management [16, 17]. Wolfert et al. [18] and Biradar et al. [19] studied smart-farm industries through surveys, while multidisciplinary IoT models are examined in [20].

It must be underlined that, to overcome some prediction stability issues [2], different ML solutions can be considered and combined to obtain a more robust prediction.

3 Methods and Dataset

The present work is part of an Industrial Research Project named Eco-Loop. The project is focused on the development of support systems to help various stakeholders (i.e. wastewater facilities and farmers) in the reuse of refined wastewater in agriculture. Wastewater refinement plants are managed by AQP SpA, while the wells, cisterns and distribution pipes are managed by a farmers' co-op, and the farmers are directly involved in the pilot with their cultivations. The system is provided with two Decision Support Systems (DSS). The first one deals with the irrigation and fertilization strategy taking into account the actual soil, the cultivation status and water chemical composition compared to the cultivation chemical requirements. The second one provides a monitoring dashboard, a warning system, data presentation facilities (e.g. laboratory analysis, weather forecasts, water availability and historical data). The first DSS provides suggestions to farmers, the second one to the plant management. Both DSSs produce (in a strictly related manner) water needs forecasting which depends (also) upon solar irradiation. The system presented in this work is used by both DSSs for the "Watering Planning" block of the Eco-Loop overall architecture reported in Fig. 1. As can be easily observed, the overall system adopts a modern IoT infrastructure.

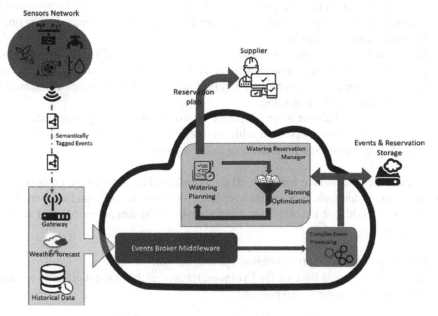

Fig. 1. Eco-Loop overall architecture [22]

Among the set of field devices, a set of weather stations has been installed near the water refining plants, near the water collection and mixing tanks and in the proximity of the cultivation fields. Weather stations are, in general, equipped with the following sensors:

- Thermometer (for air and sea surface temperature);
- Barometer (for atmospheric pressure);
- Hygrometer (for measuring humidity);
- Anemometer (for measuring wind);
- Rain gauge (for liquid precipitation);
- Pyranometer (for the solar radiation).

Acquired and processed data spans a period from July 1st, 2017 to February 28th, 2018. Considered variables are: Latitude, Longitude, Date time, Solar radiation, Rain level, Wind speed and direction, Pressure, Temperature, Relative Humidity. These data have been pre-processed: null values and poor-quality records have been filtered out, timestamps have been split for aggregation purposes. In this work the forecasting of the solar index has been performed considering various variables called predictor variables. In other words, in order to predict solar radiation, different combinations of input variables have been used. More specifically, three different use cases have been considered:

1. past values of solar radiation are used to perform forecasting of future values: it is considered as the default flat predictor. This condition assumes that the pyranometer is always available at any station, however this is not obvious because it is one of the most expansive sensors.
2. past values of max temperature and daily sun hours are used for prediction. These parameters both present a similar trend and seasonality if compared to the solar radiation. Temperature's sensors are very common and very cheap, whereas daily sun hours are a computed value.
3. future values of the solar index are predicted using the max daily temperature, daily sun hours and humidity (similar with a source of variance). Many vendors propose an integrated low-cost temperature/humidity sensor.

In the experiments here reported, all the considered stations where equipped with pyranometer, so that forecasted values have been compared to real solar radiation values. The selection of the configurations has been also driven by a maintenance cost analysis. This analysis:

- gives the possibility to offer different operational solutions,
- gives an alternative in the case the Pyranometer is available but not properly working, so that the forecasting is still possible.

Three different time horizons have been considered for forecasting aims due to different water supply periods and plants' water needs:

- one-day forecasting;
- three-day forecasting;
- week forecasting.

In order to achieve such predictions, the dataset has been modeled as follow. Raw records received from sensors had a time interval of 15 min, so that the daily mean values have been evaluated. Features have been transformed by rescaling them within a comparable range (Fig. 2).

A Multi Layer Perceptron (MLP) has been considered as the base classifier. MLP is a feed-forward neural network, wherein connections between the nodes do not form a cycle (opposite to recurrent neural networks); it is the classical implementation of a neural network and the universal approximation theorem for neural networks states that every continuous function that maps input values to an output interval can be approximated closely by a MLP with just one hidden layer, unless you consider problems about optimization and computational constraints. Despite being so powerful, MLPs have caveats as well, in fact, different random weights initializations can lead to different validation accuracy while requiring a quite substantial number of hyperparameters to

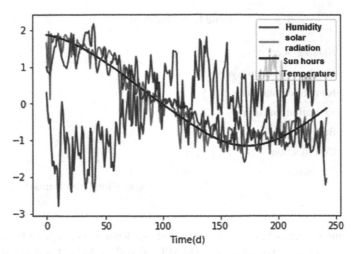

Fig. 2. A sampling of the selected features (humidity, solar radiation, sun hours and temperature) plotted in their rescaled version.

be tuned: most of the times there is not better solution than finding the best configuration empirically.

The MLP has been trained with daily data in order to achieve forecasting with the previous horizons. For each day x of the training interval, the network has been fed with the mentioned data. Successively, given actual data of a specific day, the network is able to provide the prediction of the average solar radiation expected within 1, 3 or 7 days. Figure 3 reports the forecasting pipeline.

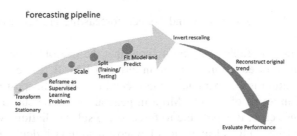

Fig. 3. Forecasting pipeline.

4 Results

The dataset has been divided into 75% (train) - 25% (test). The former contains data between 1/07/2017-2/01/2018; the latter refers to the interval 3/01/2018-28/02/2018. Training has been performed considering 10 epochs and the squared loss as cost function. Table 1 reports details.

Table 1. MLP details.

Number of epochs	10
Number of hidden layers	1
Number of neurons	3
Activation function	Tanh
Batch size	31
Cost function	Squared loss
Optimizer	Limited-memory Broyden–Fletcher–Goldfarb–Shannon

Two common metrics have been considered: the Root Mean Square Error (RMSE) and the Mean Absolute Percentage Error (MAPE). RMSE is defined as in (2), it measures differences between two time series $X_{(1,t)}$ and $X_{(2,t)}$ as the predicted value by one of the models and the observed one with variables observed over T times.

$$\text{RMSE} = \sqrt{\frac{\sum_{i=1}^{T}(X_{1,t} - X_{2,t})^2}{T}} \tag{2}$$

MAPE is defined in (3): it measures the accuracy percentage between the value and its forecast.

$$\text{MAPE} = \frac{100\%}{n} \sum_{i=1}^{n} \left| \frac{x_i - \hat{x}_i}{x_i} \right| \tag{3}$$

Results related to 1-day, 3-days and 1-week forecasting are reported, respectively, in Tables 1, 3 and 4.

Results reported in Table 2, 3 and 4 show that the use of the standard predictor reaches best results for each time horizon in terms of RMSE. Table 4 reveals that the standard configuration is comparable to the second one (the cheapest solution) for 1-week prediction in terms of MAPE. More in general, it can be stated that, given the specific project use case, it is possible to forecast the solar radiation even if its values are not available (Fig. 4) thus leaving to a low-cost field device deployment as well as to a more robust prediction in case on data unavailability. Execution time always allow real time results delivery.

Table 2. Results of 1-day horizon forecasting. The first column represents the feature combinations. RMSE metric is MJm^{-2}, MAPE is in % and time in [s].

Configuration	RMSE	MAPE	Time
1 – Solar radiation	0.0562	13.18	0.035
2 – Max temperature and sun hours	0.0653	15.41	0.039
3 – Max daily temperature, daily sun hours, humidity	0.0745	17.11	0.044

Table 3. Results of 3-day horizon forecasting. The first column represents the feature combinations. RMSE metric is MJm^{-2}, MAPE is in % and time in [s].

Configuration	RMSE	MAPE	Time
1 – Solar radiation	0.0577	9.25	0.046
2 – Max temperature and sun hours	0.0687	11.61	0.046
3 – Max daily temperature, daily sun hours, humidity	0.0731	11.94	0.031

Table 4. Results of 1-week horizon forecasting. The first column represents the feature combinations. RMSE metric is MJm^{-2}, MAPE is in % and time in [s].

Configuration	RMSE	MAPE	Time
1 – Solar radiation	0.0636	6.48	0.037
2 – Max temperature and sun hours	0.0737	6.48	0.039
3 – Max daily temperature, daily sun hours, humidity	0.0739	7.07	0.042

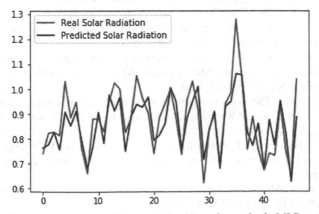

Fig. 4. One week solar radiation prediction obtained by using a single MLP model adopting as input Temperature and Sun hours; the X axis reports days, Y axis reports values in [MJ/m^2].

5 Conclusions

This work shows how to integrate solar radiation forecasting within a real complex system whose aim is to estimate water needs in terms of both production (wastewater) and usage for irrigation aims. The use of a Multi Layer Perceptron has been investigated taking into account different use cases. More specifically different input parameters have been considered taking into account data availability as well as forecasting horizons reaching predictive errors between 17% and 6%.

It emerges that the use of the standard predictor reaches best results for each time horizon in terms of RMSE. However, it has been showed that, given the specific project use case, it is possible to forecast the solar radiation even if its values are not available by using data obtained with cheap sensors (i.e. temperature and/or Humidity) thus leading to a low cost field device deployment as well as to a more robust prediction in case on data unavailability. Execution time always allow real time results delivery.

Future works will involve bagged models and other strategies.

Acknowledge. This work is within the Eco-Loop project (no. 2AT8246) funded by the Puglia POR FESR-FSE 2014-2020. Fondo Europeo Sviluppo Regionale. Azione 1.6 - Avviso Pubblico "InnoNetwork".

References

1. Diagne, M., Dvid, M., Lauret, P., Boland, J., Schmutz, N.: Review of solar irradiance forecasting methods and a proposition for small-scale insular grids. Renew. Sustain. Energy Rev. **27**, 65–76 (2013)
2. Voyant, C., et al.: Machine learning methods for solar radiation forecasting: a review. Renew. Energy **105**, 569–582 (2017)
3. Inman, R.H., Pedro, H.T.C., Coimbra, C.F.M.: Solar forecasting methods for renewable energy integration. Prog. Energy Combust. Sci. **39**(6), 535–576 (2013)
4. Balducci, F., Impedovo, D., Pirlo, G.: Machine learning applications on agricultural datasets for smart farm enhancement. Machines **6**, 38 (2018)
5. Allen, R.G., Pereira, L.S., Raes, D., Smith, M.: Crop evapotranspiration - guidelines for computing crop water 2 requirements. FAO irrigation and drainage. Paper 56 Food and Agriculture ONU (1998). ISBN 978-92-5-104219-9
6. Dobbs, A., Florita, A.: Short-term solar forecasting performance of popular machine learning algorithms. In: International Workshop on the Integration of Solar Power into Power Systems (Solar Integration Workshop), pp. 1–14 (2017)
7. Meenal, R., Selvakumar, A.I.: Assessment of SVM, empirical and ANN based solar radiation prediction models with most influencing input parameters. Renew. Energy **121**, 324–343 (2018). Elsevier
8. Pierro, M., et al.: Multi-model ensemble for day ahead prediction of photovoltaic power generation. Solar Energy **134**, 132–146 (2016). Elsevier
9. Ding, M., Wang, L., Bi, R.: An ANN-based approach for forecasting the power output of photovoltaic system. Procedia Environ. Sci. **11**, 1308–1315 (2011). Elsevier
10. Chaouachi, A., Kamel, R., Nagasaka, K.: Neural network ensemble-based solar power generation short-term forecasting. Int. J. Electr. Comput. Eng. **3**(6), 41–46 (2010). IAES

11. Chakraborty, K., Mehrotra, K., Mohan, C.K., Ranka, S.: Forecasting the behavior of multivariate time series using neural networks. Neural Netw. **5**(6), 961–970 (1992)
12. Bontempi, G., Ben Taieb, S., Le Borgne, Y.-A.: Machine learning strategies for time series forecasting. In: Aufaure, M.-A., Zimányi, E. (eds.) eBISS 2012. LNBIP, vol. 138, pp. 62–77. Springer, Heidelberg (2013). https://doi.org/10.1007/978-3-642-36318-4_3
13. Ren, Y., Suganthan, P.N., Srikanth, N.: Ensemble methods for wind and solar power forecasting - a state-of-the-art review. Renew. Sustain. Energy Rev. **50**, 82–91 (2015). https://doi.org/10.1016/j.rser.2015.04.081
14. Ren, Y., Member, S., Zhang, L., Member, S., Suganthan, P.N.: Ensemble classification and regression-recent developments, applications and future directions. IEEE Comput. Intell. Mag. **11**, 1–14 (2016)
15. Rekha, P., Rangan, V.P., Ramesh, M.V., Nibi, K.V.: High yield groundnut agronomy: An IoT based precision farming framework. In: Proceedings of the 2017 IEEE Global Humanitarian Technology Conference, San Jose, CA, USA, 19–22 October 2017, pp. 1–5 (2017)
16. Sarangdhar, A.A., Pawar, V.R.: Machine learning regression technique for cotton leaf disease detection and controlling using IoT. In: Proceedings of the 2017 International Conference of Electronics, Communication and Aerospace Technology (ICECA), Coimbatore, India, 20–22 April 2017, vol. 2, pp. 449–454 (2017)
17. Patil, S.S., Thorat, S.A.: Early detection of grapes diseases using machine learning and IoT. In: Proceedings of the 2016 2nd International Conference on Cognitive Computing and Information Processing (CCIP), Mysore, India, 12–13 August 2016, pp. 1–5 (2016)
18. Wolfert, S., Ge, L., Verdouw, C., Bogaardt, M.J.: Big data in smart farming a review. Agric. Syst. **153**, 69–80 (2017)
19. Biradarand, H.B., Shabadi, L.: Review on IoT based multidisciplinary models for smart farming. In: Proceedings of the 2nd IEEE Int. Conference on Recent Trends in Electronics, Information Communication Technology, Bangalore, India, 19–20 May 2017, pp. 1923–1926 (2017)
20. Yoon, C., Huh, M., Kang, S.G., Park, J., Lee, C.: Implement smart farm with IoT technology. In: Proceedings of the 20th International Conference on Advanced Communication Technology, Chuncheon-si Gangwon-do, Korea, 11–14 February 2018, pp. 749–752 (2018)
21. Inman, R.H., Pedro, H.T.C., Coimbra, C.F.M.: Solar forecasting methods for renewable energy integration. Progr. Energy Combust. Sci. **39**(6), 535–576 (2013)
22. Rotondi, D., Straniero, L., Saltarella, M., Balducci, F., Impedovo, D., Pirlo, G.: Semantics for wastewater reuse in agriculture*. In: 2019 IEEE International Conference on Systems, Man and Cybernetics (SMC), Bari, Italy 2019, pp. 598–603 (2019)

Spatio-Temporal Stability Analysis in Satellite Image Times Series

Mohamed Chelali[1]([✉]) [iD], Camille Kurtz[1] [iD], Anne Puissant[2] [iD], and Nicole Vincent[1] [iD]

[1] Université de Paris, LIPADE, Paris, France
{mohamed.chelali,camille.kurtz,nicole.vincent}@u-paris.fr
[2] Université de Strasbourg, LIVE, Strasbourg, France
anne.puissant@unistra.fr

Abstract. Satellite Image Time Series (SITS) provide valuable information for the study of the Earth's surface. In particular, this information may improve the comprehension, the understanding and the mapping of many phenomenons such as earthquake monitoring, urban sprawling or agricultural practices. In this article, we propose a method to define new spatio-temporal features from SITS based on the measure of the temporal stability. The proposed method is based on a compression algorithm named Run Length Encoding leading to a novel image representation from which stability features can be measured. Such features can then be used in several applications such as SITS summarizing, to make easier the interpretation of such a data-cube, or the classification of spatio-temporal patterns. The preliminary results obtained from a series of 50 Sentinel-2 optical images highlight the interest of our approach in a remote sensing application.

Keywords: Satellite Image Time Series (SITS) · Spatio-temporal features · Run Length Encoding · Temporal stability · Sentinel-2

1 Introduction

Image time series, such as Satellite Image Time Series (SITS) or MRI functional sequences in the medical domain consist of ordered sets of images taken from the same scene at different dates. Such data provide rich information with the temporal evolution of the studied areas. In the context of remote sensing, SITS provide enormous amounts of information that allow the monitoring of the surface of our planet. Recently new constellations of satellites have been launched to observe our territories, producing optical images with high spatial, spectral and temporal resolution. For example, the Sentinel-2 sensors provide SITS with a revisit time of 5 days and a spatial resolution of 10–20 m.

One of the major applications of SITS is the cartography/mapping of land covers (e.g. agricultural crops, urban areas) and the identification of land use

The authors thank the ANR for supporting this work – Grant ANR-17-CE23-0015.

changes (e.g. urbanization, deforestation). The availability of the temporal information makes possible to understand the evolution of our territory while analyzing complex patterns leading to produce or update accurate land cover maps of a particular area [11].

A main challenge for the automation of SITS analysis is to consider at the same time both the temporal and the spatial domains of the data-cube in order to take into account and benefit from all the (complementary) information carried by the data. Indeed, most of the existing methods for SITS analysis are actually solely based on temporal information [18,19]. However, for many complex tasks such as the understanding of heterogeneous agricultural practices, this may be not sufficient to get satisfactory results. The consideration of spatio-temporal features from SITS should allow the discrimination between different complex land cover classes, related to agricultural and urban land-cover practices. Note that we do not aim here to produce temporal land-cover maps or to study land use changes (e.g. urbanization) but our objective is to analyze complex land-cover classes prone to confusions when a single date image is used.

This article focuses on the specific problem of spatio-temporal features extraction from image time series that can be used for different applications. For example, such information may be used to summarize a SITS, improving the difficult photo-interpretation of the sensed scenes from a set of many images acquired at different dates, or for classification of different regions of interest into thematic classes (e.g. agricultural crops, vegetation, urban). Our contribution relies on spatio-temporal features that are based on the measure of the spatio-temporal stability of a zone, using a compression algorithm named Run Length Encoding. By compressing a data-cube with this strategy, we obtain a novel SITS representation from which stability features can be measured.

This article is organized as follow. Section 2 recalls some existing methods for SITS analysis. Section 3 presents our proposal to extract spatio-temporal features. Section 4 describes the experimental study while Sect. 5 concludes.

2   Related Works

SITS allows the observation of the Earth surface and the understanding of the evolution of our environment, where various changes can occur over the time (e.g. following natural disasters, urbanization, agricultural practices). Such environmental changes may be of different types, origins and duration [6].

Pioneer methods for SITS analysis take into account single images or stacks of images. Those that consider each image, compute different measurements per pixel as independent features and involve them in classical machine learning-based procedures. In such approaches, the date of the measurements is often ignored in the feature space. Methods designed for bi-temporal analysis, can locate and study abrupt changes occurring between two observations. These methods include image differencing [4], ratio-ing [13] or change vector analysis [14]. We also find statistical methods that consider two or more images, such as linear transformation (PCA and MAF) [16,17].

More recent methods were designed to directly deal with the specificity of image time series. This category of methods includes multi-date classification approaches, such as radiometric trajectory analysis [24], or sequential patterns techniques that group pixels that share common temporal pattern [15]. The latter strategy exploits the notion that land cover can vary through time (e.g. because of seasons, vegetation evolution [23]), and related methods take into account the order of measurements by using dedicated time series analysis methods [3]. Every pixel is viewed as a temporally ordered series of measurements, and the changes of the measurements through time are analyzed to find (temporal) patterns.

Other type of methods start by transforming the original representation of the SITS into a new one. For example, the analysis can be operated in the "frequency-domain" that includes spectral analysis or wavelet analysis [2]. Concerning the classification methods, the classical way is to measure similarity between any incoming sample and the training set and then to assign the label of the most similar class. Other methods extract more discriminative "hand-crafted" features from a new enriched space [19,21] before using the classifier.

Recently, deep learning approaches have been employed to analyze satellite images. For example, convolutional neural networks (CNN) can be applied with $2D$ convolutions to deal with the spatial domain [9]. CNNs have been also applied successfully to perform SITS classification; in this case $1D$ convolutions dealing only with the temporal domain [18] have been proposed. Other architectures of deep learning that is designed for time series are recurrent neural network (RNN) such as Long-Short Term Memory (LSTM), used successfully in [10,22]. In this context, deep learning approaches outperform traditional classifications algorithm as Random Forest [12], but they do not directly take into account the spatial dimension of the data as they consider pixels in an independent way. Although, some approaches have been proposed to consider both the temporal and the spatial dimensions of the $2D + t$ data-cube [7], the expressiveness of the underlying convolutional features is difficult to use for the interpretation of the content of the sensed scenes.

In this article, we propose an approach to define a set of spatio-temporal features from an image time series. Such spatio-temporal features can be used for different needs, for example to summarize a SITS, in order to improve and to make easier the scene interpretation, or to enrich the feature space, increasing the separability of complex land-cover classes in a classification task. The proposed approach measures the stability of a zone based on a lossless compression algorithm, here the Run Length Encoding (RLE) [8]. The notion of temporal stability has been initially proposed in a previous article [5]. In the following, we extend this strategy to now measure the spatio-temporal stability of an area, by relaxing both temporal and spatial constraints when assessing the equality between consecutive pixels through the time.

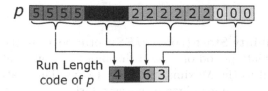

Fig. 1. Compression of a temporal pixel p with a Run Length Encoding (RLE).

3 Proposed Method

The proposed method aims to analyze the spatio-temporal stability or change of image pixels from a SITS. The spatio-temporal features are based on the measure of the spatio-temporal stability of a zone, using a compression algorithm, namely Run Length Encoding (RLE) algorithm, leading to a novel image representation from which stability features can be measured. In this Section, we first explain how the stability can be measured in the datacube induced by the SITS data. Since the notion of stability involves the study of the repetition of successive values, the notion of equality and its application level enable to relax both temporal and/or spatial constraints, leading to the definition of spatio-temporal features.

3.1 Stability Measurement

The stability is based on the repetition of the successive values. In the literature, different methods have been proposed to measure this information. In our case, we choose a compression method named Run Length Encoding (RLE) [8] that was already used for time series analysis in [1, 20]. RLE is a lossless compression algorithm. It allows the compression of a vector v of length L by storing both the number of times a value is repeated successively and its value. In our case we only consider the first information and omit the value. The resulting vector is noted $RLE(v)$ with length l. Figure 1 illustrates the computation of the RLE on a temporal pixel p.

In the following, a SITS is noted $(I_t)_{t \in [\![1,T]\!]}$ where t is the date of acquisition for each image I. The computation of the RLE is applied on a temporal pixel, noted $p(x, y) = (p_t(x, y))_{t=1}^{T}$, with $x \in [1, W]$ and $y \in [1, H]$, W and H representing respectively the width and the height of the images.

Our strategy here is to employ the RLE to change the representation of the SITS into a less voluminous structure, where stability information can be more easily measured. In this context, we define three features that can be extracted from this new representation for a given temporal pixel p:

- The **Maximal Stability** (noted MS) feature captures the longest period where the pixel intensity stays stable (i.e. without change) through the time series. It is expressed in term of number of days of the year. Such a feature value can be computed as:

$$MS(p) = \|RLE(p)\|_\infty \tag{1}$$

- The **Max Stability Start** (noted MSS) corresponds to the beginning of the maximum stability period of a pixel through time. Such information is then directly related to the Maximal Stability (MS). This feature can be informative/discriminative for different specific tasks, for example the maximum stability in artificialized areas (e.g. built-up area, impervious surfaces) starts earlier than in non-artificial zones since they do not change over time. The MSS feature value can be defined as:

$$MSS(p) = \left(\sum_{i=1}^{t_0-1} RLE(p)_i\right) \quad \text{with} \ \ t_0 \ / \ RLE(p)_{t_0} = MS(p) \tag{2}$$

- The **Number of changes** (noted NB) corresponds to the number of stability ranges, linked to the number of changes in the area covered by the studied temporal pixel. Such a feature value can be computed as:

$$NB(p) = l_p \tag{3}$$

The computation of RLE is based on the analysis of the equality between successive values. However, the images composing a SITS are not acquired at the same time. Although the images are generally corrected, the variability of the pixel intensity values can be important along the series since the distributions of the pixel values are not in the same dynamic from an image to another. This may be a consequence of the seasons and the different illumination conditions. To deal with this issue, the notion of equality has to be carefully studied in order to evaluate, in a more realistic manner, if two successive (temporal) pixel intensity values can be considered as equal or not.

3.2   Notion of Equality

The equality is a binary equivalence relation that compares two objects of a same set E. They are considered as identical if a given predicate P holds:

$$\mathfrak{E} \ : E \times E \to \qquad Bool$$
$$o_1, \ o_2 \to \begin{cases} True & \text{if} \ \ P(o_1, o_2) \\ False & \text{else} \end{cases} \tag{4}$$

When a SITS is considered, the pixel values being either continuous or discrete in interval such as $[0, 255]$ or hypercube when vectorial values are considered, the equality of values is not always significant. Then, to fix this problem, we start by applying a quantization of the pixel values. The quantization must not be applied to each image I_t, nor at temporal pixel p, it has to be done at a global level of all the pixels of the data-cube $(I_t)_t$. The quantization could be regular, fixed with respect of the usual distribution of the values or it can be adapted to the image series, to the nature of the characteristic used.

In order to perform the quantization, we apply a clustering algorithm that enables to define the significant intervals. In our case a k-Means algorithm, $k_{quantiz}$ being a parameter of the method, is used and different values have been experimented according to the precision needed in our problem. Then, the pixel values are replaced by the cluster label belonging to $\{1, 2, \dots, k_{quantiz}\}$ defining a new STIS $(J_t)_{t \in [\![1,T]\!]}$. The RLE is then actually applied on $(J_t)_t$ pixels. In this way, we define the predicate P by the strict equality between the objects, as $P(o_1, o_2) = (o_1 = o_2)$.

3.3 Towards Spatio-Temporal Stability

The features presented in Sect. 3.1 are defined for a temporal pixel and they can be naturally considered as temporal features. Since our strategy relies on the compression ability of the RLE, we also suppose that the data are as clean as possible. But with satellite images, this may not be always the case. Some images from the series can be affected by noise (e.g. salt and pepper, undetected clouds) and a SITS can be potentially affected by registration problems from one image to another. In this case, the RLE may not reflect the reality of the content of the sensed scene along time. To handle this, we are going to relax the equality definition in the temporal and/or the spatial domains, leading to an approximation of the RLE, noted $\widetilde{RLE}$, that can absorb these different noises. It is no longer a lossless compression scheme.

The RLE computation is mainly based on the notion of "runs", computed over a sequence, a run being a sub-sequence composed of successive (identical) repeated values. In the classical RLE, the predicate P is the one defined previously, leading to a "hard" equality relation. To compute the approximation of the RLE, the strategy here is to compute approximated runs over the sequence by relaxing the predicate P, used to estimate the equality between successive values. Of course the aim is to favor the shortest $\widetilde{RLE}$, that is to say, the longest runs that are assumed to be resumed by a single value. Then it is no more possible to have a linear process of the series but, recursively, the longest runs on the main series and on the sub-series when some parts of the series are already compressed.

We first define several ways to consider that a sub-series is "constant" and to compute an approximated run. This will be done through the relaxation of the predicate P in the temporal domain, in the spatial domain and finally in the spatio-temporal domain.

Let us note $s = (s_0, s_1, s_2, \dots, s_n)$ a series. Our goal here is to compute the length of the longest run in this series, starting from s_0:

- **Temporal relaxation** The first possibility is to relax the temporal domain. When $i \geq 1$, we define a new predicate as:

$$P_{s,i}^{temp} = (s_i = s_0) \vee (s_{i+1} = s_0) \tag{5}$$

where $\vee$ represents the logical or. This predicate makes it possible to skip a time value when comparing two consecutive elements of the series, according to the value of s_0.

– **Spatial relaxation** When we relax the spatial domain, the pixel value at instant s_0 is compared with the next value in $t + 1$ and its neighbors in a square window (a patch), which provides a certain spatial flexibility during the comparison. With this strategy, we can avoid noise (e.g. salt and pepper) and potential problems of image registration disturbing the comparison.
With the series s we associate the series of the neighbor of each s_i, $sN = (Ns_0, Ns_1, Ns_2, \ldots, Ns_n)$ and we define a new predicate as:

$$P_{s,i}^{spatio} = s_0 \in Ns_i \tag{6}$$

– **Spatio-temporal relaxation** A comparison completely relaxed is extracted by combining both spatial and temporal relaxations. This spatio-temporal relaxation is obtained with:

$$P_{s,i}^{spatio-temp} = P_{s,i}^{spatio} \ \lor \ P_{s,i}^{temp} \tag{7}$$

Given these predicates, we can now compute the length of the longest run of the series, starting from s_0. We first define a function c as

$$
\begin{aligned}
c: \quad &\mathbb{B} \to \quad \mathbb{Z} \\
&x \mapsto \begin{cases} 1 & \text{if } x \\ 0 & \text{otherwise} \end{cases}
\end{aligned} \tag{8}
$$

The length of the longest run (noted LR) can then be computed (using c to build a counter) as:

$$LR(s) = \max_k \left\{ k \in [1, n-1] : \sum_{i=1}^{k} c(P_{s,i}^{\star}) = k \right\} + 1 \tag{9}$$

where $P_{s,i}^{\star}$ is one of the relaxed predicates mentioned above.

Algorithmic Aspects. Now that we know how to compute an approximated run beginning at a specific position, we can compute $\widetilde{RLE}$.

Let us consider a temporal pixel $p(x, y) = (p_t(x, y))_{t=1}^{T}$ of a SITS, noted p hereinafter. We note $\widetilde{p}_t$ the sub-series of p beginning in p_t and we have

$$LR(p) = \max_t LR(\widetilde{p}_t) \quad \text{and} \quad ilr(p) = \arg\max_t LR(\widetilde{p}_t) \tag{10}$$

with $ilr(p)$ providing the position (index) of the starting point of the longest run of p. To compute all the runs composing $\widetilde{RLE}$, we implemented a greedy optimization algorithm that works as follows. The function to optimize is to obtain a RLE whose number of elements is the smallest possible. We then select the longest run using Eq. 10, keep it as an element of the final $\widetilde{RLE}$, and then re-apply this process, recursively (using a divide-and-conquer paradigm), to the left and to the right of the selected run, until we have considered all the temporal values of the pixel.

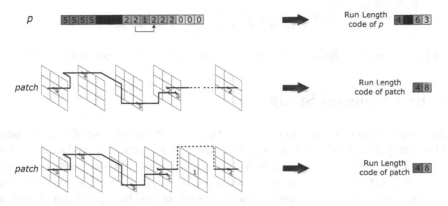

Fig. 2. Different ways for calculating an approximated Run Length Encoding ($\widetilde{RLE}$) by considering different relaxation strategies. (Top) equality with temporal relaxation; (Center) equality with spatial relaxation; (Bottom) equality with spatio-temporal relaxation.

The fundamental relation of the recursive process can be written:

$$\text{if } p = (p_t)_{t=1}^T \quad \text{with } \forall t, \ p_t = c \quad \text{then } \widetilde{RLE}(p) = T \tag{11}$$

$$\widetilde{RLE}(p) = \left(\widetilde{RLE}(p_1 \dots p_{ilr(p)}) LR(p) (\widetilde{RLE}(p_{ilr(p)+LR(p)} \dots p_T)) \right) \tag{12}$$

By considering these different relaxation strategies, we can obtain several approximations of the RLE for one temporal pixel p, noted $\widetilde{RLE}_{temp}$ when considering the temporal relaxation, $\widetilde{RLE}_{spatio}$ when considering the spatial relaxation, and $\widetilde{RLE}_{spatio-temp}$ when considering the spatio-temporal relaxation. Figure 2 illustrates the results of $\widetilde{RLE}$ with the different relaxation strategies. From these novel representations, the stability characteristics (Maximal Stability – MS, Max Stability Start – MSS and Number of changes – NB) can be computed leading to different versions of spatio-temporal features.

3.4 Stability Summarization

After extracting the three proposed features, following the hard equality strategy or a relaxed one, they can be used to summarize the SITS by combining them into one false color image, noted TS. The color composition of the summary image is following: MS in the red channel (R); NB in the green channel (G) and MSS in the blue channel (B).

In this way, instead of analyzing the whole series of images, we will only analyze a single image that summarizes all the SITS. The combination of the summary can be applied for the classical RLE (without relation) and for each approximated $\widetilde{RLE}$, we note respectively TS (from the classical RLE), TS_{temp}, TS_{spatio} and $TS_{spatio-temp}$ (from the $\widetilde{RLE}$) the resulting summary images.

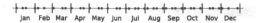

Fig. 3. Temporal distribution (over 2017) of the $T = 50$ images from the SITS.

4 Experimental Study

The proposed method has been employed in a remote sensing application related to the analysis of land-cover from a SITS. In this experimental study, we want to highlight the ability of our features to capture and to summarize spatio-temporal (stability) patterns, that cannot be handled using a single image of the SITS, the summary may be useful to assist in understanding and interpreting a terrestrial sensed scene. The experimental study is divided into two parts. In the first part, we apply the proposed method on a SITS in order to summarize it and to visualize the obtained results. In the second part, we involved the proposed spatio-temporal features in a binary classification task for the analysis of urban land-cover thematic classes.

4.1 Materials

The data used in this experimental study is an optical SITS composed of $T = 50$ images. The images, provided by the satellite Sentinel-2, have been sensed in 2017 over the same geographical area, East of France. The acquired images have been corrected and orthorectified by the French Theia program[1] to be radiometrically comparable. The images are distributed with their associated cloud, shadow and saturation masks. A pre-processing step was applied on the images with a linear interpolation on masked pixels to guarantee same size of all images of the time series. Figure 3 displays the temporal distribution of the images belonging to this SITS and Fig. 5(a) depicts two geographical areas (Strasbourg (top) and Mulhouse (bottom)) extracted from the SITS. Each image crop has a dimension of 1000^2 pixels and is composed of 4 spectral bands (Nir, R, G, B) at 10 m.

4.2 Stability Summarization for Land-Cover Analysis

In the context of our thematic study related to land-cover analysis, we chose to consider a remote sensing index, the NDVI, instead of considering all the four spectral bands. Indeed, the NDVI index is widely used in remote sensing studies to analyze land-cover from SITS since it is sensitive to the amount of vegetation. The NDVI is simply built as the multi-spectral product based on the Nir and R bands, leading to a SITS of NDVI, noted I_t^{NDVI}.

Then, according to our quantification strategy explained in Sect. 3.2, the SITS of NDVI is quantified leading to J_t^{NDVI}. In this context, we set empirically $k_{quantiz} = 4$. The proposed spatio-temporal features are then computed from J_t^{NDVI}. The extracted features are finally combined, leading to false color

---

[1] https://theia.cnes.fr/.

images, as explained in Sect. 3.4, which summarize the SITS of 50 images into unique images TS, TS_{temp}, TS_{spatio} and $TS_{spatio-temp}$. Concerning the spatial relaxations (Eq. 6), in the predicate to relax the spatial constraints, we set the spatial neighbor of a pixel to its $N = 9$ nearest neighbors.

In a comparative study, we considered as a baseline, a summarization of the STIS that consists, for each pixel of the original SITS I_t^{NDVI}, to average (through time) the values of the temporal pixels to obtain a single scalar value of NDVI. This (naive) baseline is noted $\overline{I_t^{NDVI}}$.

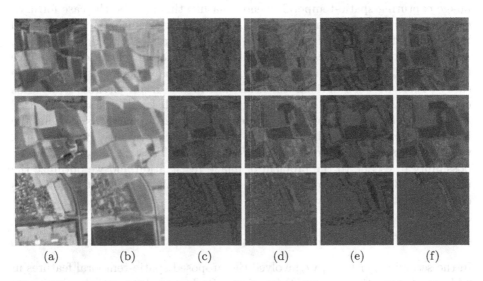

(a)   (b)   (c)   (d)   (e)   (f)

Fig. 4. Illustration of the data and results in three different geographical areas: (a) Image of the SITS at date 2017-08-26; (b) Result of the average of the SITS of NDVI $\overline{I_t^{NDVI}}$; (c), (d), (e) and (f) Obtained results of the proposed approach with the different relaxations, related respectively to TS, TS_{temp}, TS_{spatio} and $TS_{spatio-temp}$. (Color figure online)

Figure 4 presents the data of three different geographical regions sensed by the SITS and the obtained results: the two first lines are focused in agricultural areas and the third one is focused on a peri-urban area. The column (a) presents one original image of a SITS at date 2017-08-26, (b) presents the obtained results of the baseline summarization $\overline{I_t^{NDVI}}$ and (c), (d), (e), (f) are the results of the four proposed summarizations without/with the different relaxations presented, respectively TS, TS_{temp}, TS_{spatio} and $TS_{spatio-temp}$.

We remark that with $\overline{I_t^{NDVI}}$, the intensity of pixels in urban areas is very dark because the NDVI intensity increases in vegetation areas and decreases in other areas. In the agricultural areas, we visualize that we have about two noticeable gray levels that means that two types of grounds are present.

The obtained images with the proposed method allow a better visualization between all the thematic classes. The pixel colors are linked to the pixel evolution through time and a priori is a label of classes. The red color means that the region stays stable during a long time (high MS) with few changes (small NB) and its stability starts early in the year (low MSS), as illustrated in the third row in Fig. 4. The green color means that the region changes a lot through time with small MS and MSS. Such observations can easily allow a user, like a geographer, to better (and more conveniently) interpret the observed territories, by considering only a single image instead of an entire series, such an image capturing spatio-temporal phenomena like this may be the case for agricultural environments (seasons, different sowing or agricultural practices) and urban environments (constantly subject to various changes related to land-uses).

When comparing the results obtained with the different relaxations, we notice that with TS_{spatio} and $TS_{spatio-temp}$, roads, paths, parcel delineations are more easily visible between agricultural fields and in the urban environments. This can be surprising because when considering the traditional version of the RLE and the resulting TS image, these linear zones are not very visible, but the latter are elements normally showing a great temporal stability. It means that, when we relax the spatial domain with TS_{spatio} and $TS_{spatio-temp}$, we have the possibility to escape to the problem of image registration and we optimize the results by reducing noise (e.g. the salt and pepper). We notice that here there is only small differences between TS and TS_{temp}.

4.3   Classification of SITS with Spatio-Temporal Stability Features

In the second experiment, we involved the proposed spatio-temporal features in a binary classification task, for the analysis of urban land-cover thematic classes.

Reference Data. In addition to the SITS, we dispose of an imperviousness product that represents the percentage of the soil sealing. The imperviousness product is a result of a project of the European Copernicus program[2] released by the European Environment Agency. This product defines the impermeability of materials such as the urban areas (e.g. building, commercial zone, parking) and is provided at pixel level. The spatial resolution of this product is 20 m but we re-sampled it at 10 m to fit with the spatial resolution of the Sentinel-2 images. Each pixel value in this reference data estimates a degree of imperviousness (0–100%). In this thematical study, we used this product to discriminate between the natural areas (0% imperviousness) and artificialized areas (imperviousness >0%). Figure 5(b) illustrates the imperviousness reference data for two specific geographic areas. Given a SITS, the task is then to predict for each temporal pixel a binary label (i.e. artificialized vs. natural area).

---

[2] https://land.copernicus.eu/.

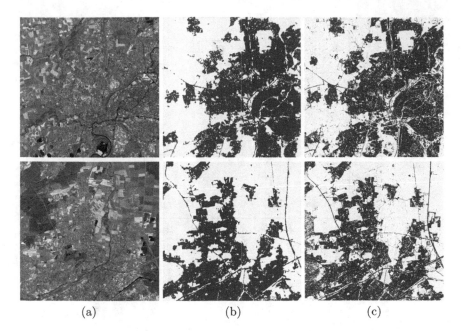

(a)   (b)   (c)

Fig. 5. Illustration of the data and results in two different geographical areas: (a) Image of the SITS at date 2017-08-26: (top) Strasbourg area and (bottom) Mulhouse area; (b) Imperviousness reference data: (red) artificialized areas and (yellow) natural areas; (c) Binary classification results from the TS features combined with *time series*. (Color figure online)

Classification Task. Our goal is to conduct a classification in order to analyze the urban land-cover using the proposed stability features. We assume they capture in very few variables the information contained in the temporal cube. We are going to evaluate the performance of these features (MS, MSS and NB, so 3 values), alone and in complement to the raw material, i.e. the temporal pixels characterized by the NDVI values ($T + 3 = 53$ values). The influence of the different relaxation strategies introduced previously will be also studied.

In order to make a comparative study, we will also consider two sets of data, where the pixels to be classified are characterized differently: (1) the pixels are characterized by their NDVI time series, noted *time series* (50 values since $T = 50$ dates) and (2) the pixels are characterized by the mean of each time series, noted $\overline{time\ series}$ (1 value) – in the same way as in Sect. 4.2.

According to the considered input size for each pixel (TS values (≤ 3) or *time series* values (≥ 50)), we have chosen to use a Decision Tree (DT) and a Random Forest (RF) for this classification task. The decision tree used is C4.5. The Random Forest classifier contains 30 trees. Each one is constructed by splitting nodes until we get pure ones. The used criterion to do the split is the classical "Gini". We also compared our approach with the use of a convolutional

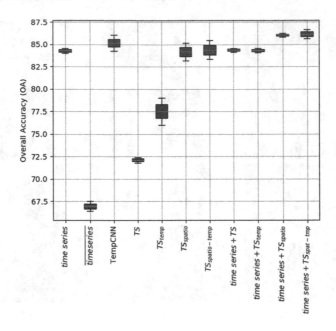

Fig. 6. Boxplot of the experiments, related to the results from Table 1.

neural network, TempCNN [18]. The convolutions are applied in the temporal domain, the input of the CNN is the NDVI *time series* (50 values).

The global process is to learn the two-class pixel classifiers on one geographical area and to test the classifier on a different area. Here, the Strasbourg and Mulhouse areas are concerned. This leads to two experiments, one learning on Strasbourg area and testing on Mulhouse area (Experience 1), the other one, on learning on Mulhouse and testing on Strasbourg (Experience 2).

Results. The evaluations provided in Table 1 and illustrated in Fig. 6 is done by computing the overall accuracy (OA) on all pixels of the test city areas.

Table 1 reports the obtained results with the different features. First we can notice that using all the *time series* gives much better results than when only the mean value is used $\overline{time\ series}$. We can also see that the use of the three features TS we have extracted from the time series provides much better results even if they are lower than when using the all *time series*. Besides we see the improvement brought by the relaxation processes proposed. The resulting spatio-temporal features enable to omit some punctual outlier values or to get rid of the registration problems we saw occurring on the different geographical limits. When processing TS features combined with *time series*, the accuracy scores are increased. The spatio and the spatio-temporal features give significantly higher results. We can also notice the TempCCN method [18], despite a longer learning phase, does not present better results than ours based on *time series* $+ TS_{spatio}$ or *time series* $+ TS_{spatio-temp}$ in this thematical study. Figure 5(c) illustrates

Table 1. Results for the two classification experiments (overall accuracy – OA).

Classifier	Features	Input size	Exp. 1	Exp. 2
RF	*time series*	50	84.56	84.02
DT	*time series*	1	66.40	67.49
CNN	TempCNN [18]	50	84.26	86.05
DT	TS	3	72.40	71.58
DT	TS_{temp}	3	76.01	79.03
DT	TS_{spatio}	3	83.17	85.18
DT	$TS_{spatio-temp}$	3	83.33	85.44
RF	*timeseries*+TS	53	84.58	84.15
RF	*timeseries*+TS_{temp}	53	84.56	84.07
RF	*timeseries*+TS_{spatio}	53	**85.85**	**86.23**
RF	*timeseries*+$TS_{spatio-temp}$	53	**85.65**	**86.65**

the visual classification results from the two experiments, by combining the TS features with *time series*.

Figure 6 enables to visualize the improvement that is globally brought using the stability features. They introduce non linear processing that is difficult to model without using recurrent studies of the time series, a temporal filtering window is often a parameter depending on the series content. As a whole, we have decreased the error rate of the initial classifier by about 10%.

5 Conclusion

We propose in this article an approach to define spatio-temporal features from SITS, based on a measure of the temporal stability. The proposal is based on a compression algorithm named Run Length Encoding (RLE), applied on the image data-cube, leading to a novel image representation from which stability features can be measured. Since the notion of stability involves the study of the repetition of successive values, we also study the notion of equality and its application level. One of our contribution relies on the definition of novel approximated versions of the RLE, by relaxing both temporal or spatial constraints in the predicates involved in the equality definition. From these new versions of RLE, we then proposed the definition of spatio-temporal stability features.

All the proposed features can be used in several applications such as SITS summarizing, to make easier the interpretation of the sensed territories from the original data-cube, or the classification of spatio-temporal patterns for land-cover analysis. The preliminary results obtained from a series of 50 Sentinel-2 images highlight the interest of our approach in a remote sensing application.

We plan to pursue our work on the notion of equality, used to decide if a pixel value is stable through time. As a limit of our work, we currently compute our spatio-temporal representation and the proposed features from mono-valued data, i.e. the temporal pixels characterized by the NDVI, that is a single scalar value. When the characteristic is vectorial, for instance all the spectral bands or a combination of remote sensing indexes, it is possible to try novel definitions of the equality, leading to a more or less constrained equality definition. Such vectorial approach is more suited when the considered scalar characteristics are independent, that is not yet the case with those involved in our application.

References

1. Aminikhanghahi, S., Cook, D.J.: A survey of methods for time series change point detection. Knowl. Inf. Syst. **51**(2), 339–367 (2016). https://doi.org/10.1007/s10115-016-0987-z
2. Andres, L., Salas, W., Skole, D.: Fourier analysis of multi-temporal AVHRR data applied to a land cover classification. IJRS **15**(5), 1115–1121 (1994)
3. Bagnall, A., Lines, J., Bostrom, A., Large, J., Keogh, E.: The great time series classification bake off: a review and experimental evaluation of recent algorithmic advances. Data Min. Knowl. Disc. **31**(3), 606–660 (2016). https://doi.org/10.1007/s10618-016-0483-9
4. Bruzzone, L., Prieto, D.: Automatic analysis of the difference image for unsupervised change detection. IEEE TGRS **38**(3), 1171–1182 (2000)
5. Chelali, M., Kurtz, C., Puissant, A., Vincent, N.: Urban land cover analysis from satellite image time series based on temporal stability. In: Proceedings of the JURSE (2019)
6. Coppin, P., Jonckheere, I., Nackaerts, K., Muys, B., Lambin, E.: Digital change detection methods in ecosystem monitoring: a review. IJRS **25**(9), 1565–1596 (2004)
7. Di Mauro, N., Vergari, A., Basile, T.M.A., Ventola, F.G., Esposito, F.: End-to-end learning of deep spatio-temporal representations for satellite image time series classification. In: Proceedings of the DC@PKDD/ECML, pp. 1–8 (2017)
8. Golomb, S.W.: Run-length encodings. IEEE TIF **12**, 399–401 (1966)
9. Huang, B., et al.: Large-scale semantic classification: outcome of the first year of inria aerial image labeling benchmark. In: Proceedings of the IGARSS, pp. 6947–6950 (2018)
10. Ienco, D., Gaetano, R., Dupaquier, C., Maurel, P.: Land cover classification via multitemporal spatial data by deep recurrent neural networks. IEEE GRSL **14**(10), 1685–1689 (2017)
11. Inglada, J., Vincent, A., Arias, M., Tardy, B., Morin, D., Rodes, I.: Operational high resolution land cover map production at the country scale using satellite image time series. RS **9**(1), 95–108 (2017)
12. Ismail Fawaz, H., Forestier, G., Weber, J., Idoumghar, L., Muller, P.-A.: Deep learning for time series classification: a review. Data Min. Knowl. Disc. **33**(4), 917–963 (2019). https://doi.org/10.1007/s10618-019-00619-1
13. Jensen, J.R.: Urban change detection mapping using Landsat digital data. CGIS **8**(21), 127–147 (1981)
14. Johnson, R., Kasischke, E.: Change vector analysis: a technique for the multispectral monitoring of land cover and condition. IJRS **19**(16), 411–426 (1998)
15. Julea, A., et al.: Unsupervised spatiotemporal mining of satellite image time series using grouped frequent sequential patterns. IEEE TGRS **49**(4), 1417–1430 (2011)

16. Millward, A.A., Piwowar, J.M., Howarth, P.J.: Time-series analysis of medium-resolution, multisensor satellite data for identifying landscape change. PERS **72**(6), 653–663 (2006)
17. Nielsen, A.A., Conradsen, K., Simpson, J.J.: Multivariate alteration detection (MAD) and MAF postprocessing in multispectral, bitemporal image data: new approaches to change detection studies. RSE **64**(1), 1–19 (1998)
18. Pelletier, C., Webb, G., Petitjean, F.: Temporal convolutional neural network for the classification of satellite image time series. RS **11**(5), 523–534 (2019)
19. Petitjean, F., Inglada, J., Gançarski, P.: Satellite image time series analysis under time warping. IEEE TGRS **50**(8), 3081–3095 (2012)
20. Ratanamahatana, C., Keogh, E., Bagnall, A.J., Lonardi, S.: A novel bit level time series representation with implication of similarity search and clustering. In: Ho, T.B., Cheung, D., Liu, H. (eds.) PAKDD 2005. LNCS (LNAI), vol. 3518, pp. 771–777. Springer, Heidelberg (2005). https://doi.org/10.1007/11430919_90
21. Ravikumar, P., Devi, V.S.: Weighted feature-based classification of time series data. In: Proceedings of the CIDM, pp. 222–228 (2014)
22. Russwurm, M., Korner, M.: Temporal vegetation modelling using long short-term memory networks for crop identification from medium-resolution multi-spectral satellite images. In: Proceedings of the EarthVision@CVPR, pp. 1496–1504 (2017)
23. Senf, C., Leitao, P., Pflugmacher, D., Van der Linden, S., Hostert, P.: Mapping land cover in complex Mediterranean landscapes using Landsat: improved classification accuracies from integrating multi-seasonal and synthetic imagery. RSE **156**, 527–536 (2015)
24. Verbesselt, J., Hyndman, R., Newnham, G., Culvenor, D.: Detecting trend and seasonal changes in satellite image time series. RSE **114**(1), 106–115 (2010)

Overview of Crowd Counting

Peizhi Zeng$^{(\boxtimes)}$ and Jun Tan

School of Mathematics, Sun Yat-sen University, Guangzhou, China
1069319173@qq.com

Abstract. Recently, counting the number of people for crowd scenes is a hot topic because of its widespread applications (e.g. video surveillance, public security). The stampede incidents frequently occur in large-scale activities at home and abroad, which have caused a lot of casualties. For example, in the Shanghai Bund stampede incident in 2015, it has reached the level of major casualties prescribed by China. Therefore, the research on the population counting problem is getting hotter and hotter. If the population density of the current scene is accurately estimated and the corresponding security measures are arranged, the occurrence of such events can be effectively reduced or avoided. For this reason, this paper reviews the history of the development of population counting in recent years and some models that are more classic and have better effects. We will also consider pros and cons of these approaches.

Keywords: Crowd counting · Density map · Density estimation · Detection · Convolutional neural networks.

1 Introduction

With the rapid pace of urbanization, crowds tend to gather more frequently, increasing requirements for effective safety monitoring, disaster relief, urban planning, and crowd management. Counting or estimating crowd size, both a priori and in real-time, is an essential element for planning and maintaining crowd safety in places of public assembly [1].

In recent years, with the rapid growth of population, group counting has been widely used in video surveillance, traffic control and sports events. Early research focused on estimating the number of people by detecting the body or head, while other methods learned the mapping from local or global features to actual quantities. Recently, the population count problem was formulated as a regression of the population density map, and then the number of people in the image was obtained by summing the values of the density map. This approach can handle serious occlusions in dense crowd images. With the success of deep learning techniques, researchers used convolutional neural networks (CNN) to generate accurate population density maps and achieved better performance than traditional methods. Good methods of crowd counting can also be extended to other domains, for instance, counting cells or bacteria from microscopic images, animal

© Springer Nature Switzerland AG 2020
Y. Lu et al. (Eds.): ICPRAI 2020, LNCS 12068, pp. 500–512, 2020.
https://doi.org/10.1007/978-3-030-59830-3_43

crowd estimates in wildlife sanctuaries, or estimating the number of vehicles at transportation hubs or traffic jams, etc.

However, population counting is still a challenging task due to large scale variations, severe occlusion, background noise, and perspective distortion. Among them, scale change is the most important issue. To better handle scale changes, researchers have proposed many multi-column or multi-branch networks. These architectures typically consist of several columns of CNN or several branches at different stages of the backbone network. These columns or branches have different receptive fields to sense changes in population size. Although these methods have been greatly improved, the scale diversity they capture is limited by the number of columns or branches.

The key challenges of scale variation lie in two aspects. First, as shown in Fig. 1 left, the people in crowd image often have very different sizes, ranging from several pixels to tens of pixels. This requires the network to be able to capture a large rang of scales. Second, as shown in Fig. 1 right, the scale usually varies continuously across the image, especially for high density images. This requires the network to be able to sample the captured scale range densely. However, non of existing methods can deal with these two challenges simultaneously.

Fig. 1. Large scale variations exist in crowd counting datasets. Left: Input image and corresponding ground truth density map from ShanghaiTech dataset [19]. Right: Input image and corresponding ground truth density map from UCF-QNRF dataset [20]

In this paper, we will compare the advantages and disadvantages of several frontier methods in the field of crowd counting, and provide some ideas for future improvement based on the existing problems.

2   Related Works

2.1   Traditional Crowd Counting Methods

A method based on detection. Early population research focused on detection-based methods. [2] Use a sliding window detector to detect people in the scene and count the number of people. Detection-based methods are mainly divided into two categories, one based on holistic detection and the other based on partial body detection. Based on the overall detection method, such as [3–6], the typical traditional method, mainly trains a classifier to detect pedestrians using features such as wavelet, HOG, and edge extracted from the whole body of the pedestrian. Learning algorithms mainly include SVM, boosting and random forest. The method based on the overall detection is mainly applied to the sparse crowd counting. As the density of the population increases, the occlusion between people becomes more and more serious. Therefore, based on partial body detection methods, it is used to deal with population counting problems [7,8], mainly by detecting part of the body structure, such as head, shoulders, etc, to count the number of people. This method has a slight improvement in effect compared to the overall based detection.

A regression-based approach. Regardless of the detection-based approach, it is difficult to deal with serious occlusion problems between people. Therefore, regression-based methods are gradually being used to solve the problem of crowd counting. Based on the regression method, the main idea is to learn the mapping of a feature to the number of people [9–11]. This method step is mainly divided into two steps. The first step is to extract low-level features such as foreground features, edge features, textures and gradient features. The second step is to learn a regression model such as linear regression, piecewise linear regression, and ridge regression. And Gaussian process regression and other methods to learn a mapping relationship between a low-level feature and the number of people.

2.2   Deep Learning-Driven Crowd Counting Deep Learning

Density estimation based methods, in particular, have received increasing research focus. These techniques have the ability to localize the crowd by generating a density estimation map using pixel-wise regression. The crowd count is then calculated as the integral of the density map. To generate maps with a retained spatial size as the inputs, deep encoder-decoder convolutional neural network (CNN) architectures are widely applied [10,12–14]. More details about this will be introduced in the second half of this article.

3   Introduction to the Current Popular Architeture

Crowd images are often captured from varying view points, resulting in a wide variety of perspectives and scale variations. People near the camera are often captured in a great level of detail i.e., their faces and at times their entire body is captured. However, in the case of people away from camera or when images are captured from an aerial viewpoint, each person is represented only as a head blob. Efficient detection of people in both these scenarios requires the model to simultaneously operate at a highly semantic level (faces/body detectors) while also recognizing the low-level head blob patterns. Although these multi-column architectures prove the ability to estimate crowd count, several disadvantages also exist in these approaches: they are hard to train caused by the multi-column architecture, and they have large amount of redundant parameters, also the speed is slow as multiple CNNs need to be run. Taking all above drawback into consideration, recent works have focused on multi-scale, single column architectures. However, all these single-column works can only capture several kinds of receptive fields, which limits the network to handle large variations in crowd images.

3.1   CrowdNet [15]

This paper [15] proposes a new learning framework for detecting population density from high-density populations. Their model achieves this using a combination of deep and shallow convolutional neural networks. An overview of the proposed architecture is shown in Fig. 2. They proposes a novel deep learning framework for estimating crowd density from static images of highly dense crowds. They use a combination of deep and shallow, fully convolutional networks to predict the density map for a given crowd image. Such a combination is used for effectively capturing both the high-level semantic information (face/body detectors) and the low-level features (blob detectors), that are necessary for crowd counting under large scale variations. As most crowd datasets have limited training samples, they also perform multiscale data augmentation.

Their deep network captures the desired high-level semantics required for crowd counting using an architectural design similar to the well-known VGG-16 [21] network. Although the VGG-16 architecture was originally trained for the purpose of object classification, the learned filters are very good generic visual descriptors and have found applications in a wide variety of vision tasks such as saliency prediction, object segmentation etc. However, crowd density estimation requires perpixel predictions unlike the problem of image classification, where a single discrete label is assigned for an entire image. They obtain these pixel-level predictions by removing the fully connected layers present in the VGG architecture, thereby making their network fully convolutional in nature.

Deep Network mainly uses to capture high-level semantics information. Here they use a structure similar to VGG network. They remove the fully connected layer and the network becomes a fully convolutional layer. At the same time, the original VGG network used 5 max-pool layers each with a stride of 2, and the final feature map size was only 1/32 of the input image size. They need to output the pixel density crowd map here, so they set the stride of the fourth max-pool layer to 1 and remove the fifth pooling layer, so that the size of the final feature map is only 1/8 of the input image size. As for the shallow Network, they aim to recognize the low-level head blob patterns, arising from people away from the camera, using a shallow convolutional network. To ensure that there is no loss of count due to max-pooling, they use average pooling layers in the shallow network. They concatenate the predictions from the deep and shallow networks, each having a spatial resolution of 1/8 times the input image, and process it using a 1×1 convolution layer. The output from this layer is upsampled to the size of the input image using bilinear interpolation to obtain the final crowd density prediction. They use Mean Absolute Error (MAE) to quantify the performance of the method. MAE computes the mean of absolute difference between the actual count and the predicted count for all the images in the dataset. The results of the proposed model is 452.5 on UCF_CC_50 Dataset.

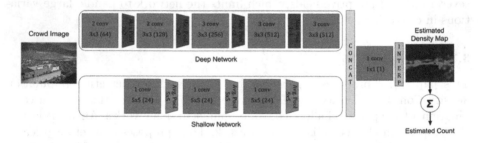

Fig. 2. Overview of the proposed architecture for crowd counting

3.2   MCNN [16]

This paper [16] aims to develop a method that can accurately estimate the crowd count from an individual image with arbitrary crowd density and arbitrary perspective. To this end, they have proposed a simple but effective Multi-column Convolutional Neural Network (MCNN) architecture to map the image to its crowd density map. The proposed MCNN allows the input image to be of arbitrary size or resolution. By utilizing filters with receptive fields of different sizes, the features learned by each column CNN are adaptive to variations in people/head size due to perspective effect or image resolution. Furthermore, the true density map is computed accurately based on geometry-adaptive kernels which do not need knowing the perspective map of the input image.

The contributions of this article are as follows: 1) A multi-column architecture is used, which has different receptive fields and can process crowd images of different sizes. 2) MCNN uses a full convolution network, and the input can be of any size. 3) A new database Shanghaitech has been built.

The structure of 3-col CNN is pretty simple and straightforward to under-stand, there are several points about the network:

- The intention of multi-col design is for addressing heads of different sizes, increasing the network's robustness to different situations.
- There is no fully-connected layer inside MCNN, which is flexible to inputs of different sizes.
- Before convolving with the final 1×1 conv layer, feature maps from 3 net-works are concatenated together. It should be noted that all feature maps before concatenating share the same size only different number of channels, so they can be concatenated.
- During training, the output density map become smaller. In order to calculate loss, the author downsampled the original input to of the same size with output, then generated corresponding ground truth density map. Finally, they use Euclidean distance as training loss.

And the model structure of that paper is shown as Fig. 3. And the MAE of the UCF_CC_50 dataset is 377.6.

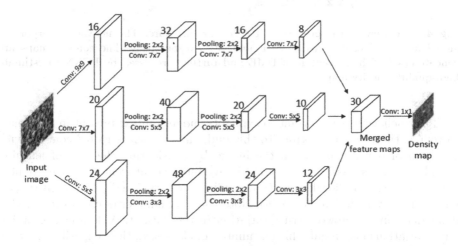

Fig. 3. The structure of the proposed multi-column convolutional neural network for crowd density map estimation

3.3   CP-CNN [17]

The proposed CP-CNN [17] method consists of a pyramid of context estimators and a Fusion-CNN. It consists of four modules: GCE, LCE, DME, and F-CNN. GCE and LCE are CNN-based networks that encode global and local context present in the input image respectively. DME is a multi-column CNN that per-forms the initial task of transforming the input image to high-dimensional fea-ture maps. Finally, F-CNN combines contextual information from GCE and

LCE with highdimensional feature maps from DME to produce highresolution and high-quality density maps. And the model structure of that paper is shown as Fig. 4. The original intention of this method is to consider the global density and local density information of the crowd in an image, and finally constrain the whole feature, so that the network can adaptively learn the characteristics of the corresponding density level for any image.

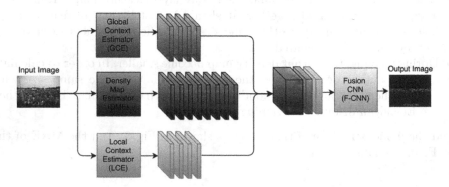

Fig. 4. Overview of the proposed CP-CNN architecture. The network incorporates global and local context using GCE and LCE respectively. The context maps are concatenated with the output of DME and further processed by FCNN to estimate high-quality density maps.

They believe it is important to explicilty model context present in the image to reduce the estimation error. To this end, they associate global context with the level of density present in the image by considering the task of learning global context as classifying the input image into five different classes: extremely low-density (ex-lo), lowdensity (lo), medium-density (med), high-density (hi) and extremely high-density (ex-hi). Note that the number of classes required is dependent on the crowd density variation in the dataset. A dataset containing large variations may require higher number of classes. In their experiments, they obtained significant improvements using five categories of density levels. As for GCE, the convolutional layers from the VGG-16 network are retained, however, the last three fully connected layers are replaced with a different configuration of fully connected layers in order to cater to our task of classification into five categories.

Existing methods for crowd density estimation have primarily focussed on achieving lower count errors rather than estimating better quality density maps. After an analysis of these results, they believe that some kind of local contextual information can aid us to achieve better quality maps. To this effect, similar to GCE, they propose to learn an image's local context by learning to classify it's local patches into one of the five classes: ex-lo, lo, med, hi, ex-hi. The local context is learned by the LCE.

The aim of DME is to transform the input image into a set of high-dimensional feature maps which will be concatenated with the contextual information provided by GCE and LCE. Estimating density maps from high-density crowd images is especially challenging due to the presence of heads with varying sizes in and across images. Inspired by the success of these methods, they use a multi-column architecture. Also, in this work, the multicolumn architecture is used to transform the input into a set of high-dimensional feature map rather than using them directly to estimate the density map.

The contextual information from GCE and LCE are combined with the high-dimensional feature maps from DME using F-CNN. Their model has also achieved a relatively good improvement on the UCF_CC_50 dataset.

3.4 CSRNet [18]

Previous analysis of crowded scenes was mainly based on multi-scale architectures. They have achieved high performance in this field, but as the network becomes deeper, the design they use also brings two significant disadvantages: a large amount of training time and invalid branching structures (for example, MCNN). This is also demonstrated in the paper by experiments. Dilated convolutional layers have been proven in segmentation tasks, their accuracy is significantly improved, and it is a good alternative to pooling layers. Although pooling layers (eg, maximum and average pooling) are widely used to maintain invariance and control overfitting, they also significantly reduce spatial resolution, which means that the spatial information of feature maps is lost. Deconvolution layers can reduce the loss of information, but the additional complexity and execution latency may not be suitable for all situations. Dilated convolution is a better choice, it uses a sparse kernel to alternately pool and convolve layers. Dilated convolution expands the receptive field without increasing the number of parameters or calculations.

This method [18] is inspired by the field of image generation. This step-by-step generation from low resolution to high resolution is quite common in the field of image generation. In this paper, they proposed a novel architecture called CSRNet for crowd counting and high-quality density map generation with an easy-trained end-to-end approach. They used the dilated convolutional layers to aggregate the multiscale contextual information in the congested scenes.

The proposed CSRNet is composed of two major components:

- a convolutional neural network (CNN) is utilized as the front-end for 2D feature extraction.
- a dilated CNN follows the front-end for the back-end, which uses dilated kernels to deliver larger reception fields and to replace pooling operations.

In this work, the front-end CNN is same as the first ten layers of VGG-16 with three pooling layers, considering the trade off between accuracy and the resource overhead. The back-end CNN is a series of dilated convolutional layers and the last layer is a $1 \times 1 \times 1$ convolutional layer producing density map. The network architecture is shown in the following Fig. 5 which is copy from [18].

Configurations of CSRNet			
A	B	C	D
input(unfixed-resolution color image)			
front-end (fine-tuned from VGG-16)			
conv3-64-1 conv3-64-1			
max-pooling			
conv3-128-1 conv3-128-1			
max-pooling			
conv3-256-1 conv3-256-1 conv3-256-1			
max-pooling			
conv3-512-1 conv3-512-1 conv3-512-1			
back-end (four different configurations)			
conv3-512-1	conv3-512-2	conv3-512-2	conv3-512-4
conv3-512-1	conv3-512-2	conv3-512-2	conv3-512-4
conv3-512-1	conv3-512-2	conv3-512-2	conv3-512-4
conv3-256-1	conv3-256-2	conv3-256-4	conv3-256-4
conv3-128-1	conv3-128-2	conv3-128-4	conv3-128-4
conv3-64-1	conv3-64-2	conv3-64-4	conv3-64-4
conv1-1-1			

Fig. 5. Configuration of CSRNet. All convolutional layers use padding to maintain the previous size. The convolutional layers' parameters are denoted as "conv-(kernel size)-(number of filters)-(dilation rate)", max-pooling layers are conducted over a 2×2 pixel window with stride 2.

The output size of this front-end network is $1/8$ of the original input size. If we continue to stack more convolutional and pooling layers (the basic components in VGG-16), the output size will be further reduced and it will be difficult to generate high-quality density maps. They try to use the dilated convolutional layer as the back end to extract deeper saliency information and maintain output resolution.

As can be seen in Fig. 5, there are four CSRNet network configurations, which have the same front-end structure but different back-end expansion rates. Regarding the front end, they use VGG-16 network (except the fully connected layer) and only use the 3×3 convolution kernels. According to VGG's paper, when using receptive fields of the same size, using more convolutional layers with small kernels is more efficient than using fewer layers with larger kernels.Finally,on UCF_CC_50 dataset, the MAE of CSRNet is 266.1.

3.5   DSNet [22]

The basic idea of this method is an end-to-end single-row CNN with more dense scale diversity to cope with large scale changes and density level differences in dense and sparse scenes. The architecture of DSNet is shown in Fig. 6. The proposed DSNet contains backbone network as feature extractor, three dense dilated convolution blocks stacked by dense residual connections that enlarges denser scale diversity, and three convolutional layers for crowd density map regression. Following CSRNet, they keep the first ten layers of VGG-16 with only three pooling layers to be their backbone network. Thus to tackle the challenges of scale variation, they propose a new dense dilated convolution block that contains three dilated convolutional layers with increasing dilation rate of 1, 2, 3. This setting preserves information from denser scales with small gap of receptive field size, i.e. 2 pixels. Another advantage of our carefully selected dilation rates is that it can overcome the gridding artifacts in DenseASPP [23].

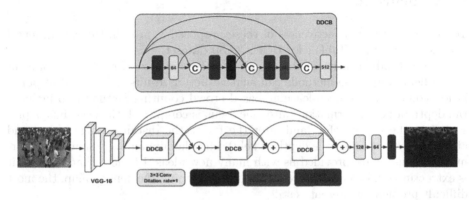

Fig. 6. The architecture of the proposed dense scale network (DSNet) for crowd counting. The DSNet consists of backbone network with the front ten layers of VGG-16, three dense dilated convolution blocks (DDCBs) with dense residual connections (DRCs), and three convolutional layers for crowd density map regression. The DDCBs with DRCs are used to enlarge scale diversity and receptive fields of features to handle large scale variations so that density maps can be estimated accurately.

This paper presents a new dense-scale single-column neural network, DSNet, for crowd counting. DSNnet consists of densely connected expansion convolution blocks, so it can output features with different receptive fields and capture crowd information at different scales. DSNet's convolution block is similar in structure to DenseASPP [23], but with different combinations of expansion rates. The author carefully chooses these ratios for the layers within the block, so that each block samples densely for continuously changing scales. At the same time, the selected expansion ratio combination can use all pixels of the receptive field to perform feature calculations to prevent gridding effects. In order to further improve the scale diversity captured by DSNet, the author stacks three dense expansion convolution blocks and uses residual connection for dense connection. The final network can sample a very large scale change range in a more dense manner, so that it can deal with the problem of large scale changes in population counting.

Most previous methods used traditional Euclidean loss to train the network, which is based on the assumption of pixel independence. This loss ignores the global and local consistency of the density map and can affect the results of population counting. In order to solve this problem, the authors propose a multi-scale density level consistency loss to ensure that the global and local density levels are consistent between the estimated population density map and the real population density map.

The authors performed extensive experiments on four challenging public population statistics sets. This method achieves the best performance compared to the most advanced methods available.

4   Conclusion

To demonstrate the effectiveness of these proposed approach, the experimental results are shown in Table 1 for UCF_CC_50 dataset.

The methods mentioned above have greatly contributed to the development of the field, and these methods can achieve better results in the field of population counting. The deep-learning-based crowd counting method also includes the depth of field information (after the focus is completed, the clear image presented in the range before and after the focus, the distance range before and after the focus is called the depth of field), and the human body structure information. These have provided us with many new ideas. I believe that whether it is extracting multi-scale features or gradually refining the density map, the most difficult problem at present is still

- how to make the model more subtle distinguishing features (such as human head overlap) in crowd-intensive areas
- How can the model "see" the local features of smaller scales in a wide-ranging image of population density.

Table 1. Comparison with state-of-the-art methods on UCF_CC_50.

Method	MAE	MSE
CrowdNet [15]	452.5	*
MCNN [16]	377.6	509.1
CP-CNN [17]	295.8	320.9
CSRNet [18]	266.1	397.5
DSNet [22]	**183.3**	**240.6**

References

1. Keith Still, G.: Crowd science and crowd counting. Impact **2019**(1), 19–23 (2019)
2. Dollar, P., Wojek, C., Schiele, B., et al.: Pedestrian detection: an evaluation of the state of the art. IEEE Trans. Pattern Anal. Mach. Intell. **34**(4), 743–761 (2011)
3. Dalal, N., Triggs, B.: Histograms of oriented gradients for human detection. In: 2005 IEEE Computer Society Conference on Computer Vision and Pattern Recognition (CVPR 2005), vol. 1, pp. 886–893. IEEE (2005)
4. Leibe, B., Seemann, E., Schiele, B.: Pedestrian detection in crowded scenes. In: 2005 IEEE Computer Society Conference on Computer Vision and Pattern Recognition (CVPR 2005), vol. 1, pp. 878–885. IEEE (2005)
5. Enzweiler, M., Gavrila, D.M.: Monocular pedestrian detection: survey and experiments. IEEE Trans. Pattern Anal. Mach. Intell. **31**(12), 2179–2195 (2008)
6. Tuzel, O., Porikli, F., Meer, P.: Pedestrian detection via classification on riemannian manifolds. IEEE Trans. Pattern Anal. Mach. Intell. **30**(10), 1713–1727 (2008)
7. Felzenszwalb, P.F., Girshick, R.B., McAllester, D., et al.: Object detection with discriminatively trained part-based models. IEEE Trans. Pattern Anal. Mach. Intell. **32**(9), 1627–1645 (2009)
8. Wu, B., Nevatia, R.: Detection and tracking of multiple, partially occluded humans by bayesian combination of edgelet based part detectors. Int. J. Comput. Vision **75**(2), 247–266 (2007)
9. Chan, A.B., Vasconcelos, N.: Bayesian poisson regression for crowd counting. In: 2009 IEEE 12th International Conference on Computer Vision, pp. 545–551. IEEE (2009)
10. Ryan, D., Denman, S., Fookes, C., et al.: Crowd counting using multiple local features. In: 2009 Digital Image Computing: Techniques and Applications, pp. 81–88. IEEE (2009)
11. Chen, K., Loy, C.C., Gong, S., et al.: Feature mining for localised crowd counting. BMVC **1**(2), 3 (2012)
12. Dong, C., Loy, C.C., Tang, X.: Accelerating the super-resolution convolutional neural network. In: Leibe, B., Matas, J., Sebe, N., Welling, M. (eds.) ECCV 2016. LNCS, vol. 9906, pp. 391–407. Springer, Cham (2016). https://doi.org/10.1007/978-3-319-46475-6_25
13. Tao, X., Gao, H., Liao, R., et al.: Detail-revealing deep video super-resolution. In: Proceedings of the IEEE International Conference on Computer Vision, pp. 4472–4480 (2017)
14. Zhao, H., Shi, J., Qi, X., et al.: Pyramid scene parsing network. In: Proceedings of the IEEE Conference on Computer Vision and Pattern Recognition, pp. 2881–2890 (2017)
15. Boominathan, L., Kruthiventi, S.S.S., Babu, R.V.: Crowdnet: a deep convolutional network for dense crowd counting. In: Proceedings of the 24th ACM International Conference on Multimedia, pp. 640–644 (2016)
16. Zhang, Y., Zhou, D., Chen, S., et al.: Single-image crowd counting via multi-column convolutional neural network. In: Proceedings of the IEEE Conference on Computer Vision and Pattern Recognition, pp. 589–597 (2016)
17. Sindagi, V.A., Patel, V.M.: Generating high-quality crowd density maps using contextual pyramid CNNs. In: Proceedings of the IEEE International Conference on Computer Vision, pp. 1861–1870 (2017)
18. Li, Y., Zhang, X., Chen, D.: CSRNet: dilated convolutional neural networks for understanding the highly congested scenes. In: Proceedings of the IEEE Conference on Computer Vision and Pattern Recognition, pp. 1091–1100 (2018)

19. Zhang, Y., Zhou, D., Chen, S., Gao, S., Ma, Y.: Singleimage crowd counting via multi-column convolutional neural network. In: Proceedings of the IEEE Conference on Computer Vision and Pattern Recognition, pp. 589–597 (2016)
20. Idrees, H., et al.: Composition loss for counting, density map estimation and localization in dense crowds. In: Ferrari, V., Hebert, M., Sminchisescu, C., Weiss, Y. (eds.) ECCV 2018. LNCS, vol. 11206, pp. 544–559. Springer, Cham (2018). https://doi.org/10.1007/978-3-030-01216-8_33
21. Simonyan, K., Zisserman, A.: Very deep convolutional networks for large-scale image recognition. arXiv preprint arXiv:1409.1556 (2014). 3.1.1
22. Dai, F., Liu, H., Ma, Y., et al.: Dense scale network for crowd counting[J]. arXiv preprint arXiv:1906.09707 (2019)
23. Yang, M., Yu, K., Zhang, C., Li, Z., Yang, K.: DenseASPP for semantic segmentation in street scenes. In: 2018 IEEE/CVF Conference on Computer Vision and Pattern Recognition, Salt Lake City, UT, 2018, pp. 3684–3692 (2008). https://doi.org/10.1109/CVPR.2018.00388

Rotation-Invariant Face Detection with Multi-task Progressive Calibration Networks

Li-Fang Zhou[1,2]($\boxtimes$) , Yu Gu[1] , Patrick S. P. Wang[3] , Fa-Yuan Liu[4] , Jie Liu[2] , and Tian-Yu Xu[2]

[1] College of Computer Science and Technology, Chongqing
University of Posts and Telecommunications, Chongqing 400065, China
zhoulf@cqupt.edu.cn
[2] College of Software Engineering, Chongqing University of Posts and Telecommunications,
Chongqing 400065, China
[3] Northeastern University, Boston, MA 02115, USA
[4] Petrochina Chongqing Marketing Company, Chongqing 400065, China

Abstract. Rotation-invariant face detection (RIFD) aims to detect faces with arbitrary rotation-in-plane (RIP) on both images and video sequences, which is challenging because of large appearance variations of rotated faces under unconstrained scenarios. The problem becomes more formidable when the speed and memory efficiency of the detectors are taken into account. To solve this problem, we propose a Multi-Task Progressive Calibration Networks (MTPCN), which not only enjoy the natural advantage of explicit geometric structure representation, but also retain important cue to guide the feature learning for precise calibration. More concretely, our framework leverages a cascaded architecture with three stages of carefully designed convolutional neural networks to predict face and landmark location with gradually decreasing RIP ranges. In addition, we propose a novel loss by further integrating geometric information into penalization, which is much more reasonable than simply measuring the differences of training samples equally. MTPCN achieves significant performance improvement on the multi-oriented FDDB dataset and the rotation WIDER face dataset. Extensive experiments are performed, and demonstrate the effectiveness of our method for each of three tasks.

Keywords: Rotation-invariant face detection · Face alignment · Multi-task learning

1 Introduction

Face detection is an important and long-standing problem in computer vision. A number of methods have been proposed in the past, including neural network based methods [1, 2, 11, 21], cascade structures [3, 8] and deformable part models (DPM) [4] detectors. Although such succinct structures have achieved remarkable results for the face detection

© Springer Nature Switzerland AG 2020
Y. Lu et al. (Eds.): ICPRAI 2020, LNCS 12068, pp. 513–524, 2020.
https://doi.org/10.1007/978-3-030-59830-3_44

task, it is still inadequate to meet the rigorous requirements of general applications (*e.g., visual surveillance system, digital equipments that need autofocus on faces, etc.*) as the face images can be captured almost from any RIP angle and undergo the large variations of face appearances.

Many effective methods were proposed to handle these challenges [1, 2, 9, 24]. Currently, the most effective RIFD method exhibits the human face with a rotated bounding box by conducting estimation of the RIP angle and then the face candidates can be rotated according to the predicted one. Such a detailed representation is more desirable for applications that require explicit face regions without other background; what is more, it does inherently model the rotation variations to guide the face detection in each face image. While those approaches just can be successfully applied to RIFD in many photos, their performance on the most challenging datasets [5] still leaves room for improvement. The following summarizes issues regarding the RIFD accuracy into two challenges:

Challenge # 1 - Semantic Ambiguity. It is not uncommon that, the presence of semantic ambiguity could deteriorate the performance of many fundamental computer vision tasks, such as object detection and tracking [17, 26, 28]. This problem emerges naturally in applications such as RIFD, where the goal is to predict RIP angle within each face candidate with the simple supervision of axis-aligned boxes along with face pose (typical pose or atypical pose) information. The semantic ambiguity results from the lack of clear definition on those pose annotations. Thus, the existing RIFD approaches could suffer from misleading training because of random annotation noises and degraded performance as pointed out by previous studies [28].

Challenge # 2 - Data Imbalance. Deep neural networks are a powerful tool for learning image representations in computer vision applications. However, training deep networks, in order to capture the large visual variations from one or few examples, is challenging [27]. Therefore, sufficient training data for data-driven approaches is also key to model performance. The recent introduction of the WIDER face dataset [5], is still limited in size and scenarios covered, due to the difficulties in data collection and annotation. Under the circumstances, the performance of general face detectors drop dramatically when the scale of training data is not sufficiently large.

To solve the problems above, we present Multi-Task Progressive Calibration Networks (MTPCN), which exploits the inherent correlation between detection and alignment to boost up their performance. MTPCN is inspired by PCN [9]. PCN casts RIFD as a multi-class classification problem and turns the classification results into regression by calculating the expected value as the RIP orientation of each face candidate. One problem with PCN is imprecise annotations. MTPCN addresses this issue by jointing face alignment learning, since it is feasible to obtain the RIP angles automatically extracted from the given ground-truth landmarks. Moreover, face alignment identifies geometry structure of human face which can be viewed as modeling highly structured output. Instead of completely relying on a standard supervisory source consists of a simple axis-aligned bounding box, such a detailed representation is more desirable for RIFD. Another problem with the regression-based RIP angles estimation approaches is ambiguous mapping between face appearance and its precise RIP angle. We turn the regression problem of RIP angles estimation into solving face alignment problem and then calculating the expected RIP angles from the predicted landmarks. By integrating all these technical innovations, we obtain a robust and accurate approach to locate face regions along with five facial landmarks for face images with full 360° RIP angles.

In short, our contributions are summarized as follows:

1. We propose a Multi-Task Progressive Calibration Networks (MTPCN) for rotation-invariant face detection. Our model provides an explicit mechanism for RIP variations modeling in the face detection problem, and is designed to integrate detection and alignment using unified cascaded CNNs by multi-task learning, which reduces the 'semantic ambiguity' intrinsically.
2. Considering the geometrical regularization, in order to mitigate the issue of data imbalance by adjusting weights of samples with great variations in RIP angles in the training set, a novel loss is designed.
3. Extensive experiments are conducted on challenging benchmarks, which show significant performance improvement of the proposed approach compared to the state-of-the-art techniques in both face detection and face alignment tasks.

2 Related Work

Rotation-invariant face detection is inherently involves two different tasks: face detection and pose estimation [12, 18]. Unifying or separating these two tasks will lead to different approaches. In the unified framework [1, 21], the detector is designed to model human faces with diverse RIP angles, whereas in separated framework the detector addressed rotation variations using multi-class classification approaches by categorizing the entire RIP range into several distinct groups according to their RIP angles [6, 12]. Though a single detectors [21] can achieve accuracy of 87% on the challenging WIDER face, learning a large neural networks for powerful image representations in a data augmentation fashion usually leads to low computational efficiency and unpractical in many applications.

There are alternative strategies for improving rotation-invariant face detection in separated framework apart from data augmentation. In [12], the part-level response signal can be generated by deformable part models (DPM) technique for inferring human faces. [6] divided the full RIP range into several groups and then developed multiple detectors in order to cater to RIFD. Unfortunately, these methods suffer from significant limitations: 1) it causes quantization issue which result from multi-class classification, and 2) it is computationally inefficient.

Currently, the most state-of-the-art RIFD method [9] casts RIFD as a multi-class classification problem and turns the classification results into regression by calculating the expected value as the RIP orientation of each face candidate. The detector, which is used to generate the RIP angle of each face candidate in a coarse-to-fine manner, is easier to achieve fast and accurate calibration by flipping original image few times. However, since the source of supervision tends to be imprecise, precise and credible rotation-invariant face detection remains challenging. Our approach MTPCN follows the high-level idea of multi-class classification by employing coarse-to-fine strategy, but MTPCN attempts to generate the semantically consistent annotations, and hence is capable of improving detectors from an explicit perspective.

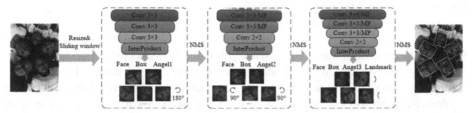

Fig. 1. The illustration of our architecture. Our MTPCN progressively calibrates the RIP orientation of each face candidate to upright for better distinguishing faces from non-faces. Specifically, MTPCN-1 first identifies face candidates and calibrates those facing down to facing up, halving the range of RIP angles from $[-180°, 180°]$ to $[-90°, 90°]$. Then the rotated face candidates are further distinguished and calibrated to an upright range of $[-45°, 45°]$ in MTPCN-2, shrinking the RIP ranges by half again. Finally, MTPCN-3 makes the accurate final decision for each face candidate to determine whether it is a face and predict the precise RIP angle. In particular, the network will output five facial landmarks' positions.

3 Multi-task Progressive Calibration Networks

We simultaneously train new cascaded CNNs on three tasks, each corresponding to a different strategy to inherently and explicitly address the face RIP variation problem. As a general pipeline of most state-of-the-art CNN-based face detection methods, face classification and bounding box regression are jointly optimized in a data-driven fashion. The CNN outputs a simple axis-aligned bounding box which only indicates the location of human face, without providing any detailed information *e.g.*, face shape and pose. To refine this information, we simultaneously learn two further tasks: the calibration based on the RIP angle prediction and facial landmark localization. The detailed configuration of our proposed MTPCN detector is illustrated in Fig. 1. The calibration process can be implemented for shrinking the RIP ranges. The task of face alignment identifies geometry structure of human face which can be viewed as modeling highly structured output. In our proposed architecture, each task is represented by a different branch departing from a shared CNN and contributes towards a final loss, which sums the three outputs together.

3.1 Problem Formulation

Our starting point is a network similar to that of [3], which we later modify in order to allow the model to be interpreted as a rotation-invariant face detector. The MTCNN [3] considers the inherent correlation between detection and alignment which potentially boosts up their performance. MTPCN uses similar strategies, but to detect and at the same time calibrate, faces. MTPCN consists of three stages, each of which not only distinguishes the faces from non-faces, but also calibrates the RIP orientation of each face candidate to upright progressively. In particular, the network will output five facial landmarks' positions. With the help of additive calibration process and explicit facial landmarks, the obtained model can easily make moderate decision for each face candidate to predict the precise RIP angle. The total loss is computed as the weighted sum of the four individual losses as follows:

$$min\ L = \sum_{i \in \{cls, reg, landmark, cal\}} \lambda_i \cdot L_i \tag{1}$$

where λ denotes c of tasks. The weight parameter is decided based on the importance of the task in the overall loss. In the following, we refer to them as *Lcls*, *Lbox*, *Llandmarks* and *Lcal*. The classification loss *Lcls* is cross-entropy loss over two classes (face *vs*▷. background) and the regression loss $\mathcal{L}_{box\ \psi}$ is the Euclidean loss defined in [3].

3.2 Stage-Wise Calibration

Previous work has turned the problem of modeling RIP variation into solving a multiclass classification problem and then directly regressing the precise RIP angles of face candidates. For example, PCN only predicts coarse RIP orientations in early stages, which is robust to the large diversity and further benefits the prediction of successive stages. Thus, the range of RIP angles is gradually decreasing, which helps distinguish faces from non-faces. Particularly, the RIP angle of a face candidate, *e.g.* θRIP, is obtained as the sum of predicted RIP angles from three stages, *e.g.* $\psi RIP_\psi = \psi1 + \psi2 + \psi3$. Given a set of candidate windows w, the coarse orientation prediction in first stage is a binary classification of the RIP angle range as follows:

$$L_{cal} = p \log g + (1 - p) \log(1 - g) \tag{2}$$

where p equals 1 if w is facing up, and equals 0 if w is facing down, and g is orientation score. To have a more accurate estimation, PCN follows the works in multi-class classification. To sum up, the first stage focuses only on identifying facing up or facing down, whereas the second stage requires predicting the coarse orientation for facing up, facing left or facing right. At the last training stage, the remainder of the RIP range is reduced to $[-45°, 45°]$, and can be directly estimated in a regression style. However, the regression-based RIP estimation approaches could suffer from overfitting because of ambiguous mapping between face appearance and its real RIP angles. On the other hand, as pointed out by previous studies, existing RIFD methods that rely on low-fidelity face representations without considerations for the global structure of faces, such as face shape and pose.

We believe the reasoning of the unique structure of facial landmarks (*e.g., eyes, nose, mouth*) help detecting faces under unconstrained environments. To this aim we show that, besides scores and bounding box coordinates, it is possible for the existing CNNs to also predict the facial landmarks in its special branch that strongly indicate the geometry structure of human face. Having these high-quality structure information, the pose of head can be determined. The proposed strategy is diagrammed in Fig. 3. Moreover, due to its multi-tasking nature, the learned CNNs achieve significant performance improvement on the multi-oriented FDDB dataset [7] and the rotation WIDER face dataset. Extensive experiments are performed, and demonstrate the effectiveness of our method for each of three tasks.

3.3 Loss Function

Pose and imaging quality are two main factors globally affecting the appearance of faces in images, which would result in poor localization of a (large) fraction of landmarks when the global structure of faces is mis-estimated. It is not uncommon that, penalizing more on errors corresponding to rare training samples than on those to rich ones achieves promising performance, as both the aforementioned concerns, say the geometric constraint and the data imbalance should all be considered. Given the face image I, d denote its distance measurement between ground-truth landmarks and the predict ones, the simplest losses arguably go to $L2$ and $L1$ losses. However, equally measuring the differences of landmark pairs is not so wise, without considering geometric/structural information. To refine this information, a novel loss is designed as follows:

$$L_{landmark} = \frac{1}{M} \sum_{m=1}^{M} \sum_{n=1}^{N} (1.5 - cos\theta) \left\| d_n^m \right\|_2^2 \tag{3}$$

in which, θ represent the RIP angles of deviation between the ground-truth and estimate RIP angles. N_ψ is the pre-defined number of landmarks per face to detect. M denotes the number of training images in each process. We note that our base loss is $L2$. It is worth mentioning that Guo *et al.* [22] considered the 3D pose information to improve the performance. They proposed an auxiliary network to obtain 3D rotation information including yaw, pitch, and roll angles and make the landmark localization stable and robust. Our approach differs in two key aspects: 1) the weighting parameter $cos\theta$ is adjusted according to the fraction of samples belonging to class c (this work simply adopts the reciprocal of fraction); 2) our detection framework aims to handle faces with full 360° RIP angles. Comparing with three Euler angles (yaw, roll, and pitch), the discrete roll deserves more efforts in our case, as it can greatly bring the large variations of face appearances.

4   Experiment

In this section, we first present a brief introduction to the experimental settings of our method. Then we evaluated our method on the FDDB dataset to verify the performances of the multi-class classification. Besides, we also have conducted experimental comparisons on rotation WIDER face dataset to evaluate the effectiveness of the proposed calibration process for estimating an arbitrary RIP angles. Finally, the face alignment method is tested on the most popular benchmark AFLW [14], which outperforms many state-of-the-art landmark detectors.

4.1   Implementation Details

Since different tasks are jointly performed, it essentially employs different types of training images during the training phase, such as face, non-face, part face and landmark face. When training, all training images of WIDER face dataset are randomly cropped to collect positives, negatives, part face and landmark face. Then, landmark faces are cropped from the same images and landmarks used by others [15]. This set contains

5,590 images from the LFW collection [16] and 7,876 images from throughout the web, all labeled for five facial landmarks. All images are randomly selected for training use. We follow [9] to generate discrete RIP angles ground truth and we use multiclass classification in early two stages. To make the model robust, we adopt online hard sample mining strategy [3] in whole three tasks which is adaptive by computing the task-specific loss and sorting them to keep the training samples with lowest scores. We follow the weight decay, learning rate and momentum of PCN. All the experiments run on a desktop computer with 1.7 GHz CPU and GTX Titan X and are implemented in Tensorflow.

4.2   Rotation-Invariant Face Detection Evaluation

To compare with other RIFD algorithms, we first evaluate our proposed model on two widely used benchmark datasets, FDDB and WIDER face. The FDDB dataset contains 2845 images and 5171 annotated faces. Since the original version of the FDDB dataset contains only a few rotated faces, it is also not suitable for evaluating RIFD, we use its extended version provided by [9] in the experiments. The WIDER face dataset, though more challenge than FDDB, is still dominated by faces with small RIP variations. For fair comparisons, we follow the same experimental settings and evaluation metrics as adopted in [9]. To this end, we manually select some images that contain rotated faces from the WIDER face test set, denoted as Rotation WIDER face, contains 400 images and 1053 rotated faces in the wild, as shown in Fig. 4.

We compare with the following three representative methods:

Divide-and-Conquer: The first kind addressed the RIFD using multi-class classification approaches by quantizing the entire face ranges into groups. In our experiments, we implement an upright face detector based on Cascade CNN [8] and run this detector four times. Then the multi-oriented detection results on test images are combined to output the final decision.

Data Augmentation: These models directly learn a mapping function from the whole training images with great variations of face appearances to the face regions. For more extensive comparison, we employ the most state-of-the-art models, such as Faster R-CNN [21], SSD300 [11] and Cascade CNN.

PCN: The designation of PCN represents rotation variation in a label-learning manner and provides a coarse-to-fine way to perform RIFD, which have demonstrated the robustness to faces with full 360 RIP angles.

Compared to PCN, Faster R-CNN (VGG16), SSD300(VGG16) and Cascade CNN, our method achieves the best accuracy on two widely used datasets including the WIDER face dataset and FDDB dataset. We respectively plot the ROC curves of these models on the Multi-Oriented FDDB dataset as shown in Fig. 2. All the results demonstrate the effectiveness of multi-class classification. As can be seen, there exist small gap between MTPCN and the PCN. It is likely caused by performing cross-task learning (Table 1).

Figure 3 compares MTPCN to different detectors mentioned above using the curves over the rotation WIDER face dataset. Our approach is significantly outperforms previous methods in this dataset, which exhibits human faces in a large range. This shows

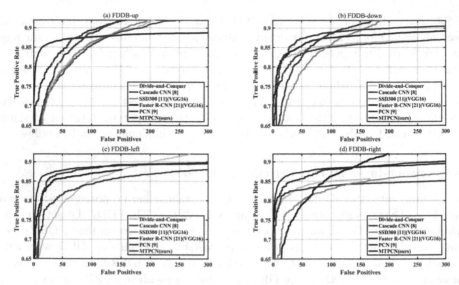

Fig. 2. Evaluation on the multi-oriented FDDB dataset.

Table 1. The FDDB recall rate (%) is at 100 false positives.

Method	Recall rate at 100 FP on FDDB					Speed		Model size
	Up	Down	Left	Right	Ave	CPU	GPU	
MTCNN [3]	89.6	–	–	–	–	28	54	1.9M
MTCNN-aug	86.2	84.1	83.6	83.5	84.4	28	54	1.9M
Divide-and-Conquer	85.5	85.2	85.5	85.6	85.5	15	20	2.2M
Cascade CNN-aug [8]	85.0	84.2	84.7	85.8	84.9	31	67	4.2M
SSD500(VGG16) [11]	86.3	86.5	85.5	86.1	86.1	1	20	95M
Faster R-CNN(VGG16) [21]	87.0	86.5	85.2	86.1	86.2	0.5	10	547M
PCN [9]	87.8	87.5	87.1	87.3	87.4	29	63	4.2M
Ours	**88.0**	**87.6**	**87.4**	**87.3**	**87.58**	25	41	5.4M

that our proposed method is capable of dealing with arbitrary RIP rotation and achieves better orientation regression accuracy. Some detection results can be viewed in Fig. 4.

4.3 Face Alignment Evaluation

In recent years, face alignment technology has developed rapidly, and more and more new challenging datasets are available for facilitating future research. But in a word, the ultimate goal is always to achieve accurate facial localization for practical purpose. Therefore, we carry out experiments on the challenging public dataset AFLW. The AFLW

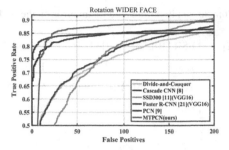

Fig. 3. Evaluation on rotation WIDER face.

Fig. 4. Our MTPCN's detection results on rotation WIDER FACE.

dataset contains 24,386 face images annotated with 21 landmarks with a large variety in head pose as shown in Fig. 2. Same as most existing methods [10, 12, 13, 19, 23, 29], we follow the same experimental settings and select 3,000 testing samples to make it comparable against the state-of-the-art approaches [3, 20, 25].

Table 2 reports the NME results obtained by the competitors. Obviously, for the general head pose, the state-of-the-art accuracy achieved with the aid of deep neural networks and data augmentation strategies, such as randomly flipping, rotating and occluded. While for face image with large rotation variations, most modern detectors often result in unpleasant results, as the unaccounted design against difficult cases. To alleviate the problems, we first distinguish faces from non-faces and calibrate it upright in early two stages. Use this progressively calibration strategy, the proposed MTPCN achieves quite promising performance.

Table 2. Comparison in normalized mean error on the AFLW dataset.

Method	TSPM [12]	ESR [10]	CMD [13]	Luxand [29]	RCPR [23]
AFLW	15.9	12.4	13.1	13.0	11.6
Method	SDM [19]	TCDCN [25]	BBFCN [20]	MTCNN [3]	**MTPCN**
AFLW	8.5	8.0	7.5	6.18	**6.06**

5 Conclusion

We present a novel Multi-Tasking Progressive Calibration Networks (MTPCN), which is advantageous because: (1) the ambiguities in the RIP angle annotations can be largely avoided, (2) specifically considers the facial geometric structure, leading to a reasonable loss function design, and (3) joint face detection, calibration and alignment task learning enables to enhance both three cross-correlated tasks. Also, experiments on the challenging benchmarks show that the above designs make a well-performing model. In the current version, MTPCN only employs the RIP variations as the geometric constraint. To further improve the accuracy of existing face detectors, more sophisticated geometric/structural information should be concerned.

Acknowledgment. This work was supported by the Science and Technology Research Program of Chongqing Municipal Education Commission (Grant No. KJZD-K201900601), by the National Natural Science Foundation of Chongqing (Grant No. cstc2019jcyj-msxmX0461), and by the Chongqing Overseas Scholars Innovation Program (Grant No. E020H2018013).

References

1. Farfade, S.S., Saberian, M.J., Li, L.J.: Multi-view face detection using deep convolutional neural networks. In: Proceedings of the 5th ACM on International Conference on Multimedia Retrieval, pp. 643–650 (2015)
2. Rowley, H.A., Baluja, S., Kanade, T.: Rotation invariant neural network-based face detection. In: Proceedings of the 1998 IEEE Computer Society Conference on Computer Vision and Pattern Recognition (Cat. No. 98CB36231), pp. 38–44. IEEE (1998)
3. Zhang, K., Zhang, Z., Li, Z., et al.: Joint face detection and alignment using multitask cascaded convolutional networks. IEEE Signal Process. Lett. 23(10), 1499–1503 (2016)
4. Felzenszwalb, P., McAllester, D., Ramanan, D.: A discriminatively trained, multiscale, deformable part model. In: 2008 IEEE Conference on Computer Vision and Pattern Recognition, pp. 1–8. IEEE (2008)
5. Yang, S., Luo, P., Loy, C.C., et al.: Wider face: a face detection benchmark. In: Proceedings of the IEEE Conference on Computer Vision and Pattern Recognition, pp. 5525–5533 (2016)
6. Huang, C., Ai, H., Li, Y., et al.: High-performance rotation invariant multiview face detection. IEEE Trans. Pattern Anal. Mach. Intell. 29(4), 671–686 (2007)
7. Jain, V., Learned-Miller, E.: FDDB: a benchmark for face detection in unconstrained settings. UMass Amherst technical report (2010)

8. Li, H., Lin, Z., Shen, X., et al.: A convolutional neural network cascade for face detection. In: Proceedings of the IEEE Conference on Computer Vision and Pattern Recognition, pp. 5325–5334 (2015)

9. Shi, X., Shan, S., Kan, M., et al.: Real-time rotation-invariant face detection with progressive calibration networks. In: Proceedings of the IEEE Conference on Computer Vision and Pattern Recognition, pp. 2295–2303 (2018)

10. Cao, X., Wei, Y., Wen, F., Sun, J.: Face alignment by explicit shape regression. Int. J. Comput. Vision 107(2), 177–190 (2013). https://doi.org/10.1007/s11263-013-0667-3

11. Zhang, Z., Qiao, S., Xie, C., et al.: Single-shot object detection with enriched semantics. In: Proceedings of the IEEE Conference on Computer Vision and Pattern Recognition, pp. 5813–5821 (2018)

12. Zhu, X., Ramanan, D.: Face detection, pose estimation, and landmark localization in the wild. In: 2012 IEEE Conference on Computer Vision and Pattern Recognition, pp. 2879–2886. IEEE (2012)

13. Yu, X., Huang, J., Zhang, S., et al.: Pose-free facial landmark fitting via optimized part mixtures and cascaded deformable shape model. In: Proceedings of the IEEE International Conference on Computer Vision, pp. 1944–1951 (2013)

14. Koestinger, M., Wohlhart, P., Roth, P.M., et al.: Annotated facial landmarks in the wild: a large-scale, real-world database for facial landmark localization. In: 2011 IEEE International Conference on Computer Vision Workshops (ICCV Workshops), pp. 2144–2151. IEEE (2011)

15. Wu, Y., Hassner, T., Kim, K.G., et al.: Facial landmark detection with tweaked convolutional neural networks. IEEE Trans. Pattern Anal. Mach. Intell. 40(12), 3067–3074 (2017)

16. Huang, G.B., Ramesh, M., Berg, T., Learned-Miller, E., et al.: Labeled faces in the wild: a database for studying face recognition in unconstrained environments. UMass, Amherst, MA, USA, Technical report 07–49, October 2007

17. Wang, Q., Zhang, L., Bertinetto, L., et al.: Fast online object tracking and segmentation: a unifying approach. In: Proceedings of the IEEE Conference on Computer Vision and Pattern Recognition, pp. 1328–1338 (2019)

18. Ranjan, R., Patel, V.M., Chellappa, R.: Hyperface: a deep multi-task learning framework for face detection, landmark localization, pose estimation, and gender recognition. IEEE Trans. Pattern Anal. Mach. Intell. 41(1), 121–135 (2017)

19. Xiong, X., De la Torre, F.: Supervised descent method and its applications to face alignment. In: Proceedings of the IEEE Conference on Computer Vision and Pattern Recognition, pp. 532–539 (2013)

20. Liang, Z., Ding, S., Lin, L.: Unconstrained facial landmark localization with backbone-branches fully-convolutional networks. arXiv preprint arXiv:1507.03409 (2015)

21. Ren, S., He, K., Girshick, R., et al.: Faster R-CNN: towards real-time object detection with region proposal networks. In: Advances in Neural Information Processing Systems, pp. 91–99 (2015)

22. Guo, X., Li, S., Zhang, J., et al.: PFLD: a practical facial landmark detector. arXiv preprint arXiv:1902.10859 (2019)

23. Burgos-Artizzu, X.P., Perona, P., Dollár, P.: Robust face landmark estimation under occlusion. In: Proceedings of the IEEE International Conference on Computer Vision, pp. 1513–1520 (2013)

24. Yang, B., Yang, C., Liu, Q., et al.: Joint rotation-invariance face detection and alignment with angle-sensitivity cascaded networks. In: Proceedings of the 27th ACM International Conference on Multimedia, pp. 1473–1480 (2019)

25. Zhang, Z., Luo, P., Loy, C.C., Tang, X.: Facial landmark detection by deep multi-task learning. In: Fleet, D., Pajdla, T., Schiele, B., Tuytelaars, T. (eds.) ECCV 2014. LNCS, vol. 8694, pp. 94–108. Springer, Cham (2014). https://doi.org/10.1007/978-3-319-10599-4_7

26. Zhou, X., Zhuo, J., Krahenbuhl, P.: Bottom-up object detection by grouping extreme and center points. In: Proceedings of the IEEE Conference on Computer Vision and Pattern Recognition, pp. 850–859 (2019)
27. Dong, X., Yu, S.I., Weng, X., et al.: Supervision-by-registration: An unsupervised approach to improve the precision of facial landmark detectors. In: Proceedings of the IEEE Conference on Computer Vision and Pattern Recognition, pp. 360–368 (2018)
28. Liu, Z., Zhu, X., Hu, G., et al.: Semantic alignment: Finding semantically consistent ground-truth for facial landmark detection. In: Proceedings of the IEEE Conference on Computer Vision and Pattern Recognition, pp. 3467–3476 (2019)
29. Luxand Incorporated: Luxand face SDK. http://www.luxand.com/

Enlacement and Interlacement Shape Descriptors

Michaël Clément[1]([✉]) [ID], Camille Kurtz[2] [ID], and Laurent Wendling[2] [ID]

[1] Univ. Bordeaux, CNRS, Bordeaux INP, LaBRI UMR 5800, 33400 Talence, France
michael.clement@labri.fr
[2] Université de Paris, LIPADE, 75006 Paris, France

Abstract. We propose a novel approach to characterize complex 2D shapes based on enlacement and interlacement directional spatial relations. This new relational concept allows to assess in a polar space how the concave parts of objects are intertwined following a set of directions. In addition, such a spatial relationship has an interesting behavior considering the common properties in pattern recognition such as translation, rotation, scale and symmetry. A shape descriptor is defined by considering the enlacement of its own shape and the disk area that surrounds it. An experimental study carried out on two datasets of binary shapes highlights the discriminating ability of these new shape descriptors.

Keywords: Shape descriptors · Spatial relations · Enlacement · Interlacement

1 Introduction

A 2D shape recognition system is usually roughly decomposed into several successive steps [1,2]. First, the shapes are extracted from their surrounding background. This step, called segmentation, relies heavily on *a priori* knowledge of the images to be processed. Then, a new representation is built from the extracted patterns. This representation aims to computationally judge the similarity or distance between two patterns, usually to perform classification. It may either be a set of measurements made on the patterns, forming a vector of features, or a symbolic description of how the pattern can be divided into basic shapes. The choice of one representation instead of another one generally relies on the application under consideration, and following underlying conditions such as robustness against noise and small distortions, invariance with regards to common geometrical transformations or tolerance to occlusions. The shape representation for object recognition has been the subject of numerous researches; and extensive surveys of shape analysis can be found in the literature, for example in [3–5]. Two main categories of shape descriptors are often encountered: those working on a shape as a whole (called region-based descriptors) and those working on the contours of the shape (called contour-based descriptors).

ⓒ Springer Nature Switzerland AG 2020
Y. Lu et al. (Eds.): ICPRAI 2020, LNCS 12068, pp. 525–537, 2020.
https://doi.org/10.1007/978-3-030-59830-3_45

The usual descriptors based on the contours include the well-known Fourier descriptors [6–8] and the curvature approaches [9,10], where a shape is described in a scale space. The shape context [11] is also a robust descriptor to small perturbations of parts of the shape. The main drawback lies in the lack of guarantee with scale-invariance. Other approaches are based on the extraction of the skeleton considering [12,13] (for instance, a graph defined from the medial axis). Shock graphs associated to both global optimization and graph matching provide a powerful tool [14] for discriminating shapes but such methods are generally highly sensitive to scale variations. Since contour descriptors are based on the boundary of a shape, they are not able to properly capture the internal structure of a shape. Furthermore, such methods are not suitable for disjoint shapes or shapes with holes because boundary information is not available. Therefore, they remain limited to certain types of applications.

Region-based descriptors work on a shape as a whole taking into account all the pixels within it. Comparative studies [15,16] have demonstrated the interest of the Zernike moments, and numerous researches have been focused on improving their invariance properties [17] and speeding up the fast computation of the Zernike moments [18]. Fourier-Mellin has been introduced to avoid the mathematical application problems of the standard transform but an approximation is required to process with numerical data [19]. To overcome the drawbacks of contour based Fourier descriptors, Zhang and Lu [8] have proposed a region-based generic Fourier descriptor based on the polar discrete transform. Experimental results show that this approach outperforms common contour-based (classical Fourier and curvature approaches) and region-based (Zernike moments) shape descriptors. These methods were often adapted to deal with the specificities of the application under consideration. On the whole, region-based methods are more suited to general applications. However, they are generally more computationally intense and most approaches need to normalize (centroid position, re-sampling) the image to be processed with geometrical properties. Such a normalization may introduce errors, sensitivity to noise, and consequently inaccuracy in the recognition/decision process.

Among region-based descriptors, relative position descriptors aim at assessing a specific measure following a set of directions. Most of the works have been focused on the notions on force histogram [20], phi-descriptor [21], meta directional histogram [22] built from fuzzy landscapes [23] and Radon transform [24]. Such methods are less sensitive to the noise and they preserve common geometrical properties. They are generally less relevant when the number of classes grows drastically due to the compact aspect of the descriptor but they offer important (complementary) information when they are combined with other descriptors.

In this paper, our contribution is the proposition of a novel region-based approach in the context of this generic shape recognition problem. The underlying strategy of our method is to capture how the parts of one object are messed or twisted from a circular background area, enabling capture more complex patterns, whose shapes are, for instance, affected by complex concavities and (deep) levels of imbrication. The remainder of this article is organized as follows. Section 2 recalls the concept of directional spatial enlacement, initially proposed in [25]. From this model, we derive novel shape descriptors, presented in Sects. 3 and 4, which are defined by considering the enlacement/interlacement of the object shape and the disk area surrounding it. An experimental study carried out on two databases of binary shapes is described in Sect. 5. Finally, conclusions and perspectives will be found in Sect. 6.

2   Directional Spatial Enlacement

The notion of directional spatial enlacement, modeled as an "enlacement" histogram, was initially proposed to assess a new spatial relation between two binary objects [25]. Such a histogram was also embedded in pattern recognition applications considering local features calculated from pairs of broad objects [26]. We show here that it can be easily used to provide a discriminate descriptor (or signature) characterizing only one shape. We recall first basic notions aiming at describing such a kind of signature that efficiently integrates both whole shape and spatial description following a set of directions. A full description of the underlying theoretical developments can be found in [25,27].

Let (A, B) be a couple of two-dimensional objects. The goal is to measure how an object A is enlaced with an object B (and reciprocally).

Handling with Points. The relative positions of individual points $a_i \in A$ and $b_i \in B$ are considered as arguments to put in favor of the proposition "A is enlaced by B". Let us consider an oriented straight line $\Delta^{(\theta,\rho)}$. The functions $f_A^{(\theta,\rho)}$ and $f_B^{(\theta,\rho)}$ designate longitudinal cuts along the oriented line $\Delta^{(\theta,\rho)}$. The quantity of A located *after* a point x on a line $\Delta^{(\theta,\rho)}$ is given by:

$$\int_x^{+\infty} f_A^{(\theta,\rho)}(y)\, \mathrm{d}y. \tag{1}$$

Handling with Segments. Let us consider a binary object A. The integral determines the cumulated length of the segments on the interval $[x, +\infty[$ of this longitudinal cut. Consequently, to capture the quantity of A located *after* points of B on this line, we have:

$$\int_{-\infty}^{+\infty} f_B^{(\theta,\rho)}(x) \int_x^{+\infty} f_A^{(\theta,\rho)}(y)\, \mathrm{d}y\, \mathrm{d}x. \tag{2}$$

Symmetrically, the quantity of A located *before* parts of B is obtained with:

$$\int_{-\infty}^{+\infty} f_A^{(\theta,\rho)}(x) \int_x^{+\infty} f_B^{(\theta,\rho)}(y)\, \mathrm{d}y\, \mathrm{d}x. \tag{3}$$

The goal is then to combine these two quantities, in order to take into account both parts of B *before* and *after* parts of A, leading us to the following general definition of the one-dimensional enlacement. Let f and g be two bounded measurable functions with compact support from $\mathbb{R}$ to $\mathbb{R}$. The enlacement of f with regards to g is defined as:

$$E(f,g) = \int_{-\infty}^{+\infty} g(x) \int_{x}^{+\infty} f(y) \int_{y}^{+\infty} g(z)\, \mathrm{d}z\, \mathrm{d}y\, \mathrm{d}x. \tag{4}$$

This generic definition between two real-valued functions can be applied to longitudinal cuts $f_A^{(\theta,\rho)}$ and $f_B^{(\theta,\rho)}$. The value $E(f_A^{(\theta,\rho)}, f_B^{(\theta,\rho)})$ represents the overall enlacement of A by B along the oriented line $\Delta^{(\theta,\rho)}$.

Handling with Objects. Let $\theta \in \mathbb{R}$ be an orientation angle and let us consider the pencil of all parallel lines $\{\Delta^{(\theta,\rho)}, \rho \in \mathbb{R}\}$ in this direction; they will slice any object into a pencil of longitudinal cuts. The enlacement of an object with regards to another in this direction θ, is obtained by aggregating all the one-dimensional enlacement values, as it follows. Let $\theta \in \mathbb{R}$ be an orientation angle, and let A and B be two objects. The enlacement of A by B in the direction θ is given by:

$$E_{AB}(\theta) = \int_{-\infty}^{+\infty} E(f_A^{(\theta,\rho)}, f_B^{(\theta,\rho)})\, \mathrm{d}\rho. \tag{5}$$

Properties. Considering the definition of the enlacement from parallel longitudinal cuts, the following properties can be easily checked [25]:

- For any two objects A and B, the directional enlacement E_{AB} is **periodic** with period π, that is $\forall \theta \in \mathbb{R}$, $E_{AB}(\theta) = E_{AB}(\theta + k\pi), k \in \mathbb{Z}$;
- For any two objects A and B, the directional enlacement E_{AB} is invariant with regards to **translations**. Let T_v by a translation by a vector $v \in \mathbb{C} \simeq \mathbb{R}^2$, we have: $E_{T_v(A)T_v(B)}(\theta) = E_{AB}(\theta)$;
- For any two objects A and B, the directional enlacement E_{AB} is quasi-invariant with regards to **rotations**. Let $\alpha \in \mathbb{R}$ be a rotation angle and let R_α be a rotation transformation, we have: $E_{R_\alpha(A)R_\alpha(B)}(\theta) = E_{AB}(\theta - \alpha)$;
- For any two objects A and B, the directional enlacement E_{AB} is quasi-invariant with regards to **scaling** transformations. Let $\lambda \in \mathbb{R}$ be a scaling factor and let S_λ be a scaling transformation, we have: $E_{S_\lambda(A)S_\lambda(B)}(\theta) = \lambda^4 E_{AB}(\theta)$.

3   Enlacement Descriptor

In its initial definition, the enlacement is calculated between two generic binary objects A and B. Considering one single shape, a longitudinal cut can split it into several segments assuming the existence of holes or concavities. The goal is to assess such internal information by calculating the enlacement between a shape and a surrounding area (including the convex hull). Let us consider a shape A and its centroid $c_A(x_a, y_a)$. Let r_A be the maximal radius of A, calculated from

c_A. The surrounding area, noted $\bar{A}$ is defined from the disk D of center c_A and radius r_A such that: $\bar{A} = D \backslash A$ (that is $D = A \cup \bar{A}$).

The scalar enlacement value in a given direction should be relative to the overall area in order to compare the enlacement of different couples of shapes having the same area (but different shapes). Let us define the area $\|A\|_1$ of a shape A by:

$$\|A\|_1 = \iint_{\mathbb{R}^2} |f_A(x,y)| \, dx \, dy. \tag{6}$$

The directional enlacement descriptor of A with regards to $\bar{A}$ is defined by the following function (with $\mathcal{E}_{A\bar{A}}$ is equal to 0 if $A = D$):

$$\mathcal{E}_{A\bar{A}} : \mathbb{R} \longrightarrow \mathbb{R}$$
$$\theta \longmapsto \frac{E_{A\bar{A}}(\theta)}{\|A\|_1 \left(\pi r_A^2 - \|A\|_1\right)}. \tag{7}$$

The enlacement descriptor $\mathcal{E}_{A\bar{A}}$ takes the form of a function, which associates to each angle θ the enlacement value of A by its surrounding area $\bar{A}$ normalized by the combination of the two respective areas. Note that both descriptors $\mathcal{E}_{A\bar{A}}$ and $\mathcal{E}_{\bar{A}A}$ provide complementary information (*i.e.* enlacement of the enclosing circle by the shape, and enlacement of the shape by the enclosing circle respectively). In this sense, they can also be concatenated into a single feature vector $\mathcal{E}_{A\bar{A}}\mathcal{E}_{\bar{A}A}$.

An overall number of elementary occurrences of enlacement is obtained by considering each longitudinal cut of the shape in direction θ. The descriptor can be interpreted as a circular histogram, which we call the $\mathcal{E}$-histogram. Using a disk area as an overlapping support instead of a rectangle area keeps consistency (in an isotropic manner), considering the pencil of longitudinal cuts following any direction, as the same amount of pairwise information is processed. Thanks to this normalization the enlacement histogram $\mathcal{E}_{A\bar{A}}$ is invariant with regards to scaling transformations, that is: let $\lambda \in \mathbb{R}$ be a scaling factor and let S_λ be the associated scaling transformation; we have: $\mathcal{E}_{S_\lambda(A)S_\lambda(\bar{A})}(\theta) = \mathcal{E}_{A\bar{A}}(\theta)$.

It is easy to show that the properties of periodicity (considering opposite directions), translation (as shapes are processed independently of their location in the frame) and rotation (a rotation of angle θ implies a circular shift of θ in the descriptor) are preserved from $E_{A\bar{A}}$ for the enlacement histogram $\mathcal{E}_{A\bar{A}}$. Then $\mathcal{E}$-histogram of a shape is invariant with regards to translations and scaling transformations, and quasi-invariant to rotations.

4   Interlacement Descriptor

The enlacement histogram $\mathcal{E}_{A\bar{A}}$ defines a shape descriptor that quantitatively characterizes how a shape A is enlaced by its surrounding background $\bar{A}$. As a relative position descriptor, the descriptor $\mathcal{E}_{A\bar{A}}$ does not completely describe the spatial configuration of A and $\bar{A}$. So the opposite descriptor $\mathcal{E}_{\bar{A}A}$ is complementary to $\mathcal{E}_{A\bar{A}}$ and can also be considered to improve the description of the shape under consideration. The combination of these two descriptors provides a description of the mutual interlacement of the shape and its surrounding area. The directional interlacement descriptor is defined by the harmonic mean as:

$$\mathcal{I}_{A\bar{A}}(\theta) = \frac{2 \, \mathcal{E}_{AA}(\theta) \times \mathcal{E}_{\bar{A}A}(\theta)}{\mathcal{E}_{A\bar{A}}(\theta) + \mathcal{E}_{\bar{A}A}(\theta)} \tag{8}$$

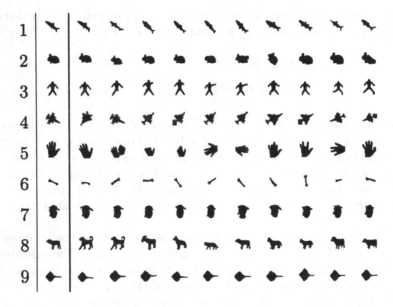

Fig. 1. Binary 2D shapes from the Sharvit B_1 dataset (99 shapes, 9 classes). Similarly, the Sharvit B_2 contains 216 shapes (12 classes) with the same kind of distorsions.

The interlacement descriptor can also be interpreted as a circular histogram, called the $\mathcal{I}$-*histogram*. The invariance properties are preserved for the interlacement histogram.

5   Experimental Study

For this experimental study, the proposed descriptors have been employed for a 2D shape classification task on two datasets of binary shapes followed by a comparative study involving other baseline shape descriptors.

5.1   Datasets

Two shape datasets have been considered to highlight the interest of our proposed enlacement and interlacement shape descriptors.

Sharvit B_1 and B_2. The first shape dataset is provided by Sharvit *et al.* [28]. This dataset is decomposed into two independent datasets, B_1 that contains 9 categories with 11 binary shapes in each category (see Fig. 1) and B_2 that contains 18 categories with 12 shapes in each category. Considering B_1, a few of the shapes are occluded (airplanes and hands) and some shapes are partially represented (rabbits, men, and hands). There are also distorted objects (tools) and heterogeneous shapes in the same cluster (animals). The dataset B_2 also contains shapes with similar distortions. These two Sharvit sub-datasets are widely used in the literature to compare feature descriptors designed to classify 2D shapes in presence of noise, artifacts and distorted parts.

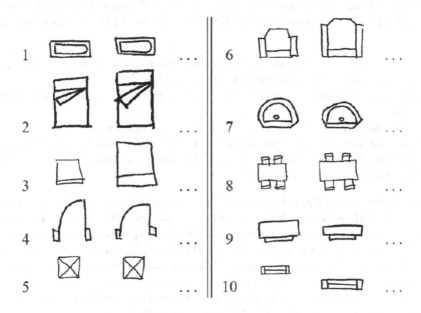

Fig. 2. Samples from the CVC dataset composed of 3000 binary 2D shapes (300 classes) representing various hand-drawn patterns.

CVC. The second dataset consists of 3000 binary shapes categorized into 10 classes, with equal number of samples in each category (see Fig. 2). It is kindly provided by CVC Barcelone. This dataset is composed of complex 2D hand-drawn patterns, with shapes having holes and disconnected parts. The patterns were drawn by ten persons using the Anoto digital pen and paper.

5.2 Experimental Protocol

We used the proposed enlacement and interlacement descriptors to classify the binary shapes from the Sharvit and CVC datasets into their corresponding categories. For the first experiment, we studied the influence of the number of directions for our proposed descriptors (as well as there potential combinations) on the Sharvit dataset, since this baseline dataset contains more generic and heterogeneous shapes. Then, we evaluated the impact of using different classifiers on the larger CVC dataset. For both datasets, we also perform a comparative study with other baseline shape descriptors.

Parameter Settings. For each 2D shape of the datasets, we computed both enlacement shape descriptors $\mathcal{E}_{A\bar{A}}$ and $\mathcal{E}_{\bar{A}A}$, as well as the interlacement descriptor $\mathcal{I}_A$. All these descriptors are computed onto a set of k discrete directions (the influence of k being studied hereinafter) equally spaced along the $[0, \pi]$ interval.

The resulting feature vectors are used to train different classifiers to recognize the different shape categories from the two datasets. In these experiments, we tested:

- k-nearest neighbors classifiers (k-NN) using the ℓ_2 norm on features. The number of neighbors has been empirically set to $k = 1$ (for both datasets, using more neighbors did not improve the results, either for majority of weighted votes);
- SVM classifiers following a *one-versus-all* strategy for multiclass classification. The soft-margin hyperparameter C was optimized by performing grid search on the training set for each tested features. We report results for SVMs using linear and gaussian kernels.

To better assess the robustness of the approach, this classification process is coupled with a cross validation strategy. For both the Sharvit B_1 and B_2 datasets, we employed a leave-one-out strategy, as these datasets are relatively small. For the CVC dataset we used a k-fold cross validation strategy with $k = 3$. That is, the training set representing 2/3 of its total size, while the remaining 1/3 is used for testing, and this process is repeated 3 times so that the whole dataset is classified. For each strategy, we evaluate the classification results using the accuracy score (*i.e.* the rate of correctly classified shapes).

5.3   Influence of the Number of Directions

In this first experiment, we studied the influence of the number of discrete directions $k \in [0, \pi]$ when computing the enlacement descriptors $\mathcal{E}_{A\bar{A}}$ and $\mathcal{E}_{\bar{A}A}$, the interlacement $\mathcal{I}_A$, as well as their combinations $\mathcal{E}_{A\bar{A}}\mathcal{E}_{\bar{A}A}$ and $\mathcal{E}_{A\bar{A}}\mathcal{E}_{\bar{A}A}\mathcal{I}_A$. Note that when combining descriptors, the size is increased accordingly: $\mathcal{E}_{A\bar{A}}\mathcal{E}_{\bar{A}A}$ has $2k$ values and $\mathcal{E}_{A\bar{A}}\mathcal{E}_{\bar{A}A}\mathcal{I}_A$ has $3k$ values. The different numbers of directions tested were $k \in \{4, 8, 16, 32, 64\}$. For all of these values, we apply the k-NN classifier with the corresponding descriptors on both Sharvit datasets B_1 and B_2, following the leave-one-out cross validation strategy.

Figure 3 presents the evolution of the accuracy scores for these different number of directions, on Sharvit (a) B_1 and (b) B_2. On B_1, we can observe relatively stable recognition rates for small number of directions, with a slight decrease in performance when using 64 directions. The instability is probably due to the

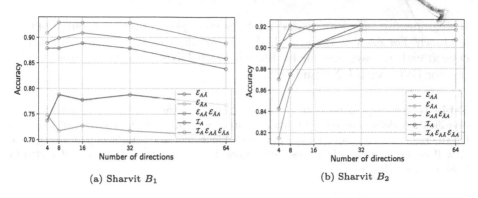

Fig. 3. Accuracy scores for the Sharvit B_1 and B_2 datasets, for varying number of directions of the enlacement and interlacement shape descriptors. Classification is performed with k-NN ($k = 1$) on the ℓ_2-norm and using leave-one-out cross validation.

small size of this dataset. For B_2, we observe a more homogeneous increase in accuracy for all features when increasing the number directions, reaching a plateau for 16–32 directions. Overall, experiments with 32 directions seem to be a good compromise between performance and size. This value will therefore be retained for the remainder of the experimental study.

From these results, we can also observe the relative performances of the different considered features. In particular, we can see that the combination $\mathcal{E}_{A\bar{A}}\mathcal{E}_{\bar{A}A}$ of both enlacement descriptors seems to always improve the results compared to their individual parts. Further combining into $\mathcal{E}_{A\bar{A}}\mathcal{E}_{\bar{A}A}\mathcal{I}_A$ again slightly improves performance. This seems to confirm our hypothesis that these descriptors represent complementary rich information, captured both from the shape itself and its background counterpart, that can be useful to better characterize complex visual patterns.

5.4   Comparative Study and Discussion

We now present a comparative study with several commonly used shape descriptors from the literature on both datasets. In these experiments, we used the following baseline descriptors:

- Angular Signature (AS) [29];
- Yang [30];
- 7 Moments of Zernike (Zernike7) [18];
- Angular Radial Transform (ART) [31];
- Generic Fourier Descriptors (GFD) [8];
- Force Histograms ($\mathcal{F}_0^{AA}$ and $\mathcal{F}_2^{AA}$) [20].

All these methods result in feature vectors for each shape, and are supposed to be relatively robust to different affine transforms such as translation, rotation, or scaling. The five first methods are global shape descriptors, while Force

Histograms are based on spatial relations concepts, similarly to our proposed approach. For each of these methods, we use the same experimental protocol defined earlier: leave-one-out for Sharvit (with k-NN), and 3-fold cross validation for CVC (with k-NN and SVMs).

Table 1. Accuracy scores obtained on the Sharvit datasets B_1 and B_2 with a 1-NN classification (leave-one-out). On the left are the results for the baseline descriptors. On the right are the results for our proposed enlacement and interlacement descriptors, as well as their combinations. Best and second best scores are respectively depicted in bold and underlined.

Baseline descriptors	B_1	B_2	Our descriptors	B_1	B_2
AS	75.76	67.59	$\mathcal{E}_{A\bar{A}}$	87.88	90.74
Yang	82.83	81.94	$\mathcal{E}_{\bar{A}A}$	71.72	<u>91.67</u>
Zernike7	<u>93.94</u>	92.59	$\mathcal{I}_A$	78.79	**92.13**
ART	<u>93.94</u>	**94.44**	$\mathcal{E}_{A\bar{A}}\mathcal{E}_{\bar{A}A}$	<u>89.90</u>	92.13
GFD	**95.96**	<u>93.06</u>	$\mathcal{I}_A\mathcal{E}_{A\bar{A}}\mathcal{E}_{\bar{A}A}$	92.93	92.13
$\mathcal{F}_0^{AA}$	77.78	74.54			
$\mathcal{F}_2^{AA}$	56.57	51.85			

In Table 1, we report the accuracy scores obtained on the Sharvit B_1 and B_2 datasets for the five baseline methods, and we compare them to our proposed enlacement and interlacement descriptors, as well as their combinations. From this Table, we can observe that our proposed shape descriptors provide competitive and comparable results with the tested baseline descriptors. On this dataset, the best accuracy scores were achieved using global descriptors such as GFD, ART or Zernike moments, which can be considered as relevant baselines. Our descriptors that combine pairwise enlacements and also interlacement outperform common directional based descriptors (*i.e.* Force Histograms $\mathcal{F}_0^{AA}$ and $\mathcal{F}_2^{AA}$). Although some shapes present concavities, most of the them are relatively compact, which can be crippling for our enlacement criterion.

Table 2. Accuracy scores obtained on the CVC dataset with different classifiers (k-NN and SVMs with linear and gaussian kernels), following a 3-fold cross validation. Best and second best scores are respectively depicted in bold and underlined.

	k-NN ($k=1$)	SVM (linear)	SVM (gaussian)
GFD	89.57	87.83	90.43
$\mathcal{E}_{A\bar{A}}$	83.13	90.17	90.57
$\mathcal{E}_{\bar{A}A}$	83.87	87.43	89.87
$\mathcal{I}_A$	85.73	88.17	90.23
$\mathcal{E}_{A\bar{A}}\mathcal{E}_{\bar{A}A}$	<u>93.83</u>	<u>96.30</u>	<u>96.87</u>
$\mathcal{I}_A\mathcal{E}_{A\bar{A}}\mathcal{E}_{\bar{A}A}$	**94.73**	**96.53**	**97.00**

In Table 2 we report the accuracy scores obtained on the larger CVC dataset for the GFD baseline descriptors, and we compare them to our proposed enlacement and interlacement descriptors. Here, we also report results for different classifiers, $i.e.$ k-NN and SVMs with linear and gaussian kernels. On this dataset, our combined enlacement and interlacement features significantly improve upon GFD descriptors (with 36 bins as suggested by [8]. We also found that increasing the number of bins for the GFD descriptors to 96 values ($i.c.$ to match the dimensionality of our combined features) results in slightly worse performance (from 0.7% to 4% lower accuracy scores depending on the classifier).

The good performance of our method on this dataset can be explained by the fact that the shapes are more complex, usually with multiple line strokes resulting in many concavities. The enlacement is specifically designed to capture the internal structure of this types of shapes, characterizing the alternating patterns between strokes in different directions. Therefore, our proposed features seem to be more efficient and powerful for this type of application.

6   Conclusion

A new shape descriptor based on the assessment of both enlacement and interlacement behavior following a set of directions has been proposed in this paper. Experimental study on three databases attests of the promising interest of such directional descriptors which can easily keep nice properties as homothety, relation and translation. Further works are dedicated to extending our descriptors by considering, for instance, the skeleton and the remaining medial axis calculated between the shape and the surrounding area. The goal will be to provide a suitable derivative approach to make tests on other complex specific datasets (such as symbols, texts) where the underlying structure is close to the skeleton (following the width). We plan also to integrate the Chanfrein distance (as in [32]) to better integrate the internal structure of the shape.

References

1. Jain, A.K., Duin, R.P.W., Mao, J.: Statistical pattern recognition: a review. IEEE Trans. Pattern Anal. Mach. Intell. **22**(1), 4–37 (2000)
2. Smeulders, A.W.M., Worring, M., Santini, S., Gupta, A., Jain, R.C.: Content-based image retrieval at the end of the early years. IEEE Trans. Pattern Anal. Mach. Intell. **22**(12), 1349–1380 (2000)
3. Loncaric, S.: A survey of shape analysis techniques. Pattern Recogn. **31**(8), 983–1001 (1998)
4. Zhang, D., Lu, G.: Review of shape representation and description techniques. Pattern Recogn. **37**(1), 1–19 (2004)
5. Chen, C.H.: Handbook of Pattern Recognition and Computer Vision, 5th edn. World Scientific, New Jersey (2016)
6. Kauppinen, H., Seppänen, T., Pietikäinen, M.: An experimental comparison of autoregressive and fourier-based descriptors in 2D shape classification. IEEE Trans. Pattern Anal. Mach. Intell. **17**(2), 201–207 (1995)
7. Persoon, E., Fu, K.S.: Shape discrimination using fourier descriptors. IEEE Trans. Pattern Anal. Mach. Intell. **8**(3), 388–397 (1986)

8. Zhang, D., Lu, G.: Shape-based image retrieval using generic fourier descriptor. Sig. Process. Image Commun. **17**(10), 825–848 (2002)
9. Mokhtarian, F., Abbasi, S.: Shape similarity retrieval under affine transforms. Pattern Recogn. **35**(1), 31–41 (2002)
10. Urdiales, C., Bandera, A., Hernández, F.S.: Non-parametric planar shape representation based on adaptive curvature functions. Pattern Recogn. **35**(1), 43–53 (2002)
11. Belongie, S.J., Malik, J., Puzicha, J.: Shape matching and object recognition using shape contexts. IEEE Trans. Pattern Anal. Mach. Intell. **24**(4), 509–522 (2002)
12. Kimia, B.B., Tannenbaum, A.R., Zucker, S.W.: Shapes, shocks, and deformations I: the components of two-dimensional shape and the reaction-diffusion space. Int. J. Comput. Vision **15**(3), 189–224 (1995)
13. Zhu, S.H., Yuille, A.L.: FORMS: a flexible object recognition and modelling system. Int. J. Comput. Vision **20**(3), 187–212 (1996)
14. Siddiqi, K., Shokoufandeh, A., Dickinson, S.J., Zucker, S.W.: Shock graphs and shape matching. Int. J. Comput. Vision **35**(1), 13–32 (1999)
15. Teh, C.H., Chin, R.T.: On image analysis by the methods of moments. IEEE Trans. Pattern Anal. Mach. Intell. **10**(4), 496–513 (1988)
16. Bailey, R.R., Srinath, M.D.: Orthogonal moment features for use with parametric and non-parametric classifiers. IEEE Trans. Pattern Anal. Mach. Intell. **18**(4), 389–399 (1996)
17. Bin, Y., Jia-Xiong, P.: Invariance analysis of improved zernike moments. J. Opt. A: Pure Appl. Opt. **4**(6), 606–614 (2002)
18. Khotanzad, A., Hong, Y.H.: Invariant image recognition by zernike moments. IEEE Trans. Pattern Anal. Mach. Intell. **12**(5), 489–497 (1990)
19. Ghorbel, F.: A complete invariant description for gray-level images by the harmonic analysis approach. Pattern Recogn. Lett. **15**(10), 1043–1051 (1994)
20. Matsakis, P., Wendling, L.: A new way to represent the relative position between areal objects. IEEE Trans. Pattern Anal. Mach. Intell. **21**(7), 634–643 (1999)
21. Matsakis, P., Naeem, M., Rahbarnia, F.: Introducing the Φ-descriptor - a most versatile relative position descriptor. In: ICPRAM, Proceedings, pp. 87–98 (2015)
22. Delaye, A., Anquetil, E.: Learning of fuzzy spatial relations between handwritten patterns. Int. J. Data Mining Modell. Manage. **6**(2), 127–147 (2014)
23. Bloch, I.: Fuzzy relative position between objects in image processing: a morphological approach. IEEE Trans. Pattern Anal. Mach. Intell. **21**(7), 657–664 (1999)
24. Tabbone, S., Wendling, L.: Binary shape normalization using the radon transform. In: DGCI, Proceedings, pp. 184–193 (2003)
25. Clément, M., Poulenard, A., Kurtz, C., Wendling, L.: Directional enlacement histograms for the description of complex spatial configurations between objects. IEEE Trans. Pattern Anal. Mach. Intell. **39**(12), 2366–2380 (2017)
26. Clément, M., Coustaty, M., Kurtz, C., Wendling, L.: Local enlacement histograms for historical drop caps style recognition. In: ICDAR, Proceedings, pp. 299–304 (2017)
27. Clément, M., Kurtz, C., Wendling, L.: Fuzzy directional enlacement landscapes. In: DGCI, Proceedings, pp. 171–182 (2017)
28. Sharvit, D., Chan, J., Tek, H., Kimia, B.B.: Symmetry-based indexing of image databases. J. Vis. Commun. Image Represent. **9**(4), 366–380 (1998)
29. Bernier, T., Landry, J.A.: A new method for representing and matching shapes of natural objects. Pattern Recogn. **36**(8), 1711–1723 (2003)

30. Yang, S.: Symbol recognition via statistical integration of pixel-level constraint histograms: a new descriptor. IEEE Trans. Pattern Anal. Mach. Intell. **27**(2), 278–281 (2005)

31. Kim, W.Y.: A new region-based shape descriptor. ISO/IEC MPEG99/M5472 (1999)

32. Tabbone, S., Wendling, L., Salmon, J.P.: A new shape descriptor defined on the radon transform. Comput. Vis. Image Underst. **102**(1), 42–51 (2006)

Image Orientation Detection Using Convolutional Neural Network

Hongjian Zhan(ID), Xiao Tu, Shujing Lyu(✉)(ID), and Yue Lu(ID)

Shanghai Key Laboratory of Multidimensional Information Processing,
East China Normal University, Shanghai 200241, China
sjlv@cs.ecnu.edu.cn

Abstract. Image orientation detection is often a prerequisite for many applications of image understanding and recognition. Recently, with the development of deep learning, mang significant Convolutional Neural Network (CNN) architectures are proposed and widely used in computer vision areas. In order to investigate the performance of CNN on image orientation detection task, in this paper, we first evaluate several famous CNN architectures, such as AlexNet, GoogleNet and VGGNet on this task, then we test a new CNN architecture by combining these networks. We collect six kinds of image, including landscape, block, indoor, human face, mail and natural images, in which the first three ones are regarded as difficult categories of orientation detection by previous work. The experiment results on these datasets indicate the effectiveness of the proposed network on image orientation detection task.

Keywords: Orientation detection · Convolutional neural network · Deep learning

1 Introduction

Automatic image orientation detection is a difficult problem, and plays an important role in many situations [2,6,8,10,12,21,22]. Nowadays, with the proliferation of digital cameras and smart phones, we can take pictures anytime and anywhere. Therefore, we need a photo management system to help us handle a mass of images. When we take photos by smart phones or cameras, we don't always make our phones upright deliberately, so the pictures we get are in different orientation. The first function of photo management system is to detecte the image orientation.

Nowadays, 3D accelerometer or G-sensor is a basic component for a smart phone. Engineers try to use the rotation information of G-sensor to estimate the rotation of the image, based on which an upside-down image can be corrected to a right orientation [3]. But it is effective only when the device has a vertical rotation, namely, we should set up our device if we want to use the automatic corrective function. Obviously, it is not friendly to users. On the other hand, the orientation information of the image is also used to estimate the camera orientation [7,11].

Y. Lu et al. (Eds.): ICPRAI 2020, LNCS 12068, pp. 538–546, 2020.
https://doi.org/10.1007/978-3-030-59830-3_46

In another situation, automatic mail sorting system, in order to recognize the address and postcode, it also needs the mail image orientation information. A mailpiece is a rectangle generally. When we feed it into an automatic sorting machine, suppose we don't trim them into a common orientation, we have to detect the orientation by analyzing the mail images before recognizing its addresses or postcode. In some practices, there are some signs on the envelope to indicate the orientation.

Many methods have been proposed to solve the orientation detection task by the content analysis [1,9,20]. These approaches used hand-crafted features, such as spatial color histograms and edges, which are fed into statistical classifiers. Vailaya et al. [20] extracted a codebook from a vector quantizer to estimate the class-conditional densities of the observed features which are used in a Bayesian learning framework to predict the orientation. Shumeet [1] combined the outputs of hundreds of classifiers trained with AdaBoost to determine the correct orientation of an image. Zhang et al. [24] also used a classifier to find whether the images are of indoor or outdoor scenes to improve the orientation detection accuracy. All these methods still cannot work well in real world, so the automatic image orientation detection is still an open issue.

Deep learning is very hot in almost all computer vision areas recent years. Many new and fantastic networks are proposed for specific or general purposes. Convolutional neural network is an important branch of deep learning and plays a significant role in image processing. There are many famous CNN models, such as AlexNet [16], the breakthrough of deep learning methods on ILSVRC2012, which contains max pooling, dropout and other new components; GoogleNet [17–19], the winner of ILSVRC2014 and now it has several derived versions; VGGNet [14], the winner on localization task and second on classification task of ILSVRC2014. These networks are applied to many other tasks and work very well with necessary modifications.

In this paper, we propose a deep-learning-based image orientation detection system. First we evaluate several existed famous networks on orientation detection task, then we create some new structures based on those networks to find an efficient network for this task. We use Caffe [5] to conduct our experiments with necessary modifications.

The remainder of this paper is organized as follows. In Sect. 2 we present the details of the proposed method. Then, the experimental setting and experimental results on six different datasets are described and analyzed in Sect. 3. Finally, in Sect. 4 we conclude this paper and discusses the future work.

2 Methods

AlexNet. AlexNet [16] is a milestone in the development of deep learning and neural network. It provides many basic ideas of designing a convolutional neural network. It consists of convolutional layers, max pooling layers, local response normalization layers and dropout layers. The most important meaning of AlexNet is that it provides a new way to handle computer vision problems.

VGGNet. VGGNet [14] is a classical network. It is similar to AlexNet in some words. In VGGNet, the typical convolutional layer is consisting of a stacked convolutional layers with small kernel size, rather than a single convolutional layer with large kernel size. With this structure it can extract better feature than AlexNet. VGGNet is deeper than AlexNet.

GoogleNet. GoogleNet [18] is also a powerful architecture. It has 22 layers (27 layers if we consider pooling layers). It shows that with deeper network we can get a better results or performance. In GoogelNet there is a new structure calls inception. It divides a layer into four parts than followed with convolutional operations with different size of kernels. After that the authors concatenate these feature maps as the output of a inception structure. There are many derived GoogleNets, such as adding batch normalization [4] and replacing $5 * 5$ kernel with two $3 * 3$ like VGGNet.

According to previous networks and the experimental results on them, we definite a new network to handle this task. With the success of CNN, researchers want to find the truth of CNN. Zeiler et al. [23] tried to visualize convolutional layers from down-level to up-level to illustrate the different function of each convolutional layer. He found that the projections from each layer of AlexNet showed the hierarchical nature of the features in CNN network. Layer 2 and 3(low level) respond corners, edge conjunctions and more complex invariance respectively, the up-level layer 4 shows significant information and layer 5 shows entire objects with significant pose variation. So in this task, after trading off the model complexity and performance, we use three convolutional-based layers in our model.

In the second convolutional layer, we use a inception structure, and the third convolutional layer we employ the stacked convolutional layer to learning more special features.

As the common layers in convolutional networks for classification, we add pooling layer after each convolutional layers and full connected layers in the end of the network. We also use dropout [15] in our model.

3 Experiments

We conduct experiments on six datasets to demonstrate the effectiveness of the proposed method. We assume that the input images are restricted to only four possible rotations. For example, given an upright image, there are other three orientations by rotating it $90°$, $180°$, $270°$. So we have four possible orientations. Our goal is to find which orientation an input image belongs to.

3.1 Data Description

We obtain the datasets from three sources: the first one is open large image datasets, the LFW and ImageNet-2012, the second one is China post mail sorting system, while the third is the Internet from Baidu's Image Search by searching

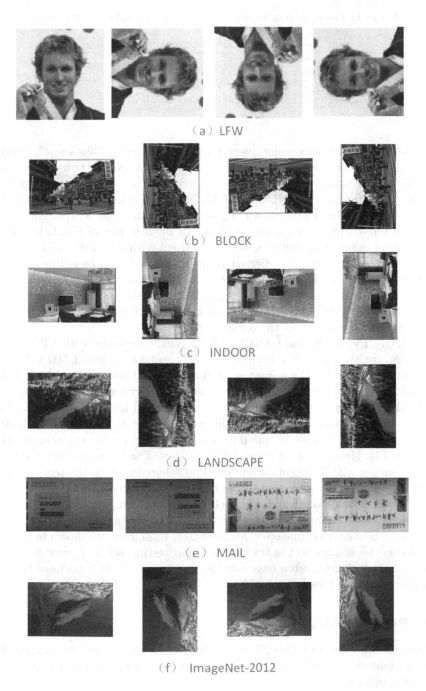

（a）LFW

（b）BLOCK

（c）INDOOR

（d）LANDSCAPE

（e）MAIL

（f）ImageNet-2012

Fig. 1. Examples of our six datasets. The images in five datasets have four orientations, but in dataset MAIL there are only two orientations under the consideration of the real application background.

Table 1. The detection accuracy of different networks on all datasets.

	LFW	MAIL	INDOOR	LANDSCAPE	ImageNet-2012	BLOCK
AlexNet	0.9993	0.9850	0.7731	0.8135	0.7761	0.9127
VGGNet	0.9997	0.9831	0.8162	0.8173	0.7923	0.9322
GoogleNet	0.9993	0.9722	0.7913	0.7622	0.8011	0.8913
The proposed	0.9995	0.9857	0.8261	0.8333	0.8079	0.9366

for images of multi-class using query terms of "indoor", "landscape", "block". The image number of them are different, ranging from several thousand to dozens of million. We name the datasets which gathered from the Internet according to the query terms, i.e., INDOOR, LANDSCAPE, BLOCK respectively. The dataset from China post mail sorting system is named as MAIL. All original images are in correct orientation, as shown in the first row of Fig. 1. In order to create other orientation samples, we rotate them with 90°, 180°, 270°.

LFW is a face dataset collected from Yahoo website, including abundant situations of real world, shown in Fig. 1(a). It contains 13,233 unique images. After rotating 90°, 180°, 270°, we get a dataset with $13233 * 4$ images in four orientations. 10,000 unique images and their rotations are used for training, while the rest are construct the testing set.

BLOCK, INDOOR, and LANDSCAPE have the same size with 3,000 images, examples are shown in Fig. 1(b)–(d). After rotating we get $2,500 * 4$ images for training set and $500 * 4$ images for testing set. These images don't have a highlighted part in a glance, and sometimes have a symmetric structure.

MAIL contains 20,000 images. In consideration of the real application background, we just rotate the mail image in two degrees, i.e., 0° and 180°. There are two types of mail images in this dataset, handwritten and typewritten, which shown in Fig. 1(e), each containing 10000 images. The training dataset contains 15,000 images with 7,500 of each class. The others build up the testing set.

All datasets above only contain a single class of images. In order to simulate the truly situation, we also use a much larger and complex dataset. ILSVRC-2012 is a subset of ImageNet with more than 1,200,000 images of 1000 categories, which is also known as ImageNet-2012 dataset. Examples are shown in Fig. 1(f). We divide the images to the training set and testing set as it was set in [13]. Images in this dataset often have complex backgrounds. It is perhaps the most complex dataset for various classification tasks.

3.2   Pre-Processing

All images are resized to $256 * 256$ before feeding into the network, and the channel information are remained, which means we don't transform colorful images into gray images.

Table 2. More details about accuracies of the proposed network on LFW, MAIL and INDOOR datasets. The left-most column is the ground truth, the top row indicates the prediction orientations.

	0°	90°	180°	270°
(a) LFW				
0°	**0.9997**	0.0000	0.0000	0.0003
90°	0.0003	**0.9991**	0.0003	0.0003
180°	0.0000	0.0003	**0.9997**	0.0000
270°	0.0003	0.0000	0.0000	**0.9997**
Average accuracy	**0.9995**			
(b) MAIL				
0°	**0.9841**	*	0.0159	*
90°	*	*	*	*
180°	0.0127	*	**0.9873**	*
270°	*	*	*	*
Average accuracy	**0.9857**			
(c) INDOOR				
0°	**0.8397**	0.0284	0.1055	0.0264
90°	0.0400	**0.8160**	0.0380	0.1060
180°	0.0907	0.0383	**0.8367**	0.0343
270°	0.0485	0.1071	0.0323	**0.8121**
Average accuracy	**0.8261**			

3.3 Experimental Settings

We train the network with batch size = 128 for 20 K iterations. We also apply weight decay and momentum in our experiments with setting values to $1 * 10^{-4}$ and 0.9, respectively. For optimization, we use the SGD to update the network parameters.

3.4 Experimental Results

We test four networks in our experiments, AlexNet, GoogleNet, VGGNet and the proposed one. The testing results on six datasets of all four networks are shown in Table 1. And in Tables 2 and 3 we present the more detailed results of the proposed network under the different orientations.

The experiments show that convolutional neural network is suitable for image orientation detection task. All architectures obtain good results on all datasets. Especially on LFW dataset, the accuracies are close to 1. The proposed method leverages the advantages of existed networks and achieves better results on this task.

Table 3. More details about accuracies of the proposed network on LANDSCAPE, ImageNet-2012 and BLOCK datasets. The left-most column is the ground truth, the top row indicates the prediction orientations.

	0°	90°	180°	270°
(a) LANDSCAPE				
0°	**0.8375**	0.0375	0.0958	0.0292
90°	0.0249	**0.8382**	0.0415	0.0954
180°	0.1000	0.0375	**0.8250**	0.0375
270°	0.0293	0.1004	0.0377	**0.8326**
Average accuracy	**0.8333**			
(b) ImageNet-2012				
0°	**0.8538**	0.0517	0.0553	0.0391
90°	0.0598	**0.7893**	0.0794	0.0715
180°	0.0552	0.0740	**0.7902**	0.0805
270°	0.0400	0.0702	0.0915	**0.7983**
Average accuracy	**0.8079**			
(c) BLOCK				
0°	**0.9386**	0.0144	0.0289	0.0180
90°	0.0147	**0.9414**	0.0183	0.0256
180°	0.0256	0.0220	**0.9304**	0.0220
270°	0.0113	0.0377	0.0151	**0.9358**
Average accuracy	**0.9366**			

More detail accuracy information on each dataset is shown is Table 2. LFW is a face dataset and in four orientations the structure of images belongs to a common pattern. We get a very high accuracy. But ImageNet-2012 is much complex than LFW that have various backgrounds and structure, it is hard to study a common pattern to represent all images.

Compared to the other methods, we archive better results in almost all datasets. Especial in the much complex dataset ImageNet-2012 our network gets a higher accuracy on orientation detection task. The experimental results have demonstrated that our network is a better structure to detect the orientation of images compared to the existed networks.

3.5   Computational Cost

Our experiments were performed on a DELL workstation. The CPU is Intel Xeon E5-1650, and the GPU is NVIDIA TITAN X. The software is the latest version of caffe with cuDNN V5 accelerated on Ubuntu 14.04 system. The average testing time is 5 ms per image with GPU.

4   Conclusions

Automatic image orientation detection is an important and difficult problem. By taking advantage of some efficient structures in the existed convolutional neural network architectures, we present a new CNN model to handle this problem. Compared to previous approaches that used hand-crafted features, our approach extracts the structured features directly from the raw images. Experimental results on several datasets indicate that the proposed approach has excellent ability on image orientation detection task. We also find that in nature images, the performance is still not good. In the future, we will improve the accuracy by using better efficient neural networks.

References

1. Baluja, S.: Automated image-orientation detection: a scalable boosting approach. Pattern Anal. Appl. **10**(3), 247–263 (2007)
2. Chen, Y.: Learning orientation-estimation convolutional neural network for building detection in optical remote sensing image. CoRR abs/1903.05862 (2019)
3. Hassan, M.A., Sazonov, E.: Orientation-based food image capture for head mounted egocentric camera. In: 41st Annual International Conference of the IEEE Engineering in Medicine and Biology Society, EMBC 2019, Berlin, Germany, 23–27 July, 2019, pp. 7145–7148 (2019)
4. Ioffe, S., Szegedy, C.: Batch normalization: accelerating deep network training by reducing internal covariate shift. arXiv preprint arXiv:1502.03167 (2015)
5. Jia, Y., et al.: Caffe: convolutional architecture for fast feature embedding. arXiv preprint arXiv:1408.5093 (2014)
6. Kumar, D., Singh, R.: Multi orientation text detection in natural imagery: a comparative survey. IJCVIP **8**(4), 41–56 (2018)
7. Lee, J., Yoon, K.: Joint estimation of camera orientation and vanishing points from an image sequence in a non-manhattan world. Int. J. Comput. Vision **127**(10), 1426–1442 (2019)
8. Li, J., Zhang, H., Wang, J., Xiao, Y., Wan, W.: Orientation-aware saliency guided JND model for robust image watermarking. IEEE Access **7**, 41261–41272 (2019)
9. Luo, J., Boutell, M.: A probabilistic approach to image orientation detection via confidence-based integration of low level and semantic cues. IEEE Trans. Pattern Anal. Mach. Intell. **27**(5), 715–726 (2005)
10. Morra, L., Famouri, S., Karakus, H.C., Lamberti, F.: Automatic detection of canonical image orientation by convolutional neural networks. In: IEEE 23rd International Symposium on Consumer Technologies, ISCT 2019, Ancona, Italy, June 19–21, 2019, pp. 118–123 (2019)
11. Olmschenk, G., Tang, H., Zhu, Z.: Pitch and roll camera orientation from a single 2D image using convolutional neural networks. In: 14th Conference on Computer and Robot Vision, CRV 2017, Edmonton, AB, Canada, 16–19 May, 2017, pp. 261–268 (2017)
12. Prince, M., Alsuhibany, S.A., Siddiqi, N.A.: A step towards the optimal estimation of image orientation. IEEE Access **7**, 185750–185759 (2019)
13. Russakovsky, O., et al.: ImageNet large scale visual recognition challenge. Int. J. Comput. Vision **115**(3), 211–252 (2015). https://doi.org/10.1007/s11263-015-0816-y

14. Simonyan, K., Zisserman, A.: Very deep convolutional networks for large-scale image recognition. arXiv preprint arXiv:1409.1556 (2014)
15. Srivastava, N., Hinton, G.E., Krizhevsky, A., Sutskever, I., Salakhutdinov, R.: Dropout: a simple way to prevent neural networks from overfitting. J. Mach. Learn. Res. **15**(1), 1929–1958 (2014)
16. Sutskever, I., Hinton, G.E., Krizhevsky, A.: Imagenet classification with deep convolutional neural networks. Adv. Neural Inf. Process. Syst. **19**, 1097–1105 (2012)
17. Szegedy, C., Ioffe, S., Vanhoucke, V., Alemi, A.: Inception-v4, inception-ResNet and the impact of residual connections on learning. arXiv preprint arXiv:1602.07261 (2016)
18. Szegedy, C.: Going deeper with convolutions. In: Computer Vision and Pattern Recognition, pp. 1–9 (2014)
19. Szegedy, C., Vanhoucke, V., Ioffe, S., Shlens, J., Wojna, Z.: Rethinking the inception architecture for computer vision. In: Proceedings of the IEEE Conference on Computer Vision and Pattern Recognition, pp. 2818–2826 (2016)
20. Vailaya, A., Zhang, H., Yang, C., Liu, F.I., Jain, A.K.: Automatic image orientation detection. IEEE Trans. Image Process. **11**(7), 746–755 (2002)
21. Wang, J., Wan, W., Li, X.X., Sun, J.D., Zhang, H.X.: Color image watermarking based on orientation diversity and color complexity. Expert Syst. Appl. **140** (2020)
22. Wu, M., Han, C., Guo, T., Zhao, T.: Registration and matching method for directed point set with orientation attributes and local information. Comput. Vis. Image Underst. **191**, 102866 (2020)
23. Zeiler, M.D., Fergus, R.: Visualizing and understanding convolutional networks. In: Fleet, D., Pajdla, T., Schiele, B., Tuytelaars, T. (eds.) ECCV 2014. LNCS, vol. 8689, pp. 818–833. Springer, Cham (2014). https://doi.org/10.1007/978-3-319-10590-1_53
24. Zhang, L., Li, M., Zhang, H.J.: Boosting image orientation detection with indoor vs. outdoor classification. In: Proceedings of the Sixth IEEE Workshop on Applications of Computer Vision, pp. 95–99. IEEE (2002)

Multi-layer Cross-domain Non-negative Matrix Factorization for Cross-scene Dimension Reduction on Hyperspectral Images

Hong Chen[1] , Kewei Gong[3] , Ling Lei[1] , Minchao Ye[1(⊠)] ,
and Yuntao Qian[2]

[1] Key Laboratory of Electromagnetic Wave Information Technology and Metrology
of Zhejiang Province, College of Information Engineering, China Jiliang University,
Hangzhou 310018, China
yeminchao@cjlu.edu.cn
[2] College of Computer Science, Zhejiang University, Hangzhou 310027, China
[3] Xiaoshan District Sub-Bureau of Hangzhou Public Security Bureau,
Hangzhou 311200, China

Abstract. Many applications on hyperspectral images (HSIs) always suffer from the high dimension low sample size problem. Hence dimension reduction (DR) is a necessary pre-process for HSIs. Most existing DR algorithms only concentrate on a single dataset. However, similar HSI scenes may share information between each other. How to utilize the shared information is an interesting research topic. This research work concentrates on cross-scene DR for HSIs. Combining the idea of dual dictionary non-negative matrix factorization (DDNMF) and stacked autoencoder (SAE), a multi-layer cross-domain non-negative matrix factorization (MLCDNMF) algorithm is proposed in this work to perform DR across different HSI scenes. MLCDNMF has following characters: 1) it provides the ability of both homogenous transfer learning and heterogeneous transfer learning; 2) it containers two DDNMF layers, and works in a way like SAE; 3) it is beyond SAE due to the graphs built based on the correlations between samples; 4) it is a flexible model which does not require strict one-to-one sample correspondences between different scenes. Experiments on cross-scene HSI datasets show good performance of the proposed MLCDNMF algorithm.

Keywords: Hyperspectral image · Cross-scene dimension reduction · Non-negative matrix factorization · Stacked autoencoder.

Supported by the National Natural Science Foundation of China (grant number 61701468), the National Key Research and Development Program of China (grant number 2018YFB0505000) and the Outstanding Student Achievement Cultivation Program of China Jiliang University (grant number 2019YW24).

Y. Lu et al. (Eds.): ICPRAI 2020, LNCS 12068, pp. 547–554, 2020.
https://doi.org/10.1007/978-3-030-59830-3_47

1   Introduction and Related Work

High feature dimensional and small sample size conditions have limited the progress of hyperspectral image (IISI) classification for decades. In recent years, various dimension reduction (DR) algorithms have been proposed, including PCA [1], discriminant analysis [4], deep learning [9], etc. However, aforementioned DR methods are designed for a single HSI scene, few methods can handle cross-scene DR. Once we take a glance at two similar HSI scenes (namely source scene and target scene), we can find the co-occurrence of land cover classes. Transfer learning is a powerful tool for exploring shared information between similar HSI scenes. For HSIs, different types of transfer learning algorithms are needed depending on whether source and target scenes are captured with the same HSI sensor. If source scene and target scene are captured by the same sensor, homogeneous transfer learning can be applied, otherwise, heterogeneous transfer learning is needed. Our earlier researches cover some algorithms for homogeneous transfer learning in HSIs [6,7], but heterogeneous transfer learning is more desired in HSI applications, since in most cases, source and target scenes are captured by different sensors.

For performing heterogeneous transfer learning, we have proposed the dual-dictionary non-negative matrix factorization (DDNMF) for cross-scene DR in our latest research [2]. Despite the effectiveness of DDNMF in cross-scene DR, the shallow model limits its performance. So in this work, we extend DDNMF to a deep model by incorporating the architecture of stacked autoencoder (SAE). The newly developed deep model is named multi-layer cross-domain non-negative matrix factorization (MLCDNMF), which has two autoencoder layers implemented by DDNMF, where the first layer aligns different input feature dimensions to a same one, and the second layer further enhances feature invariance between source and target scenes. In experiments on cross-scene HSI dataset, MLCDNMF achieves impressive results, proving its effectiveness of cross-scene DR.

2   Multi-layer Cross-domain Non-negative Matrix Factorization Model

2.1   SAE Based on NMF

SAE is a common method for unsupervised deep feature extraction, which can be traced back to 1994 [3]. It has been successfully applied to DR for HSIs [8]. In this work, non-negative matrix factorization (NMF) is adopted for building an autoencoder. NMF factorizes the non-negative input signals $\mathbf{X} \in \mathbb{R}_+^{m \times p}$ into two low-dimensional non-negative matrices $\mathbf{U} \in \mathbb{R}_+^{m \times r}$ and $\mathbf{V} \in \mathbb{R}_+^{p \times r}$:

$$\min_{\mathbf{U},\mathbf{V}} \|\mathbf{X} - \mathbf{U}\mathbf{V}^{\mathrm{T}}\|_F^2 \quad \text{s.t.} \quad \mathbf{U} \geq 0, \mathbf{V} \geq 0. \tag{1}$$

As shown in Fig. 1, the process of NMF is exactly a non-negative autoencoder.

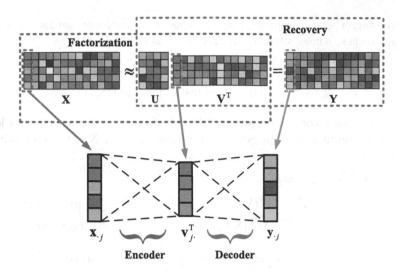

Fig. 1. Non-negative autoencoder based on NMF.

- Encoder: we denote the jth column of $\mathbf{X}$ as $\mathbf{x}_{\cdot j}$ and the jth row of $\mathbf{V}$ as $\mathbf{v}_{j\cdot}$, then NMF transforms $\mathbf{x}_{\cdot j} \in \mathbb{R}_+^{m \times 1}$ to $\mathbf{v}_{j\cdot}^{\mathrm{T}} \in \mathbb{R}_+^{r \times 1}$, and $\mathbf{v}_{j\cdot}^{\mathrm{T}}$ can be seen as the code of $\mathbf{x}_{\cdot j}$.
- Decoder: the recovery of NMF can be expressed as $\mathbf{Y} = \mathbf{U}\mathbf{V}^{\mathrm{T}}$, and we denote the jth column of $\mathbf{Y}$ as $\mathbf{y}_{\cdot j}$, then the recovery transforms $\mathbf{v}_{j\cdot}^{\mathrm{T}} \in \mathbb{R}_+^{r \times 1}$ to $\mathbf{y}_{\cdot j} \in \mathbb{R}_+^{m \times 1}$, which is the approximation of $\mathbf{x}_{\cdot j}$.

By stacking NMFs, we can build the SAE.

2.2 MLCDNMF

For cross-scene DR, there are two input matrices $\mathbf{X}^{\mathcal{S}}$ and $\mathbf{X}^{\mathcal{T}}$, where superscripts $\mathcal{S}$ and $\mathcal{T}$ are used for indicating source and target scenes, respectively. Meanwhile, the class labels of pixels are required for building connections between pixels. Three graphs are defined for maintaining within-scene and cross-scene data manifold structures.

1. source-source graph, representing the sample similarity within source scene, whose adjacent matrix $\mathbf{W}^{\mathcal{S}}$ is defined as

$$
w_{ij}^{\mathcal{S}} = \begin{cases} \frac{1}{Z^{\mathcal{S}}} \frac{\langle \mathbf{x}_{\cdot i}^{\mathcal{S}}, \mathbf{x}_{\cdot j}^{\mathcal{S}} \rangle}{\|\mathbf{x}_{\cdot i}^{\mathcal{S}}\|_2 \|\mathbf{x}_{\cdot j}^{\mathcal{S}}\|_2}, & class(\mathbf{x}_{\cdot i}^{\mathcal{S}}) = class(\mathbf{x}_{\cdot j}^{\mathcal{S}}) \\ 0, & class(\mathbf{x}_{\cdot i}^{\mathcal{S}}) \neq class(\mathbf{x}_{\cdot j}^{\mathcal{S}}) \end{cases}, \tag{2}
$$

2. target-target graph, representing the sample similarity within target scene, whose adjacent matrix $\mathbf{W}^{\mathcal{T}}$ is defined as

$$
w_{ij}^{\mathcal{T}} = \begin{cases} \frac{1}{Z^{\mathcal{T}}} \frac{\langle \mathbf{x}_{\cdot i}^{\mathcal{T}}, \mathbf{x}_{\cdot j}^{\mathcal{T}} \rangle}{\|\mathbf{x}_{\cdot i}^{\mathcal{T}}\|_2 \|\mathbf{x}_{\cdot j}^{\mathcal{T}}\|_2}, & class(\mathbf{x}_{\cdot i}^{\mathcal{T}}) = class(\mathbf{x}_{\cdot j}^{\mathcal{T}}) \\ 0, & class(\mathbf{x}_{\cdot i}^{\mathcal{T}}) \neq class(\mathbf{x}_{\cdot j}^{\mathcal{T}}) \end{cases}, \tag{3}
$$

3. source-target graph, representing the class co-occurrence across source and target scenes, whose adjacent matrix $\mathbf{W}^{\mathcal{ST}}$ is defined as

$$w_{ij}^{\mathcal{ST}} = \begin{cases} \frac{1}{Z^{\mathcal{ST}}}, & class(\mathbf{x}_{\cdot i}^{\mathcal{S}}) = class(\mathbf{x}_{\cdot j}^{\mathcal{T}}) \\ 0, & class(\mathbf{x}_{\cdot i}^{\mathcal{S}}) \neq class(\mathbf{x}_{\cdot j}^{\mathcal{T}}) \end{cases}. \tag{4}$$

In our previous work, we proposed DDNMF for heterogeneous transfer learning [2]. With input source/target data matrices $\mathbf{X}^{\mathcal{S}}$ and $\mathbf{X}^{\mathcal{T}}$, the cost function is defined as

$$
\begin{aligned}
\mathcal{C}(\mathbf{U}^{\mathcal{S}}, \mathbf{V}^{\mathcal{S}}, \mathbf{U}^{\mathcal{T}}, \mathbf{V}^{\mathcal{T}}) = & \\
& \|\mathbf{X}^{\mathcal{S}} - \mathbf{U}^{\mathcal{S}}(\mathbf{V}^{\mathcal{S}})^{\mathrm{T}}\|_F^2 && \text{(source error)} \\
& + \alpha\|\mathbf{X}^{\mathcal{T}} - \mathbf{U}^{\mathcal{T}}(\mathbf{V}^{\mathcal{T}})^{\mathrm{T}}\|_F^2 && \text{(target error)} \\
& + \frac{\lambda}{2}\sum_{i=1}^{p}\sum_{j=1}^{p} w_{ij}^{\mathcal{S}}\|\mathbf{v}_{i\cdot}^{\mathcal{S}} - \mathbf{v}_{j\cdot}^{\mathcal{S}}\|_2^2 && \text{(source-source graph)} \\
& + \frac{\lambda}{2}\sum_{i=1}^{q}\sum_{j=1}^{q} w_{ij}^{\mathcal{T}}\|\mathbf{v}_{i\cdot}^{\mathcal{T}} - \mathbf{v}_{j\cdot}^{\mathcal{T}}\|_2^2 && \text{(target-target graph)} \\
& + \frac{\lambda\mu}{2}\sum_{i=1}^{p}\sum_{j=1}^{q} w_{ij}^{\mathcal{ST}}\|\mathbf{v}_{i\cdot}^{\mathcal{S}} - \mathbf{v}_{j\cdot}^{\mathcal{T}}\|_2^2 && \text{(source-target graph)},
\end{aligned}
\tag{5}
$$

where $\mathbf{U}^{\mathcal{S}}$ and $\mathbf{U}^{\mathcal{T}}$ are source and target dictionaries, which need to be trained. They are expected to transform the source and target data into a shared low-dimensional feature subspace. $\mathbf{V}^{\mathcal{S}}$ and $\mathbf{V}^{\mathcal{T}}$ are low-dimensional features of $\mathbf{X}^{\mathcal{S}}$ and $\mathbf{X}^{\mathcal{T}}$, respectively. $\alpha = \|\mathbf{X}^{\mathcal{S}}\|_F^2/\|\mathbf{X}^{\mathcal{T}}\|_F^2$ is the balancing factor between source and target scenes. λ is the parameter controlling the strength of graph regularization, and μ is parameter balancing within-scene graphs and cross-scene graph. The detailed parameter settings can be found in [2]. Experimental results in [2] imply that DDNMF aligns the data manifolds of two domains.

By combining the cross-scene DR ability of DDNMF and the superior feature extraction performance of SAE, the MLCDNMF model is proposed, whose framework is shown in Fig. 2. MLCDNMF is built by stacking DDNMF layers in the way of SAE and adding the activation between adjacent layers. MLCDNMF works with following procedures. We use superscripts $(\cdot)^{(1)}$ and $(\cdot)^{(2)}$ to denote the variables of first and second layers, respectively.

1. Input features (spectrums of pixels) are normalized with min-max normalization to avoid numerical problems caused by the scale of input data, and the normalized data $(\mathbf{X}^{\mathcal{S}})^{(1)}$ and $(\mathbf{X}^{\mathcal{T}})^{(1)}$ are fed into first DDNMF layer. Note that $(\mathbf{X}^{\mathcal{S}})^{(1)}$ and $(\mathbf{X}^{\mathcal{T}})^{(1)}$ have different feature dimensions.
2. DDNMF of first layer is performed, which transfers $(\mathbf{X}^{\mathcal{S}})^{(1)}$ and $(\mathbf{X}^{\mathcal{T}})^{(1)}$ into $((\mathbf{V}^{\mathcal{S}})^{(1)})^{\mathrm{T}}$ and $((\mathbf{V}^{\mathcal{T}})^{(1)})^{\mathrm{T}}$, respectively. After the DDNMF, $((\mathbf{V}^{\mathcal{S}})^{(1)})^{\mathrm{T}}$ and $((\mathbf{V}^{\mathcal{T}})^{(1)})^{\mathrm{T}}$ have the same feature dimensions. The role of first DDNMF layer is to transform different input spaces into a shared one, which is a key step for

heterogeneous transfer learning. However, feature shift may still exist between source and target scenes, hence further alignments are required.

3. Activation and min-max normalization are performed on the output features of first layer $((\mathbf{V}^{\mathcal{S}})^{(1)})^{\mathrm{T}}$ and $((\mathbf{V}^{\mathcal{T}})^{(1)})^{\mathrm{T}}$, before they are fed into the second layer. Applying activation function brings the ability of non-linear feature representation, and the min-max normalization can ensure non-negativity and avoid numerical problems. With activation and min-max normalization procedures, $((\mathbf{V}^{\mathcal{S}})^{(1)})^{\mathrm{T}}$ and $((\mathbf{V}^{\mathcal{T}})^{(1)})^{\mathrm{T}}$ are transformed to $(\mathbf{X}^{\mathcal{S}})^{(2)}$ and $(\mathbf{X}^{\mathcal{T}})^{(2)}$, which are the input of the second DDNMF layer.

4. Though the input of second layer $(\mathbf{X}^{\mathcal{S}})^{(2)}$ and $(\mathbf{X}^{\mathcal{T}})^{(2)}$ have the same feature dimensions, there may exist feature shift between $(\mathbf{X}^{\mathcal{S}})^{(2)}$ and $(\mathbf{X}^{\mathcal{T}})^{(2)}$. What the second layer does is to further eliminate the feature shift. After performing DDNMF, the outputs $((\mathbf{V}^{\mathcal{S}})^{(2)})^{\mathrm{T}}$ and $((\mathbf{V}^{\mathcal{T}})^{(2)})^{\mathrm{T}}$ from second layer are regarded as the final outputs of MLCDNMF.

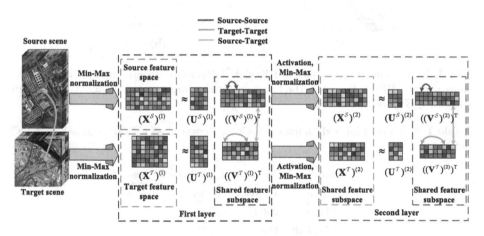

Fig. 2. The framework of MLCDNMF.

3   Experiments

To test the feature invariance achieved by the DR algorithm, classification experiments are conducted on the cross-scene HSI dataset from [2], which is named RPaviaU-DPaviaC dataset. It is composed of the source scene ROSIS Pavia University (RPaviaU) and the target scene DAIS Pavia Center (DPaviaC). RPaviaU scene was taken by ROSIS HSI sensor over the University of Pavia, Italy, and the data cube is sized $610 \times 340 \times 103$. DPaviaC scene was obtained via DAIS HSI sensor over the center of Pavia city, Italy, and the data cube is sized $400 \times 400 \times 72$. Seven land cover classes are shared between two scenes. The visualized data cubes and ground truth maps are illustrated in Fig. 3. The details of shared land cover classes can be found in Table 1.

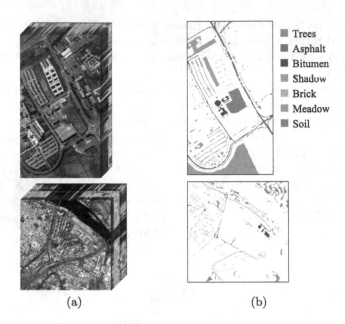

Fig. 3. Source and target scenes in RPaviaU-DPaviaC dataset. The upper one is the source scene (RPaviaU), while the lower one is the target scene (DPaviaC). (a) Data cubes. (b) Ground truth maps.

Table 1. Number of labeled samples in each land cover class within RPaviaU-DPaviaC dataset.

Class		Number of labeled samples	
#	Name	RPaviaU	DPaviaC
1	Trees	3064	2424
2	Asphalt	6631	1704
3	Bitumen	1330	685
4	Shadow	947	241
5	Brick	3682	2237
6	Meadow	18649	1251
7	Soil	5029	1475

We randomly select 200 labeled samples per class in each scene as training samples, and the remaining labeled samples are treated as test samples. To test the feature invariance, following steps are adopted for evaluation:

1. Train the DR model with both source and target training samples.
2. Perform DR on source training set and target test set, obtaining low dimensional features.
3. Train a support vector machine (SVM) classifier on source training set.

4. Perform prediction on target test set and evaluate the accuracies.

It should be noted that training set and test set are selected from different scenes in order to test the cross-scene feature invariance achieved by the DR algorithm.

Three cross-domain DR algorithms are compared: 1) Manifold Alignment (MA) [5]; 2) DDNMF [2]; 3) MLCDNMF proposed in this work. For all compared algorithms, the output dimension is set to $r \in \{5, 6, \dots, 25\}$, and especially for MLCDNMF, the output dimension of first DDNMF layer is set to $2r$, the output dimension of second DDNMF layer (final output) is set to r, and sigmoid function is selected for activation. For SVM classifier, RBF kernel is adopted, so there are two parameters: γ for the RBF kernel, and C for penalty. Five-fold cross validation is performed with $\gamma \in \{2^{-15}, 2^{-14}, \dots, 2^5\}$ and $C \in \{10^{-1}, 10^0, \dots, 10^4\}$. Overall accuracy (OA), average accuracy (AA) and kappa coefficient (κ) are utilized for accuracy evaluation. For each DR algorithm, the OA, AA and κ with each r value are collected, and the best one among different r values is selected for comparison. The results listed in Table 2 implies that the proposed MLCDNMF achieves best accuracies among all compared cross-domain DR algorithms.

Table 2. Accuracies achieved by different DR algorithms.

DR algorithm	OA	AA	κ
MA	0.2420	0.1486	0.0088
DDNMF	0.6520	0.6770	0.5764
MLCDNMF	**0.7691**	**0.7961**	**0.7212**

Though MLCDNMF achieves best results in the comparison, we still have some ideas for future work, which may provide opportunities to further promote the performance of MLCDNMF.

1. In this work, the output dimension of first layer is empirically set to $2r$. However, tuning this parameter may lead to better results.
2. Only one kind of activation function, i.e., sigmoid, is adopted in the experiments. Changing different activation functions may also benefit MLCDNMF.
3. Currently, only the outputs of the last layer are included in the final output. Maybe including the more outputs from other layers can help to obtain better accuracies.
4. Combing MLCDNMF and other cross-domain feature selection algorithms may produce better results.

4   Conclusion

In this work, we have proposed a MLCDNMF model for cross-scene DR, which is a deep extension of DDNMF. It combines the cross-scene DR ability of DDNMF and the deep feature extraction performance of SAE. In MLCDNMF, two DDNMF layers are stacked to form a cross-scene SAE. MLCDNMF conveys more cross-scene feature invariance than DDNMF. Moreover, MLCDNMF provide the opportunity of non-linear feature extraction through the activation between layers. Experiments on cross-scene HSI dataset has confirmed the superior of MLCDNMF.

References

1. Agarwal, A., El-Ghazawi, T., El-Askary, H., Le-Moigne, J.: Efficient hierarchical-PCA dimension reduction for hyperspectral imagery. In: Proceedings pf the IEEE International Symposium on Signal Processing and Information Technology, pp. 353–356, December 2007
2. Chen, H., Ye, M., Lu, H., Lei, L., Qian, Y.: Dual dictionary learning for mining a unified feature subspace between different hyperspectral image scenes. In: Proceedings of the IEEE International Geoscience and Remote Sensing Symposium, pp. 1096–1099 (2019)
3. Hinton, G.E., Zemel, R.S.: Autoencoders, minimum description length and Helmholtz free energy. In: Proceedings of the Advances in Neural Information Processing Systems, pp. 3–10 (1994)
4. Li, W., Feng, F., Li, H., Du, Q.: Discriminant analysis-based dimension reduction for hyperspectral image classification: a survey of the most recent advances and an experimental comparison of different techniques. IEEE Geosci. Remote Sens. Mag. **6**(1), 15–34 (2018)
5. Wang, C., Mahadevan, S.: A general framework for manifold alignment. In: Proceedings of the AAAI Fall Symposium Series, pp. 79–86 (2009)
6. Ye, M., Qian, Y., Zhou, J., Tang, Y.Y.: Dictionary learning-based feature-level domain adaptation for cross-scene hyperspectral image classification. IEEE Trans. Geosci. Remote Sens. **55**(3), 1544–1562 (2017)
7. Ye, M., Zheng, W., Lu, H., Zeng, X., Qian, Y.: Cross-scene hyperspectral image classification based on DWT and manifold-constrained subspace learning. Int. J. Wavelets Multiresolution Inf. Process. **15**(06), 1–16 (2017)
8. Zabalza, J., Ren, J., Zheng, J., Zhao, H., Qing, C., Yang, Z., Du, P., Marshall, S.: Novel segmented stacked autoencoder for effective dimensionality reduction and feature extraction in hyperspectral imaging. Neurocomputing **185**, 1–10 (2016)
9. Zhao, W., Du, S.: Spectral-spatial feature extraction for hyperspectral image classification: a dimension reduction and deep learning approach. IEEE Trans. Geosci. Remote Sens. **54**(8), 4544–4554 (2016)

A Comprehensive Unconstrained, License Plate Database

Nicola Nobile$^{(\boxtimes)}$ ⓘ, Hoi Kei Phocbe Chan ⓘ, and Marleah Blom ⓘ

CENPARMI, Concordia University, Montreal, Canada
nicola@cenparmi.concordia.ca, hkpc07@gmail.com,
marleah@encs.concordia.ca

Abstract. In this paper, information about a large and diverse database of license plates from countries around the world is presented. CENPARMI's growing database contains images of isolated plates, including a small percentage of vanity plates, as well as landscapes, where the vehicle is included in the image. Many images contain multiple license plates, have complex scenery, have motion blur, and contain high light and shadow contrast. Photos were taken during seasons, many are occluded by foreground objects, and some were taken through mist, fog, snow, and/or glass. In many cases, the license plate, camera, or both were in motion. In order to make training more robust, different cameras, lenses, focal lengths, shutter speeds, and ISOs were used. A summary of proposed guidelines used for license plate design are outlined, details about the database of license plates are presented, followed by suggestions for future work.

Keywords: License plates · Database · Off-line image recognition

1 Introduction

License plates from around the world come in a wide range of formats, character set, restrictions, fonts, and layouts. CENPARMI is currently building a large database of license plates from several countries. Although there are several license plate databases, they contain plates that are limited to only one region, have few points-of-view of the plates, and were taken in ideal environments. We wanted to create a more challenging database that included license plates from many countries, states, and provinces, were taken from many different angles and heights, has moving and stationary vehicles, and offers some challenging environment conditions such as snow and rain.

1.1 Limitations

Many countries issuing license plates belong to the American Association of Motor Vehicle Administrators (AAMVA). AAMVA, based in Arlington, Virginia, creates standards for road safety and researches topics related to the registration of vehicles. Members must conform to legibility guidelines established by the AAMVA. Some countries limit

Y. Lu et al. (Eds.): ICPRAI 2020, LNCS 12068, pp. 555–561, 2020.
https://doi.org/10.1007/978-3-030-59830-3_48

the character set used based on human visual identification or more recently, based on results by computerized recognition systems and cameras.

License plates are most effective when they are designed to optimize legibility to the human eye as well as for automated license plate readers (ALPRs) [1]. As a result, AAMVA imposes strict standards in order to make identification easier for both human and machine. The relevant policies are related to display and design. For display, license plates must be displayed horizontally in the space designated by the vehicle manufacturer. Some issuing authorities may use decals - stickers used to add a graphic to a license plate. Decals should not be used to identify a license plate since they have a shorter life expectancy that the license plate. They are for presentation purposes only. Some additional AAMVA proposed guidelines for license plate design are shown in Table 1 [1].

Table 1. Some AAMVA proposed guidelines

1. Characters are at least 2.5 inches in height, proportionally wide, and spaced no less than 0.25 inches apart
2. Character stroke weight (thickness of lines) are between 0.2 and 0.4 inches
3. Characters are positioned on the plate no less than 1.25 inches away from the top and bottom edges of the license plate
4. The font and spacing present each alphanumeric as a distinct and identifiable character. Standardized fonts and font sizes that clearly distinguish characters are used
5. Non-alphanumeric characters are allowed. They are considered part of the license plate number and are to be accurately displayed on the license plate
6. Graphics on license plates must not distort or interfere with the readability of the characters or with any other identifying information on the plate
7. Graphics must be to the right or left side of the license plate number
8. A background is allowed but must not interfere with the ability to read the license plate number by both a human and by an ALPR

Because of item number 4, some jurisdictions limit the characters uses to avoid confusion in identification. Only one of a group of similar looking characters will be used. For example, the letter "I" and the number "1", "B" and "8" look very similar from far away. In this case, the issuing authority may ban the letters "I" and "B" to both improve legibility and to avoid near duplicate plates.

Many fonts can be found on plates as there are Asian characters (like traditional Chinese and Japanese), special characters (such as dash and accents) and alphanumeric characters from North America and Europe. License plates in our database are either decorative plates, personalized/vanity plates, diplomatic plates and common plates. While plates at the back of a vehicle are compulsory, not all places require a front license plates like the province of Québec in Canada.

Although car owners are required to maintain good quality of their plates and obey certain restrictions, these rules may not always be adhered to due to dirt, snow, or other coverings, such as shields or frames. In many cases, information displayed on the license plates may be obstructed or may be difficult to see at night. For example, we have seen license plates so dirty that it was difficult to read the contents, especially snow covered plates, plates with tinted or smoked bubble shields in front of them. Some license plates have a frame that partially covers the letters and/or digits of the plate. Also, there were some unlit plates, which makes reading them difficult at night.

Many more issues arise for ALPR because of generous rules such as vanity plates. Although vanity plates must obey some restrictions, they do not necessarily need to follow the standard plate format of a given district. For example, in Quebec, Canada, government passenger issued driver license plates follow a "LDD LLL" or a "LLLD-DDD" pattern [9]. Where "L" is a letter and "D" is a digit. An ALPR system can no longer take advantage of this pattern. We have seen vanity plates as long as "JIMIDAR" and "VITRIOL", and as short as "7Z", "MV", "HEY", and "6888".

Additionally, more governing agencies allow for a background image or logos/icons to be displayed, which greatly affects character segmentation and recognition. Veteran plates are known to have background and logos included in them. To renew a license plate, some places still use stickers to indicate the expiration date. Finally, electric cars, in Quebec, have license plates in a light green colour, which may affect plate identification if a colour threshold is being used. Figure 1 shows some samples that challenge plate recognition such as vanity plates, logos, background image, and frames.

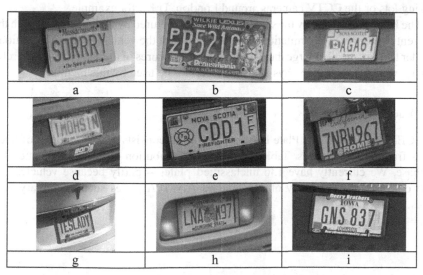

Fig. 1. Challenging License Plates Including Vanity (a, c, e, g), with Background (b, h), with Logo (c, e), with Obstructing Frame (b, f), Light Green Coloured Plate (g), with Sticker (b, h, i), with Smoked Bubble Shield (d) (Color figure online)

2 Database Overview

Our goal is to create a large vehicle license plate database. We decided to collect images of license plates from around the world in order to make ALPR systems more robust. Users may choose to focus their recognizers on one specific plate type if so desired.

Some plates are clean, legible and taken at a close distance. Meanwhile, some plates are challenging to read as they may be blurry considering they were taken either 1) from a long distance, 2) while cars are moving, 3) various angles, 4) while walking, 5) under challenging weather conditions (such as rain, slush and snow), 6) broken/bent/rusted, 7) colorful backgrounds, 8) have stickers or other markings, 9) difficult fonts, 10) varied font colors, 11) plates' composition, 12) protected by tinted or smoked cover, 13) have license plate frames from dealership or other organizations.

It is important to note that tinted covers are illegal in places like Ontario [4] and certain states like Connecticut in the United States, do not allow plates to be covered by frames [5].

Other datasets, such as the Reid dataset [6], was intentionally made of low quality images, extracted from 9.5 h of video. This dataset simulates surveillance cameras by placing the Full-HD video capture devices on bridges above eight highway locations. The Reid dataset contains 76, 412 color license plate images of different lengths, image blur and slight occlusion. Since the extracted plates were taken from the same viewpoint, we found this a limitation to our goal to have plate images from different heights, as well as distance. Many surveillance systems capture vehicle images close to the ground. Police cars mounted with ALPR systems, and CCTV cameras such as those monitoring parking lots or the CCTV cameras around London, U.K., for example.

The HDR dataset [6], by the same authors, was captured by DSLR cameras with three different exposures. The limitation with this set is that the vehicle needs to be stationary in order to dynamically merge the three different exposures of the same scene. Therefore, moving vehicles from highways and streets cannot be used by this method.

2.1 Locations

This CENPARMI License Plate Database currently consists of 2200 images of license plates from eight countries. Table 2 shows the distribution of the current state of our database. We currently have 136 unclassified plates – partly because vehicles come from out-of-town, and partly of nonstandard AAMVA plates such as military issued plates.

Table 2. License database country distribution

Country	Count
China	151
Japan	13
Canada	714
U.S.A.	181
Jordan	199
Palestine	302
Turkey	468
Taiwan	36
Unknown	136
Total:	2200

3   Data Collection

Photos of license plates were taken from several vantage points. The distance from the plates range from the macro scale (isolated plate) to a landscape view where other objects and/or vehicles may appear in the photo. Figure 2 shows some variations in our database in terms of distance from and angle of the plate, the surrounding conditions, and the camera placement in relation to the plates.

Many images contain several vehicles, and license plates, in them. Luo et al. [2] performed research on detecting multiple license plates in complex scenes. Our database is unconstrained. Ideal for research similar to Silva and Jung [3] where they identify vehicles from an unconstrained scene, locate the license plate, rectify the plate, and then perform OCR on the corrected plate image.

To maintain privacy, we took special care to remove identifiable people from photos by either cropping, blurring, or using Photoshop to remove them.

3.1   Capture Devices

In order to prevent overfitting, we used several cameras models, lenses, and resolutions. Both high-end professional single-lens reflex (SLR) cameras and wireless phone cameras were used. Table 3 lists the cameras used. For the majority of the photos, the camera was handheld. As a result, some images may be rotated by a few degrees. Both landscape and portrait orientations are represented.

Resolutions vary based on the capture device. SLR cameras, with high end lenses, provide a higher resolution and photo size. Most of the time, the white balance was set to automatic. This is mostly the case for phone camera. For the SLR cameras, sunny, cloudy, or a manual white balance may have been chosen for some photos.

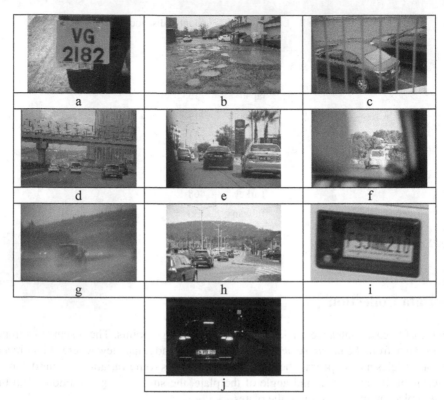

Fig. 2. Database Photo Variations: Close and isolated (a, i), Far (b, d, f), Occluded (c, f), Both vehicles and camera in motion (d, e, f, g), Multiple license plates in photo (d, e, h), Mist conditions (g), Shot through a car window (d, e, f, g), Vehicles in motion (d, e, f, g, h, j), still camera, High vantage point (c), Different lighting (i), Night (j).

Table 3. License plate capture devices

Camera maker	Camera model	Camera type
Canon	EOS REBEL T2i	SLR
Canon	EOS 6D Mark II	SLR
Sony	DSC-WX220	Point and shoot
Apple	iPhone 6	Phone
Apple	iPhone 6 s	Phone
LGE	Nexus 5X	Phone
Samsung	SGH-I337M	Phone
Sony	H8324	Phone
Vivo	X23 Magic color	Phone

4 Conclusion and Future Work

As we are in the early stages of collecting license plate images, we intend to increase the number of samples. This includes not only the number but the variety of the images, like personalized plates with emoji that is now available to vehicle owners in Australia [7]. In addition, some states are experimenting with digital license plates [8] that makes the physical change from a metal plate to a Liquid Crystal Display (LCD) or e-Ink display. Participants of digital plates had the choice of using the standard white background with black lettering, or the inverted black background with white lettering. We will also plan to use different capture devices and lenses.

4.1 Labeling/Ground Truth

We manually extracted some isolated characters from a few selected license plates. However, because we intend to drastically increase the number of samples, this will not be a feasible method. We are in the process of building a semi-automatic tool to assist in this labeling procedure. The software will locate the license plate(s) within an image, perform OCR on the plate and attempt to identify it, and extract the isolated characters from them. The user will have to ability to change the automatic proposed labels.

Information we intend to include for each image will include the capture device (camera make and model), resolution, lens, focal length, number of license plates in the image, the location of the plate(s) within the image, the content of the plates including any special symbols, and the country the plate was issued. Temporal information such as time of day and season, and physical information such as plate occlusion and the weather situation such as rain, snow, mist, or fog, that can affect recognition, will be included as well. Some photos were taken through a window and in some cases, either the vehicle, the camera, or both were in motion.

References

1. American Association of Motor Vehicle Administrators (AAMVA), License Plate Standard (2016)
2. Luo, Y., Li, Y., Huang, S., Han, F.: Multiple Chinese vehicle license plate localization in complex scenes. In: 2018 IEEE 3rd International Conference on Image, Vision and Computing (ICIVC), Chongqing, China, pp. 745–749 (2018)
3. Silva, M., Jung, C.R.: License plate detection and recognition in unconstrained scenarios. In: European Conference on Computer Vision, Munich, Germany, pp. 593–609 (2018)
4. https://toronto.citynews.ca/2006/07/17/licence-plate-covers-not-permitted/
5. http://www.wfsb.com/news/local-woman-warning-drivers-of-license-plate-frame-law-after/article_39ec11ba-414b-11e9-8eef-ef030a73f602.amp.html
6. Španhel, J., Sochor, J., Juránek, R., Herout, A., Maršík, L., Zemčík, P.: Holistic recognition of low quality license plates by CNN using track annotated data. In: 2017 14th IEEE International Conference on Advanced Video and Signal Based Surveillance (AVSS), Lecce, pp. 1–6 (2017)
7. Personalized Plates Queensland. https://www.ppq.com.au/_/media/files/pdf/downloadable-ebrochure-2020v2.pdf?la=en
8. State of California Department of Motor Vehicles. Report On Alternative Registration Products Pilot Program, p. 9, August 2019
9. Categories of Licence Plates, SAAQ (2020). https://saaq.gouv.qc.ca/en/vehicle-registration/categories-licence-plates/

An Extended Evaluation of the Impact of Different Modules in ST-VQA Systems

Viviana Beltrán[1]([⊠]) [iD], Mickaël Coustaty[1] [iD], Nicholas Journet[2] [iD],
Juan C. Caicedo[3,4] [iD], and Antoine Doucet[1] [iD]

[1] University of La Rochelle, 17000 La Rochelle, France
{vbeltran,mickael.coustaty,antoine.doucet}@univ-lr.fr
[2] University of Bordeaux, 33000 Bordeaux, France
journet@labri.fr
[3] Fundación Universitaria Konrad Lorenz, Bogotá, Colombia
[4] Broad Institute of MIT and Harvard, Cambridge, MA 02142, USA
jcaicedo@broadinstitute.org

Abstract. Scene Text VQA has been recently proposed as a new challenging task in the context of multimodal content description. The aim is to teach traditional VQA models to read text contained in natural images by performing a semantic analysis between the visual content and the textual information contained in associated questions to give the correct answer. In this work, we present results obtained after evaluating the relevance of different modules in the proposed frameworks using several experimental setups and baselines, as well as to expose some of the main drawbacks and difficulties when facing this problem. We makes use of a strong VQA architecture and explore key model components such as suitable embeddings for each modality, relevance of the dimension of the answer space, calculation of scores and appropriate selection of the number of spaces in the copy module, and the gain in improvement when additional data is sent to the system. We make emphasis and present alternative solutions to the out-of-vocabulary (OOV) problem which is one of the critical points when solving this task. For the experimental phase, we make use of the TextVQA database, which is one of the main databases targeting this problem.

Keywords: Visual question answering · Scene text recognition · Deep learning · Copy module

1 Introduction

Scene Text Visual Question Answering (ST-VQA) targets the specific task where understanding the textual information in a scene (text contained in signs, posters or ads in the image) is required in order to give the correct answer. Although deep

This work has been supported by the French region of Nouvelle Aquitaine under the ANIMONS project and by the MIRES research federation.

learning has been used with acceptable accuracy results of ≈*70%* for traditional VQA tasks, when solving the ST-VQA problem the accuracy drops to ≈*27%* demonstrating the challenge ahead. As a semantic description task, many related issues need to be addressed, we mention the most relevant ones. First, understand the type of question: although the task is well defined, current databases fail to define clean samples fitting the specifications of the task. Figure 1 presents samples of triplets *(Image, Question and Answers)* from TextVQA database [13]. Each sample contains an image, a question associated and the ground truth list of answers given for 10 human annotators (for cases where the answers given by annotators are all the same, we put the unique answer, if the answers are different, we put the entire set of answers given by annotators). We present examples for wrong annotation cases *(A-E)* where there is no need to read the text in the image to answer the question or the answer given by the annotator is not correct *(case C)*, the answer fall into the category 'Yes'/'No' *(case D)* or the case where the sample doesn't fit the requirements but it is possible to filter those samples by the answer given by the annotator *(case E)*, and correct cases when the answer is as expected, some text present in the image required in the question *(case F)*.

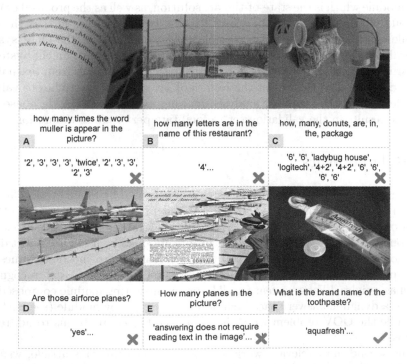

Fig. 1. Samples of triplets (Image, Question, List of answers given by 10 annotators) from TextVQA database for wrong (A-E) and correct (F) annotation cases for the problem of ST-VQA.

Other problem associated, is the detection and recognition of the text present in the visual data in the wild or from natural scenes, which remains a challenge for current models facing this problem because of all variations contained in them [9]. One of the most difficult tasks, even in traditional VQA systems [6] is the reasoning required to resolve spatial and visual references that involves understanding about the question and the visual information at the same time. Another related problem is the representation of the answer space, as this can contain unlimited words in any possibly language, which makes infeasible the establishment of a fixed pool of answers, and yields the problem of "out-of-vocabulary" answers (words not contained in the pool of answers are often called OOV words).

In this document, we present an incremental ablation study and an analysis for the modules comprised when solving the ST-VQA task. Our contributions are the following: we evaluate a strong architecture widely used in the context of media description for VQA systems applied to the problem of ST-VQA. We tested representative feature extractor models for the modalities involved in the VQA system. We also evaluate the relevance of the dimension of the answer space for the case of fixed set of words and for the case when the copy module is used as the main strategy for the OOV problem. We expose drawbacks related with the copy module which is the state-of-the-art solution, as well as the proposal of using a second metric to compute the scores for the dynamic spaces so that the copy module can take advantage of texts not 100% recognized for the OCR system. We evaluate the performance of including additional data to train the system in the form of a complementary network representing embeddings from textual and visual data. Finally, we present the results of several ablative studies to validate the relevance of the proposed analysis, making use of TextVQA database with baseline results in the validation set to facilitate performance evaluation.

2   Related Work

The very first work introducing the task was proposed by Singh et al. [13], they created a new database called TextVQA and presented a strategy (LoRRA) based on deep learning to solve the task. Their strategy contains the following components, a VQA system to process inputs obtained by using an object detection model for the visual features and GloVe vectors [11] to encode the question, a reading component to include OCRs extracted by using a text recognition model as weighted Fasttext features [5] and an answering module composed of a fixed + a dynamic answer space included by using a copy module (see Sect. 3.4) to handle the OOV problem which is one of the biggest problems to address in this task.

The second most relevant work at the time of writing this article, was presented by Biten et al. [1]. Similar to [13], they also introduced a new database (we refer to this database as ICDAR db) created for the ICDAR 2019 Robust Reading Challenge on Scene Text Visual Question Answering (see [4] for details on the competition). In this work, they presented final results for the proposed competitions from different participants addressing the task of ST-VQA under three tasks of increasing difficulty. The winning strategy, VTA, makes use of a strong architecture based on two types of attention, bottom-up (since they used an object detection model as a visual feature extractor) and top-down attention (by including OCR information extracted with an OCR recognition system). For the text, they use a pre-trained Bert model [14] to turn all the text into sentence embeddings, including object names, OCR recognition results, questions and answers (from training set). Having these embeddings for the text and for the images, they use a similar architecture as the one presented by Anderson et al. [2] to get the answer.

By analysing models and results presented for both main strategies, LoRRA [13] and VTA [4], their architectures are very similar, as well as obtained results in the target database, TextVQA and ICDAR db respectively. They both use an object detection model to extract visual features, an embedding method for the text (question, answers, OCRs), a VQA system, and an answer module. For VTA, the accuracy in task 1 of the challenge is *43.52%* which is significant better than the accuracy for tasks 2 *(17.77%)* and 3 *(18.13%)*, (for the complete definition of the tasks, see "Table II: Main Results Table" at [1], where tasks are 1: Strongly Contextualized, 2: Weakly Contextualise, and 3: Open Dictionary). In task 1, ground truth text is provided in the database as a set of possible words related to the scene, while for tasks 2 and 3, these texts are obtained by using an OCR recognition system and therefore relying on its performance to obtain ground truth texts from the images. By making a fair comparison, the results obtained in tasks 2 and 3 can be compared to LoRRA, with an accuracy of *27.63%* (see "Table 2: Evaluation on TextVQA" at [13]), as in both cases, the strategy relies on using OCRs obtained for a recognition system. The strategy in [13] performs better because of the inclusion of the copy module (see Sect. 3.4) that allow to handle the OOV problem, however, as we will discuss in Sect. 5, this solution is far from being optimal as it presents many limitations and a dubious performance.

Other works such as [3] describe the strategy used for one of the competitors, VQA-DML in [4], in which the main difference is the use of a n-gram representation for the answer space that allow to handle the OOV problem as well as giving the system the possibility to extend the answer space but not the dimension of it (i.e., the number of possible words formed from a n-gram combination increases, while the dimension of the target vector keeps a reasonable size). However, it also poses another challenge as it is required to add an additional stage for retrieving the correct answer from the n-gram predicted representation. While the low accuracy reported of VQA-DML for ICDAR db database (approx. 11%) [4] can be attributed to the straightforward architecture used for their authors, more analysis are required to determine the convenience of using this n-gram representation for the answer space. As this task is attracting attention, recent works present the task by introducing new databases, [10] introduces a new database, OCR-VQA–200K comprising images of bookcovers, [12] introduces a database containing images of business brands, movie posters and book covers.

3   Architecture Description

In order to perform comparisons with the state-of-the-art, we make use of similar frameworks for our experimental setups. Taking into account Fig. 2, we can divide the framework into modules. The following is the description of each one.

3.1   The Embeddings Module

The embeddings module represents the process of computing input features for the modalities involved, visual and textual (and other possible data such as OCRs, and localized features). For this reason, different specialized models for each modality can be studied (see Sect. 4.3 for information of the models tested during experimentation phase).

3.2   VQA Model

The VQA module represents the component in which the data is combined. We make use of a similar architecture as the one presented in [2]. It inputs the features extracted by using module A and used them to train the network and give the correct answer. We make use of attention mechanisms directed from the question network to the visual network (and the complementary network, see Table 3).

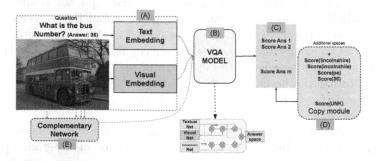

Fig. 2. Modules comprised in ST-VQA frameworks. Modules A, B and C represent the basic modules comprised in an STVQA framework. Modules D and E are added as strategies for improve the performance. A) Embedding module for input data of different modalities (Visual/Textual/OCRs), B) VQA system, C) Answer space, D) Copy Module, and E) Complementary network.

3.3   Answer Space Module

The answer module is in charge of the representation of the target vector relying on an answer space. We evaluated the usage of a fixed answer space commonly know as a bag of words (BoW), in which the score of each space in the final vector will indicate the presence or absence of the word.

3.4   Copy Module

The copy module works as a mechanism to handle the OOV problem. For this task, it is specially required, because the dimension of the answer space can growth unlimited. It works by adding a set of additional spaces to the fixed answer space (module C), filled with scores computed by using the OCRs recognized in the image. Thus, the final dimension of the answer space will be the one fixed by the set of selected answers from the training data + the set of dynamic words with a fixed number of spaces representing the OCRs.

We propose to compute the scores using two different metrics. First, the *Human Score* metric used in [13] computed as follows:

$$HS(\textbf{ans}) = min(\frac{\#\,humans\,that\,said\,\textbf{ans}}{3}, 1) \tag{1}$$

Each OCR will be taken as the *'ans'* to compute the score, this means that for *'ans'* to get a $HS = 1$, *'ans'* should be present in the set of answers given by the annotators at least three times. Figure 3 shows an example of calculation of scores using Eq. 1 for an image with two different questions associated. For the first question Qi, the answer is composed of two words, (*eddie, izzard*), which are outside the fixed answer space. The copy module could help to use the OCRs extracted from the image as an advantage, however, in this case, the *Human Score, Eq.* 1, will be zero for all the OCRs, because it seeks a perfect match between the ground truth answer *'eddie izzard'* and each one of the OCRs in a separate way *['eddie', 'izzard']*, leading to a zero vector as the target representation for this sample. For the second question associated Qj, it works as expected, as there exists an exact match between the ground truth answer and the OCRs.

Another example of the use of the copy module is when the set of human answers contains more than one answer for the same question, as it is expected to be the same for all the 10 annotators, but in some cases, they can differ and give a different answer. As an example, if the set of human answers is *[stop, emergency stop, emergency stop, stop, stop, emergency stop, emergency stop, it is an emergency stop, emergency stop, unanswerable]*, and the set of OCRs recognized is *[stop, emergency]*, although the majority of answers in the ground truth is *'emergency stop'*, it also contains another ground truth answers such as *'stop'*, which results convenient at setting a score for the OCR *'stop'*. As these two previous cases, there are others in which the copy module may or may not work because it relies in having texts 100% well recognized in the images (*'one' is completely different from 'one:' when computing the Human Score*), or when answers contain more than one token.

In order to improve the calculation of scores when partial matches are found in the OCRs of the image, we propose to use a second metric based on the Average Normalized Levenshtein Similarity (ANLS), computed as follows:

$$ANLS\,Score(ocr) = \frac{1}{M} \sum_{i=0}^{M-1} 1 - NL(ans_i, ocr)) \tag{2}$$

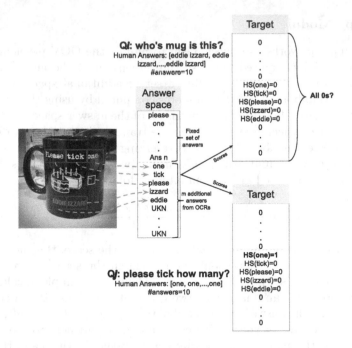

Fig. 3. Sample of assignation of scores using Human Score in the positive and negative cases of match between ground truth answers and the set of OCRs of the image.

where M = 10 is the set of answers given by 10 human annotators. Thus, for the previous example in Fig. 3, for Qi, the new scores for the target OCRs will be $Score("izzard") = 0.4166$ and $Score("eddie") = 0.5$

There are also problems related with the OCR system used, such as recognizing a single word separated in each one of their characters, or non recognizing the target text (i.e., the ground truth word from the human answer set that truly appears in the image). In Sect. 5, we discussed about the advantages and disadvantages for training models that uses the copy module as the main strategy to solve the OOV problem, as well as the dubious results when evaluating the performance in validation.

3.5   Complementary Network

The module E represents the inclusion of networks that input additional data into the VQA system. We tested three different setups: First, Fasttext embeddings [5] from OCRs recognized in the images (we use the OCRs available in the database) as in [13], however, we do not concatenate the order of the OCRs, and we input the average of the embeddings from all the OCRs available without weighting them. Second, we use global features extracted from the visual boxes containing the text recognized in the image, for training, we use the ground truth answers to filter the boxes with text matching in at least 30% one of the answers, for validation, we use the entire image. We also test the scenario when both, Fasttext and global features are sent into the VQA system together.

4 Experiments

4.1 Databases

As we mention in Sect. 2, in order to be able to explore different evaluation scenarios, we are using the TextVQA database [13], as it provides baselines results for the validation set for comparison purposes. This database contains 34,602 training samples and 5,000 validation samples, with almost 50% of the answers being unique. This shows the difficulty of using a fixed set of words in the answer space.

4.2 Evaluation Metrics

Two main metrics are used to evaluate the performance in this task: First, the accuracy computed as follows:

$$accuracy(y, \hat{y}) = \frac{1}{N_{samples}} \sum_{k=0}^{N_{samples}-1} 1(\hat{y}_i = y_i) \tag{3}$$

where $1(k)$ is the indicator function.

In the cases where the copy module is used, the calculation of the accuracy changes by using the Human Score accuracy (Eq. 1), the predicted answer is obtained by getting the index of the max value in the output vector, if the index is in the first part of the prediction (fixed space), the answer will be one of the fixed space shared among all the samples, if the index is in the additional/dynamic part of the output vector, the answer will be one of the set of OCRs recognized in the image at the index position. The second performance metric is the Average Normalized Levenshtein Similarity (ANLS) computed as follows:

$$ANLS = \frac{1}{N} \sum_{i=0}^{N} \left(max_j \, s(a_{ij}, o_{qi}) \right) \tag{4}$$

$s(a_{ij}, o_{qi}) = \begin{cases} (1 - NL(a_{ij}, o_{qi})) & if \ NL(a_{ij}, o_{qi}) < \tau \\ 0 & if \ NL(a_{ij}, o_{qi}) \geq \tau \end{cases}$ where N is the total number of questions, M is the total number of ground truth answers per question, a_{ij} the ground truth answers where i = 0, ..., N, and j = 0, ..., M, o_{qi} be the network's answer for the i-th question q_i, and τ is the threshold that determines if the answer has been correctly selected but not properly recognized, or on the contrary, the output is a wrong text selected from the options and given as an answer [4].

For this task, the second metric, ANLS (Eq. 4), could be more convenient, as the system can find partial matches among the set of words in the answer space.

On the contrary, evaluating the accuracy (Eq. 3), imposes a huge penalty if the model does not find perfect matches for the answers. This is directly related with the answer module and the OOV strategy used. However, taking into account the value of τ for the calculation of ANLS score that penalizes predictions matching in less than 50% of the characters, the performance for both metrics is expected to be similar.

4.3 Baselines and Ablations

We perform several ablation studies for the modules described in Sect. 3, where we aim to analyse the performance, drawbacks and future improvements when targeting this task. We describe the ablations performed, to see if adding/ changing/ replacing key modules of the system would lead us to obtain better results. For the first 5 models, we wanted to analyse the embedding module (see module A of Fig. 2), for the image and question data, and select the best one to test the rest of evaluation scenarios. The components included in this set of studies from Fig. 2 are modules A, B and C. For the answer space, we use the set of 3997 most frequent answers (Small Set SS, where the answers selected are those with frequencies ≥ 2) in the training database. As representative embeddings, we compare two models for the images, ResNet101[8] and Faster R-CNN (bottom-up (BU) attention) [2] with final representations of *2048-dim* and *36 (features per image with a 2048-dim each one)* respectively. And two embedding models for the text, GloVe [11] and BERT [7], with final representations of *300-dim* and *768-dim* respectively, for both models, we use a set of 15 tokens as the maximum length. The vocabulary size extracted of the questions is 9312 unique words. Therefore, the scenarios evaluated are: GloVe + ResNet, GloVe + BU, BERT + ResNet, BERT + BU.

After selecting the best set of embeddings based on the previous results, i.e., GloVe embeddings for the question, bottom-up features (BU) for the images and having fixed a small answer space (SS), we wanted to evaluate if the performance improve by increasing the size of the answer space to a large one (LS). The last row in Table 1 presents the result by using a larger set for the answer space of 7999 most frequent answers in the training database. Table 1 presents the results obtained for these first 5 models for validation samples in which the answer is contained in the selected fixed set of answers, i.e., for the answer space SS, the # of samples get reduced to 18,516 for training and 2,214 samples in validation. For the answer space LS, the # of samples get reduced to 21,183 for training and 2,290 samples in validation. Thus, to make a fair comparison, results are reported over these validation subsets.

The second set of evaluation scenarios aim to analyse the inclusion of the copy module, based on results from Table 1. The components included from Fig. 2 are modules A, B, C and D. We wanted to evaluate the appropriate number of additional spaces, for this, we test three different numbers, first, 50 spaces following the work [13], second, by taking the average of OCRs of all training samples (≈ 9.8) * 2, i.e., 20 spaces, and finally, by taking the average

Table 1. Performance for representative embedding models for visual and textual data in ST-VQA systems, with a fixed set of words in the answer space. Validation results are reported over the set of samples which answers are contained in a small fixed set SS or a larger set LS.

Model	Acc	ANLS	AVG
GloVe + ResNet101 + SS	0.1853	0.2274	0.2065
GloVe + BU + SS	**0.2005**	**0.2474**	**0.2240**
BERT + ResNet101 + SS	0.1910	0.2319	0.2115
BERT + BU + SS	0.1978	0.2366	0.2172
GloVe + BU + LS	0.1860	0.2279	0.2069

of OCRs from all training samples, i.e., 10 spaces. In this case, the data sets contain 100% of samples (34602 for training and 5000 for validation).

As we discussed in Sect. 3.4, the assignation of scores using Eq. 1, does not take advantage of text not 100% recognized, leaving many samples with zero score vectors. In this case, we wanted to change the assignation of scores by using the average ANLS score (see Eq. 2) over the set of human answers. The last row in Table 2 changes the assignation of scores using ANLS score metric. Table 2 presents the results for this set of evaluation scenarios.

Table 2. ST-VQA performance with the inclusion of the copy module with the assignation of scores using Human Score metric and by exploring the number of additional spaces for the OCRs to 50, 20 and 10. The last result changes the calculation of scores using the ANLS score metric.

Model	Acc	ANLS	AVG
50 spaces + Human Score	**0.1854**	**0.1835**	**0.1844**
20 spaces + Human Score	0.1778	0.1799	0.1788
10 spaces + Human Score	0.1792	0.1817	0.1804
50 spaces + ANLS Score	0.1705	0.1816	0.1761

To evaluate if the inclusion of more information into the VQA system could help the performance, we test the inclusion of three complementary data: the average of fasttext embeddings [5] from OCRs recognized in the images, similar as in [13], but without the addition of order and weighted information, a concatenation of global descriptors extracted from boxes containing target text, and finally, by sending into the VQA module both of them. For this evaluation scenario, the components included from Fig. 2 are modules A, B, C, D and E. We use the best model from Table 2, adding a top-down attention in the VQA system from the question towards the complementary network data. Table 3 presents the results obtained for this set of experiments.

Table 3. ST-VQA performance when complementary data is sent into the VQA module. Three types of complementary data were evaluated, Fasttext embeddings from OCRs recognized in the image, Global features extracted from the box containing the target text data and, finally, the combination of them.

Model	Acc	ANLS	AVG
OCR Fasttext	**0.1848**	**0.1942**	**0.1895**
Global Visual features (GVF)	0.1756	0.1797	0.1776
OCR Fasttext + GVF	0.1843	0.1932	0.1887

5   Discussion

The best results from Table 1 are obtained by using GloVe vectors + Fast R-CNN (or bottom-up BU) features. The slightly better performance of GloVe over BERT can be attributed to the fact that the structure and meaning of the words in the questions for this database is shared, and therefore the context does not play an important role in the discrimination of different samples. Also, as the last result in the Table showed, increasing the set of possible answers not necessarily implies an improvement of the performance (see also results of small set SA vs large set LA at "Table 2: Evaluation on TextVQA" [13] that confirm our result). This is because the set of possible answers can contain any combination of characters in different languages that are found in natural images, in the case of TextVQA database, there are more than 19,000 different answers among 34,000 samples. This makes unfeasible the establishment of a manageable fixed set of words as the answer space, and raises the question in how to handle the OOV problem?

For Tables 2 and 3, the copy module was included as a strategy to handle the OOV problem. The best number of additional spaces to include in the answer space for this database was 50, this means that the performance improves as more text data recognized in the image is sent into the system. On the contrary, the last result in the Table that tested the new metric to compute the scores, ANLS Score, did not show an improvement in the performance, which is related to the fact that for the majority of samples in the database with at least one OCR recognized, the answers are composed of only one token, and therefore the scores will be similar (for only 8.9% of the samples in TextVQA, answers contain more than one token).

Finally, regarding the evaluation of the inclusion of additional data, fasttext embeddings showed a small improvement for the performance. On the contrary, the inclusion of the global visual descriptors with the target textual data did not show any relevance, this could be the attention mechanism used, as it is the same used for both embeddings. However, a deeper analysis regarding the optimal attention mechanism is required to determine if the extra data is helping the system to learn, we leave it as a future work.

Is the copy mechanism solving the OOV problem in a suitable way?
We wanted to give final comments regarding the convenience of using the copy

module as a strategy for the OOV problem. Although, the copy module partially solves the OOV problem, each item in the dynamic space could represent as many different words exist in the OCR space of all samples, and at the end, the prediction of the correct answer over these values becomes almost a randomly choice that depends on the position of the OCR. Better solutions to handle the OOV are required as many tasks in the state-of-the-art are facing the same problem. The n-gram representation for the answer space could be a solution as with this, a larger set of answers can be represented by a fixed and manageable set of n-grams. However, it is required to perform deeper analysis of the implications of its usage.

6   Conclusions

We presented an incremental and extended study for the task of ST-VQA by performing an analysis of the modules required in any framework addressing this task. As one of the main analysed aspects, we evaluated the relevance of the dimension when a fixed set of words (BoW solution) is used as the answer space that for this problem turned out to be of little importance. We also evaluated the performance of the model when using the copy module under two different metrics for the calculation of the scores, both of them ended up with similar performance as the majority of data contains answers with only one token. Our final evaluation was the performance when including complementary data to train the system in the form of an additional network resulting in a slightly improvement of the performance. Finally, we expose some of the main drawbacks of current solutions, specially when handling the OOV problem showing us the need for better and more robust strategies. As a future work, we want to explore the performance when more data is used in the training phase, as we have noticed in the state-of-the-art, data augmentation has not been used when addressing this task. We also want to explore robust OOV strategies that do not rely on the copy mechanism, and finally, to study the mechanisms of inclusion of complementary data into the system that can help in the improvement of the performance.

References

1. Scene text visual question answering. arXiv preprint arXiv:1905.13648 (2019)
2. Anderson, P., et al.: Bottom-up and top-down attention for image captioning and visual question answering. In: Proceedings of the IEEE Conference on Computer Vision and Pattern Recognition, pp. 6077–6086 (2018)
3. Beltr, V., Journet, N., Coustaty, M., Doucet, A., et al.: Semantic text recognition via visual question answering. In: 2019 International Conference on Document Analysis and Recognition Workshops (ICDARW), vol. 5, pp. 97–102. IEEE (2019)
4. Biten, A.F., et al.: ICDAR 2019 competition on scene text visual question answering. arXiv preprint arXiv:1907.00490 (2019)
5. Bojanowski, P., Grave, E., Joulin, A., Mikolov, T.: Enriching word vectors with subword information. arXiv preprint arXiv:1607.04606 (2016)

6. Cadene, R., Ben-Younes, H., Cord, M., Thome, N.: MUREL: multimodal relational reasoning for visual question answering. In: Proceedings of the IEEE Conference on Computer Vision and Pattern Recognition, pp. 1989–1998 (2019)

7. Devlin, J., Chang, M.W., Lee, K., Toutanova, K.: BERT: pre-training of deep bidirectional transformers for language understanding. arXiv preprint arXiv:1810.04805 (2018)

8. He, K., Zhang, X., Ren, S., Sun, J.: Deep residual learning for image recognition. In: Proceedings of the IEEE Conference on Computer Vision and Pattern Recognition, pp. 770–778 (2016)

9. Liu, X., Meng, G., Pan, C.: Scene text detection and recognition with advances in deep learning: a survey. Int. J. Doc. Anal. Recognit. (IJDAR) **22**(2), 143–162 (2019). https://doi.org/10.1007/s10032-019-00320-5

10. Mishra, A., Shekhar, S., Singh, A.K., Chakraborty, A.: OCR-VQA: visual question answering by reading text in images. In: ICDAR (2019)

11. Pennington, J., Socher, R., Manning, C.: Glove: global vectors for word representation. In: Proceedings of the 2014 Conference on Empirical Methods in Natural Language Processing (EMNLP), pp. 1532–1543 (2014)

12. Singh, A.K., Mishra, A., Shekhar, S., Chakraborty, A.: From strings to things: knowledge-enabled VQA model that can read and reason. In: Proceedings of the IEEE International Conference on Computer Vision, pp. 4602–4612 (2019)

13. Singh, A., et al.: Towards VQA models that can read. arXiv preprint arXiv:1904.08920 (2019)

14. Wolf, T., et al.: Huggingface's transformers: state-of-the-art natural language processing. ArXiv abs/1910.03771 (2019)

Computer-Aided Wartegg Drawing Completion Test

Lili Liu[1]([✉])[iD], Graziella Pettinati[2][iD], and Ching Y. Suen[1][iD]

[1] Centre for Pattern Recognition and Machine Intelligence (CENPARMI), Concordia University, Montreal, QC, Canada
li_lil@encs.concordia.ca, suen@cse.concordia.ca
[2] Expertise Graziella Pettinati Inc., Montreal, Canada
info@graziellapettinati.com

Abstract. Wartegg Drawing Completion Test (WDCT) is one of the most commonly used personality analysis tools of graphology, which is the mapping from the inside world to personal qualities. It helps institutes or individuals to have a better knowledge of intrinsic personality characters. However, the WDCT evaluation of the applicants was manually performed by human experts, thus the accessibility and outcome of WDCT are heavily restricted by the availability and experience of the expert. To overcome such issues, this paper proposes the computer-aided WDCT (CA-WDCT) system, a fully-automatic WDCT system based on Digital Image Processing (DIP) and Machine Learning techniques. The CA-WDCT system extracts multimodal features and analyzes them under the Big-Five traits automatically. This CA-WDCT system can mitigate the heavy manual labour of psychologists and provide clients with flexible access.

Keywords: Personality analysis · Wartegg drawing completion test · Computer-aided analysis

1 Introduction

The Wartegg drawing completion test (WDCT) was first proposed by the Germany psychologist Ehrig Wartegg in 1939 [1]. It has been developed and utilized widely in German-speaking countries and Latin America. However, it is not so popular in English-speaking countries [2]. In particular, in North America the first WTDC manual was published by Kinget in 1952 [3]. Based on the work of Kinget, Italy researcher Crisi extended and validated the Wartegg test in1998 and 2016 subsequently [4,5]. The form of WDCT consists eight boxes as shown in Fig. 1.

© Springer Nature Switzerland AG 2020
Y. Lu et al. (Eds.): ICPRAI 2020, LNCS 12068, pp. 575–580, 2020.
https://doi.org/10.1007/978-3-030-59830-3_50

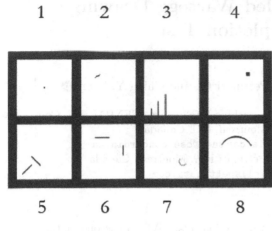

Fig. 1. The WDCT test form.

The WDCT shows its ability in a great range of application fields. It has been used in sociability evaluation by Gardziella in 1985 [6]. Gregory and Pereira have tried to imply WDCT in the personnel selection during recruitment process [7,8]. In 2003, Deinlein used WDCT for further vocational counselling [9]. It can be also used in the field of education [10,11], couple and family relationship therapy and inside condition diagnosis of patients [12,13]. The WDCT shows its ability in a great range of application fields.

However, the form evaluation during the application of Wartegg test was done by psychologists by hand rather than automatically or electronically. Only three works have tried to exploit the computer aided analysis system implication. In 2008, Campos [14], the first try of organizing wartegg test through a software, was created to analyse test results and also to provide a person's personality. It transformed the WDCT form into an electronic version, which reduces the process time by 60%. Fast Wartegg Analyzer Tool (FWAT) is another kind of web-based application designed to preparing the origin image data for further exploration faster and easier, which is created by Yuanna in 2015. The speed of evaluation process accelerated by 65% of times compared to manual operation [15]. In 2018, Yuana promoted an approach based on the FWAT to calculate the cosine-similarity between the ideal characters and candidates' characters in order to select a perfect employee [16]. Though there are some attempts in the computer-aided applications for WDCT evaluation, they had only explored to display the form digitally and generate electronic results with extracted features instead of fully automatically evaluation. In order to achieve the target of improving the evaluation process of WDCT, we propose the computer-aided WDCT (CA-WDCT), a fully-automatic WDCT framework based on Digital Image Processing (DIP) and Machine Learning techniques. This CA-WDCT system can free the labour of psychologists and make it possible for any person in the world to access the WDCT at any time.

2   Methodology

The CA-WDCT analysis process is described in four steps as shown in Fig. 2. First of all, we need to isolate the eight single boxes of the wartegg test form. Then, five predefined features are extracted automatically from each box by computer program. Once the features are extracted from drawings, character analysis can be processed for a better understanding of hand-drawings and corresponding personalities based on the Big Five personality traits. Finally, the CA-WDCT system generates personality evaluation within a professional psychology report.

Image Isolation. The first step of this project is to isolate the Wartegg test form from the scanned questionnaire and then segment it into eight boxes. We proposed a scissor algorithm for extracting the eight single boxes from questionnaires. The algorithm extracts the external contours with a modified template matching method to find out rectangle contour, which is hardly changed with the picture's affine transformation and its rotated angle. The distortion of the rectangle can be corrected with the angles. Within the cropped form, the system calculates the pixel distribution for each column and row to find out the watershed of border and drawings. Then, the "scissor" starts to work, cutting the

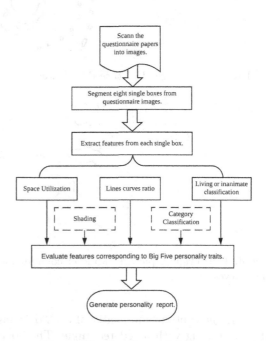

Fig. 2. The flowchart of this paper.

form aligning the watershed like a paper cutting scissor into eight single boxes and stored following their numerical order. This method keeps a 100% of cutting accuracy even with rotated and messy strokes without clear boundaries.

Feature Extraction. Feature extraction is the most important part of CA-WDCT system. It makes fully-automatic analysis possible without human intervention. Moreover, the extracted features form the key clues for personality evaluation and report generation. Once we get the single box of drawing, the data for further steps are prepared well. As shown in Fig. 2, we have five features needed to be extracted from each box. They are space utilization, line curves ratio, living or inanimate drawings classification. The first three features are implemented in this paper, and two further exploit features shading and category classification.

Space Utilization and Distribution. The first feature we have exploited is space utilization and drawings distribution. As shown in Fig. 3, the software

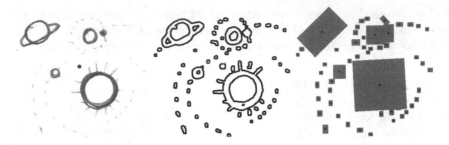

Fig. 3. Left: original picture. Middle: contoured picture. Right: utilized space.

Fig. 4. Left: original picture. Middle: colored line and curves. Right: connected line and curves.

extracts the contour and center of each drawing, aligns the center covering each abject drawing with a red rectangle. The space utilization can be obtained by calculating the ratio between the red rectangle area and the whole boxes area. The distribution is calculated with the distance of each object drawing. The distance can be calculated from the location of each center point.

Lines and Curves Ratio. The line-curve ratio feature is a kind of typical personality clue of drawings. Unlike human understanding, when two lines connect with each other, it is easier for the program to regard them as one object, which is actually should be treated as two. The system was programmed to find out the cross corner of two or more lines and split all lines apart. Then trying to connect the crossed lines back with extra heuristic information or principal component analysis. This two-stage cut-and-connect operation splits the connected lines separately with understandable meanings.

Living and Animate Objects. In this part, we extracted the drawing features of their living or inanimate category. A special convolutional neural network Lenet-5 [17] is used for classifying living and animate drawing objects. The first step is to augment the data with four different augmentations: Image Sharpening, Unsharp Mask, Adaptive Threshing and Binary. The ratio of training: validation:

testing is 6 : 2 : 2. Images are resized into 28 × 28 and normalization is applied. Lenet-5, a classical model of deep neural networks, has been commonly used in deep learning area especially in handwriting classification. The model is a 6 layer convolutional (Conv) neural network (CNN) with repeated the architecture of Conv, Relu and Pooling for twice and following a fully-connected layer and softmax function at the end. The model details are shown in Table 1. In order to improve the result, we used the Adam optimizer, Mini-batch and early stopping method while training the model. With this CNN model, we got the training accuracy of 98.46% and testing accuracy of 86.15%.

3   Discussion and Future Work

For now, we have discussed the first several steps in this CA-WDCT system research. This work has finished isolating the eight boxes from the questionnaire images and extracted three features (space utilization, living and animation classification, lines-curves ratio). In the future step, at least two more features will be exploited and explored. Once the features are extracted from drawings, character analysis can be processed for a better understanding of hand-drawings and corresponding personalities based on the Big Five personality traits. The Big The big five personality traits are Openness, Conscientiousness, Extraversion, Agreeableness and Neuroticism. The CA-WDCT will analyze the above characteristics of each box and score them. The scoring results and the distribution of the eight boxes' scores are then mapped to the big five personalities analysis. The CA-WDCT system can generate a psychology report. This CA-WDCT feature extraction and analysis method can mitigate the heavy manual labour of psychologists and benefit clients with easier access.

Table 1. Lenet-5 model used in this paper.

Layer	Filter shape	Activation function
Conv2D	(32, (7, 7))	Relu
MaxPooling2D	(2, 2)	–
Conv2D	(64, (5, 5))	Relu
MaxPooling2D	(2, 2)	–
Conv2D	(64, (3, 3))	Relu
Flatten+Dense	(64)	Relu
Dense	(2)	Sigmoid

Acknowledgement. This research was supported by the Natural Sciences and Engineering Research Council of Canada.

References

1. Wartegg, E.: Gestaltung und Charakter. Ausdrucksdeutung zeichnerischer Gestaltung und Entwurf einer charakterologischen Typologie (1939)
2. Roivanen, E.: A brief history of the Wartegg drawing test. Gestalt Theory **31**(1), 55 (2009)

3. Kinget, M.: The Drawing-Completion Test. Grune & Stratton, New York (1952)
4. Crisi, A.: Manuale del Test di Wartegg. MaGi, Roma (1998)
5. Crisi, A., Dentale, F.: The Wartegg drawing completion test: inter-rater agreement and criterion validity of three new scoring categories. Int. J. Psychol. Psychol. Ther. **16**(1), 83–90 (2016)
6. Gardziella, M.: Wartegg-piirustustesti. Tr. It. The Wartegg Drawing Test. Psykologien Kustannus Oy, Jyvaskyla (1985)
7. Pereira, F.M., Primi, R., Cobéro, C.: Validade de testes utilizados em seleção de pessoal segundo recrutadores. Revista Psicologia-Teoria e Prática **5**(2) (2003)
8. Gregory, R.J.: Psychological Testing: History, Principles, and Applications, 3rd edn. Allyn & Bacon Inc., Boston (2000)
9. Deinlein, W., Boss, M.: Befragung zum Stand der Testanwendung in der allgemeinen Berufs-Studien-und Laufbahnbereitung in der deutschen Schweiz [The use of tests in vocational counseling in German-speaking Switzerland]. Postgraduate thesis, NABB-6, Switzerland. http://www.panorama.ch/files. Accessed 31 May 2007 (2003)
10. Mellberg, K.: The Wartegg drawing completion test as a predictor of adjustment and success in industrial school. Scand. J. Psychol. **13**(1), 34–38 (1972)
11. Ceccarelli, C.: L'Uso degli Strumenti Psicodiagnostici [The use of psychodiagnostic methods]. Societa Italiana di Psicologia dei Servizi Ospedalieri e Territoriali (1999). http://www.sipsot.it. Accessed 31 May 2007
12. La Spina, C.M., et al.: The psychological projective test Wartegg-Zeichentest: a contribution to the knowledge of patients with cancer. J. Clin. Oncol. **28**(15-suppl), e19678–e19678 (2010)
13. Rizzo, A., Villa, L.D., Crisi, A.: Can the problematic internet use evolve in a pre-psychotic state? A single case study with the Wartegg. Comput. Hum. Behavi. **51**, 532–538 (2015)
14. Lizarazo, J.F.S., Rincon, J.L.S.,Higuera, B.M.R.: Wartegg Automatic Test Qualifier (2008)
15. Yuana, R.A., Harjunowibowo, D., Karyanta, N.: FWAT (Fast Wartegg Analyzer Tool) for personality identification. Adv. Sci. Lett. **21**(10), 3114–3117 (2015)
16. Yuana, R.A., Harjunowibowo, D., Karyanta, N.A., Budiyanto, C.W.: Data similarity filtering of Wartegg personality test result using cosine-similarity. Int. J. Recent Contrib. Eng. Sci. IT (iJES) **6**(3), 19–28 (2018)
17. LeCun, Y., Bottou, L., Bengio, Y., Haffner, P.: Gradient-based learning applied to document recognition. Proc. IEEE **86**(11), 2278–2324 (1998)

Medical Imaging and Applications

A Novel ECG Signal Classification Algorithm Based on Common and Specific Components Separation

Jianfeng Huang[1], Chao Huang[2], Lihua Yang[3], and Qian Zhang[2(✉)]

[1] School of Financial Mathematics and Statistics, Guangdong University of Finance, Guangzhou, China
1249422615@qq.com
[2] College of Mathematics and Statistics, Shenzhen Key Laboratory of Advanced Machine Learning and Applications, Shenzhen University, Shenzhen, China
{hchao,mazhangq}@szu.edu.cn
[3] Guangdong Province Key Laboratory of Computational Science, School of Mathematics, Sun Yat-sen University, Guangzhou, China
mcsylh@mail.sysu.edu.cn

Abstract. Electrocardiography (ECG) signal classification is a challenging task since the characteristics of ECG signals vary significantly for different patients. In this paper, we propose a new method for ECG signal classification based on the separation of common and specific components of a signal. The common components are obtained via Canonical Correlation Analysis (CCA). After removing the common components from the signal, we map the specific components to a lower dimensional feature space for classification. We first establish a basic model in the binary classification setting and then extend it to a more general version. Numerical experiments results on the MIT-BIH Arrhythmia Database are presented and discussed.

Keywords: ECG signal classification · Canonical correlation analysis · Components separation

1 Introduction

Electrocardiography (ECG) is the recording of the electrical activity of the heart by using electrodes placed on the skin. ECG signal conveys a great deal of diagnostic information about the structure of the heart and the function of its electrical conduction system. The advantages of using ECG include low cost, immediate availability and easy implementation; moreover, ECG is a non-invasive procedure. ECG signal is one of the best-recognized biomedical signals and has been a subject of studies for over 100 years [1].

This work is supported by National Natural Science Foundation of China (Nos. 11601532, 11501377, 11431015, 11601346) and Interdisciplinary Innovation Team of Shenzhen University.

© Springer Nature Switzerland AG 2020
Y. Lu et al. (Eds.): ICPRAI 2020, LNCS 12068, pp. 583–595, 2020.
https://doi.org/10.1007/978-3-030-59830-3_51

ECG signal classification is a challenging task and many algorithms have been proposed to classify ECG signals: support vector machine [2], artificial neural networks [3], convolutional neural network [4], etc. The neural network approach suffers the drawback of its slow convergence to a local minimum, and the determination of the network size or configuration is unclear. There are also lots of research on feature construction, such as wavelet transform [5], empirical mode decomposition [6], and hidden Markov models [7], and mixture-of-experts method [8]. However, handcrafted features may not represent the underlying difference among the classes, thereby limiting the classification performance.

In this paper, we propose a novel ECG classification model based on separation of the common and specific components of a signal. The main idea is as follows. Generally speaking, real-world signals consist of different basic components. The various combinations of basic components give rise to different classes of signals. Some of these basic components are common in all signal classes, meaning that they appear in the signal decompositions of every class. We call them common components. On the other hand, there are some basic components only appearing in the signal decomposition of one specific class. We call such components specific components. It is a general belief that the specific components play a dominant role in the classification process. For a concrete example, one might consider the classification of human voices. Human voices consist of harmonic components and non-harmonic (noisy) components. By common sense, it is the noisy part, which contains rich timbre textures, helps one identify the speaker. On the other hand, the harmonic components are universal among speakers and provide little information for discrimination of different speakers. This general principle can be applied to ECG signal classification as well, and that is the motivation of our work. The goal is to separate the common and specific components of a signal and to perform signal classification based on the specific components.

The rest of the paper is organized as follows. We establish a common-specific component model for ECG signal classification in Sect. 2, in which we give the mathematical definitions of common and specific components. Then we extend the model to a more general setting by using the classical results from Canonical Correlation Analysis (CCA). Numerical experiments on the public MIT-BIH Arrhythmia database are presented in Sect. 3, which show the proposed method achieves promising results. In Sect. 4 we draw a conclusion.

The following notations will be used throughout this paper. $\mathbb{R}$ denotes the set of real numbers. $\mathbb{R}^{m \times n}$ is the set of $m \times n$ real matrices. Any vector $v \in \mathbb{R}^m$ is regarded as a column vector, i.e., an $m \times 1$ matrix. The column space of a matrix M is denoted by $\mathrm{span}M$. The projection of a vector v onto $\mathrm{span}A$ is denoted by $\mathrm{proj}(v, A)$.

2   Classification Model

In this section, we aim to define a classification model for ECG signals. The essence of a classification model lies in the design of a classifier χ: a mapping from the signal space to the class label set $\{1, \ldots, K\}$ if there are K classes. The classifier χ is fed with an input signal s and outputs a class label $\chi(s)$. Usually one defines a feature mapping Φ which maps a higher-dimensional signal to a lower-dimensional feature vector, thus reducing the problem to the classification of feature vectors. If the feature mapping Φ well captures the characteristics of the signals, the feature vectors would be clustered into different groups and easily be separated by using generic classifiers like support vector machines. In the sequel, we aim to define a feature mapping Φ based on the specific components of the signal, and then construct the classifier χ by Φ.

2.1   Basic Model

At first, we consider the simple case of binary classification, i.e. there are only two signal classes. Suppose we are given training samples from two classes: $\{s_1, s_2, \ldots, s_p\}$ from class 1 and $\{s_{p+1}, s_{p+2}, \ldots, s_{p+q}\}$ from class 2 respectively. Each sample s_i is a vector in $\mathbb{R}^m$. Define matrices

$$S_1 = [s_1, \ldots, s_p] \in \mathbb{R}^{m \times p}, S_2 = [s_{p+1}, \ldots, s_{p+q}] \in \mathbb{R}^{m \times q}.$$

Next we suppose every sample of class 1 can be expressed as a linear combination of a set of basic vectors, and so we suppose for class 2. More precisely, we assume there exist three sets of vectors $\{f_1, \ldots, f_J\}$, $\{g_1, \ldots, g_K\}$ and $\{h_1, \ldots, h_L\}$ such that $s_i = \sum_{j=1}^{J} a_{ij} f_j + \sum_{k=1}^{K} b_{ik} g_k, i = 1, \ldots, p$ and $s_i = \sum_{j=1}^{J} c_{ij} f_j + \sum_{l=1}^{L} d_{il} h_l, i = p + 1, \ldots, p + q$, where the coefficients $a_{ij}, c_{ij}, b_{ik}, d_{il} \in \mathbb{R}$ and the basic vectors $f_j, g_k, h_l \in \mathbb{R}^m$. We call $\{f_j\}$ the *common components* of the two classes, $\{g_k\}$ the *specific components* of class 1 and $\{h_l\}$ the *specific components* of class 2. As mentioned earlier, we aim to classify the signals based on specific components $\{g_k\}$ and $\{h_l\}$.

In general, the choice of $\{f_j\}$, $\{g_k\}$ and $\{h_l\}$ is not unique. We rewrite the above equations in the matrix form

$$S_1 = FA + GB, \quad S_2 = FC + HD, \tag{1}$$

where $F = [f_1, \ldots, f_J] \in \mathbb{R}^{m \times J}$, $G = [g_1, \ldots, g_K] \in \mathbb{R}^{m \times K}$, $H = [h_1, \ldots, h_L] \in \mathbb{R}^{m \times L}$, and A, B, C, D represent the coefficients. There are many choices of (F, G, H) satisfying the relation (1). One can find two trivial instances immediately: $F = [S_1, S_2]$, $G = 0$, $H = 0$ or $F = 0$, $G = S_1$, $H = S_2$. Furthermore, any linear combinations of the solutions of (1) would give another solution. Therefore, in order to uniquely determine the representation (1), one must propose further restrictions on the basic vectors F, G and H. From (1), we have

$$\text{span} S_1 \subset \text{span} F + \text{span} G, \text{span} S_2 \subset \text{span} F + \text{span} H. \tag{2}$$

That means S_1 and S_2 can be linearly expressed by F, G and H. To reduce the redundancy in F, G and H, it is natural to require

$$\text{span} S_1 = \text{span} F + \text{span} G, \text{span} S_2 = \text{span} F + \text{span} H, \tag{3}$$

and the basic vectors f_j, g_k, h_l to be linearly independent. Under these requirements, one can prove the span of F is the intersection of the span of S_1 and S_2, i.e.

$$\mathrm{span}F = \mathrm{span}S_1 \cap \mathrm{span}S_2. \tag{4}$$

Since the data S_1 and S_2 are given, one can calculate $\mathrm{span}S_1 \cap \mathrm{span}S_2 = \mathrm{span}F$, where the F can be any basis of the subspace $\mathrm{span}F$. It follows from (3) and (4) that

$$\mathrm{span}G = \mathrm{span}S_1/\mathrm{span}F, \quad \mathrm{span}H = \mathrm{span}S_2/\mathrm{span}F, \tag{5}$$

where '/' means the quotient of two subspaces. That means we can take $\mathrm{span}G$ as the complement of $\mathrm{span}F$ in $\mathrm{span}S_1$, and take $\mathrm{span}H$ as the complement of $\mathrm{span}F$ in $\mathrm{span}S_2$. Thus we split the space $\mathrm{span}S_1 + \mathrm{span}S_2$ into a direct sum:

$$\mathrm{span}S_1 + \mathrm{span}S_2 = \mathrm{span}F \oplus \mathrm{span}G \oplus \mathrm{span}H. \tag{6}$$

See Fig. 1 for an illustration.

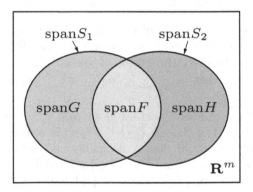

Fig. 1. Basic model

For any training sample s_i, $i = 1, \ldots, p+q$, it can be uniquely decomposed into three parts:

$$s_i = \mathrm{proj}(s_i, F) + \mathrm{proj}(s_i, G) + \mathrm{proj}(s_i, H).$$

Removing the common components part $\mathrm{proj}(s_i, F)$, we keep the remaining part for classification. Let

$$\tilde{s}_i := \mathrm{proj}(s_i, G) + \mathrm{proj}(s_i, H) = \mathrm{proj}(s, [G, H]). \tag{7}$$

Now we can calculate the angles between $\tilde{s}_i$ and the two subspaces $\mathrm{span}G$ and $\mathrm{span}H$ respectively:

$$\cos \alpha := \frac{\|\mathrm{proj}(\tilde{s}_i, G)\|}{\|\tilde{s}_i\|}, \quad \cos \beta := \frac{\|\mathrm{proj}(\tilde{s}_i, H)\|}{\|\tilde{s}_i\|}. \tag{8}$$

Here $\| \cdot \|$ denotes the Euclidean norm of a vector. The angles α and β give a measure of the closeness of the signal $\tilde{s}_i$ to the subspaces span$G$ and span$H$, respectively. See Fig. 2 for a geometric illustration.

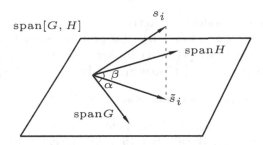

Fig. 2. Projections and angles defined by (7) and (8)

Based on these two angles, we define a simple feature mapping as follows

$$\Phi(s) := (\cos \alpha, \cos \beta). \qquad (9)$$

Therefore the feature vector $\Phi(s)$ is a point on the unit square $[0,1]^2 \subset \mathbb{R}^2$. If $\cos \alpha \geq \cos \beta$, that means s is closer to subspace span$G$ than to subspace span$H$. So a binary classifier can be defined as follows: $\chi(s) = 1$ if $\cos \alpha \geq \cos \beta$ or $\chi(s) = 2$ otherwise.

2.2   Improved Model

The definitions of F, G and H by (4) and (5) is rather simple and idealistic, and it may have problems when applied in practice. The computation of span$F$ through span$S_1 \cap$ spanS_2 might become unstable due to noise in the training data S_1 and S_2. Therefore the basic model should be improved to a more robust and stable form. Relaxing the assumption (4), we now suppose there exist two sets of basic vectors F_1 and F_2 such that

$$\text{span}S_1 = \text{span}F_1 + \text{span}G, \text{span}S_2 = \text{span}F_2 + \text{span}H, \qquad (10)$$

where F_1 and F_2 are required to be close to each other, i.e. F_1 and F_2 jointly represent the "common" components of the two classes. Like before, we require the linear independence between the basic vectors, i.e., the set of the columns of F_1, F_2, G and H are linearly independent. It is easy to see that relation (10) is a generalization of (3).

From (10), we have span$F_1 \subset$ spanS_1 and span$F_2 \subset$ spanS_2, i.e. F_1 can be linearly expressed by S_1, and F_2 can be linearly expressed by S_2. Now the problem is to find F_1 in spanS_1 and F_2 in spanS_2 with the requirement that they should be close to each other. To measure the closeness of two vectors, we adopt the definition of correlation coefficient:

$$\rho(f,g) := \frac{\langle f,g \rangle}{\|f\| \cdot \|g\|}, \quad \forall f,g \subset \mathbb{R}^m,$$

where $\langle \cdot, \cdot \rangle$ is the inner product of two vectors. In fact, $\rho(f,g)$ measures the angle between two vectors. It turns out that this problem is related to the theory of Canonical Correlation Analysis (CCA) [9]. According to classical results in CCA, there exist vectors $\{u_1, \ldots, u_r\} \subset \mathrm{span}S_1$ and $\{v_1, \ldots, v_r\} \subset \mathrm{span}S_2$ satisfying

$$u_k, v_k = \arg\max\{\rho(u,v) : u \perp U_k, v \perp V_k\},$$

where $U_k = \mathrm{span}[u_1, \ldots, u_{k-1}]$, $V_k = \mathrm{span}[v_1, \ldots, v_{k-1}]$ for $k = 1, \ldots, r$ with $r = \min\{\mathrm{rank}S_1, \mathrm{rank}S_2\}$. To calculate $\{u_k, v_k\}$, one can use the Matlab built-in routine **canoncorr**.

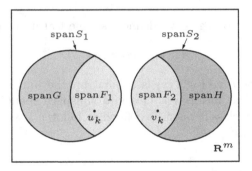

Fig. 3. Improved model

These pairs of vectors u_k and v_k obtained through CCA procedure are close to each other and could be taken as vectors of F_1 and F_2. Note the correlation coefficient $\rho_k = \rho(u_k, v_k)$ is descending over k, i.e.

$$1 \geq \rho(u_1, v_1) \geq \rho(u_2, v_2) \geq \cdots \geq \rho(u_r, v_r) \geq 0.$$

That means u_1 and v_1 is the most correlated, while u_r and v_r is the least correlated. Setting a threshold $\rho_0 \in (0, 1]$, we can determine an integer k_0 such that

$$\rho(u_{k_0}, v_{k_0}) \geq \rho_0 > \rho(u_{k_0+1}, v_{k_0+1}).$$

We can define $F_1 := [u_1, \ldots, u_{k_0}]$, $F_2 := [v_1, \ldots, v_{k_0}]$, $\mathrm{span}G := \mathrm{span}S_1/\mathrm{span}F_1$ and $\mathrm{span}H := \mathrm{span}S_2/\mathrm{span}F_2$, thus F_1, F_2, G, H satisfies the improved model (10). See Fig. 3 for an illustration. The choice of the threshold ρ_0 determines the ratio of the common and specific components of the signal, and hence affect the performance of the classification. When ρ_0 is near 0, the ratio of common to specific component increases. That means after removing the common components, the signals from the two classes have less common components. If ρ_0 gets closer to 1, the ratio of common to specific components decreases. That means the signals have more common components. However, the numerical experiments later show that classification performances are not quite sensitive to different choices of ρ_0. So we usually set ρ_0 to be 0.5.

The feature mapping Φ can be similarly defined as before. Consider the projection of a signal s onto the subspace $\text{span}[G, H]$, and then define $\Phi(s)$ based on the two angles of the projection with respect to the subspaces $\text{span}G$ and $\text{span}H$. The experiments show that the feature vectors from the two classes are well separated by the straight line $y = x$. Table 1 gives a summary of the proposed binary classification algorithm.

Table 1. Binary classification algorithm

Training Stage
Input: Training data S_1, S_2 and correlation threshold ρ_0.
Output: Specific components G, H.
1. $\{u_k, v_k, \rho_k\} := \texttt{canoncorr}\ (S_1, S_2)$.
2. $k_0 := \max\{k : \rho_k \geq \rho_0\}$.
3. $U := [u_1, \ldots, u_{k_0}]$, $V := [v_1, \ldots, v_{k_0}]$.
4. $G := \text{span}S_1/\text{span}U$, $H := \text{span}S_2/\text{span}V$.
Testing Stage
Input: Specific components G, H, and testing signal s.
Output: Class label $\chi(s)$.
1. $\tilde{s} := \text{proj}(s, [G, H])$.
2. $\Phi(s) := \left(\frac{\text{proj}(\tilde{s}, G)}{\|\tilde{s}\|}, \frac{\text{proj}(\tilde{s}, H)}{\|\tilde{s}\|} \right) =: (\cos\alpha, \cos\beta)$.
3. $\chi(s) = 1$ if $\cos\alpha \geq \cos\beta$ or $\chi(s) = 2$ otherwise.

2.3 Multiclass Classification

We proceed to consider the case of multiclass classification. We use a simple voting strategy here. Suppose there are K classes: $1, 2, \ldots, K$. For each pair of classes i and j, using algorithm in Table 1 one can define a binary classifier $\chi_{ij}(s)$ whose output is either i or j. For a signal s, there are a total of $K(K-1)/2$ outputs of such $\chi_{ij}(s)$. We then count the occurring frequency of the labels in these outputs, i.e. for $k = 1, \ldots, K$, let

$$\text{freq}(k) := \#\{(i, j) \mid \chi_{ij}(s) = k,\ 1 \leq i < j \leq K\},$$

where $\#$ means the cardinality of a set. The label that has the largest frequency is then defined as the output label:

$$\chi(s) := \arg\max_{1 \leq k \leq K} \text{freq}(k).$$

3　Numerical Experiments

3.1　ECG Signal Data and Preprocessing

We performed classification experiments on the MIT-BIH Arrhythmia Database [10]. This benchmark database contains 48 records, each containing a two-channel ECG signal for 30-min duration selected from 24-hour recordings of 47 individuals (25 men and 22 women). Continuous ECG signals are bandpass-filtered at 0.1–100 Hz and then digitized 360 Hz. The database contains annotations for both timing information and beat class information rigorously verified by independent experts. The whole database contains more than 109000 annotations of normal and 14 different types of abnormal heartbeats.

Association for the Advancement of Medical Instrumentation (AAMI) recommend the heart beats be classified into five types: **S** (Supraventricular ectopic beat), **V** (Ventricular ectopic beat), **F** (Fusion beat), **Q** (Unknown beat) and **N** (beats not in **S**, **V**, **F** or **Q** classes) [11]. Table 2 lists the correspondence between AAMI beat types and the MIT-BIH beat annotations. For each recording, MIT-BIH database provides an annotation file that contains the information of the beat types and their time positions. Based on this information, one can extract each single beat from the ECG signal recording. Before classification, the following preprocessing operations are preformed to the beats: 1) Normalize the lengths of each beat to 400 using a linear warping transformation; 2) Shift the maximum of each beat to the center of the time interval; 3) Subtract the mean of each beat so that they are zero-mean.

Table 2. Mapping table between AAMI beat types and MIT-BIH beat annotations.

AAMI type	N	S	V	F	Q
MIT-BIH ann	N, L, R, e, j	A, a, J, S	V, E	F	/, f, Q

3.2　Binary Classification Experiment

We carry out the binary classification experiment on the recording set 119 of MIT-BIH. For the purpose of binary classification, we divide the beats into two classes. The positive class contains **V** type beats. The negative class contains non-**V** type beats, i.e. types **N**, **S**, **V** and **F**. We randomly select 100 beats from each class for training purpose and 300 beats from each class for testing purpose. After the CCA procedure, we obtain 100 pairs of correlated vectors u_k and v_k. Figure 4 plots the correlation coefficient $\rho(u_k, u_k)$ which decreases from 1 to 0 as k grows. Figure 5 plots the first few pairs of correlated vectors u_k and v_k.

Figure 6 plots the distribution of feature points $\Phi(s_i)$ for different choices of the threshold ρ_0. As shown from the figures, when ρ_0 is small, the feature points of both classes have a larger spread and locate much further away from the

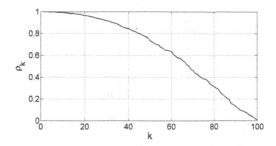

Fig. 4. Correlation coefficients $\rho(u_k, v_k)$.

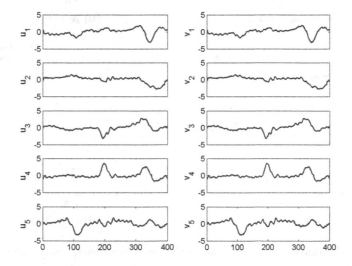

Fig. 5. Correlated vectors u_k and v_k (only showing for $k \leq 5$).

separating line $y = x$; when ρ_0 is large, the feature points are more concentrated and locate nearer to the line $y = x$.

To evaluate the performance of the classification, we use the following conventional quantities: Accu. $= \frac{TP+TN}{TP+TN+FP+FN}$, Sens. $= \frac{TP}{TP+FN}$ and Spec. $= \frac{TN}{TN+FP}$, where TP, FN, FP, TN represent the percentage of True Positive, False Negative, False Positive and True Negative.

We perform binary classification experiments on different recordings from the MIT-BIH database, with results listed in Table 3. The best performance is achieved on recording 119, with the accuracy up to 99.98%. The worst is on recording 207, with the accuracy 93.40%. On average the accuracy is 96.7%, the sensitivity is 97.2% and the specificity is 96.2%. Table 4 gives a comparison between the proposed method with other works, showing that our method achieves approximately the same level of performance with other methods. Considering that our model has a much simpler architecture and no specific preprocessing techniques, this result is rather promising.

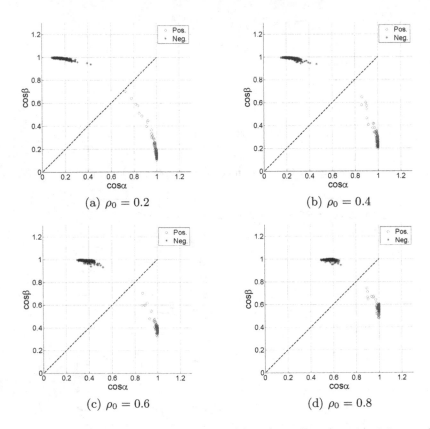

Fig. 6. Distribution of the feature points $\Phi(s) = (\cos\alpha, \cos\beta)$ under different choices of the threshold ρ_0.

Table 3. Results of binary-classification experiment on MIT-BIH database. Column 1 lists the ID number of the recordings. Columns 2 and 3 list the numbers of the **V**-type beats and non-**V**-type beats from each recording.

Rec. No	**V** type	non-**V** type	Accu	Sens	Spec
106	520	1507	97.75 ± 0.93	99.65 ± 0.48	95.85 ± 1.77
119	444	1543	$\mathbf{99.98 \pm 0.11}$	100.00 ± 0.00	99.95 ± 0.22
207	210	1650	$\mathbf{93.40 \pm 1.70}$	96.00 ± 3.79	90.80 ± 3.01
208	992	1963	96.15 ± 1.65	94.65 ± 3.29	97.65 ± 1.96
214	256	2006	96.43 ± 1.33	98.35 ± 1.11	94.50 ± 2.91
221	396	2031	99.45 ± 0.44	99.70 ± 0.46	99.20 ± 0.68
223	473	2132	93.45 ± 1.57	89.15 ± 2.76	97.75 ± 2.07
228	362	1691	95.43 ± 2.23	98.35 ± 1.77	92.50 ± 3.72
233	831	2248	98.23 ± 1.15	99.10 ± 0.89	97.35 ± 2.13
Average	498.2	1863.4	96.70 ± 1.23	97.22 ± 1.62	96.17 ± 2.05

3.3 Multiclass Classification Experiment

We use the proposed method to perform a multiclass classification experiment on MIT-BIH database. Table 5 is the confusion matrix of the experiment. Type **Q** is the most recognized type, with a correct rate of 91.7%. Type **S** is difficult to be recognized by the classifier, with only 48.50% correctly classified, 18.83% wrongly classified to **N** type and 16.00% wrongly classified to **F** type.

Table 4. Comparison of the proposed method and other works.

Reference	Method	Accu	Sens	Spec
Hu (1997, [8])	MoE	94.8	78.9	96.8
Chazal (2004, [11])	LDA	96.4	77.5	98.9
Jiang (2007, [12])	BbNN, EA	98.8	94.3	99.4
Ince (2009, [13])	MD, PSO	97.9	90.3	98.8
Kiranyaz (2016, [4])	CNN	98.9	95.9	99.4
Proposed	CCA	**96.7**	**97.2**	**96.2**

Table 5. Confusion matrix of the proposed multiclass classifier.

Input \ Output	N	S	V	F	Q
N	60.67	10.17	4.50	22.17	2.50
S	18.83	**48.50**	14.00	16.00	2.67
V	7.83	10.00	66.33	11.17	4.67
F	6.67	3.17	5.67	83.83	0.67
Q	5.33	1.33	0.67	1.50	**91.17**

4 Conclusion

In this paper, we propose a new classification model for ECG signals. To classify different classes of heart beats, we propose to decompose a signal into common and specific components. We define the common components by using classical results from Canonical Correlation Analysis. Assuming that only specific components play a dominant role in classification, we map the specific components of the signal to a two-dimensional feature space, which is then classified by a linear binary classifier. The proposed method is easily implemented and have good mathematical foundations. The design of the classifier is totally data-driven, with no prior knowledge of data or complicated handcrafted feature constructions needed. Numerical experiments on the MIT-BIH Arrhythmia Database are presented and the results are promising. The proposed model makes no assumption on the signals, so in principle it can be applied to any other types of signals.

However, it is necessary to point out several limitations of the proposed model. First, we use the two angles α and β to form the feature vector, which might be too simple to achieve a high performance. Second, we assume the linear independence between the components of ECG signals, which might be too strict a condition for the real world signals to satisfy. Third, it remains an open problem that how the choice of the threshold ρ_0 affects the outcome of classification.

To overcome these limitations, one may consider various directions of improvements. For examples, one can use the projection vector of the signal as the feature vector, which carries more information and helps to enhance the performance. Second, one could relax the linear independence restrictions and consider vectors that form a frame, then the model might become more flexible but complicated. Third, an interesting problem is to determine the best threshold ρ_0. Finally, we remark that the proposed model provides an alternative to the traditional classification scheme and it is worth further investigations.

References

1. Gacek, A., Pedrycz, W.: ECG Signal Processing, Classification and Interpretation: A Comprehensive Framework of Computational Intelligence. Springer, London (2011). https://doi.org/10.1007/978-0-85729-868-3
2. Osowski, S., Hoai, L.T., Markiewicz, T.: Support vector machine-based expert system for reliable heartbeat recognition. IEEE Trans. Biomed. Eng. **51**, 582–589 (2004)
3. Silipo, R., Marchesi, C.: Artificial neural networks for automatic ECG analysis. IEEE Trans. Signal Process. **46**, 1417–1425 (1998)
4. Kiranyaz, S., Ince, T., Gabbouj, M.: Real-time patient-specific ECG classification by 1-D convolutional neural networks. IEEE Trans. Biomed. Eng. **63**, 664–675 (2016)
5. Martinez, J.P., Almeida, R., Olmos, S., Rocha, A.P., Laguna, P.: A wavelet-based ECG delineator: evaluation on standard databases. IEEE Trans. Biomed. Eng. **51**, 570–581 (2004)
6. Kabir, M.A., Shahnaz, C.: Denoising of ECG signals based on noise reduction algorithms in EMD and wavelet domains. Biomed. Signal Process. Control **7**, 481–489 (2012)
7. Andreao, R.V., Dorizzi, B., Boudy, J.: ECG signal analysis through hidden Markov models. IEEE Trans. Biomed. Eng. **53**, 1541–1549 (2006)
8. Hu, Y., Palreddy, S., Tompkins, W.J.: A patient-adaptable ECG beat classifier using a mixture of experts approach. IEEE Trans. Biomed. Eng. **44**, 891–900 (1997)
9. Hotelling, H.: Relations between two sets of variates. Biometrika **28**(3/4), 321–377 (1936)
10. Moody, G.B., Mark, R.G.: The impact of the MIT-BIH arrhythmia database. IEEE Eng. Med. Biol. Mag. **20**, 45–50 (2001)
11. Chazal, P., O'Dwyer, M., Reilly, R.B.: Automatic classification of heartbeats using ECG morphology and heartbeat interval features. IEEE Trans. Biomed. Eng. **51**, 1196–1206 (2004)

12. Jiang, W., Kong, S.G.: Block-based neural networks for personalized ECG signal classification. IEEE Trans. Neural Networks **18**, 1750–1761 (2007)
13. Ince, T., Kiranyaz, S., Gabbouj, M.: A generic and robust system for automated patient-specific classification of ECG signals. IEEE Trans. Biomed. Eng. **56**, 1415–1426 (2009)

Sit-to-Stand Test for Neurodegenerative Diseases Video Classification

Vincenzo Dentamaro[⊠] [iD], Donato Impedovo[⊠] [iD], and Giuseppe Pirlo[⊠] [iD]

Dipartimento di Informatica, Università degli studi di Bari, 70121 Bari, Italy
vincenzo@gatech.edu, {donato.impedovo,giuseppe.pirlo}@uniba.it

Abstract. In this paper, an automatic video diagnosis system for dementia classification is presented. Starting from video recordings of patients and control subjects, performing sit-to-stand test, the designed system is capable of extracting relevant patterns for binary discern patients with dementia from healthy subjects. The proposed system achieves an accuracy 0.808 by using the rigorous inter-patient separation scheme especially suited for medical purposes. This separation scheme provides the use of some people for training and others, different, people for testing. This work is an original and pioneering work on sit-to-stand video classification for neurodegenerative diseases, thus the novelty in this study is both on phases segmentation and experimental setup.

Keywords: Sit-to-stand · TUG test · Neurodegenerative diseases classification · Image processing · Behavioral biometric

1 Introduction and Related Work

Neurodegenerative diseases are difficult both to predict and to identify [1], given their unknown etiology (in most cases) and the manifestation of symptoms common to many other pathologies. At time of writing, there is no cure to completely eradicate this type of pathology, but only treatments that slow down its course. Thus, the only option available, at the moment, is to identify the disease as quickly as possible and start treatments promptly.

In this work, patients with various kind of dementia (many of them suffer of Parkinson disease [3]), plus control subjects, are asked to perform five-time Sit-To-Stand exercise. Their exercises are recorded though three cameras. The produced system aims to predict whether a patient is suffering of dementia or not. This is a preliminary study on the subject. Sit-to-stand classification falls within behavioral biometrics techniques for neurodegenerative disease assessment. Some authors of this paper, used handwriting for the early assessment of neurodegenerative diseases [21].

At moment, there is not such big literature coverage for specifically sit-to-stand classification with cameras, thus the novelty in this study is both on phases segmentation and experimental setup.

Below is a summary of the global panorama, in which the proposed study is located, for the identification of neurodegenerative diseases through gait analysis.

© Springer Nature Switzerland AG 2020
Y. Lu et al. (Eds.): ICPRAI 2020, LNCS 12068, pp. 596–609, 2020.
https://doi.org/10.1007/978-3-030-59830-3_52

All authors of this study, have previously published a work on neurodegenerative diseases gait classification through video image processing [14] (Dentamaro et al.).

Gait Analysis consists in the study of walking and locomotor activity. It is practically the study of motion, the observation and measurement of movements, body mechanics and muscle activity. The Gait Analysis, in 1983, was approached by Gordon Rose in the clinical context. The suggested use consists in the analysis and collection of objective data, such as joint angles and walking speed, later used to express a diagnosis [4].

This study, however, focuses more on the analysis of the posture rather than on the actual gait analysis. It examines different patients during the execution of the Sit-To-Stand exercise.

The Sit-To-Stand exercise is part of the Time-Up and Go Test (TUG Test), a widely used clinical balance test used mainly for the assessment of functional mobility [11] (Podsiadlo et al.). In this scenario, the patient performs a series of movements that involve the muscles potentially affected by neurodegenerative diseases. One of the tests that can be performed is the Sit-To-Stand exercise.

The Sit-To-Stand test was used for various purposes during various tests. Among these we include the collection of indicators for the control of posture, risk of falling, weakness of the lower limbs and as a measure of disability [12] (Applebaum et al.).

The most common type of Sit-To-Stand is called "Five Times Sit-To-Stand Test" [13] (Duncan et al.), which consists in repeating the Sit-To-Stand movement five times. Specifically, each subject begins by crossing the arms over the chest while sitting on a chair. After the examiner's start command, the patient rises from the chair by levering on each of the legs, reaching a standing position with complete distention of the spine (where possible), and then reside to repeat the exercise. The exercise stopwatch starts when the examiner pronounces the first "GO" command, and ends when the patient resides for the fifth time. The patient should avoid resting on the backrest, if possible, throughout the procedure. For the classification of the phases within the execution of the Sit-To-Stand, this work refers to the classification provided by Tsukahara et al. [6]. The Sit-to-Stand, therefore, can be distinguished in 2 different phases, also shown in Fig. 1:

- Sit-to-Stand transfer phase: in which the patient stands up to reach the standing position; This is also called *stand-up* phase.
- Stand-to-Sit transfer phase: where the patient, after reaching the standing position, resides; This is also called *sit-down* phase.

Fig. 1. Sit-to-Stand task

From a pure pattern recognition perspective, there are various, but unfortunately not a high number yet, of authors who have developed various algorithms for sit-to-stand segmentation and classification tasks which depend on the kind of sensors used.

Matthew et al. [7] have developed an algorithm that combines motion capture with inertial sensor measurements for the acquisition of the complete state of a patient's "joint angles" during the execution of Sit-To-Stand exercise.

Whitney et al. [5], on the other hand, have conducted studies on the validity of the extractable features during this type of exercise, reaching a diagnosis accuracy of 65% on subjects with equilibrium disorders, of 80% on Activities-specific Balance Confidence scale (ABC) and 78% on the Dynamic Gait Index (DGI).

Li et al. [8], through video footage of patients during the execution of a TUG (Timed-Up-and-Go) exercise, proposed an algorithm for the automatic segmentation of the video in subtasks. This video is processed by a "Human Pose Estimator" to identify the set of coordinates of the key parts of the human body, normalized for subsequent analysis. After concatenating the data, a subtask segmentation is performed using a DTW algorithm (Dynamic Time Warping), in order to extract significant features corresponding to the various phases of the exercise. The study produced an average accuracy of 93%, falling to 80% in one case.

Zheng et al. [20] used inertial sensors (MEMS sensors such as accelerometers, gyroscope and so on) for mainly extracting time-based features for computing time differences in each phase of Sit-to-stand (stand-up, sit-down and total time) with respect to control subjects. Their results showed that Alzheimer [2] patients needed more time in all phases.

One of the least invasive and expensive systems currently available is certainly Image Processing [14]. This type of data collection consists in recording a patient during the execution of a specific exercise (from the simplest test such as the walk to a more demanding one like the TUG - Timed Up-and-Go exercise). Through the frames obtained from the shooting, several type of features are then extracted and used for classification. This is important, because no additional hardware needs to be wore by patients. In this way the system is more comfortable and unbiased, since the patient is free to move without worrying about the attached hardware.

The system proposed in our study, is based solely on image processing techniques through video recordings of patients performing sit-to-stand task. The paper is organized in the following way: Sect. 2 describes the dataset used. Section 3 shows the proposed classification pipeline. Section 4 presents a description of modern image processing, pose estimation and filtering techniques used in this work for the two phases segmentation: stand-up phase and sit-down phase. Section 5 describes the features extracted, experimental setup and results. Discussion of the results are presented in Sect. 6. Conclusion and future work are presented in Sect. 7.

2 Dataset Description

The dataset used deals with 7 patients with dementia in various stages and 3 subjects of the normative sample. It is known that a larger dataset is needed, but this is a preliminary study on the subject, and a bigger and better balanced dataset is currently in production.

The exercise was taken by using three cameras placed at different angles: right, left and center. Recorded videos have been preprocessed before being released into the dataset. After being synchronized, the resolution was raised to 480p and the framerate was stabilized at 20 fps.

Finally, the dataset is composed of 10 subjects; 7 patients whose diagnosis of dementia is known and 3 of the normative sample. For each patient, there are several sessions taken from multiple angles, therefore for each patient there are 4 to 6 videos due to the different exercise sessions. The normative sample, instead, were subjected to a single session, making only a maximum of 2 different videos available for subject. Furthermore, for 2 healthy subjects out of 3, one of the three cameras is not usable, due to the incorrect estimate of the Pose Estimation caused by the excessive distance of the camera with respect to the interested subject. On balance, therefore, the dataset has 38 videos of patients with dementia and only 4 videos of healthy subjects.

3 Proposed Classification Pipeline

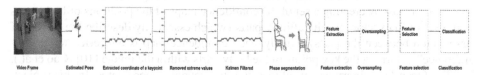

Fig. 2. Proposed classification pipeline

Figure 2 shows the proposed classification pipeline. From each frame of the video in input, the estimated pose is extracted. The coordinates of each key point of the estimated pose thorough time, are aggregated in one single vector. Extreme values are removed and subsequently, Kalman filter is applied. Depending on computed angles of different joints, phases segmentation is performed. It follows, feature extraction and oversampling with LICIC technique [17], because, as will be shown later, instances are strongly imbalanced. At the end of the pipeline there is feature selection and ultimately, classification.

4 Phases Segmentation

Phases segmentation make use of image processing techniques, such as pose estimation, for extracting the coordinates of each body part of one or more people in every single video frame. Subsequently the coordinates are filtered using Kalman Filter and angles of few, important, junctions are computed for, later, determining the final phase. The phases, as already stated, are: stand-up and sit-down.

4.1 Image Processing and Pose Estimation

The typical Image Processing system presented in [9] consists of several digital and analogue video capture cameras. Some possible data extraction methods can be:

- Threshold Filtering: conversion of black and white images;
- Pixel Count: count of light and dark pixels in the image;
- Background Segmentation: removal of the image background;
- Pose Estimation: image analysis to identify the people present;
- Depth Measurement (or Range Imaging);

The Pose Estimation deals with the location of parts of the body of one or more people in an image, providing a series of 2D points that locate the main points of the human body. However, the identification process presents several challenges:

1. Each image can contain an indefinite number of people in any position or size;
2. The interaction between people translates into the complexity of calculating the human parts due to contacts or occlusions;

Runtime complexity tends to grow depending directly on the number of people in the image;

The system takes a coloured image of variable size as input and returns the 2D coordinates of the anatomical key points of a person. Initially, a feedforward network makes a forecast by creating a set of 2D confidence maps of the positions of body parts and a set of 2D vectors for the affinity parts (which essentially describe the degree of association between the parts). Finally, each confidence map is converted to "greedy inference", returning the key points of the body parts within the image [10] (Cao et al.).

The OpenPose library was used to obtain the patient's coordinates. In this project, COCO [15] (Microsoft COCO 2014) training was used. In the following list, the indices of the vector and the part of the body they indicate, are listed:

0. Head	1. Neck	2. Shoulder Right
3. Right Elbow	4. Right Wrist	5. Left Shoulder
6. Left Elbow	7. Left Wrist	8. Right hip
9. Right knee	10. Right Ankle	11. Left hip
12. Left Knee	13. Left Ankle	

Each video, which includes the patient performing the Sit-To-Stand exercise, was processed with the OpenCV library. OpenCV routines are able to fragment the video frame by frame and obtain useful information, such as framerate of the video and total duration, just to name a few. The extracted data is composed by an array of coordinates, containing, for each frame, the x and y coordinates of the joints of each person in the image, its boolean visibilities and timestamp at which the frame is taken. One of the problems with OpenPose, is the fact that, a single person does not have a fixed index within all frames. This happens when there is more than one person in the video. This implies that the coordinates concerning our patient are placed side by side with those of the other persons present in the frame, such as for example the nurse, or the relatives that assist to the exercise.

To identify the patient's coordinates among all, a system has been defined where the first frame of the video is shown, and the user is asked to click the centre of gravity of the person whose analysis is to be performed. This centre of gravity is usually the hip. Once clicked, the coordinates within the image are saved. These coordinates will then be used to compare the various positions of people within the image. For each frame, therefore, the centre of gravity of a person is calculated with the formula in Eq. 1:

$$Cg = (Rhip + Lhip)/2 \qquad (1)$$

The nearest (in Euclidean sense) center of gravity Cg with the user selected coordinates (the clicked once) define the index of the patient. This index never changes through frames. It is necessary to note, that this solution is especially designed for Sit-To-Stand test, and it will not work for other behavioral biometric analysis, such as gait analysis.

4.2 Linear Coordinates Interpolation

Once all the coordinates of all the frames are computed by OpenPose, there is another problem to face. This problem derives from the difficulties of the pose estimation algorithm in the detection of a person's joints. In fact, if a limb is not visible, the pose estimation algorithm will mark that joint as not visible, entering the extreme coordinates (coordinates at corners of the image). Ignoring these frames would greatly reduce the frames available for feature calculation, so it was decided to use a linear interpolation algorithm for the missing points.

The algorithm takes the x and y coordinates of each joint of all the frames, finds the intervals in which there are invalid values (specifically zeros or extreme coordinate values) and replace them by linearly interpolating them, with the progressive values included between the last and the next successfully detected coordinates.

4.3 Smoothing with Kalman Filter

Joints coordinates of consecutive frames, suffer from high coordinate variance: key point of same body part in two consecutive frames, may have high variance due to the inaccuracies of the pose estimation algorithm.

For smoothing coordinates through time, it has been decided to use a linear Kalman filter [16]. Kalman Filter is an important and widely used estimation algorithm. The Kalman Filter computes estimates of hidden variables based on inaccurate and uncertain measurements. The aim of the algorithm is to predict the future state of the system by using the past estimations. The Kalman filter is used in control system, tracking, space navigation and so on.

In this case, the Kalman filter is used as a smoothing technique. Given the entire sequence of joint coordinates, the Kalman filter, through 5 iterations of the estimation operation (more iteration did not increase the quality of the output), it outputs the smoothed sequence of coordinates. This task is performed for all key points.

4.4 Angle Computation and Segmentation

Before synthetizing features, in the feature extraction phase, it was decided to segment the video into phases so as to be able to distinguish, during the execution of the exercise, whether the patient was in the stand-up phase or in the sit-down phase. Instead of synthetizing features at a predetermined number of frames, it has been decided to synthetize features every time a phase ends. To achieve this, the angle formed by the patient's shoulder, hip and knee was taken into consideration.

The calculation of the angle has been implemented through the Eq. 2.

$$\gamma = deg \left(arc\ cos \left(\frac{\overrightarrow{AB} * \overrightarrow{BC}}{|AB| * |BC|} \right) \right) \tag{2}$$

Where deg converts radians to degrees. $\vec{A}$, $\vec{B}$ and $\vec{C}$ are respectively the vectors containing the coordinates of left shoulder, left hip and left knee. Thus, the $\overrightarrow{AB}$ vector, represents the shoulder-hip vector and the $\overrightarrow{BC}$ coordinates vector, represents the hip-knee coordinates vector. The formulation for computing this vector is presented in Eq. 3. Equation 4 shows the computation of the scalar product between vectors.

$$\overrightarrow{AB} = \vec{B} - \vec{A} = \left(\overrightarrow{B_x} - \overrightarrow{A_x}, \overrightarrow{B_y} - \overrightarrow{A_x} \right) \tag{3}$$

$$\overrightarrow{AB} * \overrightarrow{BC} = \left(\overrightarrow{AB_x}, \overrightarrow{AB_y} \right) * \left(\overrightarrow{BC_x}, \overrightarrow{BC_x} \right) \tag{4}$$

From this formulation (Eq. 2) it is not automatically possible to distinguish an acute angle from an obtuse one, therefore all vectors were rotated in order to be able to easily distinguish the type of angle presented. The rotation is performed iteratively trying all degree variations, until $\vec{A}_x$ is equal to $\vec{B}_x$.

The formulation used for the rotation is presented in Eq. 5.

For each θ from $1°$ to $360°$ and until $\vec{A}_x$ is equal to $\vec{B}_x$:

$$A_{xr} = \overrightarrow{O_x} + \cos(\theta) * \left(\overrightarrow{A_x} - \overrightarrow{O_x} \right) - \sin(\theta) * \left(\overrightarrow{A_y} - \overrightarrow{O_y} \right) \tag{5}$$

$$A_{yr} = \overrightarrow{O_y} + \sin(\theta) * \left(\overrightarrow{A_x} - \overrightarrow{O_x} \right) + \cos(\theta) * \left(\overrightarrow{A_y} - \overrightarrow{O_y} \right)$$

A_{xr} and A_{yr} in Eq. 5, are the rotated coordinates for x and y coordinates of the $\vec{A}$ vector. The rotation is performed in block for vectors $\vec{A}$, $\vec{B}$ and $\vec{C}$. In this case, the rotation is performed with respect to the point $\vec{O}$ that represents the hip, in such a way to have the x coordinate of the shoulder and hip identical, therefore aligned. Once this scenario is achieved, it is enough to compare the y component of $\vec{A}$ and $\vec{C}$ (the shoulder and knee): if the y component of $\vec{C}$ vector (the knee) is greater than the y component of the $\vec{A}$ vector (the shoulder), then the angle is acute, otherwise it will be obtuse. If the calculated angle is obtuse, $90°$ will be added to γ (from Eq. 2).

The segmentation idea is simple but effective. If a person is sitting, the angle described by his body will tend to be near $90°$, while a person standing will describe an angle that is around 180°. Keeping this principle in mind, if considering the variation of the angle described by shoulder, hip and knee through time, it is possible to identify the frames in which the patient is rising or sitting. Specifically, if the difference between the average of the previous angles and the current angle is negative, then the patient is rising, if positive, the patient is sitting. The frame in which this difference changes from negative to positive and vice versa, marks one of the phases.

5 Features, Experiment and Results

5.1 Features

The preliminary features contained in Table 1 are synthetized for each phase. This means that, during the exercise, if the patient stands up and sits down five times, there will be 10 phases, 5 stand up phases and 5 sit down phases. It has been decided to synthetize features at every phase, rather than at every predefined number of frames (or seconds) because of the heterogeneity of the dataset: some patients were unable to stand up because affected of severe neurodegenerative disease.

On the other hand, some patients and the control subjects were able to stand up and sit down very fast, therefore it is not trivial to find the right threshold to which features are to be synthetized.

Concretely, features of a sample point will be computed at the end of each phase.

Features in Table 1 are synthetized for each phase and for all 14 key points. To further synthetize the resulting vectors obtained from Table 1, the final function features presented in Table 2, were used.

In practice, the features in Table 1 are applied on a temporal stream of data coordinates. Thus, the result, are vectors of computed features. In order to further synthetize those vectors, the functional features in Table 2 are applied.

In addition, function features presented in Table 2 are also applied to head-neck-hip angle, shoulder-hip-knee angle and hip-knee-ankle angle. Thus, the total number of features per sample is: 14 (joints) * 5 (function features) * 10 (preliminary features) + 3 (selected angles) * 5 (function features) = 715 features.

5.2 Experiment Design

In order to extract the relevant patterns that are able to discern subjects with dementia and subjects without, it is necessary to apply an *inter*-patient separation scheme as presented in [14]. The *inter*-patient uses extracted features of some people for training and extracted features of completely other people for testing. This separation scheme is better suited for medical purposes, this is because with an *intra*-patient separation scheme, the i.i.d. (independently and identically distributed) assumption among instances, in this specific case, is not achievable.

In addition to this separation scheme, because of the limitations of the dataset (low amount of data with strong imbalance), already described in Sect. 2, the experiment was designed in the following way: because there are only 3 subjects of the normative sample, it has been decided to keep two random subjects of the normative sample for

Table 1. Preliminaries features

Feature name	Formulation
Displacement	$d_i = \sqrt{\Delta x_i^2 + \Delta y_i^2}$
Displacement x	$\Delta x_i = x_{i+1} - x_i$
Displacement y	$\Delta y_i = y_{i+1} - y_i$
Velocity	$v_i = d_i / \Delta t_i$
Velocity x	$v_{x,i} = \Delta x_i / \Delta t_i$
Velocity y	$v_{y,i} = \Delta y_i / \Delta t_i$
Acceleration	$a = v_i / \Delta t_i$
Acceleration x	$a_{x,i} = v_{x,i} / \Delta t_i$
Acceleration y	$a_{y,i} = v_{y,i} / \Delta t_i$
Tangent angle	$\rho_i = \tan^{-1}(\Delta y_i / \Delta x_i)$

Table 2. Function features

Feature name	Formulation
Mean	$\bar{x} = \frac{x_1 + x_2 + .. + x_n}{n}$
Median	given $x = [x_1, x_2, .., x_n]$ thus $\mu = x\left[\frac{n}{2}\right]$
Standard deviation	$\sigma_X = \sqrt{\frac{\sum_{i=1}^{N}(x_i - \bar{x})^2}{N}}$ where $\bar{x} = \frac{1}{N} \sum_{i=1}^{N} x_i$
1 and 99 percentile	$n = \left[\frac{P}{100} \times N\right]$

training and one random subject of the normative sample (different from the two subjects previously used) for test. At the same manner, two random subjects among the 7 total subjects of the patients with dementia for training and one random subject, different from the two previously chosen, for testing. This process was repeated 21 times for each feature selection technique, classification technique and with or without oversampling. Results were averaged and reported in Table 3.

The produced training set suffers of the well-known "course of dimensionality" [17]. For this reason, it has been decided to perform, before feature selection, a novel oversampling technique called LICIC [17]. The oversampling technique called LICIC [17], instead, creates new instances balancing the minority classes. The synthetic instances are made preserving nonlinearities and the particular pattern present in each specific class, by copying most important components and permutating less important components among instances of the same class, and thus, create new offspring. LICIC is applied inside each traint-test separation process described previously, thus LICIC will be used 21 times. It has been avoided to use LICIC at the beginning of the pipeline for the dataset as a whole, because it would bleed specific patterns of samples used for testing, also in training, thus making results, less reliable. This kind of bleeding is not considered cheating, but it is not suitable for medical application research.

Feature selection was performed comparing accuracies of Univariate Selection with mutual information [18] and Extra Trees [19] with ordered feature importance.

Univariate selection with mutual information is a technique that chooses a predefined number of features, in this case 100, basing the choice on the result of a univariate statistical test. In this case, the univariate selection uses the mutual information criteria for ranking the features. The mutual information among two variables measures a sort of "mutual dependency" between them. Zero means no dependency and high positive value means high dependency. Extra Trees, instead, is an ensemble technique wich uses the results of multiple de-correlated trees, aggregating them and outputting the classification result. During the growth of this forest, the feature importance, which in this case is represented by the Gini index, is computed. At the end, instead of classifying samples, it is only necessary to order the features with respect to their Gini index, from higher to lower and keep the first 100 features in descending order.

The classification techniques are: feed forward neural network with "Adam" solver, "Relu" as activation function and two hidden layers with respectively 858 neurons in the first and 572 in the second. K-Nearest Neighbor using 3 nearest neighbors. Support Vector Machine with linear kernel. Random Forest classifier with 50 trees and maximum depth of 5 for conquering overfitting. Finally, AdaBoost with 10 decision trees as week learner each with 10 as max depth.

There are, ongoing, new patient recordings aimed at increasing the numerosity of the dataset.

5.3 Results

All classifiers are trained with 10-Fold stratified cross validation. Results on test sets are reported in Table 3.

Where F1 is the univariate selection with mutual information feature selection technique, and F2 is the Extra Tree technique used for feature selection.

Table 3. Results

Technique	Sensitivity	Specificity	Precision	Accuracy
SVM	0.908	0.393	0.604	0.650
SVM + F1	0.866	0.563	0.688	0.714
SVM + F1 + LICIC	0.884	0.530	0.665	0.707
SVM + F2	0.921	0.671	0.763	0.796
SVM + F2 + LICIC	0.934	0.466	0.654	0.700
AdaBoost	0.895	0.504	0.663	0.700
AdaBoost + F1	0.938	0.616	0.729	0.777
AdaBoost + F1 + LICIC	0.932	0.680	0.767	0.806
AdaBoost + F2	0.916	0.509	0.680	0.713
AdaBoost + F2 + LICIC	0.849	0.573	0.694	0.711
K-Nearest Neighbor	0.848	0.498	0.635	0.673
K-Nearest Neighbor + F1	0.912	0.544	0.677	0.728
K-Nearest Neighbor + F1 + LICIC	0.902	0.597	0.704	0.750
K-Nearest Neighbor + F2	0.897	0.577	0.698	0.737
K-Nearest Neighbor + F2 + LICIC	0.928	0.624	0.725	0.776
Neural Network	0.976	0.402	0.653	0.689
Neural Network + F1	0.937	0.521	0.672	0.729
Neural Network + F1 + LICIC	0.907	0.548	0.696	0.727
Neural Network + F2	0.940	0.594	0.724	0.767
Neural Network + F2 + LICIC	**0.944**	**0.672**	**0.780**	**0.808**
Random Forest	0.876	0.458	0.655	0.667
Random Forest + F1	0.966	0.515	0.515	0.741
Random Forest + F1 + LICIC	0.893	0.714	0.781	0.804
Random Forest + F2	0.917	0.444	0.647	0.681
Random Forest + F2 + LICIC	**0.936**	**0.680**	**0.766**	**0.808**

6   Discussion

Results from Table 3 clearly shows that the proposed pipeline in Sect. 3 is effective. Both, the Feed Forward Neural Network and Random Forest achieved the same accuracy of 0.808. Neural Network is more sensible and thus predicts positives slightly better than Random Forest. On the other hand, Random Forest is slightly better in avoiding false negatives. But the Neural Network's precision is higher compared to the Random Forest result. This means that the correctly classified elements were relevant.

Results show that the Extra Tree feature selection technique is slightly better compared to the univariate selection with mutual information. The presence of LICIC technique for balancing the dataset, improves the accuracies of 4.1% for the Neural Network and of 12.7% for Random Forest in exactly same conditions.

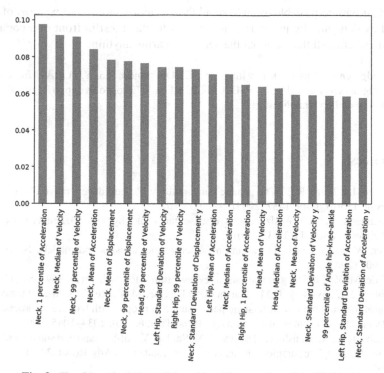

Fig. 3. Top 20 ranked Extra Tree selected features based on Gini index

In Fig. 3 it is possible to note that the 55% of top 20 meaningful features are synthetized from the neck movement, followed by head, left hip and right hip. Therefore, differently from the global medical assumption of measuring legs muscle and shoulder muscles activities, the neck and the hip are the clearest winners when comes up to sit-to-stand classification with computer vision and image processing techniques.

7   Conclusions and Future Work

In this work, an automatic video diagnosis system for sit-to-stand phases segmentation and dementia classification was developed. The proposed pipeline was effective for correctly segment the stand-up and sit-down phases. In a second step, features are extracted and classification is performed with an inter-patient separation scheme, especially designed for medical purposes. This is a preliminary work in the field. Despite the numerosity of the dataset, results are encouraging. In particular, features extracted from the neck movements are meaningful for sit-to-stand video classification. In a future work, more data will be available, but also new features, such as computational kinematic features, will be employed for capturing micro and macro parameters of body joints movements and the execution speed of the task will be investigated in more detail [22]. Also 3D data points of the pose can be built starting from data captured by the three cameras, this would increase the overall accuracy. In addition to an expected increase in accuracy, it will be possible to understand the particular movement patterns of people affected by dementia by just performing the sit-to-stand test in front of a commercial low-cost camera. All that, without the need of wearing anything.

Acknowledgments. This work is within the BESIDE project (no. YJTGRA7) funded by the Regione Puglia POR Puglia FESR - FSE 2014-2020. Fondo Europeo Sviluppo Regionale. Azione 1.6 - Avviso pubblico "InnoNetwork".

References

1. Bertram, L., Tanzi, R.E.: The genetic epidemiology of neurodegenerative disease. J. Clin. Investig. **115**(6), 1449–1457 (2005)
2. Alzheimer's, A.: 2015 Alzheimer's disease facts and figures. Alzheimer's Dement. J. Alzheimer's Assoc. **11**(3), 332 (2015)
3. Dauer, W., Przedborski, S.: Parkinson's disease: mechanisms and models. Neuron **39**(6), 889–909 (2003)
4. Whittle, M.W.: Clinical gait analysis: a review. Hum. Mov. Sci. **15**(3), 369–387 (1996)
5. Whitney, S.L., Wrisley, D.M., Marchetti, G.F., Gee, M.A., Redfern, M.S., Furman, J.M.: Clinical measurement of sit-to-stand performance in people with balance disorders: validity of data for the five-times-sit-to-stand test. Phys. Ther. **85**(10), 1034–1045 (2005)
6. Tsukahara, A., Kawanishi, R., Hasegawa, Y., Sankai, Y.: Sit-to-stand and stand-to-sit transfer support for complete paraplegic patients with robot suit HAL. Adv. Robot. **24**(11), 1615–1638 (2010)
7. Matthew, R.P., Seko, S., Bajcsy, R.: Fusing motion-capture and inertial measurements for improved joint state recovery: an application for sit-to-stand actions. In: 2017 39th Annual International Conference of the IEEE Engineering in Medicine and Biology Society (EMBC), pp. 1893–1896. IEEE, July 2017
8. Li, T., et al.: Automatic timed up-and-go sub-task segmentation for parkinson's disease patients using video-based activity classification. IEEE Trans. Neural Syst. Rehabil. Eng. **26**(11), 2189–2199 (2018)
9. Muro-De-La-Herran, A., Garcia-Zapirain, B., Mendez-Zorrilla, A.: Gait analysis methods: An overview of wearable and non-wearable systems, highlighting clinical applications. Sensors **14**(2), 3362–3394 (2014)
10. Cao, Z., Simon, T., Wei, S.E., Sheikh, Y.: Realtime multi-person 2D pose estimation using part affinity fields. In: Proceedings of the IEEE Conference on Computer Vision and Pattern Recognition, pp. 7291–7299 (2017)
11. Podsiadlo, D., Richardson, S.: The timed "Up & Go": a test of basic functional mobility for frail elderly persons. J. Am. Geriatr. Soc. **39**(2), 142–148 (1991)
12. Applebaum, E.V., et al.: Modified 30-second Sit to Stand test predicts falls in a cohort of institutionalized older veterans. PLoS ONE **12**(5), e0176946 (2017)
13. Duncan, R.P., Leddy, A.L., Earhart, G.M.: Five times sit-to-stand test performance in Parkinson's disease. Arch. Phys. Med. Rehabil. **92**(9), 1431–1436 (2011)

14. Dentamaro, V., Impedovo, D., Pirlo, G.: Real-time neurodegenerative disease video classification with severity prediction. In: Ricci, E., Rota Bulò, S., Snoek, C., Lanz, O., Messelodi, S., Sebe, N. (eds.) ICIAP 2019. LNCS, vol. 11752, pp. 618–628. Springer, Cham (2019). https://doi.org/10.1007/978-3-030-30645-8_56

15. Lin, T.-Y., et al.: Microsoft COCO: Common Objects in Context. In: Fleet, D., Pajdla, T., Schiele, B., Tuytelaars, T. (eds.) ECCV 2014. LNCS, vol. 8693, pp. 740–755. Springer, Cham (2014). https://doi.org/10.1007/978-3-319-10602-1_48

16. Welch, G., Bishop, G.: An introduction to the Kalman filter, pp. 41–95 (1995)

17. Dentamaro, V., Impedovo, D., Pirlo, G.: LICIC: less important components for imbalanced multiclass classification. Information 9(12), 317 (2018)

18. Ross, B.C.: Mutual information between discrete and continuous data sets. PLoS ONE 9(2), e87357 (2014)

19. Geurts, P., Ernst, D., Wehenkel, L.: Extremely randomized trees. Mach. Learn. 63(1), 3–42 (2006)

20. Zheng, E., Chen, B., Wang, X., Huang, Y., Wang, Q.: On the design of a wearable multi-sensor system for recognizing motion modes and sit-to-stand transition. Int. J. Adv. Rob. Syst. 11(2), 30 (2014)

21. Impedovo, D., Pirlo, G.: Dynamic handwriting analysis for the assessment of neurodegenerative diseases: a pattern recognition perspective. IEEE Rev. Biomed. Eng. 12, 209–220 (2018)

22. Impedovo, D.: Velocity-based signal features for the assessment of Parkinsonian handwriting. IEEE Signal Process. Lett. 26(4), 632–636 (2019)

A Two Stage Method for Abnormality Diagnosis of Musculoskeletal Radiographs

Yunxue Shao[1,2](✉) [iD] and Xin Wang[2](✉) [iD]

[1] School of Computer Science and Technology, Nanjing Tech University, Nanjing, China
csshyx@njtech.edu.cn
[2] College of Computer Science, Inner Monglia University, Hohhot, China
13234751492@163.com

Abstract. In this paper, a two stage method is proposed for bone image abnormality detection. The core idea of this method is based on our observation and analysis of the abnormal images. The abnormal images are divided into two categories: one is the abnormal images containing abnormal objects, the other is the abnormal images of the bone with inconspicuous lesions. The abnormal images containing abnormal objects are easy to be classified, so that we can focus on the abnormal images which are difficult to classify. The proposed two stage method enables the classifier to extract better features and learn better classification parameters for the abnormal images that are difficult to classify. We carried out experiments on a large-scale X-ray dataset MURA. SENet154 and DenseNet201 are used as the classification networks. Compare to one stage method, the proposed method can improve test accuracies by 1.71% (SENet154) and 1.43% (DenseNet201), respectively, which shows the effectiveness of the proposed method.

Keywords: Musculoskeletal Radiographs (MURA) · X-ray · Two stage method

1 Introduction

Musculoskeletal disorders are a major burden on individuals, health systems and social care systems. The United Nations and the World Health Organization have recognized this burden through the 2000–2010 Bone and Joint Decade [1]. They are the most common cause of severe long-term pain and physical disability. In the past decade, the number of fractures associated with osteoporosis has almost doubled. It is estimated that 40% of all women over the age of 50 will have osteoporotic fractures. Joint disease is particularly important in the elderly, accounting for half of the total number of chronic diseases in the elderly aged 65 and over [2]. According to GBD, 1.7 billion people worldwide are imaged for musculoskeletal disorders, and they are the second leading cause of disability [3]. Convolutional neural networks have proven to be an effective model for solving a wide range of visual problems. Through nonlinear activation function and down sampling operator, CNN can produce image representations of different modes and obtain global receptive fields by stacking the convolutional layers. Based on the advantages of DCNN on non-medical images, DCNN is beginning to be little by little

© Springer Nature Switzerland AG 2020
Y. Lu et al. (Eds.): ICPRAI 2020, LNCS 12068, pp. 610–621, 2020.
https://doi.org/10.1007/978-3-030-59830-3_53

applied to medical image classification and detection problems. For example, Spanhol et al. [4] proposed the use of DCNN to classify breast cancer pathology images. Li et al. [5] put forward a DCNN-based lung nodule classification system. Roth et al. [6] proposed an algorithm for developing a lymph node detection system using DCNN. Based on the above successful experience of DCNN applied to medical images, we tried to apply DCNN to diagnose lesions in radioactive bone studies. The Department of Computer Science, Medicine and Radiology at Stanford University published a public dataset MURA on musculoskeletal X-rays, the largest dataset with 40,561 images from 14,863 upper limb studies. On this dataset, Rajpurkar et al. [7] replaces the last layer of the DenseNet169 convolutional neural network with a single output and then replaces the Sigmoid activation function used to identify the lesion. Banga et al. [8] proposes the ensemble200 model, which consists of three CNN models. A set of trained convolutional neural network models is assembled and then superimposed into an integrated model. The output of the integrated model is the predicted average score and anomalous probability. Kaliyugarasan et al. [9] adds some additional layers to the DenseNet121 convolutional neural network and uses transfer learning to train the MURA.

In this paper, we propose a method called two stage. From a dataset perspective, we first focus on selecting cases that are clearly characterized and easily identifiable, and then group the remaining cases into one category. In this way, the dataset is manually divided into two subsets. These two subsets are independent of each other and have no intersection. The union of these two subsets is the MURA dataset. After splitting into two sub-datasets, we can solve them by using existing models, saving time and effort. In order to prove the effectiveness of the algorithm, we carried out experiments on the MURA dataset, and the experimental results also proved the effectiveness of the method.

2 Model

2.1 SENet and DenseNet

We use the SENET [10] and DenseNet [11] models for abnormality diagnosis.

SENet
The core of SENet is SE-Block, which is used to weight the features of each channel. Figure 1 shows the structure of SE-Block.

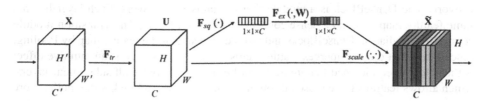

Fig. 1. SE-block structure

First, after the F_{tr} operations, that is, the conventional convolution operation, the feature map U to be processed is obtained, and then the F_{sq} and F_{scale} operations are performed on the U.

For the F_{sq} operation, the U is first subjected to a global average pooling operation to obtain a feature map Z of $1 \times 1 \times C$. The mathematical expression is:

$$Z_C = F_{sq}(U_c) = \frac{1}{H \times W} \sum_{i=1}^{H} \sum_{j=1}^{W} U_c(i, j) \tag{1}$$

Where U_c represents the Cth channel of U. Z is then used as an input to a fully connected neural network, and S represents the weighting factor of the channels of a set of feature map. Formally:

$$s = F_{ex}(z, W) = \sigma(g(z, W)) = \sigma(W_2\sigma(W_1z)) \tag{2}$$

Where $F_{ex}(., W)$ represents the calculation process of a fully connected layer, and σ represents the RELU function, $W_1 \epsilon \mathbb{R}^{\frac{C}{r} \times C}$, $W_2 \epsilon \mathbb{R}^{C \times \frac{C}{r}}$.

For the F_{scale} operation, the importance of the feature map of the difference channels is represented by the S calculated just by multiplication with the feature map U, Formally:

$$\widetilde{X}_c = F_{scale}(u_c, s_c) = s_c.u_c \tag{3}$$

Where $\widetilde{X} = [\widetilde{x}_1, \widetilde{x}_2, \ldots, \widetilde{x}_c]$.

SENet fuses the global pooling information of different feature channels into the training process of the neural network, so that the weight function of the network can be used to train the weights of different channels. SE-Block can be embedded in many mainstream convolutional neural network structures. Here we choose the SENet154 version. SENet154 is constructed by incorporating SE-Block into a modified version of 64 * 4d ResNetXt-152 which extends the original ResNeXt-101 by adopting the block stacking strategy of ResNet-152.

DenseNet

DenseNet [11] differs from ResNet [12–14], in that features are never combined by summing before passing features to a layer; instead, by connecting them to combine features, DenseNet ensures maximum information between layers flow. The structure of DenseNet is to interconnect all the layers, specifically, each layer will accept all its previous layers as its additional input.

As shown in Fig. 2. The DenseBlock + Transition structure is used in the DenseNet network. The DenseBlock is a module that contains many layers. Each layer has the same feature map size and a dense connection between layers. The Transition module connects two adjacent DenseBlocks and reduces the size of the feature map by Pooling. Since DenseNet directly merges feature maps from different layers, an intuitive effect is that each layer's learned feature map can be used directly by all subsequent layers, which allows features to be reused throughout the network and makes the model more concise.

This experiment selects DenseNet201 version, and replaces the last fully connected layer with the number of neurons by 2.

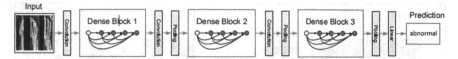

Fig. 2. There are three dense blocks in the figure, the layer between two adjacent blocks is called the transition layer, and the feature map size is changed by convolution and pooling.

2.2 Dataset

Musculoskeletal Radiographs (MURA) [7] is one of the largest X-ray databases currently collected and open by Stanford University. Each study contains one or more images, and those images are all labeled as normal or abnormal by Stanford Hospital's Board Certified Radiologists. MURA contains 40,561 musculoskeletal X-rays of 14,863 study. In more than 10,000 studies, there are 9067 normal upper extremity musculoskeletal and 5915 upper limb abnormal musculoskeletal, including shoulder, humerus, elbow, forearm, wrist, hand and finger. The Fig. 3 lists positive and negative categories distribution of study.

Study	Train		Validation		Total
	Normal	Abnormal	Normal	Abnormal	
Elbow	1094	660	92	66	1912
Finger	1280	655	92	83	2110
Hand	1497	521	101	66	2185
Humerus	321	271	68	67	727
Forearm	590	287	69	64	1010
Shoulder	1364	1457	99	95	3015
Wrist	2134	1326	140	97	3697
Total No. of Studies	8280	5177	661	538	14656

Fig. 3. Musculoskeletal radiography dataset of 14656 studies

The lesions in the MURA dataset are mainly composed of abnormal objects, fractures, injuries, arthritis, and the like. We used MURA's training set (13457 studies, 36808 images) and validation set (1199 studies, 3197 images) to conduct experiments. The test set (207 studies, 556 images) was not published, and our experimental results below are based on the validation set.

A partial picture of the dataset is shown in Fig. 4.

2.3 Two Stage Method

One or Two Stage Dataset

In order to better illustrate the effectiveness of the algorithm, we trained on all MURA datasets and we call it one stage. The difference between the one stage and the two stage is

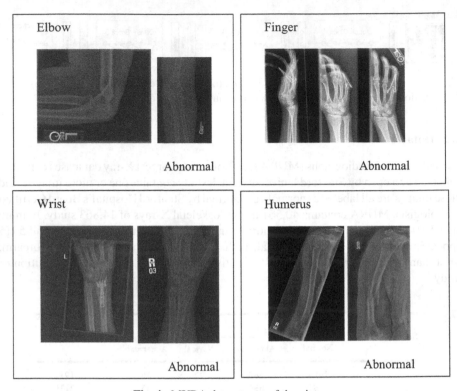

Fig. 4. MURA dataset part of the picture

whether the dataset is divided. Dividing is based on the consideration that medical image datasets are different from other fields, and the main difficulties are dataset acquisition, data annotation, etc.

1. Data acquisition

Data is the core resource required for deep learning algorithms. Only mastering algorithms and lacking data can not achieve better training results. At this stage, there is very little medical data available, and the data collected by only two or three hospitals is far from enough.

2. Data annotation

On the basis of obtaining data, deep learning combines prior knowledge to train the model, and the training set needs to be labeled in advance. Since most annotations rely on manual recognition, there is a subjective problem for doctors, that is, doctors with different qualifications for the same medical image may have inconsistent results.

Therefore, in one stage, we expand the dataset, and the accuracy of the dataset cannot be improved by more training. In the end we took a two stage algorithm. Through observation, we found that there are a large number of abnormal objects in the MURA

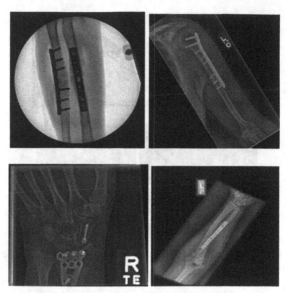

Fig. 5. Part of the dataset in the part1 of the two-stage, the lesion is the presence of an abnormal object

dataset. The existence of these abnormal objects can be understood as non-medical personnel. Therefore, we have separately sorted out all the cases of abnormal objects in MURA. The abnormal objects are shown in the Fig. 5.

The lesions of the remaining MURA datasets are mainly composed of degenerative arthritis, injuries, fractures, etc. These cases are difficult to identify for our non-medical personnel, and for deep learning models, it is difficult to learn from them. Therefore, we put abnormal images that are not easy to recognize into one category, the advantage of this is that the classifier can better focus on the difficult samples. As shown in Fig. 6.

Therefore, the two stage method first identifies whether there is an abnormal object in the image. We call it the part1 of the two stage. Secondly, it is used to identify the lesions such as fractures and injuries. We call it the part2 of the two stage. The two parts negative class are MURA negative class, due to the existence of positive and negative sample imbalance, we have carried out sample expansion on both parts, the expansion method is to rotate 10 degrees and simultaneously crop the top and left sides of the picture by 20 pixels.

Two Stage

First, the part1 is used to identify those cases with obvious characteristics. If a test case is judged to be abnormal, the algorithm outputs 1, otherwise the test case is further sent to the part2, and the part2 is often used to identify those features that are not easy to recognize. If the part2 of the algorithm judges to be abnormal, the algorithm outputs 1, otherwise it is 0. The algorithm flow chart is shown in Fig. 7.

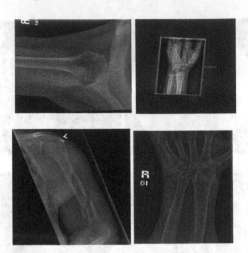

Fig. 6. Part of the dataset in the part2 of the two-stage, the main site of the lesion is fracture, degenerative arthritis, injury, etc.

3 Experiment

3.1 Preprocessing

Histogram equalization is a way to enhance image contrast. By analyzing the statistical histogram of the MURA, we choose Contrast Limited Adaptive Histogram Equalization (CLAHE) [15] for image enhancement.

CLAHE limits the enhancement of contrast by limiting the height of the local histogram, thereby limiting noise amplification and over-enhancement of local contrast.

The histogram value of the CLAHE method is

$$Hist'(i) = \begin{cases} Hist(i) + L & Hist(i) < T \\ H_{max} & Hist(i) \geq T \end{cases}$$

Where $Hist(i)$ is the derivative of the cumulative distribution function of the sliding window local histogram, and H_{max} is the maximum height of the histogram. We cut off the histogram from the threshold T, and then evenly distribute the truncated portion over the entire grayscale range to ensure that the total histogram area is constant, so that the entire histogram rises by a height L. The preprocessed image is shown in Fig. 8.

3.2 One Stage vs Two Stage

Experimental Parameters

Before we feed the image into the neural network, we normalized each image to have the same mean and standard deviation of image. The ImageNet Large Scale Visual Recognition Challenge (ILSVRC) [16] has become the standard benchmark for large-scale object recognition.ILSVRC2012 consists of 1000 object classes and 1,431,167

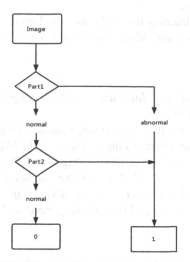

Fig. 7. Flow chart for the Two-stage method

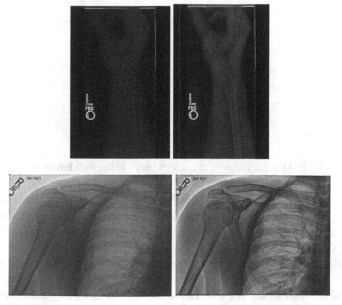

Fig. 8. After using the CLAHE method, the left side of the figure is the original picture, and the right side is the enhanced result.

annotated images. Considering that MURA is very different from the ILSVRC2012, we calculated the mean and variance of the MURA dataset, which were [0.246, 0.246, 0.246], [0.147, 0.147, 0.147]. Then resize the image to (320, 320), we augmented the data during training by random left and right flip. We use the pre-trained SENET154 and DenseNet201 parameters on ImageNet to initialize the network. The weight attenuation coefficient is 0.0001, the learning rate is 0.0001, the loss function is SoftmaxCrossEntropyLoss, the optimization algorithm is adam, beta1 = 0.9, beta2 = 0.999, batchsize selects 16.

Experimental Result

We use FineTuning technology to fine tune the two model parameters. The results of the fine tuning are shown in Table 1.

From Table 1, we can see that the two-stage method proposed by the SENet154 or DenseNet201 model further improves the accuracy of the MURA. SENet154 increased from 0.8214 of one stage to 0.8385 of two stage, and DenseNet201 increased from 0.8205 to 0.8348. Using the SENet154 model, the validation set accuracy of two-stage part1 is 0.9861, and the validation set accuracy of two-stage part2 is 0.7953. Using the DenseNet201 model, the accuracy of the two-stage part1 is 0.9848, and the accuracy of the two-stage part2 is 0.7912.

Table 1. SENet154 and DenseNet201 use the accuracy of the two-stage method on the validation set

Stage	SENet-154	DenseNet-201
One-stage	0.8214	0.8205
Two-stage	0.8385	0.8348
Two-stage part1	0.9861	0.9848
Two-stage part2	0.7953	0.7912

Next, let's look at the accuracy of the seven parts in the validation set after fine-tuning SENET154, as shown in Table 2.

From Table 2, we can see that with SENet154, our two stage method has improved in six parts, in which the finger is increased by about 2%, the hand, humerus, shoulders, and elbows are increased by 1%, and the forearm is slightly improved. But the wrist has dropped by 0.0015.

The performance of DenseNet201 in the seven parts of the validation set is shown in Table 3.

From Table 3 we can see that with DenseNet201, forearm, finger increased by 2 percentage points, shoulder increased by 1 percentage point, and wrist has slightly improved. However hand and elbow dropped by 1%.

3.3   Model Interpretation

Zhou et al. [17] Proposed that CNN not only has strong image processing and classification capabilities, but also can locate key parts of the image, which is called class activation mapping (CAM). Formally, let M(x) be CAM, then,

$$M(x) = \sum_k w_k f_k(x) \qquad (4)$$

Table 2. Using the SENet154 model, we present a two-stage approach to verifying the performance of the seven upper limb bones on the set

Bone	One-stage	Two-stage	Val_acc
Wrist	0.8756	0.8741	− 0.0015
Forearm	0.8439	0.8472	+ 0.0033
Hand	0.7957	0.8109	+ 0.0152
Humerus	0.8711	0.8850	+ 0.0139
Shoulder	0.7957	0.8064	+ 0.0107
Elbow	0.8495	0.8624	+ 0.0129
Finger	0.7787	0.7983	+ 0.0196

Table 3. Using the DenseNet201 model, we present a two-stage approach to verifying the performance of the seven upper limb bones on the set

Bone	One-stage	Two-stage	Val_acc
Wrist	0.8619	0.8695	+ 0.0076
Forearm	0.7907	0.8173	+ 0.0266
Hand	0.8196	0.8000	− 0.0196
Humerus	0.9094	0.9024	−0.0070
Shoulder	0.8082	0.8188	+ 0.0106
Elbow	0.8538	0.8409	− 0.0129
Finger	0.7766	0.8026	+ 0.0260

Where X represents the input image and $f_k(x)$ represents the Kth feature map of the set of feature images output by the last convolutional layer of the trained image. w_k represents the weight of the kth full join.

We first obtain the last set of feature maps of SENET154 trained on the two stage part2 dataset, and then multiply and add the feature maps to the fully connected weight matrix of a certain category to obtain an image in a specific category. To high light the lesions in the original radioactive image, we scaled the M(x) dimension up to its original size and fused M(x) and the original image in a 6:4 ratio.

By using CAM, we visualized the portion of the MURA partial image that contributed the most to the model anomaly prediction. Figure 9 shows the radioactive image and the corresponding cam image.

From Fig. 9, we can see that our method successfully locates its abnormal parts and highlights the area of the lesion, showing the effectiveness of our method.

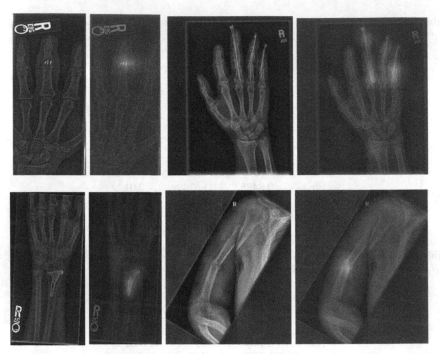

Fig. 9. The figure shows the effect diagram after using the CAMs method. The left side of the figure is the original image, and the right side is the result image

4   Conclusion

In this paper, we propose a two stage method. The method starts from the dataset perspective, divides the dataset into two sub-datasets according to the characteristics of the dataset, and then uses SENET154 and DenseNet201 models to solve the subdataset to prove that we propose. The two stage approach can be applied to any existing model. We performed experiments on the MURA dataset of one of the largest radioactive X-ray bone dataset, and the experimental results also demonstrated the effectiveness of our two stage approach. However, the algorithm we proposed does not improve significantly on individual bone parts, so we will further analyze this.

Acknowledgement. This study was supported by the National Natural Science Foundation of China (NSFC) under Grant no. 61563039.

References

1. Woolf, A.D., Pfleger, B.: Burden of major musculoskeletal conditions. Bull. World Health Organ. **81**(9), 646–656 (2003)
2. Olmarker, K.: The bone and joint decade 2000–2010. Eur. Spine J. **7**(4), 269 (1998)

3. Nafsiah, M., Indra, M.S., Indang, T., Iqbal, E., Karen, H.S., Pungkas, B.A., et al.: On the road to universal health care in indonesia, 1990–2016: a systematic analysis for the global burden of disease study 2016. The Lancet (2018). S0140673618305956

4. Spanhol, F.A., Oliveira, L.S., Petitjean, C., Heutte, L.: Breast cancer histo pathological image classification using convolutional neural networks. In: International Joint Conference on Neural Networks. IEEE (2016)

5. Wei, L., Peng, C., Dazhe, Z., Junbo, W.: Pulmonary nodule classification with deep convolutional neural networks on computed tomography images. Comput. Math. Methods Med. **2016**, 1–7 (2016)

6. Roth, H.R., Lu, L., Seff, A., Cherry, K.M., Hoffman, J., Wang, S., et al.: A new 2.5D representation for lymph node detection using random sets of deep convolutional neural network observations. Med Image Comput. Comput. Assist. Interv. (2014)

7. Rajpurkar, P., et al.: Mura: Large dataset for abnormality detection in musculoskeletal radiographs. arXiv preprint arXiv:1712.06957 (2017)

8. Banga, D., Waiganjo, P.: Abnormality detection in musculoskeletal radiographs with convolutional neural networks (Ensembles) and Performance Optimization. arXiv preprint arXiv: 1908.02170 (2019)

9. Kaliyugarasan, S.K.: Deep transfer learning in medical imaging. MS thesis. The University of Bergen (2019)

10. Hu, J., Shen, L., Sun, G.: Squeeze-and-excitation networks. In: Proceedings of the IEEE Conference on Computer Vision and Pattern Recognition, pp. 7132–7141 (2018)

11. Huang, G., Liu, Z., Van Der Maaten, L., Weinberger, K.Q.: Densely connected convolutional networks. In: Proceedings of the IEEE Conference on Computer Vision and Pattern Recognition, pp. 4700–4708 (2017)

12. He, K., Zhang, X., Ren, S., Sun, J.: Deep residual learning for image recognition. In: Proceedings of the IEEE Conference on Computer Vision and Pattern Recognition, pp. 770–778 (2016)

13. He, K., Zhang, X., Ren, S., Sun, J.: Identity mappings in deep residual networks. In: Leibe, B., Matas, J., Sebe, N., Welling, M. (eds.) ECCV 2016. LNCS, vol. 9908, pp. 630–645. Springer, Cham (2016). https://doi.org/10.1007/978-3-319-46493-0_38

14. Xie, S., Girshick, R., Dollár, P., Tu, Z., He, K.: Aggregated residual transformations for deep neural networks. In: Proceedings of the IEEE Conference on Computer Vision and Pattern Recognition, pp. 1492–1500 (2017)

15. Pizer, S.M., Amburn, E.P., Austin, J.D., Cromartie, R.: Adaptive histogram eqalization and its variations. In: Computer Vision, Graphics, and Image Processing (CVGIP), vol. 39, pp. 355–368 (1987)

16. Russakovsky, O., et al.: Imagenet large scale visual recognition challenge. Int. J. Comput. Vision **115**(3), 211–252 (2015)

17. Zhou, B., Khosla, A., Lapedriza, A., Oliva, A., Torralba, A.: Learning deep features for discriminative localization. In: Proceedings of the IEEE Conference on Computer Vision and Pattern Recognition, pp. 2921–2929 (2016)

Clinical Decision Support Systems for Predicting Patients Liable to Acquire Acute Myocardial Infarctions

Fu-Hsing Wu[2] , Hsuan-Hung Lin[2] , Po-Chou Chan[2] , Chien-Ming Tseng[2] ,
Yung-Fu Chen[1,3(✉)] , and Chih-Sheng Lin[1,4(✉)]

[1] Department of Radiology, BenQ Medical Center, The Affiliated BenQ Hospital of Nanjing
Medical University, Nanjing 210019, Jiangsu, China
Chihsheng.Lin@benqmedicalcenter.com
[2] Department of Management Information Systems, Central Taiwan University of Science
and Technology, Taichung 40601, Taiwan
[3] Department of Dental Technology and Materials Science, Central Taiwan University of
Science and Technology, Taichung 40601, Taiwan
yfchen@ctust.edu.tw
[4] Department of Medical Imaging and Radiological Sciences, Central Taiwan University of
Science and Technology, Taichung 40601, Taiwan

Abstract. Acute myocardial infarction (AMI) is a major cause of death worldwide. There are around 0.8 million persons suffered from AMI annually in the
US and the death rate reaches 27%. The risk factors of AMI were reported to
include hypertension, family history, smoking habit, diabetes, serenity, obesity,
cholesterol, alcoholism, coronary artery disease, etc. In this study, data acquired
from a subset of the National Health Insurance Research Database (NHIRD) of
Taiwan were used to develop the clinical decision support system (CDSS) for
predicting AMI. Support vector machine integrated with genetic algorithm (IGS)
was adopted to design the AMI prediction models. Data of 6087 AMI patients and
6087 non-AMI patients, each includes 50 features, were acquired for designing
the predictive models. Tenfold cross validation and three objective functions were
used for obtaining the optimal model with best prediction performance during
training. The experimental results show that the CDSSs reach a prediction performance with accuracy, sensitivity, specificity, and area under ROC curve (AUC) of
81.47–84.11%, 75.46–80.94%, 86.48–88.21%, and 0.8602–0.8935, respectively.
The IGS algorithm and comorbidity-related features are promising in designing
strong CDSS models for predicting patients who may acquire AMI in the near
future.

Keywords: Acute Myocardial Infarction (AMI) · Comorbidity ·
Comorbidity-related features · Clinical Decision Support System (CDSS) ·
Integrated genetic algorithm and support vector machine (IGS)

© Springer Nature Switzerland AG 2020
Y. Lu et al. (Eds.): ICPRAI 2020, LNCS 12068, pp. 622–634, 2020.
https://doi.org/10.1007/978-3-030-59830-3_54

1 Introduction

Cardiovascular diseases (CVDs) are ranked as the leading cause of death globally by WHO [1]. Among the CVDs, AMI is one of the major causes of death. There are 0.8 million (M) people who have experienced an AMI annually in the US and 0.2 M of them die [2]. In Taiwan, CVD, following cancer and followed by pneumonia, was listed as the second leading cause of death [3].

AMI is caused by the interruption of blood supply to the myocardium, characterized by a severe and rapid onset of symptoms which may include chest pain, dyspnea, sweating, and palpitations [4]. It is associated with poor lifestyle, family history, and comorbidity. The risk factors of AMI include stressful life, diabetes, hypertension, family history, smoking, alcohol consumption, abdominal obesity, cholesterol, cardiovascular health, body mass index, and physical activity [5–7].

CDSSs provide useful message and professional knowledge to improve the diagnosis performance, treatment outcome, and health care quality in clinical environment. They have been widely applied in disease diagnosis, disease treatment, medical alert, event reminder, drug dose, and drug prescription [8–17]. Recently, we have adopted SVM technique to design CDSSs for predicting successful ventilator weaning for patients admitted in respiratory caring centers [11] as well as for classifying different type of pap smear cells and discriminating normal from abnormal cells [18]. Furthermore, we have also constructed predictive models using the National Health Insurance Research Database (NHIRD) for predicting fracture for patients taking corticosteroids [19], readmission for patients admitted with all-cause conditions [20], and erectile dysfunctions (EDs) for men with various comorbidities [21]. A CDSS for precisely predicting AMI is expected to be useful in reducing the death of CVDs worldwide. This study aims to develop a CDSS for effectively predicting patients who are liable to acquire AMI based on the data retrieved from the NHIRD.

2 Materials and Methods

2.1 Data Acquisition

The raw data adopted in this study consist of 1 million patient data randomly sampled from the entire NHIRD containing the data of around 23 M Taiwanese citizens enrolled in the NHI program. The NHIRD includes data of medical facility registries, inpatient orders, ambulatory cares, prescription drugs, and physicians providing services, etc. Widespread researches have been conducted using the NHIRD for investigating the health status of Taiwanese citizens and the association between different diseases. Till July 2016, around 4,000 research papers have been published using the datasets of NHIRD [22].

In this study, the positive cases were patients aged 20 years old and older, diagnosed with AMI (ICD-9-CM Code 410.x) by physicians between Jan. 2002 and Dec. 2013 and confirmed for at least three outpatient visits or at least one inpatient visit. The negative cases were those without history of AMI randomly chosen from the NHIRD and were 1:1 frequency matched with the AMI patients by age and index year.

2.2   Statistical Analyses

Statistical analyses of the retrieved data were conducted using SPSS statistical software package version 17.0 (SPSS Inc.). Mean and standard deviation (SD) were calculated for describing characteristics of the continuous variables and number of events counted for describing the nominal variables. Difference of continuous and nominal variables between AMI and non-AMI groups were compared with Student's t-test and Chi-square test, respectively. The significance was defined as $p < 0.05$.

2.3   Feature Selection and CDSS Design

We considered the main risk factors and comorbidities of AMI reported in literature [5–7] and the key factors suggested in our previous studies [20, 21]. Because living habits, physiological signals, and family histories of patients are not available in the NHIRD, they were not considered in the current study. Table 1 lists the ICD-9-CM codes of 12 comorbidities (risk factors) included for analyses and to be considered for designing the CDSSs. In addition to presence of these 12 comorbidities, patient's ages when individual comorbidities were diagnosed as well as follow-up duration (in years) and annual physician visits of the comorbidities were also considered, accounting to a total of 50 features to be included for CDSS design (Table 2).

Table 1. ICD-9-CM codes of included comorbidities

Comorbidities	ICD-9-CM codes
Hypertension	401–405
Diabetes	250
Hyperlipidemia	272.0–272.4
Coronary artery disease	414.01
Shock	785.5
Stagnation heart failure	428.x
Cancer	140.0–208.9
Cerebrovascular disease	430.x–438.x
Pulmonary edema	518.4, 514.x
Acute renal failure	584.x, 586.x, 788.5
Chronic renal failure	585.x, 403.x, 404.x, 996.7, v451
Arrhythmia	427.x

Support vector machine (SVM) is regarded as one of an effective data classification and regression methods [23, 24]. The effectiveness of an SVM model depends on the selection of the kernel parameter (r) and soft margin parameter (C). The wrapper method integrating genetic algorithm (GA) and SVM, namely IGS, was demonstrated to be an effective algorithm for recursively training predictive models, selecting data features, and adjusting model parameters at each iteration until the optimal model has been obtained [25–27]. Our recent studies have demonstrated that IGS algorithm is effective in designing models to predict fractures for patients taking corticosteroids [19], hospital readmissions for those admitted with all-cause conditions [20], and ED for men with other comorbidities [21]. Figure 1 shows the chromosome pattern and flow chart of the IGS algorithm, in which GA is used for selecting the optimal SVM parameters (C and r) and significant features for constructing the SVM models.

C	r	X_1	X_2	X_3	...	X_i	...	X_p
SVM parameters		Features (Among p features, X_i =1 if the ith feature is selected, otherwise X_i =0.)						

(a)

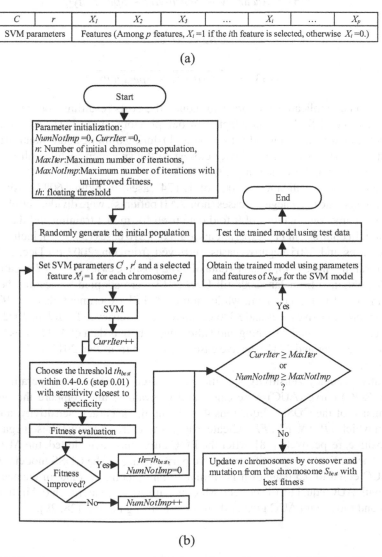

(b)

Fig. 1. (a) The chromosome pattern and (b) the software flow chart of IGS algorithm with the floating threshold method

2.4 Model Training, Testing and Evaluation

In this study, two independent training and testing (ITT) experiments were performed to verify the prediction performance of designed CDSS [18]. During the training phase, tenfold cross validation and 3 objective functions were used to train and validate the CDSSs. Instead of accuracy, as indicated in Eqs. 1–3, combination of accuracy, sensitivity, and specificity, area under the ROC curve (AUC), as well as g-mean, respectively, were used for obtaining the models with best prediction performance [20, 21].

$$OB1 = Accuracy - |Sensitivity - Specificity|. \qquad (1)$$

$$OB2 = AUC. \qquad (2)$$

$$OB3 = \sqrt{Sensitivity \times Specificity} \qquad (3)$$

For clinical applications, in order to reduce the costs resulting from false negative predictions, the IGS algorithm adopted in our previous study [21] was modified by introducing floating threshold adjustment (Fig. 1) during training and validation to obtain the optimal model with the sensitivity closest to the specificity while still keeping its fitness value best at the same time [20].

In tenfold cross-validation, a total of 12174 cases, including 6087 positive cases (AMI patients) and 6087 negative cases (non-AMI patients), were divided into 10 clusters (folds) with any combinations of 9-fold data used for model training and the rest 1 for validation. In contrast, in ITT experiments, data of 4210 cases (34.58%), including 2105 positive cases and 2105 negative cases, retrieved from Jan. 2002 till Dec. 2007 were used for training the CDSS, whereas data of 3982 positive cases and 3982 negative cases (65.42%) obtained from Jan. 2008 till Dec. 2013 were adopted for testing the trained model in the 1st ITT experiment; while, in the 2nd ITT experiment, data of 6972 cases (63.18%; 3846 positive cases and 3846 negative cases) retrieved from Jan. 2002 till Dec. 2010 were used for model training and validating, whereas data of 4482 cases (36.82%; 2241 positive cases and 2241 negative cases) retrieved from Jan. 2011 till Dec. 2013 for model testing.

Finally, accuracy, sensitivity, specificity, and area under receiver operating characteristic (ROC) curve (AUC) were calculated to evaluate and compare the prediction performance of the CDSSs. Equations 4–6 define the accuracy, sensitivity, and specificity, in which *TP*, *TN*, *FN*, *FP* indicate true positive, true negative, false negative, and false positive, respectively [18]. After the ROC curve has been plotted, the AUC can be obtained accordingly to quantify predictive performance of the CDSS models. A model with AUC equal to 0.5 denotes it is similar to an outcome of random chance while a model with AUC equal to 1 is perfect. Models with AUC greater than 0.7 indicate good models and those with AUC greater than 0.8 are strong models [28, 29].

$$Accuracy = \frac{TP + TN}{TP + FN + TN + FP}. \tag{4}$$

$$Sensitivity = \frac{TP}{TP + FN}. \tag{5}$$

$$Specificity = \frac{TN}{TN + FP}. \tag{6}$$

3 Experimental Results

Demographics, comorbidities, and comorbidity-related factors between non-AMI and AMI patients are compared in Table 2. As shown in the table, the occurrence of all the comorbidities considered in the AMI patients exhibit significant higher rates ($p < 0.01$) than the non-AMI patients. Moreover, AMI patients acquired most comorbidities, except cancer and acute renal failure, at significantly younger age ($p < 0.05$) than non-AMI patients. Furthermore, compared to the non-AMI patients, those with AMI also show significantly lower frequency ($p < 0.01$) of annual physician visits in 10 (except shock and cancer) of the 12 comorbidities. Nonetheless, regarding the follow-up durations, AMI patients exhibited significantly longer in some comorbidities (diabetes, pulmonary edema, and chronic renal failure) but significantly shorter in others (hypertension, coronary artery disease, and cancer) before AMI events occurred compared to the non-AMI patients.

The comorbidities and other comorbidity-related features selected for designing CDSS models based on three different objective functions (*OB1*, *OB2* and *OB3*) are also shown in Table 2 (columns 5–7). Notably, although age was exactly matched and had no difference between AMI and non-AMI groups, it was selected and adopted as an important factor for designing the CDSSs based on all 3 objective functions. Moreover, although the feature combinations are different among 3 models, presence of comorbidities and comorbidity-related features were mostly selected.

The prediction performances of CDSSs designed using the IGS method with floating threshold method are shown in Table 3. As indicated in the table, after cross-validation, the obtained accuracy, sensitivity, specificity, and AUC were 79.94–80.68%, 79.74–80.63%, 80.13–80.92%, and 0.8649–0.8738, respectively. The optimal threshold values (th_{best}) of the three models (*OB1*, *OB2*, and *OB3*), meeting the requirements of the sensitivity closest to the specificity and the best fitness values, were 0.42, 0.42, and 0.44, respectively.

In the 1st ITT experiment, the obtained accuracy, sensitivity, specificity, and AUC were 81.47–82.84%, 75.46–77.8%, 86.48–87.89%, and 0.8602–0.8801, respectively; whereas in the 2nd ITT experiment, the accuracy, sensitivity, specificity, and AUC were 83.19–84.11%, 78.98–80.94%, 86.74–88.21%, and 0.8734–0.8935, respectively.

Table 2. Comparisons of demographic information, presence of individual comorbidities, and comorbidity-related features between non-AMI and AMI patients

Features	Non-AMI (N − 6087)	AMI (N = 6087)	p-value	Feature selected		
				OB1	OB2	OB3
Age, mean(SD)	65.7 (14.2)	65.7 (14.2)	1	V	V	V
Gender						
Male, N (%)	4241 (69.7)	4241 (69.7)	1			
Female, N (%)	1846 (30.3)	1846 (30.3)	1			
Comorbidity, N (%)						
Hypertension[‡]	4182 (68.7)	5349 (87.9)	<0.001	V	V	V
Diabetes[‡]	2335 (38.4)	3440 (56.5)	<0.001			V
Hyperlipidemia[‡]	3094 (50.8)	4485 (73.7)	<0.001			
Coronary artery disease[‡]	499 (8.2)	4235 (69.6)	<0.001			V
Shock[‡]	208 (3.4)	1088 (17.9)	<0.001	V	V	
Stagnation heart failure[‡]	880 (14.5)	2799 (46)	<0.001	V		V
Cancer	959 (15.8)	1092 (17.9)	0.001	V	V	V
Cerebrovascular disease	1862 (30.6)	2073 (34.1)	0.001	V		
Pulmonary edema[‡]	98 (1.6)	855 (14)	<0.001	V		V
Acute renal failure[‡]	434 (7.1)	1280 (21)	<0.001	V		V
Chronic renal failure[‡]	829 (13.6)	2028 (33.3)	<0.001	V		V
Arrhythmia[‡]	1532 (25.2)	2523 (41.4)	<0.001	V		V
Age when the comorbidity diagnosed in years old, mean (SD)						
Hypertension[***]	59.5 (21.1)	57.3 (20.6)	<0.001	V	V	V
Diabetes[*]	59.7 (21.6)	58.4 (20.1)	0.018			
Hyperlipidemia[***]	58.3 (21.1)	56.3 (20.6)	<0.001	V		V
Coronary artery disease[***]	66.4 (22.1)	59.5 (21.7)	<0.001	V	V	V
Shock[*]	69.8 (24.3)	65.4 (23.1)	0.014			V
Stagnation heart failure[***]	67.9 (22.6)	62.9 (22.4)	<0.001	V		
Cancer	64.7 (21.3)	63.3 (22.0)	0.158	V		V
Cerebrovascular disease[*]	64.5 (22.3)	62.9 (21.3)	0.014	V		

(continued)

Table 2. (*continued*)

Features	Non-AMI ($N =$ 6087)	AMI ($N = 6087$)	p-value	Feature selected		
				OB1	OB2	OB3
Pulmonary edema[*]	70.2 (22.1)	65.3 (21.8)	0.037	V	V	V
Acute renal failure	66.2 (24.7)	65.4 (22.0)	0.526			
Chronic renal failure[***]	66.2 (21.6)	62.9 (21.8)	<0.001	V	V	V
Arrhythmia[**]	63.9 (22.0)	61.9 (22.0)	0.006	V		
Follow-up duration of the comorbidity in years, mean (SD)						
Hypertension[**]	7.3 (4.1)	7.0 (3.9)	0.003	V	V	V
Diabetes[***]	6.2 (4.2)	6.8 (4.1)	<0.001	V	V	V
Hyperlipidemia	5.1 (3.7)	5.1 (3.5)	0.98		V	
Coronary artery disease[*]	3.5 (2.6)	3.1 (2.6)	0.038	V	V	V
Shock	2.0 (2.2)	2.3 (2.5)	0.502	V	V	V
Stagnation heart failure	3.6 (3.3)	3.6 (3.0)	0.917	V		V
Cancer[**]	4.1 (3.7)	3.6 (3.4)	0.008			V
Cerebrovascular disease	4.5 (3.6)	4.5 (3.5)	0.892	V		
Pulmonary edema[*]	1.1 (1.2)	2.2 (2.2)	0.04			V
Acute renal failure	2.2 (2.5)	2.5 (2.5)	0.167	V		
Chronic renal failure[***]	3.3 (3.3)	3.9 (3.2)	<0.001	V		
Arrhythmia	4.3 (3.6)	4.1 (3.3)	0.069	V	V	V
Outpatient and inpatient visits per year for the comorbidity, mean (SD)						
Hypertension[***]	7.8 (5.6)	5.3 (5.3)	<0.001	V	V	V
Diabetes[***]	9.6 (7.4)	7.1 (6.5)	<0.001	V	V	V
Hyperlipidemia[***]	6.0 (5.3)	4.4 (5.1)	<0.001	V	V	
Coronary artery disease[***]	4.7 (5.4)	1.9 (3.5)	<0.001	V	V	V
Shock	1.8 (1.6)	1.5 (1.9)	0.456	V		V
Stagnation heart failure[***]	6.3 (6.0)	4.5 (5.6)	<0.001	V	V	

(*continued*)

Table 2. (*continued*)

Features	Non-AMI (N = 6087)	AMI (N = 6087)	p-value	Feature selected		
				OB1	OB2	OB3
Cancer	13.2 (12.9)	13 (15.0)	0.838	V	V	
Cerebrovascular disease***	8.9 (10)	7.3 (10.3)	<0.001	V		V
Pulmonary edema**	5.3 (7.9)	1.9 (3.2)	0.006		V	V
Acute renal failure**	6.3 (7.7)	4.1 (5.7)	0.001	V	V	V
Chronic renal failure**	10.2 (10.6)	8.3 (10.4)	0.001	V		
Arrhythmia***	6.1 (5.6)	4.2 (5.2)	<0.001	V	V	V

Chi-square test with $^{\ddagger}p < 0.001$. Student's unpaired t-test with $*p < 0.05$, $**p < 0.01$, and $***p < 0.001$.

Table 3. Results for AMI prediction using IGS algorithm with the floating threshold method

Model	Experiment	Accuracy	Sensitivity	Specificity	AUC	Threshold
OB1	10-flod cross-validation	80.63%	80.63%	80.63%	0.8672	0.42
	1st ITT	82.84%	77.80%	87.89%	0.8810	
	2nd ITT	**84.11%**	**80.94%**	87.28%	**0.8935**	
OB2	10-flod cross-validation	79.94%	79.74%	80.13%	0.8738	0.42
	1st ITT	81.94%	77.39%	*86.48%*	0.8659	
	2nd ITT	83.60%	78.98%	**88.21%**	0.8734	
OB3	10-flod cross-validation	80.68%	80.43%	80.92%	0.8649	0.44
	1st ITT	*81.47%*	*75.46%*	87.49%	*0.8602*	
	2nd ITT	83.19%	79.65%	86.74%	0.8750	

4 Discussions

As shown in Table 2, although the positive cases and negative cases were age-matched in this study, age was still chosen as a salient feature during training for designing the CDSSs using three objective functions, mimicking that age is associated with acquisition of AMI. The mean frequency of annual physician visits of most of the comorbidities for the AMI group was significantly lower than the non-AMI group, indicating that less frequent physician visits or bad disease management was associated with deterioration of comorbidities, resulting in a higher chance of acquiring AMI. This result is consistent to our previous study reporting that patients with ED exhibited significantly lower annual physician visits on relevant comorbidities than those without ED [21].

As listed in Table 2 (columns 1–3), follow-up durations of most of the 12 comorbidities present insignificant differences between AMI and non-AMI patients. However, presence of comorbidity, age when the comorbidity diagnosed, and annual physician visits of comorbidities all exhibit significant differences between AMI and non-AMI groups for most of the 12 comorbidities, which again are consistent with the models for predicting ED [21].

There are 39, 23, and 33 salient features selected for constructing *OB1*, *OB2* and *OB3* models, respectively. Although *OB2* model adopted the fewest features for model construction, its predictive performance was similar to both *OB1* and *OB3* models. Furthermore, it can be observed that, except follow-up duration of hyperlipidemia, 22 of the 23 features selected by using *OB2* were also selected by *OB1* or *OB3*. Similar phenomenon can also be observed in our previous study [21].

Our experiment results show that using the IGS algorithm with presence of comorbidities and other comorbidity-related features are able to design strong CDSS models (AUCs greater than 0.8) in predicting patients who are liable to acquire AMI in the near future. However further improvement in the predictive performance still needs to be conducted. As reported in previous literature [5–7], further improvement in predictive performance may be achieved by including features selected from physiological signals, living habits, smoking habit, alcoholism, family history, BMI, and other relevant comorbidities. Physiological signals, family history, BMI, and other lifestyle factors were not available in the NHIRD. In this pilot study, only 12 comorbidities were considered; other comorbidities which can be acquired from the NHIRD should also be considered for further improving the predictive performance. For example, ED, gout, sleep disorder, anxiety, depression, and atopic dermatitis were also reported to be associated with CVD and AMI [30–34] and should also be included for CDSS design. In literature [30, 31, 35], gout, sleep disorder, depression were shown to be associated with ED, which in turn is believed to be a precursor of CVD [32, 33]. Shen et al. demonstrated anxiety was an independent factor for predicting MI in men [34]. Atopic dermatitis were also shown to be associated with ischemic stroke, angina, coronary vascular disease, MI, congestive heart failure, and peripheral vascular disease [36–38].

The designed CDSS can be applied in the clinical setting by embedding it in the physician order entry (POE) system. During the physician visit, the CDSS can use the patient data stored in the electronic medical records to make predictions to screen person at high risk of acquiring AMI. And then, physicians may administrate effective treatments and medications for managing comorbidities and provide advices and suggestions to reduce the risk of acquiring AMI.

Non-contact vital sign sensing technologies (including ballistocardiograph BCG [39–43], and mmWave or radio frequency radar [44–47]) may be considered for monitoring the heart rate and respiration rate of the person at high risk of acquiring AMI. Once the AMI occurs, heart rate and respiration rate become abnormal. Immediately, the alert message will be sent to the care personnel for emergent medical service to reduce the sudden death.

5   Conclusions and Future Works

The designed CDSSs achieved good performance for predicting patients who have higher probability of acquiring AMI. Most comorbidities and their related features are significant in discriminating AMI from non-AMI patients. Moreover, the combination of age, presence of related comorbidities, and other comorbidity-related features, including diagnosed age and annual physician visits of individual comorbidities, is useful for designing the AMI prediction models. Future studies will focus on aggressively screening patients who are liable to acquire AMI and administrate effective m-Health interventions for promoting health education and monitoring acquired comorbidities to prevent occurrence or recurrence of AMI.

Acknowledgments. This study was partially supported by Ministry of Science and Technology, Taiwan (MOST 109-2410-H-166-001) and Central Taiwan University of Science and Technology, Taichung, Taiwan (Grant No. CTU108-P-019). Fu-Hsing Wu, Hsuan-Hung Lin, and Po-Chou Chan contributed equally to this work. Correspondence should be addressed to Yung-Fu Chen or Chih-Sheng Lin.

References

1. World Health Organization, https://www.who.int/cardiovascular_diseases/about_cvd/en/. Accessed 20 Nov 2019
2. Boateng, S., et al.: Acute myocardial infarction. Dis. Mon. **59**(3), 83–96 (2013)
3. Ministry of Health and Welfare of Taiwan, https://www.mohw.gov.tw/cp-16-48057-1.html. Accessed 20 Nov 2019
4. ICD   9,   http://www.icd9data.com/2009/Volume1/390-459/410-414/410/default.htm. Accessed 20 Nov 2019
5. Lanas, F., et al.: Risk factors for acute myocardial infarction in Latin America: The INTERHEART Latin American study. Circulation **115**(9), 1067–1074 (2007)
6. Atiq, M.: Recent Advances in Cardiovascular Risk Factors. IntechOpen, Croatia (2012)
7. Isiozor, N.M., et al.: Ideal cardiovascular health and risk of acute myocardial infarction among Finnish men. Atherosclerosis **289**, 126–131 (2019)
8. Garg, A.X., et al.: Effects of computerized clinical decision support systems on practitioner performance and patient outcomes: A systematic review. JAMA **293**(10), 1223–1238 (2005)
9. Porat, T., et al.: Eliciting user decision requirements for designing computerized diagnostic support for family physicians. J. Cognit. Eng. Decis. Mak. **10**(1), 57–73 (2016)
10. Horng, S., et al.: Creating an automated trigger for sepsis clinical decision support at emergency department triage using machine learning. PLoS ONE **12**(4), e0174708 (2017)
11. Hsu, J.-C., et al.: Clinical verification of a clinical decision support system for ventilator weaning. Biomed. Eng. Online **12**(1), S4 (2013)
12. Luo, G., et al.: A systematic review of predictive modeling for bronchiolitis. Int. J. Med. Informatics **83**(10), 691–714 (2014)
13. Dunn Lopez, K., et al.: Integrative review of clinical decision support for registered nurses in acute care settings. J. Am. Med. Inform. Assoc. **24**(2), 441–450 (2017)
14. Scheepers-Hoeks, A.-M.J., et al.: Physicians' responses to clinical decision support on an intensive care unit—comparison of four different alerting methods. Artif. Intell. Med. **59**(1), 33–38 (2013)

15. Otto, A.K., et al.: The development of a clinical decision support system for the management of pediatric food allergy. Clin. Pediatr. **56**(6), 571–578 (2017)
16. Ammenwerth, E., et al.: The effect of electronic prescribing on medication errors and adverse drug events: A systematic review. J. Am. Med. Inform. Assoc. **15**(5), 585–600 (2008)
17. Baypinar, F., et al.: Physicians' compliance with a clinical decision support system alerting during the prescribing process. J. Med. Syst. **41**(6), 96 (2017)
18. Chen, Y.-F., et al.: Semi-automatic segmentation and classification of pap smear cells. IEEE J. Biomed. Health Inform. **18**(1), 94–108 (2013)
19. Chen, Y.-F., et al.: Design of a clinical decision support system for fracture prediction using imbalanced dataset. J. Healthcare Eng. **2018**, 9621640 (2018)
20. Lai, H.-J., et al.: Designing a clinical decision support system to predict readmissions for patients admitted with all-cause conditions. J. Ambient Intell. Human. Comput. (2020). https://doi.org/10.1007/s12652-019-01579-6
21. Chen, Y.-F., et al.: Design of a Clinical Decision Support System for Predicting Erectile Dysfunction in Men Using NHIRD Dataset. IEEE J. Biomed. Health Inform. **23**(5), 2127–2137 (2018)
22. Chang, C.-C., et al.: Perioperative medicine and Taiwan National Health Insurance Research Database. Acta Anaesthesiologica Taiwanica **54**(3), 93–96 (2016)
23. Decoste, D., et al.: Training invariant support vector machines. Mach. Learn. **46**(1–3), 161–190 (2002)
24. LeCun, Y., et al.: Comparison of learning algorithms for handwritten digit recognition. In: International Conference on Artificial Neural Networks, pp. 53–60. Perth, Australia (1995)
25. Lillywhite, K., et al.: Self-tuned evolution-constructed features for general object recognition. Pattern Recogn. **45**(1), 241–251 (2012)
26. Tao, P., et al.: An improved intrusion detection algorithm based on GA and SVM. IEEE Access **6**, 13624–13631 (2018)
27. Tao, Z., et al.: GA-SVM based feature selection and parameter optimization in hospitalization expense modeling. Appl. Soft Comput. **75**, 323–332 (2019)
28. Bradley, A.P.: The use of the area under the ROC curve in the evaluation of machine learning algorithms. Pattern Recogn. **30**(7), 1145–1159 (1997)
29. Cortes, C., et al.: AUC optimization vs. error rate minimization. In: Advances in Neural Information Processing Systems, pp. 313–320 (2004)
30. Lin, H.H., et al.: Increased risk of erectile dysfunction among patients with sleep disorders: A nationwide population-based cohort study. Int. J. Clin. Pract. **69**(8), 846–852 (2015)
31. Chen, Y.-F., et al.: Gout and a subsequent increased risk of erectile dysfunction in men aged 64 and under: a nationwide cohort study in Taiwan. J. Rheumatol. **42**(10), 1898–1905 (2015)
32. Thompson, I.M., et al.: Erectile dysfunction and subsequent cardiovascular disease. JAMA **294**(23), 2996–3002 (2005)
33. Speel, T., et al.: The risk of coronary heart disease in men with erectile dysfunction. Eur. Urol. **44**(3), 366–371 (2003)
34. Shen, B.-J., et al.: Anxiety characteristics independently and prospectively predict myocardial infarction in men: the unique contribution of anxiety among psychologic factors. J. Am. Coll. Cardiol. **51**(2), 113–119 (2008)
35. Seftel, A. D., et al.: The prevalence of hypertension, hyperlipidemia, diabetes mellitus and depression in men with erectile dysfunction. J. Urology **171**(6 Part 1), 2341–2345 (2004)
36. Andersen, Y. M., et al.: Risk of myocardial infarction, ischemic stroke, and cardiovascular death in patients with atopic dermatitis. J. Allergy Clin. Immunol. **138**(1), 310–312, e3 (2016)
37. Silverberg, J.I.: Association between adult atopic dermatitis, cardiovascular disease, and increased heart attacks in three population-based studies. Allergy **70**(10), 1300–1308 (2015)
38. Su, V.Y.-F., et al.: Atopic dermatitis and risk of ischemic stroke: A nationwide population-based study. Ann. Med. **46**(2), 84–89 (2014)

39. Paalasmaa, J., et al.: Adaptive heartbeat modeling for beat-to-beat heart rate measurement in ballistocardiograms. IEEE J. Biomed. Health Inform. **19**(6), 1945–1952 (2014)
40. Sadek, I.: Ballistocardiogram signal processing: A literature review. arXiv:1807.00951 (2018)
41. Alivar, A., et al.: Motion artifact detection and reduction in bed-based ballistocardiogram. IEEE Access **7**, 13693–13703 (2019)
42. Javaid, A.Q., et al.: Quantifying and reducing posture-dependent distortion in ballistocardio-gram measurements. IEEE J. Biomed. Health Inform. **19**(5), 1549–1556 (2015)
43. Kim, C.-S., et al.: Ballistocardiogram: Mechanism and potential for unobtrusive cardiovas-cular health monitoring. Sci. Rep. **6**, 31297 (2016)
44. Rabbani, M.S., et al.: Accurate remote vital sign monitoring with 10 GHz ultra-wide patch antenna array. AEU-Int. J. Electron. Commun. **77**, 36–42 (2017)
45. Cai, W., et al.: Low power SI class E power amplifier and Rf switch for health care. arXiv: 1701.01771 (2017)
46. Adib, F., et al.: Smart homes that monitor breathing and heart rate. In: Proceedings of the 33rd Annual ACM Conference on Human Factors in Computing Systems, pp. 837–846. (2015)
47. Staderini, E.M.: UWB radars in medicine. IEEE Aerosp. Electron. Syst. Mag. **17**(1), 13–18 (2002)

A Novel Deep Learning Approach for Liver MRI Classification and HCC Detection

Rim Messaoudi[1,2](✉) ⓘ, Faouzi Jaziri[2] ⓘ, Antoine Vacavant[2] ⓘ, Achraf Mtibaa[1,3] ⓘ, and Faïez Gargouri[1,4] ⓘ

[1] MIRACL Laboratory, University of Sfax, Sfax, Tunisia
rimmessaoudii@gmail.com
[2] Institut Pascal, Université Clermont Auvergne, UMR6602 CNRS/UCA/SIGMA, 63171 Aubière, France
[3] National School of Electronic and Telecommunications, University of Sfax, Sfax, Tunisia
[4] Higher Institute of Computer Science and Multimedia, University of Sfax, Sfax, Tunisia

Abstract. This work proposes a deep learning algorithm based on the Convolutional Neural Network (CNN) architecture to detect HepatoCellular Carcinoma (HCC) from liver DCE-MRI (Dynamic Contrast-Enhanced MRI) sequences. The Deep Learning technique is an artificial intelligence technique (AI) that tries to imitate the human brain work in the training data and creating models used for decision. Actually, it is widely used for various clinical issues. To diagnose HCC, radiologists consider three different phases during contrast injection (before injection; arterial phase; portal phase for instance). This paper presents an approach that offers a parallel preprocessing algorithm. It allows HCC detection and localization in MRI images via a CNN algorithm. The created CNN model reached an accuracy level of 90% in both arterial and portal phases using MRI patches of 64 × 64 pixels. We mention also its ability to decrease false detection comparing with our previous works. The obtained good accuracy is considered to be ameliorated in our future works.

Keywords: Medical image analysis · HCC · Deep learning · CNN classification · MRI

1 Introduction

Liver cancer begins by affecting healthy cells and forming a mass presenting a tumor. Liver cancer is considered the sixth most common cancer worldwide. According to the Adrienne Wilson Liver Cancer Association[1], this year there will be over 42,030 (29,480 men and 12,550 women) new cases of liver cancer and around 31,780 (21,600 men and 10,180 women) deaths. HCC (HepatoCellular Carcinoma) is considered the most dangerous liver cancer according to the World Health Organization (WHO)[2] Diagnosis

---

[1] https://www.bluefaery.org/statistics/.
[2] https://www.who.int/fr.

© Springer Nature Switzerland AG 2020
Y. Lu et al. (Eds.): ICPRAI 2020, LNCS 12068, pp. 635–645, 2020.
https://doi.org/10.1007/978-3-030-59830-3_55

of this cancer is used through different modalities such as MRI, CT scans and Ultrasound. The follow-up of HCC is defining the high-risk patient. Male gender, older age, alcohol, smoking, family history of HCC, and the cirrhosis are also among risks factors regardless of different regions [1, 2]. The American Association for the Study of Liver Diseases (AASLD), the Asian-Pacific Association for the Study of the Liver (APASL), and the European Association for the Study of the Liver (EASL) societies declared the importance of liver cancer surveillance with abdominal imaging such as MRI and CT scans each 6 months [3].

HCC detection is a very challenging task, and has been treated in several research works. In this context [4] proposed an automatic CNN with Multi-Magnification Input Images. The algorithm is based on two main steps: (a) the extraction of cell level and structure features maps from high and low-magnifications images respectively by separating general convolutional networks, and (b) the integration of multi-magnification features by the use of fully connected network. [5] proposed a method that classifies common hepatic lesions on multi-phasic MRI. This classification approach is realized through different steps: (a) crop and resize a set of images around each tumor, (b) split randomly lesions into train and test sets, (c) create 100 copies of each image presenting a lesion through various image processing techniques, (d) using the training set to associate image patterns with the adequate lesion class, (e) validation of the model by using the test set and (f) evaluating the model performance. In addition, [6] integrated deep learning into MRI image and proposed a framework for HCC lesions detection from DCE-MRI (Dynamic Contrast-Enhanced MRI) sequences. This approach has employed the U-Net architecture in the classification process. [7] proposed a parallel Framework for HCC detection in DCE-MRI sequences with wavelet-based description and SVM (Support Vector Machine)classification. This work realized parallel patch-based processing of DCE-MRI images characterizing three contrast-enhancement phases. In addition, authors used a wavelet-based descriptor, presenting the local signal evolution of slices during DCE-MRI acquisition process, extracted in a SVM classifier. [14] presented an approach that has generated synthetic medical images applying deep learning GANs (Generative Adversarial Networks). This approach is used to improve the performance and the effectiveness of CNN for medical image classification.

All the previous cited works have improved the importance of applying classification methods to classify medical images and data. Although, they do not deeply used CNN to classify liver MRI images especially for CNN and show its effectiveness to reach high measures of accuracy. In this paper, we propose a HCC detection framework that uses CNN architecture. In fact, we have developed a novel algorithm that proposes (1) a parallel preprocessing algorithm. It takes as input MRI images and gives fragments with equal sizes as output, and (2) creates a parallel CNN algorithm that allows HCC detection and localization in MRI images. The goal here is to increase the Dice (or F-measure, or F1) compared to our previous works [6, 8] by trying to keep an interesting accuracy rate. Dice coefficient is considered as an important measure in medical imaging. It is used to gauge the similarity degree of two samples and this is efficient in our case because input classes have not the same size. The rest of the paper is organized as follows: Sect. 2 presents the material employed in this work. Section 3 explains the proposed approach.

Section 4 exposes the obtained results and discusses and Sect. 5 concludes the paper and gives an overview of the future works.

2 Material and Method

2.1 Data-Set

In this work, we have used 9 patients who underwent hepatic MRI in our department as standard of care for their cirrhosis. They all suffer from HCC diagnosed in the MRI by a radiologist, according to EASL criteria [9]: known cirrhosis (according to clinical, biological and imaging data), focal lesion of at least 1 cm, hypervascular in the arterial phase with washout in the portal venous or delayed phases. MR exams were performed on a 1.5 T Optima (General Electric Healthcare, Milwaukee, WI) with a phased array coil. It consisted of axial T2 Single Shot Fast Spin Echo (SSFSE), axial T2 Fat Sat Propeller, axial diffusion weighted imaging and multiphase 3D Fast Spoiled Gradient Echo T1 LAVA before and while injection of 0.2 mL/kg of gadobenate dimeglumine (MultiHance®, Bracco).MR images were obtained in the axial plane with a section thickness of 4 mm, a 2 mm intersection gap, matrix of 512×512 and a field of view of 420×420–500×500 mm^3, which covered the whole liver. In this study, we have applied 2D (slice by slice) data, only T1 images were used to detect HCC.

2.2 Liver Segmentation

Liver segmentation from MRI images is important in the process of tumor detection. In this work, we have applied the segmentation model developed by [7, 13]. The method consists of developing an automatic model for both modalities MRI and CT scans. In the first step, authors extracted 4 different statistical models with 68 livers segmented by clinical experts obtained from Shape 2015 [10], IRCAD [11] and SLIVER07 [12] databases. The obtained 4 models were constructed according to their variabilities from a standard shape liver. Therefore, statistical model and all patient volumes have equivalent dimensions. Then, we localize the liver on the images with the mean dimensions of a standard liver. This localization of the liver allows us to compute a threshold to isolate pixels that belong to the liver. After the thresholding on each slice, we apply a contour enhancement process. At this step we use the model the liver as a probability map to localize the liver and we then perform an active contoursegmentation method (fast marching) resulting in a binary mask. Finally, to erase errors due to over-segmentation by a process considering the global shape of the liver. The steps of the segmentation method are presented in Fig. 1.

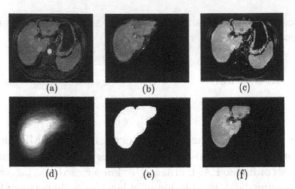

Fig. 1. Liver segmentation steps: (a) a patient slice, (b) the largest liver surface (threshold computation), (c) thresholding and contour enhancement, (d) liver model for localization and active contour method, (e) the obtained mask, and (f) the final segmentation result.

3   Deep Learning for HCC Detection by CNN

In this section, we present our deep learning approach. We will give an overview of the different steps as well as the generated contributions. Our approach is composed of five main steps; (1) Pre processing, (2) Training phase, (3) Prediction Phase, (4) Testing phase and (5) the validation phase. Figure 2 shows our Deep Learning workflow.

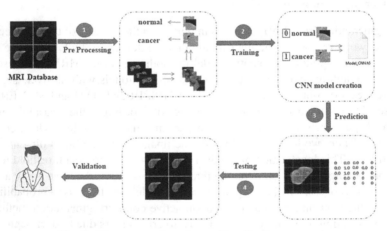

Fig. 2. Liver tumor detection and localization approach

3.1   Pre-processing Step

This step focuses on images preparation and manipulation. These images are useful later to prepare the databases. During the MRI process, the radiologist or the doctor takes three different MRI images for each patient, depending on the time after the contrast injection. This will take into account the evolution of information in these different image sequences. Figure 3 exposes an example of these images. (a) Phase 1: without injection, (b) Phase 2: short time of (30 s–1 min) after the contrast injection, and (c) Phase 3: long time (2–5 min) after the contrast injection.

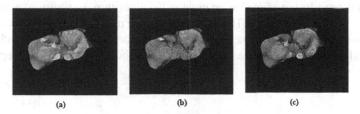

(a)   (b)   (c)

Fig. 3. MRI images in the three phases

Based on a well-defined fragment size; 64 × 64 pixels in our case defined as follows; each block size is 58 × 58 with a 3-pixel overlap in the 4 directions. We perform a division of MRI images of several patients. Figure 4 shows an example of one image division. In the case of the image size is smaller than the size of the fragment, we use resizing mechanisms. In our case, we have applied the padding technique. We add values from 0 to the end of the matrix until we wait for a new size matrix for the fragment. Moreover, our program eliminates each fragment that admits a total medium of intensity equal to 0: it is the case we have a fragment with only a black background. We have used later the different fragments resulting from this to create Deep Learning databases.

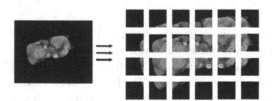

Fig. 4. MRI images fragments

3.2 Training Phase

The ability to learn in the Deep Learning algorithm has a very important feature, but this technique requires a large computing time and several training information. This algorithm follows a well-defined learning technique, it learns from the tagged information. Hence, we have input variables (x) and output variables (Y) and we use an algorithm to execute a mapping function between input and output. The goal is to approximate the mapping function so that when we have new input data (x), we can predict the output variables (Y) for that data. The proposed model is a consecutive of convolution blocks dedicated to extract and identify images features. These images are interconnected with a fully connected block that classifies fragments. In order to create the model and implement the fragment database, we have initialized the different relevant variables in our algorithm; the epoch number was fixed in the value 100. We have initialized 41,794 steps per epoch for the train and 6,000 steps for validation. We have used another variable "CallBacks" which interrupts the training if there is no change in the error value. This value shows the difference between the results obtained after the forward propagation phase and the results introduced as an input to the algorithm. In addition, we have fixed the Batch size on the value 10. This variable helps to minimize the time of calculation since it shows the number of images to use during a forward propagation cycle. Weight and bias values are initialized in a randomly. Figure 5 shows the proposed Deep Learning model.

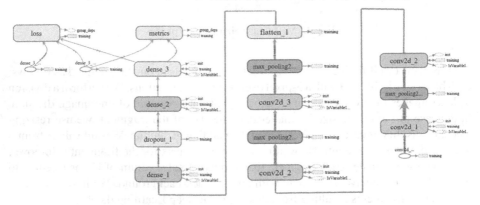

Fig. 5. The proposed Deep Learning model

We perform various convolutions on our model, where each task uses an alternative filter. After the convolution phase, we use an activation function to achieve a nonlinear output. It is the ReLU activation function. Finally, features go through a last block called Pooling which reduces the size of the map. The last step is the model registration. Indeed, the model is saved under the extension (.h5) because it characterizes structured format that supports a large mass of information.

3.3 Prediction Phase

After the training phase and the recording of the model, we have prepared new images for the prediction. Such an image is divided into non-overlapping 64 × 64 fragments and we predicted each fragment independently of the others. As shown in Fig. 7, the value 0 is attributed to a normal fragment and the value 1 indicates that the fragment potentially contains a tumor, since each fragment name has its abscissa and its ordered in the source image, we can build this matrix where each value represents the prediction of the corresponding fragment. Figure 6 presents an example illustrating the prediction phase description.

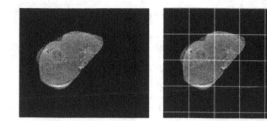 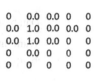

Fig. 6. The prediction phase description

3.4 Testing and Validation Phases

In these steps, we test the efficiency of our algorithm by using different images related to different real patient's cases. More details about the carried out results are mentioned in the next section.

4 Results

4.1 Applied Tools

To develop our approach, we have used Python[3] which offers a dynamic system and automatic memory management. It supports several programming paradigms, including object-oriented, imperative, functional programming and procedural styles. It promotes structured, functional and object-oriented imperative programming. We have used also various libraries and packages such as Tensor Flow[4] which is an open source platform for machine learning. It is a comprehensive, flexible ecosystem of libraries and community resources. Also, Keras[5] is applied in our Deep Learning model. It is a high-level neural networks API and enabling fast experimentation. Numpy[6] is the fundamental API for computing with Python language. It contains a powerful N-dimensional array object and various tools for integrating C/C ++ and Fortran code. OpenCV[7] is also applied in our approach. It is an open graphic library, developed by Intel and applied in real time images processing. To run the deep learning algorithm, we have used Google Colaboratory Notebook because it provides a totally free GPU and other free cloud services.

4.2 Database Description

The used database, as mentioned in Table 1, is composed of a set of fragments, divided as follows; for the training phase, we have used 26,162 normal patches and 15,632 cancerous patches. For the Testing phase, we have used 3,588 normal patches and 2,412 cancerous patches. The size of each patch is 64 × 64pixels.

---

[3] https://www.python.org/.
[4] https://www.tensorflow.org/.
[5] https://keras.io/.
[6] https://numpy.org/.
[7] https://opencv.org/.

Table 1. Deep Learning database description

Database (number of patches)	Normal	Cancer	Total
Training Data	26,162	15,632	41,794
Testing Data	3,588	2,412	6,000
Total	29,750	18,044	47,794

4.3 Performance Metrics

The most commonly used evaluation methods for a Deep Learning model are mentioned in Table 2. Hence, we have used a set of measures of performance such as Sensitivity (SEN), Accuracy (ACC), Specificity (SPC), and F1 score.

Table 2. Performance metrics

TP	True Positive
TN	True Negative
FP	False Positive
FN	False Negative
SEN	TP/TP + FN
ACC	TP/TP + FP
SPC	TN/(TN + FP)
F1 Score	(2*Recall * Precision)/(Recall + Precision)

4.4 Experimental Results

The proposed HCC liver cancer classification model was implemented with Python with system configurations such as an i7 processor with 8 GB RAM. The training of the Deep Learning algorithm requires a very important computing time, we have integrated the TenserFlow Framework to accelerate computing time. The execution of the algorithm with Google Collaboratory Notebook reached 98% (Fig. 5a) and the error rate was 0.03 (Fig. 5b). Tables 3 and 4 demonstrate respectively the performance levels of liver cancer MRI classification rates for the proposed approach. In this test, we have taken the 9 patients of our dataset and we tried to predict the existence of HCC tumors using our model. The proposed model provides better classification results.

Results obtained by our model were illustrated respectively in Tables 3 and 4 for both phases 2 (With Contrast) and 3 (After contrast).To evaluate our model, we have used the same datasets employed in works [6, 8].We have applied the SVM method in [6] and U-Net in [8] while we have applied CNN in this work. The performance of the proposed CNN model is determined by its ability to detect cancerous and normal

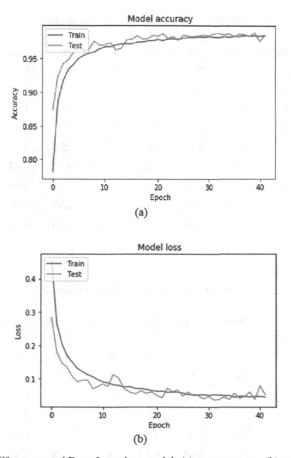

Fig. 7. The proposed Deep Learning model: (a) accuracy rate, (b) error rate

patches from MRI images. We have used the same evaluation criteria as our previous works. Hence, based on the obtained findings, the model is able to predict the medical status of the liver. The accuracy has clearly indicated that the proposed algorithm is deeply efficient in detecting HCC slices. For the phase 2, results were 80% (SEN), 71% (Acc), 92% (SPC) and 74% (F1 score). For phase 3, results were; 81% (SEN), 68% (Acc), 90% (SPC) and 73% (F1 score). The total accuracy level reached 90% in both phases. Regarding our previous studies [6] and [8], F1 score level has been increased and reached respectively 74% and 73% for phase 2 and phase 3 considering [6].This rate is the most suitable and adapted measure to evaluate our model because the used classes have not the same size. In fact, F1 score has been increased from 66.5% in [8] to 74% in our work. In addition, based on results obtained in [6], Sensitivity rate has been increased in phase 3 from 76% to 81%. Also, Specificity rates have been increased respectively in both phases 2 and 3 from 87% and 82% to 92% and 90% in our case.

Table 3. With-contrast phase evaluation

	Patient 1	Patient 2	Patient 3	Patient 4	Patient 5	Patient 6	Patient 7	Patient 8	Patient 9	Total
TP	6	3	6	30	116	5	98	27	163	454
TN	65	17	38	169	454	40	503	43	628	1957
FP	14	0	5	11	22	5	70	2	31	160
FN	4	0	6	0	10	0	65	5	15	105
SEN	0.5	1	0.5	1	0.92	1	0.6	0.84	0.91	0.80
ACC	0.5	1	0.5	0.73	0.84	0.5	0.58	0.93	0.84	0.71
SPC	0.87	1	0.88	0.93	0.95	0.88	0.87	0.95	0.95	0.92
F1 Score	0.5	1	0.5	0.84	0.87	0.66	0.59	0.88	0.87	0.74

Table 4. After-contrast phase evaluation

	Patient 1	Patient 2	Patient 3	Patient 4	Patient 5	Patient 6	Patient 7	Patient 8	Patient 9	Total
TP	6	3	6	17	111	5	109	24	159	440
TN	67	17	38	62	454	40	537	44	618	1877
FP	12	0	5	9	21	5	51	6	32	141
FN	4	0	6	1	14	0	48	3	27	103
SEN	0.6	1	0.5	0.94	0.88	1	0.69	0.88	0.85	0.81
ACC	0.33	1	0.54	0.65	0.84	0.5	0.68	0.8	0.83	0.68
SPC	0.84	1	0.88	0.87	0.95	0.88	0.91	0.88	0.95	0.90
F1 Score	0.42	1	0.52	0.77	0.86	0.66	0.68	0.84	0.84	0.73

5 Conclusion

This article discusses the topic of HCC detection through deep learning CNN architectures. It proposed a Deep Learning Approach for Liver MRI Classification and HCC Detection The proposed approach has offered a preprocessing script that provides the necessary information for the entry of the Deep Learning algorithm. Also, it creates a parallel Deep Learning Algorithm that targets the detection and localization of tumors in a medical liver image. Results have shown a better classification in case of liver MRI Images compared with others classification techniques, in term of F1 measure, but still with a high accuracy. Moreover, applying an automatic CNN classification approach achieved better precision and accuracy levels especially in the third phase which reached 90%. Moreover, according to the experimental findings, the proposed approach is effective for the classification of the human liver MRI images in terms of accuracy, sensitivity, and specificity. As future works, we aim at ameliorating the performance level of our algorithm by including other semantic techniques in the detection process.

Acknowledgement. This work was financially supported by the "PHC Utique" program of the French Ministry of Foreign Affairs and Ministry of higher education and research and the Tunisian Ministry of higher education and scientific research in the CMCU project number 18G1139 – Campus France Code 39319SM.

References

1. Bialecki, E.S., Di Bisceglie, A.M.: Diagnosis of hepatocellular carcinoma. HPB (Oxford) **7**(1), 26–34 (2005)
2. Ghouri, Y. A., Mian, I., Rowe, J. H.: Review of hepatocellular carcinoma: Epidemiology, etiology, and carcinogenesis. J. Carcinog. **16**(1) (2017)
3. Ayoub, W. S., Steggerda, J., Yang, J. D., Kuo, A., Sundaram, V., Lu, S. C.: Current status of hepatocellular carcinoma detection: screening strategies and novel biomarkers. Ther. Adv. Med. Oncol. **11** (2019)
4. Huang, W. C., et. al.: Automatic HCC detection using convolutional network with multi-magnification input images. In: IEEE International Conference on Artificial Intelligence Circuits and Systems (AICAS), pp. 194–198, IEEE, Hsinchu, Taiwan (2019)
5. Charlie, A. H., et al.: Deep learning for liver tumor diagnosis part I: Development of a convolutional neural network classifier for multi-phasic MRI. Eur. Radiol. **29**(7), 3338–3347 (2019)
6. Fabijańska, A., Vacavant, A., Lebre, M.-A., Pavan, A.L.M., de Pina, Diana R., Abergel, A., Chabrot, P., Magnin, B.: U-CatcHCC: an accurate HCC detector in hepatic DCE-MRI sequences based on an U-Net framework. In: Chmielewski, Leszek J., Kozera, R., Orłowski, A., Wojciechowski, K., Bruckstein, Alfred M., Petkov, N. (eds.) ICCVG 2018. LNCS, vol. 11114, pp. 319–328. Springer, Cham (2018). https://doi.org/10.1007/978-3-030-00692-1_28
7. Lebre, M.-A., et al.: Medical image processing and numerical simulation for digital hepatic parenchymal blood flow. In: Tsaftaris, Sotirios A., Gooya, A., Frangi, Alejandro F., Prince, Jerry L. (eds.) SASHIMI 2017. LNCS, vol. 10557, pp. 99–108. Springer, Cham (2017). https://doi.org/10.1007/978-3-319-68127-6_11
8. Pavan, A. L. M., et al.: A parallel framework for HCC detection in DCE-MRI sequences with wavelet-based description and SVM classification. In: Proceedings of the 33rd Annual ACM Symposium on Applied Computing, pp. 14–21, Pau, France (2018)
9. European Association for the Study of the Liver. 2012. EASL–EORTC Clinical Practice Guidelines: Management of hepatocellular carcinoma. J. Hepatol. **56**(4), 908–943 (2018)
10. Kistler, M., Bonaretti, S., Pfahrer, M., Niklaus, R., Büchler, P.: The virtual skeleton database: an open access repository for biomedical research and collaboration. J. Med. Internet Res. **15**(11), e245 (2013)
11. Research Institute against Digestive Cancer. IRCAD dataset, http://www.ircad.fr/research/3d-ircadb-01/
12. Heimann, T., et al.: Comparison and evaluation of methods for liver segmentation from CT datasets. IEEE Trans. Med. Imaging **28**(8), 1251–1265 (2009)
13. Lebre, M.-A., et al.: A robust multi-variability model based liver segmentation algorithm for CT-scan and MRI modalities. Comput. Med. Imaging Graph. **76**, 11 pages, (2019)
14. Frid-Adar, M., Diamant, I., Klang, E., Amitai, M., Goldberger, J., Greenspan, H.: GAN-based synthetic medical image augmentation for increased CNN performance in liver lesion classification. Neurocomputing **321**, 321–331 (2018)

An Integrated Deep Architecture for Lesion Detection in Breast MRI

Ghazal Rouhafzay[1,2] (iD), Yonggang Li[4] (iD), Haitao Guan[5] (iD), Chang Shu[1] (iD),
Rafik Goubran[3] (iD), and Pengcheng Xi[1,3(✉)] (iD)

[1] National Research Council Canada, Ottawa, ON K1A 0R6, Canada
pengcheng.xi@nrc-cnrc.gc.ca
[2] University of Ottawa, Ottawa, ON K1N 6N5, Canada
[3] Carleton University, Ottawa, ON K1S 5B6, Canada
[4] First Affiliated Hospital of Soochow University, Jiangsu 215006, China
liyonggang224@163.com
[5] Nantong No. 3 People's Hospital, Jiangsu 226000, China

Abstract. Complex nature of medical images and tedious process of data exploration calls for the development of Computer Aided Detection (CADe) methods to ease the process of lesion detection. Recent deep learning-based object detectors from computer vision are adapted to the creation of CADe lesion detectors. This research starts with state-of-the-art object detectors, namely Faster R-CNN, YOLO v2 and Grad-CAM, to determine the location of lesions in Magnetic Resonance Images of breast. A series of experiments are conducted to find the best set up for maximizing the Average Precision (AP) of each method. Consequently, AP values of 0.6993 and 0.7651 are obtained for Faster R-CNN and YOLO v2 respectively. Taking into consideration the pros and cons of each method, we propose different integration architectures in order to overcome the shortcomings of each algorithm, hence enhancing the overall lesion detection performance. The integrated architectures succeed to obtain an AP value up to 0.8097 while providing explainable reasoning that is essential for medical CADe.

Keywords: Magnetic Resonance Imaging · Breast lesion detection · Deep learning · Deep CNN · Medical CADe

1 Introduction

As the most common cancer in women worldwide, breast cancer is receiving a huge research attention in terms of revealing influential factors and searching for diagnosis and treatment. Early diagnosis of breast abnormalities is proven to play a pivotal role in raising survival rate. Despite the fact that mammography is the most effective tool for breast screening, Magnetic Resonance Imaging (MRI) of breast is recommended to assess the extent of breast cancer due to its capability of monitoring blood flow into tumors. Furthermore, breast MRI can be effective in cases where breast tissue is dense and mammogram or Ultrasound (US) imaging fails to detect lesions. However, the false positive rate of tumor diagnosis by radiologists using breast MRI is reported

© Crown 2020
Y. Lu et al. (Eds.): ICPRAI 2020, LNCS 12068, pp. 646–659, 2020.
https://doi.org/10.1007/978-3-030-59830-3_56

to be relatively high compared to other imaging modalities because MRI interpretation calls for a high level of expertise. A relevant study conveys the specificity of human judgment on this imaging modality to be between 75% and 87% [1].

Computed Aided Detection (CADe) is an open field of research for assisting radiologists on tackling with confusing cases. In particular, the advancement in deep learning-based architectures have made great strides in biomedical image processing. Once deep neural networks are properly trained, they are capable of capturing nonlinear and nonstationary nature of training data. As a result, they can extract key features and subtle structures of medical images, which are hard to spot by human eyes.

This work proposes a deep learning-based architecture through integrating state-of-the-art deep detectors in order to achieve the best possible performance. It also represents a framework leveraging global and local views for solving the problem. As part of the architecture, our approach provides explainable reasoning for lesion detection in breast MRI. Specifically, we focus on shortcomings of each method through analysing their performance to come up with an optimal integrated decision-making strategy.

The paper is organized as follows. Section 2 discusses relevant works from the literature. Further information about each method is provided in Sect. 3. Results are reported and discussed in Sect. 4. Section 5 concludes the work.

2 Literature Review

Complex nature of medical data calls for a powerful tool capable of extracting and learning biological patterns. Accordingly, it has received great interest in using deep Convolutional Neural Networks (CNN) for processing medical images. CNNs have been successfully applied to radiotherapy [2], attenuation correction for MRI and Positron Emission Tomography (PET) [3], tissues segmentation [4] and lesion detection [5, 6]. Moreover, the high rate of breast cancer in women worldwide has attracted an enormous research attention. As a result, machine learning solutions are applied to different breast imaging modalities, including mammograms [5], ultrasound [6], microwave breast Imaging [7] and MRI [8–10] to provide better treatment planning for this disease.

Among all the breast screening techniques, MRI provides decisive information about the nature of breast tissue and tumor characterizations; however, this imaging modality is prone to human error since it requires high level of expertise in interpretation. Ha et al. [8] designs a CNN network with ten convolutional layers to classify breast tumors through MRI to see if the tumor will respond completely, partially or does not respond to chemotherapy for treatment. They succeed in predicting up to 88% about the outcome of the treatment. In another research [9], the authors design a deeper network with residual feedbacks to determine the subtype of tumor, which is a crucial step in treatment planning. Herent et al. [10] take advantage of a Resnet50 [11] network to extract features from breast MRI. Extracted features are fed into an attention block for local prediction as well as into a logistic regression module for possible lesion types. Authors in [12] compare the performance of a VGG [13] network pre-trained on images from another domain and fine-tuned on breast MRIs and conclude the superiority of the latter. In a recent research, Whitney et al. [14] conduct a study on the success of transfer learning on different imaging modalities of breast.

Besides the classification of lesions in breast tissue, determining their exact location in a full image can be performed using deep CNNs. Al-Mansi et al. [15] apply a YOLO [16] object detector to breast mammograms to determine mass locations in a full mammogram. Detected masses are then fed into a binary classifier to determine if the mass is malignant or benign. Chiao et al. [17] leverage a mask R-CNN [18] for tumor detection in breast sonograms. Faster R-CNN [19] is applied to digital breast tomo-synthesis to detect mass [20].

In this work we apply three different object detectors, namely YOLO v2 [21], faster R-CNN [19] and Grad-CAM [22], to breast MRI for lesion detection. In each detector, three different CNN architectures are experimented with and results are compared. Hence six different decision-making frameworks are proposed and tested to boost the overall reliability of lesion detections. The YOLO v2 and faster R-CNN conduct local detections while the Grad-CAM provides a global view for the detections. Moreover, Grad-CAM provides explainable reasoning as part of the solution.

3   Deep Learning for Abnormality Detection

A variety of applications have benefitted from the development of large annotated image databases together with recent successes in transfer learning. It uses pre-trained deep CNNs by integrating different computational modules. Object detection is one of such applications targeting both classifying and locating existing objects in a scene. This framework can be leveraged to generate a network exploring in a medical image to locate and classify abnormalities. This section discusses three state-of-the-art methods for object detection.

3.1   Faster R-CNN

Faster R-CNN is the second successor of Region-based CNN [23], where the Selective Search algorithm [24] is replaced by a Region Proposal Network (RPN), guiding the process of object detection and hence accelerating the performance. A CNN is firstly trained on full images whose last shared convolutional layer outputs a feature map. The RPN applies a sliding window of size 3×3 to the feature map. The sliding window is convolved with the feature map to produce inputs for a classification as well as a regression network. It also accompanies 9 anchors as it passes the feature map and repeatedly measures the Intersection over Union (IoU) between each anchor and the ground-truth bounding box. The regression network is to learn the offset between the anchor and the ground-truth. If the IoU value is larger than 0.7, the anchor is positively weighted; if the IoU is less than 0.3, the anchor is negatively weighted in the loss function. Any anchor with IoU value in between is assigned a zero weight. Figure 1 summarizes the overall framework of applying the faster R-CNN to abnormality detection in breast MRI.

The training of a faster R-CNN model includes alternative steps, including training an RPN and reusing the convolution part for training the detector, after which the RPN is fine-tuned. The last step is to re-train the detector.

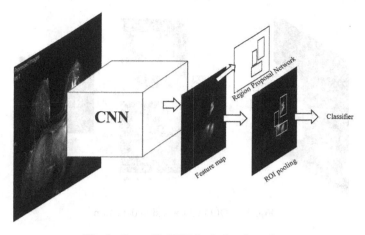

Fig. 1. Faster R-CNN for lesion detection

3.2 YOLO v2

YOLO is another deep learning-based object detector recognized for its real time applicability. It first divides a given image into an $S \times S$ grid where each grid is responsible for detecting the object whose center falls into that grid cell. The image passes through a deep CNN which is supposed to learn a tensor of size $(S, S, B \times 5 + C)$, where B is the number of bounding boxes and C is the number of possible classes. The number 5 corresponds to the five parameters predicted for each cell of the grid, i.e. (x, y, w, h, P_c), where x and y are the coordinates of the upper left corner of the bounding box, and w and h are its width and height. P_c represents the confidence score of the bounding box. A threshold value is applied to remove boxes with low confidence values. The overall class confidence value is computed as *Confidence Score* $\times$ *conditional class probability*. In YOLO v2, a series of refinements are applied to the model to increase the precision while reducing the running time. This includes adding batch normalization to convolutional layers, conducting multi-scale training, moving the class prediction from the cell level to bounding box level and picking better priors for anchors through a k-mean clustering algorithm. Instead of learning the offset between the anchors and ground truth, the model learns the offsets with respect to the grid cells.

Figure 2 illustrates the overall framework of YOLO v2. Training a YOLO v2 model requires training one single model and testing the model on a new image requires only one forward-pass computation.

3.3 Grad-CAM

Using a Grad-CAM model for object detection starts from building a classifier. In this work, we start using two sets of MRI images, one set labelled as normal and the other labelled as abnormal. These images are used to fine tune existing deep CNN models in order to build a binary classifier (see Fig. 3). The output layer is then removed from the binary classifier in preparation for building a Grad-CAM model.

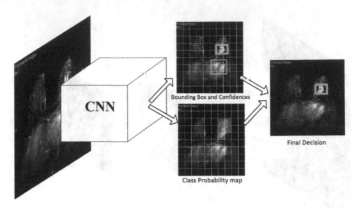

Fig. 2. YOLO v2 for lesion detection

The idea of Grad-CAM is to compute the gradient of the final classification score with respect to the final convolutional layer in the network. Given a test image, if it is classified as abnormal by the binary classifier built above, the classification score will be used to compute gradients in regard to the last convolutional layer in the network. The gradients along with the convolutional feature maps are then linearly combined for computing a color map, which highlights the class-specific regions in the input image. The last step in this approach is to convert the abnormality map into a binary image and compute bounding boxes. They are then used to compare with ground-truth bounding boxes for verifications.

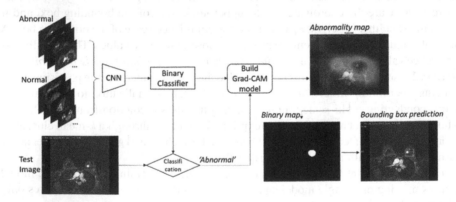

Fig. 3. Grad-CAM for lesion detection

Figure 3 introduces the abnormality detection approach using Grad-CAM in details: first, two sets of training images are fed to a deep CNN for training a binary classifier, which is then configured for building a Grad-CAM model. Given a test image, if it is classified by the binary classifier as "abnormal", it will be fed to the Grad-CAM model for computing a color-coded map highlighting the abnormalities. The abnormality map is then converted to a binary map using a threshold value (=0.9), which is then used to compute a bounding box as the final detection.

3.4 Integrated Detection Approaches

In this section, we propose six different integration approaches through combining results from the implemented lesion detectors to boost the overall performance of lesion detection in terms of decreasing the missing rate and enhancing the average precision. The overall integration method is formulated as follows to minimize the number of missed detections. It capitalizes on results from the three detectors while improving the bounding box boundaries by combining the results from YOLO v2 and Faster R-CNN detectors. To achieve this, for each test sample the detected bounding boxes by Faster R-CNN and YOLO v2 are computed and the overlapping ratio between all detected bounding boxes is computed. If the overlapping ratio is more than 50%, the two detections are considered to refer to the same abnormality. In this case, a Linear Regression model is used to determine the final bounding box of such abnormality. If the overlapping ratio is less than 50%, the integrated solution takes in both detections as a positive case. Subsequently, the detected bounding boxes are compared with the boxes achieved by Grad-Cam method and the overlapping ratios are computed. Again, if the overlapping ratio between the detection by Grad-CAM and YOLO v2/Faster R-CNN is less than 50%, the Grad-Cam detection is added as a new detection case. The overlap ratio is obtained as follows:

$$overlap\ ratio = \frac{Faster\ R - CNN\ bounding\ box \cap YOLO\ v2\ bounding\ box}{\min(Faster\ R - CNN\ bounding\ box,\ YOLO\ v2\ bounding\ box)}$$

Here is the pseudocode used for the proposed integration approach:

```
for all detected bounding boxes by YOLO v2 and Faster R-CNN
    compute the overlap ratio
    if overlap ratio ≥ 0.5
        Train linear regression models to learn a mapping of the
        four parameters of the Faster R-CNN and YOLO v2 bounding
        boxes into the ground truth.
    else
        add all detected bounding boxes

    for all bounding boxes in previous step and in Grad-CAM
        compute the overlap ratio
        if overlap ratio < 0.5
            add the detection by Grad-CAM to the results
```

In order to train the linear regression models, we take advantage of Matlab Regression Learner App and train the regression models on bounding boxes both detected by Faster R-CNN and YOLO v2 object detectors for which there is a correspondence in ground truth (an overlapping ratio larger than 0.5). As such four linear regression models are trained to map the four representing parameters of the bounding boxes (i.e. the pixel coordinates of the upper left corner as well as the width and the length) detected by Faster R-CNN and YOLO v2 into a final solution. They are tested on the 20% initial split for testing. Results of the overall integrated solution is presented in Sect. 4 as "All Methods". Five other combination strategies are implemented as follows:

- **Faster R-CNN ∪ YOLO v2:** This method excludes the Grad-CAM detection from the overall integration method.

- **Faster R-CNN ∩ YOLO v2:** This method only includes linear regression output of matched detections between Faster R-CNN and YOLO v2.
- **Faster R-CNN priority:** This method includes linear regression output of matched detections between Faster R CNN and YOLO v2 as well as detections by Faster R-CNN which are not detected by YOLO v2.
- **YOLO v2 priority:** This method includes linear regression output of matched detections between Faster R-CNN and YOLO v2 as well as detections by YOLO v2 which are not detected by Faster R-CNN.
- **Faster R-CNN ∩ YOLO v2 + Grad-CAM:** This method includes linear regression output of matched detections between Faster R-CNN and YOLO v2 as well as detections achieved by Grad-CAM.

4   Results and Discussion

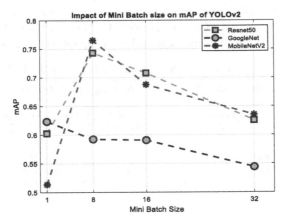

Fig. 4. Influence of the minimum batch size on AP for YOLO v2 object detector.

We have leveraged the previously explained deep object detection frameworks to determine abnormality locations in a dataset which is composed of 1,988 abnormal breast MR images. The dataset is acquired by multiple MRI scanners and from different viewpoints. A data cleaning procedure is then applied to crop images with large margins. In order to train the binary classifiers in Grad-CAM, 1,173 normal MR images are added to the dataset. Equal number of abnormal cases are randomly selected to produce a balanced dataset for training. All networks are trained on a machine equipped with NVIDIA RTX 2080 graphics card with a split of 80/20 for training and testing. The available GPU memory supports training YOLO models with a batch size up to 32, allowing us to monitor the influence of batch size on the performance. Figure 4 illustrates the results of this comparison. However, in the case of Faster R-CNN, the available GPU memory can only afford training on a single image in a batch. Learning rate is also tuned for all networks to achieve the best performance.

Table 1 compares the obtained Average Precision values for the three methods with different CNNs. For YOLO v2, the best results are achieved while optimizing batch sizes as mentioned in these tables. The threshold value here is the ratio of IoU between detected bounding box and ground truth, in order to be considered as a successful detection. One can notice that for a threshold of 0.5, MobileNetV2 achieves the best average precision compared to other architectures in YOLO v2 while in Faster R-CNN, Resnet50 outperforms other architecture. For all CNNs, YOLO v2 achieves a better performance compared to Faster R-CNN. Grad-CAM, as expected, results in a lower average precision since it does not directly rely on ground truth bounding boxes during training. Instead, this method looks at the image in a global scale. Nonetheless, when the IoU threshold is reduced, Grad-CAM has certain level of detection, confirming that the color mapping does point toward the location of abnormalities.

Table 1. Average Precisions for different methods for Th = 0.1 and Th = 0.5

Method	Resnet50		GoogLeNet		MobileNetV2	
	Th = 0.1	Th = 0.5	Th = 0.1	Th = 0.5	Th = 0.1	Th = 0.5
YOLO v2	0.8056	0.7431	0.6712	0.6232	**0.8228**	**0.7651**
Faster R-CNN	**0.8131**	**0.6993**	0.7663	0.6172	0.7859	0.6521
Grad-CAM	**0.2383**	**0.003**	0.0684	0.0000907	0.0407	0.000057

Table 2 reports the Average Precision values achieved with different integration approaches. The Faster R-CNN ∩ YOLO v2+ Grad-CAM described in Sect. 3.4 achieves the highest AP value.

Table 2. Average Precisions for different integration strategies

	All methods	Faster R-CNN ∪ YOLO v2	Faster R-CNN priority	YOLO v2 priority	Faster R-CNN ∩ YOLO v2	Faster R-CNN ∩ YOLO v2 + Grad-CAM
AP	0.7259	0.6717	0.7387	0.7987	0.7586	**0.8097**

Figure 5 depict the Precision-Recall curves for all implemented methods. The graph confirms that YOLO v2 trained on Resnet50 and MobileNetV2 are the two leading models. However, the highest possible recall value, i.e. highest True Positive detection with lowest False Negative, is achieved by the integrated solution.

Figure 6 plots the Precision-Recall curves for the proposed integration strategies. One can notice that Faster R-CNN ∩ YOLO v2, i.e., the approach relying on the output of linear regression model for matching detection results by Faster R-CNN and YOLO v2 has the highest accuracy for lower recall values; however, it doesn't succeed to get higher recall values. In other words, there remain False Negative cases. This is confirmed by Fig. 8 where the minimum missing rate for this method is higher than other methods.

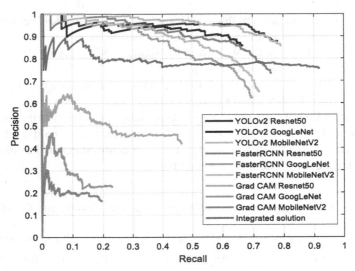

Fig. 5. Comparison of different techniques in terms of Precision-Recall

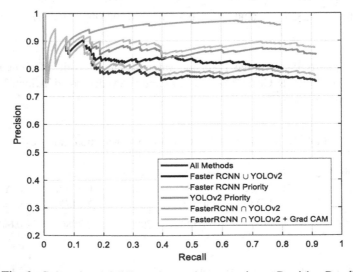

Fig. 6. Comparison of different integration strategies on Precision-Recall

When training an architecture for lesion detections, a crucial goal is to reduce the number of missed lesions. Therefore, in this section we plot the Miss Rate against the False Positive per Image (Fig. 7) to find out which method can achieve the minimum Miss Rate. Evidently, allowing more false positive detections can reduce the number of missed detections. Again, YOLO v2 with MobilenetV2 and Resnet50 succeed to achieve the minimum miss-rate among the implemented solutions. However, the integrated approach considerably outperforms other models in term of minimizing the missing rate.

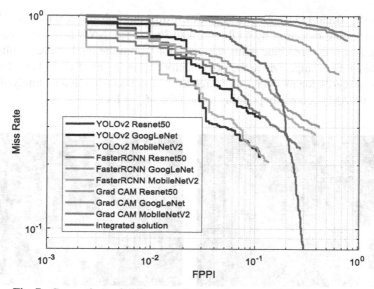

Fig. 7. Comparison of different techniques in terms of Miss Rate – FPPI

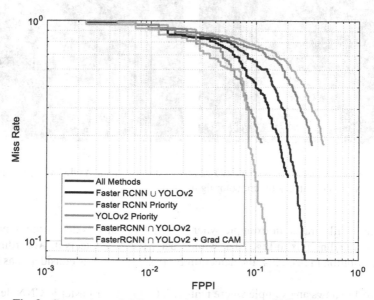

Fig. 8. Comparison of different techniques in terms of Miss Rate – FPPI

Figure 8 plots the Miss Rate - FPPI for different integration approaches. The overall integration approach (All Methods) capitalizing on all possible detections succeeds to achieve the minimum number of False Negatives at the expense of higher False positives.

Figure 9, 10 and 11 visualize the detection results through different methods. Figure 9 a and b illustrate detection by YOLO v2 and Faster R-CNN where the ground truth bounding box is colored blue while the detections from the models are highlighted in green. This is an example where the two detectors agree in detection and thus the output of the linear regression model is considered as the final decision. Interestingly, the result from Grad-CAM is correctly pointing toward the lesion location.

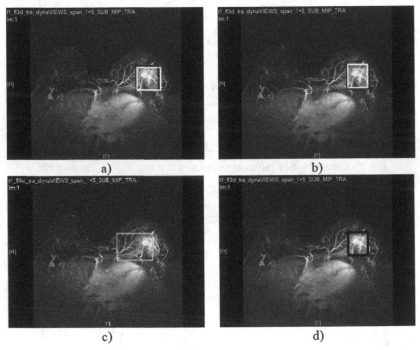

Fig. 9. Example of success of both YOLO v2 and Faster R-CNN a) YOLO v2 b) Faster R-CNN c) Grad-CAM d) integrated solution.(color figure online)

Figure 10 illustrates an example where YOLO v2 fails to detect the abnormality (no green bounding box in Fig. 10 a) while Faster R-CNN detects it. According to the proposed solution in Sect. 3.4, the detection by Faster R-CNN is considered as the final decision.

Figure 11 shows an example where neither YOLO v2 nor Faster R-CNN detects the lesion while Grad-CAM points toward the location of abnormality and thus is chosen as the output of the integrated solution.

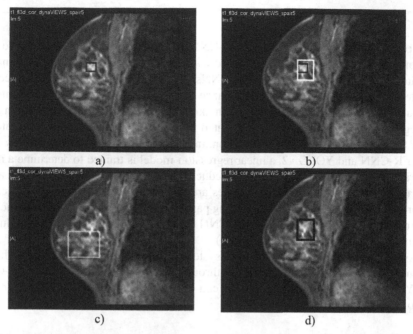

Fig. 10. An Example of Failure of YOLO v2: a) YOLO v2 b) Faster R-CNN c) Grad-CAM d) integrated solution.(color figure online)

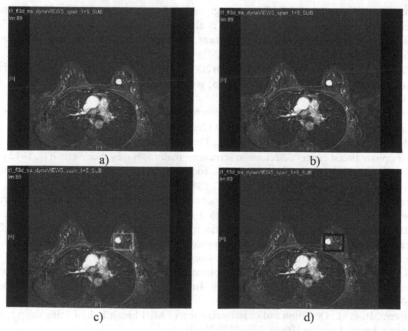

Fig. 11. An example of failure of both YOLO v2 and Faster R-CNN: a) YOLO v2 b) Faster R-CNN c) Grad-CAM d) integrated solution.(color figure online)

5 Conclusion

This research adopts YOLO v2, Faster R-CNN and Grad-CAM and trains them for the task of lesion detection in breast MRI. Grad-CAM shows success in terms of explainability, clarifying the decision taken by deep CNNs. Once the best set up for each architecture is determined to achieve the best performance on the dataset, a series of experiments are conducted to develop integrated decision-making approaches. Six different combination strategies are proposed, tested and evaluated in terms of precision and missed rate. In order to increase the precision of detection, in cases where a lesion is detected by both Faster R-CNN and YOLO v2, a linear regression model is trained to determine a more accurate bounding box for the lesion. To reduce the missed rate, all possible lesion areas detected by any of the studied approaches are merged to the overall solution. Results confirm that an integrated decision-making paradigm can both outperform each method in terms of average precision (Faster R-CNN ∩YOLO v2 + Grad-CAM) and reducing the number of missed cases (All methods).

As such, a priority-aware diagnosis system can be developed to detect high-risk lesions with more precise bounding boxes through matching detections by Faster R-CNN and YOLO v2. Other suspicious lesions can be determined by the overall integration approach for further test and analysis.

References

1. Lehman, C.D., et al.: BREAST imaging: national performance benchmarks for modern screening digital mammography Lehman et al. Radiology **283**(1), 49–58 (2017)
2. Meyer, P., Noblet, V., Mazzara, C., Lallement, A.: Survey on deep learning for radiotherapy. Comput. Biol. Med. **98**(May), 126–146 (2018)
3. Mehranian, A., Arabi, H., Zaidi, H.: Vision 20/20: magnetic resonance imaging-guided attenuation correction in PET/MRI: challenges, solutions, and opportunities. Med. Phys. **43**(3), 1130–1155 (2016)
4. Zhou, T., Ruan, S., Canu, S.: A review: deep learning for medical image segmentation using multi-modality fusion. Array **3–4**, 100004 (2019)
5. Shen, L., Margolies, L.R., Rothstein, J.H., Fluder, E., McBride, R., Sieh, W.: Deep learning to improve breast cancer detection on screening mammography. Sci. Rep. **9**(1), 1–12 (2019)
6. Han, S., et al.: A deep learning framework for supporting the classification of breast lesions in ultrasound images. Phys. Med. Biol. **62**(19), 7714–7728 (2017)
7. Rana, S.P., et al.: Machine learning approaches for automated lesion detection in microwave breast imaging clinical data. Sci. Rep. **9**(1), 1–12 (2019)
8. Ha, R., et al.: Prior to initiation of chemotherapy, can we predict breast tumor response? deep learning convolutional neural networks approach using a breast mri tumor dataset. J. Digit. Imaging **32**(5), 693–701 (2018). https://doi.org/10.1007/s10278-018-0144-1
9. Ha, R., et al.: Predicting breast cancer molecular subtype with MRI dataset utilizing convolutional neural network algorithm. J. Digit. Imaging **32**(2), 276–282 (2019). https://doi.org/10.1007/s10278-019-00179-2
10. Herent, P., et al.: Detection and characterization of MRI breast lesions using deep learning. Diagn. Interv. Imaging **100**(4), 219–225 (2019)
11. He, K., Zhang, X., Ren, S., Sun, J.: Deep residual learning for image recognition. In: Proceedings of IEEE Computer Society Conference on Computer Vision and Pattern Recognition, vol. 2016, pp. 770–778 (2016)

12. Amit, G., Ben-Ari, R., Hadad, O., Monovich, E., Granot, N., Hashoul, S.: Classification of breast MRI lesions using small-size training sets: comparison of deep learning approaches. In: Proceedings, Medical Imaging 2017: Computer-Aided Diagnosis, vol. 101234, p. 101341H (2017)

13. Simonyan, K., Zisserman, A.: Very deep convolutional networks for large-scale image recognition. In: 3rd International Conference on Learning Representations, ICLR 2015, Conference Track Proceedings, pp. 1–14 (2015)

14. Whitney, H., Li, H., Yu, J., Liu, P., Giger, M.L.: Comparison of breast MRI tumor classification using radiomics, transfer learning from deep convolutional neural networks, and fusion methods, pp. 1–15 (2019)

15. Al-masni, M.A., et al.: Simultaneous detection and classification of breast masses in digital mammograms via a deep learning YOLO-based CAD system. Comput. Methods Programs Biomed. **157**, 85–94 (2018)

16. Redmon, J., Divvala, S., Girshick, R., Farhadi, A.: You only look once: unified, real-time object detection. In: Proceedings of the IEEE Computer Society Conference on Computer Vision and Pattern Recognition, December 2016, pp. 779–788 (2016)

17. Chiao, J.Y., Chen, K.Y., Liao, K.Y.K., Hsieh, P.H., Zhang, G., Huang, T.C.: Detection and classification the breast tumors using mask R-CNN on sonograms. Med. (Baltimore) **98**(19), e15200 (2019)

18. He, K., Gkioxari, G., Dollar, P., Girshick, R.: Mask R-CNN. In: Proceedings of the IEEE International Conference on Computer Vision, October 2017, pp. 2980–2988 (2017)

19. Ren, S., He, K., Girshick, R., Sun, J.: Faster R-CNN: towards real-time object detection with region proposal networks. IEEE Trans. Pattern Anal. Mach. Intell. **39**(6), 1137–1149 (2017)

20. Fan, M., Li, Y., Zheng, S., Peng, W., Tang, W., Li, L.: Computer-aided detection of mass in digital breast tomosynthesis using a faster region-based convolutional neural network. Methods **166**, 103–111 (2019)

21. Redmon, J., Farhadi, A.: YOLO9000: better, faster, stronger. In: Proceedings of 30th IEEE Conference on Computer Vision and Pattern Recognition, CVPR 2017, January 2017, pp. 6517–6525 (2017)

22. Selvaraju, R.R., Cogswell, M., Das, A., Vedantam, R., Parikh, D., Batra, D.: Grad-CAM: visual explanations from deep networks via gradient-based localization. Int. J. Comput. Vis. **128**(2), 336–359 (2019). https://doi.org/10.1007/s11263-019-01228-7

23. Girshick, R., Donahue, J., Darrell, T., Malik, J.: Rich feature hierarchies for accurate object detection and semantic segmentation. In: Proceedings of the IEEE Computer Society Conference on Computer Vision and Pattern Recognition, pp. 580–587 (2014)

24. Uijlings, J.R.R., Van De Sande, K.E.A., Gevers, T., Smeulders, A.W.M.: Selective search for object recognition. Int. J. Comput. Vis. **104**(2), 154–171 (2013)

Single Image Super-Resolution for Medical Image Applications

Tamarafinide V. Dittimi$^{(\boxtimes)}$ and Ching Y. Suen

CENPARMI, Concordia University, Montreal, QC, Canada
{t_dittim,suen}@encs.concordia.ca

Abstract. In medical imaging, high-resolution images are expected to have the ability to deliver a more precise diagnosis with the practical application of high-resolution displays. This research proposes a deep learning method for single image super-resolution that learns an end-to-end mapping between the low and high-resolution images. It redesigns the SRGAN, using VGG19 network for feature extraction, setting discriminator network's working space as feature space, and adding the loss function based on the mean square error of pixel space, gaining more details by incorporating SRCNN layers to increase the PSNR in the reconstruction at the same time. To thoroughly investigate the system, we compared the performance with other architectures on MNIST and CIFAR-10 dataset with a further evaluation conducted on Chest x-ray.

Keywords: Chest radiography · Deep learning · Generative adversarial network · Super resolution

1 Introduction

Recently, CNN's have widely been implemented for image Super Resolution (SR), [12] developed a trained a three-layer deep fully convolutional network that used a bicubic interpolation to upscale an input image end-to-end to achieve state-of-the-art SR performance. It presented a super-resolution convolutional neural network. It is a novel deep learning approach for the super-resolution of a single image by restructuring a deep convolutional neural network from the traditional sparse-coding-based SR techniques [4]. The method learns an end-to-end mapping between the low- and high-resolution images and jointly optimizes all layers in the network, unlike conventional systems that analyze each component separately. The technique is simple, robust, and could be applied to other low-level vision problems achieving state-of-the-art restoration quality, and speed for practical on-line usage. It proves that enabling the network to learn the upscaling filters directly can further increase performance both in accuracy and speed [7].

The first experiment was conducted using MNIST and CIFAR-10 Dataset. The MNIST database includes handwritten digits that have been normalized and centered in a fixed-size format; the training had 60,000 samples, and the testing set included 10,000 samples [3], while the CIFAR-10 dataset comprised 60,000 images of 32×32 pixels

© Springer Nature Switzerland AG 2020
Y. Lu et al. (Eds.): ICPRAI 2020, LNCS 12068, pp. 660–666, 2020.
https://doi.org/10.1007/978-3-030-59830-3_57

with ten classes using 50,000 training and 10,000 for the testing set [11]. A further experimental evaluation was conducted using medical images from the Chest x-ray dataset. The Chest x-ray database was gotten from patients' routine clinical checks of pediatric patients between one to five years old in Guangzhou Women and Children's Medical Center located in Guangzhou; China made up of 5,863 images grouped into two patient categories [9].

The model was coded in Python 3.6.7, Conda 4.5.11, Tensorflow 1.12.0, and Keras 2.1.6. The experiment was run on Intel® Core™ i7-3770 CPU @ 3.60 GHz × 8, 64 GB RAM, 64-bit, and Nvidia® TITAN XP 12 GB. For the evaluation, we divided the data into 70% for training and 30% for testing. Each model was trained for 30,000 epochs, and at every 500 epochs, the validation set was evaluated on 20% of random samples removed from the training set. The test parameters used was an Adam optimizer, beta1 (momentum of Adam optimizer) = 0.5, learning rate = 0.0003. Mini-batch sizes range from one, 32, to 128 (i.e., each epoch contains only one batch).

The rest of the paper is organized as follows: Sect. 2 discusses related literature. While in Sect. 3 the proposed system including its architecture, design, dataset, experimental setup and stages of the approach is studied. Explanations of the results and discussion of the proposed system are presented in Sect. 4. Finally, Sect. 5 concludes and provides insight into future work that can be conducted.

2   Literature Review

Several researchers have addressed the image super-resolution problem; prior algorithms utilized filter techniques like bilinear, bicubic, and Lanczos filtering. However, these approaches were only able to generate smooth output without recovering any high-frequency information. [1] suggested edge features that generate a high-resolution edge map using a rectangular center-on-surround-off filter before performing piecewise linear interpolation of the zero crossings in the output of the filter. [13] exploited the interpolation constraint using the geometric duality between the high-resolution covariance and its corresponding low-resolution covariance. The accuracy was limited due to the oversimplification of the SR problem, but the approach was computationally efficient. [18] developed a technique to enhance visual effects by using a multi-scale dictionary to capture and upscale samples of similar image patches at different scales. [16] precomputed the corresponding embedding matrix and anchored the neighborhood embedding of a low-resolution patch to the nearest atom in the dictionary while enhancing the approach by using the training data in the testing procedure.

[5] introduced a technique that took into consideration self-similarity and employed example-based SR and classic multi-image SR utilizing patch redundancies across image scales. [7] employed a geometric variation to expand the internal patch search while explicitly localizing planes in the scene and applying the detected perspective geometry in guiding the patch search process. [17] utilized a feed-forward network based on the learned iterative shrinkage and thresholding algorithm integrated with a sparse representation prior. [10] applied a deep Convolutional Neural Network (CNN) based on recursive structures and skip connections that have 20 layers in addition to a small filter with a high learning rate and an adjustable gradient-clipping. The approach used

a limited model parameter to identify that for long-range pixel dependencies using a highly performant architecture. [6] proposed a method that reconstructs realistic texture detail while avoiding edge artifacts using a convolutional sparse coding approach that improves consistency by processing the whole image rather than using overlapping patches. Meanwhile [8] employed loss function closer to the perceptual similarity to regain visually more finite HR images. [14] introduced a PixelGAN generative autoencoder that combines a generative PixelCNN conditioned on a latent code with a generative adversarial network (GAN) inference network that can impose arbitrary prior to the latent code distribution. The technique proved that different priors result in different decompositions of information between the latent code and the autoregressive decoder thus showing that the PixelGAN autoencoder with a categorical prior can be directly employed in semi-supervised sceneries with competitive results on MNIST, SVHN, and NORB dataset.

[2] proposed a new deep network architecture. It is a pixel recursive super-resolution model that extends the PixelCNN. The technique is a fully probabilistic approach that tackles super-resolution with small inputs, demonstrating that images can be enlarged with improved resolution. The model produces a diversity of plausible high-resolution images at large magnification factors and shows how high-resolution images sampled from this network can fool a naive human observer a significant fraction of the time. [12] applied a deep residual network for image super-resolution (SR); the SRGAN technique could infer photo-realistic original images four times upscaling factors. They proposed a perceptual loss function which comprises a content loss and an adversarial loss. The content loss is driven by perceptual similarity rather than the similarity in pixel space; while the adversarial loss uses a discriminator network that is trained to differentiate between the super-resolved images and original photo-realistic images to push our solution to the natural image manifold. The method could recover photo-realistic textures from heavily downsampled images on public benchmarks with more photo-realistic than reconstructions obtained with state-of-the-art reference methods.

Lastly, [15] proposed a cascade training approach to deep learning to improve the accuracy of the neural networks using SRCNN while gradually increasing the number of network layers using the cascade trimming approach. The method gradually reduces the layer of the network layer by layer without significant loss on its discriminative ability.

3 Proposed Method

This research proposes an extension to Super-resolution Convolutional Neural Network (SRCNN) by updating the filter size and feature map of the low and high-resolution images before breaking up the images into patches for training to reduce memory consumption and handling big images of the medical dataset. It also involved applying some modification to the SRGAN to handle grayscale medical dataset by stacking the input on the three layers as if it was an RGB before converting the output to grayscale. In addition to determining the best layer in the VGG network to compute the loss function and selecting the best patch size in the discriminator. Lastly, combining the SRGAN, using VGG19 network for feature extraction, setting discriminator network's working space as feature space, and adding the loss function based on the mean square error of pixel space, then gain more details by incorporating SRCNN layers to increase the PSNR in the reconstruction at the same time.

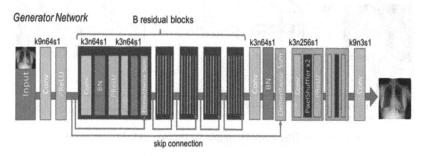

Fig. 1. SRGAN generator architecture

Figure 1 shows the generator architecture consisting of an input layer which is convolutional and takes the low-resolution image with a scale of 0.25 to the real image (Ground truth). It also included 19 residual blocks with skip connections to overcome the degrading problem with deeper layers. We added both the output of the residual blocks with the output of the first convolutional layer to keep the low feature information such as edges. Lastly, it included a final layer which is a deconvolutional layer to produce our output image. While the discriminator comprises of the Input layer that includes both high-resolution images (produced from a generator or a real image) and validity which indicates if it is fake or real. It also comprises of 10 blocks of convolutional layers and a Final dense layer to predict if it is fake or real as shown in the discriminator architecture in Fig. 2. Both showing corresponding kernel size (k), number of feature maps (n) and stride (s) indicated for each convolutional layer.

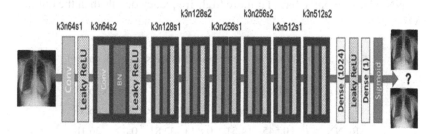

Fig. 2. SRGAN discriminator architecture

In the SRCNN model, there are only three parts: patch extraction and representation, non-linear mapping, and reconstruction. The patch extraction and representation involve upscaling the low-resolution input to the desired size using bicubic interpolation. After that, a non-linear mapping is performed to map a low-resolution vector to a high-resolution vector. Finally, the image is reconstructed in the last phase as shown in Fig. 3.

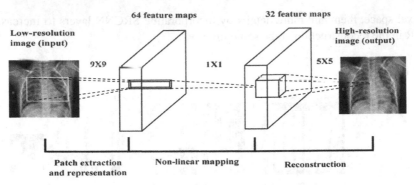

Fig. 3. The filter maps from low resolution to high resolution

4   Result and Discussion

Table 1 shows the SSIM and PSNR on the different techniques for MNIST, CIFAR-10, and Chest x-ray datasets. This SSIM and PSNR score was calculated on a random sample from the test set, and the SRCNN_GAN method was able to give a better SSIM score than the SRCNN and SRGAN methods while the PSNR of the SRCNN_GAN method was also higher than the SRGAN and SRCNN methods for all dataset. In this case, the quality was measured concerning edge fidelity. Interestingly, the model's output perceptual accuracy is reflected in the PSNR and SSIM metrics. This was achievable the outputs from the SRCNN_GAN were visually closer to the original HR ground truth images. The SRGAN output images contain high-frequency information that is like the HR ground truth images while the SRCNN and SRGAN models tend to smooth and brighten the images out to achieve a high PSNR however due to the deblurring function it failed to outperform SRCNN_GAN. Lastly, the discriminator network seeks out the high-frequency information that differentiates HR and LR images, thus forcing the SRCNN_GAN output to have far more high-frequency details than the output of the SRGAN.

Table 1. Comparison of metrics on all models using all dataset

	MNIST		CIFAR-10		Chest x-ray	
	SSIM	PSNR	SSIM	PSNR	SSIM	PSNR
SRCNN	0.545	14.51	0.890	32.81	0.832	26.18
SRGAN	0.579	14.49	0.888	35.83	0.895	31.26
SRCNN-GAN	0.917	25.87	0.992	41.93	0.991	38.36

Figure 4 shows an illustration of sample i) ground truth, ii) SRCNN, iii) SRGAN, and iv) SRCNN_GAN model trained on Chest x-ray image. The SRCNN_GAN trains end-to-end on all datasets using a 64×64 input image, upscaled four times using the extended SRGAN to the 256×256 image before deblurring based on the extended

SRCNN. Although the SRCNN and SRGAN enhanced the image with high resolution, the change in brightness caused a problem in the dataset as the brightness preservation is necessary for the resolution optimization. However, in comparison with the ground truth, the SRCNN_GAN method results produced a sharper image.

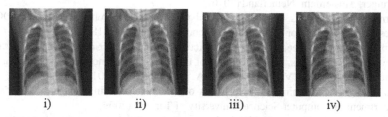

i)   ii)   iii)   iv)

Fig. 4. The Sample i) ground truth ii) SRCNN iii) SRGAN and iv) SRCNN_GAN model trained on Chest x-ray database.

5   Conclusion

This paper proposes a variant of the Single Image Super-resolution for medical analysis by redesigning the SRGAN. By using the VGG19 network for feature extraction, setting the discriminator network's working space as feature space, and adding the loss function based on the mean square error of pixel space, gaining more details by incorporating SRCNN layers to increase the PSNR in the reconstruction at the same time. Future research will perform an in-depth survey of the literature in order to register the corresponding already published work including more technical papers. We will also perform a comparison of the complexity of implementation with existing methods in order to add more value to the results presented and discuss the limitations imposed by the proposed configuration. Lastly, several tradeoffs will be explored using different network structures, parameter settings and finetuning to achieve the best performance and speed.

References

1. Allebach, J., Wong, P.W.: Edge-directed interpolation. In: Proceedings of the International Conference on Image Processing (ICIP), Lausanne, Switzerland, pp. 707–710 (1996)
2. Dahl, R., Norouzi, M., Shlens, J.: Pixel recursive super resolution. In: Proceedings of the International Conference on Computer Vision (ICCV), Venice, Italy, vol. 1, no. 2 (2017)
3. Deng, L.: The MNIST database of handwritten digit images for machine learning research [best of the web]. IEEE Sig. Process. Mag. **29**(6), 141–142 (2012)
4. Dong, C., Loy, C., He, K., Tang, X.: Image super-resolution using deep convolutional networks. IEEE Trans. Pattern Anal. Mach. Intell. **38**(2), 295–307 (2016)
5. Glasner, D., Bagon, S., Irani, M.: Super-resolution from a single image. In: Proceeding of the 12th International Conference in Computer Vision, Kyoto, Japan, pp. 349–356 (2009)
6. Gu, S., Zuo, W., Xie, Q., Meng, D., Feng, X., Zhang, L.: Convolutional sparse coding for image super-resolution. In: Proceeding of the International Conference on Computer Vision (ICCV), Las Condes, Chile, pp. 1823–1831 (2015)

7. Huang, J., Singh, A., Ahuja, N.: Single image super-resolution from transformed self-exemplars. In: Proceedings of the IEEE Conference on Computer Vision and Pattern Recognition (CVPR), Boston, Massachusetts, USA, pp. 5197–5206 (2015)
8. Johnson, J., Alahi, A., Li, F : Perceptual losses for real-time style transfer and super-resolution. In: Proceedings of the European Conference on Computer Vision (ECCV), pp. 694–711. Springer, Amsterdam, Netherland (2016)
9. Kermany, D., Zhang, K., Goldbaum, M.: Labeled Optical Coherence Tomography (OCT) and Chest X-Ray Images for Classification, Mendeley Data, v2 (2018)
10. Kim, J., Kwon, L., Lee, J., Lee, K.M.: Accurate image super-resolution using very deep convolutional networks. In: Proceedings of the IEEE Conference on Computer Vision and Pattern Recognition (CVPR), Nevada, USA, pp. 1646–1654 (2016)
11. Krizhevsky, A.: Learning multiple layers of features from tiny images. Master's thesis, Department of Computer Science, University of Toronto (2009)
12. Ledig, C., et. al.: Photo-realistic single image super-resolution using a generative adversarial network. In: Proceedings of the IEEE Conference on Computer Vision and Pattern Recognition (CVPR), Hawai, USA, vol. 2, no. 3, p. 4 (2017)
13. Li, X., Orchard, M.: New edge-directed interpolation. IEEE Trans. Image Process. (TIP) **10**, 1521–1527 (2001)
14. Makhzani, A., Frey, B.J.: Pixelgan autoencoders, In: Advances in Neural Information Processing Systems, pp. 1972–1982 (2017)
15. Rahman, S., Banik, P., Naha, S.: LDA based paper currency recognition system using edge histogram descriptor. In: 17th International Conference on IEEE Computer and Information Technology (ICCIT), pp. 326–331 (2014)
16. Timofte, R., De Smet, V., Van Gool, L.: Anchored neighborhood regression for fast example-based super-resolution. In: Proceedings of the IEEE International Conference on Computer Vision (ACCV), Darling Harbour, Sydney (2014)
17. Wang, Z., Liu, D., Yang, J., Hannand W., Huang, T.: Deep networks for image super-resolution with sparse prior. In: Proceedings of the IEEE International Conference on Computer Vision (ICCV), Las Condes, Chile, pp. 370–378 (2015)
18. Zhang, X., Gao, X., Tao, D., Li, X.: Multi-scale dictionary for single image super-resolution, In: Proceedings of the IEEE Conference on Computer Vision and Pattern Recognition, Providence, Rhode Island, pp. 1114–1121 (2012)

Forensic Studies and Medical Diagnosis

A Blob Detector Images-Based Method for Counterfeit Coin Detection by Fuzzy Association Rules Mining

Maryam Sharifi Rad[1] , Saeed Khazaee[1]([✉]) , Li Liu[1] , and Ching Y. Suen[2]

[1] School of Information Engineering, Nanchang University, Nanchang, China
{sey_shar,s_khaza}@encs.concordia.ca, liuli_033@163.com
[2] CENPARMI, Concordia University, Montreal, Canada
suen@encs.concordia.ca

Abstract. Image processing techniques using the knowledge obtained from known historical data has become recently one of the most intensively studied topics in decision science and computer science. This paper presents an automatic system for fake coins detection based on image content. In this study, a blob detector image-based method by fuzzy association rules mining is proposed to detect counterfeit coins. This method consists of two-stages. In the first stage, the original image dataset is preprocessed by a blob detector. This provides all frequent features that must be mined in the next stage. In the second stage, fuzzy association rules mining extracts the effective fuzzy rules and classifies automatically the coin image data. The performance of the proposed method has been compared with some other methods and we demonstrate that our framework surpasses in terms of classification accuracy, which is a desirable level when compared with recent studies in this field. This research demonstrates the proposed framework is a reliable intelligent detection system and can be utilized for other applications based on image content.

Keywords: Image mining · Preprocessing · Blob detection · Fuzzy association rules · Fake coin

1 Introduction

One of the most important research areas of analyzing image database and pattern recognition is designing the automated detection systems. Thanks to modern facilities, very large databases can be collected nowadays. These databases require special approaches for processing, analyzing, and efficient use of them. With the widespread use of automatic systems, such as vending machines, parking meters, banks and so on, coin recognition plays a vital role in all aspects of our daily life. Moreover, museums have increased the demand for automatic systems to classify historical coins [1]. However, in recent years, a lot of illegal counterfeiting rings manufacture and sell fake coins, which have

© Springer Nature Switzerland AG 2020
Y. Lu et al. (Eds.): ICPRAI 2020, LNCS 12068, pp. 669–684, 2020.
https://doi.org/10.1007/978-3-030-59830-3_58

caused great loss and damage to society [2]. Although forensic experts can be employed to detect the suspected coins, having a human label every possible object in a vast collection of coin images is a daunting task. Accordingly, an automatic counterfeit coin detection system can converge toward the desired point in an effective manner. Basically, counterfeit coin detection is a difficult process because of widely varying input patterns, cluttered images, and various rotations which are the great challenges. As a significant topic of security, counterfeit coin detection has become the focus of research in the field of numismatics. There is a growing interest in counterfeit coin detection community toward the application of image mining techniques in this field. The image mining concept deals with implicit knowledge extraction, image data relationship and other patterns that are not clearly stored in the images. For the past few years, a very large number of digital images have been generated on multimedia applications. As a result, building a computer-aided system to mine valuable information from the large digital image datasets is becoming a priority for many researchers. This trend has affected our daily life in every way, and advances in image acquisition have led to tremendous growth in significantly large image datasets. Although in recent years several studies have been developed based on image mining, still it is a challenging task. Basically, an image may contain several significant objects that form the concept of the image together. In fact, the semantic of a whole image cannot be represented only by its low-level features. To get closer to human perception, fuzzy association rules mining can be applied as an effective way to map low-level features into high-level semantics. Hence, the feasibility of applying fuzzy association rules mining to detect counterfeit coins will be demonstrated in this research. It is worthy to note that the combination of association rules mining and fuzzy sets theory discussed in this study presents a new approach for the better counterfeit coin detection with flexible prediction power.

This paper presents an automatic system for fake coin detection based on image content using fuzzy association rules mining. In this study, a blob detection method is used for the feature extraction of image datasets and fuzzy association rules mining is applied for intelligent classification. The proposed system performance is compared with some other methods in this field. This research demonstrates the proposed model is a reliable intelligent detection system and can be utilized even for other applications based on image content.

The remainder of this paper is organized as follows. Section 2 revisits the related studies in the field of counterfeit coin detection as well as the image mining approaches. In Sect. 3, we propose our methodology for counterfeit coin detection. The experimental validations are presented in Sect. 4, while the conclusions and the directions for the future works are listed in Sect. 5.

2   Research and Related Background

Image processing techniques using the knowledge obtained from known historical data has become recently one of the most intensively studied topics in decision science and computer science. During the last years, image-based approaches to detect fake coins have been extensively studied in the literature. Furthermore, a few studies based on pattern recognition techniques and classification algorithms have been proposed that exploit images to detect fake coins. Related work to mining image content, we can mention the following. In [2], an image-based approach to detect the fake coins based on the characteristics of coin images has been proposed. The dimension of coin images is determined by the number of prototypes. They have computed the dissimilarity between the coin images by the local key points on each image using the DOG detector and the SIFT descriptor. In [3], an algorithm for coin Recognition using Circular Hough Transform (CHT) has been proposed. The proposed system first uses canny edge detection to generate an edge map, then uses CHT to recognize the coins and further find their radius. The basic limitation arises when the image is captured from a distance. In [4], a system for Indian coin recognition by Heuristic approach and Hough Transform (HT) has been proposed. The proposed method has been limited to recognizing only the Indian Coins. In [5], a method to detect two-euro fake coins based on the coin images captured by an optical mouse sensor has been developed. In this study, the basic limitation arises during the coin rotation and it is vulnerable to distortions. In [6], a counterfeit coin detection method to detect fake Danish coins based on their image characteristics has been proposed. Despite the promising results achieved, the dataset used was extremely small which consisted of only 16 coins. Furthermore, counterfeit coin detection can be developed based on different methods and techniques [7, 8]. It is worthy to note that many studies have been conducted on coins' colors and radius-based features to detect fake coins. However, there is growing evidence that merging image mining approaches and advanced storage technology together can produce more efficient and accurate image content-based systems than traditional systems. Several interesting studies involving image mining and the concepts of association rules using object-based features have been extended in [1, 9–12]. Although most of them have been developed based on different approaches for mining of images to extract strong association rules, still it is a challenging task.

On the other hand, association rule mining is one of the most significant approaches for pattern detection in which the interesting relationships among items in image datasets will be determined. The main purpose of image mining is to extract useful and non-explicit information from images stored in large repositories. By extracting rules from an image dataset, the content of images will be analyzed and the information for image classification will be obtained. Although the current image mining approaches are far from maturity and integrity, they open a vast room and promising research direction. Since the presentation of the image mining technique, this area remained one of the current hot research topics in knowledge discovery.

In this research, we develop an image mining system on top of the concept of fuzzy association rules that helps us to discover the implicit information from the images in the way closer to the human's viewpoint. In general, fuzzy logic can be used to identify complex patterns or structural variations in image datasets. Therefore, in this study, a new framework based on fuzzy concepts to rely on semi-automatic mining on the image datasets will be proposed.

2.1 Foundation of Rules Mining

Agrawal et al. [13] have proposed the Apriori algorithm which is the best-known fundamental algorithm to find Boolean association rules. During the past years, several studies have been conducted on extracting association rules from datasets. In spite of the novelties of the proposed methods, it is clear that the methods were not robust enough to distinguish the sharp boundary problem. In other words, they either ignore or over-emphasize the elements near the boundary of intervals in the mining process. As a remedy to the sharp boundary problem, the fuzzy set concept, introduced by Zadeh [14] has been used more frequently in mining association rules. This approach is better than the partitioning method because fuzzy sets provide a smooth transition between members and non-members of a set and increase the flexibility of systems [15]. The extraction algorithm of association rules based on the precise concept has been widely used. In recent years, many researchers have made a thorough study of fuzzy sets concepts. Despite the wide use of the association rules mining, the research on the fuzzy-based association rules mining is relatively scarce. It is worthy to note that the fuzzy concept is an excellent tool to extract association rules. In this research, fuzzy association rules mining is used for counterfeit coin detection and it is considered as the key component of the proposed method because of its affinity with the human knowledge representation.

2.2 Definitions

A fuzzy association rule is demonstrated by LHS[1]⇒RHS[2] form with both LHS and RHS allowed to contain multiple items. The measures for establishing a fuzzy association rule's fitness are *Support* and *Confidence*. So, a fuzzy association rule can be represented as $X \Rightarrow Y$ (s, c), where s is called the *Support* and c is called the *Confidence* of rule:

$$\text{Support } (X \Rightarrow Y) = N_{X \& Y}/N_{\text{Database}} \tag{1}$$

$$\text{Confidence } (X \Rightarrow Y) = N_{X \& Y}/N_X \tag{2}$$

where $N_{X \& Y}$ represents the number of transactions, which contain X and Y; N_X represents the number of transactions, which contain X; and $N_{Database}$ represents the number of transactions in the database [16]. As mentioned above, *Support* of a fuzzy association rule is defined as the percentage of transactions that contain all items (both LHS and RHS) and *Confidence* of a fuzzy association rule is defined as the percentage of LHS items that also contains RHS. A fuzzy association rule holds if its *Support* is greater than minimum *Support* (*Min_S*) and its *Confidence* is greater than minimum *Confidence* (*Min_C*), where *Min_S* and *Min_C* are configurable. The problem of mining fuzzy association rules is to extract all fuzzy rules that have *Support* and *Confidence* greater than some user-specified minimum *Support* and minimum *Confidence* thresholds, respectively. Therefore, the problem of finding fuzzy association rules is decomposed into sub-problems of discovering all itemset with at least *Min_S* (also called large itemset)

---

[1] Informal Shorthand for the Left-Hand Side of an Equation.
[2] Informal Shorthand for the Right-Hand Side of an Equation.

and using these large itemsets to generate the desired fuzzy rules (tested for *Min_C*). Large itemset generation is achieved by generating candidate itemset and keeping the ones with *Min_S*. A fuzzy association rule in the image database is a fuzzy rule that associates visual object features and the relationship among objects in images. The crucial part of image mining is to separate similar objects in different images. One premise behind supporting objects is to get rid of the demand for manual indexing of image content. The problem of mining fuzzy association rules is to generate all rules that have *Support* and *Confidence* greater than some user-specified minimum *Support* and minimum *Confidence* thresholds, respectively. In this study, the FGBRMA[3] algorithm is used to mine fuzzy association rules [17]. This algorithm consists of two phases: large fuzzy grids generation and fuzzy association rule extraction. In the first phase, this algorithm uses a fuzzy partition method for each feature so that it divided into K different linguistic values ($K = 2, 3, \ldots$). In fact, each feature is viewed as a linguistic variable and the variables are divided into various linguistic values. Each linguistic value can be used to represent a candidate 1-dimensional fuzzy grid. Moreover, to generate a candidate high-dimensional fuzzy grid, we can apply two large 1-dimensional fuzzy grids to constrict a candidate 2-dimensional fuzzy grid. To check whether this fuzzy grid is large or not, the fuzzy *Support* is computed (Eq. (1)). When its fuzzy *Support* is larger than or equal to the predetermined minimum fuzzy *Support* (*Min_FS*), it can be said that it is a large k-dimensional fuzzy grid. When all the large fuzzy grids have been generated, the next phase will be started. In the second phase, each rule R is generated by two large fuzzy grids. To check whether this rule is effective or not, its fuzzy *Confidence* is computed (Eq. (2)). When its fuzzy *Confidence* is larger than or equal to the predetermined minimum fuzzy *Confidence* (*Min_FC*), the rule is considered as an effective rule. The FGBRMA is an effective algorithm since it scans the dataset only once and applies Boolean operations on tables to generate fuzzy grids and fuzzy association rules. In this algorithm, a table structure, called FGTTFS, is implemented to generate large fuzzy grids. This table consists of the following substructures:

- Fuzzy grids substructure (FG): each row represents a fuzzy grid, and each column represents a linguistic value.
- Transaction substructure (TT): each column represents a tuple t_p, while each element records the membership degree of t_p which belongs to the corresponding fuzzy grid.
- Fuzzy *Support* substructure (FS): stores the fuzzy *Support* corresponding to the fuzzy grid.

In this study, FGBRMA is employed as the FGBRM (Dataset, *Min_FS*, *Min_FC*) function in the proposed framework.

2.3 Fuzzy Set Theory

In the objective world, the information is mostly vague and indefinite, so the research on the fuzzy concept has important practical significance [18]. In recent years, many researchers have made a thorough study of fuzzy sets concepts. Despite the wide use of

---

[3] Fuzzy Grids-Based Rule Mining Algorithm.

the association rules mining, the research on the fuzzy-based association rules mining is relatively few. It is worthy to note that the fuzzy concept is an excellent tool to extract association rules. In the information society, the data collected in many real systems are inaccurate. If this uncertain information cannot be properly analyzed, it will probably lead to a great error between the inference result and the objective fact. Many researchers have been trying to find a reasonable way to deal with uncertain data. In 1965, Zadeh proposed the fuzzy sets as an extension of the classical set theory [14]. According to the definition in mathematics, the elements in fuzzy sets have membership value. Fuzzy sets theory is now widely used in different fields, such as linguistics, decision systems, and cluster analysis. The fuzzy sets theory uses a membership function to describe the fuzzy relation. The range of the membership function value is the unit interval [0,1] [18]. Fuzzy sets can be represented as (U, m), wherein the domain U is a set, and m: $U \to [0,1]$ is a membership function. For each $x \in U$, the value of $m(x)$ is called the membership degree of x in (U, m). for the finite set $U = \{x_1, ..., x_n\}$, the fuzzy sets (U, m) is usually represented by $\{m(x_1)/x_1, ..., m(x_n)/x_n\}$. For a given domain U, a mapping $m_A: U \to [0,1]$ can determine a fuzzy subset A, whose representation is as follows:

$$A = \begin{cases} \sum_{x_i \in U} \frac{m_A(x_i)}{x_i}, & U \text{ is a finite set } \{x_1, \ldots, x_n\} \\ \int \frac{m_A(x)}{x}, & U \text{ is an infinite set, and } x \in U \end{cases} \tag{3}$$

where m_A is the membership function, and $m_A(x)$ indicates the membership degree of x to A. Note that $\sum_{x_i \in U} \frac{m_A(x_i)}{x_i}$ is not an expression of the sum of fractions, but merely a sign [18].

3 Methodology

In this study, a blob detector image-based method by fuzzy association rules mining is proposed to use in counterfeit coin detection. This method consists of two-stages. In the first stage, the original image dataset is preprocessed by using a blob detector. This provides all frequent features that should be mined in the next stage. In the second stage, fuzzy association rules mining extracts the effective fuzzy rules and classifies the coin image data. The main goal in the proposed framework is to discover knowledge in image datasets. This is an important attempt to combine fuzzy association rules and the coin image dataset, although there has been remarkable research in the image domain. In this study, a new framework to discover the frequent objects and to extract the interesting fuzzy association rules for counterfeit coin detection will be proposed. The method proposed here is a fuzzy association rules-based image mining approach by using the blob detection method and it extends the techniques that we proposed in [16]. The fundamental component in image mining is to identify similar objects in different images. We transform the images with a set of transactions, each transaction representing one image with the features extracted as well as other given characteristics along with the class label. The result of this phase is a transactional dataset to be mined in the next phase of our proposed framework. In this way, the unclassified images can be automatically classified by generated fuzzy association rules. From another point of view, the most delicate part of the classification with fuzzy association rules mining is the construction of the classifier itself. Figure 1 demonstrates a schematic view of the proposed framework. In the first module, the digital coin images are segmented for finding the region of interest (ROI). In the second module, the frequent objects in the images using fuzzy association rules mining are extracted. The fuzzy association rules can discover relationships among objects in image datasets. So, each input digital coin image is associated with a keyword i.e. fake or genuine. For a new image to be classified, the extracted fuzzy association rules are deployed to determine the most relevant class for the coin image efficiently. Therefore, the class of coin images can be predicted by using the classification fuzzy rules mined. The method proposed in this study classifies the digital coin images into two categories: genuine and fake. The details of the proposed framework are described in the following subsections.

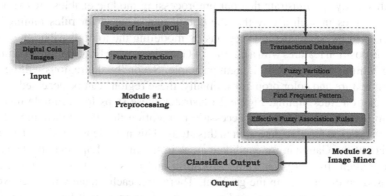

Fig. 1. Schematic view of the proposed framework.

3.1 Data Preprocessing

Data preprocessing is necessary to enhance the quality of coin images and develop the feature extraction method more reliably. In this study, the goal of the preprocessing is to bridge the gap from the low-level image properties to the objects in which can be ultimately considered as an object recognition problem. In the preprocessing step, a method will be applied for image retrieval based on segmentation into regions and querying using properties of these regions. It is necessary to consider that image segmentation is basically an effortful issue. Several algorithms have been conducted on image segmentation which some of them inevitably make inaccurate segmentation, causing some degradation in the efficiency of any model that utilizes the segmentation results. Feature extraction is the most significant component of designing an intelligent system based on image pattern recognition since even the best classifier will run inaccurately if the features are not chosen well.

During the preprocessing step, we use a method based on blob detection which is structured around a sequence of increasingly specialized grouping activities that produce a "Blobworld" representation of an image. This preprocessing is a transformation from the raw pixel data to a small set of localized coherent regions in color and textual space.

"Blobworld" technique has been remarked as a noticeable method for object recognition that is based on image segmentation using the Expectation-Maximization algorithm on combined color and texture features [19]. In this step, we detect objects in images and extend the concept of image mining using objects in the image dataset. By extracting ROIs that roughly correspond to the objects, we can access the images at the level of objects rather than global image properties. The preprocessing module of the proposed framework produces a "Blobworld" representation of each image and each "blob" is considered as a 2-D ellipse that possesses several attributes. The pixels will be grouped into regions by modeling the joint distribution of color, texture, and position features. In this manner, the EM[4] algorithm is applied to estimate the parameters of a mixture of Gaussians model of the joint distribution of pixel color and texture features. This approach is related to the MDL[5] principle to perform segmentation based on motion.

In this study, we illustrate that our preprocessing module enables the extraction of images features in such a way that the output can be used for rules mining. Firstly, all pixels will be subdivided into regions by modeling the joint distribution of texture, color, and position features with a mixture of Gaussians using Expectation-Maximization and the Minimum Description Length principle. Secondly, the regions will be defined based on texture and color properties. Finally, these regions can be accessed as an item for association rules mining. Figure 2 illustrates these steps for a sample image from pixels to feature extraction. It is necessary to mention that the python module called *skimage* was used for blob detection in this study. This module is based on detecting the Laplacian of the Gaussian for candidate areas in the image. Laplacian of the Gaussian finds a gradient from a given central location and tries to determine an edge of the blob based on differences in the gradient. Therefore, each image will be converted to greyscale and then can be displayed as an array. This array will be then forwarded as input to the Laplacian of Gaussian function in *skimage*. The preprocessing result of this study is a list of (x, y, r) coordinates, where (x, y) shows the center of the blob and r indicates the radius of the blob. After blob detection, we create a new dataset, such that each transaction represents one image with the visual blobs extracted along with the class label. The new transactional dataset is submitted to the image miner module as input for rules mining step. Figure 3 demonstrates a sample of the preprocessing step in finding the blobs of an image.

3.2   Image Mining

In this study, the fuzzy grids-based rule mining algorithm is used to mine fuzzy association rules. In this algorithm, each object feature is viewed as a linguistic variable, and the variables are divided into various linguistic terms. Fuzzy sets provide a smooth transition between members and non-members of a set and increase the flexibility of systems. To define fuzzy membership functions, three linguistic values are determined for every feature and triangular membership functions are used for each linguistic value.

A predefined membership function is assigned to each feature, and the linguistic terms can be expressed by the membership function shown in Fig. 4. The parameters

---

[4] Expectation-Maximization
[5] Minimum Description Length

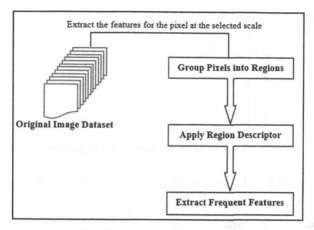

Fig. 2. Preprocessing flowchart.

(a)   (b)

Fig. 3. Preprocessing phase on an example image (one Dollar US coin): (a) original image; (b) preprocessed image after blob detection.

α, β, and γ in the fuzzy membership function for feature Fi are set as follows [20]: β: average value of feature Fi in the dataset; γ: The largest value of feature Fi in the dataset; $\alpha = 2\beta - \gamma$.

After the definition of the fuzzy membership function, the images transaction datasets are ready to be mined. At this point, the proposed framework is built on top of the fuzzy data set and the preprocessed images are considered as a transaction of features. The problem of fuzzy rule generation is to extract all fuzzy rules with *Support* and *Confidence* greater than a threshold parameter established by the user. In this step, the frequent itemsets are found by computing the Fuzzy Support counts of candidate itemsets. To check whether each candidate itemset is large or not, its Fuzzy Support is computed. When its Fuzzy Support is larger than or equal to the pre-determined Minimum Fuzzy Support (called *Min_FS*), it is considered as a frequent itemset. After finding all frequent itemsets, fuzzy association rules can be generated. To check whether each fuzzy rule R is effective or not, its Fuzzy Confidence is computed. When its Fuzzy Confidence is larger than or equal to the pre-determined Minimum Fuzzy Confidence (called *Min_FC*), the fuzzy rule is considered as an acceptable rule. These fuzzy rules will be then used to classify the digital coin images into two categories: genuine, and fake.

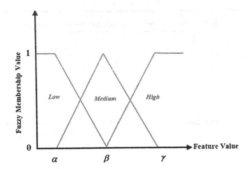

Fig. 4. Definition of the fuzzy membership function.

3.3 Classifying a New Image

In this subsection, we briefly describe how to classify a new image by using the classification rules obtained through the fuzzy association rules mining. The set of fuzzy rules represent the actual classifier. This categorizer is applied to predict to which classes new images are attached. Given a new image, the classification process searches in this set of fuzzy rules to discover the class that is the closest to be attached to the new image presented for categorization. To classify a new image, the class that has the most rules matching will be attached to the new image as a label. Consider that a new image should be classified: First, the blobs discussed in subsection *A* are extracted and then these blobs would yield a list of applicable fuzzy rules in the limit given by the *Min_FC* threshold. If the applicable fuzzy rules are grouped by category in their consequent part and the groups are ordered by the sum of fuzzy rules' confidences, the ordered groups would indicate the most significant category (i.e. a fake or genuine) that should be attached to the new image to be classified. It is noted that the length of the large itemset and the fuzzy confidence value are both critical factors to find the most possible class that the image should belong to.

4   Experiments

The experiments of this study were conducted in the environment of Microsoft Windows 8.1-64 bit and hardware of the test environment consisted of an i7-4500U 4.2 GHz CPU (only one core was used), DDR3 6 GB RAM.

4.1 Evaluation Metrics

In this subsection, the performance of the proposed framework in terms of the most common metrics has been evaluated. In this research, we used a total of 1050 genuine and 1000 fake coins for training and evaluating the system respectively. Several experiments were conducted to evaluate the combination of ROIs method and fuzzy association rules mining aiming to counterfeit coin detection. In this research, the performance of the

FAR-based classifier, as a tool for the detection of the fake coin, has been evaluated in terms of sensitivity, specificity, and accuracy. These measures are calculated as follows:

$$Sensitivity = TP/(TP + FN) \tag{4}$$

$$Specificity = TN(TN + FP) \tag{5}$$

$$Accuracy = (TP + TN)/(TP + TN + FP + FN) \tag{6}$$

where, TP and TN are the number of true positives and true negatives, respectively. Also, FP and FN are the number of false positives and false negatives, respectively.

4.2 Dataset

The dataset applied in this study consists of real data. The main reason why we study real coin images instead of any simulative data is to avoid insignificance and we try to focus on the reliability of the discovered knowledge. It is important to note that to have access to fake coins is a very difficult undertaking due to the legal issue and the access to more fake coins is usually restricted. In this research, we scanned two types of datasets including Chinese and US coins to evaluate the performance of the proposed method. To this end, a precise 3-D scanner has been used to scan the Chinese and US one Dollar coins. Statistical details of two types of Chinese coins and also US one Dollar are summarized in Table 1. In Fig. 5, samples of genuine and fake coins of years, Chinese half Yuan 1942, Chinese One Yuan 1997 and US One Dollar 1908 are indicated.

Table 1. Characteristics of dataset components used for training and testing.

Coin specification	Number of genuine coins	Number of fake coins	Training set		Test set	
			Genuine	Fake	Genuine	Fake
Chinese half Yuan 1942	550	500	340	300	210	200
Chinese One Yuan 1997	400	400	250	250	150	150
US One Dollar	100	100	60	60	40	40

4.3 Results

The proposed framework consists of two modules. First, the automated segmentation of coin images is performed to obtain the region of interest (ROI). Next, the transaction database has been generated and fuzzy association rules mining by using the FGBRMA algorithm have been extracted. With the application of fuzzy logic, the linguistic features have been extracted, and all of them have been added to the final feature subset. Our transaction database has a record for every input coin image and each record has a keyword i.e. genuine or fake along with unique labels. The labels of the input coin's image i.e. F or G and the extracted blobs are used to build the transactional dataset. Each record of the transactional dataset is submitted to the image miner module to extract fuzzy association rules. In the preprocessing step, the median filtering is used to remove digitization noise. Furthermore, thresholding operation and contrast enhancement will be applied to the segmentation of coin images. It is noted that extracting fuzzy rules at the highest *Min_FS* and *Min_FC* percentage yields the best results in terms of sensitivity, specificity, and accuracy. When the proposed framework is implemented with *Min_FS* = 85%, and *Min_FC* = 90%, a total of 1084 rules have been discovered. From these, there are 132 rules consist of two elements, 437 rules with three elements, 301 rules with four elements, and 214 rules with five elements. In this way, some of the mined rules are listed in Table 2. As mentioned earlier, an image is made up of a collection of blobs and each blob represents a region of the image which is relatively homogeneous with respect to color and texture. Oi, j is a blob in the transaction dataset where ($1 \leq i \leq N$; N shows the number of all blobs in the dataset) and ($j = Low, Medium,$ or $High$) and also integers behind the blobs show quantifying the occurrence of the blob. We use G as a keyword for genuine images and F as a keyword for fake images.

Table 3 illustrates the performance of the proposed framework in terms of precision, sensitivity, F-Value, and accuracy. It is interesting to note that, as expected, the capabilities of the proposed framework reveal the effectiveness of data mining techniques. The accuracy of the proposed method in classifying both classes fake and genuine comparing with some other methods reported in [16, 21, 22], has been shown in Table 4, as well.

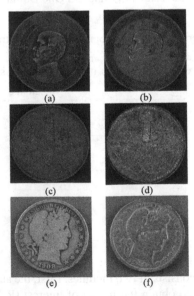

Fig. 5. Examples of genuine and fake coins: (a) genuine Chinese half Yuan 1942, (b) fake Chinese half Yuan 1942, (c) genuine Chinese One Yuan 1997, (d) fake Chinese One Yuan 1997, (e) genuine US One Dollar 1908, (f) fake US One Dollar 1908.

To make an accurate comparison, the same dataset and experimental conditions have been applied to all methods. Furthermore, the parameters have been selected according to the guidelines provided by the authors in the papers in which each method has been presented. As we can see clearly, the proposed method performs better than all the other recent methods. The results illustrate that the proposed method has reached a remarkable improvement in the classification of the coins, especially for One Yuan Chinese 1997, and One Dollar US 1908.

In addition, the accuracy of the proposed framework depends steadily on the precision of the parameters such as *Min_FS* and *Min_FC* in this research. Therefore, determining the optimum values of these parameters can improve the performance of the proposed framework, effectively. Experimental results indicate that the optimized parameters can converge faster, and it is very important to enhance the image mining performance in the future works.

Table 2. Some of the fuzzy association rules with fuzzy support (FS) $\geq 85\%$ and fuzzy confidence (FC) $\geq 90\%$.

Fuzzy rules	FS value	FC value
$2O_{1, Low} \wedge 1O_{2, Medium} \Rightarrow 1O_{1, Medium} \wedge G$	95%	100%
$1O_{2, Medium} \wedge 1O_{2, Low} \Rightarrow 2O_{1, High} \wedge F$	90%	95%
$1O_{1, High} \wedge 2O_{2, Medium} \Rightarrow 2O_{2, Medium} \wedge G \wedge 1O_{3, Medium}$	95%	100%
$2O_{1, High} \wedge 2O_{2, Medium} \wedge G \wedge 1O_{3, High} \Rightarrow 1O_{4, Medium} \wedge 2O_{5, High}$	85%	95%
$1O_{2, Medium} \wedge F \wedge 2O_{3, High} \wedge 2O_{4, Low} \Rightarrow 1O_{4, Medium} \wedge 1O_{5, High} \wedge 2O_{6, Medium}$	95%	100%

Table 3. Performance of the proposed method in terms of precision, sensitivity, F-Value, and accuracy.

Dataset	Class	Precision (%)	Sensitivity (%)	F-value (%)	Accuracy (%)
Chinese half Yuan 1942	Genuine	0.985	0.960	0.972	95.6
	Fake	0.855	0.940	0.895	
Chinese One Yuan 1997	Genuine	0.985	0.975	0.980	96.8
	Fake	0.904	0.940	0.922	
US One Dollar 1908	Genuine	0.973	0.888	0.928	89
	Fake	0.667	0.900	0.766	

Table 4. Comparison of the proposed framework with some other counterfeit coin detection systems in terms of accuracy in recognizing both genuine and fake class.

Datasets	Sharifi et al. [16]	Khazaee et al. [21]	Ali et al. [22]	Proposed framework
Half Yuan Chinese 1942	91.2	94.5	90.0	95.6
One Yuan Chinese 1997	88.5	86.7	88.9	96.8
One Dollar US 1908	77	80	71	89
Training All Coins together	63.4	66.8	62.4	73.6

5   Conclusion and Future Works

In this paper, we have presented a method based on the fuzzy association rules mining for counterfeit coin detection. The main concern of the proposed framework was to construct a fake coin detector by using the fuzzy association relations among the coin image objects. Fuzzy association rules mining can adequately utilize a large amount of data and extract important relationships among items. In the proposed framework, an image mining method using the blob detection technique has been introduced for finding important information from the coin images. It is interesting to note that by using the fuzzy association rules mining technique equipped by the blob detection method, the accuracy of the proposed framework is an acceptable level when compared with recent research in this field. In this study, the problem with a general form has been described to provide a common framework for other problems appearing in other domains. The research activities proposed here will significantly advance the field of counterfeit coin detection and image classification. These contributions will enable researchers to focus on the conceptual fuzzy associations among the objects in the images. They are innovative in the sense that they will lead to a new breed of feature extraction to discover the implicit information from the images in the way closer to the human's viewpoint.

Direction for future work can be stated as follows:

- It could be interesting to consider the similar behaving features in the mining of fuzzy association rules for image classification. The key strength of fuzzy association rule mining is its completeness. This strength, however, comes with a major drawback. It often produces a huge number of fuzzy associations rules. This is particularly true for datasets whose attributes are highly correlated. In other words, rules having a high value of support and confidence give redundant or irrelevant information, which makes them uninteresting rules. In several previous algorithms for fuzzy association rules mining, the objective is to find frequent fuzzy itemsets by expanding techniques presented for the binary form and the problems existing in the fuzzy form, remain yet. To address this limitation, we need a measure of the dependency among features. Correlation measures can be used to augment the support and confidence. Based on our research, another interesting idea for testing independence and/or correlation can be Chi-square test statistics. Therefore, a novel technique to overcome these problems by focusing on similar behaving attributes for future studies in this direction could be performed.

- The efficiency of the proposed model could be improved also. The accuracy of the proposed framework depends strongly on the precision of the parameters of the image miner module (*Min-FS* and *Min-FC*). To find the threshold values for minimum fuzzy support and confidence, we can apply the genetic algorithm (GA) and particle swarm optimization (PSO) algorithm to determine the optimum values of them. Therefore, finding the optimum values of parameters with the help of GA and PSO for future studies could be performed.

References

1. Van Der Maaten, L.J., Postma, E.O.: Towards automatic coin classification. In: Proceedings of the EVA-Vienna 2006, Vienna, Austria, pp. 19–26 (2006)
2. Liu, L., Lu, Y., Suen, C.Y.: An image-based approach to detection of fake coins. IEEE Trans. Inf. Forensics Secur. **12**(5), 1227–1239 (2017)
3. Jain, N., Jain, N.: Coin recognition using circular hough transform. Int. J. Electron. Commun. Comput. Technol. (IJECCT) **2**(3), 2249–7838 (2012)
4. Velu, C.M., Vivekanandan, P.: Indian coin recognition system of image segmentation by heuristic approach and hough transform. Int. J. Open Probl. Compt. Math. **2**(2), 224–271 (2009)
5. Tresanchez, M., Palleja, T., Teixido, M., Palacin, J.: Using the optical mouse sensor as a two-euro counterfeit coin detector. Sensors **9**(9), 7083–7096 (2009)
6. Sun, K., et al.: Detection of counterfeit coins based on shape and lettering features. In: Proceeding of International Conference on Computer Applications in Industry and Engineering, pp. 165–170 (2015)
7. Kim, S., Lee, S.H., Ro, Y.M.: Image-based coin recognition using rotation-invariant region binary patterns based on gradient magnitudes. J. Vis. Commun. Image Represent. **32**(C), 217–223 (2015)
8. Shen, L., Jia, S., Ji, Z., Chen, W.S.: Extracting local texture features for image-based coin recognition. IET Image Process. **5**(5), 394–401 (2011)
9. Deshmukh, J., Bhosle, U.: Image mining using association rule for medical image dataset. In: International Conference on Computational Modeling and Security, pp. 117–124 (2016)
10. Lee, A.J.T., Hong, R.W., Ko, W.M., Tsao, W.K., Lin, H.H.: Mining spatial association rules in image databases. Inf. Sci. **177**(7), 1593–1608 (2007)
11. Rajendran, P., Madheswaran, M.: Novel fuzzy association rule image mining algorithm for medical decision support system. Int. J. Comput. Appl. **1**(20), 87–94 (2010)
12. Ribeiro, M.X., Traina, A.J.M., Traina, C., Marques, P.M.A.: An association rule-based method to support medical image diagnosis with efficiency. IEEE Trans. Multimed. **10**(2), 277–285 (2008)
13. Agrawal, R., Imielinski, T., Swami, A.: Mining association rules between sets of items in large databases. In: Proceedings of the ACM SIGMOD ICMD, Washington DC, pp. 207–216 (1993)
14. Zadeh, L.: Fuzzy sets. Proc. Inf. Control **8**(3), 338–353 (1965)

15. Sheikhan, M., Sharifi Rad, M.: Gravitational search algorithm-optimized neural misuse detector with selected features by fuzzy grids-based association rules mining. Neural Comput. Appl. **23**(7–8), 2451–2463 (2013)
16. Sharifi Rad, M., Khazaee, S., Suen, C.: Counterfeit coin detection based on image content by fuzzy association rules mining. In: Proceeding of ICPRAI 2018, Montreal, Canada, Center for Pattern Recognition and Machine Intelligence, pp. 285–289 (2018)
17. Hu, Yc., Chen, Rs., Tzeng, Gh.: Discovering fuzzy association rules using fuzzy partition methods. Knowl. Based Syst. **16**, 137–147 (2003)
18. Zou, C., Deng, H., Wan, J., Wang, Z., Deng, P.: Mining and updating association rules based on fuzzy concept Lattice. Future Gener. Comput. Syst. **82**, 698–706 (2018)
19. Carson, C., Belongie, S., Greenspan, H., Malik, J.: Blobworld: image segmentation using expectation-maximization and its application to image querying. IEEE Trans. Pattern Anal. Mach. Intell. **24**(8), 1026–1038 (2002)
20. Mabu, S., Chen, C., Lu, N., Shimada, K., Hirasawa, K.: An intrusion-detection model based on fuzzy class-association-rule mining using genetic network programming. IEEE Trans. Syst. Man Cybern. Part C Appl. Rev. **41**(1), 130–139 (2011)
21. Khazaee, S., Sharifi Rad, M., Suen, C.: Restoring height-map images of shiny coins using spline approximation to detect counterfeit coins. In: Proceeding of ICPRAI 2018, Montreal, Canada, Center for Pattern Recognition and Machine Intelligence, pp. 383–387 (2018)
22. Hmood, A., Suen, C.: An ensemble of character features and fine-tuned convolutional neural network for spurious coin detection. In: Language Processing, Pattern Recognition, and Intelligent Systems Frontiers in Pattern Recognition and Artificial Intelligence, pp. 169–187 (2019)

Interpreting Deep Glucose Predictive Models for Diabetic People Using RETAIN

Maxime De Bois[1](✉)(iD), Mounîm A. El Yacoubi[2](✉)(iD), and Mehdi Ammi[3](✉)(iD)

[1] CNRS-LIMSI and Université Paris Saclay, Orsay, France
maxime.debois@limsi.fr
[2] Samovar, CNRS, Télécom SudParis, Institut Polytechnique de Paris, Évry, France
mounim.el_yacoubi@telecom-sudparis.eu
[3] Université Paris 8, Saint-Denis, France
ammi@ai.univ-paris8.fr

Abstract. Progress in the biomedical field through the use of deep learning is hindered by the lack of interpretability of the models. In this paper, we study the RETAIN architecture for the forecasting of future glucose values for diabetic people. Thanks to its two-level attention mechanism, the RETAIN model is interpretable while remaining as efficient as standard neural networks.

We evaluate the model on a real-world type-2 diabetic population and we compare it to a random forest model and a LSTM-based recurrent neural network. Our results show that the RETAIN model outperforms the former and equals the latter on common accuracy metrics and clinical acceptability metrics, thereby proving its legitimacy in the context of glucose level forecasting. Furthermore, we propose tools to take advantage of the RETAIN interpretable nature. As informative for the patients as for the practitioners, it can enhance the understanding of the predictions made by the model and improve the design of future glucose predictive models.

Keywords: Deep learning · Glucose prediction · Diabetes · Neural networks · Attention · Interpretability.

1 Introduction

Diabetes is undoubtedly one of the major diseases of the modern world as it has been inputed a total of 1.5 million deaths in 2012 [16]. The every day challenge faced by diabetic people is the regulation of their blood glucose level which is troubled by either the non-production of insulin (type-1 diabetes) or the increasing body resistance to its action (type-2 diabetes). Diabetic people are at risk of facing short terms complications (e.g., coma, death) due to their glycemia falling

This work is supported by the "IDI 2017" project funded by the IDEX Paris-Saclay, ANR-11-IDEX-0003-02.

too low (hypoglycemia) and also long-term complications (e.g., cardiovascular diseases, blindness) when it gets to high (hyperglycemia).

To help the patients coping with their disease, a lot of technological efforts have been made in the recent years. For instance, by enabling the diabetic patient to forgo the use of lancets to get his or her glucose level, continuous glucose monitoring (CGM) devices (e.g., FreeStyle Libre [10]) are getting more and more common. Besides, we are witnessing the rise of coaching applications specifically made for diabetic people (e.g., mySugr [12]). From a research perspective, current endeavors are focused towards the building of glucose predictive models. Using past glucose values, carbohydrate (CHO) intakes, insulin infusions, and more, the models forecast the future glucose values at horizons varying from 30 min (short-term) to 120 min (long-term) [11].

Thanks to the increasing availability of data and the access to more computing power, the glucose predictive models are shifting from rather simple models (e.g., autoregressive models [13]), to more complex algorithms from the machine learning and deep learning field. Daskalaki et al. have demonstrated the superiority of feed-forward neural networks over the autoregressive models in the context of short-term glucose forecasting [2]. Georga et al. explored the usability of extreme learning machines for short-term glucose prediction as well [5]. Recurrent neural networks have recently generated a lot of interest because of their temporal nature, making them particularly suitable for the task of predicting future glucose values [3,9]. As time-series can be seen as one-dimension images, convolutional neural networks, which are very popular in the image recognition community, have also been tried out for the forecasting of future glucose values [7].

Even though deep models can be effective for the task of glucose prediction, they have a sizable downside: the deeper the model, the more difficult it is to understand its behavior. This is especially an issue for biomedical applications for which it is important to be able to interpret the models in order to understand why a prediction is being made. To address this issue, Georga et al. showed that Random Forests (RF), while being highly interpretable, can achieve good performances for the task of glucose prediction [4].

Recently, Choi et al. proposed a neural network, called RETAIN, specifically designed for healthcare applications dealing with temporal inputs. Featuring a two-level attention mechanism, the model is meant to be as performant as standard neural networks while being interpretable. This property is highly valuable for the prediction of future glucose values. On one hand, it would help the practitioner design better and safer models by providing a better error analysis tool. On the other hand, for the patient, it would help him or her understand his or her desease better.

In this work, we study the use of the RETAIN architecture for the challenging task of the forecast of future glucose values for diabetic people. In particular, we adapt its interpretability feature to regression problems and propose several analysis and visualization tools to interpret the predictions made by the model.

The rest of the paper is structured as follows. First, we describe the RETAIN architecture and how the predictions are interpreted from it. Then, we describe the overall experimental methodology. Finally, we provide the results and analysis of the experiments before concluding.

2   RETAIN

This section presents the RETAIN architecture that has been previously introduced in [1] and its interpretation for time-series forecasting, and in particular for glucose prediction.

2.1   Architecture

Most of the efficiency of the RETAIN model comes from its two levels of attention: the *time-level* attention (also called *visit-level* attention [1]), and the *variable-level* attention. The general attention mechanism comes from the natural language processing field where it enables the model to understand relationships between words in a sentence [15]. Here, when dealing with temporal inputs (e.g., time-series, events), while the time-level attention makes the network focus on specific time-steps, the variable-level attention enphasizes specific input within the time-steps.

The predictions of the RETAIN model are made following five different steps for which Fig. 1 provides a graphical representation. In the following, t refers to the current time-step the prediction is made, r to the number of different input variables, H to the size of the history (the number of past values for every input variable), PH to the prediction horizon, the subscript $i \in [t - H, t]$ to the i-th time-step, and the subscript $j \in [1, r]$ to the j-th input variable.

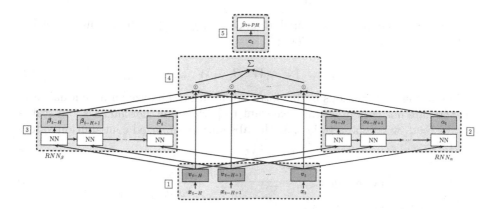

Fig. 1. Graphical overview of the RETAIN model. **Step 1**: The input signals are transformed into embeddings. **Step 2**: Time-level attention weights are computed from the embeddings. **Step 3**: Variable-level attention weights are also computed from the embeddings. **Step 4**: Using the attention weights, the context vector is computed. **Step 5**: The prediction is made from the context vector.

- **Step 1**: Each time-step input vector x_i is linearly transformed into a learnable embedding v_i following: $v_i = W_{emb}x_i$.
- **Step 2**: These embeddings are given as inputs to a recurrent neural network, RNN_α, which outputs the time-level attention weights α_i (see [1] for more details).
- **Step 3**: Similarly, the embeddings are fed into a second recurrent neural network, RNN_β, which computes the variable-level attention weights β_i (see [1] for more details).
- **Step 4**: Using both attention weights, the context vector c_t is computed following: $c_t = \sum_{i=t-H}^{t} \alpha_i \beta_i \odot v_i$.
- **Step 5**: The predictions of the model are made by linearly transforming the context vectors: $\hat{y_{t+PH}} = W c_t + b$.

The only difference between our architecture and the original one is that we do not compute the attention weights in reverse time-order (Steps 2 and 3) [1], but rather in forward order, as the latter yielded better performances for our application.

2.2 Interpretation

In their original paper, the authors of RETAIN propose a way to interpret the outputs of the RETAIN model in the context of multiclass classification. We propose here an adaptation of the methodology to regression problems.

By going through the different operations made in the model, we can express the prediction $\hat{y}_{t+PH}$ in this form:

$$\hat{y}_{t+PH} = \sum_{i=t-H}^{t} \sum_{j=1}^{r} x_{i,j} \alpha_i W (\beta_i \odot W_{emb}[:,j]) + b \tag{1}$$

We can then express the contribution $\omega(\hat{y}_{t+PH}, x_{i,j})$ of the $x_{i,j}$ input feature on the prediction $\hat{y}_{t+PH}$ as follows:

$$\omega(\hat{y}_{t+PH}, x_{i,j}) = \alpha_i W (\beta_i \odot W_{emb}[:,j]) \, x_{i,j} \tag{2}$$

While the contributions in this form are useful to analyze an individual sample, they are not very practical if we want to perform further analysis and statistics. Instead, we propose to look at the absolute normalized contribution values $\omega_{AN}(\hat{y}_{t+PH}, x_{i,j})$:

$$\omega_{AN}(\hat{y}_{t+PH}, x_{i,j}) = \frac{|\omega(\hat{y}_{t+PH}, x_{i,j})|}{\sum_{i=t-H}^{t} \sum_{j=1}^{r} |\omega(\hat{y}_{t+PH}, x_{i,j})|} \tag{3}$$

Taking the absolute values makes the computation of the mean contribution across the samples more representative of the overall contributions, preventing positive and negative contributions from canceling each other. Normalizing the contributions makes the contributions independent from the prediction value itself, enabling a better comparison between samples.

3   Methods

3.1   Experimental Data

In this study, we use the IDIAB dataset whose collection has been approved by the french ethical committee (ID RCB 2018-A00312-53). It is made of data coming from 5 type-2 diabetic patients (4F/1M, age 58.8 ± 8.28 years old, BMI 30.76 ± 5.14 kg/m^2, HbA1c 6.8 ± 0.71%) that have been monitored for 31.8 ± 1.17 days in free-living conditions. Whereas their glucose level (in mg/dL) was recorded through the use of the FreeStyle Libre continuous glucose monitoring device, data related to CHO intakes (in g) and insulin (in units) infusions were manually reported through the mySugr (mySugr GmbH) smartphone coaching app for diabetes.

3.2   Models

In this study, we build global glucose predictive models. Whereas personalized glucode predictive models are often more accurate, global models have the advantage of being easier to train by avoiding overfitting thanks to more training data.

We describe here the preprocessing and training steps of the different models used in this study.

3.3   Data Preprocessing

After splitting the patients into four training patients and one testing patient, we have splitted each training patient's data into a training set and a validation set following a 75%/25% distribution.

To predict the future glucose values at an horizon of 30 min, the models are given as inputs the histories of glucose values, insulin infusions, and CHO intakes of the past 3 h. For every patient, these inputs are standardized (zero mean and unit variance) w.r.t their respective training set.

3.4   Model Training

The Random Forest (RF) model [14] is one of the two baseline models used in this study. Its main strength is that it provides generally good performances while being easily interpretable. Here, a forest of size 100 is fitted using the mean-squared error (MSE) criterion. The minimum number of samples per leaf has been set to 25 to reduce the overfitting of the model to the training set.

Our second baseline, the LSTM model, has been implemented with an architecture that matches the computational complexity of the RETAIN model described below. In particular, every time-step input variables are embedded into a learnable vector of size 64. These embeddings are then given to a 2-layer LSTM model with 128 units per layer. The latter has been trained to minimize the MSE loss function with the Adam optimizer (learning rate of 10^{-3}, mini-batch size of 50). To prevent the overfitting of the network to the training set, the early stopping methodology (patience of 25) has been used.

As for the LSTM model, the RETAIN model has an embedding size of 64. Both RNN_α and RNN_β are made of one layer of 128 LSTM units. Similarly, the Adam optimizer (same learning rate and mini-batch size) with the early stopping methodology was used to fit the model.

All the hyperparameters have been optimized by grid search on the validation set on a subspace delimited by manual search.

3.5   Evaluation

The models have been evaluated with a 4-fold cross-validation on the training patients followed by a leave-one-(patient)-out cross-validation.

Four different metrics have been used: the Root-Mean-Squared Error (RMSE), the Mean Percentage Absolute Error (MAPE), the Time Lag (TL), and the Continuous Glucose-Error Grid Analysis (CG-EGA).

Both the RMSE and MAPE metrics give a measure of the accuracy of the prediction. The TL metric provides an estimate of the time gained by doing the prediction and is computed as the time-shift (in minutes) that maximizes the correlation between the true and the predicted glucose values. Finally, the CG-EGA measures the clinical acceptability of the predictions [6]. By analyzing both the prediction accuracy and the accuracy of the variation between two consecutive predictions, the CG-EGA classifies the prediction either as an accurate prediction (AP), a benign error (BE), or an erroneous prediction (EP). For a model to be clinically acceptable, it needs to have high AP and low EP rates.

4   Results and Discussion

4.1   Experimental Results

The performances of the three models are shown in Table 1 and Table 2. With an average deterioration of 1.4% in RMSE/MAPE/TL when compared to the LSTM model, the RETAIN model displays a comparable prediction accuracy. Its clinical acceptability is also very similar to the LSTM model.

When compared to the RF model, the RETAIN model shows an improvement of 8.5%, 8.4%, 20.7% in the RMSE, MAPE, and TL metrics respectively. It also has a better clinical acceptability with a lower EP rate (−9.7%) which comes at the cost of a slightly lower AP rate (−3.3% of the remaining room for improvement).

Overall, these results are showing that the RETAIN model is a legitimate model for the task of glucose prediction.

Table 1. Performances of the models with mean ± standard deviation, averaged on the population.

Model	RMSE	MAPE	TL
RF	19.23 ± 6.73	9.37 ± 1.58	15.31 ± 3.38
LSTM	**17.52 ± 5.52**	**8.35 ± 1.30**	**12.01 ± 2.36**
RETAIN	17.60 ± 4.90	8.58 ± 0.84	12.14 ± 2.53

Table 2. Clinical acceptability of the models with mean ± standard deviation, averaged on the population.

Model	CG-EGA		
	AP	BE	EP
RF	**86.00 ± 4.37**	**10.79 ± 3.59**	3.21 ± 0.84
LSTM	85.67 ± 3.28	11.46 ± 2.47	**2.87 ± 0.95**
RETAIN	85.54 ± 5.41	11.56 ± 4.50	2.90 ± 0.95

4.2 Interpreting the RETAIN Model

The real strength of the RETAIN model, however, lies in its interpretability. We propose here several different visualization tools for the analysis of the behavior of the RETAIN model. To ease the reading, we will refer to the contribution as the absolute normalized contribution, presented in Sect. 2.2.

First, by looking at the individual maximum contribution of the input variables, we can see if each of them has ever contributed significantly to the prediction. Figure 2 plots the maximum contribution of the model inputs related to the 3-h histories of glucose values, insulin infusions, and CHO intakes. We can see that the older an input value is, the less contribution it has. The decrease in the contribution is faster for the insulin and CHO signals (close to zero after 30 min) than for the glucose signal (close to zero after 60 min). This suggests that it is not useful in this context to use histories that are longer than one hour. Reducing the number of past values inputed to the model should increase the performances by making it harder to overfit and should reduce the training time. Such an analysis is not possible with a standard LSTM model.

From a different perspective, we can look at the behavior of the model when an event occurs. Figure 3 depicts the behavior of the model following the occurrence of two different events: insulin infusions and CHO intakes. We can compare these plots to the mean contribution when no event has occured in the last hour with Fig. 4.

When either one of the events occurs, we can see that the glucose value that has the most importance is not the current glucose value, but the previous one (which is the value 5 min before the event). This specific value keeps a relative high importance as the time moves on. This shows that, when an event occurs, the model uses the last glucose value before the event as a value of reference. On

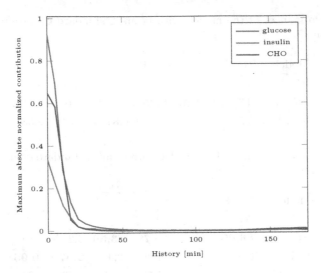

Fig. 2. Maximum absolute normalized contribution of the input signals (history of glucose, insulin, and CHO).

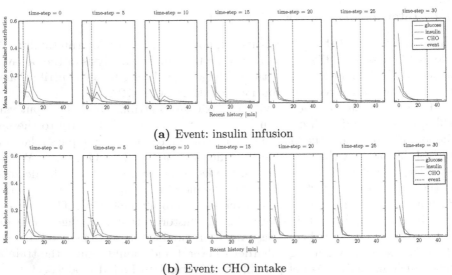

(a) Event: insulin infusion

(b) Event: CHO intake

Fig. 3. Mean evolution through time of the absolute normalized contribution of the input signals (history of glucose, insulin, and CHO) after the occurrence of an event: Fig. 3a, insulin infusion; and Fig. 3b, CHO intake.

the other hand, for both insulin and CHO signals, when their respective event occurs, the contribution of the value of the event is relatively high for the next 20 min. However, after this time, the contribution of the event is close to zero and the mean contribution profile becomes similar to the one for which no event has occured in the past hour, depicted by Fig. 4.

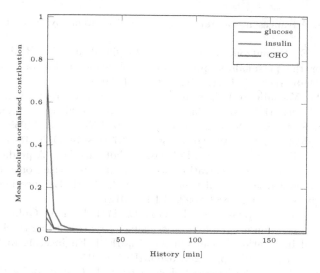

Fig. 4. Mean absolute normalized contribution of the input signals when no event (CHO intake or insulin infusion) occured in the last hour.

5   Conclusion

In this study, we have presented the application of the RETAIN model proposed by Choi *et al.* [1] to the challenging task of 30-min ahead-of-time glucose prediction for diabetic people. Using a two-level attention mechanism, the RETAIN model is able to produce interpretable predictions, which is highly valuable in the context of a biomedical field.

We have evaluated the model on a type-2 diabetic population of 5 patients and compared it against a Random Forest and a LSTM-based recurrent neural network. By being interpetable while respectively equalling and outperforming the LSTM and the RF models, we show that the RETAIN model is very promising.

In the future, we plan to extend the study to another dataset, namely the Ohio T1DM dataset [8]. In particular, this dataset comprises 6 type-1 diabetic patients with similar data. Also, thanks to the interpretability of the RETAIN model, we plan to explore variants of its architecture and input data (e.g., physical activity measures).

Acknowledgment. We would like to thank the diabetes health network Revesdiab and Dr. Sylvie JOANNIDIS for their help in building the IDIAB dataset used in this study.

References

1. Choi, E., Bahadori, M.T., Sun, J., Kulas, J., Schuetz, A., Stewart, W.: Retain: an interpretable predictive model for healthcare using reverse time attention mech-

anism. In: Advances in Neural Information Processing Systems, pp. 3504–3512 (2016)

2. Daskalaki, E., Prountzou, A., Diem, P., Mougiakakou, S.G.: Real-time adaptive models for the personalized prediction of glycemic profile in type 1 diabetes patients. Diab. Technol. Ther. **14**(2), 168–174 (2012)

3. De Bois, M., Yacoubi, M.A.E., Ammi, M.: Prediction-coherent LSTM-based recurrent neural network for safer glucose predictions in diabetic people. In: Gedeon, T., Wong, K.W., Lee, M. (eds.) ICONIP 2019. LNCS, vol. 11955, pp. 510–521. Springer, Cham (2019). https://doi.org/10.1007/978-3-030-36718-3_43

4. Georga, E.I., Protopappas, V.C., Polyzos, D., Fotiadis, D.I.: A predictive model of subcutaneous glucose concentration in type 1 diabetes based on random forests. In: 2012 Annual International Conference of the IEEE Engineering in Medicine and Biology Society, pp. 2889–2892. IEEE (2012)

5. Georga, E.I., Protopappas, V.C., Polyzos, D., Fotiadis, D.I.: Online prediction of glucose concentration in type 1 diabetes using extreme learning machines. In: 2015 37th Annual International Conference of the IEEE Engineering in Medicine and Biology Society (EMBC), pp. 3262–3265. IEEE (2015)

6. Kovatchev, B.P., Gonder-Frederick, L.A., Cox, D.J., Clarke, W.L.: Evaluating the accuracy of continuous glucose-monitoring sensors: continuous glucose-error grid analysis illustrated by therasense freestyle navigator data. Diabetes Care **27**(8), 1922–1928 (2004)

7. Li, K., Daniels, J., Liu, C., Herrero-Vinas, P., Georgiou, P.: Convolutional recurrent neural networks for glucose prediction. IEEE J. Biomed. Health Inf. **24**(2), 603–613 (2019)

8. Marling, C., Bunescu, R.: The OhioT1DM dataset for blood glucose level prediction. In: The 3rd International Workshop on Knowledge Discovery in Healthcare Data, Stockholm, Sweden (2018)

9. Mirshekarian, S., Bunescu, R., Marling, C., Schwartz, F.: Using LSTMs to learn physiological models of blood glucose behavior. In: 2017 39th Annual International Conference of the IEEE Engineering in Medicine and Biology Society (EMBC), pp. 2887–2891. IEEE (2017)

10. Ólafsdóttir, A.F., et al.: A clinical trial of the accuracy and treatment experience of the flash glucose monitor freestyle libre in adults with type 1 diabetes. Diab. Technol. Therapeutics **19**(3), 164–172 (2017)

11. Oviedo, S., Vehí, J., Calm, R., Armengol, J.: A review of personalized blood glucose prediction strategies for T1DM patients. Int. J. Numer. Meth. biomed. Eng. **33**(6), e2833 (2017)

12. Rose, K., Koenig, M., Wiesbauer, F.: Evaluating success for behavioral change in diabetes via mhealth and gamification: Mysugr's keys to retention and patient engagement. Diab. Technol. Ther. **15**, A114 (2013)

13. Sparacino, G., Zanderigo, F., Corazza, S., Maran, A., Facchinetti, A., Cobelli, C.: Glucose concentration can be predicted ahead in time from continuous glucose monitoring sensor time-series. IEEE Trans. Biomed. Eng. **54**(5), 931–937 (2007)

14. Svetnik, V., Liaw, A., Tong, C., Culberson, J.C., Sheridan, R.P., Feuston, B.P.: Random forest: a classification and regression tool for compound classification and QSAR modeling. J. Chem. Inf. Comput. Sci. **43**(6), 1947–1958 (2003)

15. Vaswani, A., et al.: Attention is all you need. In: Advances in Neural Information Processing Systems, pp. 5998–6008 (2017)

16. World Health Organization, et al.: Global report on diabetes. World Health Organization (2016)

Abusive Language Detection Using BERT Pre-trained Embedding

Nasrin Baratalipour$^{(\boxtimes)}$ ⓘ, Ching Y. Suen ⓘ, and Olga Ormandjieva

Concordia University, Montreal, Canada
k_barata@encs.concordia.ca, {suen,ormandj}@cse.concordia.ca

Abstract. The rapid growth in social communication increases the importance of detecting toxic languages. However, detecting toxic language is difficult because of deliberately noisy words and lack of labeled data. These issues cause a low recall in toxic language detection. To address these, we utilized pre-trained BERT models for toxic language detection. We hypothesize pre-trained sub-words embeddings allow BERT models to quickly learn the meaning of obfuscation words and, hence improve the recall of the models on toxic language detection. Our results confirm this hypothesis and show that fine-tuned BERT models perform on a par with the state-of-the-art on the Twitter dataset and outperform the state-of-the-art on the Wikipedia dataset.

Keywords: BERT · Pre-trained word embedding · Abusive language in social media

1 Introduction

Usage of social networks is growing rapidly for sharing private and/or intimate information that assist users to get in close contact with others. However, sharing such information may lead to social consequences. For example, anytime one engages online, whether on message board forums, comments, or social media, there is always a serious risk that he or she may be the target of ridicule and even harassment. In addition, message posts can contain the sharing of some kinds of toxic (abusive or offensive) contents which can yield threats like cyber-bullying. Recent studies show the prevalence of toxic language in social networks and it raises the necessities to detect toxic language [2].

Detecting toxic language is often more difficult than one expects for two reasons:

1. Deliberately noisy inputs: The intentional usage of obfuscation of words and phrases (such as fcukk, w0m3n, banislam), misspelling, and rare words makes toxic language detection difficult [6,8]. As models consider all these noisy words as unknown words, and lose the ability to distinguish between actual rare words and deliberately noisy words.

© Springer Nature Switzerland AG 2020
Y. Lu et al. (Eds.): ICPRAI 2020, LNCS 12068, pp. 695–701, 2020.
https://doi.org/10.1007/978-3-030-59830-3_60

2. Data deficiency: Another obstacle in abusive language detection is the shortage of labeled data. For example, the only available dataset in Twitter has a few hundred thousand (16K) human-labeled examples. However, this quantity is not sufficient to train a significant number of parameters in deep learning models.

As a result of these two issues, machine learning models suffer from a low recall in toxic language detection. For example, HybridCNN [9], one of the early deep learning models for toxic language detection, has achieved a recall of %76.6 and %67.9 in detecting racism and sexism tweets, respectively. To improve the performance of deep learning models several approaches have been proposed. For example, [7] has proposed to use neural character-based models to address noisy inputs. [5] has addressed the lack of annotated data by using pre-trained embeddings to improve recall of their model.

To the best of our knowledge, there is no work which addresses both noisy inputs and data deficiency at the same time. In this paper, we propose to utilize BERT [3] to simultaneously deal with these two problems. BERT uses Wordpieces tokenizer [14] which gives a good balance between the flexibility of single characters and the efficiency of full words. Therefore, it allows encoding of noisy inputs as a sequence of sub-word tokens (which include single characters, if no larger tokens are appropriate), and it also preserves the semantic of common subwords in the representation. Moreover, BERT can use masked language model [3] to train pre-trained embedding for sub-word tokens only by using not-labeled data. [3] have shown that pre-trained BERT models can efficiently be fine-tuned in another task even with low amount of labeled data.

In this work, we use pre-trained BERT models and fine-tuned them on toxic language detection. Our experiment on the Wikipedia dataset and Twitter dataset show that this approach can outperform or achieve on par results with the state-of-the-art approaches in these datasets.

Our contributions are three-fold: first, we improved the state-of-the-art results toxic language detection on Wikipedia dataset, second, we empirically demonstrate that Wordpieces tokenizer can handle obfuscation words/rare words. And lastly, we show how little resources in this domain can be addressed by using pre-trained language model.

2   Related Work

Many efforts have been made recently to detect toxic language in social media [1,4,9]. The obfuscation words, or/and rare words is one challenge in the field as these words are intentionally used to avoid being captured with the abusive detection models [6,8,10,13,15]. The problem of not explicitly dealing with deliberately noisy input is that a single generic out-of-vocabulary embedding is assigned to all un-seen words. Using a single embedding for all unseen words leads to lose the ability to distinguish between obfuscated and non-obfuscated or rare words. [8] addressed this issue by defining the character n-grams features to represent the meaning of a text (containing obfuscation words), and used this character-based representation for the abusive language prediction. [7] have extended this trend by using character-based word composition models to obtain an individual embedding for each word including the un-seen words.

Data deficiency is another challenge of abusive language detection. With the lack of data, the network cannot fully capture the meaning of each word in the data. Transfer learning helps to learn a generic representation of abusive language that is useful for detecting various types of abusive language [11]. The recent works utilize this idea by employing transfer learning through pre-trained language model, and word embedding. [5] have utilized a modified SWEM-concat [12] architecture, incorporated with glove embedding to better handle infrequent and unknown words.

In this work, we use sub-tokens instead of words and train an embedding for sub-tokens. This approach allows us to handle both noisy inputs and data deficiency at the same time.

3 Model

In this work, we utilized BERT to obtain a fixed-dimensional representation of the inputs. BERT (Bidirectional Encoder Representations from Transformers) is an attention based mechanism that learns contextual relations between words (or sub-words) in a text [3]. As Fig. 1 shows the input of the model is a sequence of tokens (each token obtained from Wordpiece [14]), and the output is a sequence of vectors, in which each vector corresponds to an input token with the same index. In BERT an additional token is added to input, which is referred as [CLS]. The final hidden state of [CLS] is assumed to represent the aggregate sequence input. This token is used to predict the classes.

To detect toxic language, we first take a pre-trained BERT model, and fine-tune the parameters end-to-end. To do so, we use [CLS] outputs and pass it to a Softmax layer to get predictions. Then we used a cross-entropy loss function to fine-tune the parameters of the network.

4 Experiments and Results

4.1 Dataset

In this paper, we use two data sets from the literature to train and evaluate our own classifier. To address all the categories of hateful speech, they used different strategies of labeling the collected data. The characteristics of the datasets is listed below.

- Twitter dataset
  [13] released the dataset as a list of 16,907 tweet IDs and their corresponding annotations. At the time of the experiment, we could not extract most of racism Tweets. Hence, we obtain the dataset used in [5]. This dataset contains 15,908 tweets, 1,924 (12%) are racism, 3,086 (19%) are sexism, and the remaining 10,898 (68%) are neither; Note that [13]'s dataset has fairly different distribution, i.e., 11.7% racism, 20.0% sexism, and 68.3% neither.

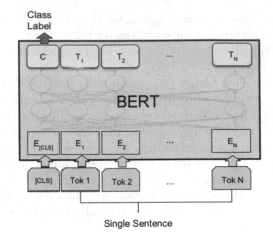

Fig. 1. Toxic language detection model formed by incorporating BERT with one additional output layer. Figure adapted from [3]

– Wikipedia comment dataset
   We utilize the toxicity and personal attack datasets of comments collected
   from the English Wikipedia Talk page [15]. As [7] we use the majority annotation of each comment to resolve its gold label: if a comment is deemed
   toxic (alternatively, attacking) by more than half of the annotators, we label
   it as abusive; otherwise, as non-abusive. 13,590 (11.7%) of the 115,864 comments in W-ATT and 15,362 (9.6%) of the 159,686 comments in W-TOX are
   abusive.

4.2   Experimental Setup

We normalize the input by lowercasing all words, then used BertTokenizer (containing Wordpieces tokenization [14] to split words into sub-word units). We
selected BERT_base as the pre-trained BERT model. The model consists of 12
Transformer blocks, 12 self-attention heads, and 768 hidden dimension with total
parameters of 110M [3].

The classifier is trained with a batch size of 8. The dropout probability was
set to 0.1. Adam optimizer was used with the starting learning rate of 2e−5.
We specifically use torch.optim.lr_scheduler.ReduceLROnPlateau to divide the
learning rate by 0.1 when loss is not decreased for 10 iterations in a row. We
stop the training when the learning rate reaches 1e−10. By this approach we
imitate the early stopping behaviour in our training process.

For the Wikipedia dataset, following previous work [15], we conduct a standard 60:40 train test split experiment on the dataset, in which 95,692 comments
(10.0% abusive) are used for training and 63,994 (9.1% abusive) for testing.
For the Twitter dataset, like [7], we report the macro precision, recall, and F1
averaged over 10 folds of stratified CV.

4.3 Results

Table 1 reports our result along with the state-of-the-art results on Wikipedia and Twitter datasets. As the table shows, we outperform the state-of-the-art on Wikipedia dataset, and achieved comparable results on Twitter dataset. One primary reason for outperforming the state-of-the-art on Wikipedia is the use of a pre-trained model which is trained on Wikipedia articles. It decreases the number of unknown words (included noisy and rare words) exposed the model and helps it to better train the toxic comments in this domain.

In addition to the above experiments, we conducted experiments with and without employing BERT pre-trained language model on Twitter and Wikipedia dataset, to show the importance of using pre-trained embedding, and simply fine-tuning the pre-trained parameters on our downstream task. As it is shown in the Table 2, pre-trained embedding significantly improves the abusive language prediction. For both Wikipedia and Twitter datasets, the BERT model with pre-training provides with similar precision as the model trained from scratch. However, we can see a significant improvement in recall in both dataset. The reason for this improvement is that we do not need to train all the model parameters from scratch. Therefore, with our current datasets (not large enough to train a model from scratch) we still can learn a model which predicts abusive language without losing the model generalization.

Table 1. The F1 comparison of our approach (Pre-trained BERT + fine tuning), and recent approaches on toxic language prediction on Twitter dataset [13], and Wikipedia dataset [15].

Method	Twitter dataset	Wikipedia dataset
Augmented WS + CNG[7]	79.8	87
Augmented HS + CNG [7]	79.21	89.35
TWEM [5]	86	–
Pre-trained BERT + fine tuning (our approach)	84.01	91.27

Table 2. Experiments of our approach with and without pre-trained models. As it is shown the recall significantly improves with employing a pre-trained model.

Method	Twitter dataset			Wikipedia dataset		
	F1	Precision	Recall	F1	Precision	Recall
BERT (from scratch)	0.69	0.86	0.62	0.80	0.82	0.78
Pre-trained BERT + fine tuning	0.84	0.83	0.84	0.91	0.87	0.96

5 Result Analysis and Discussion

To better understand the model behaviour we looked through some of the test data results and recorded the following observations:

– Once pre-training is used the model does not need to learn from scratch to capture the meaning of each word in the corpus, and the linguistic phenomena. Therefore, the model better understands the meaning of a text, and accordingly better learns to detect abusive/sexism comments. For example, the following tweet may do not contain any toxic word, but the language model used in this tweet is toxic:
  any real world examples of women actually being jailed for it?
– Bert tokenizer is trained on Wikipedia. So, the vocabulary used in this tokenizer, does not contain specific domain related words in Twitter, such as "ffs", "fk" etc. For example this tweet was not recognized as abusive language, as the word "fk" is not in current tokenizer vocabulary:
  fk you kat and andre! #mkr
  One idea to improve the model is to re-train Bert tokenizer in order to include more domain specific words in the tokenizer vocabulary and hence less unknown words. Another idea is to provide a list of domain-specific words and explicitly feed it to the network [2].
– Many of the false negatives are references to characters in the TV show "My Kitchen Rules", rather than something about women in general. Such examples may be harmless in isolation but could potentially be sexist or racist in context. For example:
  "classy and elegant" is one way of putting it... #mkr
– We noticed a large number of miss-annotation in the Twitter dataset: *@xxx what is that?* labeled as *sexism*, but does not seem to be.

6 Conclusion and Future Work

In this work we demonstrate that BERT model with the capability of encoding a deep sense of language context, can be successfully utilized for the toxic language detection task. We fine-tuned BERT model on Wikipedia dataset [15] and Twitter dataset [13], two common datasets in the toxic language detection task, and achieved a better performance compared to the state-of-the-art results. For the future work we suggest re-training Wordpieces tokenizer [14] for a large domain-specific unlabeled data (e.g. a large collection of tweets in different topics), and then pre-training BERT Language Model on this dataset. Therefore, the model will be able to better understand the domain specific language, and accordingly predict the toxic comments in that domain.

References

1. Davidson, T., Warmsley, D., Macy, M., Weber, I.: Automated hate speech detection and the problem of offensive language. In: Eleventh International AAAI Conference on Web and Social Media (2017)

2. Delisle, L., Kalaitzis, A., Majewski, K., de Berker, A., Marin, M., Cornebise, J.: A large-scale crowdsourced analysis of abuse against women journalists and politicians on twitter. arXiv preprint arXiv:1902.03093 (2019)
3. Devlin, J., Chang, M.W., Lee, K., Toutanova, K.: Bert: Pre-training of deep bidirectional transformers for language understanding. arXiv preprint arXiv:1810.04805 (2018)
4. Founta, A.M., Chatzakou, D., Kourtellis, N., Blackburn, J., Vakali, A., Leontiadis, I.: A unified deep learning architecture for abuse detection. arXiv preprint arXiv:1802.00385 (2018)
5. Kshirsagar, R., Cukuvac, T., McKeown, K., McGregor, S.: Predictive embeddings for hate speech detection on twitter. In: Proceedings of the 2nd Workshop on Abusive Language Online (ALW2), Brussels, Belgium, pp. 26–32. Association for Computational Linguistics, October 2018
6. Mishra, P., Del Tredici, M., Yannakoudakis, H., Shutova, E.: Author profiling for abuse detection. In: Proceedings of the 27th International Conference on Computational Linguistics, pp. 1088–1098 (2018)
7. Mishra, P., Yannakoudakis, H., Shutova, E.: Neural character-based composition models for abuse detection. arXiv preprint arXiv:1809.00378 (2018)
8. Nobata, C., Tetreault, J., Thomas, A., Mehdad, Y., Chang, Y.: Abusive language detection in online user content. In: Proceedings of the 25th International Conference on World Wide Web, pp. 145–153. International World Wide Web Conferences Steering Committee (2016)
9. Park, J.H., Fung, P.: One-step and two-step classification for abusive language detection on Twitter. arXiv preprint arXiv:1706.01206 (2017)
10. Qian, J., ElSherief, M., Belding, E.M., Wang, W.Y.: Leveraging intra-user and inter-user representation learning for automated hate speech detection. arXiv preprint arXiv:1804.03124 (2018)
11. Sahlgren, M., Isbister, T., Olsson, F.: Learning representations for detecting abusive language. In: Proceedings of the 2nd Workshop on Abusive Language Online (ALW2), pp. 115–123 (2018)
12. Shen, D., et al.: On the use of word embeddings alone to represent natural language sequences (2018)
13. Waseem, Z., Hovy, D.: Hateful symbols or hateful people? predictive features for hate speech detection on Twitter. In: Proceedings of the NAACL Student Research Workshop, San Diego, California, pp. 88–93. Association for Computational Linguistics, June 2016. https://doi.org/10.18653/v1/N16-2013
14. Wu, Y., et al.: Google's neural machine translation system: bridging the gap between human and machine translation. arXiv preprint arXiv:1609.08144 (2016)
15. Wulczyn, E., Thain, N., Dixon, L.: Ex machina: personal attacks seen at scale. In: Proceedings of the 26th International Conference on World Wide Web, pp. 1391–1399. International World Wide Web Conferences Steering Committee (2017)

Classification of Criminal News Over Time Using Bidirectional LSTM

Mireya Tovar Vidal[1]([⊠]) [iD], Emmanuel Santos Rodríguez[1] [iD],
and José A. Reyes-Ortiz[2] [iD]

[1] Faculty of Computer Science, Benemerita Universidad Autonoma de Puebla,
14 sur y Av, C.U., San Claudio, Puebla, Mexico
`mtovar@cs.buap.mx, e.ss.rdz@gmail.com`
[2] Universidad Autonoma Metropolitana, Av. San Pablo Xalpa 180, 02200 Azcapotzalco,
Mexico City, Mexico
`jaro@azc.uam.mx`

Abstract. With the rapid expansion of digital newspapers, readers have an overwhelming amount of news available daily. However, it is difficult to keep track of the news that is only of interest to the reader. Because of this, this research discusses the use of deep learning for the classification of news, especially crime related, published by Mexican digital newspapers as well as an analysis of the predictions obtained through the proposed model. According to the experimental results, the proposed system achieves 98.87% of accuracy.

Keywords: Natural language processing · News classification · Machine learning · Long-Short term memory · Deep learning

1 Introduction

Currently, we have a large amount of data, which increases every second and this trend will only increase in the future. However, most of this data is not structured. Text processing is a very important task because most of the information available, not only on the Internet but in various media is found in written form. Because of this, finding a set of tools to exploit the knowledge contained in text is indispensable. Text classification is the task of assigning labels to a text according to its content. This is one of the most important tasks of processing natural language, because it allows us to apply text classification in different areas such as spam detection, sentiment analysis, content detection inappropriate, social network monitoring, etc. [22]. However, there are different approaches to perform this task as rule-based systems, machine learning based or hybrid systems

In this paper, an approach for crime news classification, based on the headlines, collected from Twitter using deep learning is proposed. A baseline is provided to compare the performance of our approach in terms of accuracy, recall and F_1.

The document is structured as follows: Sect. 2 presents a description of works related to news classification through different methods; in Sect. 3 the solution methodology is

Y. Lu et al. (Eds.): ICPRAI 2020, LNCS 12068, pp. 702–713, 2020.
https://doi.org/10.1007/978-3-030-59830-3_61

shown, and its components are detailed; Sect. 4 discusses the obtained results and Sect. 5 presents the conclusions.

2   Related Work

There are different proposals for text classification, especially for news classification. However, most of these are focused in English news. Therefore, below, various solutions are shown for the problem mentioned above.

Cerviño et al. [1] propose a comparison of three machine learning techniques (ANN, MultiBinomial Naive Bayes and kNN) to tackle Spanish news classification of the newspaper La Capital. The proposed approach relies on TF IDF and dimensionality reduction to weight the documents approximately equal. The best technique for this corpus is, according to the authors, MultiBinomial Naive Bayes, however they didn't perform extensive tests, nor did they provide a benchmark.

The work of Liliana et al. [3] consists in a three-phase system (preprocessing, learning and classification) built upon a Support Vector Machine for classifying Indonesian news. As there are four categories for the news the classification is approached as a one against one (pairwise classification). This system achieves a good accuracy (85%), however the authors suggest trying another multiclass mechanism or diversifying news sources.

Mangal and Goyal [2] describe a news classification system for an Indian language, Punjabi, using Naive Bayes classifier. This system classifies the news into four categories: terror attack news, murder related news, accidental news and suicide news. The news set is obtained from different news websites written in Punjabi. The performance of this proposal is measured through precision, recall and F1 and it achieves a reasonable success even though punjabi has a poor resource pool for preprocessing tools.

Krishnalal et al. [4] propose a system for classifying online news based on Hidden Markov Model for feature extraction and multiclass Support Vector Machines to perform classification. A comparison provided by the authors shows that this approach outperforms kNN and SVM in terms of accuracy. Nevertheless, there is ambiguity in the features that may lead to misclassification. It is important to highlight that the preprocessing phase for this system is crucial because it reduces computing time and noise.

Dilrukshi et al. [5] introduce a method to classify news as from the headlines shared on Twitter by Sri Lankan news groups. To extract the features, they used Bag-of-Words approach, but they reduced data dimensionality by removing the most and the least frequent words, as it's likely that these don't carry important information. For the classification task, a Support Vector Machine is used as it supports high dimensional data. The system performance is measured through effectiveness which is defined as the system ability to satisfy the user in terms of the relevance of the retrieved messages. At the same time, the effectiveness is measured by precision and recall. The overall effectiveness of the proposed system is good, except for one of the classes.

Nowak et al. [6] demonstrate how to classify text using RNN variants such as LSTM, Bi LSTM and GRU. These tests are performed on three datasets. We have special interest in the amazon books review dataset because the authors only used review titles to perform sentiment analysis. Also, they compare this approach to a Bag-Of-Words one. The provided results show that the all neural networks models, especially Bi LSTM, outperform the Bag-Of-Words approach in terms of accuracy by a considerable margin. However, it should be considered using full reviews on the Amazon dataset.

Fauzi et al. [21] discuss the high dimensionality problem associated created by selecting irrelevant features for classification, resulting in a declining accuracy. They address this problem using Information Gain for feature reduction in a first phase and Maximal Marginal Relevance for Feature Selection to select features in a second phase. The classification in done using Naive Bayes classifier. This two-phase system is proven to achieve good accuracy with reduced computational cost.

García-Mendoza and Gambino Juárez [23] proposed a system which classifies news into sections for each newspaper. The news used were gathered in a span of six months from twitter accounts of three Mexican newspapers. This approach relies in a vector representation of the news articles content to subsequently apply different classifiers such as decision trees, Naive Bayes, SVM and logistic regression. The results show that the best classifiers for this task are SVM and logistic regression.

3 Proposed Approach

This section is divided into two parts. First, the model architecture of this proposal is discussed, then the key parts of the model are presented.

The model proposed in this paper consists of the following components, as presented in Fig. 1:

- Input layer
- Embedding layer
- LSTM layer
- Max pooling layer
- Fully connected layer
- Output

3.1 Word Embeddings

The input layer for this model is a dense vector representation of the words in the document fed into the input, where each word is represented by a real value. This representation, called word embeddings, provides some advantages over sparse word representations, such as One-Hot Encoding, because it requires less dimensions and it preserves better the semantic of the words. Given the relatively small dataset, we decided to use pre-trained embeddings rather than training the word embeddings. Therefore, we employ Spanish word embeddings, which are trained on the Spanish Billion Word Corpus [7] using Word2Vec.

3.2 Long Short-Term Memory

Recurrent Neural Networks (RNN) can use their memory to process input sequences, thus they can deal with sequential data such as text. However, they suffer from van-ishing or exploding gradient problems [9]. To overcome this problem, a new architecture was designed: Long Short-Term Memory (LSTM) which can effectively deal with long term dependencies through its gate mechanism. LSTM units are mainly composed of three elements: input gate, output gate and forget gate. The input gate controls the information added to the cells, the forget gate decides what information should be kept or removed and the output gate decides what parts of the cell state will be output. However, an in-depth look of this architecture can be found in [8, 10].

For this approach, we want to exploit both future and past information, therefore we propose a bidirectional Long Short-Term Memory. As stated in [11], this architecture is more effective than unidirectional LSTMs and is appropriate for tasks where context is crucial. Also, it's convenient to have access to future and past context for sequence modeling tasks [15].

3.3 Regularization

Dropout [12] is a regularization technique where the core idea is to randomly drop units as well as their connections during training to prevent overfitting, improving the network performance in a variety of domains. We use dropout in the penultimate layer of the model.

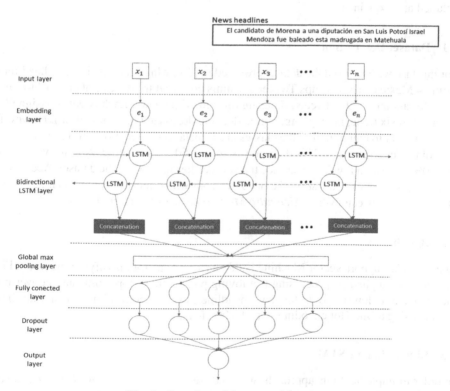

Fig. 1. Overview of the proposed system

On the other hand, we apply L2 Regularization for the LSTM layers, which is defined as follows.

$$|\mathrm{w}|_2^2 = \sum_{i=1}^{m} |w_i|^2 \qquad (1)$$

3.4 Pooling

We employ a one-dimension global max pooling layer after the Bidirectional LSTM layer to extract representative information. This reduces the dimensionality but keeps the key information. However, this operation loses information about locality of the words in a text. In [13], the impact of the pooling strategy is discussed, and it is stated that 1-max pooling outperforms other pooling strategies for sentence classification.

4 Results and Discussion

In this section, a description of the used data is analyzed, and the results obtained are detailed and explained.

4.1 Dataset Description

For this task we have used the dataset provided in [20], which consists in news headlines from 24 Mexican news groups Twitter accounts collected from November 2017 to June 2019 for a total of 7988 tweets, focusing only in crime news. Each tweet is assigned a label out of six (rape, kidnapping, homicide, suicide, assault or exploitation). However, it is important to highlight that the classes are imbalanced since most crimes committed are homicides and assault. This dataset was split in 60% for training, 20% for validation and 20% for test. In Fig. 2, we can see the sources for the news in the dataset. According to Fig. 2, *Noticieros Televisa* is the account from which most tweets percentage were obtained (1639 tweets) while *Alejandro Marti* represents the least (1 tweet).

4.2 Baseline

In this work, Support Vector Machine is taken as baseline for classifying the news. For this, first we applied preprocessing (removing punctuation, stop words and converting the sentences to lowercase) to the news headlines, then we converted the headlines into vectors using a One-Hot encoding with 10,000 features.

4.3 Bidirectional LSTM

In order to implement our approach, we used Keras with TensorFlow 2.0 as backend, which in turn allowed us to train the network using a GPU. The experiments were run on a GeForce GTX 1050Ti GPU and an Intel I5 CPU. The network is trained to minimize the binary cross-entropy loss, using ADAM optimization with batch size 64 during 50

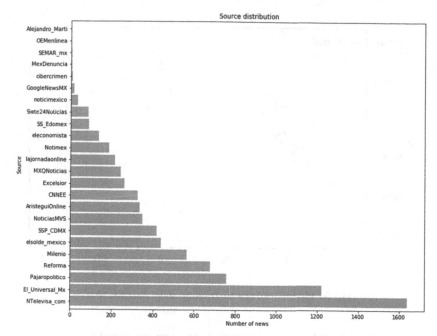

Fig. 2. Source distribution for the dataset

epochs. We performed Random Search [19] using Keras tuner to determine the number of units for hidden layers and the dropout rate. We used 190 units for LSTM and fully connected layers. The L2 regularization used is on the order of 10^{-3} and the dropout was set to 0.15. For the embedding layer, we employ the 300-dimensional word vectors trained by Cardellino [7].

In Fig. 3 we can observe the plot of the model history for training and validation sets. According to the figure, it might be better to stop the training before epoch 25 using a callback function.

4.4 Classification Results

The results obtained by the proposed model and the baseline were measured using Scikit-Learn metrics. We employ accuracy, precision and balanced accuracy as well as classification report to get a detailed insight.

After training our model for 50 epochs, we obtained the classifications results which are shown in Table 1. According to Table 1, it is observed that the class with lowest scores is class *Suicide*, even though it has more support than class *Exploitation*.

With respect to the baseline, the scores for precision are similar in Kidnapping and Rape classes, but for the rest are lower than the proposed model. This can be observed specially in the Exploitation class where our model scores 100% precision while the baseline obtains 60%. However, the baseline recall and F1 scores are worse than the proposal for the classes with the lowest support (Table 2).

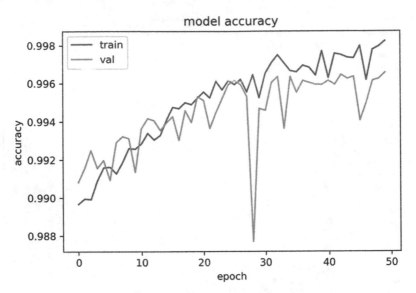

Fig. 3. Plot of accuracy for training and validation sets

Table 1. Classification report for the proposed model

Class	Precision	Recall	F_1	Support
Assault	99	100	100	542
Exploitation	100	100	100	26
Homicide	99	99	99	768
Kidnapping	97	97	97	76
Suicide	92	92	92	63
Rape	98	99	99	123

Table 2. Classification report for the baseline

Class	Precision	Recall	F_1	Support
Assault	93	92	92	542
Exploitation	60	23	33	26
Homicide	86	95	91	768
Kidnapping	99	91	95	76
Suicide	83	40	54	63
Rape	96	82	89	123

Nevertheless, when comparing the models, it can be seen in Table 3 that our proposal outperforms the baseline in all the metrics used.

Table 3. Comparison of the baseline and proposed model using accuracy, precision and balanced precision

Model	Accuracy	Precision	Balanced precision
Bi-LSTM	98.87	97.86	97.74
Baseline	89.54	86.23	89.54

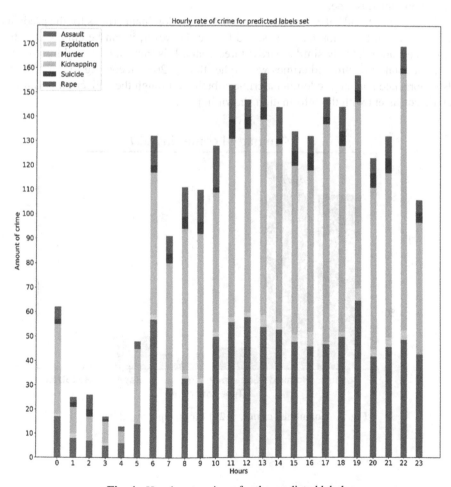

Fig. 4. Hourly rate crimes for the predicted labels

4.5 Analysis

Once we obtained the predictions for the test set, we perform a brief analysis of these predictions to determine the most committed crimes through time. We specifically focus on hourly and annually crimes.

In Fig. 4, it is observed the hourly rate crimes for the predicted labels of the tests set. Each color represents a type of crime: red for assault, yellow for exploitation, green for murder, cyan for kidnapping, dark blue for suicide and magenta for rape. It is easy to see that most common crimes are assault which reaches its peak at 7:00 p.m. while murder reaches its at 10:00 p.m. Also, it is important to note that the hour with less committed crimes is at 4:00 am (13 crimes) and the hour with most committed crimes is at 10:00 p.m. (169 crimes).

It is interesting to note that at 4:00 am, besides the predominant crimes, the only crime committed is rape.

From Fig. 5, 6 and 7 it can be perceived that the predominant crimes for the predicted labels are assault and homicide, as stated before. However, it can be appreciated that crime rates are relatively similar across three years. It is important to highlight that in 2018 the total of committed crimes was higher than in 2017 according to [14]. We can also appreciate that the predominant crimes obtained through the collection of our data set are consistent with the information shown in [14].

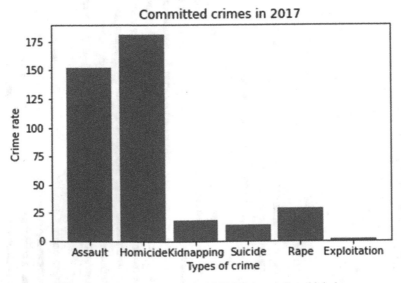

Fig. 5. Committed crimes in 2017 for the predicted labels

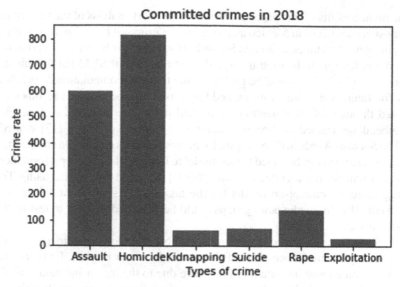

Fig. 6. Committed crimes in 2018 for the predicted labels

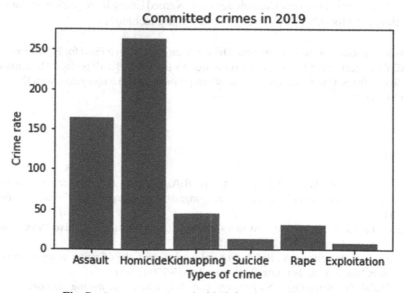

Fig. 7. Committed crimes in 2019 for the predicted labels

5   Conclusions

In this paper, a model has been proposed to classify news headlines from crimes. In this aspect, a good percentage of correctly classified instances was reached while outperforming the baseline in all the metrics used, but especially in terms of recall and F_1.

The main benefits of this paper are to provide a short analysis of the time of the day when most crimes occur in México and to present a proposal based on deep learning for classifying news headlines written in Spanish. However, it is important to point out that this model is feasible to be used in large datasets instead of SVM but for this dataset, Support Vector Machine should be preferred due to lower computational complexity.

As for future work, the word embeddings used can be changed to other models calculated through different methods or trained on another corpora, such as FastText [16] embeddings trained on Spanish Unannoted Corpora or Glove [17] embeddings trained on Spanish Words Billion Corpus, to improve metrics results. On another hand, an attention mechanism can be added to this model to improve the overall performance [15]. However, a transformer architecture like BERT [18] could be applied, using Transfer Learning, to reach state-of-art results for the task here described. Additionally, more tweets from different media news groups could be collected in order to test further the proposed model.

In addition, the importance of work related to Natural Language Processing in Spanish must be highlighted since most of the works focus on English. Furthermore, this topic will become more important in the future due to the increasing data available. In this light, a system capable of classifying news according to some specific criteria such as crime news only could be useful for designing recommendation systems to digital newspaper readers or to extract knowledge (e.g. Named Entity Recognition, Relationship Extraction, etc.) for other tasks through data mining techniques.

Acknowledgment. This work is supported by the Sectoral Research Fund for Education with the CONACYT project 257357 and partially supported by the VIEP-BUAP project. The authors also would like to thank Universidad Autonoma Metropolitana, unit Azcapotzalco, with the research project SI001-18.

References

1. Cerviño Beresi, U., García Adeva, J.J., Calvo, R.A., Ceccatto, H.A.: Automatic classification of news articles in Spanish. In: Actas del Congreso Argentina de Ciencias de Computación (2004)
2. Bajaj Mangal, S., Goyal, V.: Text news classification system using Naïve Bayes classifier. Res. Cell Int. J. Eng. Sci. **3**, 209–2013 (2014)
3. Liliana, D.Y., Hardianto, A., Ridok, M.: Indonesian news classification using support vector machine. World Acad. Sci. Eng. Technol. **81**, 767–770 (2011)
4. Krishnalal, G., Rengarajan, S., Srinivasagan, K.: A new text mining approachl based on HMM-SVM for web news classification. Int. J. Comput. Appl. **1**, 98–104 (2010)
5. Dilrukshi, K., De Zoysa, K., Caldera, A.: Twitter news classification using SVM. In: 2013 8th International Conference on Computer Science & Education, Colombo, pp. 287–291 (2013)
6. Nowak, J., Taspinar, A., Scherer, R.: LSTM recurrent neural networks for short text and sentiment classification. In: Rutkowski, L., Korytkowski, M., Scherer, R., Tadeusiewicz, R., Zadeh, Lotfi A., Zurada, Jacek M. (eds.) ICAISC 2017. LNCS (LNAI), vol. 10246, pp. 553–562. Springer, Cham (2017). https://doi.org/10.1007/978-3-319-59060-8_50
7. Cardellino, C.: Spanish Billion Words Corpus and Embeddings, March 2016. https://crscardellino.github.io/SBWCE/

8. Hochreiter, S., Schmidhuber, J.: Long short-term memory. Neural Comput. **9**(8), 1735–1780 (1997)
9. Hochreiter, S., Bengio, Y., Frasconi, P., Schmidhuber, J.: Gradient flow in recurrent nets: the difficulty of learning long-term dependencies (2001)
10. Gers, F., Schraudolph, N., Schmidhuber, J.: Learning precise timing with LSTM recurrent networks. J. Mach. Learn. Res. **3**, 115–143 (2002)
11. Graves, A., Schmidhuber, J.: Framewise phoneme classification with bidirectional LSTM and other neural network architectures. Neural Netw. Official J. Int. Neural Netw. Soc. **18**(5), 602–610 (2005)
12. Srivastava, N., Hinton, G.E., Krizhevsky, A., Sutskever, I., Salakhutdinov, R.: Dropout: a simple way to prevent neural networks from overfitting. J. Mach. Learn. Res. **15**, 1929–1958 (2014)
13. Zhang, Y., Wallace, B.C.: A sensitivity analysis of (and practitioners' guide to) convolutional neural networks for sentence classification. In: Proceedings of International Joint Conference Natural Language Processing, pp. 253–263 (2017)
14. Incidencia delictiva del Fuero Común. https://www.gob.mx/sesnsp/acciones-y-programas/inc idencia-delictiva-del-fuero-comun-nueva-metodologia. Accessed 11 Dec 2019
15. Jing, R.: A self-attention based LSTM network for text classification. J. Phys: Conf. Ser. **1207**, 012008 (2019)
16. Bojanowski, P., Grave, E., Joulin, A., Mikolov, T.: Enriching word vectors with subword information. Trans. Assoc. Comput. Linguist. **5**, 135–146 (2016)
17. Pennington, J., Socher, R., Manning, C.: Glove: global vectors for word representation. In: Proceedings of Conference on Empirical Methods in Natural Language Processing, vol. 14, pp. 1532–1543 (2014)
18. Devlin, J., Chang, M., Lee, K., Toutanova, K.: BERT: pre-training of deep bidirectional transformers for language understanding. NAACL-HLT **1**, 4171–4186 (2019)
19. Bergstra, J., Bengio, Y.: Random search for hyper-parameter optimization. J. Mach. Learn. Res. **13**, 281–305 (2012)
20. Reyes-Ortiz, J.A., Bravo, M.: Enhancing patterns with linguistic information for criminal event recognition. J. Intell. Fuzzy Syst. **34**(5), 3027–3036 (2018)
21. Fauzi, M.A., Arifin, A.Z., Gosaria, S.C., Prabowo, I.S.: Indonesian news classification using naïve bayes and two-phase feature selection model. Indonesian J. Electr. Eng. Comput. Sci. **8**(3), 610–615 (2017)
22. Aggarwal, C.C., Zhai, C.: A survey of text classification algorithms. In: Aggarwal, C., Zhai, C. (eds.) Mining text data, pp. 163–222. Springer, Boston (2012). https://doi.org/10.1007/ 978-1-4614-3223-4_6
23. García-Mendoza, C.-V., Gambino Juárez, O.: News article classification of Mexican newspapers. In: Mata-Rivera, M.F., Zagal-Flores, R. (eds.) WITCOM 2018. CCIS, vol. 944, pp. 101–109. Springer, Cham (2018). https://doi.org/10.1007/978-3-030-03763-5_9

End-to-End Generative Adversarial Network for Palm-Vein Recognition

Huafeng Qin[1]([⊠]) [iD] and Mounîm A. El Yacoubi[2] [iD]

[1] Chongqing Engineering Laboratory of Detection Control and Integrated System, Chongqing Technology and Business University, Chongqing, China
qinhuafengfeng@163.com
[2] Telecom SudParis, Institut Polytechnique de Paris, Courcouronnes, France
mounim.el_yacoubi@telecom-sudparis.eu

Abstract. Palm-vein recognition has received increasing researchers' attention in recent years. However, palm-vein recognition faces various challenges in practical applications, one of which is the lack of robustness against image quality degradation, resulting in reduction of the verification accuracy. To address this problem, this paper proposes an end-to-end convolutional neural network to automatically extract vein network features, thus without resorting to any hand-crafted features. Firstly, we label the palm-vein pixels based on several handcraft-based segmentation methods and reconstruct a training set accordingly. Secondly, an end-to-end vein segmentation model is proposed based on a generative adversarial network. After training, this model outputs a map where each value is the probability that the corresponding pixel belongs to a vein pattern. The resulting map is then subject to binarization by thresholding and stored in a binary image, used subsequently for verification matching. The experimental results on the public CASIA palm-vein dataset demonstrate the effectiveness of our proposed method.

Keywords: Palm-vein · Generative adversarial network · U-Net · Convolutional neural network

1 Introduction

With the rapid development of digital economy and internet technology, traditional verification techniques based on passwords or smart cards can hardly meet the requirements of convenience, reliability, and security in practical applications. In this context, automatic personal verification using physiological and/or behavioral characteristics has received increasing attention and has become one of the most critical and challenging tasks in information security. Currently, the various modalities employed for personal verification are divided into two categories: (1) Extrinsic modalities: fingerprint [1], iris [2], gait [3], face [4], hand shape [5] and signature [6], and (2) Intrinsic modalities such as finger-vein [7], palm-vein [8], dorsal hand-vein [9]. The extrinsic biometric traits are prone to attack, causing some concerns on privacy and security in practice applications. For example, face and fingerprint are easy to acquire and their fake versions have

© Springer Nature Switzerland AG 2020
Y. Lu et al. (Eds.): ICPRAI 2020, LNCS 12068, pp. 714–724, 2020.
https://doi.org/10.1007/978-3-030-59830-3_62

been successfully employed to fool the recognition systems. Similarly, iris recognition systems are also prone to be attacked by the iris fake versions [10, 11].

Unlike the extrinsic biometric features, vein traits are concealed in our body and thus are not easy to copy, which results in high security and privacy. Compared with the finger-vein, the palm-vein has more complex lines and structures, that are beneficial to improve the verification rate. Therefore, the palm-vein verification technology has received more and more attention from researchers and industry.

2 Related Works

The veins are concealed beneath the skin and are therefore difficult to observe in visible light. However, they can be captured by infrared illumination with wavelength of 850 nm [7–9]. Some medical research works [12, 13] have shown that the structure of blood vessels has high uniqueness for each individual [13], and can distinguish even identical twins [12, 14]. Therefore, the vein authentication technology has been widely investigated in the past years [7–9, 15–27].

Palm-vein recognition is still a challenging task because the acquisition process is inherently affected by many factors such as temperature, equipment, user habits, illumination, etc. As a result, the captured image includes noise and irregular shadow, which will ultimately degrade the performance of palm-vein verification systems. To solve these problems, various methods are proposed to extract the palm-vein texture. In general, they can be broadly classified into two categories:

(1) Handcraft-based segmentation approaches: Some researchers assume that the cross-profile of vein patterns show a valley shape and build a mathematical model to detect valleys for vein segmentation. The representative approaches include line tracking methods [15, 16], and curvature-based measures [17–20]. Other studies observe that the vein patterns show a line-like texture in a predefined neighborhood region, so the Gabor filters [7, 21], matched filters [22], wide line detector [23] and neural networks [24] are proposed to extract vein textures.

(2) Deep learning-based segmentation methods: Convolutional neural networks have shown to outperform the state of the art in the computer vision field. Different from handcraft-based segmentation methods, deep learning based segmentation methods are an end-to-end architecture without the manual attribute distribution assumption. Some researchers have applied it for vein segmentation. For example, Qin et al. [25] proposed a CNN to segment vein patterns for verification. To improve the performance, an iterative deep neural network [26] is proposed for hand-vein verification. To address the wrong label problem, a generative adversarial network (GAN) [27] is employed to extract the finger-vein texture.

The handcrafted methods [7–9, 15–24] described above depend on the assumption distributions such as valleys and line segments. However, these assumptions may not be always effective because the vein pixel values may create different distributions. Therefore, these existing methods do not always perform well in practical applications. The deep learning based methods [25–27] are capable of extracting the vein texture without requiring any assumption and have shown better performance than handcrafted approaches. However, they still have two critical issues. First, they divide an image into various patches and build a patch-based dataset to train a deep neural network for feature extraction. Therefore, they do not account for global correlations when processing individual patches, which may lead to failures caused by noise and local irregular shadow regions. In fact, there is a tradeoff between localization accuracy and the use of global context. Large patches require more max-pooling layers which may degrade the localization accuracy, while small patches prevent the network from capturing the global context. Second, the classification strategy for each patch is computationally intensive for both the training and testing phases.

The adversarial learning framework allows us to model the underlying distribution of plausible samples only from training data, without manually interacting with parameters controlling complex mathematical models. This framework was successfully applied for various computer vision tasks. Some researchers brought it into medical image segmentation field [28, 29] and harnessed it for retina image segmentation, brain segmentation, and neuronal membranes segmentation.

Inspired by this idea, in this paper, we propose an end-to-end vein image segmentation model for hand-vein verification. The main contributions of this paper can be summarized as follows: 1) this work makes the first attempt to accommodate a generative adversarial network (GAN) on the hand-vein verification task. First, different from existing approaches based on manual labeling schemes, we employ an existing hand-crafted image segmentation approach to extract the vein network from an image and the resulting binary image is used to automatically label each pixel. Secondly, unlike exiting vein segmentation approaches [25–27] which divide an image into several patches and construct a patch-based set for training, the entire hand-vein images and theirs corresponding ground truths (binary images) are directly input into GAN for training. In the testing phase, the entire image is taken as input of GAN to predict the probability of each of its pixels to belong to a vein pattern. Therefore, our approach takes into account localization and the use of global context at the same time. Moreover, it can directly predict binary vein patterns from vein images in one forward propagation without preprocessing and post-processing, which results in low-time cost. to perform training of the hand-vein generation module and the hand image to vein network mapping, the adversarial loss and the binary cross entropy loss are combined in our GAN model. The U-net allows to generate realistic vein networks by minimizing an adversarial loss, while the cross entropy loss guarantees that the output of U-net is globally consistent. Therefore, the proposed framework provides an effective end-to-end hand vein image segmented tool, capable of extracting vein networks from an input hand vein image. 2) We carry out rigorous experiments to investigate the capacity of our vein segmentation model. Our experimental results show that the proposed model is capable of extracting the vein patterns from raw hand vein images and achieve better performance compared to existing approaches.

3 The Proposed Method

In this paper, we propose a generative adversarial network to extract palm vein patterns for verification. Firstly, given an image, various handcraft-based methods are employed to segment the vein network and the resulting binary images are combined to label the vein pixels. Then, a generative adversarial network with a U-Net structure is built and trained to segment the vein patterns. Finally, we match the vein patterns for verification.

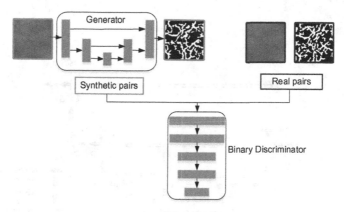

Fig. 1. Overview of our approach

3.1 Labeling Vein Pattern

Similar to work [25], we employ the Repeated line tracking [15], Gabor filters [7], and Hessian phase [9] techniques to segment the vein patterns in a given image I. We get in this way three corresponding binary vein images L_1, L_2, and L_3, where 1 and 0 denote the vein pixel and background pixel respectively. Finally, the three binary images are combined to obtain a labeled image L as follows.

$$L(i,j) = \begin{cases} 1 & \sum_{k=1}^{K} l_k(i,j) \geq \frac{K}{2} \\ 0 & \sum_{k=1}^{K} l_k(i,j) < \frac{K}{2} \end{cases} \tag{1}$$

where K is the number of segmentation approaches. We use the labeled map $L(i,j)$. (0 and 1 denote background and vein pixels respectively) as the ground truth of the corresponding grayscale palm-vein image I and construct accordingly the training set.

4 GAN Framework Structure

GANs [30] is a specific framework of a generative model aiming at implicitly learning the data distribution p_{data} from a set of samples $(x_1, x_2, \ldots, x_m)$ (e.g. images) in order to further generate new ones drawn from the learned distribution. In our work, we have explored GAN for hand-vein recognition. The GAN consists of two modules i.e. generator and discriminator. The generator takes a hand image as input and generates a vein probability map with the same size. The values in the probability map, ranging from 0 to 1, indicate the probability of each pixel to belong to a vein pattern. The discriminator takes a hand image and a vein image for training and determine whether the vein image is the ground truth (real) or rather the output of the generator (fake). The framework of our GAN is depicted in Fig. 1.

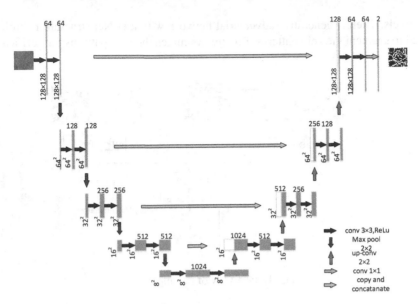

Fig. 2. The architecture of U-Net

Generator Architecture. For the generator, we follow the spirit of U-Net [31] where the initial convolutional feature maps are skip-connected to upsampled layers from bottleneck layers. This skip-connection is crucial to segmentation tasks as the initial feature maps maintain low-level features such as edges and blobs that can be properly exploited for accurate segmentation. As shown in Fig. 2, the U-Net architecture consists of a contracting path and an expansive path. In the contracting path, down-sampling is a classical convolutional neural network, which consists of five layers. In each layer, there are two 3×3 convolution layers, followed by a Leaky ReLU activation function $f(x) = max(x, leak \times x)$ and a 2×2 max pooling layer with stride 2. Here, the first derivative of Leaky ReLU activation function is 1 for $x > 0$ and *leak* for $x \leq 0$. So, the leaky rectifier allows for a small, non-zero gradient when the units are saturated and not active. In other words, the Leaky ReLU network still learns slowly when training traditional ReLU networks with constant 0 gradients. The number of convolution kernels in the next layer is twice times than the number of convolution kernels in the previous layer, and finally the number of convolution kernels in the fifth layer is 1024. In the expansive path, each layer has two 2×2 up-convolution layers, followed by two 3×3 convolution layers and a Leaky ReLU activation function.

Discriminator Architecture. The discriminator network has a typical CNN architecture that takes the input image of size 128×128 and outputs one decision: is this a real pair (ground truth) or is it a fake pair (output of generator)? In this network, there are four convolutional layers with a kernel size of 3×3 and one fully connection layer. Strided convolutions are applied to each convolution layer to reduce spatial dimensionality instead of using pooling layers. Batch-normalization is applied to each layer of the network, except for the input and output layers. The Leaky ReLU activation functions is applied to all layers except the output layer which uses the Sigmoid function for the likelihood probability score of the image.

Objective Function. Let the generator G be a mapping from a hand image x to a vein image y. Then, taking a pair of (x, y) as input, the discriminator D makes a binary decision $\{0, 1\}$, where 0 or 1 represent that y is produced by the generator or is the ground truth, respectively. Adversarial networks are trained by optimizing the following loss function of a two-player minimax game.

$$L_{adv} = \mathbb{E}_{x,y \sim p_{data}(x,y)}\big[\log D(x, y)\big] + \mathbb{E}_{x \sim p_{data}(x)}\big[\log(1 - D(x, G(x)))\big] \qquad (2)$$

where $\mathbb{E}_{x,y \sim p_{data}(x,y)}$ is the expectation over the pairs (x, y) sampled from the joint data distribution of real pairs $p_{data}(x, y)$ and $\mathbb{E}_{x \sim p_{data}(x)}$ is the expectation over the x sampled from the real vein network distribution $p_{data}(x)$.

Although optimizing the above loss function induces G to generate visually sharp results, recent work in [28] and [29] has shown that considering some global loss such as L_1 provides more consistent results for image synthesis. Inspired by works [28] and [29], as the binary cross entropy loss [31] has shown good performance for image segmentation, we employ it to penalize the distance between the ground truth and the output of generator and formulate accordingly the following objective function.

$$L_{seg} = L_{adv}(G, D) + \lambda L_{binary} \qquad (3)$$

where $L_{binary} = \mathbb{E}_{x,y \sim p_{data}(x,y)} - y.\log(G(x)) - (1 - y)\big[\log(1 - G(x)\big]$ and λ balances the contribution of the two losses. To train the model, we use the Adam optimizer with a fixed learning rate of $2e^{-4}$ and an exponential decay rate for the 1st moment estimates of $\beta_1 = 0.5$, and we set the trade-off coefficient in Eq. (3) to $\lambda = 500$.

5 Experiment and Results

Our methods are implemented in Matlab and conducted on a high performance computer with 8 Core E3- 1270v3 3.5 GHz processor, 16 GB of RAM, and a NVIDIA Quadro GTX1080ti graphics card. To test the performance of our approach, we have carried out several experiments on a public hand-vein database. Some existing approaches (i.e. Repeated line tracking [15], Gabor filters [7], and Hessian phase [9]) have been employed to segment the hand-vein network and the resulting segmentation images are used to label the vein pixels in each image by Eq. (1). Then, a data set is constructed to train the proposed model. For testing, our model takes a hand image as input and outputs an image map of the same size that is then subject to binarization by a threshold of 0.5. We match the resulting binary vein image for verification. In our tests, the Maximum principle curvature [20] and CNN [25] techniques also show good performance for vein verification. In the experiments, we compare all approaches mentioned above to get more insights into the problem of hand-vein verification. The performances are shown in the following experiments.

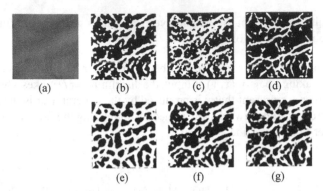

Fig. 3. Segmented results. (a) Original hand-vein image, (b) vein patterns extracted from (a) using Hessian phase, (c) vein patterns extracted from (a) using Repeated line tracking, (d) vein patterns extracted from (a) using Maximum principle curvature, (e) vein patterns extracted from (a) using Gabor filter, (f) vein patterns extracted from (a) using CNN, and (f) vein patterns extracted from (a) using the proposed approach.

5.1 CASIA Datasets

The CASIA Multi-Spectral Palm-print Image Database includes 7200 palm images that are collected by a self-designed multiple spectral imaging device. All images with a resolution of 8 bit are collected from 100 different people. These images are taken from six different wavelength bands which are 460 nm, 630 nm, 700 nm, 850 nm, 940 nm and white light respectively. All images are captured in two separate sessions with a time interval of more than one month. In each session, the left hand and right hand of each subject provide 3 image samples respectively. In this paper, we only employ the images acquired under the 850 nm wavelength. Totally, there are 1200 images (100 subjects × 2 hands × 2 sessions × 3 images). In our experiments, a pre-processing method [26] is employed to extract the region of interest (ROI) image and to resize the resulting image to 128 × 128.

5.2 Verification Results

There are 200 hands in the CASIA Dataset. In our experiments, we split it into three datasets: 70 hands with 420 (70 hands × 6 samples) samples in the training dataset, 30 hands with 180 (30 hands × 6 samples) samples in the validation dataset, and 100 hands with 600 (100 hands × 6 samples) samples in the test dataset. The training dataset is used to train our approach and the validation dataset is employed for hyper-parameter selection. After training, our model can be used to segment the vein network. For comparison, we also apply existing approaches to extract the vein patterns. The segmentation results of the compared approaches are shown in Fig. 3. From Fig. 3, we observe that the Repeated line tracking-based approach and the Gabor filter-based approach are prone to over-segmentation (generate additional non-vein patterns). By contrast, the Maximum principle curvature descriptor fails to extract some actual vein patterns. Hessian phase and CNN show better performance, but these techniques generate noise and broken vein patterns. Compared to existing approaches, the proposed approach is robust to noise and can extract connected and a smooth vein network, as shown in Fig. 3(g).

Table 1. EER of various approaches

Methods	EER (%)	Computation time (s)
Repeated line tracking [15]	4.00	8.35
Maximum principle curvature [20]	2.33	1.43
Gabor filters [7]	1.00	2.96
Hessian phase [9]	1.33	1.25
CNN[25]	0.74	5.58
The proposed approach	0.33	0.03

From the quantitative viewpoint, our experiments aim at assessing the verification performance of our approach w.r.t existing approaches. The test dataset consists of 100 hands with 600 (100 hands × 6 samples) samples, captured in two sessions. We select the 3 hand-vein images acquired during the first session as training data while the corresponding 3 images acquired during the second session are employed as testing data. Thus, 300 (100 × 3) genuine scores are produced by matching images from the same hand. The impostor matching score computation is time costly as there are 178, 200 (6 × 6 × 100 × 99/2) matching groups. To reduce time cost, all hands are randomly split into 10 groups and then the impostor matching scores are computed for each group. For example, we divide the 100 hands from the test dataset into 10 groups where each group includes 60 (6 × 10) images from 10 hands. For each group, matching the i-th sample at different sessions from different hands ($i = 1, 2, 3, 4, 5, 6$) produces 270 (10 × 9 × 3) impostor matching scores. Hence, there are totally 2, 700 (270 × 10 groups) matching scores for the 10 groups of test dataset. The False Rejection Rate (FRR) is computed according to the genuine scores while the False Acceptance Rate is computed according to the impostor scores. The Equal Error Rate (EER) is the error rate when FAR is equal to FRR. The experimental results obtained with the various approaches mentioned previously are listed in Table 1 and the corresponding receiver operating characteristics (ROC) curves (the FAR against the FAR) are depicted in Fig. 4.

The experimental results (Table 1 and Fig. 4) show that the proposed approach outperforms the existing approaches mentioned in this paper in terms of reducing the verification error and the average computation cost. For example, the Repeated line tracking, Gabor filters and Hessian phase methods achieve 4.00% EER, 1.00% EER and 1.33 EER, respectively. By taking the labeled data from these three baselines as input of our model, the verification error is reduced to 0.33%. Overall, the deep learning based approaches (e.g. CNN and the proposed approach) achieve lower EER than the handcrafted features engineering based approaches. This can be explained by the fact that the handcrafted feature engineering-based segmentation methods extract explicitly some image processing-based features (low level features) that might discard relevant information for vein pattern classification. By contrast, the deep learning based methods automatically learn high-level features that are directly related to vein patterns. Also, we can observe that the proposed model achieves lower EER than CNN. Such a good performance may be attributed to the fact that the CNN takes a patch as input. For small patches, more detailed vein patterns are extracted but including more noise. This noise can generate mismatch errors, which reduces verification accuracy. On the contrary, if patches are too large, the CNN will take into account more global information, which results in missing detailed vein textures and inaccurate localization. Unlike CNN, the proposed approach takes the entire image as input instead of patch, which enables it to consider the localization and global context at the same time, thereby achieving higher verification accuracy.

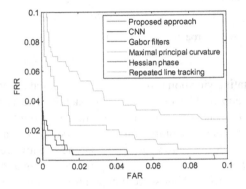

Fig. 4. Receiver operating characteristics

As shown in Table 1, we also compare the proposed GAN model with existing approaches in terms of inference time. It can be observed that the repeated line tracking method has the highest time cost for single image segmentation, followed by the CNN model. Compared to existing methods, including CNN, the proposed approach achieves the lowest time cost (i.e. 0.03 s) to segment a hand-vein image. This may be explained by the fact that the proposed approach takes the entire image as input instead of one patch at a time. So, it can predict the vein patterns of an image in one forward pass, which reduces the computation time substantially.

6 Conclusions

In this paper, we present a GAN model that robustly extracts palm vein patterns for person verification. Firstly, the vein pixels are labeled by the combination of several handcraft-based segmentation methods. Secondly, the proposed model is trained adversarially and the U-Net model is employed to predict the probability of the palm vein pixels. To perform matching, the output of U-Net is subject to binarization by thresholding. Finally, we match the resulting binary images for verification. Experiment results on a large public dataset demonstrate the ability of the proposed approach to significantly reduce the verification error.

As the vein capturing system is affected by many factors, many images include low quality regions where the separability between vein and non-vein patterns is poor. Matching such regions degrades, as a result, the verification accuracy. In the future, we intend to improve our model by learning the distribution of patches at different scales and orientations so that more robust vein patterns are extracted from the image, especially low quality regions.

Acknowledgement. This work was supported by Institut Mines Telecom/Telecom SudParis, the National Natural Science Foundation of China (Grant No. 61976030, 61402063), the Natural Science Foundation Project of Chongqing (Grant No. cstc2017jcyjAX0002, Grant No. cstc2018jcyjAX0095, Grant No. cstc2017zdcy-zdyfX0067), and the Natural Science Foundation of Chongqing Municipal Education Commission (KJQN201900848).

References

1. Jain, A.K., Feng, J.J.: Latent fingerprint matching. IEEE Trans. Pattern Anal. Mach. Intell. **33**(1), 88–100 (2010)
2. Al-Waisy, A.S., Qahwaji, R., Ipson, S., Al-Fahdawi, S., Nagem, T.A.M.: A multi-biometric iris recognition system based on a deep learning approach. Pattern Anal. Appl. **21**(3), 783–802 (2017). https://doi.org/10.1007/s10044-017-0656-1
3. Zou, Q., Ni, L., Wang, Q., Li, Q., Wang, S.: Robust gait recognition by integrating inertial and RGBD sensors. IEEE Trans. Cybern. **48**(4), 1136–1150 (2017)
4. Ding, C.X., Tao, D.C.: Trunk-branch ensemble convolutional neural networks for video-based face recognition. IEEE Trans. Pattern Anal. Mach. Intell. **40**(4), 1002–1014 (2017)
5. Boukhayma, A., Bem, R.D., Torr, P.H.: 3D hand shape and pose from images in the wild. In: Proceedings of the IEEE Conference on Computer Vision and Pattern Recognition, pp. 10843–10852. IEEE Computer Society, Los Alamitos (2019)
6. Diaz, M., Fischer, A., Ferrer, M.A., Plamondon, R.: Dynamic signature verification system based on one real signature. IEEE Trans. Cybern. **48**(1), 228–239 (2016)
7. Kumar, A., Zhou, Y.: Human identification using finger images. IEEE Trans. Image Process. **21**(4), 2228–2244 (2012)
8. Zhou, Y., Kumar, A.: Human identification using palm-vein images. IEEE Trans. Inf. Forensics Secur. **6**(4), 1259–1274 (2011)
9. Kumar, A., Prathyusha, K.V.: Personal authentication using hand vein triangulation and knuckle shape. IEEE Trans. Image Process. **18**(9), 2127–2136 (2009)
10. Ruiz-Albacete, V., Tome-Gonzalez, P., Alonso-Fernandez, F., Galbally, J., Fierrez, J., Ortega-Garcia, J.: Direct attacks using fake images in iris verification. In: Schouten, B., Juul, N.C., Drygajlo, A., Tistarelli, M. (eds.) BioID 2008. LNCS, vol. 5372, pp. 181–190. Springer, Heidelberg (2008). https://doi.org/10.1007/978-3-540-89991-4_19
11. Menotti, D., Chiachia, G., Pinto, A., et al.: Deep representations for iris, face, and fingerprint spoofing detection. IEEE Trans. Inf. Forensics Secur. **10**(4), 864–879 (2015)
12. Tanaka, T., Kubo, N.: Biometric authentication by hand vein patterns. In: Sice 2004 Conference, pp. 249–253. IEEE Computer Society, Los Alamitos (2004)
13. Shahin, M., Badawi, A., Kamel, M.: Biometric authentication using fast correlation of near infrared hand vein patterns. Int. J. Biomed. Sci. **2**(1), 141–148 (2007)
14. Jain, A.K., Ross, A., Prabhakar, S.: An introduction to biometric recognition. IEEE Trans. Circ. Syst. Video Technol. **14**(1), 4–20 (2004)

15. Miura, N., Nagasaka, A., Miyatake, T.: Feature extraction of finger-vein patterns based on repeated line tracking and its application to personal identification. Mach. Vis. Appl. **15**(4), 194–203 (2004)
16. Qin, H., Qin, L., Yu, C.: Region growth–based feature extraction method for finger-vein recognition. Opt. Eng. **50**(5), 057208 (2011)
17. Qin, H., He, X., Yao, X., Li, H.: Finger-vein verification based on the curvature in radon space. Exp. Syst. Appl. **82**, 151–161 (2017)
18. Yang, L., Yang, G., Yin, Y., Xi, X.: Finger vein recognition with anatomy structure analysis. IEEE Trans. Circ. Syst. Video Technol. **28**, 1892–1905 (2017)
19. Qin, H., Qin, L., Xue, L., He, X., Yu, C., Liang, X.: Finger-vein verification based on multifeatures fusion. Sensors **13**(11), 15048–15067 (2013)
20. Miura, N., Nagasaka, A., Miyatake, T.: Extraction of finger-vein patterns using maximum curvature points in image profiles. IEICE-Trans. Inf. Syst. **90**(8), 1185–1194 (2007)
21. Yu C, B., Qin, H, F., Cui, Y, Z., Hu, X, Q.: Finger-vein image recognition combining modified hausdorff distance with minutiae feature matching. Interdisc. Sci. Comput. Life Sci. **1**(4), 280–289 (2009)
22. Chaudhuri, S., Chatterjee, S., Katz, N., Nelson, M., Goldbaum, M.: Detection of blood vessels in retinal images using two-dimensional matched filters. IEEE Trans. Med. Imaging **8**(3), 263–269 (1989)
23. Huang, B., Dai, Y., Li, R., Tang, D., Li, W.: Finger-vein authentication based on wide line detector and pattern normalization. In: International Conference on Pattern Recognition, pp. 1269–1272. IEEE Computer Society, Los Alamitos (2010)
24. Zhang, Z., Ma, S., Han, X.: "Multiscale feature extraction of finger-vein patterns based on curvelets and local interconnection structure neural network. In: International Conference on Pattern Recognition, pp. 145–148. IEEE Computer Society, Los Alamitos (2006)
25. Qin, H., El-Yacoubi, M, A.: Deep representation-based feature extraction and recovering for finger-vein verification. IEEE Trans. Inf. Forensics Secur. **12**(8), 1816–1829 (2017)
26. Qin, H., El-Yacoubi, M.A., Lin, J., et al.: An iterative deep neural network for hand-vein verification. IEEE Access **7**, 34823–34837 (2019)
27. Yang, W., Hui, C., Chen, Z., et al.: FV-GAN: finger vein representation using generative adversarial networks. IEEE Trans. Inf. Forensics Secur. **14**(9), 2512–2524 (2019)
28. Zhang, Y., Yang, L., Chen, J., Fredericksen, M., Hughes, David P., Chen, Danny Z.: Deep adversarial networks for biomedical image segmentation utilizing unannotated images. In: Descoteaux, M., Maier-Hein, L., Franz, A., Jannin, P., Collins, D.Louis, Duchesne, S. (eds.) MICCAI 2017. LNCS, vol. 10435, pp. 408–416. Springer, Cham (2017). https://doi.org/10.1007/978-3-319-66179-7_47
29. Dou, Q., Ouyang, C., Chen, C., Chen, H., Heng, P, A.: Unsupervised cross-modality domain adaptation of convnets for biomedical image segmentations with adversarial loss (2018). arXiv preprint arXiv:1804.10916
30. Goodfellow, I., et al.: Generative adversarial nets. In Advances in Neural Information Processing Systems, pp. 2672–2680. Curran Associates, New York (2014)
31. Ronneberger, O., Fischer, P., Brox, T.: U-Net: convolutional networks for biomedical image segmentation. In: Navab, N., Hornegger, J., Wells, William M., Frangi, Alejandro F. (eds.) MICCAI 2015. LNCS, vol. 9351, pp. 234–241. Springer, Cham (2015). https://doi.org/10.1007/978-3-319-24574-4_28

A Survey on Peripheral Blood Smear Analysis Using Deep Learning

Rabiah Al-qudah$^{(\boxtimes)}$ and Ching Y. Suen

Concordia University, Montreal, QC H3G 1M8, Canada
r_alquda@encs.concordia.ca, suen@cs.concordia.ca

Abstract. Peripheral Blood Smear (PBS) analysis is a routine test carried out in specialized medical laboratories by specialists to assess some aspects of health status that are measured and assessed through blood. PBS analysis is prone to human errors and the usage of computer-based analysis can greatly enhance this process in terms of accuracy and cost. Despite the challenges, Deep Learning neural networks have shown impressive performance in this context. In this study the recent contributions are summarized along with the main challenges and future directions in this context.

Keywords: Deep learning · Blood smears · Malaria · Leukemia · Cell classification.

1 Introduction

Blood cell analysis is a vital source of information for Medical doctors to diagnose patients for certain diseases. They provide important indicators of our health status. A blood test must be performed in a specialized Medical laboratory by specialists. There are three major types of blood cells: Red Blood Cells (RBCs), White Blood Cells (WBCs), and Platelets. RBCs, also known as erythrocytes, are the most common blood cell type and are responsible primarily for carrying oxygen and carbon dioxide to the entire body. WBCs, also known as leukocytes, are the primary defense system against infectious disease and foreign invaders. WBCs are much less common than RBCs. For example, the number of WBCs in adult males ranges from 4.5 to 11.5 thousand in 1 microliter, where the number of RBCs in adult males ranges from 4.6 to 6 million in 1 microlitre [32]. Platelets, also known as thrombocytes, are non-nucleated entities and are 2 to 4 μm in diameter, and is responsible for repairing blood vessels in case of injury. A Peripheral Blood Smears (PBS), also known as a blood films, is the result of dispersing and staining a thin layer of blood on a microscope slide that is made of glass. Typically, it is hard, even for experts, to categorise some abnormal blood cells on the slide. PBSs are used to verify the results obtained by automated analyzer tools, to identify abnormal, immature, and/or atypical cells, and to recognize morphological abnormalities that

© Springer Nature Switzerland AG 2020
Y. Lu et al. (Eds.): ICPRAI 2020, LNCS 12068, pp. 725–738, 2020.
https://doi.org/10.1007/978-3-030-59830-3_63

are beyond the capabilities of automated analyzers. Blood smear analysis is a daily, time consuming procedure for lab specialists, hence, the automation of blood smear analysis has attracted the attention of researchers in recent years, and despite the good results achieved so far, many challenges still arise. Image processing techniques [12,13,37], Machine Learning techniques [2,3,9], and Deep Learning techniques have been widely employed in analyzing blood samples and diagnosing abnormalities. In this work, a comprehensive survey of the recent work done in PBS analysis using deep learning networks is presented. The rest of this paper is organized as follows: Sect. 2 presents the recent contributions in the area of blood smear analysis using deep learning and highlights the main findings and conclusions of them. Section 3 concludes the paper, discusses and suggests open research areas for future investigations. Finally, in Sect. 4 this work is concluded.

2   Related Works on Peripheral Blood Smear Analysis

As mentioned above, blood smear analysis is an important diagnostic tool for medical doctors. Recent advancements in machine and deep learning paved the way for researcher to utilize these learning networks for blood smear analysis. The researchers focused mainly on three distinct directions in this context:

1. Malaria Detection
2. Blood Cell Detection and Classification
3. Leukemia Diagnosis

Now, we present the recent state-of-the-art techniques used for each of these directions in greater details.

2.1   Malaria Detection

Malaria is a life-threatening disease caused by parasites that are transmitted to people through the bites of infected female Anopheles mosquitoes, The estimated number of malaria deaths stood at 435,000 in 2017. One of the most accurate methods to diagnose malaria is by examining a PBS to look for Plasmodium falciparum parasites infecting some RBCs. The National Institute of Health (NIH) dataset [11,29] is a widely used dataset that was published in 2018. It consists of segmented cells from thin blood smear slide in which images were collected and photographed at Chittagong Medical College Hospital, Bangladesh. The dataset contains 27,558 light microscopic cell images, with equal instances of Plasmodium parasitized and uninfected segmented red blood cell images. Many studies used this dataset to train their models [16,17,28–30].

In [17], a Convolution Neural Network (CNN) architecture that is comprised of 12 layers was trained with the NIH dataset after applying a set of augmentation operations such as horizontal flip, vertical flip, width shift, height shift, fill mode, zoom range, and rotational range. The dataset was preprocessed by applying normalization, gamma correction, and logarithmic correction to improve brightness and adjust contrast operations. This model achieved an accuracy of 98.23% and F1 score of 97.74%. In [30], the NIH dataset instances were mean normalized, and several augmentation techniques including rotations, translation, shearing, zooming, and flipping were performed. A CNN that consists of 3 blocks was trained for Malaria detection. Moreover Visual Geometry Group-19 (VGG-19), SqueezeNet, InceptionResNet-V2 were customized by truncating them at their deepest convolutional layer and adding a Global Average Pooling (GAP) and dense layers. Several combinations of the listed models were ensembled by taking the average of the predictions, VGG-19 and SqueezeNet; combination outperformed the individual models and other ensembles in all performance metrics with an accuracy of 99.51%. The work in [16] utilized the same NIH set to train a CNN of 3 convolutional layers, one hidden layer, input, flatten and output layers. This shallow CNN achieved a good accuracy of 95% with no augmentation or preprocessing. In [28], the NIH dataset was preprocessed by stain normalization, Min-Max Normalization, and Standardization. Many augmentation techniques have been applied: horizontal and vertical flips, Gaussian blur, rotation, horizontal and vertical shifting, darkening and lightening, ZCA whitening, and feature wise standardization, and Change of color space and Gaussian Blur. The dataset size was extended to 137,940 after augmentation. This work proposed a CNN that consists of 8 convolution layers. A VGG16 deep network was also customized by removing the pre-trained fully convolution layers, and adding a dense layer, a dropout layer, and a fully connected layer. A third architecture called CNNEx-SVM was trained by emitting the customised VGG16 features to an Support Vector Machine (SVM). Finally, all models were ensembled by taking a weighted average of all predictions. The customised VGG16 achieve an accuracy of 97.6%, and the ensemble one achieved an accuracy of 97.7%.

In [29], a multi-scale Laplacian of Gaussian (LoG) filter was applied to whole-slide images from the NIH dataset to detect RBCs centroids. The detected cells were then segmented. Morphology opening operation is then applied to remove artifacts. F1-score of 0.952 was achieved in the detection phase. For cell classification a custom CNN of three convolutional layers and two fully connected layers and other pre-trained CNNs (AlexNet, VGG-16, Xception, DenseNet-121, and ResNet-50) were employed. ResNet outperformed the other networks with 95.7% accuracy. The accuracy was improved to 95.9% by evaluating the optimal layer for feature extraction, as the final layer isn't necessarily the optimal one. Some general conclusions can be drawn from [16,17,28,30], and [29] since it was all trained using the same dataset:

1. Preprocessing and augmenting the dataset has helped in improving the testing results.
2. Ensemble Learning when accompanied with preprocessing and augmentation can be a very effective and outperforms individual deep models.

Other works considered different datasets for training and testing [5,7,8,22,24, 34,43]. For example, the authors in [43] developed an Android smartphone application using a dataset of 1819 whole slide thick smear images from 150 patients. The parasite in this work is detected using a pipeline that starts by applying an intensity-based Iterative Global Minimum Screening (IGMS) procedure to reduce the size of the initial search space and limit the number of regions of interest which are fed to a CNN model consisting of seven convolutional layers. The classification accuracy was 93.46%. The work in [8] annotated over 92k objects of the four major malaria species. Quality control is applied to each instance. Firstly, by checking the standard deviation of grayscale pixel values and the dynamic range of the gradient of the grayscale against preset thresholds. If either of these values is lower than the threshold it will be rejected. Secondly, focus metrics are calculated on the instance and on a corresponding artificially blurred version. Next, RBCs are detected and counted by applying binary grayscale clustering; candidate objects are chosen by finding connected-components in the threshold image. The pipeline then splits into two branches, one for rings (for quantitation), and one for late stages (for species ID). Each branch has two parts: First, a Gradient-boosted Tree (GBT) classifier was trained to act as a high-sensitivity distractor. Detected late stage parasites also pass through a species ID module. Finally, the various outputs are combined to deliver patient-level quantitation and species ID predictions. Two CNNs were proposed for classification, one for the ring branch and a second one for the late stage. The ring branch CNN had 3 convolutional layers, followed by two Inception modules and one fully connected layer where the late stage branch CNN was a fully convolutional neural network with 7 convolutional layers. The work in [5] combined two datasets, a one collected by the authors and another public dataset from the Institute for Molecular Medicine Finland (FIMM). It has a collection of digital images infected by P.falciparum. The overall collection has a total of 1030 images of infected cells, and 1520 images of non-infected ones. The VGG network proposed in [5] was customized by removing the last three layers. The features obtained from the network are then fed to an SVM, and the model is called VGG19-SVM. Accuracy of 93.13% and F-score 91.66% were achieved. In [34], a dataset of 1000 instances, multi-wavelength was utilized to increase the sample size, 45°, and 135° rotations were also applied. The authors opted to utilize AlexNet, VGG-16, ResNet50, GoogLeNet, and a customized CNN network of 5 convolutional layers and 2 fully connected layers. ResNet outperformed with an accuracy of 97.6% in classifying the test set as healthy or infected.

A CNN framework is presented in [22] that is able to perform the extended depth of field images from z-stacks of thick blood films for automated malaria diagnosis. Two deep architectures were proposed, EDoF-CNN-3D, and EDoF-CNN-Max. In EDoF-CNN-3D, the encoder part of the network was modified by replacing the two-dimensional convolutions with three-dimensional ones. The output tensor was flattened on the z-axis before the residual layers of the network by average pooling. Where in EDoF-CNN-Max combines the idea behind the Siamese networks and the one behind the wavelet-based MSD EDoF. Each focal plane is passed through the encoder part of the network, and the maximum of the activation values are selected before going through the residual layers. The detection recall of the EDoF-CNN-3D method is 73%. While the work in [24] used other whole slides dataset which consists of 800 infected cells and 2000 healthy cells. Two augmentation techniques were applied on half of the dataset to produce two augmented sub-datasets: Image interpolation in the spatial domain (For any two images A and B in the dataset, we can generate a new image by finding a weighted average C), and image interpolation in the feature domain (For any two images A and B in the dataset, two 30-point feature vectors FA and FB can be obtained by the stacked autoencoders that have been trained we can generate a new 30-point vector FC by finding a weighted average). The authors in [7] used 4100 whole slide peripheral blood smear images to train a Deep Belief Network (DBN) to classify the objects to either parasites or non-parasites. The objects are extracted from peripheral blood smear images using the level set method. A concatenated feature of color (histogram-based features and color coherence vector) and texture (Haralick features, LBP features, and gray level run length matrix feature) were used to initialize the visible layer of the 4 hidden layer DBN. The deep network achieved an F-score of 89.66.

Although the models in studies [43] to [7] were trained with different models, however, a general finding is that using very deep neural networks, such as ResNet, can significantly improve the overall detection results.

2.2   Blood Cell Detection and Classification

The work in [26] proposed an architecture for microcytic hypochromia. The target features were a combination of blood smear image features extracted by AlexNet deep convolutional neural network and clinical features from (Red blood cell count (RBC), Haemoglobin concentration (HB), Red blood cell distribution width (RDW)). Samples were collected especially for this research from twenty patients. Both Principal Component Analysis (PCA), and Linear Discriminant Analysis (LDA) were used to reduce the feature set with minimal loss of information. k-Nearest Neighbors (k-NN), SVM, and Neural Network (NN), were employed for the classification phase. Each model was trained with three different feature sets: the clinical features, image features and fused features. The neural network and the SVM classifier scored 99% accuracy at testing when trained with the fused features, which shows the superiority of the proposed fusion model.

The work in [27,41] classifies WBCs not only to their main types but also to some morphological abnormalities. In [41], a total of 14,700 annotated whole-slide images that include 11 categories of leukocytes were considered. Cells recognition was performed using Single Shot Detector (SSD) and YOLO3, Different variations of SSD and YOLOv3 were examined, 0.931 MaP and accuracy of 90.09% score were reported for the SSD 300 × 300, and the highest MaP scored with YOLOv3 320 × 320 is 0.92. In [27], the authors collected a private dataset that contains a total of 92480 leukocytes belonging to 40 categories, with one object of interest in each instance. To handle the dataset imbalance, many augmentation techniques were applied such as, horizontal and vertical flips, and adding random noises and color changes to the original images. The architecture of the proposed deep residual neural network consists of 7 convolutional layers, 2 fully connected layers and three residual blocks to improve its performance.

The authors examined 7 different schemes by using different activation functions to train the network. The average classification accuracy was 76.84%. The work in [14] studied blood-cell classification in medical hyperspectral imaging (MHSI). It utilized four different architectures: SVM, typical VGG16, CNN without Gabor wavelet, CNN with Gabor wavelet and a combination of modulated Gabor wavelet and CNN kernels, named as MGCNN. In MGCNN, each convolutional layer performs a dot product between multi-scale and orientation Gabor operators and the initial CNN kernels, to transform the convolutional kernels into the frequency domain in order to extract the features. Three datasets were utilized for testing, (1) Bloodcells1-3: CNN achieved the lowest overall accuracy (OA) of 81.35% and the highest OA is achieved using the proposed model with a score of 94.03%, (2) Bloodcells2-2 where 88.70%, and 94.40% were scored by CNN and the proposed model, respectively, (3) white blood cells dataset: SVM with PCA scored the lowest with overall accuracy of 90.83% and the proposed model achieved the highest accuracy score of 97.65%.

In [35], a subset of the All-IDB1 whole slide image dataset was used, as the authors selected 42 images and performed a pixel wise annotation on them. The training set size was increased from 29 to 145 images by performing random reflection and translation augmentation techniques. The class weighting technique was used to handle the dataset imbalance caused by having RBCs appear seven times more then WBCs. The set was then fed to a SegNet for semantic segmentation purposes. The highest accuracy was 94%, scored for the WBCs class.

In [40], a dataset available on GitHub of WBC images was used. Augmentation techniques such as random rotation, scaling, reflection, and shearing were performed. Gaussian noise was also applied to a subset of the training and testing set to train the network on poor quality images. The resulting 12,500 instances set is then fed to 3 sets of experiments that are made up of 10, 20, 30 CNNs where each CNN is constructed by generating random numbers of convolution blocks and layer sizes from preset ranges. The feature maps of each experiment were then concatenated and emitted to a PatternNet deep network to ensemble the strongest features to contribute in the final decision. The 30-CNNs experiment outperformed the 10 and 20 CNNs experiments with an accuracy score of 99.37%. In [4] the input images are acquired from Pinterest online open source haematology database. RBCs are cropped from the blood smear to generate the dataset of normal, acanthocyte, sickle cell, teardrop and elliptocyte cell. The authors utilized SVM and AlexNet deep network. SVM model outperformed the AlexNet model, the authors referred this to the small dataset size. It is noted that there's a noticeable difference between the results of models, for example, the SVM model achieved 100% accuracy in classifying Achantocyte where the deep learning model achieved 0%. The work in [42] combines Fourier Ptychographic Microscopy (FPM) and an improved version of You Only Look Once (YOLO) networks for WCB detection. In order to improve the detection of the microscopic WBCs, the feature maps of the last three layers are concatenated and passed to a final convolution layer. The proposed model was trained and tested on a 1000 whole slide image set. For the work in [19], the authors try to address the problem caused by the lack of some deep networks ability to fully exploit the long-term dependence relationship between certain key features of images and image labels. A combination of a CNN and Recurrent Neural Network (RNN) is employed to deepen the understanding of the image context. A dataset comprising 12,444 augmented and rotated images of blood cells were collected from Kaggle and BCCD public datasets. The proposed network consists of Pre-trained convolutional neural network layer, RNN layer, Merge layer, and fully connected layer with Softmax output. The proposed model achieved an accuracy of 90.79%. In [23], a dataset of blood cell images was augmented by performing rotation, reflection and translation, three pipelines one for each blood cell were implemented to perform the classification using CNNs and UNet deep networks. The 64000 blood cells dataset used in [31] is a combination of all-idb, DPDx, ASH image bank and other images available on Google. The authors of [31] proposed a two-stage solution, In the first phase, a contour aware CNN was used for the segmentation of individual cells. In order to classify WBCs into five subtypes, features were extracted by a CNN and forwarded to ELM for classification. In order to classify RBCs into normal and identify abnormalities, several features were extracted such as centroid, medial axis ratio, and cell deform ratio, extracted features are forwarded to ELM for classification. Overall RBC classification accuracy was 90.10%, the highest WBC subtype accuracy was for Monocyte 98.68%.

2.3 Leukemia Diagnosis

Leukemia is a fatal malignancy and has two main types: Acute and chronic which depending on how fast it progresses. Moreover, there are two subtypes of each leukemia main types depending on the size and the shape of the WBC: Lymphoid and myeloid. Acute leukemia is usually diagnosed after having clinical signs and symptoms that need to be confirmed by laboratory investigations. Complete blood count for WBCs, RBCs, platelets, and a peripheral blood smear are the initial tests. In many cases they will not be enough to confirm the diagnosis, which is why the clinical practice is to do a bone marrow smear and biopsy. A bone marrow specimen will have a smear and a biopsy. This specimen is usually good enough to confirm the diagnosis of leukemia, but more testing is mandatory for subtyping the leukemia into lymphoid or myeloid, and then subclassifying each subtype. In bone marrow specimen, number and shape of WBCs are the key point in diagnosing leukemia. Findings will be different between acute and chronic, myeloid and lymphoid, and each entity will have its own criteria of diagnosis. The work in [1,33] performs Leukemia diagnosis and classifies the result into its subtypes. In [1], two public leukemia datasets (ALL-IDB and ASH Image Bank) were used to train the network to classify the samples into one of the four main Leukemia types. The number of samples increased to 8 times for both datasets by applying shifting, rotation, and flipping. A CNN of 2 convolution layers, a Flatten layer, followed by a fully connected layer was proposed in this work. The accuracy of classification obtained was 81.74%. The work in [21, 25, 36, 38, 39] aims to train a well generalized model to detect Leukemia. In [25] segmented white blood cell images of the C-NMC dataset were augmented by performing horizontal and vertical flips, and random translations. A Squeeze-and-Excitation-ResNeXt50 network achieved a weighted F1-score of 88.91%. In [21] only one object of interest appears in each instance all images were transferred to grayscale and the cell region was then binarized using the threshold estimated by Otsu's method followed by the erosion operation. The authors opted to train the model using a ResNet with two fully connected layers and utilize bagging ensemble training strategy. The model achieved an F1-score of 0.84. The work in [39] trained three deep architectures (AlexNet, CaffeNet, Vgg-f) to generate features. The features space was then reduced by applying the gain ratio algorithm, before emitting the features to an SVM classifier. An accuracy score of 100% when the classification was performed by concatenating and reducing the features obtained by all models. In [36], CNN architecture comprising of 5 convolutional layers and 2 layers (fully connected and softmax) was trained on ALL-IDB1 dataset after applying many augmentation operations: histogram equalization, translation, reflection, rotation, shearing, conversion to grayscale, and blurring. The proposed method achieved an accuracy of 96.6%. In [38], feature maps were extracted from the All-IDB1 dataset using AlexNet, CaffeNet, Vgg-f deep networks before being classified using SVM, Multilayer Perceptron (MLP) and Random Forest (RF). The authors also experimented the concatenation of feature maps obtained by all deep networks, the feature space was reduced by utilizing PCA technique and the majority voting rule to combine the outcomes obtained by each classifier. An accuracy score of 100% was achieved. A general conclusion can be highlighted from the literature in this subsection is that extracting features from an ensemble of deep networks and concatenating them before classification can be a very effective approach in detecting Leukemia. In Table 1 we summarize all the literature mentioned in this section.

3   Discussion and Future Directions

Despite the vast advancement in artificial intelligence and the robust frameworks presented in PBS analysis context, many challenges remain to be tackled. One of the biggest challenges is data scarcity because of the limited number of datasets that are publicly available, where most of them are kept private which makes it harder to reproduce and improve the work presented in the literature. One the other hand, annotating PBS datasets is tedious and can only be done by medical experts. And due to the high diversity in blood cell shapes and morphology, sometimes even experts have different annotations for the same cell, which leads to taking the annotation with the highest number of experts votes. Furthermore, PBS datasets suffer from data imbalance due to the natural distribution of WBCs and RBCs in blood. Imbalance could lead to a model that exhibits bias towards the majority class, and in some extreme cases would lead to ignoring the minority class altogether [15]. Imbalance in this context has been handled

Table 1. Summary of the reviewed literature

Reference/year	Method	Dataset	Results
Malaria detection			
[17], (2019)	CNN	Augmentation, NIH	Accuracy: 98.23%, F1 score: 97.74%
[30], (2019)	CNN, VGG19, SqueezeNet, InceptionResNet-V2	Augmentation, NIH	Accuracy: 99.51%
[16], (2019)	CNN	NIH	Accuracy: 95%
[28], (2019)	CNN, VGG16, SVM	Augmentation, NIH	Accuracy: 97.7%
[29], (2018)	CNN, AlexNet, VGG-16, Xception, DenseNet-121, ResNet-50	NIH	Accuracy: 95.9%
[43], (2019)	CNN	Private dataset	Accuracy: 93.46%
[8], (2019)	CNN	Private dataset	Accuracy: ring: 94.8%, late: 96.6%
[5], (2019)	VGG, SVM	Private dataset + FIMM	Accuracy: 93.13%, F-score 91.66%
[34], (2019)	AlexNet, VGG-16, ResNet50, GoogLeNet, CNN	Augmentation, Private dataset	Accuracy: 97.6%
[22], (2019)	CNN	Private dataset	Recall: 73%
[24], (2018)	CNN	Augmentation, Private dataset	Accuracy 99%
[7], (2017)	Deep Belief Network (DBN)	Private dataset	F-score: 89.66%
Blood cell detection and classification			
[26], (2019)	AlexNet, SVM	Private dataset	Accuracy: 99%
[41], (2019)	YOLO, SSD	Private dataset	Mean accuracy: 90.09%
[27], (2018)	Residual Network	Augmentation, Private dataset	Accuracy: 76.84%
[14], (2019)	SVM, VGG16, CNN, CNN with Gabor wavelet	Private dataset	Accuracy: 97.65%
[35], (2017)	SegNet	Augmentation, ALL-IDB	Accuracy: 89.45%
[40], (2018)	CNN, PatterNet	Private dataset	Accuracy: 99.37%
[4], (2018)	SVM, AlexNet	Pinterest online open source haematology database	Acanthocyte Accuracy: 100%
[42], (2018)	YOLO	Private dataset	Precision: 100%
[19], (2018)	CNN, RNN	Augmentation, Kaggle and BCCD	Accuracy: 90.79%
[23], (2018)	CNN, U-Net	Augmentation, Private dataset	Specificity: 99.11%, Sensitivity: 100%
[31], (2017)	Contour aware CNN, ELM	ALL-IDB, DPDx, ASH, Google	RBC accuracy: 94.71%, WBC Accuracy: 98.68%
Leukemia detection			
[1], (2019)	CNN	Augmentation, ALL-IDB and ASH Image Bank	Accuracy: 81.74%
[25], (2019)	Squeeze-and-Excitation-ResNeXt50	Augmented, C-NMC dataset	F1-score: 88.91%
[21], (2019)	ResNet	Private dataset	F1-score: 0.84
[39], (2018)	AlexNet, CaffeNet, VGG-f, SVM	Private dataset	Accuracy: 100 %
[36], (2018)	CNN	Augmentation, ALL-IDB1	Accuracy: 96.6%
[38], (2017)	AlexNet, CaffeNet, VGG-f, SVM	All-IDB1	Accuracy: 100%

mostly by dynamic sampling approaches [20] and augmentation techniques such as rotation and translation. A drawback of these augmentation techniques is that it is performed by trial and error, and there is no formal method that can determine if an augmentation strategy will improve results until after training, which might lead to a time consuming training process [18]. A promising solution to the data imbalance and scarcity issues is the utilization of neural networks that generate synthetic data. For example the Generative Adversarial Networks (GAN) [6] can be used to generate the synthetic data efficiently, the work in [18] also employed a novel deep network to perform smart augmentations that can guarantee the effectiveness of the synthetic images.

In the context of PBS analysis research two types of datasets can be found: light microscopic images and whole-slide images, as shown in Fig. 2 and 1, respectively. The first type is being extensively investigated compared to the second one despite the fact that the whole-slide sets poses more realistic and challenging scenarios, as the entities in the blood appear microscopic, touching and crowded instead of the simple scenario represented in the light microscopic images where only one object of interest appears. Many PBS analysis application areas have not been completely addressed using whole-slide images and there is still room for improvement in this context.

Stains are very important diagnostic tools, they help in confirming and subtyping diseases in general. Staining even in laboratory automated devices have not reached the ideal status yet because of chances of procedure errors or human error. Laboratory services in diagnosis will step hugely forward if computers manage to do virtual staining accurately. This will save time, money and will eliminate chances of procedure errors. We recognize that more work is still needed for the analysis of unstained slides as this direction is lacking investigation by computer researchers. The work in [10] is an example of a recent work in this direction. It is worth mentioning that many computer researchers still compare their work results to the manual methods of classifying and counting blood cells, where these manual methods are not in use anymore and are already replaced by certified automatic analyzers that can perform the task in seconds. Hence, counting and classifying normal blood components is not a current problem anymore, but counting and classifying abnormal cells is the problem that needs automated tools. If computers can read a slide and highlight number and shape and structure of all abnormal cells, this will make diagnosis and classification of leukemia for example an easy process in most cases. Currently medical specialists need many stains and tools such as Flowcytometry to determine the type, number, shape of cells which is time, money and effort consuming.

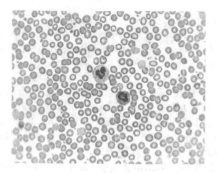

Fig. 1. Whole-slide images

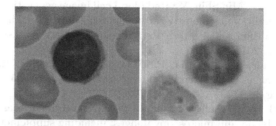

Fig. 2. Light microscopic images

4 Conclusion

A Peripheral Blood Smear (PBS), also known as a blood film, is the result of dispersing and staining a thin layer of blood on a microscope slide to assess some aspects of health status. Despite the several deep learning techniques and architectures have been proposed to automate this procedure, many areas are not tackled enough. In this study we present the recent contributions and discuss the promising future directions in the context of PBS analysis.

Acknowledgment. We thank Dr. Mohammad Al-Qudah from the pathology department at Jordan University of Science and Technology for his help in the medical background and for useful discussions about the possible future directions in PBS analysis.

References

1. Ahmed, A.N., Yiğit, A., Isik, Z., Alpkocak, A.: Identification of leukemia subtypes from microscopic images using convolutional neural network. Diagnostics **9**(3), 104 (2019). https://doi.org/10.3390/diagnostics9030104
2. Alam, M.M., Islam, M.T.: Machine learning approach of automatic identification and counting of blood cells. Healthcare Technol. Lett. **6**(4), 103–108 (2019). https://doi.org/10.1049/htl.2018.5098

3. Aliyu, H.A., Sudirman, R., Abdul Razak, M.A., Abd Wahab, M.A.: Red blood cell classification: deep learning architecture versus support vector machine. In: 2nd International Conference on BioSignal Analysis, Processing and Systems (ICBAPS), Kuching, Malaysia, pp. 142–147, July 2018. https://doi.org/10.1109/ICBAPS.2018.8527398

4. Aliyu, H.A., Sudirman, R., Abdul Razak, M.A., Abd Wahab, M.A.: Red blood cell classification: deep learning architecture versus support vector machine. In: 2018 2nd International Conference on BioSignal Analysis, Processing and Systems (ICBAPS), Kuching, Malaysia, pp. 142–147, July 2018. https://doi.org/10.1109/ICBAPS.2018.8527398

5. Arunagiri, V., B, R.: Deep learning approach to detect malaria from microscopic images. Multimedia Tools Appl., 1–21 (2019). https://doi.org/10.1007/s11042-019-7162-y

6. Bailo, O., Ham, D., Min Shin, Y.: Red blood cell image generation for data augmentation using conditional generative adversarial networks. In: The IEEE Conference on Computer Vision and Pattern Recognition (CVPR) Workshops, Long Beach, CA, USA, June 2019

7. Bibin, D.S., Nair, M., Punitha, P.: Malaria parasite detection from peripheral blood smear images using deep belief networks. IEEE Access 5, 9099–9108 (2017). https://doi.org/10.1109/ACCESS.2017.2705642

8. Delahunt, C.B., et al.: Fully-automated patient-level malaria assessment on field-prepared thin blood film microscopy images, including supplementary information. CoRR abs/1908.01901 (2019). arXiv:1908.01901v1

9. Elsalamony, H.A.: Detection of anaemia disease in human red blood cells using cell signature, neural networks and SVM. Multimedia Tools Appl. 77(12), 15047–15074 (2018). https://doi.org/10.1007/s11042-017-5088-9

10. Go, T., Kim, J.H., Byeon, H., Lee, S.J.: Machine learning-based in-line holographic sensing of unstained malaria-infected red blood cells. J. Biophotonics 11(9), e201800101 (2018). https://doi.org/10.1002/jbio.201800101

11. National Institutes of Health: Nih malaria dataset (2018). https://lhncbc.nlm.nih.gov/publication/pub9932

12. Hegde, R., Prasad, K., Hebbar, H., Singh, B.M.: Comparison of traditional image processing and deep learning approaches for classification of white blood cells in peripheral blood smear images. Biocybernetics Biomed. Eng. 39(2), 382–392 (2019). https://doi.org/10.1016/j.bbe.2019.01.005

13. Hegde, R., Prasad, K., Hebbar, H., Singh, B.M.: Image processing approach for detection of leukocytes in peripheral blood smears. J. Med. Syst. 43(5) (2019). https://doi.org/10.1007/s10916-019-1219-3

14. Huang, Q., Li, W., Zhang, B., Li, Q., Tao, R., Lovell, N.H.: Blood cell classification based on hyperspectral imaging with modulated Gabor and CNN. IEEE J. Biomed. Health Inform., 1 (2019). https://doi.org/10.1109/JBHI.2019.2905623

15. Johnson, J.M., Khoshgoftaar, T.M.: Survey on deep learning with class imbalance. J. Big Data 6(1), 27 (2019). https://doi.org/10.1186/s40537-019-0192-5

16. Kalkan, S.C., Sahingoz, O.K.: Deep learning based classification of malaria from slide images. In: 2019 Scientific Meeting on Electrical-Electronics Biomedical Engineering and Computer Science (EBBT), Istanbul, Turkey, pp. 1–4, April 2019. https://doi.org/10.1109/EBBT.2019.8741702

17. Kumar, R., Singh, S.K., Khamparia, A.: Malaria detection using custom convolutional neural network model on blood smear slide images. In: Luhach, A.K., Jat, D.S., Hawari, K.B.G., Gao, X.-Z., Lingras, P. (eds.) ICAICR 2019. CCIS, vol. 1075, pp. 20–28. Springer, Singapore (2019). https://doi.org/10.1007/978-981-15-0108-1_3

18. Lemley, J., Bazrafkan, S., Corcoran, P.: Smart augmentation learning an optimal data augmentation strategy. IEEE Access **5**, 5858–5869 (2017). https://doi.org/10.1109/ACCESS.2017.2696121

19. Liang, G., Hong, H., Xie, W., Zheng, L.: Combining convolutional neural network with recursive neural network for blood cell image classification. IEEE Access **6**, 36188–36197 (2018). https://doi.org/10.1109/ACCESS.2018.2846685

20. Lin, M., Tang, K., Yao, X.: Dynamic sampling approach to training neural networks for multiclass imbalance classification. IEEE Trans. Neural Netw. Learn. Syst. **24**(4), 647–660 (2013). https://doi.org/10.1109/TNNLS.2012.2228231

21. Liu, Y., Long, F.: Acute lymphoblastic leukemia cells image analysis with deep bagging ensemble learning. bioRxiv (2019). https://doi.org/10.1101/580852

22. Manescu, P., et al.: Deep learning enhanced extended depth-of-field for thick blood-film malaria high-throughput microscopy. CoRR abs/1906.07496 (2019). arXiv:1903.06056v1

23. Mundhra, D., Cheluvaraju, B., Rampure, J., Rai Dastidar, T.: Analyzing microscopic images of peripheral blood smear using deep learning. In: Cardoso, M.J., et al. (eds.) DLMIA/ML-CDS -2017. LNCS, vol. 10553, pp. 178–185. Springer, Cham (2017). https://doi.org/10.1007/978-3-319-67558-9_21

24. Pan, W., Dong, Y., Wu, D.: Classification of malaria-infected cells using deep convolutional neural networks, pp. 159–172. IntechOpen, September 2018. https://doi.org/10.5772/intechopen.72426

25. Prellberg, J., Kramer, O.: Acute lymphoblastic leukemia classification from microscopic images using convolutional neural networks. CoRR abs/1906.09020 (2019)

26. Purwar, S., Tripathi, R.K., Ranjan, R., Saxena, R.: Detection of microcytic hypochromia using CBC and blood film features extracted from convolution neural network by different classifiers. Multimedia Tools Appl. (2019). https://doi.org/10.1007/s11042-019-07927-0

27. Qin, F., Gao, N., Peng, Y., Wu, Z., Shen, S., Grudtsin, A.: Fine-grained leukocyte classification with deep residual learning for microscopic images. Comput. Methods Programs Biomed. **162**(8), 243–252 (2018). https://doi.org/10.1016/j.cmpb.2018.05.024

28. Rahman, A., et al.: Improving malaria parasite detection from red blood cell using deep convolutional neural networks. CoRR abs/1907.10418 (2019). arXiv:1907.10418v1

29. Rajaraman, S., et al.: Pre-trained convolutional neural networks as feature extractors toward improved malaria parasite detection in thin blood smear images. PeerJ **6**(e4568) (2018). https://doi.org/10.7717/peerj.4568

30. Rajaraman, S., Jaeger, S., Antani, S.K.: Performance evaluation of deep neural ensembles toward malaria parasite detection in thin-blood smear images. PeerJ **7**(e6977) (2019). https://doi.org/10.7717/peerj.6977

31. Razzak, M.I., Naz, S.: Microscopic blood smear segmentation and classification using deep contour aware CNN and extreme machine learning. In: 2017 IEEE Conference on Computer Vision and Pattern Recognition Workshops (CVPRW), pp. 801–807, July 2017. https://doi.org/10.1109/CVPRW.2017.111

32. Rodak, B.F., Fritsma, G.A., Doig, K.: Hematology: Clinical Principles and Applications. Elsevier Health Sciences (2007). https://books.google.com/books?id=6sfacydDNsUC

33. Shafique, S., Tehsin, S.: Acute lymphoblastic leukemia detection and classification of its subtypes using pretrained deep convolutional neural networks. Technol. Cancer Res. Treat. **17**, 1–7 (2018). https://doi.org/10.1177/1533033818802789

34. Singla, N., Srivastava, V.: Deep learning enabled multi-wavelength spatial coherence microscope for the classification of malaria-infected stages with limited labelled data size. CoRR abs/1903.06056 (2019). arXiv:1903.06056v1

35. Thanh, T., Kwon, O.H., Kwon, K.R., Lee, S.H., Kang, K.W.: Blood cell images segmentation using deep learning semantic segmentation. In: 2018 IEEE International Conference on Electronics and Communication Engineering (ICECE), Xian, China, pp. 13–16, December 2018. https://doi.org/10.1109/ICECOME.2018.8644754

36. Thanh, T., Vununu, C., Atoev, S., Lee, S.H., Kwon, K.R.: Leukemia blood cell image classification using convolutional neural network. Int. J. Comput. Theory Eng. **10**(2), 54–58 (2018). https://doi.org/10.7763/IJCTE.2018.V10.1198

37. Varma, S.L., Chavan, S.S.: Detection of malaria parasite based on thick and thin blood smear images using local binary pattern. In: Iyer, B., Nalbalwar, S.L., Pathak, N.P. (eds.) Computing, Communication and Signal Processing. AISC, vol. 810, pp. 967–975. Springer, Singapore (2019). https://doi.org/10.1007/978-981-13-1513-8_98

38. Vogado, L.H.S., Veras, R.D.M.S., Andrade, A.R., Araujo, F.H.D.D., Silva, R.R.V., Aires, K.R.T.: Diagnosing leukemia in blood smear images using an ensemble of classifiers and pre-trained convolutional neural networks. In: 2017 30th SIBGRAPI Conference on Graphics, Patterns and Images (SIBGRAPI), Campo Grande, Brazil, pp. 367–373, October 2017. https://doi.org/10.1109/SIBGRAPI.2017.55

39. Vogado, L.H., Veras, R.M., Araujo, F.H., Silva, R.R., Aires, K.R.: Leukemia diagnosis in blood slides using transfer learning in CNNs and SVM for classification. Eng. Appl. Artif. Intell. **72**(C), 415–422 (2018). https://doi.org/10.1016/j.engappai.2018.04.024

40. Wang, J.L., Li, A.Y., Huang, M., Ibrahim, A.K., Zhuang, H., Ali, A.M.: Classification of white blood cells with pattern net-fused ensemble of convolutional neural networks (PECNN). In: 2018 IEEE International Symposium on Signal Processing and Information Technology (ISSPIT), Louisville, KY, USA, pp. 325–330, December 2018. https://doi.org/10.1109/ISSPIT.2018.8642630

41. Wang, Q., Bi, S., Sun, M., Wang, Y., Wang, D., Yang, S.: Deep learning approach to peripheral leukocyte recognition. PloS One **14**(6) (2019). https://doi.org/10.1371/journal.pone.0218808

42. Wang, X., Xu, T., Zhang, J., Chen, S., Zhang, Y.: SO-YOLO based WBC detection with Fourier ptychographic microscopy. IEEE Access **6**, 51566–51576 (2018). https://doi.org/10.1109/ACCESS.2018.2865541

43. Yang, F., et al.: Deep learning for smartphone-based malaria parasite detection in thick blood smears. IEEE J. Biomed. Health Inform. **24**(5), 1427–1438 (2019). https://doi.org/10.1109/JBHI.2019.2939121

Author Index

Printed in the United States
By Bookmasters